HOLT INTRODUCTORY ALGEBRA 2

Russell F. Jacobs

Holt, Rinehart and Winston
Harcourt Brace Jovanovich
HBJ
Austin • Orlando • San Diego • Chicago • Dallas • Toronto

ABOUT THE AUTHOR

RUSSELL F. JACOBS

Formerly Mathematics Supervisor for the Phoenix Union High School System
Phoenix, Arizona

EDITORIAL ADVISORS

Printed in the United States of America

ISBN 0-03-076984-1

5 6 7 040 97

Contents

Operations on Numbers

CHAPTER 1

The history of computers began thousands of years ago with the *abacus.* Blaize Pascal (1623–1662) built the first mechanical adding machine, the *Pascaline.* Charles Babbage (1791–1871) tried to develop the *analytical engine* to perform arithmetic calculations.

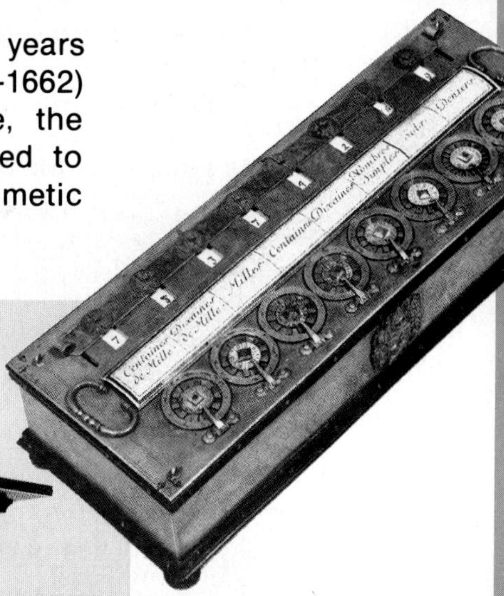

1.1 Opposites and Absolute Value

Note on this number line that 3 and -3 (negative 3) are the same distance from 0 and are in opposite directions from 0. Thus, 3 and -3 are ***opposites.***

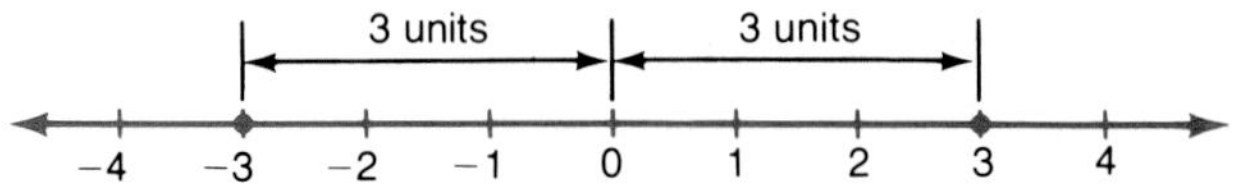

0 is its own opposite.

The symbol for the opposite of a number is the dash, $-$.

Table 1

Number	Opposite	Simplest Form of the Opposite
-5	$-(-5)$	5
2	-2	-2
0	-0	0

The opposite of a negative number is positive. $-(-7) = 7$
The opposite of a positive number is negative. $-7 = -7$
The opposite of 0 is 0. $-0 = 0$

To indicate distance, but not direction, from zero, you use absolute value. Since 3 and -3 are the same distance from 0, they have the same ***absolute value.***

The absolute value of 3 is 3. The absolute value of -3 is 3.

The symbol $|\ |$ is used to represent absolute value.

Table 2

Absolute Value	Meaning	Value
$\lvert -5 \rvert$	Distance of -5 from 0	5
$\lvert 2 \rvert$	Distance of 2 from 0	2
$\lvert 0 \rvert$	Distance of 0 from 0	0
$-\lvert 4 \rvert$	Opposite of the distance of 4 from 0	-4

The absolute value of a positive number equals the number. The absolute value of a negative number equals the opposite of the number. The absolute value of 0 equals 0. $|7| = 7$ $|-7| = 7$ $|0| = 0$

$|x| = x$ if $x \geq 0$; $|x| = -x$ if $x < 0$.

EXAMPLE Which number has the greater absolute value?

a. -15 or 9 **b.** -8 or 13

Solutions:

a. $|-15| = 15$; $|9| = 9$ **Answer:** -15

b. $|-8| = 8$; $|13| = 13$ **Answer:** 13

CLASSROOM EXERCISES

Write the opposite of each number in simplest form. (Table 1)

1. -25 **2.** 0 **3.** -0.001 **4.** 0.001 **5.** $-\frac{2}{3}$
6. $-4\frac{1}{2}$ **7.** $8\frac{7}{8}$ **8.** $-(-1\frac{2}{3})$ **9.** $-(-8.5)$ **10.** $-(-12)$

Simplify. (Table 2)

11. $|9|$ **12.** $|-13|$ **13.** $|-0|$ **14.** $|-2.38|$ **15.** $|\frac{4}{5}|$
16. $|-3\frac{5}{8}|$ **17.** $|0.053|$ **18.** $-|-47|$ **19.** $|-(-\frac{1}{4})|$ **20.** $|-100|$
21. $-|0.01|$ **22.** $-|8.7|$ **23.** $-|-2.5|$ **24.** $-|-\frac{1}{2}|$ **25.** $-|-2\frac{1}{2}|$

Which number has the greater absolute value? (Example)

26. 42 or -37 **27.** 12 or -3 **28.** 7 or -2 **29.** -5 or 4 **30.** -3 or -8

WRITTEN EXERCISES

Goal: To write the absolute value of a number
Sample Problems: $|8|$; $|-7|$ **Answers:** 8; 7

Write the opposite of each number in simplest form. (Table 1)

1. 23 **2.** -41 **3.** -0.12 **4.** 0.03 **5.** 0
6. -1 **7.** $\frac{3}{5}$ **8.** $1\frac{3}{8}$ **9.** $-5\frac{6}{7}$ **10.** $-\frac{2}{7}$

Simplify. (Table 2)

11. $|-18|$ **12.** $|-35|$ **13.** $-|227|$ **14.** $-|508|$ **15.** $|\frac{7}{8}|$
16. $|\frac{2}{3}|$ **17.** $|-5\frac{1}{8}|$ **18.** $|-3\frac{3}{4}|$ **19.** $|0.083|$ **20.** $|0.006|$
21. $|-2.73|$ **22.** $-|-5.18|$ **23.** $|-(-23)|$ **24.** $|-(-72)|$ **25.** $-|-6.8|$

Which number has the greater absolute value? (Example)

26. -23 or 21 **27.** 12 or 7 **28.** -23 or 6 **29.** 23 or -6
30. -8 or -7 **31.** -13 or -8 **32.** 7 or -3 **33.** 13 or -5

1.2 Addition

You can use this rule for adding on the number line.

> ***Rule for Adding on the Number Line***
>
> 1. Graph the first addend.
> 2. From this point draw an arrow for the second addend. If this addend is negative, draw the arrow to the left. If this addend is positive, draw the arrow to the right.
> 3. Read the coordinate of the point where the arrow ends.

Table

<table>
<tr><th>Problems</th><th>Solution</th><th></th></tr>
<tr><td>1. $-2 + (-7)$</td><td>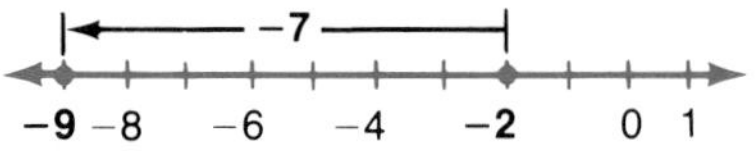
</td><td>$-2 + (-7) = -9$</td></tr>
<tr><td>2. $-7 + 9$
NOTE: $|9| > |-7|$</td><td>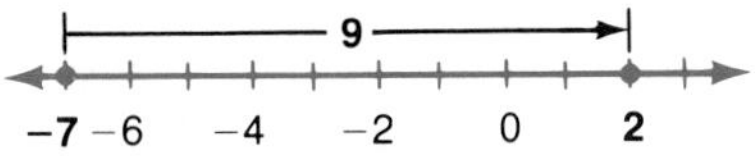
</td><td>$-7 + 9 = 2$</td></tr>
<tr><td>3. $-8 + 5$
NOTE: $|-8| > |5|$</td><td>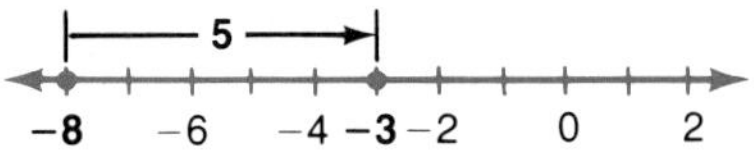
</td><td>$-8 + 5 = -3$</td></tr>
<tr><td>4. $-6 + 6$
NOTE: $|-6| = |6|$</td><td>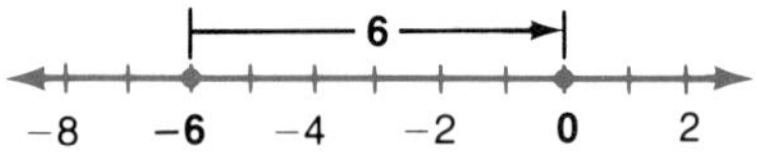
</td><td>$-6 + 6 = 0$</td></tr>
<tr><td>5. $4 + (-4)$
NOTE: $|4| = |-4|$</td><td>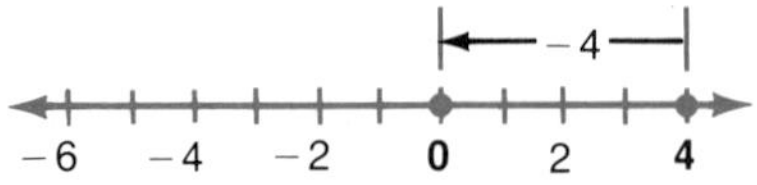
</td><td>$4 + (-4) = 0$</td></tr>
</table>

Problem 1 above shows that the sum of two negative numbers is a negative number. The following rule tells how to find the sum of two negative numbers without using the number line.

> ***Rule for Adding Two Negative Numbers***
>
> 1. Add the absolute values.
> 2. Write the opposite of the result.

EXAMPLE 1 Add: $-6 + (-3)$

Solution: [1] Add the absolute values. ⟶ $|-6| + |-3| = 6 + 3 = 9$

[2] Write the opposite of the result. ⟶ $-6 + (-3) = -9$

P–1 **Add.**

a. $-2 + (-7)$ **b.** $-1 + -5$ **c.** $-4.7 + 0$

The sum of any number n and 0 is n.

Problems 2–5 in the table show that the sum of a positive number and a negative number may be a positive number, a negative number, or zero. The following rules tell how to use absolute value to find the the sum of a positive and a negative number without using the number line.

Rule for Adding a Negative Number and a Positive Number

1. Subtract the lesser absolute value from the greater absolute value.
2. The answer is positive if the positive number has the greater absolute value. $8 + (-3) = 5$

   The answer is negative if the negative number has the greater absolute value. $5 + (-7) = -2$

   The answer is zero if the two absolute values are equal. $9 + (-9) = 0$

EXAMPLE 2 Add: $-9 + 15$

$|\mathbf{15}| > |\mathbf{-9}|$

Solution:

[1] Subtract the lesser absolute value from the greater. ⟶ $|15| - |-9| = 15 - 9$
$= 6$

[2] The answer is positive because the positive number has the greater absolute value. ⟶ $-9 + 15 = 6$

P–2 **Add.**

a. $-7 + 9$ **b.** $5\frac{3}{4} + (-3\frac{1}{4})$ **c.** $-23.1 + (25.7)$

EXAMPLE 3 A liquefied gas had a temperature of −13°C. It was slowly heated until its temperature increased 8°C. What was its temperature after being heated?

Solution: $-13 + 8 = \underline{\ ?\ }$

1. Subtract the lesser absolute value from the greater. ⟶ $|-13| - |8| = 13 - 8 = 5$
2. The answer is negative because the negative number has the greater absolute value. ⟶ $-13 + 8 = -5$

The temperature of the gas was −5°C after being heated.

P–3 **Add.**

a. $-11 + 8$ **b.** $-6\frac{2}{3} + 5\frac{1}{3}$ **c.** $-8.5 + 1.1$

Since the absolute values of opposites are equal, the sum of opposites is 0.

$-3 + 3 = 0$

$1.8 + (-1.8) = 0$

CLASSROOM EXERCISES

Match each of Exercises 1–4 with one of the graphs **a–d.** (Table)

1. $1 + (-6)$ **2.** $-7 + 11$ **3.** $4 + (-7)$ **4.** $-5 + 6$

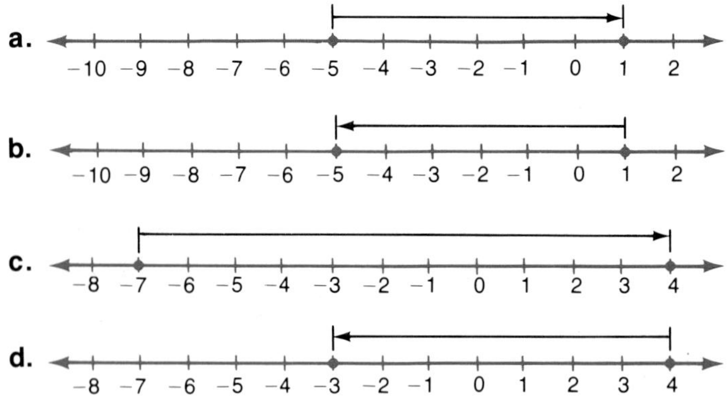

Add. (Example 1)

5. $-28 + (-12)$ **6.** $-19 + (-13)$ **7.** $-3\frac{1}{2} + (-2\frac{1}{2})$

8. $-6\frac{1}{5} + (-3\frac{4}{5})$ **9.** $-0.5 + (-0.2)$ **10.** $-4.3 + (-1.5)$

(Example 2)

11. $28 + (-12)$ 12. $-16 + 32$ 13. $-2.4 + 5.8$

14. $0.28 + (-0.13)$ 15. $-\frac{2}{5} + \frac{3}{5}$ 16. $-\frac{1}{3} + \frac{2}{3}$

(Example 3)

17. $-36 + 30$ 18. $52 + (-61)$ 19. $22.6 + (-25.9)$

20. $-16.8 + 9.3$ 21. $-12\frac{5}{8} + 5\frac{1}{8}$ 22. $3\frac{1}{4} + (-6\frac{3}{4})$

23. $-56 + 56$ 24. $38\frac{1}{9} + (-38\frac{1}{9})$ 25. $2.18 + (-2.18)$

WRITTEN EXERCISES

Goal: To add positive and negative numbers
Sample Problems: **a.** $(-49) + (52)$ **b.** $(-32) + (-12)$
Answers: **a.** 3 **b.** -44

Add. (Example 1)

1. $-23 + (-19)$ 2. $-38 + (-17)$ 3. $-47 + (-59)$

4. $-66 + (-48)$ 5. $0 + (-0.08)$ 6. $-17.8 + (-9.5)$

(Example 2)

7. $26 + (-18)$ 8. $-24 + 32$ 9. $14 + (-10)$

10. $12 + (-8)$ 11. $-0.12 + 0.25$ 12. $-2.3 + 8.6$

(Example 3)

13. $-39 + 14$ 14. $-24 + 15$ 15. $23 + (-55)$

16. $25 + (-44)$ 17. $-28.7 + 19.8$ 18. $-43.3 + 29.7$

MIXED PRACTICE

19. $-32 + (-21)$ 20. $-20 + (-38)$ 21. $-31 + 47$

22. $-24 + 14$ 23. $-29.6 + (-15.9)$ 24. $-0.38 + (-0.45)$

25. $4.67 + (-4.67)$ 26. $81.3 + (-81.3)$ 27. $-42 + 42$

28. $-26 + 26$ 29. $-\frac{2}{3} + \frac{5}{3}$ 30. $-\frac{9}{13} + \frac{7}{13}$

APPLICATIONS (Example 3)

31. The thermometer reading at 6:00 A.M. was -4°C. By noon, it had risen 9°C. Find the thermometer reading at noon.

32. On Monday, the change in the price of a stock was $-2\frac{1}{4}$ points. On Tuesday, the change was $-3\frac{3}{4}$ points. Find the total change in price for the two days.

REVIEW CAPSULE FOR SECTION 1.3

*Add or subtract as indicated. Compare the answers in **a** and **b**.*

1. a. $8 - 4$ **b.** $8 + (-4)$ **2. a.** $12 - 8$ **b.** $12 + (-8)$ **3. a.** $16 - 9$ **b.** $16 + (-9)$

4. a. $21 - 11$ **b.** $21 + (-11)$ **5. a.** $33 - 12$ **b.** $33 + (-12)$ **6. a.** $1.5 - 1.5$ **b.** $1.5 + (-1.5)$

1.3 Subtraction

You can think of subtraction as meaning "add the opposite."

Subtraction	Addition	Answer
$5 - 2$	$5 + (-2)$	3
$3 - 8$	$3 + (-8)$	-5
$7 - (-3)$	$7 + 3$	?
$-2.7 - (-3.5)$	$-2.7 + 3.5$	?
$0 - 37$	$0 + (-37)$	?
$-10.8 - (-10.8)$	$-10.8 + 10.8$	?

P–1 **Find the missing answers in the table above.**

To subtract a number, add its opposite. $7 - 5 = 7 + (-5)$

$a - b = a + (-b)$ $-8 - (-3) = -8 + 3$

EXAMPLE 1 Subtract: $12 - 25$

Solution: [1] Write a sum for the difference. → $12 - 25 = 12 + (-25)$

[2] Add. → $= -13$

EXAMPLE 2 Subtract: $-13 - (-9)$

Solution: [1] Write a sum for the difference. → $-13 - (-9) = -13 + 9$

[2] Add. → $= -4$

P–2 **Subtract.**

a. $30 - 12$ **b.** $12 - (-8)$ **c.** $-18 - 7$ **d.** $-5.6 - (-9.8)$

EXAMPLE 3 The thermometer reading at noon on a January day was $-12°C$. The wind chill temperature was $-17°C$. How much lower was the wind chill temperature than the thermometer reading?

Solution: $-12-(-17)=\underline{\ ?\ }$

[1] Write a sum for the difference. ⟶ $-12-(-17)=-12+17$

[2] Add. ⟶ $=5$

The wind chill temperature was 5°C lower than the thermometer reading.

CLASSROOM EXERCISES

Write a sum for each difference. (Step 1, Examples 1 and 2)

1. $20-15$ **2.** $6-11$ **3.** $0-12$ **4.** $18-(-5)$

5. $(-12)-6$ **6.** $18-(-7)$ **7.** $-20-(-18)$ **8.** $-13-(-19)$

9. $5-(-23)$ **10.** $7-12$ **11.** $-3.3-6.2$ **12.** $-2.6-3.4$

Subtract. (Example 1)

13. $12-16$ **14.** $0-18$ **15.** $-15-8$ **16.** $-15-15$

17. $-4\frac{1}{6}-1\frac{1}{6}$ **18.** $-7\frac{1}{4}-2\frac{1}{4}$ **19.** $-7.3-2.7$ **20.** $-2.7-7.3$

(Example 2)

21. $6-(-10)$ **22.** $18-(-3)$ **23.** $-24-(-10)$ **24.** $0-(-32)$

WRITTEN EXERCISES

Goal: To subtract positive and negative numbers

Sample Problems: **a.** $-11-(-11)$ **b.** $-11-11$

Answers: **a.** 0 **b.** -22

Subtract. (Example 1)

1. $9-5$ **2.** $18-6$ **3.** $24-18$ **4.** $16-9$

5. $0-5$ **6.** $6-13$ **7.** $14-23$ **8.** $28-35$

9. $0-22$ **10.** $0-36$ **11.** $-9.3-1.5$ **12.** $-7.6-18.1$

13. $-18.9-5.8$ **14.** $-0.14-0.58$ **15.** $\frac{1}{8}-\frac{7}{8}$ **16.** $-\frac{5}{6}-\frac{1}{6}$

(Example 2)

17. $25 - (-6)$ **18.** $19 - (-7)$ **19.** $13 - (-19)$ **20.** $26 - (-11)$

21. $-16 - (-18)$ **22.** $-14 - (-20)$ **23.** $0 - (-42)$ **24.** $0 - (-96)$

25. $2.7 - (-5.6)$ **26.** $4.3 - (-8.9)$ **27.** $-8\frac{5}{6} - (-4\frac{5}{6})$ **28.** $-10\frac{3}{8} - (-4\frac{3}{8})$

MIXED PRACTICE

29. $11 - 24$ **30.** $11 - (-24)$ **31.** $0 - 18$ **32.** $0 - (-18)$

33. $-4 - (-12)$ **34.** $-4 - 12$ **35.** $7.8 - 3.4$ **36.** $3.4 - 7.8$

37. $-3.2 - (-4.5)$ **38.** $-0.71 - 0.82$ **39.** $\frac{5}{7} - \frac{6}{7}$ **40.** $-\frac{1}{3} - (-\frac{4}{3})$

APPLICATIONS

(Example 3)

41. When mixed equally with water in an automobile's radiator, antifreeze has a freezing point of −17°F. By itself, antifreeze has a freezing point of −37°F. How much lower is the freezing point for the pure antifreeze than for the mixture?

42. The record low temperature for Houston, Texas, is 5°F. The record low temperature for Buffalo, New York, is −21°F. How much lower is the record low temperature for Buffalo than for Houston?

43. A new business had a net income of −\$582 in August. Its net income for September was −\$620. How much lower was the business' September net income than its August net income?

44. On May 31, Jill's checking account balance was \$265. On June 31, her checking account balance was −\$14. How much lower was Jill's account balance on June 31 than on May 31?

REVIEW CAPSULE FOR SECTION 1.4

Write a fraction for each of the following.

1. $1\frac{1}{4}$ **2.** 9 **3.** $1\frac{5}{6}$ **4.** 13 **5.** $5\frac{2}{5}$

Multiply.

6. $(10)(3)$ **7.** $(4)(1.3)$ **8.** $(\frac{1}{2})(18)$ **9.** $(\frac{1}{2})(\frac{2}{3})$

10. $(1.5)(1.5)$ **11.** $(0.8)(0.7)$ **12.** $(0.9)(1.6)$ **13.** $(5)(0)$

14. $(0)(28)$ **15.** $1\frac{1}{2} \times \frac{2}{3}$ **16.** $\frac{3}{4} \times \frac{8}{9}$ **17.** $2\frac{1}{4} \times 3\frac{1}{5}$

1.4 Multiplication

In the following multiplication problems, one factor, 8, stays the same. The other factor decreases by 1.

$8(3) = 24$
$8(2) = 16$
$8(1) = 8$
$8(0) = \underline{\ ?\ }$
$8(-1) = \underline{\ ?\ }$
$8(-2) = \underline{\ ?\ }$
$8(-3) = \underline{\ ?\ }$

Each product is eight less than the preceding one.

The product of a positive number and a negative number is a negative number.

P–1 **If the pattern continues, what is the answer to 8(0)?**

> ***Multiplication Property of Zero***
>
> For any number a, $\mathbf{a \cdot 0}$ or $\mathbf{0 \cdot a = 0}$.
>
> $3 \cdot 0 = 0$
>
> $0 \cdot \frac{1}{2} = 0$

P–2 **If the pattern above continues, what are the other missing numbers?**

The following rule tells how to find the product of a positive and a negative number by using absolute value.

> ***Rule for Multiplying a Positive Number and a Negative Number***
>
> 1. Multiply the absolute values. $(3)(-7) = -21$
> 2. Write the opposite of the result. $(-8)(4) = -32$

EXAMPLE 1 Multiply: $(9)(-8)$

Solution: [1] Multiply the absolute values. ⟶ $|9| \cdot |-8| = 9 \cdot 8$
$= 72$

[2] Write the opposite of the result. ⟶ $(9)(-8) = -72$

P–3 **Multiply.**

a. $(-6)(7)$ **b.** $(4)(-12)$ **c.** $(-\frac{1}{2})(0)$ **d.** $(-\frac{3}{4})(\frac{4}{3})$

In the following multiplication problems, one factor, −6, stays the same. The other factor decreases by 1.

$-6(3) = -18$
$-6(2) = -12$
$-6(1) = -6$
$-6(0) = 0$
$-6(-1) = \underline{\ ?\ }$
$-6(-2) = \underline{\ ?\ }$
$-6(-3) = \underline{\ ?\ }$

Each product is greater (less negative) than the answer that precedes it.

The product of two negative numbers is a positive number.

P–4 **If the pattern continues, what are the missing numbers?**

The following rule tells you how to find the product of two negative numbers by using absolute value.

> ***Rule for Multiplying Two Negative Numbers***
>
> 1. Multiply the absolute values.
> 2. Write the result. $(-7)(-6) = 42$

EXAMPLE 2 Multiply: $(-8)(-5)$

Solution: [1] Multiply the absolute values. ⟶ $|-8| \cdot |-5| = 8 \cdot 5$
$= 40$

[2] Write the result. ⟶ $(-8)(-5) = 40$

P–5 **Multiply.**

a. $(-5)(-9)$ **b.** $(-7)(-10)$ **c.** $(-2)(-0.13)$ **d.** $(-6)(-0.5)$

EXAMPLE 3 At 7:00 P.M. on a winter day, the wind chill temperature was six times the thermometer reading of −5°F. What was the wind chill temperature?

Solution: $(6)(-5) = \underline{\ ?\ }$

[1] Multiply the absolute values. ⟶ $|6| \cdot |-5| = 6 \cdot 5$
$= 30$

[2] Write the opposite of the result. ⟶ $(6)(-5) = -30$

The wind chill temperature at 7:00 P.M. was −30°F.

CLASSROOM EXERCISES

Multiply. (Example 1)

1. $(-4)(9)$
2. $(10)(-6)$
3. $(-12)(8)$
4. $(-4)(12)$
5. $(-\frac{1}{3})(9)$
6. $(-15)(\frac{1}{5})$
7. $(-37)(0)$
8. $(0)(-75)$
9. $(4)(-2.4)$
10. $(-2)(1.8)$
11. $(-3)(0.7)$
12. $(-10)(2.6)$

Multiply. (Example 2)

13. $(-11)(-7)$
14. $(-5)(-9)$
15. $(-10)(-15)$
16. $(-14)(-6)$
17. $(-7)(-21)$
18. $(-43)(-3)$
19. $(-5)(-1.2)$
20. $(-0.7)(-30)$
21. $(-\frac{1}{4})(-4)$
22. $(-\frac{1}{8})(-8)$
23. $(-\frac{3}{8})(-\frac{8}{3})$
24. $(-\frac{2}{5})(-\frac{3}{5})$

WRITTEN EXERCISES

Goal: To multiply positive and negative numbers

Sample Problems: a. $(-7)(-8)$ **b.** $(-9)(12)$

Answers: a. 56 **b.** -108

Multiply. (Example 1)

1. $(-14)(5)$
2. $(-16)(6)$
3. $(0)(-63)$
4. $(-30)(0)$
5. $(22)(-3)$
6. $(18)(-7)$
7. $(-2)(24)$
8. $(-4)(32)$
9. $(15)(-\frac{2}{3})$
10. $(20)(-\frac{3}{4})$
11. $(\frac{3}{5})(-\frac{4}{9})$
12. $(-\frac{2}{9})(\frac{3}{4})$
13. $(-12)(1.7)$
14. $(-9)(3.4)$
15. $(-1)(8.36)$
16. $(9.01)(-1)$

(Example 2)

17. $(-24)(-3)$
18. $(-16)(-8)$
19. $(-\frac{4}{5})(-\frac{5}{4})$
20. $(-\frac{7}{4})(-\frac{4}{7})$
21. $(-15)(-5.8)$
22. $(-24)(-2.7)$
23. $(-160)(-20)$
24. $(-180)(-99)$
25. $(-1.65)(-0.2)$
26. $(-1.25)(-0.8)$
27. $(-1)(-99)$
28. $(-86)(-1)$

MIXED PRACTICE

29. $(36)(-12)$
30. $(-42)(8)$
31. $(-18)(-11)$
32. $(-23)(-5)$
33. $(-240)(15)$
34. $(-325)(7)$
35. $(817)(0)$
36. $(0)(-341)$
37. $(-1.2)(-1.3)$
38. $(-2.4)(-5.1)$
39. $(-1\frac{2}{3})(2\frac{4}{5})$
40. $(-1\frac{3}{5})(-3\frac{1}{3})$

APPLICATIONS (Example 3)

41. The net change in the price of a stock for five successive days was $-\frac{3}{4}$ of a point for each day. Find the net change for the five days.

42. On a winter day, the wind chill temperature was eight times the thermometer reading of −4°C. Find the wind chill temperature.

MORE CHALLENGING EXERCISES

Let $*$ mean: "Double the number with the larger absolute value. Then multiply by the remaining number." Compute each value.

EXAMPLE $3 * (-5)$ **SOLUTION** $3 * (-5) = 3 \cdot 2(-5)$
$= 3 \cdot (-10)$
$= -30$

43. $-6 * 2$ **44.** $-4 * (-8)$ **45.** $1.2 * 3.6$ **46.** $\frac{1}{2} * (-\frac{3}{8})$

47. Compute the value for $2 * (-4)$ and for $-4 * 2$. Is $*$ a commutative operation? Explain.

MID-CHAPTER REVIEW

Write the opposite of each number in simplest form. (Section 1.1)

1. 17 **2.** 32 **3.** $-\frac{1}{2}$ **4.** $-1\frac{1}{2}$ **5.** −0.06

Simplify. (Section 1.1)

6. $|20|$ **7.** $|0|$ **8.** $|-3.7|$ **9.** $|2.5|$ **10.** $|-1\frac{3}{5}|$

Which number has the greater absolute value? (Section 1.1)

11. −17 or 12 **12.** −8 or −12 **13.** 10 or −5 **14.** 9 or −38

Add. (Section 1.2)

15. $-20 + (-13)$ **16.** $-15 + 35$ **17.** $-18.1 + 19.1$

18. $-6.8 + 6.8$ **19.** $-12\frac{1}{2} + (-13\frac{1}{2})$ **20.** $8\frac{1}{4} + (-12\frac{3}{4})$

Solve. (Section 1.2)

21. The temperature of a solution was −5°C. It was heated until its temperature rose 11°C. What was its temperature after being heated?

22. The thermometer reading at 4:00 A.M. was −8°F. By 1:00 P.M., it had risen 17°F. Find the thermometer reading at 1:00 P.M.

Write a sum for each difference. (Section 1.3)

23. $11 - 7$ **24.** $-8.7 - 9.6$ **25.** $12.1 - (-7.9)$
26. $3.6 - 5.4$ **27.** $-7 - 4$ **28.** $-8 - (-10)$

Subtract. (Section 1.3)

29. $13 - 28$ **30.** $-4 - (-12)$ **31.** $0 - (-9.25)$
32. $2.6 - (-3.3)$ **33.** $\frac{1}{2} - 7$ **34.** $-6\frac{1}{2} - 6\frac{1}{2}$

Solve. (Section 1.3)

35. The high temperature on a winter day was 14°F. The low temperature was −11°F. How much higher was the high temperature than the low temperature?

36. The wind chill temperature on a winter day was −17°F. The thermometer reading was 6°F. How much lower was the wind chill temperature than the thermometer reading?

Multiply. (Section 1.4)

37. $(-7)(8)$ **38.** $(-4)(13)$ **39.** $(9)(-\frac{1}{3})$ **40.** $(12)(-\frac{1}{4})$
41. $(-1)(-1)$ **42.** $(-4)(-5)$ **43.** $(-4)(-8.4)$ **44.** $(-10)(-4.5)$

Solve. (Section 1.4)

45. On a winter day, the low temperature was −2°F. The high temperature was −11 times the low temperature. Find the day's high temperature.

46. The net change in the price of a stock for four successive days was $-1\frac{1}{2}$ points each day. Find the total change in the price of the stock for the four days.

REVIEW CAPSULE FOR SECTION 1.5

Write a fraction for each of the following.

1. $1\frac{1}{2}$ **2.** 12 **3.** $1\frac{3}{5}$ **4.** $2\frac{1}{3}$ **5.** $3\frac{1}{7}$
6. $4\frac{2}{5}$ **7.** $5\frac{2}{3}$ **8.** $2\frac{1}{6}$ **9.** 7 **10.** $3\frac{3}{4}$

Multiply or divide as indicated.

11. $\frac{7}{2} \times \frac{2}{7}$ **12.** $-\frac{3}{5} \times (-\frac{5}{3})$ **13.** $1\frac{2}{3} \times \frac{3}{5}$ **14.** $(-6) \times (-\frac{1}{6})$
15. $24 \div 8$ **16.** $\frac{24}{8}$ **17.** $24 \times \frac{1}{8}$ **18.** $4\frac{1}{2} \div 6$

1.5 Division

Each of the following products equals 1.

$$\frac{7}{2} \times \frac{2}{7} = 1 \qquad -\frac{3}{5} \times \left(-\frac{5}{3}\right) = 1 \qquad \frac{5}{3} \times \frac{3}{5} = 1$$

Definition

Two numbers are ***reciprocals*** of each other if their product is 1.

$\frac{7}{2}$ and $\frac{2}{7}$

$a \cdot \frac{1}{a} = 1$ (a is not zero.)

$-\frac{5}{3}$ and $-\frac{3}{5}$

P–1 **What is the reciprocal of each number below?**

a. $\frac{3}{10}$ **b.** -3 **c.** $1\frac{1}{3}$ **d.** 1 **e.** -1

The following illustrates that

dividing by a number is the same as multiplying by its reciprocal.

$$36 \div 9 = 4 \qquad 36 \times \frac{1}{9} = 4$$

9 and $\frac{1}{9}$ are reciprocals.

To divide by a number, multiply by its reciprocal.

$a \div b = a \cdot \frac{1}{b}$, $b \neq 0$ (b is not zero.)

$$3\tfrac{1}{2} \div 1\tfrac{3}{4} = \frac{7}{2} \times \frac{4}{7} = \frac{28}{14} = 2$$

Any number except zero can be a divisor. Thus, "$\frac{6}{2}$" is permitted because "$6 = 3 \cdot 2$" is true. However, "$\frac{6}{0}$" is not permitted because "$6 = a \cdot 0$" is false no matter what number a represents. $a \cdot 0$ always equals 0.

EXAMPLE 1 Divide: $(-12) \div (-\frac{1}{3})$

Solution: The reciprocal of $-\frac{1}{3}$ is $-\frac{3}{1}$, or -3.

1 Write a product for the quotient. ——→ $(-12) \div (-\frac{1}{3}) = (-12) \times (-3)$

2 Multiply. ——→ $= 36$

P–2 **Divide.**

a. $(-16) \div (-\frac{1}{4})$ **b.** $(-\frac{5}{9}) \div (-5)$ **c.** $0 \div (-\frac{1}{8})$

$0 \cdot a = 0$

EXAMPLE 2 Divide: $4\frac{1}{2} \div (-6)$

Solution: The reciprocal of -6 is $-\frac{1}{6}$.

[1] Write a product for the quotient. ⟶ $4\frac{1}{2} \div (-6) = \frac{9}{2} \times (-\frac{1}{6})$

[2] Multiply. ⟶ $= -\frac{9}{12}$

[3] Write the answer in lowest terms. ⟶ $= -\frac{3}{4}$

P–3 **Divide.**

a. $5\frac{1}{6} \div (-3)$ **b.** $(-18) \div 1\frac{1}{6}$ **c.** $\frac{-2.4}{8}$

This represents the quotient of −2.4 and 8.

Rules for Division

1. If a positive number is divided by a positive number, the quotient is a positive number.	$30 \div 3 = 10$
2. If a negative number is divided by a negative number, the quotient is a positive number.	$-4\frac{1}{2} \div -2\frac{1}{4} = 2$
3. If a negative number is divided by a positive number, the quotient is a negative number.	$-25 \div 5 = -5$
4. If a positive number is divided by a negative number, the quotient is a negative number.	$35 \div -7 = -5$
5. If zero is divided by a positive or a negative number, the quotient is zero.	$0 \div -6 = 0$
6. Zero can never be a divisor.	$-5 \div 0$ is undefined.

EXAMPLE 3 A city's deficiency in rainfall for one year (12 months) was −36 centimeters. Find the average deficiency per month.

Solution:

[1] Write the division expression. ⟶ $-36 \div 12 = \underline{\ ?\ }$

[2] Divide. ⟶ $-36 \div 12 = -3$

The average deficiency in rainfall per month was −3 centimeters.

CLASSROOM EXERCISES

Write a product for each quotient. (Step 1 in Examples 1 and 2)

1. $-8 \div \frac{1}{4}$
2. $-12 \div 3$
3. $28 \div (-7)$
4. $\frac{-3.2}{4}$

Divide. (Example 1)

5. $-35 \div (-\frac{1}{5})$
6. $-32 \div (-\frac{1}{2})$
7. $-30 \div (-10)$
8. $\frac{-44}{-11}$

(Example 2)

9. $4\frac{1}{2} \div (-9)$
10. $3\frac{1}{5} \div (-8)$
11. $0 \div (-10\frac{3}{4})$
12. $0 \div (-\frac{1}{5})$

WRITTEN EXERCISES

Goal: To divide positive and negative numbers
Sample Problems: a. $125 \div (-5)$ **b.** $(-1\frac{1}{2}) \div (-6\frac{1}{4})$
Answers: a. -25 **b.** $\frac{6}{25}$

Divide. (Example 1)

1. $-480 \div (-20)$
2. $(-325) \div (-25)$
3. $\frac{-27}{-9}$
4. $\frac{-132}{-12}$
5. $-16 \div (-1\frac{1}{2})$
6. $-25 \div (-1\frac{1}{4})$
7. $-1\frac{1}{4} \div (-7\frac{1}{2})$
8. $-4\frac{1}{2} \div (-\frac{3}{4})$

(Example 2)

9. $16 \div (-1\frac{1}{2})$
10. $10 \div (-\frac{1}{3})$
11. $3\frac{3}{4} \div (-\frac{3}{4})$
12. $2\frac{2}{5} \div (-\frac{2}{5})$
13. $\frac{24}{-8}$
14. $\frac{18}{-9}$
15. $108 \div (-12)$
16. $176 \div (-16)$
17. $0 \div 17$
18. $0 \div (-81)$
19. $-121 \div 11$
20. $-216 \div 12$
21. $-6.3 \div 7$
22. $-1.8 \div 0.3$
23. $-2\frac{1}{2} \div \frac{1}{4}$
24. $-3\frac{1}{6} \div \frac{1}{3}$

MIXED PRACTICE

25. $\frac{36}{-9}$
26. $\frac{-54}{-6}$
27. $-25 \div 5$
28. $72 \div (-12)$
29. $-13 \div (-\frac{1}{4})$
30. $-18 \div \frac{2}{3}$
31. $\frac{0}{-3}$
32. $\frac{0}{-21.2}$
33. $\frac{-39}{13}$
34. $\frac{17}{-2}$
35. $-\frac{7}{9} \div 2\frac{1}{3}$
36. $-\frac{1}{2} \div 1\frac{1}{8}$
37. $\frac{110}{-11}$
38. $\frac{-23}{10}$
39. $20.5 \div (-0.5)$
40. $-3.6 \div (-1.2)$

APPLICATIONS (Example 3)

41. A company had a net loss for the year of $-\$28,200$. Find the average net loss per month.

42. Over a period of ten days, a stock showed a net change in price of $-3\frac{3}{4}$ points. Find the average change per day.

1.6 Order of Operations

An ***expression,*** such as $12 - 18$ or $2 \cdot 5 + 7$, includes at least one of the operations of addition, subtraction, multiplication, or division.

P–1 **Which of the following answers are correct?**

a. $2 \cdot 5 + 7 = 10 + 7$ **Ans. 17**

b. $2 \cdot 5 + 7 = 2 \cdot 12$ **Ans. 24**

c. $10 \div 5 - 3 = 2 - 3$ **Ans. −1**

d. $10 \div 5 - 3 = 10 \div 2$ **Ans. 5**

Rules are needed so that a given expression has only one value.

> If only addition and subtraction are involved, the operations are performed from left to right.
> $8 - 2 - 1 = 5$
> $1 - 2 + 8 = 7$

EXAMPLE 1 Evaluate: **a.** $12 + 5 - 4$ **b.** $4 - 9 + 3$

Evaluate means "find the value of."

Solutions: **a.** $12 + 5 - 4 = 17 - 4 = 13$ **b.** $4 - 9 + 3 = -5 + 3 = -2$

> If only multiplication and division are involved, the operations are performed from left to right.
> $8 \times 2 \div 4 = 4$
> $27 \div 9 \div 3 = 1$

EXAMPLE 2 Evaluate: **a.** $28 \div 7(-9)$ **b.** $(-2)(-13) \div (-4)$

Solutions: **a.** $28 \div 7(-9) = 4(-9) = -36$ **b.** $(-2)(-13) \div (-4) = 26 \div (-4) = -6\frac{1}{2}$

> Multiplication or division is performed before addition or subtraction.
> $2 \cdot 5 + 7 = 17$
> $10 \div 5 - 3 = -1$

EXAMPLE 3 Evaluate.

a. $2 \cdot 5 + 7$ **b.** $4 + 3 \cdot 6$ **c.** $10 - 6 \cdot 2$ **d.** $10 \div 5 - 3$

Solutions: **a.** $2 \cdot 5 + 7 = 10 + 7 = 17$ **b.** $4 + 3 \cdot 6 = 4 + 18 = 22$

c. $10 - 6 \cdot 2 = 10 - 12 = -2$ **d.** $10 \div 5 - 3 = 2 - 3 = -1$

CLASSROOM EXERCISES

Evaluate. (Example 1)

1. $12 - 3 + 7$ **2.** $14 + 1 - 9$ **3.** $7 - 10 + 3$ **4.** $5 - 8 - 11$

(Example 2)

5. $2 \cdot 15 \div 6$ **6.** $42 \div 7 \cdot 3$ **7.** $-8 \div 2(-3)$ **8.** $-10 \div (-2) \div (-5)$

(Example 3)

9. $3 + 10 \cdot 4$ **10.** $20 - 4 \cdot 2$ **11.** $21 - 27 \div 3$ **12.** $3 + 5(-7)$

13. $(-3)(5) - (-2)(2)$ **14.** $-9 \div 3 - 7$ **15.** $11 \div (-\frac{1}{2}) + 6$ **16.** $-8 + (-4) \div (-\frac{1}{4})$

WRITTEN EXERCISES

Goal: To evaluate an expression that involves order of operations
Sample Problems: a. Evaluate: $5 + 8 \cdot 3$ **b.** Evaluate: $-2(-3) + 3 \div (-1)$
Answers: a. 29 **b.** 3

Evaluate. (Example 1)

1. $5 - 8 + 12$ **2.** $10 - 14 + 7$ **3.** $5 - 10 - 7$ **4.** $-6 + 7 + 8$

5. $-0.4 + 0.7 - 0.8$ **6.** $1.2 - 2.3 + 3.6$ **7.** $8 + 9 - 10 + 6$ **8.** $17 - 13 - 2 + 10$

(Example 2)

9. $8 \cdot 2 \div 4$ **10.** $20 \div 5 \cdot 4$ **11.** $-240 \div 8 \cdot 2$ **12.** $9(-8) \div (-12)$

13. $-3 \cdot 12 \div (-4)$ **14.** $-6(-8) \div 3(4)$ **15.** $15 \div \frac{1}{2}(-3)$ **16.** $(-27) \div \frac{1}{3}(-5)$

(Example 3)

17. $6 - 5 \cdot 4$ **18.** $12 - 3 \cdot 9$ **19.** $-17 + (-9)2$ **20.** $-21 + 5(-3)$

21. $12 - 32 \div (-4)$ **22.** $15 - 39 \div (-3)$ **23.** $6.3 + 4.8 \div (-8)$ **24.** $1\frac{1}{8} + 3\frac{3}{8} \div (-18)$

MIXED PRACTICE

25. $8 - 5 + 9(-2)$ **26.** $9 + 2 - 15$ **27.** $20 - 6 \cdot 7$ **28.** $-3 \cdot 5 - 21$

29. $16 + 14 \div (-2)$ **30.** $21 \div 7 + 4$ **31.** $-1(8) - 2(-3)$ **32.** $4(3) + 9(8)$

33. $-3 + 18 \div (-9)$ **34.** $-5 - 12 \div 6$ **35.** $-66 \div (-2)(-4)$ **36.** $-200 \div (-2)(-3)$

NON-ROUTINE PROBLEMS

37. Suppose a person was born on the 25th day of the year 35 B.C. and died on the 25th day of the year 15 A.D. How many years did the person live?

38. Use four 4's and one or more of the operation symbols, +, −, ÷, or ×, to express the whole numbers from 1 to 4.
Example: $(4 \div 4) + 4 - 4 = 1$

1.7 Powers

An expression such as $(-2)^3$ is called a ***power.*** The 3 is the ***exponent*** and the -2 is the ***base.*** The exponent tells how many times to use the base as a factor.

EXAMPLE 1 Evaluate: $(-2)^3$

Solution: Use -2 as a factor 3 times ⟶ $(-2)^3 = (-2)(-2)(-2) = -8$

Powers are evaluated before multiplication and division are performed and before addition and subtraction are performed.

EXAMPLE 2 Evaluate.

a. $2 \cdot 3^2$ **b.** $-8 \div 4^3$ **c.** $-6 + 2^3$ **d.** $(-5)^2 - 6(-2)$

Solutions: **a.** $2 \cdot 3^2 = 2 \cdot 9 = 18$

b. $-8 \div 4^3 = (-8) \div 64 = \frac{-8}{64} = -\frac{1}{8}$

c. $-6 + 2^3 = -6 + 8$
$= 2$

d. $(-5)^2 - 6(-2) = 25 - (-12)$
$= 25 + 12 = 37$

P–1 **Evaluate:** **a.** $3 \cdot 2^2$ **b.** $(-4)^2 \div 2^3$ **c.** $-10 - 5(-\frac{1}{2})$

A ***variable*** is a letter such as x that represents one or more numbers. An ***algebraic expression*** is an expression with one or more variables, such as the following.

$$x + 2 \qquad 5n \qquad p^2 - 7y \qquad z^3$$

To evaluate an algebraic expression, replace each variable with its value. Then perform the operations.

EXAMPLE 3 Evaluate $3n^2 - 4$ when $n = 5$.

Solution:
1. Replace n with 5. ⟶ $3n^2 - 4 = 3(5)^2 - 4$
2. Evaluate the power. Then multiply. ⟶ $= 3(25) - 4$
3. Subtract. ⟶ $= 75 - 4 = 71$

P–2 **Evaluate each expression when $p = 3$.**

a. $p^2 - 5$ **b.** $5 - p^2$ **c.** $4p^2 - p$

EXAMPLE 4 Evaluate $x^2 - 3x$ when $x = -2$.

Solution:
1. Replace x with -2. ⟶ $x^2 - 3x = (-2)^2 - 3(-2)$
2. Evaluate the power. Then multiply. ⟶ $= 4 - 3(-2)$
3. Subtract. ⟶ $= 4 - (-6) = 10$

CLASSROOM EXERCISES

Evaluate each expression in Exercises 1–14. (Example 1)

1. 6^2 **2.** 2^3 **3.** 5^3 **4.** $(-3)^2$ **5.** $(-4)^3$ **6.** $(-1)^4$

(Example 2)

7. $12 - 4^2$ **8.** $3 \cdot 7^2$ **9.** $2 \cdot 3^2 - 10$ **10.** $4 - 3 \cdot 2^2$

11. $8^2 \div (-4)$ **12.** $-9 \div 3^3$ **13.** $-10 - 10^2$ **14.** $(-4)^2 - 3(-4)$

Evaluate each expression when $t = 4$ (Example 3)

15. $t^2 - 5$ **16.** $2 - t^2$ **17.** $-4t - 7$ **18.** $-13 + 7t$

Evaluate each expression when $y = -3$. (Example 3)

19. $2y^2$ **20.** $y^3 + 6$ **21.** $-8 - y^2$ **22.** $-2y + 13$

Evaluate each expression when $n = 5$. (Example 4)

23. $n^2 + n$ **24.** $n^3 - 7n$ **25.** $2n^2 - 5n$ **26.** $-2n^2 - 4n + 5$

Evaluate each expression when $x = -2$. (Example 4)

27. $3x - x^2$ **28.** $x^3 - 7x$ **29.** $3x^2 + 6x$ **30.** $5x^3 - x^2 - x$

WRITTEN EXERCISES

Goal: To evaluate an expression
Sample Problems: Evaluate: **a.** $5 + 9 \cdot 3^2$ **b.** $n^2 - 6$ when $n = 3$
Answers: **a.** 86 **b.** 3

Evaluate each expression in Exercises 1–14. (Example 1)

1. 11^2 **2.** 6^3 **3.** 10^3 **4.** $(-5)^2$ **5.** $(-7)^3$ **6.** $(-3)^3$

(Example 2)

7. $6(-3)^2$ **8.** $-4 \cdot (-1)^3$ **9.** $5 + 2 \cdot 4^2$ **10.** $9 + 3 \cdot 5^2$

11. $4 \cdot 2^3 - 50$ **12.** $(-5)3^2 + 37$ **13.** $5^2 - 4^2 + 3^2$ **14.** $5^2 + 2^2 - 3^2$

Evaluate each expression when $n = 3$. (Example 3)

15. $n^2 + 4$ **16.** $n^2 - 7$ **17.** $8n^2 - 12$ **18.** $-4n^2 + 9$

Evaluate each expression when $r = -1$. (Example 3)

19. $7 + r^2$ **20.** $13 - r^3$ **21.** $4r^2 - 6$ **22.** $-5r^2 + 10$

Evaluate each expression when $s = 4$ or when $w = -10$. (Example 4)

23. $s^2 - 4s$ **24.** $10s^2 - s$ **25.** $-5s^2 - s$ **26.** $s^2 + s + 1$

27. $3s^2 - 5s + 2$ **28.** $6s^2 + 3s - 5$ **29.** $s^3 + 2s^2 - 3s - 1$ **30.** $3s^3 - 4s^2 + 2s - 5$

31. $3w^2$ **32.** $-4w^2$ **33.** $2w^2 + w$ **34.** $5w - w^2$

35. $w^2 + w - 30$ **36.** $2w^2 - 4w + 20$ **37.** $w^3 + 2w^2 - w$ **38.** $3w^3 - 4w^2 - w + 60$

MIXED PRACTICE *Evaluate each expression.*

39. 4^4 **40.** $3 \cdot 2^3$ **41.** $(-6)^3$ **42.** $-2(-4)^2$

43. $2^2 + 3^2$ **44.** $4 \cdot 7 - 5^2$ **45.** $-6 \div 4^2$ **46.** $(-1)^5 - 4(-3)$

Evaluate each expression when $m = -5$ or when $t = 6$.

47. $1 - 5m$ **48.** $12m \div (-8)$ **49.** $t^3 - 2t^2 \div (-4)$ **50.** $-8t \div 4(-7)$

51. $-3t^2 - 4t$ **52.** $-4m - 6$ **53.** $2m - m^2$ **54.** $-t^2 - 5t$

55. $2t^2$ **56.** $-3t^3$ **57.** $2m^2 \div (-4)$ **58.** $-t^3 - 10t$

59. $m^2 + 3m - m$ **60.** $-5 + m^2$ **61.** $8(-4t^2)$ **62.** $m^2 - m + (-6)$

APPLICATIONS

The Example below shows how to use a calculator to evaluate a power of a number. Note that you press "[=]" once for the second power, twice for the third power, three times for the fourth power, and so on.

EXAMPLE Evaluate 16^3.

SOLUTION [1] [6] [×] [=] [=] → 4096.

On some calculators, press "[×]" twice here.

Since the power is three, press "[=]" twice.

Use a calculator to do the following exercises.

63. 7^2 **64.** 9^3 **65.** 5^4 **66.** 21^4 **67.** 7^6 **68.** $(-4)^3$ (HINT: First press "[4] [+/-].")

Methods of Computation

Making a Choice

In everyday situations, you often have to consider whether you need an **exact answer** to a problem or whether an **estimate** will give an answer that is close enough. Then you have to decide whether to use **mental computation,** a **calculator, paper and pencil,** or some combination of these to solve the problem efficiently.

EXERCISES

1. Robert went to the hardware store to buy several items for a project. He was on his way to the check-out counter when he discovered that he had exactly \$16.55 in his wallet. Robert thought: "I'll have to find the total cost to be sure I have enough money to pay for everything."

   a. Does Robert need to find the exact total or will an estimate be close enough? Explain.

   b. Which would be the more efficient way to estimate the sum, using a calculator or using mental computation? Explain.

   Robert used **front-end estimation** in this way.

   [1] Add the dollars first. $\$4 + \$1 + \$1 + \$2 + \$4 = \$12$

   [2] Estimate the cents.

   0.95 ← ***About \$1***
   0.90 ← ***About \$1***
   0.05, 0.19, 0.49 ← ***About 75¢***

   Estimate: $\$12 + \$2 + 75¢ =$ **\$14.75**

   c. How far was Robert's estimate from the actual cost?

   d. Did Robert have enough money to purchase all the items?

2. The manager of Bargain Books listed one week's sales of the store's top selling books. The books are to be ranked in order, starting with the greatest number of sales.

Book Title	Sales	Rank
Yesterday	59	?
Disbelief	43	?
Of the Heart	60	?
My Life	48	?
Remember the Day	51	?
Last Boat Home	52	?
Faithful	49	?
Springtime	44	?

a. Is an exact sales listing or an approximate sales listing appropriate? Why?

b. Do you think the manager will use paper and pencil or mental computation to do the rankings? Explain your answer.

3. The table at the right gives the scores awarded by five judges for the final dives of the top scorers at Uptown High School's swimming and diving meet. The judges use this formula to award points.

$$\text{Number of Points} = \text{Sum of Middle Three Scores} \times \text{Degree of Difficulty}$$

Diver	Degree of Difficulty	Judges' Scores	Points
Jones	2.3	7 7 8 8 9	?
Lang	2.1	7 7 8 8 8	?
Grimm	2.4	6 6 7 7 8	?
Mills	2.6	7 8 8 8 9	?
Curry	2.5	8 8 8 9 9	?

a. Should an exact answer be computed or is an estimate sufficient? Why?

b. Would it be more efficient to use paper and pencil or a calculator to award points for each dive? Why?

c. Rank the divers according to points, from highest to lowest.

d. What computation method did you use to arrange the rankings? Why?

4. Maria Stern must compute one week's earnings for five of her employees. This table shows the number of hours worked and the hourly pay rate for each employee.

Employee	Hours	Rate	Earnings
Wayne	20.5	$4.30	?
Eva	24.75	$5.20	?
Gloria	17.25	$4.20	?
Bill	18.5	$4.10	?
Carl	16.75	$3.80	?

a. Should Maria estimate each employee's earnings or is an exact answer needed? Why?

b. What method or methods of computation would you use to compute the earnings? Give reasons for your answer.

c. Find the earnings for each employee.

CHAPTER SUMMARY

IMPORTANT TERMS

Opposite *(p. 2)*
Absolute value *(p. 2)*
Reciprocal *(p. 16)*
Power *(p. 21)*
Exponent *(p. 21)*
Base *(p. 21)*
Variable *(p. 21)*
Algebraic expression *(p. 21)*

IMPORTANT IDEAS

1. The absolute value of a positive number equals that number. The absolute value of a negative number equals the opposite of that number. The absolute value of 0 equals 0.
2. Rules for Adding Positive and Negative Numbers: See pages 4 and 5.
3. The sum of opposites is 0.
4. To subtract a number, add its opposite.
5. *Multiplication Property of Zero:* $a \cdot 0 = 0 \cdot a = 0$.
6. Rules for Multiplying Positive and Negative Numbers: See pages 11–12.
7. To divide by a number, multiply by its reciprocal.
8. If a is any number, $\frac{a}{0}$ does not represent a number.
9. Rules for Dividing Positive and Negative Numbers: See page 17.
10. Rules for Order of Operations
    - **a.** Multiplication or division is performed before addition or subtraction.
    - **b.** If only addition and subtraction are involved, the operations are performed from left to right.
    - **c.** If only multiplication and division are involved, the operations are performed from left to right.
    - **d.** Powers are evaluated before multiplication and division are performed *and* before addition and subtraction are performed.

CHAPTER REVIEW

SECTION 1.1

Write the opposite of each number in simplest form.

1. -1.5 **2.** 27 **3.** $-\frac{3}{8}$ **4.** $-2\frac{1}{2}$ **5.** 0

Simplify.

6. $|24|$ **7.** $|-2.6|$ **8.** $-|-\frac{3}{8}|$ **9.** $|-35|$ **10.** $-|-9|$

Which number has the greater absolute value?

11. 0 or -3 **12.** 0 or 1 **13.** -2 or 4 **14.** -5 or 3

SECTION 1.2

Add.

15. $-13 + 24$
16. $53 + (-29)$
17. $28 + (-42)$
18. $-61 + 39$
19. $-5.8 + (-7.7)$
20. $-0.29 + (-0.58)$

21. On Thursday, the change in a stock's price was $-1\frac{7}{8}$ points. On Friday, the change was $-1\frac{3}{8}$ points. Find the total change in price for the two days.

22. In February, two new record low temperatures were set in a city. The new records were -12°F and -19°F. What is the sum of these two new record low temperatures?

SECTION 1.3

Write each difference as a sum. Then compute.

23. $17 - 29$
24. $42 - 68$
25. $-27 - (-18)$
26. $-53 - (-26)$
27. $19.2 - (-4.7)$
28. $42.9 - (-18.5)$

29. The freezing point of the coolant in an automobile's radiator was -8°F. More antifreeze was added which made the freezing point -27°F. How much was the freezing point lowered?

30. The high temperature on a December day was 4°C. The low temperature was -12°C. How much higher was the high temperature than the low temperature?

SECTION 1.4

Multiply.

31. $(12)(-17)$
32. $(-15)(21)$
33. $(-18)(-25)$
34. $(-52)(-14)$
35. $(-\frac{1}{5})(45)$
36. $(\frac{1}{4})(-28)$

37. On a winter day, the low temperature for Helena, Montana, was -4 times the low temperature of 3°F for Boise, Idaho. What was the low temperature for Helena?

38. At 11:00 P.M., the thermometer reading was -2°F. The wind chill temperature was $9\frac{1}{2}$ times the thermometer reading. What was the wind chill temperature?

SECTION 1.5

Divide.

39. $\frac{-63}{9}$
40. $\frac{84}{-7}$
41. $\frac{-144}{-12}$
42. $\frac{-128}{-8}$
43. $13 \div (-\frac{1}{4})$
44. $\frac{2}{3} \div (-12)$
45. $(-1\frac{1}{4}) \div (-1\frac{7}{8})$
46. $(-3\frac{1}{2}) \div (-2\frac{5}{8})$

Solve.

47. The deficiency in rainfall in a farming region for January through June was -8.4 inches. Find the average deficiency per month for that period.

48. Over a period of five days, the net change in the price of a company's stock was $-3\frac{1}{8}$ points. Find the average change in the stock's price per day.

SECTION 1.6

Evaluate.

49. $12 - 27 + 8$

50. $23 - 38 + 6$

51. $16 + 4(-8)$

52. $(-13) + 12 \cdot 3$

53. $15 - 42 \div (-6)$

54. $24 - (-16) \div 4$

SECTION 1.7

Evaluate.

55. 4^3

56. $(-2)^3$

57. $(6)^2 - 3(-5)$

58. $(-12)(8) \div 24 + 3^2$

59. $4(-3)^2 \div (-12) - 25$

60. $-4(3)^2 \div 12 - (-15)$

Evaluate each expression when $n = 3$.

61. $n^2 - 13$

62. $4n^2 - 45$

63. $-2n^2 - n$

64. $-n - 5n^2$

65. $n^3 - n^2 + n$

66. $n^3 - 3n^2 + 1$

67. $-n^3 + 3n^2 - 1$

68. $-n^3 + n^2 - n$

69. $-3n^3 + 2n^2 - 2n$

CHAPTER 2

Real Numbers

In this chapter you will study properties of numbers. Scientists study objects to determine if they exhibit mathematical properties. Crystals, for example, exhibit the mathematical property called *symmetry*. Similarities between crystals are revealed by their symmetry.

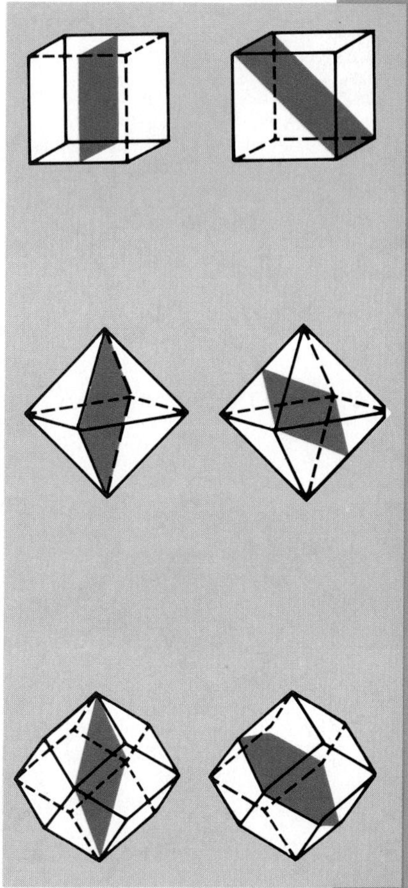

2.1 Real Numbers

The ***set of whole numbers*** consists of zero and the counting numbers.

Counting Numbers	Whole Numbers
$\{1, 2, 3, 4, 5, \cdots\}$	$\{0, 1, 2, 3, 4, 5, \cdots\}$

The three dots mean "and so on."

The ***set of integers*** consists of the set of whole numbers and their opposites. For example, 2 and −2 are opposites.

$$\{\cdots, -3, -2, -1, 0, 1, 2, 3, \cdots\}$$

P–1 **What is the opposite of each of the following?**

a. −2 **b.** 6 **c.** 0 **d.** 213

The number $-2\frac{1}{2}$ is a rational number. Here are other names for $-2\frac{1}{2}$.

$$-\frac{5}{2} \qquad \frac{-5}{2} \qquad \frac{5}{-2} \qquad \frac{-10}{4} \qquad \frac{20}{-8} \qquad -2.5 \qquad -2\frac{5}{10}$$

Definition

A ***rational number*** is a number that can be expressed in the form $\frac{a}{b}$, where a and b are integers, but b is not 0.

$-5 = -\frac{5}{1}$

$3\frac{1}{3} = \frac{10}{3}$

EXAMPLE 1 Write each number as a quotient of two integers in two ways.

a. 4 **b.** 0 **c.** $3\frac{3}{4}$ **d.** $-\frac{5}{8}$ **e.** 0.3

Solutions: **a.** $\frac{4}{1}; \frac{8}{2}$ **b.** $\frac{0}{1}; \frac{0}{5}$ **c.** $\frac{15}{4}; \frac{45}{12}$ **d.** $\frac{-5}{8}$ or $\frac{5}{-8}; \frac{-10}{16}$ **e.** $\frac{3}{10}; \frac{30}{100}$

(Other solutions are possible.)

Rational numbers can also be expressed as repeating decimals. Also, all repeating decimals are names for rational numbers.

EXAMPLE 2 Express $\frac{3}{11}$ as a repeating decimal.

Solution: $\frac{3}{11}$ means $3 \div 11$.

$$11\overline{)3.0000\cdots} = 0.2727\cdots$$

A repeating decimal has a digit or sequence of digits that is infinitely repeated.

P–2 **Express each rational number as a repeating decimal.**

a. $\frac{2}{9}$ **b.** $\frac{4}{11}$ **c.** $\frac{1}{6}$

The repeating decimal 0.2727 · · · can be abbreviated as $0.\overline{27}$. The bar indicates the digits that repeat.

The decimal form of many rational numbers is a terminating decimal. For example, $\frac{1}{4} = 0.25$, $-3\frac{1}{2} = -3.5$, and $1\frac{7}{8} = 1.875$. However, you can consider 0 as the repeating digit for these numbers.

$$\frac{1}{4} = 0.25000 \cdots \qquad -3\frac{1}{2} = -3.5000 \cdots \qquad 1\frac{7}{8} = 1.875000 \cdots$$

EXAMPLE 3

Choose from the box the set of numbers that corresponds to each word description.

a. The attendance at a concert

b. The scale on a weather thermometer

c. The rise and fall of stock prices

Rational numbers
Counting numbers
Integers

Solutions:

a. Counting numbers	As their name implies, they are used for counting.
b. Integers	A weather thermometer's scale includes 0, negative numbers, and positive numbers.
c. Rational numbers	The changes in stock prices are expressed as positive and negative fractions.

Positive numbers such as 1, 4, 9, 16, 25, and 36 are ***perfect squares,*** because each is the square of a counting number.

$$1^2 = 1 \qquad 2^2 = 4 \qquad 3^2 = 9 \qquad 4^2 = 16 \qquad 5^2 = 25 \qquad 6^2 = 36$$

Thus, square roots of perfect squares are rational.

$$\sqrt{1} = 1 \qquad \sqrt{4} = 2 \qquad \sqrt{9} = 3 \qquad \sqrt{16} = 4 \qquad \sqrt{25} = 5 \qquad \cdots$$

Square roots of counting numbers that are not perfect squares are irrational.

An ***irrational number*** cannot be expressed as a quotient of integers. Thus, the decimal form of an irrational number is nonrepeating.

$$\sqrt{2} = 1.4142135 \cdots \qquad \sqrt{3} = 1.7320508 \cdots \qquad \pi = 3.1415926 \cdots$$

EXAMPLE 4 Write Rational or Irrational to describe each number.

a. 12.326326 · · · b. 0.050050005 · · ·
c. $\sqrt{14}$ d. $-\sqrt{9}$

Solutions: **a.** Rational, because 326 repeats.
b. Irrational, because the decimal is nonrepeating.
c. Irrational, because 14 is not a perfect square.
d. Rational, because 9 is a perfect square: $-\sqrt{9}=-3$.

P–3 **Describe each number as Rational or Irrational.**

a. 0.121314 · · · **b.** $\sqrt{24}$ **c.** 4.4444 · · ·

Definition The ***set of real numbers*** contains all the rational numbers and all the irrational numbers.

EXAMPLE 5 Refer to the set below to list the numbers named.

$$\{-\frac{13}{4}, -\frac{12}{4}, -2.7, -\sqrt{2}, 0, \sqrt{3}, \pi, 4.9, \sqrt{49}\}$$

a. Integers **b.** Rational numbers
c. Irrational numbers **d.** Real numbers

Solutions: **a.** $-\frac{12}{4}, 0, \sqrt{49}$ **b.** $-\frac{13}{4}, -\frac{12}{4}, -2.7, 0, 4.9, \sqrt{49}$
c. $-\sqrt{2}, \sqrt{3}, \pi$ **d.** All are real numbers.

CLASSROOM EXERCISES

Write each number as a quotient of two integers in two ways. (Example 1)

1. $3\frac{1}{2}$ **2.** -15 **3.** 0 **4.** -0.9 **5.** $-\frac{1}{4}$ **6.** $\sqrt{9}$ **7.** $-\sqrt{25}$

Write each rational number as a repeating decimal. (Example 2)

8. 5 **9.** $\frac{1}{2}$ **10.** $\frac{1}{3}$ **11.** $\frac{4}{7}$ **12.** $\frac{1}{9}$ **13.** $-\frac{3}{8}$ **14.** $-1\frac{2}{5}$

For Exercises 15–17, choose from the box the set of numbers that corresponds to each word description. (Example 3)

Positive rational numbers
Counting numbers
Integers

15. The profit or loss in dollars of a business during one year
16. The counter on a service station pump showing the number of gallons purchased
17. The page numbers in your textbook

Describe each number as Rational or Irrational. (Example 4)

18. 1.123123 · · · **19.** 6.121231234 · · · **20.** 19 **21.** -16

Refer to the set of numbers below. Then list the numbers named in each exercise. (Example 5)

$\{28, -1.8, 3\frac{1}{4}, 0.1010010001\cdots, -124, 17, -\frac{22}{7}, 5.436436436\cdots\}$

22. Whole numbers **23.** Integers **24.** Negative numbers
25. Real numbers **26.** Rational numbers **27.** Irrational numbers

WRITTEN EXERCISES

Goal: To identify rational and irrational numbers
Sample Problem: List the rational and irrational numbers in

$\{\frac{4}{2}, \sqrt{16}, -3.5, -\sqrt{27}, 5.222\cdots, \sqrt{33}\}$.

Answer: Rationals: $\frac{4}{2}, \sqrt{16}, -3.5, 5.222\cdots$; Irrationals: $-\sqrt{27}, \sqrt{33}$

Write each number as a quotient of two integers in two ways. (Example 1)

1. $-\frac{7}{8}$ **2.** $-\frac{13}{16}$ **3.** $4\frac{3}{8}$ **4.** $-5\frac{3}{4}$ **5.** -1.83
6. 3.47 **7.** 20 **8.** -46 **9.** $-\sqrt{9}$ **10.** $-\sqrt{64}$

Write each rational number as a repeating decimal. (Example 2)

11. -24 **12.** $\frac{5}{9}$ **13.** $\frac{7}{11}$ **14.** $-3\frac{5}{8}$ **15.** $\frac{3}{7}$ **16.** $\frac{11}{12}$

For Exercises 17–19, choose from the box the set of numbers that corresponds to each word description. (Example 3)

Positive rational numbers
Whole numbers
Integers

17. A table of record high and low temperatures expressed to the nearest degree
18. The odometer on a car
19. The attendance at a basketball game

Write Rational or Irrational to describe each number. (Example 4)

20. -3.7 **21.** 7.9 **22.** $\sqrt{29}$ **23.** $-\sqrt{13}$

24. $-1\frac{5}{8}$ **25.** $3\frac{3}{4}$ **26.** $\sqrt{49}$ **27.** $-\sqrt{36}$

28. $-\pi$ **29.** $-7.1000\cdots$ **30.** $0.24681012\cdots$ **31.** $-0.3691215\cdots$

Refer to the set of numbers below. Then list the numbers named in Exercises 32–37. (Example 5)

$$\{\frac{10}{2}, -2, 4.7, 0, \pi, -1\frac{3}{4}, \frac{7}{8}, 19, -\frac{17}{4}, -6, -0.9\}$$

32. Rational numbers **33.** Irrational numbers **34.** Negative numbers

35. Positive integers **36.** Positive irrational numbers **37.** Real numbers

Select one or more letters below to describe each of the following numbers. (Examples 4 and 5)

a. Whole numbers **b.** Integers **c.** Rational numbers
d. Irrational numbers **e.** Real numbers **f.** Counting numbers

38. 7 **39.** -4 **40.** $\frac{5}{8}$ **41.** -20 **42.** -1.7 **43.** $7\frac{1}{5}$

44. $-6\frac{1}{9}$ **45.** $\frac{22}{7}$ **46.** $1.4238\cdots$ **47.** $-\frac{28}{7}$ **48.** 101 **49.** -8.0

REVIEW CAPSULE FOR SECTION 2.2

Rewrite the following as sums. (Pages 8–10)

1. $6-4$ **2.** $7-6$ **3.** $15-8$ **4.** $4-9$

5. $5-12$ **6.** $23-19$ **7.** $14-18$ **8.** $2-25$

Add. (Pages 4–7)

9. $4+13$ **10.** $8+9$ **11.** $\frac{5}{16}+(-\frac{1}{4})$ **12.** $\frac{1}{12}+(-\frac{2}{5})$

13. $1.3+(-0.2)$ **14.** $2.3+(-1.1)$ **15.** $-4+(-5)$ **16.** $-8+(-1)$

17. $-\frac{1}{3}+\frac{2}{9}$ **18.** $\frac{1}{2}+3\frac{2}{3}$ **19.** $-1.5+(-5.6)$ **20.** $0.8+(-5.9)$

Subtract. (Pages 8–10)

21. $5-3$ **22.** $12-9$ **23.** $\frac{1}{3}-\frac{1}{4}$ **24.** $\frac{3}{4}-\frac{2}{15}$

25. $-0.6-1.2$ **26.** $-0.2-0.7$ **27.** $-4-(-5)$ **28.** $-6-(-1)$

29. $1\frac{2}{3}-\frac{3}{4}$ **30.** $-\frac{5}{6}-(-\frac{11}{12})$ **31.** $0.5-3.1$ **32.** $2.4-(-8.3)$

2.2 Addition Properties

The following properties apply to addition of real numbers.

Table

Addition Property of Zero

For any real number *a*,

$$a + 0 = a \text{ and } 0 + a = a.$$

$-3.4 + 0 = -3.4$

$0 + 6\frac{1}{2} = 6\frac{1}{2}$

Addition Property of Opposites

For any real number *a*,

$$a + (-a) = 0.$$

$9.1 + (-9.1) = 0$

$8\frac{1}{3} + (-8\frac{1}{3}) = 0$

Commutative Property of Addition

Any two real numbers can be added in either order.
For any real numbers *a* and *b*,

$$a + b = b + a.$$

$\left(-\frac{3}{4}\right) + 7 = 7 + \left(-\frac{3}{4}\right)$

$y + 5 = 5 + y$

Associative Property of Addition

The way real numbers are grouped for addition does not affect their sum.
For any real numbers *a*, *b*, and *c*,

$$(a + b) + c = a + (b + c).$$

$\left(9 + \frac{7}{10}\right) + \frac{3}{10} = 9 + \left(\frac{7}{10} + \frac{3}{10}\right)$

$(6 + x) + x = 6 + (x + x)$

EXAMPLE 1

An equation is a sentence that contains the equality symbol "=."

Find a value of the variable that will make each equation true. Name the property that gives the reason for your choice.

a. $2.9 + x = 0$ **b.** $3 + n = 8\frac{1}{2} + 3$

c. $y + 0 = 4$ **d.** $(35 + 67) + 72 = 35 + (67 + a)$

Solutions:

	Equation	Value	Property
a.	$2.9 + x = 0$	$x = -2.9$	Addition Property of Opposites
b.	$3 + n = 8\frac{1}{2} + 3$	$n = 8\frac{1}{2}$	Commutative Property
c.	$y + 0 = 4$	$y = 4$	Addition Property of Zero
d.	$(35 + 67) + 72 = 35 + (67 + a)$	$a = 72$	Associative Property

P–1 **What value of y makes each equation true? by which property?**

a. $8.2 + 7.1 = 7.1 + y$ **b.** $(7 + 3\frac{1}{2}) + y = 7 + (3\frac{1}{2} + 6)$

The addition properties can be used to simplify expressions. The first step is to write the expression as a sum.

EXAMPLE 2 Simplify: $-12 + 19 - 13 + 15$

Solution: Use the Commutative and Associative Properties.

1 Write as a sum. ⟶ $-12 + 19 - 13 + 15 = -12 + 19 + (-13) + 15$

2 Regroup. ⟶ $= [-12 + (-13)] + (19 + 15)$

3 Add. ⟶ $= -25 + 34$

$= 9$

EXAMPLE 3 Simplify: $12 + t - 17$

Solution:

1 Write as a sum. ⟶ $12 + t - 17 = 12 + t + (-17)$

2 Regroup. ⟶ $= t + [12 + (-17)]$ ◀ ***By the Commutative and Associative Properties***

3 Simplify. ⟶ $= t + (-5)$

$= t - 5$

EXAMPLE 4 Simplify: $5.2 + x - 8.3 + y$

Solution: $5.2 + x - 8.3 + y = [5.2 + (-8.3)] + x + y$

$= -3.1 + x + y$

P–2 **Simplify each of the following.**

a. $3 + 4 - 7 + 9$ **b.** $9 - 2\frac{1}{2} - 7\frac{3}{8}$ **c.** $1\frac{1}{2} + y - \frac{2}{3}$ **d.** $3.2 + b - 4.0$

CLASSROOM EXERCISES

Add. (Table)

1. $6 + (-6)$ **2.** $-12 + 0$ **3.** $4\frac{1}{3} + 0$ **4.** $(-\frac{1}{3}) + \frac{1}{3}$

5. $0 + (-0.7)$ **6.** $(-0.4) + 0.4$ **7.** $1\frac{1}{2} + (-1\frac{1}{2})$ **8.** $0 + \frac{3}{8}$

Write the value of the variable that makes each equation true. (Example 1)

9. $k = 4 + (-4)$

10. $3 + (4 + b) = (3 + 4) + 5$

11. $3\frac{1}{2} + (4 + \frac{1}{3}) = (a + 4) + \frac{1}{3}$

12. $6\frac{1}{2} + 0 = n$

13. $1.4 + x = 0.6 + 1.4$

14. $4.3 + d = 4.3$

Simplify. (Example 2)

15. $4 - 5 - 2 + 6$

16. $-11 + 2 - 7 + 3$

17. $-3\frac{1}{2} + 1\frac{2}{3} - \frac{1}{6} + 2$

18. $4\frac{1}{5} - 5\frac{8}{10} + 2\frac{3}{10} - \frac{1}{2}$

(Example 3)

19. $31 - a + 23$

20. $-d + 21 - 32$

21. $2.6 + x - 0.5$

22. $y - 3.2 + 1.7$

(Example 4)

23. $x - 4 + y - 2$

24. $-z + 6 - 9 - p$

25. $12.0 - a - b - 6.9$

26. $-3.9 - 9.3 - m - k$

WRITTEN EXERCISES

Goal: To simplify expressions using the addition properties
Sample Problem: $-3\frac{1}{3} - a + \frac{2}{3} + c$ **Answer:** $-a + c - 2\frac{2}{3}$

Write the value of the variable that makes each equation true. Name the property that gives the reason for your choice. (Example 1)

1. $8 + r = 0$

2. $0 + 6 = t$

3. $15 + 7 = x + 15$

4. $y + 0 = -12$

5. $4\frac{1}{2} + (-4\frac{1}{2}) = b$

6. $(a + \frac{3}{8}) + 1\frac{1}{2} = \frac{2}{3} + (\frac{3}{8} + 1\frac{1}{2})$

Simplify. (Example 2)

7. $7 + 5 - 13 + 6$

8. $12 - 10 + 23 - 15$

9. $\frac{1}{2} - \frac{1}{4} + 2 - \frac{3}{4}$

10. $-3\frac{5}{8} + \frac{1}{4} - \frac{1}{8} - \frac{3}{8}$

11. $2.5 - 1.2 - 0.8 + 2.0$

12. $-0.3 - 5.9 + 2.8 + 12.9$

13. $2.7 + 3.6 - 4.8 + 0.2$

14. $-8.9 - 2.6 + 11.2 - 1.2$

(Example 3)

15. $5 - 1 - 2 - u$

16. $s - 16 - 0 + 9$

17. $1\frac{7}{8} + p - \frac{1}{2} - \frac{3}{8}$

18. $4\frac{1}{4} - y - 7\frac{1}{3} - 5\frac{3}{4}$

19. $-a - 0.5 + 0.2$

20. $3.8 + b - 6.4$

21. $-3.9 + f - 0.5$

22. $-r + 6.8 - 8.1$

(Example 4)

23. $6 + k - 23 - b$

24. $-d + 56 - 39 + w$

25. $d - 2\frac{1}{2} + \frac{3}{8} - f$

26. $h - 0.7 + 2.4 - m$

27. $a - 5 + 14 - b$

28. $r - 14 - g - 16$

29. $s + 5.9 - t - 5.9$

30. $-19.1 + k + 19.1 - x$

MIXED PRACTICE

31. $16 + 34 - 21 + 10$

32. $16 - 35 + 17 + 35$

33. $3\frac{1}{2} - 5\frac{3}{4} + 2\frac{1}{4} - 4\frac{1}{2}$

34. $4\frac{1}{4} - 6\frac{1}{2} - 2\frac{3}{4} + 1\frac{1}{4}$

35. $3 + 6.1 - 0.8 + 2.3$

36. $0.9 - 1.3 - 4.8 + 3.7$

37. $-18 - y + 18$

38. $31 + x - 17$

39. $-63 + 15 + q - 34$

40. $44 + j - 17 + w$

41. $c - 7\frac{5}{6} - d - 3\frac{2}{3}$

42. $7\frac{5}{8} + a - b - 13\frac{1}{4}$

NON-ROUTINE PROBLEMS

43. Steve forgot his locker number at school. The school secretary reminds him that the digits in his locker number are all odd numbers and that, when added, they total 11. There are 120 lockers. What is Steve's locker number?

44. In the addition problem below, the letters P, K, and N represent three different digits. Identify the digits.

$$\begin{array}{r} P\ K \\ +\ \ K\ N \\ \hline K\ N\ K \end{array}$$

REVIEW CAPSULE FOR SECTION 2.3

Multiply. (Pages 11–14)

1. $(5)(-6)$
2. $3 \cdot 15$
3. $(\frac{1}{3})(-\frac{3}{4})$
4. $\frac{1}{5} \cdot \frac{6}{5}$
5. $-0.4 \cdot 6.2$
6. $(-1.8)(-3.9)$
7. $(-14)(5)$
8. $(-8)(-11)$
9. $(-\frac{3}{8})(-1\frac{2}{3})$
10. $-1\frac{1}{2} \cdot 3\frac{1}{5}$
11. $1.9 \cdot 2.6$
12. $6.8(-5.3)$

Write the reciprocal of the following. (Pages 16–18)

13. 5
14. $\frac{2}{5}$
15. -4
16. $\frac{1}{3}$
17. $-\frac{3}{8}$
18. 10
19. 1
20. $-\frac{1}{12}$
21. -13
22. 11
23. -1
24. $\frac{8}{5}$

2.3 Multiplication Properties

The following properties apply to multiplication of real numbers.

Table

Multiplication Property of One

For any real number *a*,

$$a \cdot 1 = a \text{ and } 1 \cdot a = a.$$

$(-4)(1) = -4$

$1(0.3x) = 0.3x$

Multiplication Property of Reciprocals

The product of any nonzero real number and its reciprocal is 1.

$$a \cdot \frac{1}{a} = 1, a \neq 0$$

$\frac{3}{5} \cdot \frac{5}{3} = 1$

$(-7)(-\frac{1}{7}) = 1$

Commutative Property of Multiplication

Any two real numbers can be multiplied in either order.
For any real numbers *a* and *b*,

$$a \cdot b = b \cdot a.$$

$(-3)(2.5) = (2.5)(-3)$

$(4\frac{1}{5})(-8) = (-8)(4\frac{1}{5})$

Associative Property of Multiplication

The way real numbers are grouped for multiplication does not affect their product.
For any real numbers *a*, *b*, and *c*,

$$(ab)c = a(bc).$$

$(9 \cdot \frac{2}{3})5 = 9(\frac{2}{3} \cdot 5)$

$5.2(3x) = (5.2 \cdot 3)x$

EXAMPLE 1 Find the value of the variable that makes each equation true. Name the property that gives the reason for your choice.

a. $\frac{1}{3} \cdot y = 1$ **b.** $(4.1) \cdot m = (-3.5)(4.1)$

c. $n \cdot 1 = 6$ **d.** $(\frac{2}{3}d)4\frac{1}{2} = \frac{2}{3}(5 \cdot 4\frac{1}{2})$

Solutions:

	Equation	Value	Property
a.	$\frac{1}{3} \cdot y = 1$	$y = \frac{3}{1}$, or 3	Multiplication Property of Reciprocals
b.	$(4.1) \cdot m = (-3.5)(4.1)$	$m = -3.5$	Commutative Property
c.	$n \cdot 1 = 6$	$n = 6$	Multiplication Property of One
d.	$(\frac{2}{3}d)4\frac{1}{2} = \frac{2}{3}(5 \cdot 4\frac{1}{2})$	$d = 5$	Associative Property

P–1 **What value of x makes each equation true? by which property?**

a. $(8.1)(6.3) = (6.3)x$ **b.** $(7 \cdot 9)\frac{1}{2} = x(9 \cdot \frac{1}{2})$ **c.** $(-9)x = 1$

The multiplication properties can be used to simplify expressions.

EXAMPLE 2 Multiply: $(-\frac{1}{2})(\frac{1}{4})(-4)(-12)$

Solution: Use the Associative and Commutative Properties.

1 Rewrite the expression ⟶ $(-\frac{1}{2})(\frac{1}{4})(-4)(-12) = (-\frac{1}{2})(-12)(\frac{1}{4})(-4)$

2 Multiply. ⟶ $(-\frac{1}{2})(-12)(\frac{1}{4})(-4) = (6)(-1)$

$= -6$

EXAMPLE 3 Multiply: $(9)(x)(-\frac{1}{3})(y)$.

Solution:

1 $(9)(x)(-\frac{1}{3})(y) = (9)(-\frac{1}{3})(x)(y)$ ◀ ***By the Commutative and Associative Properties***

2 $(9)(-\frac{1}{3})(x)(y) = (-3)(x \cdot y)$

$= -3xy$

P–2 **Multiply in each expression.**

a. $(0.6)(3.2)(-5)$ **b.** $(\frac{1}{8})(n)(24)(16)$ **c.** $(-25)(p)(q)(-4)$

CLASSROOM EXERCISES

Multiply. (Table)

1. $(\frac{1}{4})(4)$ **2.** $-16(-\frac{1}{16})$ **3.** $62 \cdot 1$ **4.** $(1)(-23)$

5. $(\frac{10}{9})(\frac{9}{10})$ **6.** $(-\frac{2}{3})(-\frac{3}{2})$ **7.** $1(17.8)$ **8.** $(0.03)(1)$

Write the value of the variable that makes each equation true. (Example 1)

9. $(12.8)(3.04) = n(12.8)$ **10.** $(1)(-0.24) = d$ **11.** $7(16) = y(7)$

12. $(17 \cdot 3)b = 17(3 \cdot 5)$ **13.** $54 \cdot 5 = 5q$ **14.** $(4.1 \cdot 0.2)a = 4.1(0.2 \cdot 5)$

15. $\frac{3}{8}(\frac{1}{3} \cdot k) = (\frac{3}{8} \cdot \frac{1}{3})\frac{4}{5}$ **16.** $1 = \frac{5}{12} \cdot z$ **17.** $\frac{2}{5}(\frac{5}{7} \cdot m) = (\frac{2}{5} \cdot \frac{5}{7})\frac{1}{2}$

Multiply. (Examples 2 and 3)

18. $(4)(17)(5)(-3)$

19. $(-3)(19)(5)(-2)$

20. $(\frac{1}{3})(\frac{4}{5})(-6)(-\frac{5}{4})$

21. $(-\frac{3}{2})(-\frac{1}{8})(48)(-\frac{4}{3})$

22. $(14)(-4)(9)(-11)$

23. $(35)(\frac{1}{7})(10)(\frac{2}{5})$

24. $(a)(0.25)(b)(0.4)$

25. $(1.6)(y)(-0.5)(z)$

26. $(-j)(-3)(6)(k)$

27. $(-12)(m)(-n)(-10)$

28. $(x)(-6.1)(1.3)(y)$

29. $(-b)(1.8)(h)(0.6)$

WRITTEN EXERCISES

Goal: To multiply in expressions using the multiplication properties
Sample Problem: $(-\frac{5}{8})(y)(\frac{8}{5})(-z)$ **Answer:** yz

Write the value of the variable that makes each equation true. Name the property that gives the reason for your choice. (Example 1)

1. $-\frac{5}{7} = f \cdot 1$

2. $\frac{6}{11} \cdot \frac{11}{6} = h$

3. $(-2.3 \cdot 6.4)8 = -2.3(6.4 \cdot s)$

4. $1 \cdot (-9) = x$

5. $a \cdot 3 = 3 \cdot (-\frac{1}{2})$

6. $(-6.8)(3.5) = m(-6.8)$

Multiply. (Examples 2 and 3)

7. $(-2)(-5)4$

8. $(-1)(-8)(-3)$

9. $(-\frac{1}{3})(12)(-9)$

10. $(-4)(-\frac{3}{2})(-\frac{2}{3})(-\frac{1}{4})$

11. $(0.6)(-p)(0.7)(-q)$

12. $(-3.5)(c)(-d)(-40)$

13. $(-8)(18)(m)(-2)$

14. $(d)(4)(39)(f)(2)$

15. $(-3)(-3)(t)(-3)$

16. $(5)(p)(-6)(-2)$

17. $(1.8)(m)(p)(-3)$

18. $(-0.3)(a)(-0.7)(5)(b)$

19. $(-5)(x)(9)(y)(-5)$

20. $(b)(4)(-3)(c)(-5)$

21. $(1\frac{3}{10})(\frac{5}{6})(-r)(\frac{6}{5})(d)$

22. $(-\frac{1}{8})(-y)(\frac{3}{5})(-8)(m)$

23. $(w)(16)(-5)(\frac{1}{8})(z)$

24. $(-k)(14)(12)(-\frac{1}{7})(t)$

MORE CHALLENGING EXERCISES

For Exercises 25–28, a represents a positive integer and b represents a negative integer. Identify whether the value of each expression is positive, p, or negative, n.

25. $(-5)(b)(a)(a)$

26. $(-3)(a)(b)(-2)$

27. $(b)(-4)(-1)(b)(-3)$

28. $(2)(a)(b)(b)(b)$

MID-CHAPTER REVIEW

Write each number as a quotient of two integers in two ways. (Section 2.1)

1. $-\frac{5}{6}$ **2.** $-\frac{14}{17}$ **3.** $2\frac{4}{5}$ **4.** $-3\frac{1}{7}$ **5.** 3.5 **6.** −4.8

Write each rational number as a repeating decimal. (Section 2.1)

7. $\frac{4}{9}$ **8.** $\frac{2}{3}$ **9.** $\frac{8}{11}$ **10.** $-\frac{4}{7}$ **11.** $5\frac{1}{4}$ **12.** $-6\frac{1}{7}$

Write Rational or Irrational to describe each number. (Section 2.1)

13. −2.8 **14.** 8.1 **15.** 31 **16.** $\sqrt{81}$

17. $-4\frac{2}{3}$ **18.** $\frac{\pi}{4}$ **19.** $6.13000\cdots$ **20.** $2.323323332\cdots$

Refer to the set of numbers below. Then list the numbers named in each exercise. (Section 2.1)

$$\{\tfrac{2}{3}, -\sqrt{5}, -1\tfrac{2}{7}, \sqrt{36}, -9.3\}$$

21. Integers **22.** Irrational numbers **23.** Real numbers

Simplify. (Section 2.2)

24. $18 + x - 18$ **25.** $-32 - y + 32$ **26.** $-43 + w - 82 + 37$

27. $y - 12 + 8 + z$ **28.** $-12\frac{3}{4} + k + 7\frac{1}{2} - j$ **29.** $5\frac{3}{8} + r - 9\frac{3}{4} - s$

30. $t - 0.25 - w - 0.75$ **31.** $-6.3 - a + b - 9.2$ **32.** $-8.4 - 2.5 - 0.6 - 1.3$

Multiply. (Section 2.3)

33. $(6)(-2)(-\frac{1}{3})$ **34.** $(-\frac{1}{4})(-24)(8)$ **35.** $(-2)(-3)(h)(-\frac{1}{2})$

36. $(5)(-3)(x)(\frac{1}{5})$ **37.** $(-40)(-k)(-10)(4)$ **38.** $(n)(-2)(-2)(-2)(-2)$

39. $(6.3)(a)(-1.4)(y)$ **40.** $(2.2)(-0.4)(d)(-s)$ **41.** $(p)(8)(-3)(f)$.

REVIEW CAPSULE FOR SECTION 2.4

Simplify.

1. $(-3 \cdot 4)(x \cdot y \cdot y)$ **2.** $(5 \cdot 7)(a \cdot a \cdot b \cdot b)$ **3.** $(9 \cdot \frac{1}{3})(p^2 \cdot q \cdot r)$

4. $(-\frac{1}{5} \cdot 10)(m \cdot n^2 \cdot s)$ **5.** $x + (-8)$ **6.** $-t + (-1)$

7. $(1.3 \cdot 5)(f^2 \cdot g \cdot t^2)$ **8.** $w + (-16)$ **9.** $-22 + (-p)$

2.4 Special Properties

The following special properties are helpful when simplifying an expression that contains parentheses.

Opposite of a Sum

$$-(a+b)=(-a)+(-b) \qquad -(4+x)=(-4)+(-x)$$
$$=-a-b \qquad =-4-x$$

Multiplication Property of −1

$$-1(a)=-a \qquad -1(7.2)=-7.2;\ -1(-y)=y$$
$$-1(a+b)=-a-b \qquad -1(4+x)=-4-x;\ -1(-t-s)=t+s$$

Opposites and Products

$$(-a)b=-ab \qquad (-6)(m)=-6m;\ (8)(-z)=-8z$$
$$(-a)(-b)=ab \qquad (-4)(-y)=4y;\ -(5x)(-y)=5xy$$

Subtraction of a Sum or Difference

$$a-(b+c)=a-b-c \qquad 5-(m+q)=5-m-q$$
$$a-(b-c)=a-b+c \qquad 6-(c-2)=6-c+2$$

Note that the first two special properties above are equivalent. That is,

$$-(a+b)=-1(a+b).$$

Finding the opposite of a sum or difference is the same as multiplying the sum or difference by −1.

P–1 **Simplify.**

a. $-1(x+2)$ **b.** $-(4-t)$ **c.** $-1(\sqrt{2})$ **d.** $(-m)(-k)$

EXAMPLE 1 Simplify: $(-rs)(5rt)$

Solution:

1 Use the Multiplication Property of −1. ⟶ $(-rs)(5rt)=(-1rs)(5rt)$

2 Use the Commutative and Associative Properties. ⟶ $=(-1\cdot 5)(r\cdot r\cdot s\cdot t)$

3 Multiply. ⟶ $=(-5)(r^2st)$

$=-5r^2st$

EXAMPLE 2 Simplify: $(-2a)(-a)(-x)(-\frac{1}{2}x)$

Solution:

1 Multiply from the left to right. ⟶ $(-2a)(-a)(-x)(-\frac{1}{2})x = (2a^2)(\frac{1}{2}x^2)$

2 Use the Commutative and Associative Properties. ⟶ $= (2 \cdot \frac{1}{2})(a^2 \cdot x^2)$

3 Use the Multiplication Property of 1. ⟶ $= 1 \cdot a^2x^2$

$= a^2x^2$

In the following special properties, note that no divisor is 0.

Special Properties of Division

$\frac{a}{1} = a$ $\qquad \frac{7qr}{1} = 7qr$

$\frac{a}{a} = 1, a \neq 0$ $\qquad \frac{-3s+4t}{-3s+4t} = 1$

Opposites and Quotients

$\frac{-a}{b} = -\frac{a}{b}; \frac{a}{-b} = -\frac{a}{b}; b \neq 0$ $\qquad \frac{-2k}{m} = -\frac{2k}{m}; \frac{7y}{-9x} = -\frac{7y}{9x}$

$\frac{-a}{-b} = \frac{a}{b}, b \neq 0$ $\qquad \frac{-3c}{-5d^2} = \frac{3c}{5d^2}$

P–2 Simplify.

a. $\frac{-5k}{7}$ **b.** $\frac{-3cd}{-3cd}$ **c.** $\frac{6t+4m}{6t+4m}$ **d.** $\frac{2xy}{-2xy}$

EXAMPLE 3 Simplify: $\frac{-(4-t)}{t-4}, t \neq 4$

If t is not 4, then t − 4 is not zero.

Solution:

1 Simplify the numerator. ⟶ $\frac{-(4-t)}{t-4} = \frac{-4+t}{t-4}$

2 Use the Commutative Property. ⟶ $= \frac{t+(-4)}{t-4}$

3 Simplify. ⟶ $= \frac{t-4}{t-4} = 1$

CLASSROOM EXERCISES

Simplify. (Examples 1–3)

1. $3(-k)$
2. $-1(-x)$
3. $(-8)(-6y)$
4. $(-5p)(-3)$
5. $-(x + y)$
6. $-(-x - y)$
7. $(x - 2) - (y - a)$
8. $a - (6 - b)$
9. $(-20h)(\frac{1}{5})$
10. $(-\frac{1}{2})(-a)$
11. $(-x)(2y)$
12. $(xy)(-7)$
13. $(-xy)(3x)$
14. $(-a)(-a)$
15. $(-y)(-y)(z)(-z)$
16. $(-a)(-b)(-c)$
17. $(-\frac{2}{3}c)(-3ac)(-a)$
18. $(\frac{1}{5})(-\frac{7}{8}g)(10h)(-4g)$
19. $(-m)(-4n)(-4n)(m)$
20. $(2r)(-3s)(-r)(-6)$
21. $(-9) \div f,\ f \neq 0$
22. $\frac{a}{-b}$
23. $\frac{-cd}{-cd}$
24. $\frac{9a}{-9a}$
25. $\frac{x-7}{x-7},\ x \neq 7$
26. $\frac{-(a-b)}{b-a},\ b \neq a$
27. $\frac{16-c}{-(c-16)},\ c \neq 16$

WRITTEN EXERCISES

Goal: To simplify expressions using special properties

Sample Problems. **a.** $-6r - (r + 3)$ **b.** $(-3.2)(-x)(y)$

c. $\frac{2a-5}{-(5-2a)},\ a \neq 2\frac{1}{2}$

Answers: **a.** $-7r - 3$ **b.** $3.2xy$ **c.** 1

Simplify. (Examples 1–3)

1. $(-a)(-c)$
2. $x - (y + 10)$
3. $2t - (r - 1)$
4. $(-1)(f)$
5. $(-d)(4)$
6. $-(z + 11)$
7. $(m + n)(-1)$
8. $5d - (6c - 7)$
9. $2a - (11 + b)$
10. $(8a)(-ab)$
11. $(-4c)(-4d)$
12. $(-\frac{1}{3}t)(-15)$
13. $(-16)(\frac{1}{4}q)$
14. $(-2.3x)(-2y)$
15. $(-5a)(1.8a)$
16. $(-3s)(-t)(s)(2t)$
17. $(2x)(-y)(-x)(3y)$
18. $(-\frac{1}{2}xy)(-\frac{1}{3}xy)(3)$
19. $(4b)(-\frac{2}{3}ab)(-\frac{1}{2}a)$
20. $(-5ab)(1.8ac)(-\frac{1}{5}b)$
21. $(1.3r)(-0.6s)(2rs)$
22. $\frac{-7}{a}$
23. $\frac{3}{-p}$
24. $\frac{-4f}{-4f}$
25. $\frac{bc}{bc}$
26. $\frac{-(h+3)}{k}$
27. $\frac{-(2+y)}{x}$

Simplify. (Examples 1–3)

28. $\frac{-(2-x)}{x-2}, x \neq 2$

29. $\frac{f-g}{f-g}, f \neq g$

30. $\frac{a-7}{-(a+7)}, a \neq 7$

31. $\frac{-(k-g)}{-(k-g)}, k \neq g$

32. $(\frac{1}{2}xy)(-4x)(-y)$

33. $(-7)(-fg)$

34. $\frac{abc}{abc}$

35. $\frac{mp}{-(mp)}$

36. $(-1)(0.3c+4d)$

37. $(1.3r)(-0.6r)$

38. $(-3a)(5b)(\frac{1}{10}a)(-4b)$

39. $(\frac{1}{5}x)(\frac{2}{3}xy)$

40. $(p-q)-(r-s)$

41. $12-(r-3s+15)$

42. $-(18+t)$

43. $-(-r+2)$

44. $\frac{-2b+3}{-(-3+2b)}, b \neq \frac{3}{2}$

45. $\frac{-(y-2)}{y-2}, y \neq 2$

NON-ROUTINE PROBLEMS

46. Sarah doubled the amount in her checking account during the month of January. On the last day of the month, she withdrew $16. Each month after that, she repeated this procedure. However, after withdrawing $16 on the 30th of April, Sarah had no money left in her account. How much did Sarah have in her account at the beginning of January?

47. How many triangles are shown in the figure below?

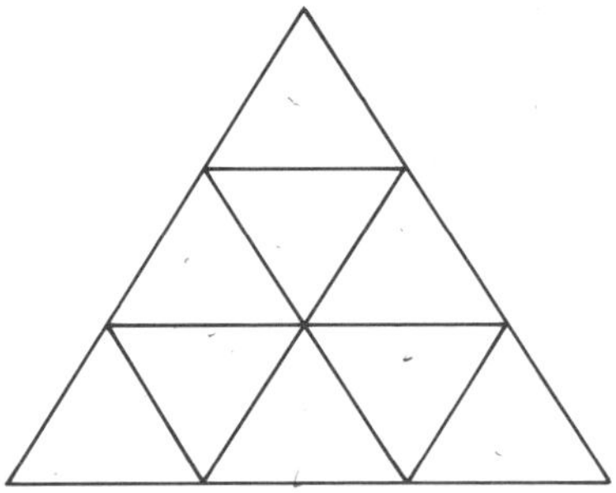

REVIEW CAPSULE FOR SECTION 2.5

Simplify.

1. $(2 \cdot 7)a + (2)(8)$

2. $3 \cdot 7 + (3 \cdot 8)x$

3. $(-5 \cdot 2) + (-5 \cdot 6)b$

4. $(-8)y + (-8 \cdot 4)$

5. $(4p)(3q) + (4p)(-2)$

6. $(\frac{1}{2}x)(-y) + (8x^2)(-y)$

Write Yes or No to indicate whether 3xy is a factor of both given expressions.

7. $-6xy, 3x^2y$

8. $2xy, 6xy^2$

9. $-18x^3y, 3xy$

10. $4x^2y^2, 6xy$

11. $9x^2y^2, -13xy$

12. $21yx, -15x^2y^2$

2.5 Multiplying and Factoring

The Distributive Property can be used to express a product as a sum.

> ***Distributive Property***
>
> For any real numbers a, b, and c:
>
> 1. $\mathbf{a(b + c) = ab + ac}$ $\quad 3(2c + 5b) = 3(2c) + 3(5b)$
> 2. $\mathbf{(b + c)a = ba + ca}$ $\quad (-5a - 7b)2c = (-5a)2c + (-7b)2c$

EXAMPLE 1 Multiply and simplify: $-3(4x + 7)$

Solution:

1. Use the Distributive Property. → $-3(4x + 7) = -3(4x) + (-3)(7)$
2. Use the Associative Property. → $= (-3 \cdot 4)x + (-3)(7)$
   $= -12x + (-21)$
3. Simplify. → $= -12x - 21$

P–1 **Multiply and simplify.**

a. $5(-3a + 2)$ **b.** $(6m + 4)(-\frac{1}{2})$ **c.** $-3r(-2s + 5)$

Note in Example 2 that $(3x - 2y)$ is written as the sum $[3x + (-2y)]$ before multiplying.

EXAMPLE 2 Multiply and simplify: $(3x - 2y)5x$

Solution: $(3x - 2y)5x = [3x + (-2y)]5x$

$= (3x)(5x) + (-2y)(5x)$

$= 15x^2 + (-10xy)$

$= 15x^2 - 10xy$

The Distributive Property can be used to express a sum as a product.

$$7x + 14 = 7(x + 2)$$

This process is called ***factoring.*** The number 7 is the common factor of $7x$ and 14 because each is divisible by 7.

P–2 **Find the common factors of each pair.**

a. $3x^2$ and $3xy$ **b.** $12ac$ and $4bc$ **c.** $-21rt$ and $-7t$

EXAMPLE 3 Factor: $3x^2 + 3xy$

Solution: [1] Find the common factor. ⟶ $3x^2 + 3xy = (3x)(x) + (3x)(y)$

[2] Rewrite using the common factor as one factor of the product. ⟶ $= (3x)(x + y)$

Note in Example 4 that $12ac - 4bc$ and $-21rt - 7t$ are written as sums before factoring.

EXAMPLE 4 Factor: **a.** $12ac - 4bc$ **b.** $-21rt - 7t$

Solutions:

a. $12ac - 4bc = 12ac + (-4bc)$

$= 3a(4c) + (-b)(4c)$

$= [3a + (-b)]4c$

$= (3a - b)4c$

b. $-21rt - 7t = -21rt + (-7t)$

$= (-7t)(3r) + (-7t)(1)$

$= (-7t)(3r + 1)$

$= -7t(3r + 1)$

P–3 **What is the factored form of each of the following?**

a. $6y + 24$ **b.** $10mn - 4n$ **c.** $15x^2 - 3x$ **d.** $-9v - 3vw$

CLASSROOM EXERCISES

Multiply and simplify. (Examples 1 and 2)

1. $10(p + q)$ **2.** $12(k - 1)$ **3.** $\frac{1}{2}(4k - 12)$ **4.** $(r + s)8$

5. $-4x(-5x - 7)$ **6.** $(a - b)14a$ **7.** $(9m - 3)(-\frac{1}{3}m)$ **8.** $(3m - 5n)(-m)$

Factor. (Examples 3 and 4)

9. $8m + 8n$ **10.** $3ax + 5ay$ **11.** $6x + 10y$ **12.** $2a + 10a^2$

13. $5x - 5y$ **14.** $4 - 6b$ **15.** $4mt - 9nt$ **16.** $x^2 - 3x$

17. $-3a - 3b$ **18.** $-5x - 10y$ **19.** $-6r^2 - r$ **20.** $-4st + 8s^2t$

WRITTEN EXERCISES

Goal: To use the Distributive Property to multiply and factor expressions
Sample Problem: a. $-5t(3r-4s)$ **b.** $21mt-9nt$
Answers: a. $-15tr+20ts$ **b.** $3t(7m-3n)$

Multiply and simplify. (Examples 1 and 2)

1. $10(a+b)$	**2.** $5(x+y)$	**3.** $-3(x+2)$	**4.** $-5(t+4)$
5. $7(r-8)$	**6.** $12(s-5)$	**7.** $8(3p+2)$	**8.** $6(4r+5)$
9. $(r+3)4r$	**10.** $(m+6)3m$	**11.** $(y-6)(-2y)$	**12.** $(n-10)(-4n)$
13. $(6t-5u)4t$	**14.** $(8k-7j)5k$	**15.** $(-8t+12)(\frac{1}{4}t)$	**16.** $(2.8b-1.4c)(-0.2ab)$

Factor. (Examples 3 and 4)

17. $5a+5b$	**18.** $14m+7n$	**19.** $4r+20s$	**20.** $3ax+6ay$
21. $2km+4kn$	**22.** $y+13y^2$	**23.** p^2q+pq^2	**24.** $6m+18n$
25. $3r-3s$	**26.** $4a-4$	**27.** $8x-8$	**28.** $12rk-18sk$
29. $8kx^2-16xy^2$	**30.** $13x^2-26xy^2$	**31.** $\frac{3}{4}rx-\frac{3}{4}ry$	**32.** $\frac{1}{3}am-\frac{1}{3}an$
33. $-14ab-7b$	**34.** $-24rs-8s$	**35.** $-3x^2+6x$	**36.** $-5n^2+15n$
37. $-2xy+5xy$	**38.** $-10a-2b$	**39.** $-4x^2-x$	**40.** $-11rs^2+33r^2s$

MORE CHALLENGING EXERCISES

Factor.

EXAMPLE: $6a^3+9a^2-3a$

SOLUTION: $6a^3+9a^2-3a=(3a)(2a^2)+(3a)(3a)+(3a)(-1)$
$=3a(2a^2+3a-1)$

41. $5a-15b+10c$	**42.** $4r+8s-24t$	**43.** $6x^3+3x^2-18x$
44. $7y^3-14y^2+21y$	**45.** $30r^2-36x-138x^2$	**46.** $-2x^2-4x^2y-8x^2y^2$

In the expression $(ax+by)$, the value of a is less than the value of b. The common factor of ax and by is an integer c.

47. Can $c=a$? Explain, using examples to illustrate your answer.

48. Can $c=b$? Explain, using examples to illustrate your answer.

REVIEW CAPSULE FOR SECTION 2.6

Add or subtract, as indicated. (Pages 4–10)

1. $12+(-5)$	**2.** $-8+(-2)$	**3.** $7+(-13)$	**4.** $-4+(-9)$
5. $8-14$	**6.** $3-(-2)$	**7.** $-5-15$	**8.** $-1-(-6)$

2.6 Combining Like Terms

The first and third terms in $3x + 4x^2 - 2x$ are like terms.

Definition

> ***Like terms*** have exactly the same variables and the same powers of these variables.

P–1 **Which of the following are like terms?**

a. $3x$ and $-5x$ **b.** $5y$ and $2y^2$ **c.** a^2b and $-2ab^2$

You can add and subtract only like terms. When you add or subtract like terms, you are ***combining*** like terms.

EXAMPLE 1 Simplify: $x^2 - 8x^2$

Solution:

[1] Factor. ⟶ $x^2 - 8x^2 = (1 - 8)x^2$

[2] Add. ⟶ $= (-7)x^2$ ◀ ***1 − 8 = 1 + (−8)***

$= -7x^2$

It may be helpful to write the expression as a sum.

EXAMPLE 2 Simplify: $x - 5 - 11x + 7$

Solution:

[1] Write as a sum. ⟶ $x - 5 - 11x + 7 = 1x + (-5) + (-11x) + 7$

[2] Group like terms. ⟶ $= [1x + (-11x)] + (-5 + 7)$

[3] Factor. ⟶ $= [1 + (-11)]x + (-5 + 7)$

$= -10x + 2$

It is not necessary to show detailed steps as in Examples 1 and 2.

EXAMPLE 3 Simplify: $-2x^2 - 3x + 5 - 5x + 11x^2$

Solution: $-2x^2 - 3x + 5 - 5x + 11x^2 = 9x^2 - 8x + 5$ ◀ $-2x^2 + 11x^2 = 9x^2$; $-3x - 5x = -8x$

CLASSROOM EXERCISES

Simplify. (Examples 1–3)

1. $3y + 8y$
2. $2ax + 7ax$
3. $7x - 5x$
4. $2r - 2r$
5. $ab + ab$
6. $2z + 10z$
7. $6y^2 - 9y^2$
8. $-3b^2 + b^2$
9. $-t^2 - t^2$
10. $2a^2b + 3ba^2$
11. $3x - 1 + 5x$
12. $5a - a + 2$
13. $-7 + 2x - 2x$
14. $-3t + t + 5$
15. $-x^2 - 5x^2 - 3$
16. $2a^2 - 3a - 7a^2$
17. $a^2 - 3a + 2a + 5$
18. $d^2 + 9 + d + 3d^2$
19. $x^2 + x + 1 - x + x^2$
20. $5d - 16f + 23d - 41 - 12f$
21. $\frac{1}{9}a^2 + \frac{1}{3} + \frac{1}{2}a + \frac{1}{6} + 2a$
22. $\frac{1}{4}x + 2 - 4 - \frac{1}{2}x + 2x^2 + 1$
23. $1.5m + 0.9n + 2.1m^2 + 4.2m + 0.1n$
24. $0.6y^2 - 0.5y - 0.3y + y^2 + 1$
25. $15p^2 - 31 - 81p^2 - 42 - p$
26. $14n + 5n^2 + 5n^2 + 16m^2 - 16n^2$

WRITTEN EXERCISES

Goal: To combine like terms

Sample Problems: **a.** $-3y + 5y$ **b.** $-3x^2 + 6 - x^2 - 4$

Answers: **a.** $2y$ **b.** $2 - 4x^2$

Simplify. (Examples 1–3)

1. $13y + 7y$
2. $2x - x - 6x$
3. $17t + 2t$
4. $5a - 11a$
5. $-2x - 15x$
6. $-m + 5m$
7. $3t^2 - 5t^2$
8. $2r^2 - 11r^2$
9. $-4z^2 + 11z^2$
10. $-11d^2 + 20d^2$
11. $-3a - 6 + 9a$
12. $9m + 2n - 3m - 5n$
13. $0.3t^2 + 2t - 0.5t^2$
14. $3c^2 - 3x - 3c + 5$
15. $-2.1g + 5.3 - 8.7 - 0.9g$
16. $5.4d - 0.6 - 8.9d - 8.5$
17. $\frac{1}{4}a - b + \frac{3}{5}a - \frac{1}{2}b$
18. $\frac{2}{3}r + \frac{5}{3}s - \frac{1}{3}s + \frac{2}{3}r$
19. $50b + 200z + 5bz - 300bz - 112b$
20. $15m - 17 - 17m - 4n - 11n$
21. $0.9t + 1.6t^2 + 0.3 + 1.7t^2 + 1.9t$
22. $0.05 + 13n^2 + 3.15 - 2.1n^2 - 3.4n$
23. $\frac{1}{3}r^2 - \frac{1}{5}s^2 - \frac{1}{3}s^2 + \frac{1}{8}r^2 + \frac{1}{3}$
24. $\frac{2}{3} - \frac{4}{5}b - \frac{3}{4}b^2 - \frac{7}{8}b^2 - \frac{3}{10}$
25. $12f^2 - 15f - 62f^2 - 45 - 2f$
26. $a^2 - 6 - 3a^2 - 12 - 5a$
27. $-0.3t + 0.7t + 0.1t$
28. $1.8ab - 0.7ab + 2.1ab$
29. $ab^2 - a^2b + ab - a^2b^2 - ba^2$
30. $-5y^2 - 3y + y^2 - y$
31. $8v^2 + 8w^2 + 2w + 2w^2 - 6w$
32. $12k^2 + 9k - 12k + 6k^2 + 6$
33. $77a^2 + 35 - 2a^2 - 10$
34. $17a^2 + 19a^2$
35. $\frac{1}{2}x - x^2 - \frac{3}{4}x - 3x^2 - \frac{1}{3}x^2 + \frac{1}{5}$
36. $\frac{1}{3}d + \frac{1}{8}d^2 - \frac{1}{2}d$

Comparing Averages

The field of statistics involves collecting, organizing, and interpreting data. **Quality control engineers** collect statistics on the quality of a product by devising and carrying out testing procedures.

Acme Quality Control, Inc. conducts gasoline mileage tests on automobiles. Each of three cars was tested five times. The data is shown in the table.

GASOLINE MILEAGE IN MILES PER GALLON (mpg)			
	Car A	*Car B*	*Car C*
Test 1	26.7	30.4	30.1
Test 2	28.4	27.5	28.2
Test 3	28.3	30.4	28.2
Test 4	28.6	27.9	28.1
Test 5	26.7	27.4	31.3

EXAMPLE: Find the mean, median, and mode of the gasoline mileages for Car A.

SOLUTION:

1 The **mean** is the sum of the mileages divided by the number of mileages.

$$\frac{26.7 + 28.4 + 28.3 + 28.6 + 26.7}{5} = \frac{138.7}{5} = 27.74, \text{ or } 27.7$$

Rounded to the nearest tenth

2 Write the mileages in order from least to greatest.
The **median** is the middle number.

26.7 26.7 (*2 numbers*) 28.3 (*median*) 28.4 28.6 (*2 numbers*)

If there are two middle numbers, the median is the mean of these two numbers.

3 The **mode** is the mileage which occurs most often. → 26.7

There can be no mode, or more than one mode.

The mean is **27.7 mpg;** the median is **28.3 mpg;** and the mode is **26.7 mpg.**

EXERCISES

1. Which statistic gives the greatest average gasoline mileage for Car A, the mean, median, or mode?

2. If you were in charge of the advertising for Car A, would you give the mean, median, or mode as the *average* mileage? Explain.

3. Find the mean, median, and mode for Car B.
4. Which average would you use to advertise Car B? Explain.
5. Find the mean, median, and mode for Car C.
6. Which average would you use to advertise Car C? Explain.

The manufacturer of each car would like to claim that the test results show that its car got the best average gasoline mileage.

7. Which manufacturer do you think would use the mean to support a claim that its car gets the best average mileage? Explain.
8. Do you think the manufacturer of Car B would use the median or the mode to support its claim for the best average mileage? Explain.
9. Which average do you think the manufacturer of Car A would use to claim the best average mileage? Explain.

*The **mode** is a useful average when analyzing data on clothing or package sizes. It indicates the most frequently used size or sizes. The **median** is useful when discussing salaries or home prices. It is not affected by a few extremely large or small values in the group. The **mean** is affected by extremely large or small values in a group. The mean is useful when discussing scores or sports performances, where all the values are equally important.*

Choose the average (mean, median, or mode) you feel would be the best to use in each case. Give a reason for your choice.

10. A dress manufacturer wants to know the "average" size that women will buy.
11. A real estate developer wants to know the "average" family income in a community.
12. A teacher wants to know the "average" score on an exam.
13. A homebuyer wants to know the "average" selling price for houses in a neighborhood.
14. A canning company wants to know the "average" can size that shoppers will buy.

CHAPTER SUMMARY

IMPORTANT TERMS

Counting numbers *(p. 30)*
Whole numbers *(p. 30)*
Integers *(p. 30)*
Rational numbers *(p. 30)*
Repeating decimal *(p. 30)*
Terminating decimal *(p. 31)*
Perfect squares *(p. 31)*
Irrational numbers *(p. 31)*
Nonrepeating decimal *(p. 31)*
Real numbers *(p. 32)*
Equation *(p. 35)*
Factoring *(p. 47)*
Like terms *(p. 50)*
Combining like terms *(p. 50)*

IMPORTANT IDEAS

1. Rational numbers can be expressed as repeating decimals.
2. The decimal forms of irrational numbers are nonrepeating.
3. The following addition properties are true for any real numbers.
   a. *Addition Property of Zero:* $a + 0 = a$
   b. *Addition Property of Opposites:* $a + (-a) = 0$
   c. *Commutative Property of Addition:* $a + b = b + a$
   d. *Associative Property of Addition:* $(a + b) + c = a + (b + c)$
4. The following multiplication properties are true for any real numbers.
   a. *Multiplication Property of One:* $a \cdot 1 = a$
   b. *Multiplication Property of Reciprocals:* $a \cdot \frac{1}{a} = 1, a \neq 0$
   c. *Commutative Property of Multiplication:* $a \cdot b = b \cdot a$
   d. *Associative Property of Multiplication:* $(ab)c = a(bc)$
5. The following special properties are true for any real numbers.
   a. *Property of the Opposite of a Sum:* $-(a + b) = (-a) + (-b)$
   b. *Multiplication Property of −1:* $-1(a) = -a; -1(a + b) = -a - b$
   c. *Opposites and Products:* $(-a)(b) = -ab; (-a)(-b) = ab$
   d. *Subtraction of a Sum or Difference:*
   $a - (b + c) = a - b - c; a - (b - c) = a - b + c$
   e. *Special Division Properties:* $\frac{a}{1} = a; \frac{a}{a} = 1, a \neq 0$
   f. *Opposites and Division:* $\frac{-a}{b} = -\frac{a}{b}; \frac{a}{-b} = -\frac{a}{b}; \frac{-a}{-b} = \frac{a}{b}; b \neq 0$
6. The opposite of a sum or difference is the same as multiplying the sum or difference by −1.
7. *Distributive Property:* For any real numbers *a*, *b*, and *c*,
   a. $a(b + c) = ab + ac$, and
   b. $(b + c)a = ba + ca$.
8. Steps for Combining Like Terms
   a. Add or subtract the numerical factors.
   b. Use the same variable(s) and exponent(s) to write one term.

CHAPTER REVIEW

SECTION 2.1

Write each number as a quotient of two integers in two ways.

1. $-\frac{13}{16}$ **2.** $-\frac{4}{5}$ **3.** $3\frac{7}{8}$ **4.** $-2\frac{3}{4}$ **5.** -37 **6.** 16

Write each rational number as a repeating decimal.

7. -21 **8.** $\frac{4}{5}$ **9.** $-\frac{1}{3}$ **10.** $\frac{7}{12}$ **11.** $4\frac{1}{2}$ **12.** $8\frac{1}{7}$

For Exercises 13–15, choose from the box the set of numbers that corresponds to each word description.

13. The counter on a service station pump showing the cost of the gasoline dispensed

14. An electronic traffic counter

15. The numbers used by a TV weatherperson to describe the excess or deficiency of rainfall

Rational numbers
Whole numbers
Positive rational numbers

Refer to the set of numbers below. Then list the numbers named in Exercises 16–21.

$$\{13\frac{1}{2}, -5, 0, -\sqrt{5}, 4.1, 13, -\sqrt{9}, \frac{5}{8}, \pi, -0.5\}$$

16. Rational numbers **17.** Irrational numbers **18.** Positive real numbers

19. Negative real numbers **20.** Integers **21.** Whole numbers

SECTION 2.2

Write the value of the variable that makes each equation true. Name the property that gives the reason for your choice.

22. $44 + (63 + 2) = (44 + 63) + d$ **23.** $h + 0 = 5\frac{1}{2}$

24. $x + 5.7 = 5.7 + 3.9$ **25.** $14 + m = 0$

Simplify.

26. $23 - 15 - 18 + 5$ **27.** $-37 + 14 - 9 + 21$

28. $-12.9 + 8.6 + 13.7 - 18.3$ **29.** $16.8 - 13.4 - 25.7 + 17.8$

30. $f - 3\frac{1}{4} + 2\frac{1}{2} - d$ **31.** $9\frac{5}{8} - 6\frac{1}{2} - a + 2\frac{7}{8}$

32. $56 + r - 92 + 26 - s - 18$ **33.** $-73 - r + 39 + w - 14 + 58$

SECTION 2.3

Multiply.

34. $(-4)(7)(-3)$
35. $(-5)(20)(-4)$
36. $(-30)(-2)(-4)$
37. $(-10)(-40)(-5)$
38. $(12)(-3)(\frac{1}{4})(-\frac{1}{9})(-17)$
39. $(\frac{1}{6})(-8)(3)(-\frac{1}{4})(-63)$
40. $(-2)(-2)(-3)(-1)$
41. $(-1)(-2)(-2)(-3)(-3)$
42. $(-6)(12)(x)(-4)$
43. $(\frac{1}{4})(y)(8)(-5)$
44. $(7)(-c)(-\frac{1}{3})(21)$
45. $(-18)(5)(-d)(-\frac{1}{9})$

SECTION 2.4

Simplify.

46. $-(s-5)$
47. $-(12m-2n-p)$
48. $-5a-(-3b+c)$
49. $w-(k+12)$
50. $(m-n)-(a+b)$
51. $(r+s)-(p-q)$
52. $(-4s)(-t)$
53. $(-k)(-12n)$
54. $(\frac{1}{4}x)(-12x)$
55. $(-\frac{1}{3}t)(-21t)$
56. $(-10ab)(5b)(-2ac)$
57. $(-4qr)(6q)(-3qr)$
58. $\frac{-xy}{-xy}$
59. $\frac{-8t}{8t}$
60. $\frac{5-s}{5-s},\ s \neq 5$
61. $\frac{a-4}{a-4},\ a \neq 4$
62. $\frac{t-1}{1-t},\ t \neq 1$
63. $\frac{-(7-x)}{x-7},\ x \neq 7$

SECTION 2.5

Multiply and simplify.

64. $-8(b+9)$
65. $(r+10)(-7)$
66. $(-5+6t)(-3)$
67. $-9(2a+1)$
68. $3r(-2r+5)$
69. $6m(-3m+5)$
70. $(-9t-12)(\frac{1}{4}t)$
71. $(3.4a+2.6)(-0.3b)$

Factor.

72. $10t-10w$
73. $\frac{1}{2}a-\frac{1}{2}b$
74. $2x+6y$
75. $15r+10s$
76. $6x^2+3x$
77. $-4y^2+2y$
78. $-12x-3y$
79. $-22xy^2+44xy$

SECTION 2.6

Simplify.

80. $5k-8k$
81. $-12t^2-15t^2-5$
82. $6.3n-4.7n-3.2-3.9n$
83. $7.2q-5.9q-12.1-2.7q$
84. $r^2s-4s^2+r^2s+s^2$
85. $-p^2q+2pq^2-qp^2-pq^2$

Equations and Problem Solving

CHAPTER 3

Many events in nature can be examined using mathematics. For example, mathematics can be used to determine the effects of a drop of liquid falling into a liquid, the speed of a flash of lightning, and the size and shape of the hexagonal-patterned snowflake.

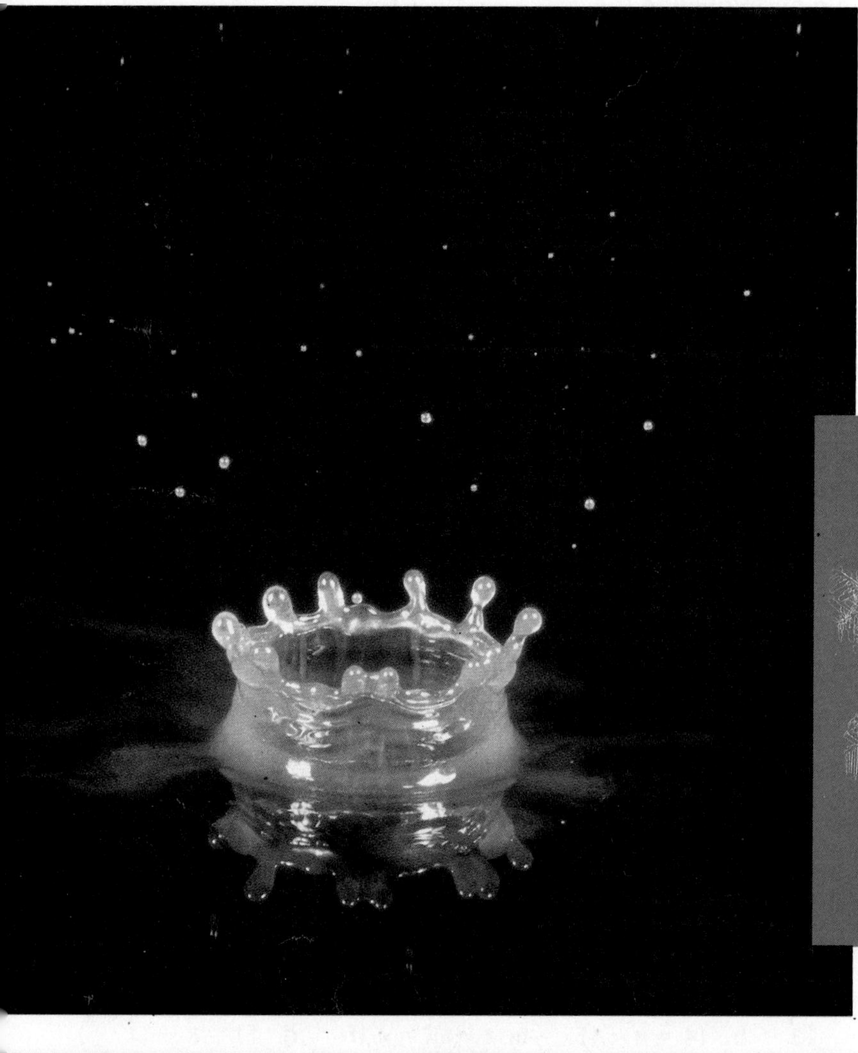

3.1 Addition and Subtraction

Recall from Section 2.2 that an ***equation*** is a sentence that contains the equality symbol "=." A ***solution*** or ***root*** of an equation is a number that makes the equation true.

P–1 **Which equations are true when x is replaced by 19?**

a. $x - 9 = 10$ **b.** $x - 9 + 9 = 10 + 9$ **c.** $x = 19$

Equations **a, b,** and **c** are equivalent. ***Equivalent equations*** have the same solution. Note that 9 was added to each side of Equation **a** to get Equation **b** and Equation **c.**

> ***Addition Property for Equations***
>
> Adding the same real number to each side of an equation forms an equivalent equation.
>
> $$x - 6 = 10$$
> $$x - 6 + \mathbf{6} = 10 + \mathbf{6}$$
> $$x = 16$$

The goal in finding the solution of an equation is to get the variable alone on one side of the equation. To ***solve*** an equation means to find its solution.

EXAMPLE 1 Solve: **a.** $n - 8 = -15$ **b.** $73 = t - 9$

Solution: a.

$$n - 8 = -15$$

[1] Add 8 to each side. → $n - 8 + \mathbf{8} = -15 + \mathbf{8}$

$$n = -7$$

Check: $n - 8 = -15$

$-7 - 8$

-15

Solution: b.

$$73 = t - 9$$

[1] Add 9 to each side. → $73 + \mathbf{9} = t - 9 + \mathbf{9}$

$$82 = t$$

Check: $73 = t - 9$

$82 - 9$

73

As Example 1 shows, you check the answer by replacing the variable with your answer to see if it is the solution.

P–2 **Solve.**

a. $y - 12 = -13$ **b.** $14 = x - 15$ **c.** $t - 18 = -18$ **d.** $11 = k - 11$

As you would expect, there is a subtraction property for equations.

Subtraction Property for Equations

Subtracting the same real number from each side of an equation forms an equivalent equation.

$$x + 7 = -2$$
$$x + 7 - \mathbf{7} = -2 - \mathbf{7}$$
$$x = -9$$

The checks are left for you to do in Example 2.

EXAMPLE 2 Solve: **a.** $3.4 + x = -7.9$. **b.** $\frac{2}{5} = y + \frac{3}{5}$

Solution:

$$3.4 + x = -7.9$$

[1] Subtract 3.4 from each side. ⟶ $3.4 + x - \mathbf{3.4} = -7.9 - \mathbf{3.4}$ ◀ $-7.9 - 3.4 = -7.9 + (-3.4)$

$$x = -11.3$$

Solution:

$$\frac{2}{5} = y + \frac{3}{5}$$

[1] Subtract $\frac{3}{5}$ from each side. ⟶ $\frac{2}{5} - \mathbf{\frac{3}{5}} = y + \frac{3}{5} - \mathbf{\frac{3}{5}}$ ◀ $\frac{2}{5} - \frac{3}{5} = \frac{2}{5} + (-\frac{3}{5})$

$$-\frac{1}{5} = y$$

P–3 **Solve.**

a. $n + 13 = 12$ **b.** $17 + k = 18$ **c.** $24 = x + 24$

CLASSROOM EXERCISES

What number must be added to or subtracted from each side of each equation? (Step 1, Examples 1 and 2)

1. $k + 12 = -9$ **2.** $x - 8 = 14$ **3.** $8 + t = 29$

4. $m - 12.9 = -4.7$ **5.** $9.2 = w - 13.7$ **6.** $8\frac{3}{4} + r = 1\frac{1}{8}$

7. $\frac{13}{8} = p - \frac{5}{2}$ **8.** $8 + m = 3$ **9.** $14 = 23 + s$

Solve. Check each answer. (Examples 1 and 2)

10. $x - 7 = 14$ **11.** $y - 12 = -7$ **12.** $p - 23 = -18$

13. $17 = q - 9$ **14.** $-0.8 = r - 0.8$ **15.** $2.5 = b - 1.5$

16. $f + \frac{3}{8} = 2\frac{1}{4}$ **17.** $h + 6\frac{1}{2} = 3\frac{2}{3}$ **18.** $12 + v = -9$

19. $32 = y + 15$ **20.** $-12 = 20 + x$ **21.** $-6 = 11 + c$

WRITTEN EXERCISES

Goal: To solve equations by using the Addition and Subtraction Properties

Sample Problems: a. $x - 19 = 12$ **b.** $-30 = 27 + n$

Answers: a. $x = 31$ **b.** $-57 = n$

Solve. Check each answer. (Example 1)

1. $x - 9 = 24$ **2.** $y - 15 = 27$ **3.** $z - 21 = -9$

4. $a - 32 = -15$ **5.** $b - 1.3 = -5.5$ **6.** $c - 27.1 = 19.3$

7. $23 = d - 15$ **8.** $-46 = r - 31$ **9.** $23 = f - 1$

10. $-236 = p - 407$ **11.** $\frac{1}{2} = g - \frac{1}{2}$ **12.** $-5\frac{1}{6} = h - 2\frac{5}{6}$

(Example 2)

13. $j + 11 = 28$ **14.** $14 + k = 33$ **15.** $n + 15 = -9$

16. $38 + q = 26$ **17.** $0.6 + x = -0.4$ **18.** $t + 0.25 = 0.75$

19. $10.0 = b + 5.2$ **20.** $30.7 = x + 27.6$ **21.** $9 = 7 + k$

22. $413 = w + 198$ **23.** $-\frac{4}{5} = \frac{2}{5} + y$ **24.** $9\frac{3}{4} = z + 2\frac{1}{2}$

MIXED PRACTICE

25. $x + 20 = -13$ **26.** $a + 2.75 = -7.48$ **27.** $7.5 = b - 2.6$

28. $c - 59.2 = -34.5$ **29.** $3.4 = d + 6.3$ **30.** $\frac{5}{6} = t - \frac{1}{6}$

31. $10.5 + f = 19.3$ **32.** $-59.6 = 80.7 + g$ **33.** $1\frac{3}{4} + x = -\frac{7}{4}$

34. $h + 26 = 405$ **35.** $k - 60.2 = -47.4$ **36.** $-7.3 = 4.1 + m$

37. $16.2 = p + 14.9$ **38.** $s - 59 = -102$ **39.** $74 + z = 12$

40. $3\frac{3}{4} = v - \frac{7}{8}$ **41.** $9.3 = -24.1 + b$ **42.** $k - 12\frac{1}{5} = 17\frac{1}{2}$

REVIEW CAPSULE FOR SECTION 3.2

Perform the indicated operations. (Pages 11–14 and 16–18)

1. $3(\frac{1}{3})$ **2.** $\frac{1}{8} \cdot 8$ **3.** $(-\frac{1}{7})(-7)$ **4.** $(-16)(-\frac{1}{16})$

5. $(\frac{3}{4})(\frac{4}{3})$ **6.** $(-\frac{7}{2})(-\frac{2}{7})$ **7.** $(-\frac{5}{16})(-\frac{16}{5})$ **8.** $(\frac{8}{3})(\frac{3}{8})$

9. $33 \div 33$ **10.** $(-4.5) \div (-4.5)$ **11.** $\frac{5}{8} \div \frac{5}{8}$ **12.** $312 \div 312$

13. $15 \cdot \frac{3}{5}$ **14.** $(-1.8)(3.9)$ **15.** $(\frac{5}{8})(-12)$ **16.** $(-9)(-13)$

17. $(-\frac{3}{4})(-\frac{7}{2})$ **18.** $(13)(-76)$ **19.** $(58)(8)$ **20.** $-12.5 \cdot 8.2$

21. $-7.2 \div -0.9$ **22.** $\frac{1}{6} \div \frac{2}{3}$ **23.** $135 \div 9$ **24.** $-13 \div \frac{1}{4}$

25. $-15.5 \div 0.6$ **26.** $0.1 \div 0.25$ **27.** $(-\frac{2}{3}) \div (\frac{4}{15})$ **28.** $-\frac{4}{5} \div 16$

3.2 Multiplication and Division

Equations involving multiplication or division are solved somewhat like equations that involve addition or subtraction.

P–1 **Which equations are true when n is replaced by 27?**

a. $\frac{n}{9} = 3$ **b.** $\frac{n}{9} \cdot 9 = 3 \cdot 9$ **c.** $n = 27$

Equations **a, b,** and **c** are equivalent. Note that each side of Equation **a** was multiplied by 9 to get Equation **b** and Equation **c.**

Multiplication Property for Equations

Multiplying each side of an equation by the same nonzero real number forms an equivalent equation.

$$\frac{x}{5} = 1.6$$

$$\frac{x}{5}(\mathbf{5}) = 1.6(\mathbf{5})$$

$$x = 8$$

Recall that $\frac{x}{4}$ means $\frac{1}{4}x$. Thus, in the example below,

$\frac{x}{4}$ (4) means $(\frac{1}{4}x)$ (4) or, $1 \cdot x$.

EXAMPLE 1 Solve: **a.** $\frac{x}{4} = -12$ **b.** $40 = \frac{y}{-8}$

Solution: a. $\frac{x}{4} = -12$

1 Multiply each side by 4. ⟶ $\frac{x}{4}(\mathbf{4}) = -12(\mathbf{4})$

$x = -48$

Check: $\frac{x}{4} = -12$; $\frac{-48}{4} = -12$

Solution: b. $40 = \frac{y}{-8}$

1 Multiply each side by −8. ⟶ $40(\mathbf{-8}) = \frac{y}{-8}(\mathbf{-8})$

$-320 = y$

Check: $40 = \frac{y}{-8}$; $\frac{-320}{-8} = 40$

P–2 **Solve.**

a. $\frac{b}{6} = -6$ **b.** $20 = \frac{x}{-5}$ **c.** $\frac{n}{3} = \frac{1}{3}$ **d.** $-32 = \frac{b}{-8}$

As you would expect, there is a division property for equations.

> ***Division Property for Equations***
>
> Dividing each side of an equation by the same nonzero real number forms an equivalent equation.
>
> $$7x = -3$$
> $$\frac{7x}{\mathbf{7}} = \frac{-3}{\mathbf{7}}$$
> $$x = -\frac{3}{7}$$

EXAMPLE 2

Solve: $-6x = -8.4$

Solution: $-6x = -8.4$

[1] Divide each side by -6. ⟶ $\frac{-6x}{\mathbf{-6}} = \frac{-8.4}{\mathbf{-6}}$

$x = 1.4$

Check: $-6x = -8.4$

$-6(1.4)$

-8.4 ←

P–3 **Solve.**

a. $3y = 0.3$ **b.** $21 = -7x$ **c.** $-5n = -5$ **d.** $-12 = 4t$

Recall that dividing by a number is the same as multiplying by its reciprocal. This idea is used in the next example.

EXAMPLE 3

Solve: $12 = -\frac{3}{5}y$

Solution: $12 = -\frac{3}{5}y$

[1] Multiply each side by $-\frac{5}{3}$. ⟶ $12(-\mathbf{\frac{5}{3}}) = (-\frac{3}{5}y)(-\mathbf{\frac{5}{3}})$ ◀ ***Since $-\frac{5}{3} \cdot (-\frac{3}{5}) = 1$, they are reciprocals.***

[2] Rewrite the equation. ⟶ $12(-\frac{5}{3}) = (-\frac{3}{5} \cdot -\frac{5}{3})y$

[3] Multiply. ⟶ $(\frac{\overset{4}{\cancel{12}}}{1})(-\frac{5}{\underset{1}{\cancel{3}}}) = 1 \cdot y$

$-20 = y$ ◀ ***The check is left for you.***

P–4 **What number should you multiply each side of the equation by to get the variable alone?**

a. $\frac{1}{2}a = 19$ **b.** $12 = -\frac{3}{4}m$ **c.** $-1\frac{2}{3}x = 16$ **d.** $-9 = \frac{4}{5}y$

CLASSROOM EXERCISES

What number should you multiply by or divide by to get the variable alone? (Step 1, Examples 1, 2, and 3)

1. $\frac{n}{12} = -89$
2. $\frac{r}{-1.9} = 3.4$
3. $14a = -72$
4. $-23m = -59$
5. $5.6 = -8.9k$
6. $-8.65 = 0.29q$
7. $127 = -\frac{2}{3}w$
8. $-28 = \frac{15}{4}x$

Solve. Check each answer (Examples 1, 2, and 3)

9. $\frac{j}{7} = 4$
10. $\frac{y}{-6} = -11$
11. $\frac{z}{7} = 20$
12. $\frac{a}{12} = -5$
13. $-4 = \frac{r}{1.2}$
14. $8 = \frac{s}{-7}$
15. $-9 = \frac{t}{-9}$
16. $0 = \frac{w}{32}$
17. $5t = -30$
18. $-8y = 56$
19. $8a = -2.4$
20. $-0.6x = -5.4$
21. $-24 = \frac{3}{2}k$
22. $-15 = -\frac{1}{2}x$
23. $-24 = -\frac{2}{3}k$
24. $-27 = \frac{3}{4}p$

WRITTEN EXERCISES

Goal: To solve equations using the Multiplication and Division Properties

Sample Problems: a. $\frac{w}{5} = -7.2$ **b.** $4\frac{2}{3} = -3x$

Answers: a. $w = -36$ **b.** $-\frac{14}{9} = x$

Solve. Check each answer. (Examples 1, 2, and 3)

1. $\frac{x}{-5} = 13$
2. $\frac{g}{-3} = 30$
3. $\frac{y}{8} = -12$
4. $\frac{k}{12} = -9$
5. $\frac{x}{8.3} = -14$
6. $\frac{s}{0.2} = 1000$
7. $\frac{t}{-4} = -15$
8. $\frac{m}{-5} = 75$
9. $23 = \frac{v}{8}$
10. $-14 = \frac{w}{-7}$
11. $-16 = \frac{x}{5.2}$
12. $-2.08 = \frac{m}{-12}$
13. $8\frac{1}{4} = \frac{s}{-4}$
14. $5\frac{1}{8} = \frac{z}{16}$
15. $0.34 = \frac{r}{-1.6}$
16. $-70 = \frac{b}{0.5}$
17. $4a = -28$
18. $6c = -54$
19. $-9d = 72$
20. $-11e = 77$
21. $-1.8x = 18.36$
22. $-5b = 3.5$
23. $-108 = 12x$
24. $-72 = -0.9y$
25. $-140 = -16x$
26. $-y = 99$
27. $-62r = 0$
28. $-90 = -6y$
29. $\frac{1}{2}z = -44$
30. $\frac{2}{3}r = -54$
31. $30 = \frac{2}{5}k$
32. $15 = \frac{3}{4}c$
33. $-\frac{2}{3}w = 20$
34. $-\frac{1}{4}a = 8$
35. $36 = \frac{3}{4}m$
36. $24 = \frac{5}{6}e$
37. $\frac{2}{5}x = -14$
38. $-100 = \frac{1}{5}y$
39. $-\frac{7}{8}p = -49$
40. $56 = -\frac{1}{4}s$

MORE CHALLENGING EXERCISES

41. $6|x| = 1.2$
42. $-\frac{1}{3}|p| = -5$
43. $\frac{|-n|}{-7} = -3$
44. $\frac{|-t|}{5} = 3.2$

REVIEW CAPSULE FOR SECTION 3.3

Simplify. (Pages 56–60)

1. $5n + 10 - 10$ **2.** $3r - 7 + 7$ **3.** $6b \div 6$

4. $(12t + 1.8 - 1.8) \div 12$ **5.** $(18x - 3\frac{1}{3} + 3\frac{1}{3}) \div 18$ **6.** $(\frac{y}{4} + 4 - 4)4$

3.3 Two or More Operations

To solve an equation that involves more than one operation, use the Addition or Subtraction Property for Equations before using the Multiplication or Division Property.

EXAMPLE 1 Solve: $5n - 7 = 14$

Solution:

$$5n - 7 = 14$$

1 Add 7 to each side. ⟶ $5n - 7 + \mathbf{7} = 14 + \mathbf{7}$

$$5n = 21$$

2 Divide each side by 5. ⟶ $\frac{5n}{\mathbf{5}} = \frac{21}{\mathbf{5}}$

$$n = 4.2$$

Check: $5n - 7 = 14$

$5(4.2) - 7$

$21 - 7$

14

Sometimes the variable is on the right side of an equation. The steps for solving are still the same.

EXAMPLE 2 Solve: $-40 = 8 - 4y$

Solution:

$$-40 = 8 - 4y$$

1 Subtract 8 from each side. ⟶ $-40 - \mathbf{8} = 8 - 4y - \mathbf{8}$

$$-48 = -4y$$

2 Divide each side by −4. ⟶ $\frac{-48}{\mathbf{-4}} = \frac{-4y}{\mathbf{-4}}$

$$12 = y$$

Check: $-40 = 8 - 4y$

$8 - 4(12)$

$8 - 48$

-40

P–1 **Solve.**

a. $3t - 7 = 14$ **b.** $-k + 16 = 8$ **c.** $29 = -3 - 8r$

Recall that $\frac{x}{12}$ means $\frac{1}{12}x$. Thus, in the example below,

$\frac{x}{12}(12)$ means $(\frac{1}{12}x)(12)$ or, $1 \cdot x$.

EXAMPLE 3 Solve: $\frac{x}{12} - 14 = -8$

Solution: $\frac{x}{12} - 14 = -8$

[1] Add 14 to each side. ⟶ $\frac{x}{12} - 14 + \mathbf{14} = -8 + \mathbf{14}$

$\frac{x}{12} = 6$

[2] Multiply each side by 12. ⟶ $\frac{x}{12}(\mathbf{12}) = 6(\mathbf{12})$ ◀ $\frac{x}{12}(12) = (\frac{1}{12})(\frac{12}{1})x$

$x = 72$

The check is left for you.

P-2 **Solve.**

a. $\frac{b}{7} + 6 = 5$ **b.** $11 + \frac{s}{5} = 16$ **c.** $-33 = -3 - \frac{d}{2}$

CLASSROOM EXERCISES

What number must be added to or subtracted from each side of each equation? (Step 1, Examples 1-3)

1. $2n + 5 = 8$ **2.** $7w - 12 = -3$ **3.** $6p - 5.6 = 7.3$ **4.** $1.4 + 0.9k = 0.5$

5. $3 - 2t = -9$ **6.** $12 = 8w + 7$ **7.** $-25 = 5 + 7q$ **8.** $16 = 2 - 12x$

9. $\frac{a}{4} - 6 = 23$ **10.** $\frac{m}{-5} + 3 = -7$ **11.** $6.5 - 1.3r = -3.7$ **12.** $0.5 + 0.8w = -2.7$

Write Yes or No after checking whether the solution of each equation below is correct. (Checks in Examples 1-3)

13. $4x - 9 = -21$; $x = -3$

14. $-3y + 7 = -6$; $y = 4$

15. $12 - 3c = -15$; $c = 1$

16. $16 = -8 + 3f$; $f = 8$

17. $25 - b = 17$; $b = -8$

18. $12 + \frac{1}{2}n = 18$; $n = \mathbf{3}$

19. $-\frac{1}{3}r + 5 = -2$; $r = 21$

20. $-19 + \frac{a}{5} = -23$; $a = -20$

21. $-5.9 = 1.3p - 2$; $p = 3$

WRITTEN EXERCISES

Goal: To solve equations using the Addition, Subtraction, Multiplication, and Division Properties for Equations

Sample Problems: **a.** $-8 + 4r = -20$ **b.** $8.5 = 5x + 6$

Answers: **a.** $r = -3$ **b.** $0.5 = x$

Solve. Show all steps. Check each answer. (Example 1)

1. $3x - 8 = 13$ **2.** $4a - 12 = 16$ **3.** $2b + 9 = -17$

4. $6x + 5 = 25$ **5.** $16 + 3c = -50$ **6.** $-56 + 24d = -254$

(Example 2)

7. $-31 = 2e + 7$ **8.** $17 = -4x + 3$ **9.** $-52 = 12 + 8y$

10. $-16\frac{1}{2} = -5\frac{1}{2} + 3x$ **11.** $-17\frac{1}{4} = 4y - 2\frac{1}{4}$ **12.** $250 = 5f - 100$

(Example 3)

13. $\frac{x}{7} - 4 = -12$ **14.** $\frac{h}{10} - 6 = -15$ **15.** $-26 + \frac{x}{1.2} = -34$

16. $-19 + \frac{k}{2.7} = -33$ **17.** $20 = \frac{a}{5} - 10$ **18.** $70 = 8 + \frac{b}{2}$

MIXED PRACTICE

19. $3x - 4.8 = -21.3$ **20.** $-32 + 30t = -287$ **21.** $\frac{h}{4} + 15 = -60$

22. $-2.4 = 10.6 - 2k$ **23.** $\frac{2}{3}g - \frac{1}{3} = \frac{2}{3}$ **24.** $50 = \frac{1}{2}d + 30$

25. $12 - 5z = 48$ **26.** $1.5f + 8.5 = 11.5$ **27.** $125 = 10c - 25$

28. $\frac{n}{6} - 13 = 20$ **29.** $4x - 12.3 = -18.5$ **30.** $12 - \frac{3}{4}m = 15$

31. $\frac{j}{3} - 4 = 7$ **32.** $2a - 9.3 = -6.7$ **33.** $\frac{1}{5}b + 15 = 25$

34. $-10.4 - p = 12.1$ **35.** $\frac{r}{8} + 15 = 12$ **36.** $\frac{2}{7}w + \frac{1}{7} = -\frac{5}{7}$

37. $86 = 9 + 3x$ **38.** $12 - \frac{t}{5} = 7$ **39.** $-36 = -4x + 6$

40. $-\frac{p}{6} + 7 = -2.5$ **41.** $5\frac{1}{3} = 12 - 8k$ **42.** $6.2 = -2x - 5.4$

NON-ROUTINE PROBLEMS

43. A customer wanted a 25-foot piece of wire. The clerk incorrectly measured the wire with a yardstick that was 2 inches too short. How many inches were missing from the customer's length of wire?

44. A three-digit number is between 300 and 500. The sum of the digits is 12. The sum of the hundred's digit and the one's digit equals the ten's digit. Find the number.

PROBLEM SOLVING AND APPLICATIONS

3.4 Using Formulas

The following word rule and formula relate distance, rate, and time.

Word Rule: Distance equals rate multiplied by time.

Formula: $d = rt$

In a formula, variables and symbols are used to represent words. If you know the values of two of the variables in $d = rt$, you use the techniques for solving equations to find the third value.

EXAMPLE 1

In 1903, Orville Wright flew his airplane a distance of 37 meters at an average speed (rate) of 3.08 meters per second. To the nearest tenth, how long did the flight last?

d = distance
r = rate
t = time

Solution:

1. Write the formula. → $d = rt$
2. Identify the known values. → $d = 37$ meters; $r = 3.08$ meters per second
3. Replace the variables. → $37 = 3.08t$
4. Solve the equation. → $\frac{37}{3.08} = t$ or, $12.01 = t$

To the nearest tenth, the flight lasted 12.0 seconds.

Word Rule: The perimeter of a rectangle equals the sum of the lengths of the sides.

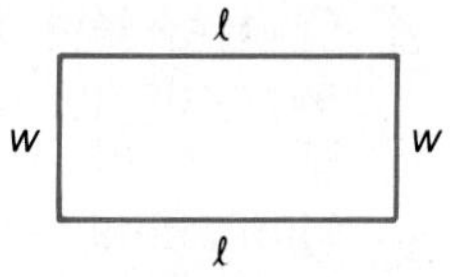

Formula: $p = \ell + \ell + w + w$

$p = 2\ell + 2w$ or, $p = 2(\ell + w)$

Combine like terms. Then factor.

EXAMPLE 2

The perimeter of this postage stamp is 9.4 centimeters (cm). The width is 2.2 centimeters. Find the length.

Solution:

1. Write the formula. → $p = 2(\ell + w)$
2. Identify the known values. → $p = 9.4$ cm; $w = 2.2$ cm
3. Replace the variables. → $9.4 = 2(\ell + 2.2)$
4. Solve the equation. → $9.4 = 2\ell + 4.4$

$5.0 = 2\ell$

$2.5 = \ell$ The length is 2.5 cm.

CLASSROOM EXERCISES

Find the value of the unknown variable. (Example 1)

1. $d = 1200$ kilometers; $r = 400$ kilometers per hour; $t = \underline{?}$
2. $d = \underline{?}$; $r = 25$ meters per minute; $t = 30$ minutes
3. $d = 280$ feet; $r = \underline{?}$; $t = 4$ seconds

(Example 2)

4. $p = \underline{?}$; $\ell = 18$ meters; $w = 12$ meters
5. $p = 150$ kilometers; $\ell = \underline{?}$; $w = 30$ kilometers
6. $p = 29.2$ millimeters; $\ell = 6.4$ millimeters; $w = \underline{?}$

WRITTEN EXERCISES

Sample Problems: **a.** Given: $d = 60.2$ m and $r = 7$ meters per second. Find t.
b. Given: $p = 37.0$ mm and $\ell = 12.7$ mm. Find w.

Answers: **a.** 8.6 seconds **b.** $w = 5.8$ mm

For each problem, find the rate, r, or the time, t. Round your answer to the nearest tenth. (Example 1)

1. On a certain day, it took 1.2 seconds for light to travel from the moon to the earth 360,000 kilometers away. Find the speed of light in kilometers per second.
2. During a thunderstorm, lightning was observed 5100 feet away. The thunderclap followed 4.5 seconds later. Find the speed of sound in feet per second.
3. A cheetah ran a distance of 80 yards traveling at an average speed of 35 yards per second. How long did it take the cheetah to run that distance?
4. In 1932, Amelia Earhart flew a plane from Newfoundland to Ireland, a distance of 3260 kilometers, at an average speed of 217 kilometers per hour. How long did her trip take?

For each problem, find the width, w, or length, ℓ. Round your answer to the nearest tenth. (Example 2)

5. The perimeter of a picture frame is 3.8 meters. The length of the frame is 1.2 meters. Find the width.
6. The perimeter of a hand calculator is 12.6 inches. The width is 2.5 inches. Find the length.

APPLICATIONS

You can use a calculator to check an equation. However, some calculators do not automatically follow the rules for the order of operations. You may have to rewrite the equation for the calculator.

EXAMPLE Check the answer to Sample Problem **a** on page 66.

SOLUTION Equation: $-8 + 4r = -20$ Answer to be checked: -3

First rewrite the left side for the calculator: $4r - 8$

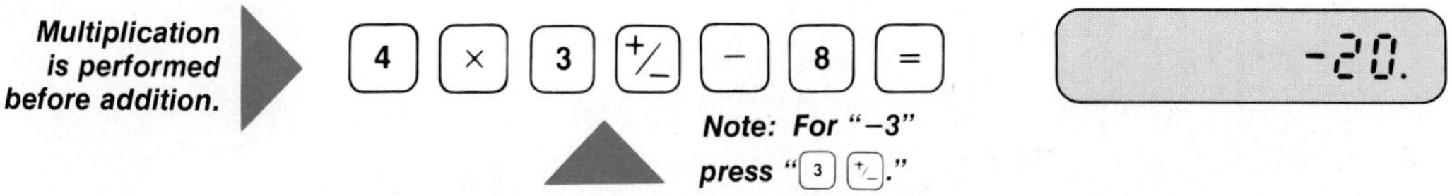

Since both sides have the same value, −20, the answer, −3, is correct.

Check the given "answer" to each equation below.

7. $5n - 7 = 14$; 4.2

8. $24 = 6 + 2x$; 3

9. $4 + 3x = -11$; −5

MID-CHAPTER REVIEW

Solve. Check each answer. (Section 3.1)

1. $b + 9 = -3$

2. $10 + w = 40$

3. $y - 1.5 = 3$

4. $-3\frac{1}{8} = k - 2\frac{1}{4}$

5. $t - 67 = 83$

6. $0.75 = a + 0.25$

Solve. Check each answer. (Section 3.2)

7. $-39 = 13a$

8. $\frac{m}{-2} = -3$

9. $-9z = 30$

10. $\frac{r}{0.5} = 16.5$

11. $6\frac{1}{2} = \frac{x}{2}$

12. $-42 = -13h$

Solve. Check each answer. (Section 3.3)

13. $3y + 7 = 28$

14. $40 = \frac{h}{3} - 10$

15. $15 = 7\frac{1}{2} - \frac{1}{2}t$

16. $-20 = 15 + 15n$

17. $1.7 + \frac{j}{6} = 3.7$

18. $8v - 17 = -33$

Solve. Round your answer to the nearest tenth. (Section 3.4)

19. A coyote ran a distance of 50 yards traveling at an average speed of 20 yards per second. How long did it take the coyote to run that distance?

20. The perimeter of a park is 196 yards. The width is 46 yards. Find the length.

Using a Formula — Engineering

It is important for an **engineer** to know the greatest safe load that a beam can bear. The following formula applies to a steel beam with a rectangular cross section. Its load is in pounds, distributed along the beam.

Greatest Safe Load

$$S = \frac{1780Ad}{l}$$

A = cross-sectional area (in²)
d = depth of beam (in)
l = distance between supports (ft)

EXAMPLE: Compute the greatest safe load in pounds for a solid steel beam 20 feet long with a cross section 8 inches wide and 6 inches deep.

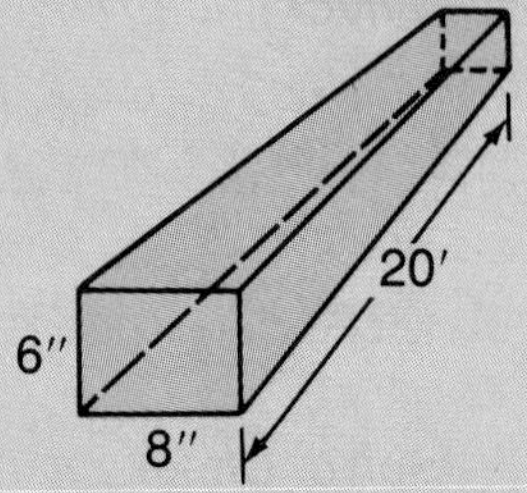

SOLUTION: $S = \frac{1780Ad}{l}$

$A = 48 \text{ in}^2 \ (6'' \times 8''),\ l = 20 \text{ ft},\ d = 6 \text{ in}$

$S = \frac{(1780)(48)(6)}{20}$

1 7 8 0 [×] 4 8 [×] 6 [÷] 2 0 [=] → 25632.

The greatest safe load is **25,632 pounds.**

EXERCISES

1. The distance between supports of a solid rectangular steel beam is 16 feet. The width of the beam is 2 inches and the depth is 4 inches. Find the greatest safe load.

2. The width of a solid rectangular steel beam is 6 inches and the depth is 6 inches. The distance between supports is 18 feet. Find the greatest safe load.

3. If the distance between the supports of a solid steel rectangular beam is increased, does the greatest safe load increase or decrease?

4. Two solid rectangular steel beams have the same depth and distance between supports. The width of one beam is twice that of the other. Compare the greatest safe loads.

5. The formula $S = \frac{1795Ad}{l}$ is used to find the greatest safe load of a steel I-beam. A steel I-beam has a length of 24 feet, a depth of 4 inches, and a cross-sectional area of 16 square inches. Find the greatest safe load.

REVIEW CAPSULE FOR SECTION 3.5

Simplify. (Pages 47–51)

1. $5n + 3 - 2n$ **2.** $2.4t - t + 1.3t$ **3.** $k + 1\frac{1}{4} - 3\frac{1}{2}k - 1\frac{3}{4}$

4. $16m - 14 - 6 - 3m + m$ **5.** $5(6 - x)$ **6.** $15(3 + 2n)$

7. $-8(k - 6)$ **8.** $-12(-5 - 3x)$ **9.** $\frac{1}{4}(2y + 3)$

3.5 Like Terms in Equations

In solving some equations, you should combine like terms first.

EXAMPLE 1 Solve and check: $5x - 7 - 3x = -19$

Solution:

$$5x - 7 - 3x = -19$$

[1] Combine like terms. → $5x - 7 - 3x = -19$

$$2x - 7 = -19$$

[2] Add 7 to each side. → $2x - 7 + \mathbf{7} = -19 + \mathbf{7}$

$$2x = -12$$

[3] Divide each side by 2. → $\frac{2x}{\mathbf{2}} = \frac{-12}{\mathbf{2}}$

$$x = -6$$

Check: $5x - 7 - 3x = -19$

$$5(-6) - 7 - 3(-6)$$
$$-30 - 7 + 18$$
$$-37 + 18$$
$$-19$$

P–1 **Solve.**

a. $7m - 3 + 4m = 30$ **b.** $9k + 6 - 3k = -36$

In solving some equations, you have to use the Distributive Property first to remove parentheses. The check in the following examples is left for you to do.

EXAMPLE 2 Solve and check: $26 = 5 - 3(2t + 3)$

Solution: [1] Use the Distributive Property. → $26 = 5 - 3(2t + 3)$

[2] Simplify. → $26 = 5 - 6t - 9$

$$26 = -4 - 6t$$

[3] Add 4 to each side. ⟶ $26 + \mathbf{4} = -4 - 6t + \mathbf{4}$

$30 = -6t$

[4] Divide each side by −6. ⟶ $\frac{30}{\mathbf{-6}} = \frac{-6t}{\mathbf{-6}}$

$-5 = t$

P–2 **Solve.**

a. $24 = 8 - 4(3y - 1)$ **b.** $17 = 12 + 5(r - 4)$

EXAMPLE 3 Solve: $2(3x - 4) - 4(x + 1) = -15$

Solution:

[1] Use the Distributive Property twice. ⟶ $2(3x - 4) - 4(x + 1) = -15$

[2] Combine like terms. ⟶ $6x - 8 - 4x - 4 = -15$

$2x - 12 = -15$

[3] Add 12 to each side. ⟶ $2x - 12 + \mathbf{12} = -15 + \mathbf{12}$

$2x = -3$

[4] Divide each side by 2. ⟶ $\frac{2x}{\mathbf{2}} = \frac{-3}{\mathbf{2}}$

$x = -1\frac{1}{2}$

CLASSROOM EXERCISES

Write the equation that results after combining like terms. (Step 1, Example 1)

1. $5y + 6 - y = 13$

2. $-10 = 8t - 3t - 7$

3. $-4r + 8 - r = 17$

4. $10q - 6 - 2q - 3q = 7$

5. $-9.7 = 5.8m - 2.7m - 4.8$

6. $-3.2a - 1.6 + 4.1a = 9.3$

a. *Write the equation that results after removing parentheses.*
b. *Then write the equation that results after combining like terms.*
(Steps, 1 and 2 in Examples 2 and 3)

7. $3(4x - 1) + 6 = 10$

8. $-5(2x + 3) - 7 = 12$

9. $\frac{1}{3} - 2(q - 5) + \frac{1}{4}q = 6$

10. $-7 = -\frac{1}{2}(4r - 6) + 13$

11. $\frac{3}{4}(x - 2) + 2(3x + \frac{1}{4}) = -11$

12. $5(k - \frac{1}{3}) - 3(2k + \frac{1}{3}) = 19$

13. $-6 = 4(3 - 2t) + 5(t - 4)$

14. $-6 + 5w - 3(4 - w) = 25$

Evaluate each expression for the given value of the variable. (Check, Example 1)

15. $3y - 8$ when $y = 5$

16. $12 - 2p$ when $p = -3$

17. $z - 5z - 3$ when $z = 2$

18. $3r - 5 - 5r$ when $r = -1$

19. $2(d - 3) - 5$ when $d = 4$

20. $-3(c - 2) - 4$ when $c = -2$

WRITTEN EXERCISES

Goal: To solve equations having like terms
Sample Problem: $-5(2x + 1) - 7x = 29$
Answer: $x = -2$

Solve. Check each answer. (Examples 1, 2, and 3)

1. $3x + 5 + x = 33$

2. $2y - 8 - 5y = -47$

3. $-z + 1.2 - 3z = 2.8$

4. $4.6a - 5.8a - 30 = -12$

5. $37 = 8x - 19 - 15x$

6. $-26 = 22 - c - 5c$

7. $\frac{1}{4}(b + 3) = 2\frac{1}{2}$

8. $15 = 3(\frac{2}{3}d - \frac{1}{3})$

9. $-2(x + 1) - 7 = 15$

10. $-3(4n + 1) + 8n = -35$

11. $-32 = -5(f - 2) + 2f$

12. $15 - 4(2g - 3) + 4g = 39$

13. $(4k - 3) - (7k + 8) = 6\frac{1}{2}$

14. $17\frac{1}{2} = \frac{1}{3}(y + 1) - (2y - \frac{1}{2})$

15. $92 = (m + 4) - (3m - 12)$

16. $2(3 - 2x) - 4(x + 4) = -118$

17. $-0.5(3n + 3) - 2(n + 0.6) = 0.8$

18. $-7 = 6(4 + 0.3) + 0.5(2r - 5)$

MIXED PRACTICE

19. $\frac{1}{2}x + 6 + 4x = 5\frac{1}{2}$

20. $7(n + 1) = 21$

21. $-12 + 2(5 - h) - 3h = -57$

22. $-3.6b + 26 - 0.8b = -84$

23. $19.2 - 23.8 = 5d - 14.8 - 7d$

24. $-2(5 - x) - 6(x + 3) = -102$

25. $0.9 - 5.7 = 10.4 - e - 3e$

26. $16 = 2r - 4(r - 6) + 6r$

27. $-85 = 5t - 9t - (14 - 2t)$

28. $4z + 6z - (20 - 8z) = 25$

29. $2\frac{1}{2}y - 3(y + \frac{1}{4}) + 6y = 13$

30. $70 = 6(2t - 1) - 3(2t + 1)$

MORE CHALLENGING EXERCISES

Find each answer without solving the equation.

31. If $2x + 3x = 15$, what is the value of $10x$?

32. If $3(x - 4) - (x - 4) = -3$, what is the value of $6(x - 4)$?

3.6 From Words to Symbols

In order to solve word problems by algebra it is necessary to translate key words of the problem to algebraic symbols. Table 1 shows some word expressions and the corresponding algebraic expressions.

Table 1

Operation	Word Expression	Algebraic Expression
Addition	The sum of the number of years, y, and 12	$y + 12$
	Thirteen plus the number of centimeters, c	$13 + c$
	\$6 more than the cost of a bus ticket, b	$b + 6$
	Some number, n, increased by $3\frac{2}{3}$	$n + 3\frac{2}{3}$
Subtraction	The number of hours, h, decreased by $5\frac{1}{2}$	$h - 5\frac{1}{2}$
	The difference between some number, x, and 9.4	$x - 9.4$
	\$54 minus the amount earned, a	$54 - a$
	The number of students, s, less 5	$s - 5$
	Four less than the number of lemons, ℓ	$\ell - 4$
Multiplication	The product of the number of feet, f, and 4.9	$4.9f$
	Some number, q, multiplied by $\frac{2}{3}$	$\frac{2}{3}q$
	Twenty-four times the number of meters, m	$24m$
	The number of people, p, doubled	$2p$
Division	The quotient of some number, k, and 15	$\frac{k}{15}$
	The number of miles, m, divided by 2.9	$\frac{m}{2.9}$

EXAMPLE 1 Write an algebraic expression for this word expression.

The record number scored decreased by 14 points

Solution: [1] Choose a variable. ⟶ Let n = the record number scored

[2] Identify the operation. ⟶ subtraction ("decreased by")

[3] Write an algebraic expression.

Think: **The record number scored decreased by 14 points**

Translate: n $-$ 14

Table 2 shows some sentences and their corresponding equations.

Table 2

Word Sentence	Equation
19.8 **is** the sum of some number and 4.3.	$19.8 = t + 4.3$
An unknown number of cars decreased by 37 **equals** 196.	$n - 37 = 196$
The product of the number of hours worked and the rate per hour, \$5.25, **is** \$183.75	$5.25w = 183.75$
The quotient of the total cost and the number of payments, 36, **equals** \$173.50.	$\frac{c}{36} = 173.50$

EXAMPLE 2 Write an equation for this sentence. Then solve the equation.
The list price of a lamp less a discount of \$3.00 equals the sale price of \$26.95.

Solution: [1] Choose a variable. ⟶ Let p = the list price.

[2] Identify the operation. ⟶ subtraction ("less")

[3] Write an equation.

Think: The list price less the discount equals the sale price.

Translate: $p \quad - \quad 3.00 \quad = \quad 26.95$

[4] Solve the equation. ⟶ $p - 3.00 + 3.00 = 26.95 + 3.00$

$p = 29.95$

The price of the lamp is \$29.95.

Note carefully how these two subtraction expressions differ.

y less 5 **is not the same as** y less than 5

$y - 5$ **is not the same as** $5 - y$.

CLASSROOM EXERCISES

For Exercises 1–10, write an algebraic expression for each word expression. Use the variable n. (Table 1 and Example 1)

1. Twelve more than a number of stamps
2. Four dollars less than the cost of two tickets
3. The total number of points divided by 8
4. A savings account balance increased by \$1200

5. Eighteen hundred entries in the race <u>times</u> the entry fee

6. The number of swimmers <u>decreased by</u> eighteen

7. The <u>quotient</u> of 1200 miles and the number of gallons of fuel

8. The number of shoppers over the last year <u>tripled</u>

9. The <u>difference</u> between twenty-six and the number of fish caught

10. Six thousand <u>plus</u> the number of records sold this year

Choose a variable and write an equation for each word sentence. Then solve the equation. (Table 2 and Example 2)

11. The low temperature for the month increased by 27° is 50°.

12. The high temperature for the day decreased by 18° is 78°.

13. The total number of points divided by 20 games equals 18 points per game.

14. The cost of an unknown number of tickets multiplied by 4 dollars equals 940 dollars.

15. The total number of persons on the flight, 104, is the number of passengers plus a flight crew of eleven.

16. The number of kilometers driven divided by 75 liters equals 10.2 kilometers per liter.

17. The average number of centimeters of rainfall less 9.4 centimeters is 69.3 centimeters, the amount of rainfall for the past year.

18. The total amount of this year's sales tripled would represent $372,000 in sales next year.

WRITTEN EXERCISES

Goal: To write and solve an algebraic equation for a word sentence involving one operation

Sample Problem: The low temperature for one day of −35° equals the high temperature decreased by 12°.

Answer: $-35 = h - 12$; $h = -23°$

Choose a variable and write an expression for each word expression. (Example 1)

1. Twice the number of taxicabs

2. Three times the number of musicians

3. The number of TV viewers increased by 25,000

4. Twenty-nine less than the number of pro golfers

5. The number of meters decreased by 15.8

6. Five hundred sixty dollars more than the amount of a loan

7. A bowler's total score divided by 28 games

8. The quotient of the number of kilometers sailed and 3.2 hours

9. The difference between the number of accidents reported and 512

10. The product of the number of cartons shipped and 14.3 kilograms

Choose a variable and write an equation for each word sentence. Then solve the equation. (Example 2)

11. The quotient of the total cost and 24 payments is 86 dollars.

12. The product of the number of tickets sold and $6.50 equals $6916.

13. The width of a rectangle, 14.2 meters, is 5.8 meters less than the length.

14. The quotient of the total cost and 288 cans of tennis balls equals the cost per can of $1.75.

15. The low temperature of $-5.3°C$ is 7.9 degrees less than the high temperature for the day.

16. The high temperature of $-12.6°F$ is 8.7 degrees more than the low temperature for the day.

17. The amount of the sale multiplied by 0.06 equals the sales tax of $33.60.

18. Fifty-four dollars is $7.75 more than the cost of the items one year ago.

19. The sum of Tim's weekly salary and his commission of $283 is $561.

20. The difference between the number of balloons purchased and the 283 balloons sold is 104.

MIXED PRACTICE

Match each word expression or word sentence in Exercises 21–28 with the corresponding expression or equation from ***a–t.***

21. Thirty-five more than the number of horses

22. The number of shoppers increased by 4500 equals 12,700.

23. The number of books in a collection less 485 books is 714.

24. The paper carrier's daily supply of papers less 17, the number not sold

25. Three times the number of state senators is 72.

26. The number of players on the squad who played plus 24, the number who did not play

27. Twice the number of planes landing each day

28. A golfer's score of 86 is 3 strokes less than his handicap score.

a. $17 - w$

b. $2p = 58$

c. $28.6 + 2.3 = m$

d. $4500n = 12{,}700$

e. $n + 24 = 56$

f. $86 - 3$

g. $485 - r = 714$

h. $3 + n = 72$

i. $r + 2.3 = 28.6$

j. $24 + q$

k. $2(y)$

l. $t + 35$

m. $s - 3 = 86$

n. $714 = k - 485$

o. $72 = 3s$

p. $12{,}700 = 4500 + x$

q. $r + 2\frac{2}{3} = 84\frac{1}{8}$

r. $p - 17$

s. $2\frac{3}{4} + b$

t. $76\frac{5}{8} + 2\frac{3}{4}$

PROBLEM SOLVING AND APPLICATIONS

3.7 Using Real Numbers

Thus far in the text, you have solved word problems involving positive and negative numbers and by replacing variables in formulas. Many word problems in algebra can be solved by writing an equation. The following steps for doing this are illustrated in the examples.

1. **Understand** the problem: Read the problem carefully to determine what information is given and what you are asked to find. Use this information to choose a variable to represent the unknown(s).
2. Make a **plan:** In algebra, the plan usually involves writing an equation.
3. **Solve** the problem.
4. **Check** your results with the statements in the problem. Write the answer in a complete sentence.

EXAMPLE 1 A meteorologist reports that the high temperature for yesterday increased by 6.2 degrees equals the high temperature for today. The high temperature for today is −3°C. Find the high temperature for yesterday.

Solution:

[1] Choose a variable to represent the unknown. ⟶ Let n = yesterday's high temperature.

[2] Write an equation.

Think:	**Yesterday's high**	**increased by**	**6.2°**	**equals**	**today's high.**
Translate:	n	$+$	6.2	$=$	-3

[3] Solve the equation. ⟶ $n + 6.2 - 6.2 = -3 - 6.2$

$$n = -9.2$$

[4] **Check:** When −9.2 is increased by 6.2, is the sum −3?

Does $-9.2 + 6.2 = -3$? Yes ✓

The high temperature for yesterday was −9.2°C.

Sometimes you use more than one operation to solve a word problem.

EXAMPLE 2 Twice the cost of a pair of skis decreased by \$40 equals \$246. Find the cost of the skis.

Solution:

[1] Choose a variable to represent the unknown. ⟶ Let $s =$ the cost of the skis.

[2] Write an equation.

Think: **Twice the cost of the skis** **decreased by** **\$40** **equals** **\$246.**

Translate: $2s - 40 = 246$

[3] Solve the equation. ⟶ $2s - 40 + 40 = 246 + 40$

$$2s = 286$$

$$\frac{2s}{2} = \frac{286}{2}$$

$$s = 143$$

[4] The check is left for you.

The cost of the skis is \$143.

Using Algebra to Solve Word Problems

1. Choose a variable. Use the variable to represent the unknowns.
2. Write an equation for the problem.
3. Solve the equation.
4. Check your answer with the statements in the original problem. Answer the question.

CLASSROOM EXERCISES

For Exercises 1–6:

a. *Choose a variable to represent the unknowns.*

b. *Write an equation for each problem.* (Example 1)

1. On a certain weekend, the Waldron family drove 12 miles less on Sunday than on Saturday. They drove 75 miles on Sunday. How many miles did they drive on Saturday?

2. Apex stock's net change on Tuesday was $\frac{3}{4}$ of a point more than its net change on Monday. Its net change on Tuesday was $-\frac{1}{8}$ of a point. Find the net change on Monday.

3. A company's loss (−) for August divided by −1.2 equals the company's profit of \$5000 for September. Find the loss for August.

4. The record high temperature for Moline, Illinois, multiplied by −0.125 equals the record low temperature of −12°F. Find the record high temperature.

(Example 2)

5. The number of letters mailed on Tuesday was 16 less than twice the number of letters mailed on Monday. There were 132 letters mailed on Tuesday. How many letters were mailed on Monday?

6. The number of tourists visiting Cosmic Caves this year is 5 more than one-half the number of tourists that visited last year. The number of tourists this year is 2410. How many tourists visited last year?

WRITTEN EXERCISES

Goal: To use equations to solve word problems

Sample Problem: A new car dealer sold 8 more cars in April than in March. The dealer sold 34 cars in April. Find the number of cars sold in March.

Answer: $n + 8 = 34$; $n = 26$ cars

For Exercises 1–11:

a. *Choose a variable to represent the unknowns.*

b. *Write an equation for each problem.*

c. *Solve the problem.* (Example 1)

1. On Saturday, the members of the Math Club washed 7 more cars in the afternoon than in the morning. They washed 18 cars in the afternoon. Find the number of cars they washed in the morning.

2. Mt. McKinley in Alaska has an elevation of 20,320 feet above sea level. The difference in elevation between Mt. McKinley and Death Valley in California is 20,602 feet. Find the elevation of Death Valley.

3. The record low temperature for Denver, Colorado, multiplied by -3.5 equals the record high temperature of 105°F. Find the record low temperature.

4. The quotient of the number of tourists and 9 tour buses equals the number of tourists on each bus. There are 35 tourists on each bus. How many tourists are there in all?

(Example 2)

5. On a winter day, four times the reading on a thermometer increased by 5° equals the wind chill temperature. The wind chill temperature is -23°F. What is the reading on the thermometer?

6. The number of books in the main library is 15 less than 3 times the number of books in the branch library. The main library has 12,000 books. Find the number of books in the branch library.

MIXED PRACTICE

7. The number of people attending graduation this year is 12 less than twice the number that attended last year. This year, there are 326 people. Find how many people attended last year.

8. On Tuesday, a nursery received a shipment of 42 new trees. The number of trees the nursery had increased by the number of new trees equals a total of 74 trees. How many trees did the nursery have before the shipment?

9. The number of hours needed to complete a job divided by 6 employees equals an average workday of 7.5 hours for each employee. Find the number of hours needed to complete the job.

10. One half the cost of a new car decreased by $1027 equals the cost of the same car after three years. The car will cost $3134 after three years. What is the cost of the new car?

11. A company's loss (−) for 1985 multiplied by -2.5 equals the company's profit of $30,000 for 1986. Find the loss for 1985.

Focus on Reasoning: Logic Tables

Logical reasoning is used to solve many types of problems.

EXAMPLE Jeff, Cathy, and Amy work on the school newspaper. One is the editor, one is a reporter, and one is a word processor. Use the clues below to find each person's job.

CLUE A Amy's only exercise is jogging.
CLUE B Jeff and the editor play tennis together.
CLUE C Amy and the reporter are cousins.

Solution: Make a table to show all the possibilities. Use an X to show that a possibility cannot be true. Use a ✓ when you are certain that a possibility is true.

[1] Read Clue A. Can you reach any conclusion about Amy's job?

[2] Read Clue B. Is Jeff the editor?
Place an X next to Jeff's name in the editor column.
Read Clue A and Clue B. Is Amy the editor?
Place an X next to Amy's name in the editor column.
Who must be the editor?
Place a ✓ next to Cathy's name in the editor column. Place X's next to Cathy's name in the reporter and word processor columns.

	Editor	Reporter	Word Processor
Jeff	X		
Cathy	✓	X	X
Amy	X		

[3] Read Clue C. Is Amy the reporter?
Place an X in the reporter column and a ✓ in the word processor column next to Amy's name.
Who must be the reporter?

Jeff: reporter **Cathy: editor**
Amy: word processor

Check the answer with the clues.

	Editor	Reporter	Word Processor
Jeff	X	✓	X
Cathy	✓	X	X
Amy	X	X	✓

EXERCISES

Copy the table. Use the clues to complete the table to solve the problem.

1. Alice, Nathan, and Marie play in the school band. One plays the drum, one plays the saxophone, and one plays the flute.

CLUE A Alice is a senior.

CLUE B Alice and the saxophone player practice together after school.

CLUE C Nathan and the flute player are sophomores.

	Drum	Saxophone	Flute
Alice			
Nathan			
Marie			

Who plays each instrument?

2. Brenda has three cats named Tiki, Moby, and Copper. One is a Persian, one is a Siamese, and one is a Himalayan.

CLUE A Copper's favorite food is Fish Treats.

CLUE B The Siamese will only eat Liver Bites.

CLUE C Tiki will not eat with the Himalayan.

	Persian	Siamese	Himalayan
Tiki	X	✓	X
Moby	X	X	✓
Copper	✓	X	X

CLUE D The Himalayan does not have a favorite food.

What is the name of Brenda's Siamese cat?

Solve. Make a table to show all the possibilities.

3. Alex, Dee, Rod, and Sue are athletes. One is on the baseball team, one is on the soccer team, one is on the track team, and one is on the golf team.

CLUE A Alex is taller than the soccer player.

CLUE B Dee and Rod do not know how to play golf.

CLUE C Neither Alex nor Dee has time to play baseball.

CLUE D Sue's sport does not use a ball.

Who plays on the soccer team?

4. Chuck, Lillian, Denise, and Brad each have a different hobby. The hobbies are stamp collecting, oil painting, kite flying, and photography.

CLUE A Chuck and Denise are older than the stamp collector.

CLUE B Denise and Brad do not paint or take photographs.

CLUE C Chuck and the photographer are cousins.

What is Lillian's hobby?

IMPORTANT TERMS

Equation *(p. 58)*
Solution *(p. 58)*
Root *(p. 58)*
Equivalent equations *(p. 58)*

IMPORTANT IDEAS

1. *Addition Property for Equations:* Adding the same real number to each side of an equation forms an equivalent equation.
2. *Subtraction Property for Equations:* Subtracting the same real number from each side of an equation forms an equivalent equation.
3. *Multiplication Property for Equations:* Multiplying each side of an equation by the same nonzero real number forms an equivalent equation.
4. *Division Property for Equations:* Dividing each side of an equation by the same nonzero real number forms an equivalent equation.
5. To solve an equation that involves more than one operation, the Addition or Subtraction Property for Equations is generally used before the Multiplication or Division Property.
6. In equations that contain parentheses, use the Distributive Property first.
7. The words sum, plus, more than, and increased by suggest the operation of addition.
8. The words decreased by, difference, minus, less, and less than suggest the operation of subtraction.
9. The words product, multiplied by, times, and doubled suggest the operation of multiplication.
10. The words quotient and divided by suggest the operation of division.
11. Using Algebra to Solve Word Problems
    - **a.** Choose a variable. Use the variable to represent the unknowns.
    - **b.** Write an equation for the problem.
    - **c.** Solve the equation.
    - **d.** Check your answer with the statements in the original problem. Answer the question.

CHAPTER REVIEW

For Exercises 1–24, solve each equation. Check each answer.

SECTION 3.1

1. $x + 23 = 16$ **2.** $y + 29 = -18$ **3.** $b - 47 = -28$ **4.** $d - 37 = 49$

5. $-28 = 14 + f$ **6.** $76 = -32 + g$ **7.** $5.6 = t - 24.7$ **8.** $-12.3 = x + 19.8$

SECTION 3.2

9. $-7k = 63$ **10.** $-96 = 12n$ **11.** $-8 = \frac{m}{14}$ **12.** $\frac{x}{-8} = 19$

13. $-\frac{3}{4}t = -27$ **14.** $60 = -\frac{12}{4}z$ **15.** $12y = -99.6$ **16.** $131.4 = -9r$

SECTION 3.3

17. $4z - 3 = 42$ **18.** $6t - 14 = 38$ **19.** $-68 = 8m + 12$ **20.** $-54 = 18 + 12p$

21. $78 - 6x = 34$ **22.** $69 = 102 - 8n$ **23.** $-22 + \frac{v}{2.4} = -46$ **24.** $-9 = \frac{w}{1.9} + 48$

SECTION 3.4

For each problem, find the rate, r, or the time, t. Round your answer to the nearest tenth.

25. A wildebeest was observed racing a distance of 86.4 meters in 4.5 seconds. Find the speed of the wildebeest in meters per second.

26. The speed of a radio signal is 186,000 miles per second. How long does it take for a radio signal to travel from the earth to Mars, which is about 35,000,000 miles away?

For each problem, find the width, w, or length, ℓ. Round your answer to the nearest tenth.

27. The perimeter of a wall mirror is 204 inches. The width is 18.1 inches. Find the length of the mirror.

28. A steel ribbon 3.8 meters long is exactly long enough to wrap around a packing carton once. The base of the carton is 1.2 meters long. How wide is the base of the carton?

SECTION 3.5

Solve each equation. Check each answer.

29. $8x + 6 - 5x = 39$

30. $56 = 12k - 8k$

31. $-19.3y - 16 + 5.9y = -83$

32. $-55 = 2.4n + 74 - 15.3n$

33. $35 = 4(t - 2) - 6t$

34. $7z - 3(z + 7) = -67$

35. $-28 = (3p - 2) - 2(5p + 12)$

36. $(18 - x) - (3x + 4) = -48$

SECTION 3.6

For Exercises 37–40, choose a variable and write an equation for the word sentence. Then solve the equation.

37. The total distance traveled less 25.6 kilometers equals 278.3 kilometers.

38. Twenty-four times the cost of each box equals $30.72.

39. A bowler's score divided by 5 games equals her average game score of 168.

40. The low temperature for one day increased by 24° equals the high temperature of 30.5°.

SECTION 3.7

For Exercises 41–44:

a. *Choose a variable to represent the unknowns.*

b. *Write an equation for each problem.*

c. *Solve.*

41. The number of rainy days last year increased by 6 equals the previous recorded high of 82 rainy days in a year. Find the number of rainy days last year.

42. The record low temperature for Raleigh, North Carolina, divided by 0.04 equals the record high temperature of 100°F. Find the record low temperature.

43. Three times the reading on a thermometer decreased by 7 degrees equals the wind chill temperature of −31°F. Find the thermometer reading.

44. Three less than two times the number of pages in the first edition of a book equals the number of pages in the second edition of the book. The second edition contains 210 pages. Find the number of pages in the first edition.

CUMULATIVE REVIEW: CHAPTERS 1–3

Simplify. (Section 1.1)

1. $|-0.07|$
2. $-|2\frac{1}{5}|$
3. $-|-12|$
4. $|-(-134)|$

Add. (Section 1.2)

5. $-8 + (-14)$
6. $-17 + (-35)$
7. $-139 + 39$
8. $3.7 + (-1.6)$

Subtract. (Section 1.3)

9. $16 - 28$
10. $-10 - 10$
11. $5\frac{2}{3} - (-1\frac{1}{3})$
12. $-19 - (-21)$

Multiply. (Section 1.4)

13. $(-4)(7)$
14. $(2)(-26)$
15. $(-1.7)(-100)$
16. $(-5)(-16)$

Divide. (Section 1.5)

17. $\frac{-20}{-5}$
18. $\frac{-72}{9}$
19. $-330 \div 22$
20. $-18 \div (-1\frac{1}{2})$

Evaluate. (Section 1.6)

21. $4 - 7 + 5$
22. $-32 \div 2(-4)$
23. $-11 + (-5)(-2)$
24. $12 \div (-3) - 18$

Evaluate each expression. (Section 1.7)

25. $2n^2 - 3$ when $n = 4$
26. $y^2 + 7y$ when $y = -3$
27. $-2x + 3x^2$ when $x = -5$

Refer to the set of numbers below. Then list the numbers named in Exercises 28–30. (Section 2.1)

$$\{-5, -\sqrt{3}, -1\frac{2}{3}, 0, 4, \pi, \frac{18}{3}\}$$

28. Rational numbers
29. Integers
30. Real numbers

Simplify. (Section 2.2)

31. $-11 + 18 - 19 + 12$
32. $6.4 - 8.2 + p - 0.4$
33. $-44 + k - w + 62$

Multiply. (Section 2.3)

34. $(-3)(6)(-2)$
35. $(-2)(-2)(x)(-2)$
36. $(-\frac{2}{3})(t)(18)(p)$

Simplify. (Section 2.4)

37. $(8xy)(-4x)$
38. $4x - (y - 3)$
39. $(2a)(-3ab)(-5b)$

Multiply and simplify. (Section 2.5)

40. $3(x+y)$ **41.** $-6(8a-1)$ **42.** $(-2x+7)5x$ **43.** $(-16x+8y)(-\frac{1}{2})$

Factor. (Section 2.5)

44. $5x-5y$ **45.** $12r-9$ **46.** $21ab+7a$ **47.** $-6x^2-18x$

Simplify. (Section 2.6)

48. $23x^2-x^2$ **49.** $9a-12+7a+4$ **50.** $3y^2-7y+2+y^2-y$

Solve. Check each answer. (Sections 3.1 through 3.3)

51. $x-3=-7$ **52.** $0.2=k+0.9$ **53.** $-2=\frac{h}{-2}$ **54.** $-8x=32$

55. $-36=\frac{2}{3}y$ **56.** $5a-3=-7$ **57.** $-4-3x=13$ **58.** $\frac{m}{6}+19=-2$

Solve each problem. (Section 3.4)

59. The perimeter of a picture frame is 37 inches. The length of the frame is 11 inches. Find the width.

60. A certain car traveled a distance of 1120 miles at an average speed of 40 miles per hour. How long did it take the car to travel that distance?

Solve. Check each answer. (Section 3.5)

61. $-n+8+4n=10$

62. $18=5x-(2x+6)$

63. $10=4-2(3-2x)$

64. $3(x-5)+2(3-2x)=18$

Choose a variable and write an equation for each word sentence. Then solve the equation. (Section 3.6)

65. The quotient of the number of apples and 17 is 5.

66. The product of 9 and the width of a rectangle is equal to 23.4 units.

Choose a variable to represent the unknowns. Write an equation for each problem. Solve. (Section 3.7)

67. Lincoln High School Cheerleaders sold 54 fewer football programs this week than they sold last week. They sold 192 football programs this week. How many programs did they sell last week?

68. The number of trees planted by C. M. Grow's Nursery in April was 3 more than twice the number of trees planted by C. M. Grow's Nursery in March. Seventy-one trees were planted in April. Find the number of trees planted in March.

Problem Solving: One Variable

CHAPTER 4

In nature, patterns can be found in the waves made by drops of rain in a still pool, in the veins of a leaf, and in the rings of a tree trunk. In mathematics, finding patterns is one method used in problem solving.

4.1 Using Conditions

To solve word problems, it is helpful to identify the conditions in the problem. One condition is used to represent the unknowns of the problem. Call this Condition 1. Another condition is used to write the equation. Call this Condition 2. Then solve the equation.

EXAMPLE 1 One model of subcompact car has 71 cubic feet more passenger space than luggage space (Condition 1). The total amount of passenger and luggage space is 87 cubic feet (Condition 2). Find the amount of both types of space.

Solution:

[1] Use Condition 1 to represent the unknowns. → Let s = the luggage space. Then $s + 71$ = the passenger space.

[2] Use Condition 2 to write an equation.

Think: Luggage space + Passenger space = Total space

Translate: $s + s + 71 = 87$

[3] Solve the equation. →

$$2s + 71 = 87$$
$$2s = 16$$
$$s = 8$$
$$s + 71 = 79$$

Don't forget to find s + 71.

[4] **Check:** Condition 1 Are there 71 more cubic feet of passenger space than luggage space? Does $79 = 8 + 71$? Yes ✓

Condition 2 Does the total amount of space equal 87 cubic feet? Does $8 + 79 = 87$? Yes ✓

The passenger space is 79 cubic feet, and the luggage space is 8 cubic feet.

Steps for Solving Word Problems

1. Identify Condition 1. Use Condition 1 to represent the unknowns.
2. Identify Condition 2. Use Condition 2 to write an equation.
3. Solve the equation.
4. Check the results with the conditions. Answer the question.

EXAMPLE 2

The number of eleventh grade members of the Ski Club this year is 5 fewer than twice the number of twelfth grade members (Condition 1). The total membership is 82 (Condition 2). How many eleventh grade members are there?

Solution:

1 Use Condition 1 to represent the unknowns. → Let n = the number of twelfth graders. Then $2n - 5$ = the number of eleventh graders.

2 Use Condition 2 to write an equation.

Think:	**Number of twelfth graders**	+	**Number of eleventh graders**	=	**Total membership**
Translate:	n	+	$2n - 5$	=	82

3 Solve the equation. →

$$3n - 5 = 82$$
$$3n = 87$$
$$n = 29$$
$$2n - 5 = 53$$

4 **Check:** Condition 1 Is the number of eleventh grade members 5 less than twice the number of twelfth grade members?

Does $53 = 2(29) - 5$? Yes ✓

Condition 2 Is the total membership 82?

Does $29 + 53 = 82$? Yes ✓

There are 29 twelfth grade members and 53 eleventh grade members.

CLASSROOM EXERCISES

For Exercises 1–10, Condition 1 is underscored once. Condition 2 is underscored twice. (Example 1)

a. *Use Condition 1 to represent the unknowns. Use p as the variable.*

b. *Use Condition 2 to write an equation for the problem.*

1. There were 186 more seats for tourist class than for first class on one flight. The total number of seats was 234.

2. The number of parking spaces for small cars in a lot is 75 less than the number for standard cars. The total number of spaces is 425.

3. The parts for a repair job on a car cost \$3.58 more than the labor. The total bill was \$75.62.

4. One model of compact car has 69 cubic feet more passenger space than cargo space. The total amount of space is 101 cubic feet.

(Example 2)

5. A city's rainfall in June was $4\frac{3}{5}$ inches less than twice the amount in May. The total rainfall for the two months was $14\frac{9}{10}$ inches.

6. A family's electric bill one month was \$0.35 less than twice the gas bill. The total amount of the two bills was \$76.30.

7. A professional golfer earned \$5,000 less than twice the amount earned the previous year. The total amount earned in two years was \$93,000.

8. The number of students in school is 153 more than ten times the number of teachers. The total number of students and teachers is 3244.

9. The number of multiple–choice questions on a test was seven less than three times the number of true–false questions. The total number of questions was 105.

10. The mass of a machine is 7.6 kilograms less than 20 times the mass of its shipping crate. The total mass of the shipment is 239.4 kilograms.

WRITTEN EXERCISES

Goal: To represent conditions of a word problem and to solve the problem

Sample Problem: The attendance at the first performance of an orchestra was 560 more than at the second performance (Condition 1). The total attendance for the two performances was 5320 (Condition 2). Find the number attending each performance.

Answers: 1st performance: 2940; 2nd performance: 2380

For some problems, Condition 1 is underscored once. Condition 2 is underscored twice.

a. *Use Condition 1 to represent the unknowns.*

b. *Use Condition 2 to write an equation for the problem.*

c. *Solve.* (Example 1)

1. A marathon race had 448 more local participants than visiting runners. The total number in the race was 3640. Find the number of local runners and the number of visiting runners.

2. A coach's teams during his career won 121 more games than they lost. The total number of games was 531. Find the number of wins and the number of losses.

3. A tennis match ending in two sets had 4 more games in the second set than in the first. The total number of games was 20. Find the number played in each set.

4. A collector has paintings and sculptures. The paintings are valued at $350,000 more than the sculptures. The total collection is valued at $1,260,000. Find the value of the paintings and the value of the sculptures.

(Example 2)

5. In one year, a retired couple spent $150 less on housing and transportation than twice the amount spent on food. The total amount spent on housing, transportation, and food was $8500. Find the amount spent for food.

6. At one point of a season the leading scorer in pro basketball had scored four fewer points on field goals than four times the number of points on free throws. Her total number of points was 1776. Find the number of points scored on field goals and the number scored on free throws.

7. A family's large car holds 2.3 gallons less fuel than twice the capacity of its compact car. The total fuel capacity of the two cars is 35.2 gallons. Find the fuel capacity of each car.

8. An art book contains 16 fewer pages of illustrations than three times the number of printed pages. The total number of pages is 512. Find the number of pages of illustrations.

MIXED PRACTICE

9. The passenger space in one car contains 5 cubic feet more than five times the cargo space. The total passenger and cargo space is 131 cubic feet. Find the size of each space.

10. A bedroom requires 120 square meters more carpeting than a living room. The total amount of carpeting needed is 968 square meters. Find the amount needed for the bedroom and the living room.

11. Simon has a part-time job. In one two-week period he worked $2\frac{3}{4}$ hours more the second week than the first. His total time for the two weeks was $45\frac{3}{4}$ hours. How many hours did he work each week?

12. A trip from a family's home to the coast required 4.5 gallons more fuel returning than going. The total amount needed for the round-trip was 180.3 gallons. How much fuel was needed for each leg of the trip?

13. Alfredo has a coin collection. He has 26 fewer U.S. coins than three times the number of foreign coins. The collection consists of 998 coins. Find the number of each type of coin.

14. The average snowfall for a city is 5.4 inches less than twice the amount of rainfall. The total amount of precipitation is 100.9 inches. Find the average amount of snowfall.

15. Mary Ann enters customer records on a computer. In a two-day period she entered 12 more records the second day than the first day. In the two days she entered 256 records. Find the number of records entered each day.

16. An insurance analyst set a goal of doubling her sales from the previous year. Her sales actually were $50,000 less than twice the previous year's sales. The total for the two years was $980,000. Find her sales for the second year.

NON-ROUTINE PROBLEMS

17. What is the smallest number of coins that a cashier needs in order to make change for a purchase of less than one dollar?

18. At a banquet, four people are to be seated on one side of the head table. In how many different ways can these four people be seated?

19. A pet store owner wanted to separate his fish so that the same number of fish were in each tank. When he separated the fish by twos, by threes, by fours, by fives, or by sixes, there was always one fish left over. What is the least number of fish the pet owner could have had?

20. Two books, Math I and Math II, are placed side by side on a bookshelf so that Math I is to the left of Math II. Each book cover is $\frac{1}{8}$ inch thick, and the pages, not counting the covers, are 2 inches thick. What is the distance from the first page of Math I to the last page of Math II?

REVIEW CAPSULE FOR SECTION 4.2

Solve and check each equation. (Pages 71–73)

1. $2t + 2(t - 10) = 84$
2. $256 = 2w + 2(w + 5)$
3. $2(5n) + 2(13n) = 151.2$
4. $143 = 2(7p) + 2(15p)$
5. $3x + 8x + 13x = 163.2$
6. $185.4 = 4r + 5r + 9r$

Represent the unknowns in each problem. Use the variable d.

EXAMPLE: The ratio of rainy days to sunny days is 1:6. **ANSWER:** $1d$; $6d$

7. The ratio of tin to copper in an alloy is 1:4.

8. The ratio of gaining stocks to losing stocks is 7:2.

9. The ratio of orange, grapefruit, and pineapple sections in a fruit cup is 5:4:2.

10. Stainless steel contains steel, chromium, and nickel in the ratio 37:9:4.

4.2 Using Geometry Formulas: Perimeter

The ***perimeter*** of a geometric figure such as a triangle, square, or rectangle is the sum of the lengths of its sides.

P–1 **What is the perimeter of each figure below?**

a.

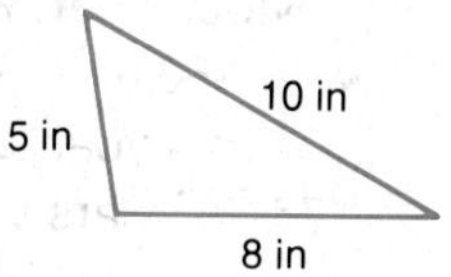

b.

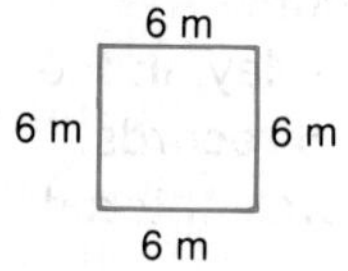

c.

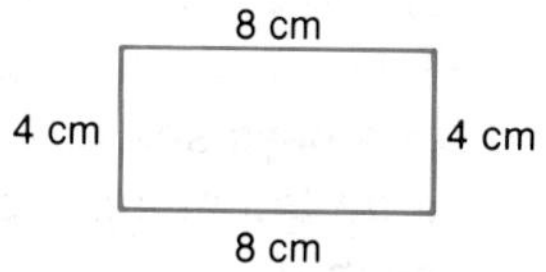

EXAMPLE 1

The length of a drawing board is 5 inches more than the width (Condition 1). The perimeter is 94 inches (Condition 2). Find the length and width.

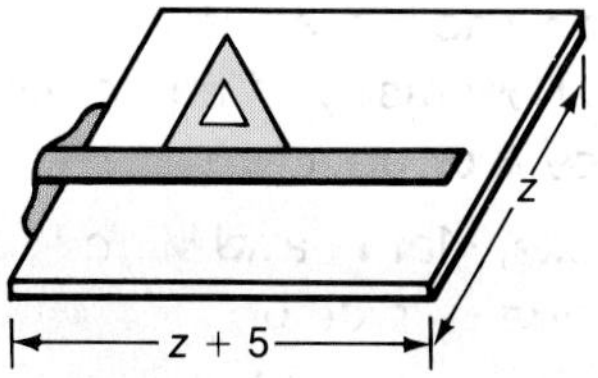

Solution:

1 Use Condition 1 to represent the unknowns. ⟶ Let $z =$ the width. Then $z + 5 =$ the length.

2 Use Condition 2 to write an equation.

$$\begin{array}{ccccc} p & = & 2\ell & + & 2w \\ \downarrow & & \downarrow & & \downarrow \\ 94 & = & 2(z+5) & + & 2z \end{array}$$

◀ ***Formula for the perimeter of a rectangle***

3 Solve the equation. ⟶

$$94 = 2z + 10 + 2z$$
$$94 = 4z + 10$$
$$84 = 4z$$
$$21 = z$$
$$z + 5 = 26$$

◀ ***Don't forget to find $z + 5$.***

4 **Check:** Condition 1 Is the length 5 inches more than the width? Does $26 = 21 + 5$? Yes ✓

Condition 2 Is the perimeter 94 inches? Does $2(26) + 2(21) = 94$? Yes ✓

The length is 26 inches and the width is 21 inches.

Instead of the lengths of the sides of a figure, sometimes the ratio of the lengths of the sides is given. For example, if the lengths of a particular triangle are in the ratio 5:8:11, the lengths can be represented by $5x$, $8x$, and $11x$.

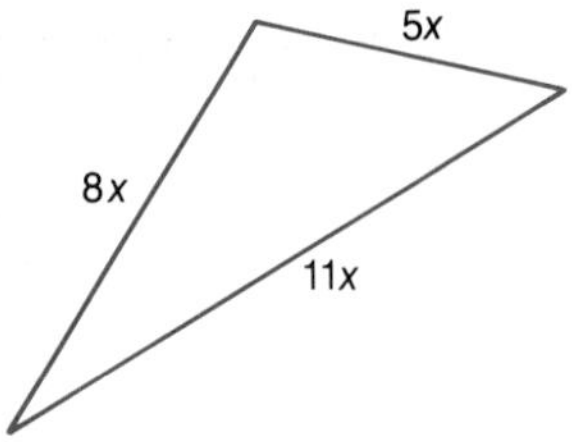

EXAMPLE 2

A cross country ski course is along a triangular route. The three legs of the course in order from the starting point have lengths in the ratio 2:3:4 (Condition 1). The length of the entire course is 45 kilometers (Condition 2). Find the length of each leg.

Solution:

[1] Use Condition 1 to represent the unknowns. → Let $2x$ = the length of the first leg.
Then $3x$ = the length of the second leg.
And $4x$ = the length of the third leg.

[2] Use Condition 2 to write an equation.

Think: **Length of first leg** + **Length of second leg** + **Length of third leg** = **Total length**

Translate: $2x + 3x + 4x = 45$

[3] Solve the equation. →

$$9x = 45$$
$$x = 5$$
$$2x = 10$$
$$3x = 15$$
$$4x = 20$$

[4] **Check:** Condition 1 Are lengths of the legs in the ratio 2:3:4?

Does $\frac{10}{15} = \frac{2}{3}$? (Does $10 \cdot 3 = 15 \cdot 2$?) Yes ✓

Does $\frac{15}{20} = \frac{3}{4}$? Yes ✓

Does $\frac{10}{20} = \frac{2}{4}$? Yes ✓

Condition 2 Does the sum of the lengths of the legs equal 45 km?

Does $10 + 15 + 20 = 45$? Yes ✓

The lengths of the legs of the course are 10 kilometers, 15 kilometers, and 20 kilometers.

CLASSROOM EXERCISES

Find the perimeter of each figure below. (P-1)

1.

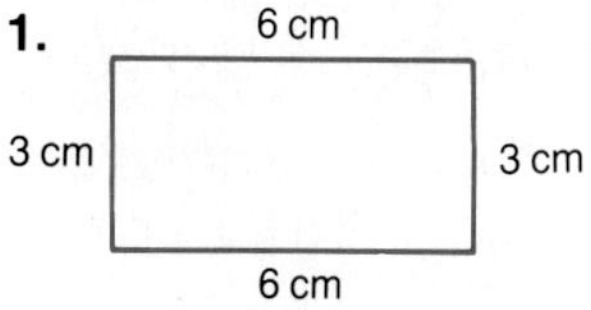

2.

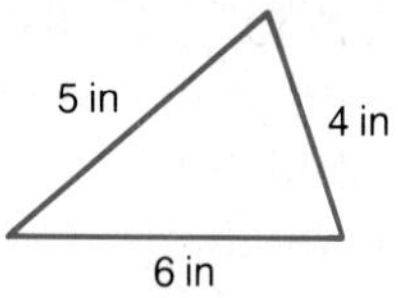

3.

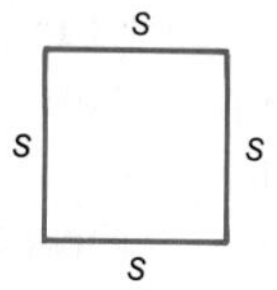

For Exercises 4–15, Condition 1 is underscored once. Condition 2 is underscored twice.

a. *Use Condition 1 to represent the unknowns. Use the variable m.*

b. *Use Condition 2 to write an equation for the problem.* (Example 1)

4. The width of a screen for viewing slides is 10 inches less than the length. The perimeter is 140 inches.

5. The length of a flag is 27 inches more than its width. The perimeter is 174 inches.

6. The top of a stereo cabinet is a rectangle with its length 16 inches more than its width. The perimeter is 102 inches.

7. The length of one size of photo slide is 1.1 centimeters more than the width. The perimeter is 11.4 centimeters.

8. A scout troop hikes along a triangular route. The first leg is twice as long as the second leg. The third leg is 17 miles long. The entire hike is for 50 miles.

9. A sailboat race is along a triangular route. The first leg is 25.3 kilometers. The third leg is three times as long as the second leg. The race has a length of 67.3 kilometers.

(Example 2)

10. One special triangle has the lengths of its sides in the ratio 3:4:5. The perimeter is 38.4 centimeters.

11. Three cities form a triangle and the distances between the cities are in the ratio 4:5:6. The perimeter of the triangle is 184.5 kilometers.

12. The ratio of the width to the length of a tennis court is 6:13. The perimeter is 228 feet.

13. The ratio of the length to the width of a door is 20:9. The perimeter is 232 inches.

14. A surveyor working on a highway locates three stakes to form a triangle. One side of the triangle measures 34.3 meters. The ratio of the lengths of the other sides is 5:9. The perimeter is 70.7 meters.

15. A guy wire supports a pole and is anchored 9.8 meters from the base of the pole. The ratio of the length of the guy wire to the height of the pole is 7:5. The perimeter of the triangle formed is 33.8 meters.

WRITTEN EXERCISES

Goal: To solve word problems involving perimeters of triangles and rectangles.

Sample Problem: The ratio of the length to the width of a flag is 19:10 (Condition 1). Find the length and width of a flag having a perimeter of 174 feet (Condition 2).

Answer: 57 feet by 30 feet

For some problems, Condition 1 is underscored once. Condition 2 is underscored twice.

a. *Use Condition 1 to represent the unknowns.*

b. *Use Condition 2 to write an equation for the problem.*

c. *Solve.* (Example 1)

1. A sign is rectangular in shape. Its length is 6 feet greater than its height. Its perimeter is 84 feet. Find its length and height.

2. A mailing carton has a rectangular base. The length is 3 inches greater than the width. The perimeter is 36 inches. Find the length and width.

3. A sculptor designs a monument that includes a triangle with a base of 8.5 meters. The longer of the other two sides has a length 1.7 meters greater than the other. The perimeter of the triangle is 51.2 meters. Find the two unknown lengths.

4. A company's trademark is triangular. One side of the original design is 12.5 centimeters long. The shorter of the other two sides has a length 6.6 centimeters less than the length of the other. The perimeter of the triangle is 36.1 centimeters. Find the two unknown lengths.

(Example 2)

5. One special triangle has sides in the ratio 5:12:13. Find the lengths of the sides if the perimeter is 204.0 meters.

6. The ratio of the width to the length of a soccer field is 15:22. The perimeter is 370 meters. Find the length and width.

7. A pilot flies a plane on a triangular route. The ratio of the lengths of the three legs of the flight is 4:7:9. The entire route is 684 kilometers long. Find the lengths of the legs of the route.

8. A picture is to be framed with its length and width in the ratio 50:31. The length of the frame needed (perimeter) is 243 centimeters. Find the length and width of the picture.

MIXED PRACTICE

9. A house has a rectangular floor plan. The length is 4 feet less than twice the width. The perimeter is 148 feet. Find the length and width.

10. An official U.S. flag must have a ratio of length to width of 19:10. Find the length and width of a flag with a perimeter of 87 inches.

11. The width of a computer terminal screen is $2\frac{1}{2}$ inches more than the height. The perimeter of the screen is 44 inches. Find the height and width.

12. A computer printout sheet is 9.9 centimeters longer than it is wide. The perimeter is 131.4 centimeters. Find the length and width.

13. A cable attached to the top of a tower is anchored 35.6 meters from the base of the tower. The perimeter of the triangle formed is 108.1 meters. The ratio of the cable's length to the tower's height is 18:11. Find the tower's height.

14. A punch press stamps a rectangular opening in a metal part. The length of the opening is 4.5 centimeters more than the width. The perimeter is 37.8 centimeters. Find the length and width of the opening.

MORE CHALLENGING EXERCISES

15. A triangle has sides in the ratio 3:4:5. A square has sides equal in length to the smallest side of the triangle. The sum of the perimeters of the two figures is 96 centimeters. Find the lengths of the sides in each figure.

16. Lot A is rectangular in shape. It is 8 meters longer than it is wide. Lot B is in the shape of a square. Each side of Lot B has the same measure as the width of Lot A. The sum of the perimeters of the two lots is 296 meters. Find the length and width of each lot.

REVIEW CAPSULE FOR SECTION 4.3

For Exercises 1–6, evaluate both expressions for the given value of the variable. (Pages 21–23)

1. $\left.\begin{matrix}5x+3\\4x-9\end{matrix}\right\}$ $x=6$

2. $\left.\begin{matrix}20-2y\\56-5y\end{matrix}\right\}$ $y=12$

3. $\left.\begin{matrix}4-3z\\z+22\end{matrix}\right\}$ $z=-4\frac{1}{2}$

4. $\left.\begin{matrix}21+m\\-19-7m\end{matrix}\right\}$ $m=-5$

5. $\left.\begin{matrix}5r+2(r-1)\\3(r+1)\end{matrix}\right\}$ $r=1\frac{1}{4}$

6. $\left.\begin{matrix}8p-2(2p-1)\\6(p+2)\end{matrix}\right\}$ $p=-5$

4.3 Variable on Both Sides

When the variable appears on both sides of an equation, the first step is to eliminate the variable from one side.

EXAMPLE 1 Solve: $5x + 12 = 3x - 6$

Solution: $5x + 12 = 3x - 6$

[1] Subtract $3x$ from each side. → $5x + 12 - \mathbf{3x} = 3x - 6 - \mathbf{3x}$

$2x + 12 = -6$

[2] Subtract 12 from each side. → $2x + 12 - \mathbf{12} = -6 - \mathbf{12}$ ◀ $-6 - 12 = -6 + (-12)$

$2x = -18$

[3] Divide each side by 2. → $\frac{2x}{\mathbf{2}} = \frac{-18}{\mathbf{2}}$

$x = -9$

Check: $5x + 12 = 3x - 6$

$5(-9) + 12$	$3(-9) - 6$
$-45 + 12$	$-27 - 6$
-33	-33

◀ $-27 - 6 = -27 + (-6)$

P–1 **Solve.**

a. $11t + 16 = 4t + 2$ **b.** $9j + 6 = 4j - 14$

EXAMPLE 2 Solve: $12 - 7r = 42 - 3r$

Solution: $12 - 7r = 42 - 3r$

[1] Add $7r$ to each side. → $12 - 7r + \mathbf{7r} = 42 - 3r + \mathbf{7r}$

$12 = 42 + 4r$

[2] Subtract 42 from each side. → $12 - \mathbf{42} = 42 + 4r - \mathbf{42}$

$-30 = 4r$

[3] Divide each side by 4. → $\frac{-30}{\mathbf{4}} = \frac{4r}{\mathbf{4}}$

$-7\frac{1}{2} = r$ ◀ *The check is left for you.*

P–2 **Solve.**

a. $3 - 2p = 17 - p$ **b.** $15 - 3f = 3 + 6f$

In equations that contain parentheses, use the Distributive Property first. Then eliminate the variable from one side of the equation.

EXAMPLE 3 Solve: $5 + 2x = 2(x - 3) + x$

Solution:

[1] Use the Distributive Property. ⟶ $5 + 2x = 2(x - 3) + x$

[2] Combine like terms. ⟶ $5 + 2x = 2x - 6 + x$

$5 + 2x = 3x - 6$

[3] Subtract 2x from each side. ⟶ $5 + 2x - \mathbf{2x} = 3x - 6 - \mathbf{2x}$

$5 = x - 6$

[4] Add 6 to each side. ⟶ $5 + \mathbf{6} = x - 6 + \mathbf{6}$

$11 = x$

Check: $5 + 2x = 2(x - 3) + x$

$5 + 2(11)$	$2(11 - 3) + 11$
$5 + 22$	$2(8) + 11$
27	27

P–3 **Solve.**

a. $9 + 4y = 3(y + 4) - y$ **b.** $2 + 10g = 6(2g + 3) + 2g$

CLASSROOM EXERCISES

Write the two equations that are formed in eliminating the variable from each side of the following equations. (Step 1, Examples 1–2)

1. $3x - 2 = 2x + 5$ **2.** $3 - 4y = 7 - 5y$ **3.** $8 + a = -3a + 6$

4. $10 + 5g = -8 + 3g$ **5.** $-4d - 6 = 12 - d$ **6.** $10b - 7 = 9 - 4b$

For each equation, write an equivalent equation that has the variable on one side. (Steps 1 and 2, Examples 1–2)

EXAMPLE: $3x - 7 = 2x + 8$ **ANSWER:** $x - 7 = 8$ or $-7 = -x + 8$

7. $4p + 2 = 6 - p$ **8.** $5 - 2r = 3r - 2$ **9.** $-3z - 8 = z + 5$

10. $3m + 2 = -2m - 8$ **11.** $10.8 - 3.4j = 9.2 + 7.1j$ **12.** $8.2 - h = 3h - 3.4$

WRITTEN EXERCISES

Goal: To solve equations that have a variable on both sides
Sample Problem: Solve: $5 + 4x = 3x - 10 - x$
Answer: $x = -7\frac{1}{2}$

Solve. Check each answer. (Example 1)

1. $3x - 5 = 2x + 12$

2. $5x - 3 = 4x + 9$

3. $12a - 9 = 10a + 17$

4. $8b - 14 = 6b + 28$

5. $1.6c - 7.5 = 14.1 + 3.4c$

6. $0.5d - 3 - d = -1.8 + d$

7. $3\frac{1}{2}k + 17 = 2\frac{3}{4}k + 12$

8. $\frac{1}{20} + \frac{1}{5}s - 11 = \frac{3}{5}s - 11$

(Example 2)

9. $12 - x = 26 - 3x$

10. $20 - 2y = 56 - 5y$

11. $2 - 3z = z + 22$

12. $5 - 2m = -22 - 4m$

13. $2\frac{1}{4} - 1\frac{4}{5}p - 19 = \frac{2}{3}p + 38 - 2\frac{1}{3}$

14. $9 - 5\frac{1}{4}x = 27 - 2\frac{1}{4}x - 2\frac{3}{4}x$

15. $2t + 1.2 - 5t = 2.8 - t$

16. $3.2 + z = -6.7 - 1.2z$

(Example 3)

17. $3(4 - x) - 22 = x + 46$

18. $-4(d - 2) - 14 = 36 + 2d$

19. $14 - 3y = -3(4 + 2y) - 7$

20. $-6g - 8 = 12 - 4(3g + 9)$

21. $0.2(n + 1) = 0.6(n - 1) - 0.7$

22. $3(y + 0.5) = 0.5y + 1.5$

23. $\frac{4}{5}y - \frac{1}{2}(y - 1) = \frac{3}{4}(y + 1)$

24. $\frac{1}{2}(z + 1) - 2z = 6z + \frac{1}{2}$

MIXED PRACTICE

25. $10 - 5y = y + 61$

26. $14 - 3b = -16 - 7b$

27. $6 - 15q = 11q - 46$

28. $21 + m = -19 - 7m$

29. $3(2 - k) - 7 = 4(2k + 8)$

30. $5z + 2(z - 1) = 3(z + 1)$

31. $-10 - 18d = 26 - 12d$

32. $-4f - 19 = 17 + f$

33. $4.4 + 2.1d = 5.7d - 17.2$

34. $2.3x - 31.5 = 14.9 + 5.2x$

35. $8p - 2(2p - 1) = 6(p + 2)$

36. $5(2r + 3) - 4r = 20r + 1$

MORE CHALLENGING EXERCISES

For Exercises 37–38, a represents an even integer and b represents an odd integer. Solve for x. Identify x as an even integer or an odd integer.

37. $2a + 2x = x + b$

38. $4x - a = 3\frac{1}{2}x - 3b$

MID-CHAPTER REVIEW

For Exercises 1–8, Condition 1 is underscored once. Condition 2 is underscored twice. Use Condition 1 to represent unknowns. Use Condition 2 to write an equation for the problem. Solve. (Section 4.1)

1. A tumbling class for 5-6 year olds at the Y had seven more girls than boys. The total number enrolled for the class was 45. Find the number of girls and the number of boys.

2. Joel bought a sports coat for $15 less than five times the cost of a pair of slacks. The total cost of the coat and slacks was $273. Find the cost of each item.

3. A traffic count between 6:00 a.m. and 8:00 a.m. on a highway showed 2900 fewer cars the first hour than the second hour. The total number of cars for the two-hour period was 24,500. Find the number for each hour.

4. A school club's receipts from special projects was $5.00 more than three times the receipts from dues. The amount of total receipts was $485. Find the amounts received from dues and from special projects.

(Section 4.2)

5. A rectangular lot has a length that is 6.2 meters greater than twice its width. The perimeter is 106.6 meters. Find the length and width.

6. An Olympic-size swimming pool has a length and width in the ratio 25:12. The pool's perimeter is 148 meters. Find the length and width.

7. A stone monument has a triangular base. The lengths of the sides of the base are in the ratio 5:4:2. The perimeter is 264 inches. Find the length of each side.

8. A rectangular desk top has a length that is 2 centimeters less than twice its width. The perimeter is 460 centimeters. Find the length and width of the desk top.

For Exercises 9–14, solve each equation. Check each answer. (Section 4.3)

9. $6d - 5 = 9d + 37$
10. $4r + 11 = 8r + 43$
11. $16 - 3p = 32 - 7p$
12. $23 - 5y = 89 - 11y$
13. $-3(x - 2) + x = 10 - 6x$
14. $14 + 5c = 2c - 7(c + 8)$

REVIEW CAPSULE FOR SECTION 4.4

Compute.

EXAMPLE: 8.5% of 350 **ANSWER:** $8.5\% = 0.085$; $0.085 \cdot 350 = 29.75$

1. 12% of $960
2. 8% of $87.50
3. 5% of 2780
4. 40% of $29.95
5. 75% of 56
6. 17% of 250
7. 4.5% of $750
8. $33\frac{1}{3}\%$ of 915

4.4 Using Per Cent

Merchandise is often offered on sale to consumers at a reduction in price. The price before the reduction is called the ***list price*** or ***regular price.*** The amount that the item is reduced is called the ***discount.*** The reduced price is called the ***sale price.***

The following word rule and formula relate sale price, list price, and discount.

Word Rule: The sale price equals the list price less the discount.

Formula: $s = p - d$ ◀ *s = sale price*, *p = list price*, *d = discount*

Discount expressed as a per cent of the list price is called the ***rate of discount.***

The following word rule and formula relate discount, rate of discount, and list price.

Word Rule: Discount equals the rate of discount times the list price.

Formula: $d = r \cdot p$ ◀ *r = rate of discount*

List Price	Rate of Discount	Discount
\$60	20%	(0.20) (60) = \$12.00
\$150	10%	(0.10) (150) = 15.00
\$225	8%	(0.08) (225) = 18.00

EXAMPLE 1 A lamp was on sale at a rate of discount of 20%. The sale price was \$25.00. Find the list price of the lamp and the discount.

Solution:

[1] Represent the list price and the discount. → Let $p =$ the list price. Then $0.20p =$ the discount.

[2] Use the formula to write an equation. →

$$s = p - d$$
$$25.00 = p - 0.20p$$

[3] Solve the equation. ⟶ $25.00 = 0.80p$

$31.25 = p$

$0.20p = 6.25$

Don't forget to find 0.20p.

[4] **Check:** Was the discount 20% of the list price?
Does $6.25 \div 31.25 = 0.2$ or 20%? Yes ✓
Did the list price less the discount equal the sale price?
Does $31.25 - 6.25 = 25.00$? Yes ✓

The list price was \$31.25 and the discount was \$6.25.

In Example 2, you are asked to organize the information given so that you can answer the question that is asked. Making a table will help you to do this and to write an equation for the problem.

EXAMPLE 2

During a softball tournament, Team B played 2 more games than Team A (Condition 1). Team A won 60% of the games it played, and Team B won 50% of the games it played. The two teams won the same number of games (Condition 2). Find the number of games played by each team.

Solution:	**Games Played (Condition 1)**	**Per Cent Won**	**Games Won (Condition 2)**
[1] Team A	x	60%	$0.60x$
Team B	$x + 2$	50%	$0.50(x + 2)$

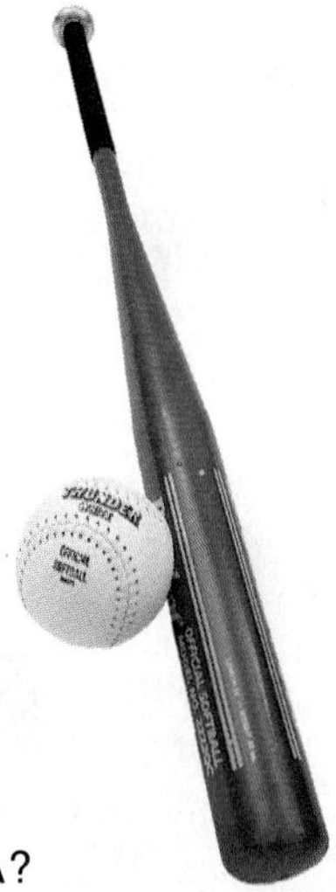

[2] Use Condition 2 to write an equation.

Think: **Games won by Team A** = **Games won by Team B**

Translate: $0.60x = 0.50(x + 2)$

[3] Solve the equation. ⟶ $0.60x = 0.50x + 1$

$0.10x = 1$

$x = 10$

$x + 2 = 12$

[4] **Check:** Condition 1 Did Team B play 2 more games than Team A?
Does $12 = 10 + 2$? Yes ✓

Condition 2 Did Team A and Team B win the same number of games?
Does $0.60(10) = 0.50(12)$? Yes ✓

Team A played 10 games and Team B played 12 games.

CLASSROOM EXERCISES

Find the discount on each item. (Table)

1. A dining table listed at \$640 with a discount rate of 20%
2. A shirt listed at \$36 with a discount rate of 25%
3. A record album listed at \$9.50 with a discount rate of 40%
4. A pick-up truck listed at \$8500 with a discount rate of 6%

Solve. Refer to the formulas on page 104. (Step 2, Example 1)

5. The rate of discount of a radio on sale was 35%. The discount was \$31.50. Find the list price.
6. The rate of discount of a pair of skis on sale was 20%. The discount was \$75. Find the list price.

(Example 1)

7. A typewriter was on sale with a discount rate of 12%. The sale price was \$396. Find the list price and the discount.
8. A sweater was on sale with a discount rate of 35%. The sale price was \$39. Find the list price and the discount.

For Exercises 9–10, Condition 1 is underscored once. Condition 2 is underscored twice.

a. *Use the given information (Condition 1 and Condition 2) to complete the table.*

b. *Use Condition 2 to write an equation for the problem.* (Example 2)

9. Julia's monthly income is \$240 greater than Emily's monthly income. Emily spends 30% of her monthly income on rent, and Julia spends 25% of her monthly income on rent. Julia and Emily spend the same amount each month on rent. Find their monthly incomes.

	Monthly Income	Per Cent Spent	Amount Spent
Emily	x	30%	?
Julia	?	25%	?

10. Before a special sale, the Sports Shop had 20 more baseball gloves in stock than baseball bats. During the sale, the shop sold 70% of its stock of bats and 50% of its stock of gloves. The same number of bats and gloves were sold. Find the number of bats and gloves in stock before the sale.

Item	Number in Stock	Per Cent Sold	Number Sold
Bats	n	70%	?
Gloves	?	50%	?

WRITTEN EXERCISES

Goal: To solve problems involving per cent

Sample Problem: A stereo was on sale for \$328 at a discount of 18%. Find the list price and the discount.

Answers: List Price: \$400; Discount: \$72

Solve. Refer to the formulas on page 104. (Example 1)

1. The rate of discount of a refrigerator on sale was 20%. The discount was \$115. Find the list price.

2. A color TV was on sale for \$331.50. The discount rate was 15%. Find the list price and the discount.

3. A woman's warmup suit was on sale at a rate of discount of 20%. The sale price was \$48. Find the list price and the discount.

4. A store offers customers a 2% discount if bills are paid within 15 days. This discount would save one customer \$5.12. Find the original amount of the customer's bill.

For Exercises 5–7, Condition 1 is underscored once. Condition 2 is underscored twice.

a. *Use Condition 1 to represent the unknowns.*

b. *Use Condition 2 to write an equation for the problem.*

c. *Solve.*

(Example 2)

5. A basketball team played 4 more games this season than last season. Last season, the team won 55% of its games and this season it won 50% of its games. The team won the same number of games this season as last season. Find the number of games played each season.

6. In August, U-Fix Auto Parts sold 40% of its stock of batteries and 30% of its stock of oil filters. At the beginning of August, there were 40 more oil filters in stock than batteries. The same number of batteries and oil filters were sold. How many of each item were in stock at the beginning of August?

7. The number of players selected for the girls' tennis team was 75% of the number who tried out. There were 4 more who tried out for the boys' team than for the girls' team, but only 60% of the boys were selected. The two teams had the same number of members. Find how many tried out for each team.

4.5 Using Distance Formulas

The following word rule and formula are used to solve time/rate/distance problems.

Word Rule: Distance equals rate multiplied by time.

Formula: $d = r \cdot t$ or $d = rt$ ◀ *d = distance, r = rate, t = time*

In solving these problems, make a table showing the rate, time and distance for each trip. For round trips,

distance going equals distance returning.

This is Condition 2 in the problems for this section. Note that Condition 2 is sometimes stated before Condition 1.

EXAMPLE 1

A salesperson took a round-trip flight to another city. The first half of the trip took 5 hours and the return trip took 4.5 hours (Condition 2). The average rate returning was 54 miles per hour (abbreviated mi/hr) more than the average rate going (Condition 1). Find the average speed returning.

[1] **Solution:** Make a table.

	Rate (Condition 1)	Time	Distance (Condition 2) $d = rt$
Going	x mi/hr	5 hr	$x(5)$, or $5x$ mi
Returning	$(x + 54)$ mi/hr	4.5 hr	$(x + 54)4.5$, or $4.5(x + 54)$ mi

[2] Use Condition 2 to write an equation.

Think: Distance going = Distance returning

Translate: $5x = 4.5(x + 54)$

[3] Solve the equation. ⟶

$$5x = 4.5x + 243$$

$$0.5x = 243$$

$$\frac{0.5x}{0.5} = \frac{243}{0.5}$$

$$x = 486$$

$$x + 54 = 540$$

◀ ***Don't forget to find x + 54.***

4 **Check:** Condition 1 Was the average rate returning 54 miles per hour more than the average rate going?

Does $540 = 486 + 54$? Yes ✓

Condition 2 Does the distance going equal the distance returning?

Does $486(5) = 540(4.5)$? Yes ✓

The average rate returning was 540 miles per hour.

The following word rule and formula are similar to the distance formula. The ***cruising range*** of a car is the distance it can travel on a full tank of fuel. ***Mileage*** means miles per gallon (mi/gal) or kilometers per liter (km/L).

Word Rule: The cruising range (distance) of a car equals the fuel tank capacity multiplied by the mileage.

Formula: $r = m \cdot c$ or $r = mc$

r = cruising range
m = mileage
c = fuel tank capacity

EXAMPLE 2

The fuel tank capacity of Car B is 36 liters less than that of Car A (Condition 1). Car A averages 7 kilometers per liter of fuel. Car B averages 9 kilometers per liter. The range of Car B is 132 kilometers less than that of Car A (Condition 2). Find the fuel tank capacity of each car.

Solution:

		Mileage	Fuel Tank Capacity (Condition 1)	Range (Condition 2) $r = mc$
1	Car A	7 km/L	x L	$7x$ km
	Car B	9 km/L	$(x - 36)$ L	$9(x - 36)$ km

2 **Think:** Range of Car B = Range of Car A − 132

Translate: $9(x - 36) = 7x - 132$

Condition 2

3

$$9x - 324 = 7x - 132$$
$$9x - 7x - 324 = 7x - 132 - 7x$$
$$2x - 324 + 324 = -132 + 324$$
$$2x = 192$$
$$x = 96$$
$$x - 36 = 60$$

Tank capacity of Car A: 96 L
Tank capacity of Car B: 60 L

4 **Check:** Condition 1 Is the fuel tank capacity of Car B 36 liters less than that of Car A? Does $60 = 96 - 36$? Yes ✓

Condition 2 Is the range of Car B 132 kilometers less than that of Car A?

Does $9(60) = 7(96) - 132$ Yes ✓

The fuel tank capacity of Car A is 96 liters.
The fuel tank capacity of Car B is 60 liters.

CLASSROOM EXERCISES

For Exercises 1–3, Condition 1 is underscored once. Condition 2 is underscored twice. (Examples 1 and 2)

a. *Use the given information to complete the table.*

b. *Use Condition 2 to write an equation for the problem.*

1. Sarah traveled to New Orleans and back by train. The trip to New Orleans took 6 hours and the trip returning took 6.4 hours. The average rate returning was 6 kilometers per hour less than the average rate going.

	Rate (Condition 1)	Time	Distance (Condition 2) $d = rt$
Going	x km/hr	6 hr	?
Returning	?	6.4 hr	?

2. On a round trip to another city, José spent 3 hours going and $2\frac{3}{4}$ hours returning. The average rate returning was 4 miles per hour faster than the average rate going.

	Rate (Condition 1)	Time	Distance (Condition 2) $d = rt$
Going	r mi/hr	?	?
Returning	?	?	?

3. The fuel tank capacity of Car B is 2 liters less than that of Car A. Car A averages 10 km/L of fuel and Car B averages 12 km/L. The range of Car B is 112 kilometers more than that of Car A.

	Mileage	Fuel Tank Capacity (Condition 1)	Range (Condition 2) $r = mc$
Car A	10 km/L	x L	?
Car B	12 km/L	?	?

WRITTEN EXERCISES

Goal: To solve distance/rate/time problems

Sample Problem: A round-trip jet flight takes 2.25 hours to reach Miami and 2 hours to return (Condition 2). The rate returning is 80 kilometers per hour greater than the rate going (Condition 1). Find the average rate going and the average rate returning.

Answers: Average rate going: 640 km/hr; Average rate returning: 720 km/hr

For Exercises 1–7, Condition 1 is underscored once. Condition 2 is underscored twice.

a. *Use Condition 1 and Condition 2 to make a table.*

b. *Use Condition 2 to write an equation for the problem.*

c. *Solve.*

(Examples 1 and 2)

1. A bus makes a round-trip run. The average rate going is 15 miles per hour more than the average rate returning. The first half of the trip takes $2\frac{4}{5}$ hours and the return trip takes 4 hours. Find the average speed on the first half of the trip.

2. An excursion boat makes a round-trip to a park. The return trip is 6 kilometers per hour slower than the rate going to the park. The trip to the park takes 4 hours and the return trip takes 5 hours. Find the average speed returning.

3. In going back and forth to work every day, Marge finds that the average rate returning is 7 miles per hour less than the rate going. It takes Marge 1 hour to go to work and $1\frac{1}{4}$ hours to return home. Find the average rate going and the average rate returning.

4. A jet makes a round-trip to San Francisco. The first half of the trip takes 5 hours and the return trip takes 4 hours. The average rate returning is 80 kilometers per hour more than the average rate going. Find the average rate going and returning.

5. Two cars have the same cruising range. Car B's fuel tank holds 35 fewer liters than Car A. Car B averages 8.5 kilometers per liter of fuel. Car A averages 5 kilometers per liter. Find the fuel tank capacity of each car.

6. The Lee family decided to drive to the beach and back. The trip going took 4 hours. The trip returning took 5 hours. The average rate returning was 10 miles per hour slower than the rate going. Find the average rate returning.

Making Tables

Surface Area/Volume

An important goal in industrial production is to keep the amount of raw materials to a minimum. For example, a food processor wants containers that provide the desired **capacity** but with dimensions that require the least amount of material **(surface area).** The formula below gives S, the surface area of a cylindrical container in square centimeters.

$$S = \frac{(3.14r^3 + c)2}{r}$$

c = capacity in milliliters
r = radius in centimeters

EXAMPLE: Find, to the nearest whole number, the value of r that will require a minimum amount of material for a cylindrical can with a capacity of 354 milliliters.

SOLUTION: $S = \frac{(3.14r^3 + 354)2}{r}$

Use a calculator to complete the table shown at the right.

r	3.0	3.5	4.0	4.5
S	?	?	?	?

By calculator: When $r = 3.0$, $S = (3^3 \times 3.14 + 354) \times 2 \div 3$.

Solve also for r = 3.5, 4.0, and 4.5.

3 × = = × 3 . 1 4

\+ 3 5 4 × 2 ÷ 3 =   292.52

In the table, the smallest value for S is 277.5. The corresponding value of r is about **4 centimeters.**

r	3.0	3.5	4.0	4.5
S	292.5	279.2	277.5	284.5

EXERCISES

1. In the Example, show that 3.8 is the value of r to the nearest *tenth*. (HINT: Find the values of S for $r = 3.7$, 3.8, and 3.9)

2. In the Example, the formula for the height of the can is $h = \frac{c}{3.14r^2}$. Find, using $r = 3.8$, the value of h that requires a minimum amount of material.

CHAPTER SUMMARY

IMPORTANT TERMS	Perimeter *(p. 95)* List price *(p. 104)* Discount *(p. 104)* Sale price *(p. 104)*	Rate of discount *(p. 104)* Mileage *(p. 109)* Cruising range *(p. 109)*

IMPORTANT IDEAS

1. Steps for Solving Word Problems
   - **a.** Identify Condition 1. Use Condition 1 to represent the unknowns.
   - **b.** Identify Condition 2. Use Condition 2 to write an equation.
   - **c.** Solve the equation.
   - **d.** Check the results with the two conditions. Answer the question.
2. In equations that contain parentheses, use the Distributive Property first.

CHAPTER REVIEW

For Exercises 1–8, Condition 1 is underscored once. Condition 2 is underscored twice. Use Condition 1 to represent the unknowns. Use Condition 2 to write an equation for the problem. Solve.

SECTION 4.1

1. A librarian issued 82 more fiction books in one day than nonfiction books. The total number of books issued was 372. How many fiction books and how many non-fiction books were issued?

2. In a 2-day period the amount of rainfall the first day was 3.7 centimeters less than the amount the second day. In two days it rained 15.3 centimeters. Find the amount it rained each day.

3. On a family trip Alice drove 100 miles more than twice the number of miles her father drove. The total number of miles was 1150. Find how many miles Alice and her father each drove.

4. The labor cost of producing a machine part is \$0.05 less than twice the cost of materials. The total unit production cost of the part is \$1.72. Find the cost of materials and the cost of labor.

SECTION 4.2

5. A helicopter pad on top of a building is rectangular in shape. The length is 11.8 meters more than the width. The perimeter is 80.0 meters. Find the length and width.

6. A ceiling of a room is covered with rectangular panels. The length of each panel is twice the width. The perimeter is 365.76 centimeters. Find the length and width.

7. The ratio of the length to the width of a rectangle is 7:3. The perimeter is 166.0 millimeters. Find the length and width.

8. A triangular plane route has legs in the ratio 5:9:13. The total route has a length of 3681 kilometers. Find the length of each leg.

SECTION 4.3

For Exercises 9–16, solve each equation. Check each answer.

9. $8h + 3 = 5h - 18$

10. $2x + 5 = 6x + 29$

11. $28 - 7y = y - 48$

12. $44 - a = -18 + 2a$

13. $\frac{1}{2}(4t - 6) - 3t = 2t - 7$

14. $-2v + 6 = 8 - \frac{1}{3}(9 - 3v)$

15. $4.3g - 28 = -1.9g + 65$

16. $62 - 3.7x = 1.9x + 34$

SECTION 4.4

Solve. Use the formulas $d = r \cdot p$ and $s = p - d$.

17. A tape deck is on sale at a discount rate of 40%. The sale price is \$274.50. Find the list price.

18. An airline offered a special fare at a discount of 18%. The special fare was \$479.70. Find the regular fare for this flight.

For Exercises 19–22, Condition 1 is underscored once. Condition 2 is underscored twice. Use Condition 1 to represent the unknowns. Use Condition 2 to write an equation for the problem. Solve.

19. Brad and Russell spend the same amount each month for rent. Brad's monthly income is \$200 greater than Russell's monthly income. Brad spends 28% of his monthly income on rent and Russell spends 32% of his monthly income on rent. Find their monthly incomes.

20. The first and second place teams in a division won 90% and 85% of their games respectively. The first place team won one more game than the second place team. They played the same number of games. Find the number of games played by each team.

SECTION 4.5

21. Juan and Maria computed the average speed of the family car on a vacation trip. The trip to their vacation spot took 4.25 hours and the return trip took 4 hours. The average speed going was 5 kilometers per hour less than the speed returning. Find the average speed each direction.

22. The fuel tank capacity of Car B is 5 liters less than that of Car A. Car A averages 11.9 kilometers per liter and Car B averages 9.8 kilometers per liter. The range of Car A is 154 kilometers more than that of Car B. Find the fuel tank capacity of each car.

CHAPTER 5

Inequalities

Water will freeze when the temperature is less than or equal to 32°F, or 0°C. To prevent the water in a car radiator from freezing, antifreeze is added to the water because it lowers the freezing point.

5.1 Graphs of Inequalities

The sentence "$x = 2$" means

"x is equal to two".

The sentence "$x < 2$" means "x is less than two". The solutions of these sentences are illustrated in the graphs below. The set of replacements for x is {integers}.

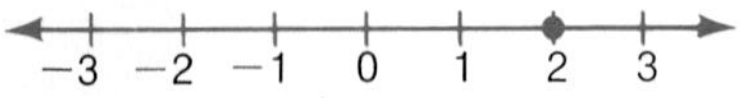

$x = 2$

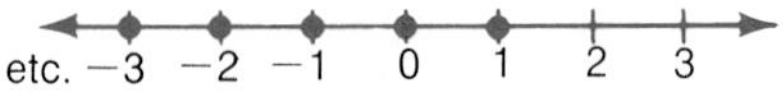

$x < 2$

P–1 **Why is "etc." necessary in the second graph?**

The set of replacements for a variable is the ***domain.*** In the second graph above the domain of x is {integers}. If the domain is {real numbers}, then the graph of $x < 2$ is as shown below.

The point corresponding to 2 is circled in order to show that 2 is not a solution.

Sentences such as $x < 2$ are called ***inequalities.*** Other kinds of inequalities are shown below.

Inequality	**Meaning**
$x > 2$	"x is greater than two."
$x \neq 2$	"x is not equal to two."
$x \leq 2$	"x is less than or equal to two."

P–2 **Why is 2 a solution of $x \leq 2$?**

P–3 **Why is 3 a solution of $x \neq 2$?**

EXAMPLE 1 Draw a graph of the solutions of $x < 3$ on a number line. Domain = {counting numbers}.

Solution:

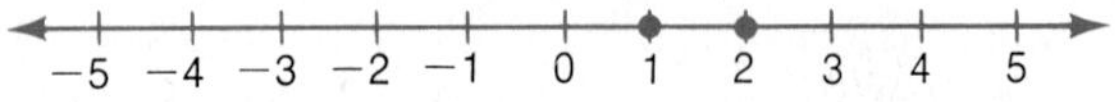

The solutions are 1 and 2. The solution set is {1, 2}.

EXAMPLE 2

Draw a graph of the solution set of $x < 3$.
Domain = {integers}.

Solution:

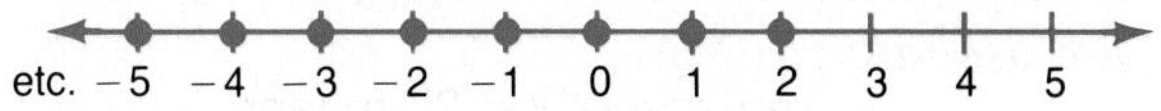

Solution set: {all integers less than 3}

If no domain is stated, it is always assumed in this text to be {real numbers}.

EXAMPLE 3

Draw a graph of $x \leq 3$.

"Graph of $x \leq 3$" means "graph of the solution set of $x \leq 3$."

Solution:

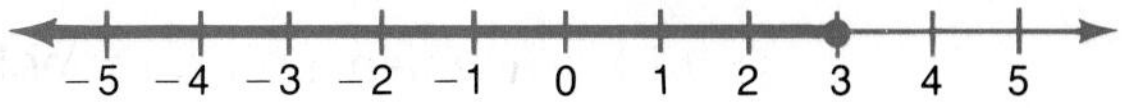

Solution set: {all real numbers less than or equal to 3}

P–4 **Why is the point corresponding to 3 darkened?**

EXAMPLE 4

Graph $x > -\frac{1}{2}$.

"Graph" means "draw a graph."

Solution:

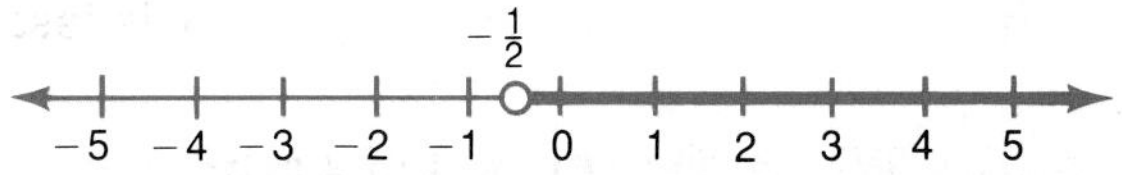

Solution set: {all real numbers greater than $-\frac{1}{2}$}

P–5 **What solution sets are graphed below?**
What domain is assumed?

a.

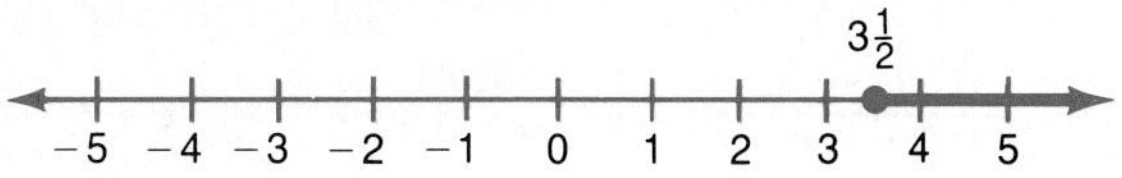

b.

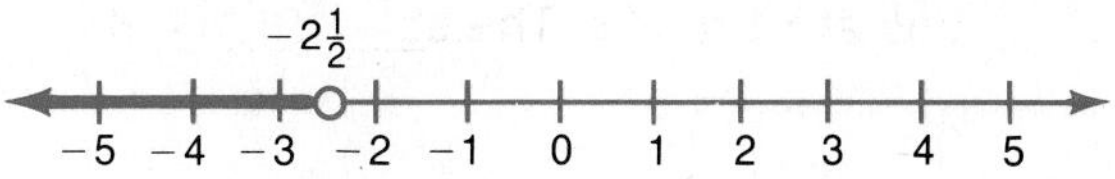

CLASSROOM EXERCISES

Find the solution set that is graphed. If necessary, name the domain that you are assuming. (P-5)

1.

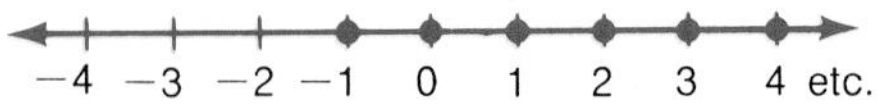

2.

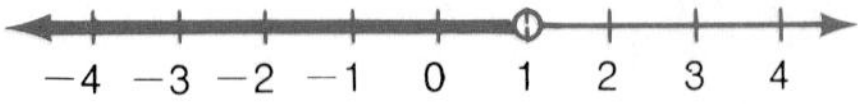

3.

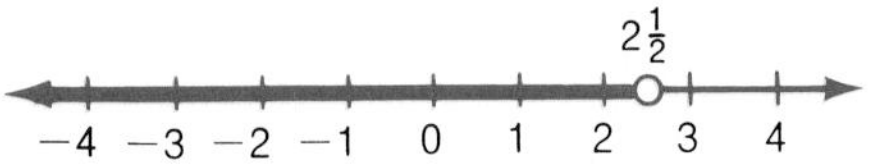

4.

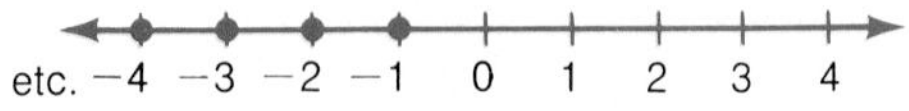

Name three solutions of each inequality. (Examples 1–4)

5. $x \leq 10$; Domain = {whole numbers}

6. $x < -9$; Domain = {integers}

7. $x \leq 3.9$

8. $x > -\pi$

9. $x \geq -5$; Domain = {rational numbers}

10. $x \geq -2$; Domain = {integers}

WRITTEN EXERCISES

Goals: To write the solution set of an inequality from a graph and to graph an inequality

Sample Problem: Graph $x < 6$. Domain = {integers}.

Answer:

Write the solution set of each graph. If necessary, write the domain you are assuming. (P-5)

1.

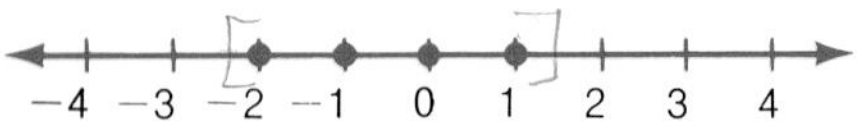

2.

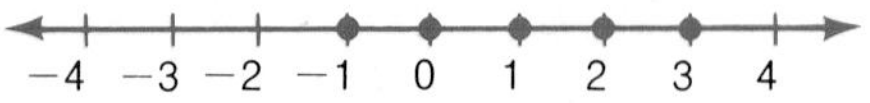

3.

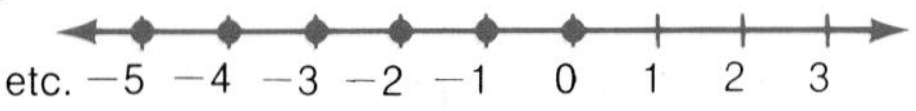

4.

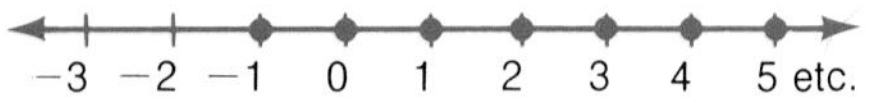

5.

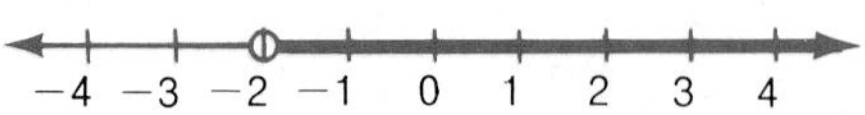

6.

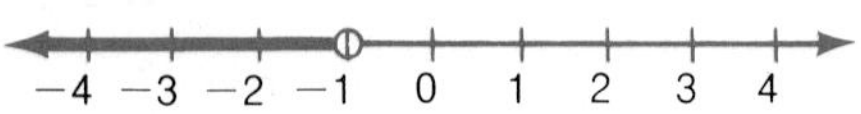

7. 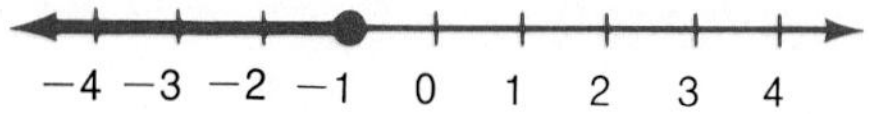

8. 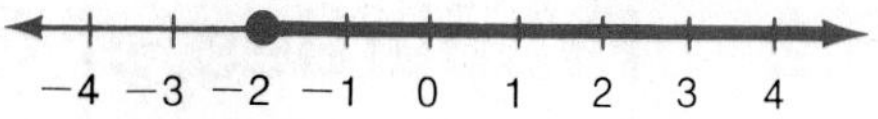

9. 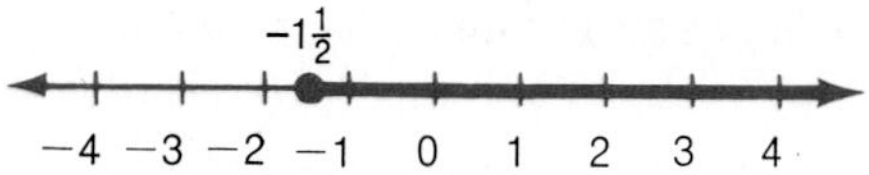

10. 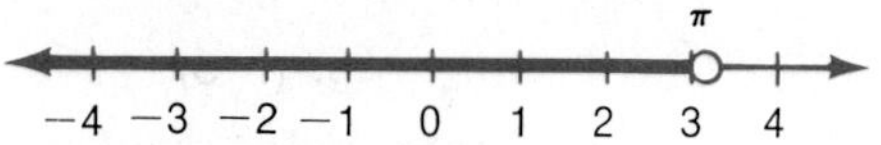

MIXED PRACTICE

Draw a graph of each inequality.

11. $x \geq 2$; Domain = {whole numbers}

12. $x < 6$; Domain = {whole numbers}

13. $x < -2$; Domain = {integers}

14. $x \geq 5$; Domain = {integers}

15. $x < -\frac{3}{4}$

16. $x \geq -3\frac{1}{2}$

17. $x \geq 2.7$

18. $x < -1.9$

MORE CHALLENGING EXERCISES

Write Yes or No to show whether each graph below describes the solution set of the sentence. Domain = {real numbers}.

19. $-x = 2$ 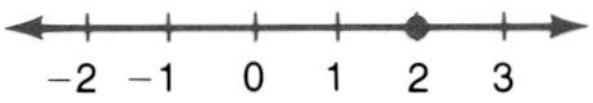

20. $-x = -3$ 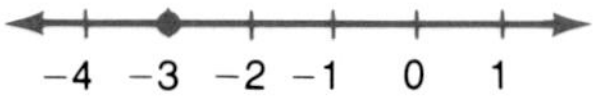

21. $-x > -3$ 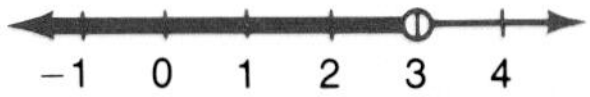

22. $-x \geq -2\frac{1}{2}$ 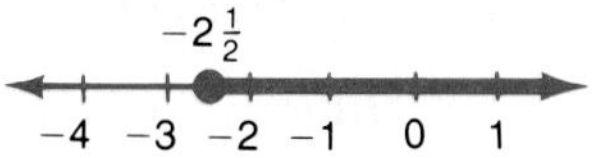

Does the graph of $x < a$ contain the point 0? Domain = {real numbers}.

23. The variable *a* represents a positive integer.

24. The variable *a* represents a negative integer.

25. The variable *a* represents a real number greater than 100.

26. The variable *a* represents zero.

REVIEW CAPSULE FOR SECTION 5.2

Change each fraction to decimal form. Then identify the greater number of each pair.

EXAMPLES: **a.** $\frac{2}{5}$ and $\frac{3}{5}$ **b.** $-\frac{7}{10}$ and $-\frac{8}{10}$

SOLUTIONS: **a.** $\frac{2}{5} = 0.4$; $\frac{3}{5} = 0.6$; $0.6 > 0.4$ Thus, $\frac{3}{5} > \frac{2}{5}$.

b. $-\frac{7}{10} = -0.7$; $-\frac{8}{10} = -0.8$; $-0.7 > -0.8$ Thus, $-\frac{7}{10} > -\frac{8}{10}$.

1. $\frac{3}{10}$ and $\frac{5}{10}$

2. $\frac{3}{4}$ and $\frac{1}{4}$

3. $\frac{2}{5}$ and $-\frac{3}{5}$

4. $\frac{6}{8}$ and $\frac{7}{8}$

5. $-\frac{4}{10}$ and $-\frac{5}{10}$

6. $-\frac{3}{8}$ and $-\frac{5}{8}$

7. $\frac{4}{6}$ and $\frac{5}{6}$

8. $-\frac{1}{3}$ and $-\frac{2}{3}$

5.2 Properties of Order

Comparison Property

If a and b are real numbers, then exactly one of the following is true.

1. $a > b$ 2. $a = b$ 3. $a < b$

P–1 **If $a = -1$ and $b = -2$, what inequality can you form by use of the Comparison Property?**

P–2 **If $a = -1.3$ and $b = \frac{-13}{-10}$, what inequality can you form by use of the Comparison Property?**

EXAMPLE 1 Write an inequality by taking the opposite of each side of $-3 < 5$.

Solution:

$$-3 < 5$$
$$-(-3) \underline{\ ?\ } -(5)$$
$$3 > -5$$

Taking the opposite of each side reverses the order.

Order Property of Opposites

If a and b are real numbers and $a < b$, then $-a > -b$.

$-5.2 < -4.7$
$5.2 > 4.7$

You can write $-3 < 5$ as $5 > -3$ and $3 > -5$ as $-5 < 3$. Similarly, you can write $a < b$ as $b > a$ and $-a > -b$ as $-b < -a$.

P–3 **What two inequalities can you form by applying the Order Property of Opposites to each of the following inequalities?**

a. $y > -6$ **b.** $-13 < t$ **c.** $-r < s$

P–4 **What two inequalities relate r and 2 in the graph below?**

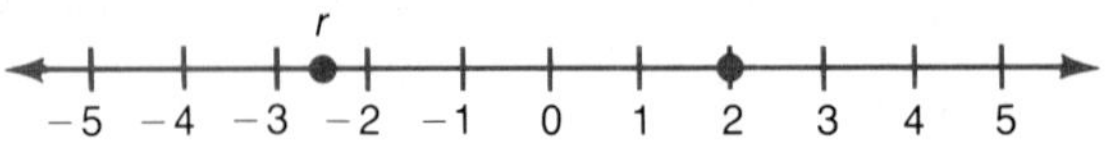

P–5 **What two inequalities relate 2 and t in the graph below?**

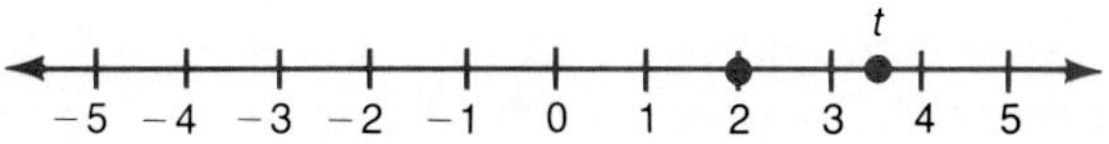

P–6 **What two inequalities relate r and t in P-4 and P-5?**

> ***Transitive Property of Order***
>
> If $a < b$ and $b < c$, then $a < c$.
>
> $-3.4 < 0$ and $0 < 1.9$. Then $-3.4 < 1.9$.

P–7 **If $x < -2$ and $-2 < y$, what two inequalities relate x and y?**

EXAMPLE 2 Suppose that "$-3 > q$" and "$-3 < p$" are both true. Write an inequality for p and q. Use "<."

Solution: $q < -3$ and $-3 < p$ ◀ ***Transitive Property of Order***

$q < p$

You know that you can write $a < b$ as $b > a$, $b < c$ as $c > b$, and $a < c$ as $c > a$.

P–8 **How would you state a Transitive Property of Order using ">"?**

CLASSROOM EXERCISES

Write an inequality for each pair of numbers. (P-1, P-2)

1. -5; 2 **2.** 15; -13 **3.** -12; -9 **4.** 14; -14

5. -16; 6 **6.** -21; -20 **7.** 1.3; 1.03 **8.** -1.6; -1.7

Complete each statement. Use <, >, or =. a, b, and c are real numbers. (Example 1, P-7)

9. If $-c > -1$, then c __?__ 1.

10. If $-a < 4$, then a __?__ -4.

11. If $a > b$ and $b > 5$, then a __?__ 5.

12. If $c < -5$ and $a > -5$, then c __?__ a.

WRITTEN EXERCISES

Goal: To write inequalities using the Comparison Property, the Order Property of Opposites, and the Transitive Property of Order

Sample Problem: Suppose that "$x < -31$" and "$y > -31$" are both true. Write an inequality for x and y. Use "<".

Answer: $x < y$

Write an inequality for the two numbers. Use "<". (P-1, P-2)

1. $-5; 4$ **2.** $17; -19$ **3.** $-12; -11$ **4.** $-5; -15$

5. $1.7; 1.6$ **6.** $0.7; -0.9$ **7.** $-\frac{3}{4}; -\frac{7}{8}$ **8.** $-\frac{1}{2}; -\frac{1}{3}$

Write a new inequality by applying the Order Property of Opposites. (Example 1)

9. $3 < -b$ **10.** $-a < 5$ **11.** $-a < -2$ **12.** $-7 > -b$

13. $-(a+b) < 2$ **14.** $-10 < -(b+a)$ **15.** $-b > a$ **16.** $-a < b$

Write True or False for each sentence. a, b, and c are real numbers. (Example 2)

17. If $c < -2$ and $-2 < a$, then $c < a$.

18. If $b > 5$ and $5 > c$, then $b > c$.

19. If $a < -2$, then $a < -5$ is true.

20. If $c > -3$, then $-1 < c$ is true.

21. If $a < 4$ and $b < 4$, then $a < b$.

22. If $a > 5$ and $b > 5$, then $b > a$.

23. If $b < c$ and $a < b$, then $a < c$.

24. If $7 > b$ and $a > 7$, then $a > b$.

Write an inequality for x and y. Use "<". (Example 2)

25. $y < -2$ and $-2 < x$

26. $y < 5$ and $5 < x$

27. $x < -3.8$ and $y > -3.8$

28. $y < 0.9$ and $x > 0.9$

29. $213 > -x$ and $y < -213$

30. $-56 > y$ and $-x < 56$

31. $-x > 19.3$ and $19.3 > -y$

32. $-y < -10.3$ and $-x > -10.3$

REVIEW CAPSULE FOR SECTION 5.3

Solve each equation. (Pages 58–60)

1. $64 = t - 71$ **2.** $x + 7 = -8$ **3.** $y - 5 = 8$

4. $3.7 = w - 4.2$ **5.** $t + 0.9 = -5.2$ **6.** $-\frac{3}{4} = \frac{7}{8} - r$

7. $-d - 1\frac{2}{3} = -\frac{5}{6}$ **8.** $-2.07 = p - 5.93$ **9.** $s + 0.87 = 0.34$

5.3 Addition and Subtraction

Compare the solution sets and graphs for each pair of inequalities in the table below. The domain is $\{-1, 0, 1, 2, 3, 4, 5, 6, 7, 8\}$.

Inequalities	Solution Sets	Graphs
$x < 5$ $x + 3 < 5 + 3$ or $x + 3 < 8$	$\{-1, 0, 1, 2, 3, 4\}$	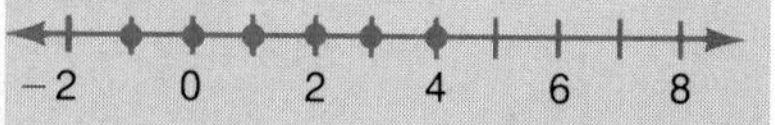
$x > 5$ $x + 3 > 5 + 3$ or $x + 3 > 8$	$\{6, 7, 8\}$	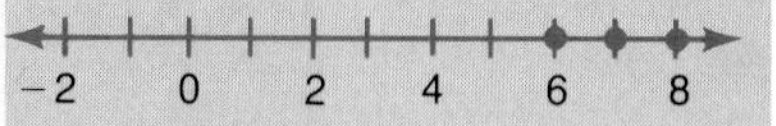

In the table above,

$x < 5$ is *equivalent* to $x + 3 < 8$
and
$x > 5$ is *equivalent* to $x + 3 > 8$,

because each pair has the ***same solution set.*** This suggests that adding the same number to each side of an inequality results in a new inequality that is equivalent to the first.

EXAMPLE 1 Solve: $x - 2 > -1$

Solve means to find the solution set.

Solution:

$x - 2 > -1$

[1] Add 2 to each side. ⟶ $x - 2 + \mathbf{2} > -1 + \mathbf{2}$

[2] Simplify. ⟶ $x > 1$

Solution set: {all real numbers greater than 1}

Addition Property for Inequalities

Adding the same real number to each side of an inequality does not change the order of the inequality. The new inequality is equivalent to the first.

$$x - 12 > 3$$
$$x - 12 + \mathbf{12} > 3 + \mathbf{12}$$
$$x > 15$$

EXAMPLE 2 Solve: $x - 5 < -1$

Solution: $x - 5 < -1$

[1] Add 5 to each side. ⟶ $x - 5 + \mathbf{5} < -1 + \mathbf{5}$

[2] Simplify. ⟶ $x < 4$

Solution set: {all real numbers less than 4}

The inequality $x + 7 + \mathbf{(-7)} < 12 + \mathbf{(-7)}$ is equivalent to $x + 7 - \mathbf{7} < 12 - \mathbf{7}$. This suggests a subtraction property.

> ***Subtraction Property for Inequalities***
>
> Subtracting the same real number from each side of an inequality does not change the order of the inequality. The new inequality is equivalent to the first.
>
> $x + \frac{1}{2} < \frac{3}{4}$
>
> $x + \frac{1}{2} - \frac{1}{2} < \frac{3}{4} - \frac{1}{2}$
>
> $x < \frac{1}{4}$

EXAMPLE 3 Solve: $x + 5.3 < 2.9$

Solution: $x + 5.3 < 2.9$

[1] Subtract 5.3 from each side. ⟶ $x + 5.3 - \mathbf{5.3} < 2.9 - \mathbf{5.3}$

[2] Simplify. ⟶ $x < -2.4$

Solution set: {all real numbers less than −2.4}

CLASSROOM EXERCISES

Write what number to add to, or subtract from, each side of the inequality to get the variable by itself. (Step 1 in Examples 1–3)

1. $r + 5 < 3$ **2.** $w - \frac{1}{2} > 2$ **3.** $-13 > a + 7$

4. $-5.6 < 1.8 + c$ **5.** $-0.8 + x < 5.9$ **6.** $-3\frac{1}{4} > t - 5\frac{1}{8}$

Solve. (Examples 1–3)

7. $x - 1 < 10$ **8.** $y + 2 > 8$ **9.** $a - 3 < 5$

10. $t + 7 > -2$
11. $w - 5 < -1$
12. $10 > -3 + s$
13. $-5 + r < -1$
14. $p - 2 > -3$
15. $-0.5 + x > -8.5$
16. $3.6 < -1.4 + y$
17. $2x - 3 - x < 5$
18. $-7 > 3a + 5 - 2a$

WRITTEN EXERCISES

Goal: To solve an inequality using the Addition or Subtraction Property for Inequalities

Sample Problems: Solve: **a.** $x - 5.3 < 2.9$ **b.** $x + 5.3 < 2.9$

Answers: a. {all real numbers less than 8.2}
b. {all real numbers less than −2.4}

Write Yes or No to show whether the two inequalities are equivalent. (Examples 1–3)

1. $x + 2 < 3;\ x < 5$
2. $-5 < y - 2;\ -7 < y$
3. $2a - 7 < -2;\ 2a < 5$
4. $5r + 6 > -1;\ 5r > -7$
5. $x + 2 < 5;\ x + 1 < 6$
6. $y - 3 < 4;\ y - 2 < 5$
7. $2x + 2 < 5x;\ 2x < 5x - 2$
8. $3t < 4t - 1;\ 3t - 1 < 4t$
9. $3w - 1 > w + 7;\ 3w - 2 > w + 8$
10. $9g - 5 > g + 3;\ 9g - 3 > g + 5$

MIXED PRACTICE

Solve.

11. $y + 2 > -3$
12. $x + 5 < 8$
13. $-3 + t < -10$
14. $-7 + r > -13$
15. $x - 1 < -2$
16. $p - 3 > -6$
17. $-3.4 > -0.3 + q$
18. $-4.5 < b - 1.2$
19. $-2 + w > 4$
20. $-5 + y > -12$
21. $4.2 < y - 3.7$
22. $3.9 < y + 1.6$

MORE CHALLENGING EXERCISES

Solve. Domain = {integers}.

23. $2 - x < 3$
24. $-y - 7 > -2$
25. $-12 > -5 - a$
26. $15 < -b + 12$
27. $-p + \frac{1}{2} \le 5\frac{1}{2}$
28. $\frac{2}{3} - q \ge -4\frac{1}{3}$

For Exercises 29–30, use the given inequalities to write an inequality for a and b. Use "<."

29. $a - y > b$ and $y > 0$
30. $y - a > c$ and $y - b < c$

NON-ROUTINE PROBLEMS

31. Dan has only a 9-quart pail, a 3-quart pail, and a 2-quart pail to carry water. How can Dan measure exactly eight quarts of water if he completely fills each of the pails just once?

32. Two shirts and one jacket cost the same amount as two sweaters. One sweater and two shirts cost the same amount as one jacket. Which costs the least—a shirt, a jacket, or a sweater?

33. Susan dug up 100 flower bulbs and separated them into five paper bags. The number of bulbs put in the first and second bags totaled 48. In the second and third bags, the bulbs totaled 32; in the third and fourth bags, the bulbs totaled 27; and in the fourth and fifth bags, the bulbs totaled 41. How many bulbs were put in each bag?

34. Trace the figure below. Connect all the points in the figure by drawing exactly four line segments without lifting your pencil off the paper. Do not draw through any one point more than once.

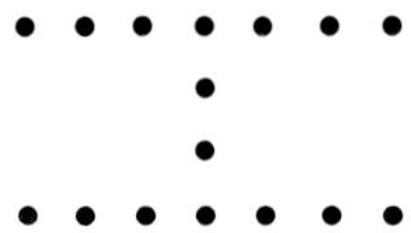

MID-CHAPTER REVIEW

Draw a graph of each inequality. (Section 5.1)

1. $x > -4$; Domain = {real numbers}
2. $x \leq 3$; Domain = {real numbers}
3. $x < 5$; Domain = {integers}
4. $x \geq -2$; Domain = {integers}
5. $x \leq 6$; Domain = {whole numbers}
6. $x > 4$; Domain = {real numbers}

Write an inequality for a and b. Use "<." (Section 5.2)

7. $a < 3$ and $b > 5$
8. $a > 2$ and $b < -3$
9. $-3.8 < a$ and $-4.3 > b$
10. $4 < -a$ and $b > -4$
11. $a > -3\frac{1}{2}$ and $b < -4$
12. $-a < 0$ and $-5 > b$

Solve. (Section 5.3)

13. $t - 8 < -3$
14. $-2 + w > 5$
15. $8.2 < x + 9.4$
16. $-4.7 > -2.3 + q$
17. $r + 1.5 < 1.2$
18. $-1.7 < y - 8.3$

REVIEW CAPSULE FOR SECTION 5.4

Solve each equation. (Pages 71–73 and 100–102)

1. $6n + 5 - n = 14$
2. $-12t + 16 = -8 - 3t$
3. $4(w - 8) = 9w + 3$
4. $1 - 5y = -3(y - 7)$
5. $-6(2 - p) = 2(-3p + 4)$
6. $-\frac{1}{4}(8m - 4) = -\frac{1}{3}(12 + 3m)$

5.4 Solving Inequalities

EXAMPLE 1 Solve: $4x + 2 - 5x < -12$

Solution: $4x + 2 - 5x < -12$

1. Combine like terms. → $-x + 2 < -12$
2. Subtract 2 from each side. → $-x + 2 \mathbf{- 2} < -12 \mathbf{- 2}$
3. Simplify. → $-x < -14$
4. Order Property of Opposites → $x > 14$

***Solution set:* {*all real numbers greater than 14*}**

The variable often appears on both sides of an inequality.

EXAMPLE 2 Solve: $2x - 5 > 3x + 2$

Solution:

1. Subtract $2x$ from each side. → $2x - 5 \mathbf{- 2x} > 3x + 2 \mathbf{- 2x}$
2. Simplify. → $-5 > x + 2$
3. Subtract 2 from each side. → $-5 \mathbf{- 2} > x + 2 \mathbf{- 2}$
4. Simplify. → $-7 > x$ or $x < -7$

Solution set: {all real numbers less than -7}

EXAMPLE 3 Solve: $3(x - 3) < 2x + 1$

Solution:

1. Distributive Property → $3x - 9 < 2x + 1$
2. Subtract $2x$ from each side. → $3x - 9 \mathbf{- 2x} < 2x + 1 \mathbf{- 2x}$
3. Simplify. → $x - 9 < 1$
4. Add 9 to each side. → $x - 9 \mathbf{+ 9} < 1 \mathbf{+ 9}$
5. Simplify. → $x < 10$

***Solution set:* {*all real numbers less than 10*}**

EXAMPLE 4 Solve: $2x - 5 > 2(x + 3)$

Solution:

1. Distributive Property → $2x - 5 > 2x + 6$
2. Subtract $2x$ from each side. → $2x - 5 \mathbf{- 2x} > 2x + 6 \mathbf{- 2x}$
3. Simplify. → $-5 > 6$

"$-5 > 6$" is false. The solution set is empty.

Since the inequality $-5 > 6$ is false, there is no replacement for x that will make $2x - 5 > 2(x + 3)$ true. The solution set is empty. The symbol for the empty set is "ϕ".

EXAMPLE 5 Solve: $3x + 7 - 2x > x - 3$

Solution:

[1] Combine like terms. ⟶ $x + 7 > x - 3$

[2] Subtract x from each side. ⟶ $x + 7 - \mathbf{x} > x - 3 - \mathbf{x}$

[3] Simplify. ⟶ $7 > -3$

The inequality $7 > -3$ is true. Thus, $3x + 7 - 2x > x - 3$ is true for any number that replaces x. The solution set is {real numbers}.

CLASSROOM EXERCISES

Write the inequality you get as the first step in solving each inequality in the Written Exercises. (Step 1 of Examples 1–5)

WRITTEN EXERCISES

Goal: To solve an inequality in which the variable appears more than once
Sample Problem: Solve: $4x - 5 < 3(x + 2)$
Answer: {all real numbers less than 11}

Solve each inequality. (Examples 1–5)

1. $-3 < -7x + 2 + 8x$
2. $8 > 4x - 3 - 3x$
3. $3\frac{1}{2}a - 2 - 2\frac{1}{2}a < -7$
4. $-5\frac{1}{4}t + 7 + 6\frac{1}{4}t < -9$
5. $12y + \frac{1}{2} - 13y > -12\frac{1}{4}$
6. $3\frac{1}{8} > -7r - \frac{1}{4} + 6r$
7. $3w - 5 < 2w + 12$
8. $s - 2 < 2s - 8$
9. $3\frac{1}{2}x - 2 > 2\frac{1}{2}x + 10$
10. $\frac{2}{3}x - 5 > -\frac{1}{3}x - 2$
11. $0.6m - 0.7 < -0.4m + 0.2$
12. $1.2n - 3.6 < 0.2n + 5.1$
13. $3(-x + 2) < -5 - 2x$
14. $x - 6 > 2(x - 5)$
15. $5 - x < -2(x - 3) - 2$
16. $\frac{1}{2}(4x - 3) + \frac{1}{2} < x - 6$
17. $5y - 2 - 3y < \frac{1}{2}(6y - 8)$
18. $5(2 - a) > -3a - 7 - a$
19. $3a - 5 - 2a < a + 2$
20. $3(n - 4) > n - 2 + 2n$
21. $5(x - 1) < 2x - 6 + 3x$

REVIEW CAPSULE FOR SECTION 5.5

Solve each equation. (Pages 61–63)

1. $\frac{3}{4}x = 27$
2. $\frac{8}{7}n = -96$
3. $-12 = -\frac{1}{5}t$
4. $\frac{r}{-6} = 2\frac{2}{3}$
5. $-0.8 = \frac{w}{7}$
6. $4p = -22$
7. $-3q = -12\frac{3}{5}$
8. $0.3s = -7.8$

5.5 Multiplication and Division

The table below shows that multiplying each side of an inequality by the same positive number does not change the direction of the inequality.

Multiplying each side by a negative number, however, reverses the direction of the inequality.

Table

Inequality	Multiply by 1.	Multiply by 2.	Multiply by −1.	Multiply by −2.
$4 < 9$	$4 < 9$	$8 < 18$	$-4 > -9$	$-8 > -18$
$-3 < 6$	$-3 < 6$	$-6 < 12$	$3 > -6$	$6 > -12$
$5 > -7$	$5 > -7$	$10 > -14$	$-5 < 7$	$-10 < 14$

This table suggests the following property.

Multiplication Property for Inequalities

If each side of an inequality is multiplied by the same nonzero real number, an equivalent inequality is obtained as follows.

1. The direction of the inequality is unchanged when each side is multiplied by the same positive number.
2. The direction of the inequality is reversed when each side is multiplied by the same negative number.

$$\frac{1}{2}x > 12$$
$$\mathbf{2}\left(\frac{1}{2}x\right) > \mathbf{2}(12)$$
$$x > 24$$

$$-\frac{2}{3}x < 6$$
$$-\frac{3}{2}\left(-\frac{2}{3}x\right) > -\frac{3}{2}(6)$$
$$x > -9$$

P–1 **What do you multiply each side of these inequalities by to get the variable alone?**

a. $\frac{1}{2}x > 3$ **b.** $-\frac{1}{3}x < 8$ **c.** $\frac{4}{5}x > 9$ **d.** $-\frac{2}{3}x > 10$

EXAMPLE 1 Solve: $-\frac{2}{3}x > 10$

Solution:

1 Multiply each side by $-\frac{3}{2}$. ⟶ $-\frac{3}{2}\left(-\frac{2}{3}x\right) < -\frac{3}{2}(10)$ ◀ ***Direction of the inequality is reversed.***

2 Simplify. ⟶ $x < -15$

Solution set: {all real numbers less than −15}

P–2 **For each inequality below, what inequality do you get if each side is divided by 3? by −3?**

a. $3x < 9$ **b.** $-6x > 6$ **c.** $12x < -3$ **d.** $-15x < 1$

Division Property for Inequalities

If each side of an inequality is divided by the same nonzero real number, an equivalent inequality is obtained as follows.

1. The direction of the inequality is unchanged when each side is divided by the same positive number.
2. The direction of the inequality is reversed when each side is divided by the same negative number.

$$5x < 15$$
$$\frac{5x}{\mathbf{5}} < \frac{15}{\mathbf{5}}$$
$$x < 3$$

$$-3x > 9$$
$$\frac{-3x}{\mathbf{-3}} < \frac{9}{\mathbf{-3}}$$
$$x < -3$$

EXAMPLE 2 Solve: $3x < 15$

Solution:

[1] Divide each side by 3. ⟶ $\frac{3x}{\mathbf{3}} < \frac{15}{\mathbf{3}}$

[2] Simplify. ⟶ $x < 5$

Solution set: {all real numbers less than 5}

EXAMPLE 3 Solve: $-5x < 20$

Solution:

[1] Divide each side by −5. ⟶ $\frac{-5x}{\mathbf{-5}} > \frac{20}{\mathbf{-5}}$ ◀ ***Dividing by a negative number reverses the order.***

[2] Simplify. ⟶ $x > -4$

Solution set: {all real numbers greater than −4}

CLASSROOM EXERCISES

Multiply each side by 5. Then write an equivalent inequality. (Table, Multiplication Property for Inequalities)

1. $x > 2$ **2.** $y < -3$ **3.** $2a > 10$ **4.** $\frac{1}{5}t < \frac{3}{5}$

Multiply each side by −3. Then write an equivalent inequality. (Step 1 of Example 1, Multiplication Property for Inequalities)

5. $y > 2$ **6.** $s < -5$ **7.** $-\frac{1}{3}r < 2$ **8.** $-\frac{1}{3}y > -10$

Write the number you would multiply or divide each side by to get the variable by itself. (Step 1 of Examples 1–3)

9. $2x < 1$ **10.** $-\frac{1}{2}x > \frac{2}{3}$ **11.** $-\frac{1}{2}x < -3$ **12.** $-\frac{1}{3}y > -2$

To form pairs of equivalent inequalities, indicate if < or > should be used as the missing symbol. (Table, P-2, Multiplication or Division Property for Inequalities)

13. $5y > 35$; y __?__ 7

14. $\frac{1}{2}t < -1$; -2 __?__ t

15. $-3 > -\frac{1}{3}x$; 9 __?__ x

16. $2 < -0.1y$; y __?__ -20

WRITTEN EXERCISES

Goal: To solve an inequality using the Multiplication or Division Property for Inequalities

Sample Problems: **a.** $-\frac{3}{4}x > 12$ **b.** $4x < -8$

Answers: **a.** {all real numbers less than −16}

b. {all real numbers less than −2}

Solve. (Examples 1–3)

1. $3y < 15$ **2.** $5y > 30$ **3.** $\frac{1}{4}t > -5$ **4.** $-10t > -20$

5. $-7r < 21$ **6.** $\frac{1}{3} < -\frac{1}{6}y$ **7.** $\frac{1}{2} > -\frac{1}{4}x$ **8.** $\frac{1}{2}x < -8$

9. $\frac{2}{3}w < -4$ **10.** $\frac{3}{5}p > -9$ **11.** $|-\frac{1}{2}|x < |-3|$ **12.** $|-\frac{1}{3}|y > |-2|$

13. $-\frac{4}{3}q > 12$ **14.** $-\frac{8}{7}b < 16$ **15.** $-3x > 11$ **16.** $-5t < -12$

17. $0.6n < -6$ **18.** $-0.9m > 18$ **19.** $-4.5 < 1.5x$ **20.** $3.4 > -17y$

NON-ROUTINE PROBLEM

21. Each of the three boxes shown in the figure at the right contains socks. The socks are either gray, G, or black, B. One box contains two gray socks; one box contains two black socks; and one box contains one gray sock and one black sock. None of the boxes is labeled correctly. From which box can you draw exactly one sock to determine how all three boxes should be labeled?

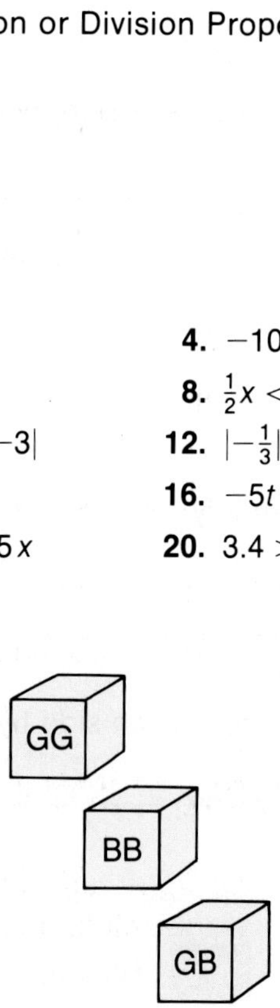

5.6 Inequalities: More Than One Step

When you multiply each side of an inequality by a positive number to get the variable alone, you are using the same steps as you used to solve an equation.

EXAMPLE 1 Solve: $3x - 2 < 10$

Solution:

[1] Add 2 to each side. ⟶ $3x - 2 + \mathbf{2} < 10 + \mathbf{2}$

[2] Simplify. ⟶ $3x < 12$

[3] Multiply by the reciprocal of 3. ⟶ $\mathbf{\frac{1}{3}}(3x) < \mathbf{\frac{1}{3}}(12)$

[4] Simplify. ⟶ $x < 4$

Solution set: {all real numbers less than 4}

P–1 **Why is the order symbol "<" not reversed in any of the steps in Example 1?**

In solving inequalities, use the Addition and Subtraction Properties for Inequalities before the Multiplication and Division Properties.

EXAMPLE 2 Solve: $-4x - 11 > 5 - 2x$

Solution:

[1] Add $2x$ to each side. ⟶ $-4x - 11 + \mathbf{2x} > 5 - 2x + \mathbf{2x}$

[2] Simplify. ⟶ $-2x - 11 > 5$

[3] Add 11 to each side. ⟶ $-2x - 11 + \mathbf{11} > 5 + \mathbf{11}$

[4] Simplify. ⟶ $-2x > 16$

[5] Multiply each side by $-\frac{1}{2}$. ⟶ $-\mathbf{\frac{1}{2}}(-2x) < -\mathbf{\frac{1}{2}}(16)$ ◀ ***The order of the inequality is reversed.***

[6] Simplify. ⟶ $x < -8$

Solution set: {all real numbers less than -8}

P–2 **Why must the order be reversed in Step [5] ?**

In inequalities that contain parentheses, use the Distributive Property first. Then eliminate the variable from one side of the inequality.

EXAMPLE 3 Solve: $-2x + 3(x + 3) > -4x - 6$

Solution:

1. Use the Distributive Property. ⟶ $-2x + 3x + 9 > -4x - 6$
2. Combine like terms. ⟶ $x + 9 > -4x - 6$
3. Add $4x$ to each side. ⟶ $x + 9 + \mathbf{4x} > -4x - 6 + \mathbf{4x}$
4. Simplify. ⟶ $5x + 9 > -6$
5. Subtract 9 from each side. ⟶ $5x + 9 - \mathbf{9} > -6 - \mathbf{9}$
6. Simplify. ⟶ $5x > -15$
7. Multiply each side by $\frac{1}{5}$. ⟶ $\mathbf{\frac{1}{5}}(5x) > \mathbf{\frac{1}{5}}(-15)$
8. Simplify. ⟶ $x > -3$

***Solution set:* {*all real numbers greater than* -3}**

P-3 **Solve: $3y - 4 < 2(-2y + 4) + 5y$**

P-4 **What is the solution set of the inequality in a below? in b?**

a	b
$2(3 - 2x) > x - 1 - 5x$	$-\frac{1}{2}x - \frac{1}{3} + \frac{3}{4}x > \frac{3}{8}x + 1 - \frac{1}{8}x$
$6 - 4x > -1 - 4x$	$-\frac{1}{3} + \frac{1}{4}x > \frac{2}{8}x + 1$
$6 - 4x + 4x > -1 - 4x + 4x$	$-\frac{1}{3} + \frac{1}{4}x - \frac{1}{4}x > \frac{1}{4}x + 1 - \frac{1}{4}x$
$6 > -1$	$-\frac{1}{3} > 1$

The steps in **a** lead to a true sentence. The solution set is {real numbers}. The steps in **b** lead to a false sentence. Its solution set is the empty set, ϕ.

CLASSROOM EXERCISES

Write the equivalent sentence you get as the first step in solving each inequality. (Step 1 of Examples 1, 2, and 3)

1. $2x + 1 > 3$ **2.** $-3x + 4 < -2$ **3.** $\frac{1}{2}x - 3 < -7$

4. $3x + 2 < x - 3$ **5.** $5 - 2x > 4x - 3$ **6.** $\frac{1}{2}x - 6 > \frac{3}{2}x + 4$

7. $5(2x - 6) < -10$ **8.** $3(x - 1) - 5 < x + 4$ **9.** $-5x + 3 > -2(x + 3) - 6$

Solve. (Example 1)

10. $2x - 2 < -4$ **11.** $3x - 6 > -12$ **12.** $3x - 1 > 5$

13. $2x + 3 < 5$ **14.** $-\frac{1}{2}x + 2 < 5$ **15.** $\frac{1}{3}x + 1 > -2$

(Example 2)

16. $1 - 2x < -x + 3$ **17.** $6x - 1 > 4x - 3$ **18.** $2.2 + 1.4x > 0.4x - 3.4$

19. $-6x + 2 > 14 - 2x$ **20.** $3x + 4 < 3x - 2$ **21.** $2.3x - 3 < 5 + 2.3x$

(Example 3)

22. $1 + 2(x - 1) > -11$ **23.** $9 > -7(x + 2) - 5$

24. $-1 + n < 3(1 - n) + 2n$ **25.** $5(x + 3) - 3x > -3 + 5x$

WRITTEN EXERCISES

Goal: To solve an inequality using at least two of the Properties of Inequalities

Sample Problem: $2x - 1 < 5$

Answer: {all real numbers less than 3}

Solve. Show the steps. (Examples 1, 2, and 3)

1. $3x - 1 < 7$ **2.** $5x + 3 < 28$

3. $-10 > 2x + 3$ **4.** $-12 > 2x - 5$

5. $-\frac{1}{2}y - 3 < -7$ **6.** $-\frac{1}{3}a + 1 > -10$

7. $3b + 7 < 5b$ **8.** $2 - 2n < -n$

9. $2m - 5 > 5m + 1$ **10.** $5t + 3 < 3t - 4$

11. $3x + 2 - 8x < 3$ **12.** $-p - 2 - 3p > -1$

13. $\frac{4}{3}q - \frac{1}{2} > \frac{7}{2} + \frac{2}{3}q$ **14.** $\frac{3}{5}r + \frac{5}{3} < -\frac{1}{3} + \frac{1}{5}r$

15. $-2(3x - 3) < 6$ **16.** $-3 > \frac{1}{2}(-4x - 6)$

17. $-4(x - 1) + 2x > 3x - 6$ **18.** $x - 6 < 3(4 - x) - 2x$

19. $-3x + \frac{1}{2} + x < -\frac{1}{2}(4x + 1)$ **20.** $-5(2x - 3) > 3 - 9x - x$

21. $-2(\frac{1}{2}x - \frac{1}{4}) < -\frac{1}{4}x + 1 - \frac{3}{4}x$ **22.** $6(3 - \frac{1}{3}x) < -\frac{1}{4}(8x + 1)$

PROBLEM SOLVING AND APPLICATIONS

5.7 Using Inequalities

The steps for solving word problems involving inequalities are similar to those for solving word problems involving equations.

EXAMPLE 1

Rosa plans to spend less than \$50 for a pair of shoes and a sweater (Condition 2). The cost of the pair of shoes she wants to buy is \$5 more than twice the cost of a sweater (Condition 1). What is the greatest amount she can spend for the sweater?

Solution:

1 Let $c =$ the cost of the sweater.
Then $2c + 5 =$ the cost of the shoes. ◀ ***Condition 1***

2 Write an inequality for the problem.

Think:	**Cost of the sweater**	+	**Cost of the shoes**	**is less than**	**\$50.**
Translate:	c	+	$2c + 5$	$<$	50

◀ ***Condition 2***

3 Solve the inequality. ⟶

$$3c + 5 < 50$$
$$3c + 5 - \mathbf{5} < 50 - \mathbf{5}$$
$$3c < 45$$
$$\frac{3c}{\mathbf{3}} < \frac{45}{\mathbf{3}}$$
$$c < 15$$

Since $c < 15$, the greatest amount Rosa can spend for a sweater is \$14.99.

4 **Check:** Is \$14.99 + 2(\$14.99) + \$5 < 50?
Is \$49.97 < \$50? Yes ✓

Recall that the *mean* of two or more scores is the sum of the scores divided by the number of scores.

In Example 2, Condition 1 is underscored once. Condition 2 is underscored twice.

EXAMPLE 2

Felipe's scores on three tests were 86, 92, and 72. He wants his mean score on four tests to be greater than 85. Find the lowest score Felipe can get on the fourth test.

Solution:

[1] Let x = Felipe's score on the fourth test.

[2] Write an inequality for the problem.

Think: **The mean of the four tests** **must be greater than** **85.**

Translate: $\frac{86 + 92 + 72 + x}{4} > 85$

[3] Solve the inequality. ⟶ $\mathbf{4}\left(\frac{86 + 92 + 72 + x}{4}\right) > \mathbf{4}(85)$

$$86 + 92 + 72 + x > 340$$

$$250 + x > 340$$

$$250 + x - \mathbf{250} > 340 - \mathbf{250}$$

$$x > 90$$

Since $x > 90$, Felipe's score must be at least 91.

[4] **Check:** Is $\frac{86 + 92 + 72 + 91}{4} > 85$? Is $\frac{341}{4} > 85$? Yes ✓

Steps for Solving Word Problems Involving Inequalities

1. Identify Condition 1. Use Condition 1 to represent the unknowns.
2. Identify Condition 2. Use Condition 2 to write an inequality.
3. Solve the inequality. Answer the question.

CLASSROOM EXERCISES

Write an inequality for each sentence. (Step 2, Examples 1 and 2)

1. The cost of a radio, c, is less than \$60.

2. The number of tickets, t, increased by 12 is greater than 70.

3. Twice the number of days, d, decreased by 3 is greater than 20.

4. Eight plus 3 times the number of meters, m, is less than 40.

5. The mean of 32, 45, and x is less than 42.

6. The mean of 82, 87, 91, and y is greater than 86.

7. The mean of 65, 70, 75, and x is greater than 72.

8. The mean of 92, 94, 98, 100, and y is less than 95.

WRITTEN EXERCISES

Goal: To solve word problems involving inequalities

Sample Problem: The Stinson family plans to spend less than $1000 for a new sofa and chair (Condition 2). They estimate that the sofa will cost $300 more than the chair (Condition 1). What is the greatest amount they can spend for the chair?

Answer: $349.99

For Exercises 1–8, Condition 1 is underscored once. Condition 2 is underscored twice.

a. *Use Condition 1 to represent the unknowns.*

b. *Use Condition 2 to write an inequality for the problem.*

c. *Solve.* (Example 1)

1. Brian estimates that a pair of tennis shoes he wants to buy costs $50 less than a certain tennis racket. He plans to spend less than $120 for both items. Find the greatest amount he can spend for the racket.

2. The Williams family plans to spend less than $800 each month for rent and food. The amount budgeted for rent is $65 more than twice the amount budgeted for food. What is the greatest amount that can be budgeted for food?

3. Andrew Klein spent more than $525 for a television and a radio. The cost of a television was $30 more than four times the cost of the radio. Is it possible that the radio cost $100?

4. At a certain concert, the number of orchestra seat tickets sold was 20 less than three times the number of balcony tickets sold. More than 400 tickets were sold in all. Is it possible that only 90 balcony tickets were sold?

(Example 2)

5. Flora's scores on three math quizzes were 85, 72, and 81. She wants her mean score on four quizzes to be greater than 80. Find the lowest score she can get on the fourth quiz.

6. Cara wants her mean score on five tests to be greater than 85. Her scores on four tests are 85, 92, 70, and 86. What is the lowest score she can get on the fifth test?

7. Ryan's scores for three games of bowling were 95, 130, and 125. In order to have a mean score greater than 110, what is the lowest score he can bowl on the fourth game?

8. Tyrone's mean bowling score for five games is greater than 120. His scores for 4 games of bowling were 125, 100, 115, and 130. Is it possible that his score on the fifth game was 132?

Using Statistics

Making Diagrams

One important task of **statisticians** is to display data so that it can be interpreted more easily. For example, the median and the mode can often be read directly from a **stem–leaf plot.**

The Bright Light Bulb Company conducts tests on the light bulbs it manufactures. Thirty-five sample light bulbs are tested to find the number of hours each one lasts before burning out. The results of one test are shown at the right.

Lifetime of Thirty-six Light Bulbs (to the nearest hour)

52	81	50	62	71	70
59	53	65	75	76	92
93	77	95	68	83	73
53	88	80	74	57	87
68	73	61	71	66	89
71	79	66	65	88	85

EXAMPLE: Construct a stem–leaf plot for the data in the table.

SOLUTION:

[1] Since the smallest entry is 50 and the largest is 95, the **stems** can be the ten's digits, 5 through 9.

Write the stems vertically. Draw a vertical line to the right.

The **leaves** are the one's digits of the 36 entries. Write each one's digit to the right of its ten's digit stem.

5	2, 9, 3, 3, 0, 7
6	8, 5, 1, 6, 2, 8, 5, 6
7	1, 7, 3, 9, 5, 4, 1, 1, 6, 0, 3
8	1, 8, 0, 3, 8, 7, 9, 5
9	3, 5, 2

5 | 2 represents the entry 52.

[2] Rewrite the leaves for each stem in increasing order from left to right. This is an **ordered stem-leaf plot.**

5	0, 2, 3, 3, 7, 9
6	1, 2, 5, 5, 6, 6, 8, 8
7	0, 1, 1, 1, 3, 3, 4, 5, 6, 7, 9
8	0, 1, 3, 5, 7, 8, 8, 9
9	2, 3, 5

EXERCISES

For Exercises 1–2, draw a stem–leaf plot for the given data.

1. ***Number of Students on Fifteen School Buses***

40	36	35	28	40
32	35	32	27	25
27	31	40	31	33

2. ***Number of Points Scored in Eighteen Basketball Games***

70	57	49	80	74	77
66	52	84	63	76	83
72	64	74	76	66	59

For Exercises 3–6, use the ordered stem-leaf plot in the Example.

3. Find the median of the data. (HINT: The mean of the 18th and 19th leaves is $\frac{1 + 3}{2}$, or 2.)

4. Find the mode of the data.

5. The mean of the data is 72.7 hours.

- **a.** Which statement would you use if you were in charge of the company's advertising? *Lasts Up to 95 Hours,* or *Lasts an Average of 70 Hours*
- **b.** Give a reason for your choice.

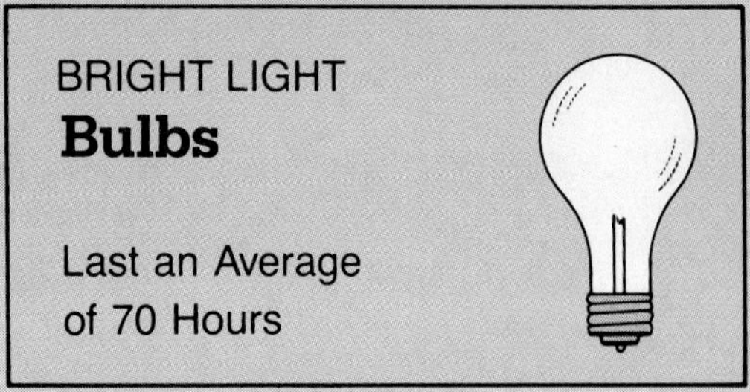

6. You purchase a Bright Light bulb.

- **a.** Is it more likely to last 75 hours or 90 hours?
- **b.** Could it last 105 hours? 45 hours? Explain.

*A **two-sided stem-leaf plot** can be used to compare two sets of data. The leaves of the second set of data are written to the left of the stems, and they are read from right to left.*

*Two local movie theaters have both shown the same movie for the past 15 days. The diagram at the right shows each theater's daily attendance for that movie. The **stems** are the hundred's and ten's digits. The **leaves** are the one's digits.*

Theater A		***Theater B***
4, 3, 0	15	
4, 2	16	9
5	17	2, 3, 3, 5, 6, 7
8, 1	18	1, 2, 5, 5, 5, 9
7, 6, 6, 0	19	6, 7
5, 4, 0	20	

0 | 20 represents the number 200.

7. Find the median and mode for the attendance at Theater A.

8. Find the median and mode for the attendance at Theater B.

9. Which theater had the greater median? the greater mode?

10. Which theater do you think had the greater mean attendance? Find the mean attendance for each theater. Was your guess correct?

CHAPTER SUMMARY

IMPORTANT TERMS

Graph *(p. 116)*
Solution set *(p. 116)*
Domain of a variable *(p. 116)*
Inequality *(p. 116)*
Empty set *(p. 128)*

IMPORTANT IDEAS

1. *Comparison Property:* If a and b are real numbers, then exactly one of the following is true: $a > b$, $a = b$, or $a < b$.
2. *Order Property of Opposites:* If a and b are real numbers and $a < b$, then $-a > -b$.
3. *Transitive Property of Order:* If $a < b$ and $b < c$, then $a < c$.
4. *Addition Property for Inequalities:* Adding the same real number to each side of an inequality does not change the direction of the inequality. The new inequality is equivalent to the first.
5. *Subtraction Property for Inequalities:* Subtracting the same real number from each side of an inequality does not change the direction of the inequality. The new inequality is equivalent to the first.
6. *Multiplication Property for Inequalities:* If each side of an inequality is multiplied by the same nonzero real number, an equivalent inequality is obtained as follows.
   a. The direction of the inequality is unchanged if each side is multiplied by the same positive number.
   b. The direction of the inequality is reversed if each side is multiplied by the same negative number.
7. *Division Property for Inequalities:* If each side of an inequality is divided by the same nonzero real number, an equivalent inequality is obtained as follows.
   a. The direction of the inequality is unchanged if each side is divided by the same positive number.
   b. The direction of the inequality is reversed if each side is divided by the same negative number.
8. Steps for Solving Word Problems Involving Inequalities
   See page 136.

CHAPTER REVIEW

SECTION 5.1

Write the solution set of each graph. If necessary, write the domain you are assuming.

1.

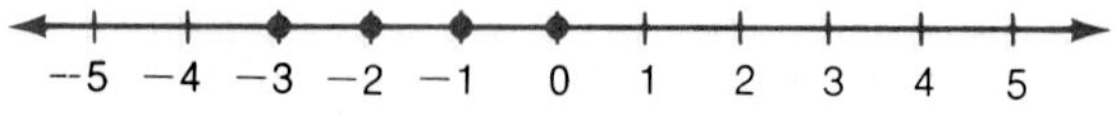

2. 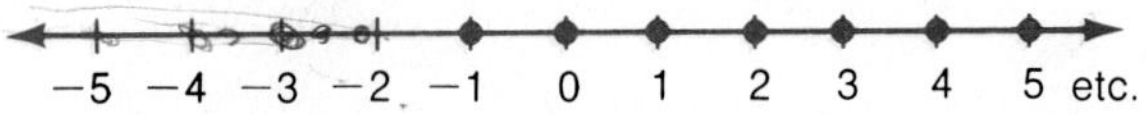

3. 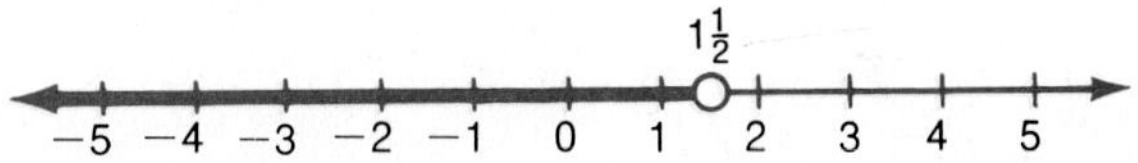

4. 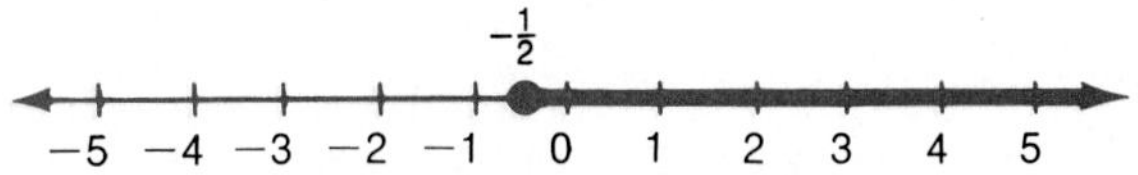

Draw a graph of each inequality.

5. $x < 5$; Domain = {whole numbers}

6. $x < -2$; Domain = {integers}

7. $x > -3$

8. $x > 1$

9. $x < 4$

10. $x < -1$

SECTION 5.2

Write a new inequality by applying the Order Property of Opposites.

11. $r < -12$

12. $x > -8$

13. $a > -3$

14. $-4.5 < t$

15. $-x > -4.2$

16. $3\frac{1}{4} > -t$

17. $-1 > -x$

18. $k < -n$

19. If $a > -5$ and $-5 > c$, what two inequalities relate a and c?

20. If $-b < c$ and $-c > a$, what two inequalities relate a and b?

Write True or False for each sentence.

21. If $a < -5$ and $-5 < b$, then $a < b$.

22. If $a > 6$ and $6 > b$, then $a < b$.

SECTION 5.3

Write Yes or No to show whether the two inequalities are equivalent.

23. $x + 3 < 5$; $x + 1 < 7$

24. $x - 4 < 10$; $x - 5 < 9$

25. $-5 > 2x - 2$; $-7 > 2x$

26. $5x - 8 > 2$; $5x > 10$

27. $2a - 7 > -2$; $2a > 5$

28. $3b - 1 < b + 7$; $3b - 2 < b + 8$

Solve.

29. $x - 3 < -2$

30. $x + 9 > -3$

31. $-4 + x < -6$

32. $-8 + x > -10$

33. $-8.3 > 1.4 + x$

34. $5.6 < x - 3.9$

SECTION 5.4

Solve.

35. $4x + 3 - 3x < -5$

36. $3 > 5x + 7 - 4x$

37. $3x - 5 > 2(x + 3)$

38. $3(2x - 1) < x - 2 + 4x$

SECTION 5.5

Solve.

39. $4x < 30$

40. $8x > -20$

41. $-9x > 45$

42. $-5x < -60$

43. $-6 < -\frac{3}{2}x$

44. $\frac{1}{2} > -\frac{1}{4}x$

SECTION 5.6

Solve.

45. $0.3 - 0.6x < 1.4x - 2.9$

46. $3.2x - 1.1 > 0.2x + 1.3$

47. $x - 19 > 12 - 2x$

48. $2x - 5 + 3x < -x + 7$

49. $-2 + 3x > 3(1 + x)$

50. $2x - 4 + x < -1 + 3x$

SECTION 5.7

For Exercises 51–54, Condition 1 is underscored once. Condition 2 is underscored twice.

Use Condition 1 to represent the unknowns. Use Condition 2 to write an inequality for the problem. Solve.

51. The cost of a coat Emilio would like to buy is \$20 more than three times the cost of a pair of pants. He plans to spend less than \$105 for both items. What is the greatest amount he can spend for the pants?

52. Julia spent more than \$25 for a book and a compact disc. The cost of the book was \$7 less than the cost of the compact disc. Is it possible that she spent only \$15 for the compact disc?

53. Suzanne's scores on three tests were 85, 93, and 78. She wants her mean score on four tests to be greater than 85. Find the lowest score she can get on the fourth test.

54. Alan's mean bowling score for five games is greater than 110. His scores for 4 games of bowling were 105, 120, 88, and 108. Is it possible that his score on the fifth game was 135?

CHAPTER 6

Powers

The *spiral galaxy* shown below is about 10^7 light-years away from Earth. In scientific notation, this is about 5.86×10^9 miles. The microscopic *diatom,* a one-celled plant, ranges in size from 5×10^{-4} inch to 8×10^{-2} inch.

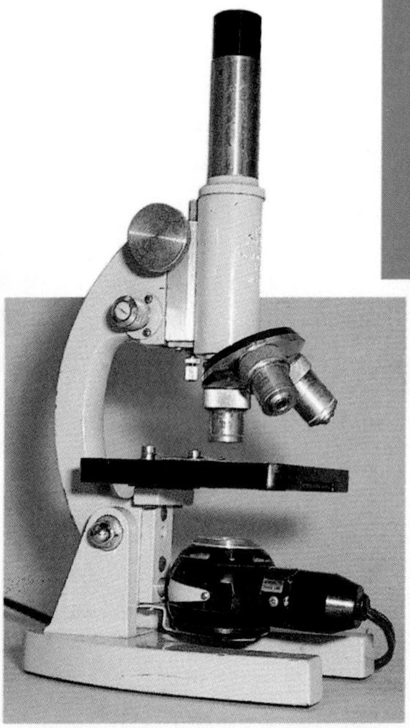

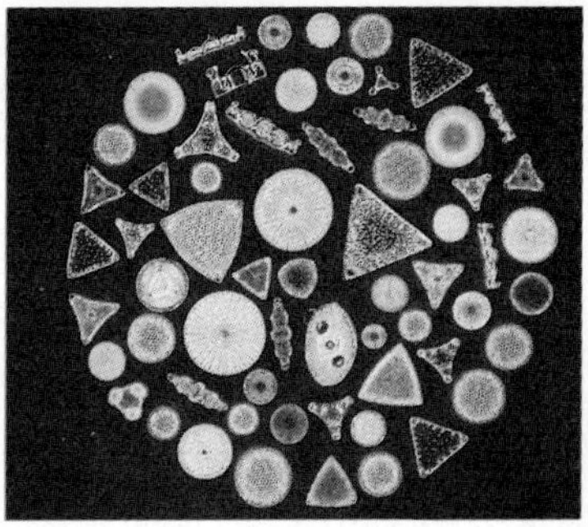

6.1 Products of Powers

P–1 **Write each power below as a product.**

a. 2^4 **b.** x^3 **c.** y^5 **d.** 4^3 **e.** z^2

2 is a factor of 2^4 four times. x is a factor of x^3 three times.

P–2 **How many times is y a factor of y^5?**

If a is a positive integer, then x^a means that x is a factor a times.

$$x^a = \underbrace{x \cdot x \cdot x \cdots x}_{a \text{ factors}}$$

EXAMPLE 1 Write $2^3 \cdot 2^5$ as one power.

Solution:

$$\begin{aligned} 2^3 \cdot 2^5 &= (2 \cdot 2 \cdot 2)(2 \cdot 2 \cdot 2 \cdot 2 \cdot 2) \\ &= (2 \cdot 2 \cdot 2 \cdot 2 \cdot 2 \cdot 2 \cdot 2 \cdot 2) \\ &= 2^8 \end{aligned}$$

P–3 **How can you express $(-3)^2$ as a product? $(-3)^4$ as a product?**

EXAMPLE 2 Write $(-3)^2(-3)^4$ as one power.

Solution:

$$\begin{aligned} (-3)^2(-3)^4 &= (-3 \cdot -3)(-3 \cdot -3 \cdot -3 \cdot -3) \\ &= (-3 \cdot -3 \cdot -3 \cdot -3 \cdot -3 \cdot -3) \\ &= (-3)^6 \end{aligned}$$

P–4 **How many times is x used as a factor in $x^a \cdot x^b$ if a and b are positive integers?**

$$x^a \cdot x^b = \underbrace{(x \cdot x \cdot x \cdots x)}_{a \text{ factors}}\underbrace{(x \cdot x \cdot x \cdots x)}_{b \text{ factors}}$$

Product Property of Powers

If a and b are integers and x is any nonzero real number, then $\mathbf{x^a \cdot x^b = x^{a+b}}$.

$13^4 \cdot 13^5 = 13^9$

$(-5.6)^2(-5.6) = (-5.6)^3$

P–5 **Write each product below as one power.**

a. $(-1.9)^2(-1.9)^3$ **b.** $(\frac{1}{2})^4(\frac{1}{2})^3$ **c.** $t^2 \cdot t^7$

Keep the same base. Add the exponents.

EXAMPLE 3 Write the product as a sum: $r^3(r^2 + r)$

Solution:

1. Distributive Property ⟶ $r^3(r^2 + r) = r^3 \cdot r^2 + r^3 \cdot r$
2. Product Property of Powers ⟶ $= r^5 + r^4$

P–6 **Write each product below expressed as a sum or difference.**

a. $t^2(t^4 + t^3)$ **b.** $k^3(3k^2 - 2k + 3)$

EXAMPLE 4 Solve: $12^4 \cdot 12^x = 12^9$

Solution:

$$12^4 \cdot 12^x = 12^9$$
$$12^{4+x} = 12^9$$
$$4 + x = 9$$
$$x = 5$$

Since the bases are equal, the exponents are also equal.

P–7 **Solve: $(-4.8)^3 \cdot (-4.8)^x = (-4.8)^7$**

CLASSROOM EXERCISES

Write each product as one power. (Examples 1–2)

1. $2^3 \cdot 2^2$ **2.** $a^5 \cdot a^2$ **3.** $(-5)^2(-5)^4$ **4.** $(-1.8)^5(-1.8)^2$

5. $(\frac{1}{2})^2(\frac{1}{2})^3$ **6.** $n \cdot n^4$ **7.** $k^{10} \cdot k^{12}$ **8.** $a^3 \cdot a \cdot a^2$

9. $7 \cdot 7^2 \cdot 7^3$ **10.** $(-1)^3(-1)^5$ **11.** $2^4 \cdot 2^3$ **12.** $m^5 \cdot m^4$

Solve each equation. (Example 4)

13. $10^x = 10^5$ **14.** $10^x \cdot 10^2 = 10^9$ **15.** $3^x \cdot 3^5 = 3^{11}$

16. $(\frac{1}{2})^3(\frac{1}{2})^5 = (\frac{1}{2})^x$ **17.** $3^x \cdot 3^2 = 3^6$ **18.** $5^3 \cdot 5^x = 5^4$

WRITTEN EXERCISES

Goal: To write a product of powers as one power

Sample Problem: $r^4 \cdot r \cdot r^3$ **Answer:** r^8

Write each product as one power. (Examples 1–2)

1. $12^3 \cdot 12^7$ **2.** $10^5 \cdot 10^4$ **3.** $5 \cdot 5^3$ **4.** $6 \cdot 6^4$

5. $(3.2)^5(3.2)^7$ **6.** $(3.14)^3(3.14)^2$ **7.** $(-15)^4(-15)^5$ **8.** $(-10)^4(-10)^2$

9. $(-6)^2(-6)^9$ **10.** $(-4)^3(-4)^5$ **11.** $(-\frac{2}{3})^2(-\frac{2}{3})^5$ **12.** $(-\frac{1}{4})^3(-\frac{1}{4})^2$

13. $a^4 \cdot a^3$ **14.** $x^2 \cdot x^5$ **15.** $r^8 \cdot r$ **16.** $y^6 \cdot y^3$

17. $a^3 \cdot a^2 \cdot a^6$ **18.** $b^4 \cdot b \cdot b^3$ **19.** $5^2 \cdot 5^7$ **20.** $12^3 \cdot 12^3$

Write each product as a sum or difference. (Example 3)

21. $y^2(y + y^2)$ **22.** $a^3(a^2 + a)$ **23.** $x^4(x^5 + x^3)$ **24.** $m^5(m^2 + m^3)$

25. $(2^3 - 2^5)2^2$ **26.** $(10^4 - 10^2)10^3$ **27.** $n(n^3 - n^2 + n)$ **28.** $s^3(s^5 - s^3 - s)$

Solve each equation. (Example 4)

29. $10^5 \cdot 10^6 = 10^x$ **30.** $8^3 \cdot 8^9 = 8^x$

31. $(0.5)^x(0.5)^2 = (0.5)^7$ **32.** $(1.2)^5(1.2)^x = (1.2)^9$

33. $3^x \cdot 3^5 = 3^{10}$ **34.** $5^5 \cdot 5^x = 5^{12}$

35. $(\pi)^x(\pi)^3 = (\pi)^9$ **36.** $(-\frac{2}{3})^2(-\frac{2}{3})^x = (-\frac{2}{3})^{11}$

REVIEW CAPSULE FOR SECTION 6.2

Evaluate both sides of each equation. Then write True or False for the equation. (Pages 21–23)

1. $\frac{2^5}{2^2} = 2^3$ **2.** $\frac{3^5}{3^3} = 3^2$ **3.** $\frac{2^6}{2^2} = 2^3$ **4.** $\frac{x^5}{y^3} = \left(\frac{x}{y}\right)^2$ when $x = -2$, $y = 1$

6.2 Quotients of Powers and Zero Exponents

P–1 **How can you express x^5 as a product? x^3 as a product?**

EXAMPLE 1 Write $\frac{x^5}{x^3}$ as one power.

Solution:

1 Meaning of a power ⟶ $\frac{x^5}{x^3} = \frac{x \cdot x \cdot x \cdot x \cdot x}{x \cdot x \cdot x}$

2 Product Rule for Fractions ⟶ $= \frac{x \cdot x \cdot x}{x \cdot x \cdot x} \cdot \frac{x \cdot x}{1}$

Remember that $\frac{x \cdot x \cdot x}{x \cdot x \cdot x} = 1$.

3 Multiplication Property of 1 ⟶ $= \frac{x \cdot x}{1}$

4 Simplify. ⟶ $= x^2$

Quotient Property of Powers

If a and b are integers and x is a nonzero real number, then $\frac{x^a}{x^b} = x^{a-b}$.

$$\frac{5^{12}}{5^3} = 5^{12-3}$$
$$= 5^9$$

P–2 **Write each quotient below as one power.**

a. $\frac{17^7}{17^4}$ **b.** $(1.8)^8 \div (1.8)^3$ **c.** $\frac{m^{12}}{m^5}$

Keep the same base. Subtract the exponents.

EXAMPLE 2 Solve: $\frac{12^t}{12^3} = 12^4$

Solution:

$$\frac{12^t}{12^3} = 12^4$$
$$12^{t-3} = 12^4$$
$$t - 3 = 4$$
$$t = 7$$

Since the bases are equal, the exponents are also equal.

P–3 **How can you express $\frac{113^5}{113^5}$ as one power by using the Quotient Property of Powers?**

P–4 **Simplify $\frac{a}{a}$ if *a* is any nonzero real number.**

$$\frac{113^5}{113^5} = 113^{5-5}$$ ◀ ***By the Quotient Property of Powers***

$$= 113^0$$

$$\frac{113^5}{113^5} = 1$$ ◀ ***By the special property: $\frac{a}{a} = 1$***

Thus, it is clear that 113^0 equals 1.

P–5 **What is the value of any nonzero real number raised to the 0 power?**

Consider $\frac{x^m}{x^m}$, in which *m* is any integer and *x* is any nonzero real number.

$$\frac{x^m}{x^m} = x^{m-m}$$ ◀ ***By the Quotient Property of Powers***

$$= x^0$$

$$\frac{x^m}{x^m} = 1$$ ◀ ***By the special property: $\frac{a}{a} = 1$***

This shows that x^0 equals 1. Thus, the following property has been proved.

> ***Property of a Zero Exponent***
>
> If *x* is any nonzero real number, then $\mathbf{x^0 = 1}$.
>
> $(-27)^0 = 1$
>
> $(\sqrt{5})^0 = 1$

P–6 **Simplify each expression below.**

a. $\left(\frac{3}{4}\right)^0 + 12^0$ **b.** $(-5.2)^0 - (\sqrt{3})^0$ **c.** $\left(\frac{t}{25}\right)^0$

CLASSROOM EXERCISES

Write each quotient as one power. No divisor equals 0. (Example 1)

1. $\frac{3^5}{3^2}$
2. $\frac{10^8}{10^3}$
3. $\frac{(-3)^6}{(-3)^2}$
4. $\frac{x^{10}}{x^7}$
5. $\frac{a^7}{a^5}$
6. $\frac{b^4}{b}$
7. $2^7 \div 2^4$
8. $y^{12} \div y^7$
9. $\left(\frac{1}{2}\right)^{10} \div \left(\frac{1}{2}\right)^3$
10. $12^3 \div 12^k$
11. $\frac{(-5)^t}{(-5)^4}$
12. $\frac{(0.5)^p}{(0.5)^q}$

Write True or False for each sentence. (Example 1)

13. $\frac{6^4}{3^2} = 2^2$
14. $\frac{10^{10}}{10^5} = 1^5$
15. $\frac{2^7}{3^2} = \left(\frac{2}{3}\right)^5$
16. $\frac{5^4}{5^3} = \frac{1}{5}$
17. $\frac{2^4}{2^2} = 2^2$
18. $\frac{13}{7^2} = (13 - 7)^3$

Simplify. (P-5, P-6)

19. 2^0
20. 100^0
21. $(-5)^0$
22. $\left(-\frac{2}{3}\right)^0$
23. $(0.5)^0$
24. $(-\pi)^0$
25. w^0; $w \neq 0$
26. $\left(\frac{1}{t}\right)^0$; $t \neq 0$

WRITTEN EXERCISES

Goal: To write a quotient of powers as one power
Sample Problem: $y^{12} \div y^4$
Answer: y^8

Write each quotient as one power. (Example 1)

1. $\frac{10^5}{10^3}$
2. $\frac{8^6}{8^3}$
3. $\left(\frac{1}{2}\right)^3 \div \left(\frac{1}{2}\right)^2$
4. $\left(\frac{1}{4}\right)^5 \div \left(\frac{1}{4}\right)^3$
5. $\frac{x^{12}}{x^8}$
6. $\frac{a^{15}}{a^{12}}$
7. $\frac{2^{12}}{2^9}$
8. $\frac{10^7}{10^3}$
9. $\frac{(0.6)^3}{(0.6)^2}$
10. $\frac{(1.3)^6}{(1.3)^5}$
11. $\frac{\pi^5}{\pi^2}$
12. $\frac{(-\pi)^5}{(-\pi)^3}$

Write True or False for each equation. (Example 1)

13. $\frac{10^{11}}{10^{10}} = 10$

14. $\frac{100^{98}}{100^{96}} = 10{,}000$

15. $\frac{10^5}{5^2} = 2^3$

16. $\frac{12^7}{3^4} = 4^3$

17. $\frac{3^8}{3^4} = 3^2$

18. $\frac{5^6}{5^2} = 5^3$

19. $\frac{(\sqrt{2})^7}{(\sqrt{2})^2} = (\sqrt{2})^5$

20. $\frac{(-\sqrt{3})^{10}}{(-\sqrt{3})^6} = (-\sqrt{3})^4$

21. $\frac{10^3 \cdot 10^5}{10^4} = 10^4$

22. $\frac{10^9}{10^2 \cdot 10^3} = 10{,}000$

Solve each equation. (Example 2)

23. $\frac{2^x}{2^2} = 2^3$

24. $\frac{3^y}{3^5} = 3^2$

25. $\frac{x^5}{10^3} = 10^2$

26. $\frac{5^9}{a^3} = 5^6$

27. $\frac{10^7}{10^2} = 10^x$

28. $\frac{10^{10}}{10^3} = 10^y$

29. $\frac{11^{2x}}{11^x} = 11^4$

30. $\frac{5^{3x}}{5^x} = 5^{12}$

31. $10^0 = x$

32. $7^0 = n$

33. $\left(\frac{1}{2}\right)^0 = y$

34. $\left(-\frac{1}{4}\right)^0 = a$

35. $(0.7)^a = 1$

36. $(-1.3)^x = 1$

37. $7^x - 5^x = 0$

38. $\left(\frac{\sqrt{2}}{\sqrt{7}}\right)^x = 1$

39. $1 = p^0$

40. $y^0 = 1$

Simplify. (P-6)

41. $93^0 - 2$

42. $(-114)^0 + 5$

43. $3x^0 + 2y^0$

44. $5m^0 + 3n^0$

45. $-2r^0 + t^0$

46. $p^0 - 4q^0$

NON-ROUTINE PROBLEM

47. Copy the addition problem at the right and find the sum. Then replace six of the digits in the problem with a zero so that the sum is 1,111.

$$\begin{array}{r} 1\ 1\ 1 \\ 3\ 3\ 3 \\ 5\ 5\ 5 \\ 7\ 7\ 7 \\ +\ 9\ 9\ 9 \\ \hline \end{array}$$

6.3 Negative Exponents

The Product and Quotient Properties of Powers apply to negative exponents as well as to positive and zero exponents.

EXAMPLE 1 Simplify $\frac{x^3}{x^6}$.

Solution:

1. Meaning of a power ⟶ $\frac{x^3}{x^6} = \frac{x \cdot x \cdot x}{x \cdot x \cdot x \cdot x \cdot x \cdot x}$
2. Product Rule for Fractions ⟶ $= \frac{x \cdot x \cdot x}{x \cdot x \cdot x} \cdot \frac{1}{x \cdot x \cdot x}$
3. Multiplication Property of 1 ⟶ $= \frac{1}{x \cdot x \cdot x}$
4. Simplify. ⟶ $= \frac{1}{x^3}$

The Quotient Property of Powers is used for the same expression.

$$\frac{x^3}{x^6} = x^{3-6}$$

$$= x^{-3}$$

◀ ***Exponent of numerator less the exponent of denominator***

Thus, the following sentence is true for all nonzero real numbers.

$$\frac{1}{x^3} = x^{-3}$$

◀ ***$\frac{1}{x^3}$ is considered to be the simplest form.***

Property of Negative Exponents

If $-a$ represents a negative integer and x is a nonzero real number, then $\mathbf{x^{-a} = \frac{1}{x^a}}$.

$$3^{-2} = \frac{1}{3^2}$$

$$(-1.3)^{-5} = \frac{1}{(-1.3)^5}$$

P–1 **Simplify.**

a. r^{-3} **b.** m^{-6} **c.** 3^{-2} **d.** 2^{-4} **e.** 10^{-3}

P–2 **What is the rule for multiplying powers of the same base?**

EXAMPLE 2

Write $y^{-5} \cdot y^3$ as one power. Then simplify.

Solution:

[1] Product Property of Powers ⟶ $y^{-5} \cdot y^3 = y^{-5+3}$

[2] Add the exponents. ⟶ $= y^{-2}$

[3] Simplify. ⟶ $= \frac{1}{y^2}$

P–3 **Write each product below as one power.**

a. $17^{-2} \cdot 17$ **b.** $(1.2)^{-2}(1.2)^6$ **c.** $r^{-1} \cdot r^3 \cdot r^{-5}$

P–4 **What is the rule for dividing powers of the same base?**

EXAMPLE 3

Write $\frac{x^{-5}}{x^{-2}}$ as one power.

Solution:

[1] Quotient Property of Powers ⟶ $\frac{x^{-5}}{x^{-2}} = x^{-5-(-2)}$

[2] Subtract the exponents. ⟶ $= x^{-3}$

The simplest form is $\frac{1}{x^3}$.

P–5 **Write each quotient below as one power.**

a. $\frac{5^{-1}}{5^{-3}}$ **b.** $\frac{(3.7)^2}{(3.7)^5}$ **c.** $\frac{y^0}{y^{-2}}$ **d.** $\frac{x^0}{x^{-5}}$ **e.** $\frac{r^{-4}}{r^0}$

EXAMPLE 4

Write as one power and simplify: $\frac{m^6 \cdot m^{-3}}{m^4}$

Solution:

[1] Product Property of Powers ⟶ $\frac{m^6 \cdot m^{-3}}{m^4} = \frac{m^3}{m^4}$

[2] Quotient Property of Powers ⟶ $= m^{-1}$

[3] Simplify. ⟶ $= \frac{1}{m}$

CLASSROOM EXERCISES

Write each expression as a power with a positive exponent. (Property of Negative Exponents)

1. 5^{-10}
2. $(-3)^{-4}$
3. $(5.2)^{-2}$
4. $(325)^{-5}$
5. r^{-8}
6. t^{-1}

Write each product or quotient as one power. (Examples 2–3)

7. $x^2 \cdot x^5$
8. $(2^{-3})(2^5)$
9. $\frac{2^5}{2^3}$
10. $\frac{10^{-2}}{10^{-1}}$

WRITTEN EXERCISES

Goal: To simplify powers and products and quotients of powers involving positive and negative exponents

Sample Problem: $\frac{a^2 \cdot a^{-7}}{a^{-10}}$ **Answer:** a^5

Simplify each expression. (Examples 2–3)

1. r^{-4}
2. t^{-8}
3. $2y^{-5}$
4. $6m^{-2}$
5. 4^{-2}
6. 3^{-3}
7. $(-3)^{-3}$
8. $(-2)^{-3}$
9. ab^{-3}
10. rs^{-2}
11. $\frac{2^3}{2^{-2}}$
12. $\frac{10^{-1}}{10^{-3}}$
13. $\frac{a^2}{a^5}$
14. $\frac{y}{y^3}$
15. $(2^{50})(2^{-49})$
16. $(10^{100})(10^{-99})$

Write each product or quotient as one power. Then simplify where necessary. (Examples 2–4)

17. $a^5 \cdot a^{-3}$
18. $b^{-2} \cdot b^7$
19. $x^{-3} \cdot x^{-2}$
20. $y^{-1} \cdot y^{-2}$
21. $\frac{n^2}{n^3}$
22. $\frac{m^5}{m^7}$
23. $\frac{r^{-2}}{r^{-5}}$
24. $\frac{s^{-6}}{s^{-3}}$
25. $\frac{t^0}{t^{-3}}$
26. $\frac{w^0}{w^{-5}}$
27. $\frac{p^{-2}}{p^0}$
28. $\frac{g^5}{g^0}$
29. $(10^2)(10^7)(10^{-5})$
30. $(5^{-2})(5^4)(5^{-3})$
31. $(y^{-5})(y^6)(y^{-3})$
32. $(b^7)(b^{-3})(b^{-2})$
33. $\frac{r^3 \cdot r^{-5}}{r^{-2}}$
34. $\frac{s^{-2} \cdot s^{-3}}{s^{-7}}$

APPLICATIONS

You can use a calculator to evaluate a fraction with a power in the denominator.

EXAMPLE Evaluate $\frac{1}{(2.9)^3}$

SOLUTION [2] [.] [9] [×] [=] [=] [÷] [=] [=] → 0.041002

On some calculators press "=" twice here.

Evaluate.

35. $\frac{1}{2^{10}}$
36. $\frac{1}{3^{12}}$
37. $\frac{1}{(4.8)^5}$
38. $\frac{1}{(0.9)^4}$
39. $\frac{1}{(0.86)^{12}}$
40. $\frac{1}{(0.92)^{15}}$
41. $\frac{1}{(3^8)(3.14)}$
42. $\frac{1}{(0.6)^7(3140)}$

MORE CHALLENGING EXERCISES

Let a be a positive integer and let b be a negative integer. Determine whether each expression is positive or negative.

43. ab^{-2}
44. $a^{-2} \cdot b$
45. $\frac{a}{a^2}$
46. $\frac{b^{-3}}{b^{-1}}$

MID-CHAPTER REVIEW

Write each product as one power. (Section 6.1)

1. $(12)^5(12)^8$
2. $(4)^3(4)^7$
3. $(-6)^4(-6)^{10}$
4. $(-3)^2(-3)^6$
5. $(\frac{3}{4})^8 \cdot (\frac{3}{4})^5$
6. $(-\frac{5}{4})^4 \cdot (-\frac{5}{4})^7$
7. $(1.4)^2(1.4)^5$
8. $(3.7)^4(3.7)^{24}$

Write each quotient as one power (Section 6.2)

9. $\frac{(21)^5}{(21)^3}$
10. $\frac{(-7)^8}{(-7)^3}$
11. $\left(\frac{3}{4}\right)^6 \div \left(\frac{3}{4}\right)^3$
12. $\frac{(5.3)^{10}}{(5.3)^5}$

Simplify.

13. $(0.9)^0$
14. $(-\frac{5}{8})^0$
15. $5^0 - 144^0$
16. $-7t^0$, $t \neq 0$

Write each product or quotient as one power. Then simplify. (Section 6.3)

17. $(0.6)^{-7}(0.6)^4$
18. $(8.1)^{-2}(8.1)^{-3}$
19. $w^{-1} \cdot w^{-5}$
20. $p^8 \cdot p^{-3}$
21. $(12)^3 \div (12)^5$
22. $(59)^{-2} \div (59)^4$
23. $\frac{m^2}{m^5}$, $m \neq 0$
24. $\frac{t^{-1}}{t^4}$, $t \neq 0$

Using Formulas Distance/Rate/Time

Speeds and distances in space are so great that **scientific notation** is useful in representing them.

Minimum Distance from Earth to:
Mars: 56,000,000 (5.6×10^7) kilometers
Sun: 147,000,000 (1.47×10^8) kilometers
Saturn: 1,200,000,000 (1.2×10^9) kilometers
Pluto: 4,270,000,000 (4.27×10^9) kilometers
Proxima Centauri (nearest star):
40,000,000,000,000 (4.0×10^{13}) kilometers

Speed of Light or of a Radio Signal:
300,000 (3.0×10^5) kilometers per second

Voyager 2 took this photograph of Saturn on July 21, 1981, at a distance of 33.9 million kilometers from the planet.

EXAMPLE: Find the approximate time in seconds needed for a radio signal to reach the earth from Saturn.

SOLUTION: Use $d = rt$.

$$d = r \times t$$

$$1.2 \times 10^9 = (3.0 \times 10^5) \times t \text{ or,}$$

$$(3.0 \times 10^5)t = 1.2 \times 10^9$$

$d = 1.2 \times 10^9$ kilometers
$r = 3.0 \times 10^5$ km per second

$$t = \frac{1.2 \times 10^9}{3.0 \times 10^5}$$

Use mental math.
$1.2 \div 0.3 = 0.4$
$10^9 \div 10^5 = 10^4$

$$= 0.4 \times 10^4, \text{ or } \mathbf{4000 \text{ seconds}}$$

EXERCISES

For Exercises 1–4, refer to the speed and distances given above to find each of the following. Round your answer to the nearest whole number.

1. The approximate time in seconds for light from the sun to reach the earth

2. The approximate time in minutes (60 seconds) for a radio signal from Mars to reach the earth

3. The approximate time in hours for a radio signal from Pluto to reach the earth

4. The approximate time in years (365 days) for light to reach the earth from the nearest star

5. Which is the greatest distance?
a. 1.48×10^6 meters
b. 14.8×10^3 kilometers
c. 1.48×10^5 kilometers
d. 1480×10^2 meters

6. Which is the fastest speed?
a. 2.74×10^5 kilometers per hour
b. 2.74×10^6 meters per hour
c. 27.4×10^3 kilometers per hour
d. 27.4×10^4 meters per hour

6.4 Powers of Products and Quotients

P–1 **What is the value of $2x^3$ if x equals 1?**

Raise 1 to the third power. Then multiply by 2.

P–2 **What is the value of $(2x)^3$ if x equals 1?**

Multiply $2 \cdot 1$. Then raise to the third power.

You can see that $2x^3$ does not equal $(2x)^3$.

EXAMPLE 1 Simplify $(2x)^3$.

Solution:

1. Meaning of a power ⟶ $(2x)^3 = (2x)(2x)(2x)$
2. Commutative and Associative Properties of Multiplication ⟶ $= (2 \cdot 2 \cdot 2)(x \cdot x \cdot x)$
3. Simplify. ⟶ $= 8x^3$

Thus, $(2x)^3 = 2^3x^3$.

Property of the Power of a Product

If x and y are nonzero real numbers and a is any integer, then $(xy)^a = x^ay^a$.

$(-3t)^3 = (-3)^3t^3$
$= -27t^3$

$(pq)^{-2} = p^{-2}q^{-2}$

P–3 **Simplify.**

a. $(xy)^2$ **b.** $(3a)^2$ **c.** $(2r)^3$ **d.** $(-3r)^2$

EXAMPLE 2 Simplify $(rs)^{-3}$.

Solution:

1. Property of the Power of a Product ⟶ $(rs)^{-3} = r^{-3}s^{-3}$
2. Property of Negative Exponents ⟶ $= \frac{1}{r^3} \cdot \frac{1}{s^3}$
3. Product Rule for Fractions ⟶ $= \frac{1}{r^3s^3}$

EXAMPLE 3 Simplify $\left(\frac{x}{3}\right)^3$.

Solution:

1 Meaning of a power ⟶ $\left(\frac{x}{3}\right)^3 = \left(\frac{x}{3}\right)\left(\frac{x}{3}\right)\left(\frac{x}{3}\right)$

2 Product Rule of Fractions ⟶ $= \frac{x \cdot x \cdot x}{3 \cdot 3 \cdot 3}$

3 Meaning of a power ⟶ $= \frac{x^3}{3^3}$, or $\frac{x^3}{27}$ ◀ ***Simplest form***

> ***Property of the Power of a Quotient***
>
> If x and y are nonzero real numbers and a is any integer, then $\left(\frac{x}{y}\right)^a = \frac{x^a}{y^a}$.
>
> $\left(\frac{2}{3}\right)^2 = \frac{2^2}{3^2}$
>
> $\left(\frac{r}{s}\right)^{-3} = \frac{r^{-3}}{s^{-3}}$

P–4 **Simplify.**

a. $\left(\frac{3}{t}\right)^2$ **b.** $\left(\frac{p}{q}\right)^4$ **c.** $\left(\frac{2r}{s}\right)^3$ **d.** $\left(\frac{1}{r}\right)^4$

Expression **c** above involves both properties.

$$\left(\frac{2r}{s}\right)^3 = \frac{(2r)^3}{s^3} = \frac{2^3r^3}{s^3} = \frac{8r^3}{s^3}$$

EXAMPLE 4 Simplify $\left(\frac{2}{x}\right)^{-3}$.

Solution:

1 Property of the Power of a Quotient ⟶ $\left(\frac{2}{x}\right)^{-3} = \frac{2^{-3}}{x^{-3}}$

2 Property of a Negative Exponent ⟶ $= \frac{1}{2^3} \div \frac{1}{x^3}$

3 Meaning of division ⟶ $= \frac{1}{2^3} \cdot \frac{x^3}{1}$

4 Simplify. ⟶ $= \frac{x^3}{8}$

CLASSROOM EXERCISES

Simplify. (Examples 1–3)

1. $(2a)^2$ **2.** $(3x)^2$ **3.** $(\frac{1}{2}y)^2$ **4.** $(2y)^3$

5. $(5n)^2$ **6.** $(4t)^2$ **7.** $(-2b)^2$ **8.** $(-x)^2$

9. $(0.5a)^2$ **10.** $(-3m)^2$ **11.** $(-2r)^3$ **12.** $(-2a)^4$

13. $(xy)^2$ **14.** $(mn)^3$ **15.** $(ab)^4$ **16.** $(st)^5$

17. $\left(\frac{3}{x}\right)^2$ **18.** $\left(\frac{y}{2}\right)^3$ **19.** $\left(\frac{m}{n}\right)^5$ **20.** $\left(\frac{2a}{b}\right)^2$

21. $\left(\frac{-x}{y}\right)^2$ **22.** $\left(\frac{r}{-3s}\right)^2$ **23.** $\left(\frac{-2}{x}\right)^3$ **24.** $\left(\frac{-1}{y}\right)^4$

WRITTEN EXERCISES

Goal: To simplify a power of a product or a power of a quotient

Sample Problems: a. $(4xy)^3$ **b.** $\left(\frac{4x}{y}\right)^3$ **Answers: a.** $64x^3y^3$ **b.** $\frac{64x^3}{y^3}$

Simplify. (Examples 1–4)

1. $(3r)^3$ **2.** $(4t)^3$ **3.** $(-2a)^2$ **4.** $(-3y)^2$ **5.** $(-3p)^3$

6. $(-2q)^3$ **7.** $(-ab)^2$ **8.** $(-xy)^2$ **9.** $(-rs)^3$ **10.** $(-mn)^3$

11. $(2xy)^2$ **12.** $(3ab)^2$ **13.** $(-3pq)^2$ **14.** $(-2mn)^2$ **15.** $(-2ab)^3$

16. $(-4cd)^3$ **17.** $\left(\frac{2}{x}\right)^3$ **18.** $\left(\frac{3}{y}\right)^2$ **19.** $\left(\frac{a}{-2}\right)^3$ **20.** $\left(\frac{b}{-3}\right)^2$

21. $\left(\frac{-x}{-y}\right)^2$ **22.** $\left(\frac{-2a}{-2b}\right)^3$ **23.** $\left(\frac{2}{y}\right)^{-1}$ **24.** $\left(\frac{3}{r}\right)^{-2}$ **25.** $\left(\frac{pq}{4}\right)^{-1}$

26. $\left(\frac{cd}{3}\right)^{-3}$ **27.** $(-3t)^{-1}$ **28.** $(-4k)^{-2}$ **29.** $\left(\frac{3a}{b}\right)^3$ **30.** $\left(\frac{2r}{3s}\right)^3$

REVIEW CAPSULE FOR SECTION 6.5

Write each product as one power. (Pages 144–146)

1. $m^4 \cdot m^4 \cdot m^4$ **2.** $(-3)^2 \cdot (-3)^2 \cdot (-3)^2$ **3.** $t^{-3} \cdot t^{-3} \cdot t^{-3}$

4. $(2a)^{-2} \cdot (2a)^{-2} \cdot (2a)^{-2}$ **5.** $g^{-5} \cdot g^{-5} \cdot g^{-5} \cdot g^{-5}$ **6.** $(-0.5)^3 \cdot (-0.5)^3$

6.5 Powers of Powers

P–1 **Write $(x^2)^3$ as a product of three factors.**

EXAMPLE 1 Simplify $(x^2)^3$.

Solution:

1 Meaning of a positive integral power ⟶ $(x^2)^3 = x^2 \cdot x^2 \cdot x^2$

2 Product Property of Powers ⟶ $= x^6$

P–2 **How is the exponent 6 related to the exponents 2 and 3?**

EXAMPLE 2 Simplify $(y^{-4})^{-3}$.

Solution:

1 Property of a Negative Exponent ⟶ $(y^{-4})^{-3} = \dfrac{1}{(y^{-4})^3}$

2 Meaning of a positive integral power ⟶ $= \dfrac{1}{(y^{-4})(y^{-4})(y^{-4})}$

3 Product Property of Powers ⟶ $= \dfrac{1}{y^{-12}}$

4 Property of a Negative Exponent ⟶ $= y^{12}$

P–3 **How is the exponent 12 related to the exponents -4 and -3?**

Property of the Power of a Power

If x is a nonzero real number and a and b are integers, then $\mathbf{(x^a)^b = x^{ab}}$

$(r^{-2})^4 = r^{(-2)(4)}$
$= r^{-8}$

P–4 **Simplify.**

a. $(m^3)^5$ **b.** $(k^{-4})^{-2}$ **c.** $(t^4)^{-1}$

Keep the same base. Multiply the exponents.

EXAMPLE 3 Simplify $(2x^2y^3)^{-2}$

Solution:

[1] Property of the Power of a Product ⟶ $(2x^2y^3)^{-2} = 2^{-2}(x^2)^{-2}(y^3)^{-2}$

[2] Property of the Power of a Power ⟶ $= 2^{-2}x^{-4}y^{-6}$

[3] Property of a Negative Exponent ⟶ $= \frac{1}{2^2} \cdot \frac{1}{x^4} \cdot \frac{1}{y^6}$

[4] Product Rule for Fractions ⟶ $= \frac{1}{4x^4y^6}$

The table below shows powers of 2.

Base	Exponent	Power
2	12	4096
2	11	2048
2	10	1024
2	9	512
2	8	256
2	7	128
2	6	64
2	5	32
2	4	16
2	3	8
2	2	4
2	1	2
2	0	1
2	−1	$\frac{1}{2}$
2	−2	$\frac{1}{4}$
2	−3	$\frac{1}{8}$
2	−4	$\frac{1}{16}$
2	−5	$\frac{1}{32}$
2	−6	$\frac{1}{64}$
2	−7	$\frac{1}{128}$
2	−8	$\frac{1}{256}$
2	−9	$\frac{1}{512}$

P–5 **Simplify.**

a. 2^9 **b.** 2^{-11} **c.** $(2^{-5})^2$ **d.** $(m^{-1})^{-5}$

EXAMPLE 4

Simplify $\left(\frac{2^4s^3t^2}{r^4}\right)^3$

Solution:

[1] Property of the Power of a Quotient ⟶ $\left(\frac{2^4s^3t^2}{r^4}\right)^3 = \frac{(2^4s^3t^2)^3}{(r^4)^3}$

[2] Property of the Power of a Product ⟶ $= \frac{(2^4)^3(s^3)^3(t^2)^3}{(r^4)^3}$

[3] Property of the Power of a Power ⟶ $= \frac{2^{12}s^9t^6}{r^{12}}$

[4] Simplify. ⟶ $= \frac{4096s^9t^6}{r^{12}}$

Use the table of powers of 2.

CLASSROOM EXERCISES

Simplify. (Examples 1–2)

1. $(3^2)^3$ **2.** $(10^4)^3$ **3.** $(a^{-1})^3$

4. $(b^2)^{-3}$ **5.** $(y^{-2})^{-3}$ **6.** $(t^{-5})^0$

7. $(0.5^2)^4$ **8.** $(100^{-2})^{10}$ **9.** $(-3^4)^6$

Simplify. (Examples 2–4)

10. $(r^2)^{-1}$ **11.** $(t^{-3})^2$ **12.** $(m^2n^{-1})^3$

13. $(3q^{-2})^{-2}$ **14.** $(r^{-10})^{-2}$ **15.** $\left(\frac{x^2}{x^3}\right)^2$

16. $\left(\frac{a^3}{b^4}\right)^3$ **17.** $\left(\frac{r^{-2}}{s^3}\right)^3$ **18.** $\left(\frac{m^{-1}}{m^{-2}}\right)^{-5}$

19. $\left(\frac{x^2}{y^{-3}}\right)^{-3}$ **20.** $\left(\frac{10^{-2}}{y^{-3}}\right)^{-1}$ **21.** $\left(\frac{-2^3}{-5^2}\right)^0$

22. $\left(\frac{6^{-2}}{8^{-3}}\right)^0$ **23.** $\left(\frac{s^{-4}}{m^{-2}}\right)^{-2}$ **24.** $\left(\frac{-x^2}{y^{-2}}\right)^{-3}$

WRITTEN EXERCISES

Goal: To simplify powers of powers
Sample Problem: $(t^{-2})^{-5}$ **Answer:** t^{10}

Simplify. (Examples 1–4)

1. $(a^2)^4$
2. $(p^3)^2$
3. $(y^{-2})^5$
4. $(r^2)^{-4}$
5. $(m^{-1})^{-5}$
6. $(n^{-2})^{-5}$
7. $(t^2)^{-2}$
8. $(w^{-3})^4$
9. $(3r^3)^2$
10. $(4s^5)^2$
11. $(2a^2b)^3$
12. $(2ab^3)^2$
13. $(2xy^2)^5$
14. $(2x^3y^2)^4$
15. $\left(\frac{x^3}{2}\right)^5$
16. $\left(\frac{y^2}{2}\right)^6$
17. $\left(\frac{2a^2b^3}{c^{21}}\right)^5$
18. $\left(\frac{2xy^3}{r^4}\right)^6$
19. $(-2t^3)^3$
20. $(-2m^{-2}n^3)^{-1}$
21. $(2xy^2)^{-3}$
22. $(3r^2s^3)^{-2}$
23. $(4m^{-2}n^3)^{-1}$
24. $(10a^{-4}b^2)^{-1}$
25. $(2^{-3}r^{-1}s^3)^{-2}$
26. $(2^{-4}p^4q^{-2})^{-3}$
27. $(-2x^2y^{-1}z^3)^{-5}$
28. $(-2r^{-2}s^4t^3)^{-6}$
29. $(0.2m^4n^5)^7$
30. $(-0.2x^5y^2)^5$

NON-ROUTINE PROBLEMS

31. Train A leaves Pittsburgh for Boston traveling 70 miles per hour. At the same time, Train B leaves Boston for Pittsburgh traveling 50 miles per hour. How far apart are the trains 1 hour before they pass each other?

32. Mr. Thomas plans to drive his car 45,000 miles. Each of the tires he buys will last a maximum of 30,000 miles. Find the fewest number of tires he will have to buy to travel 45,000 miles.

REVIEW CAPSULE FOR SECTION 6.6

*Multiply each number **(a)** by 1000 and **(b)** by 100,000.*

1. 0.0015678
2. 0.826605027
3. 48.2783
4. 5280.

*Divide each number **(a)** by 10,000 and **(b)** by 1,000,000.*

5. 675,928
6. 5,026,500
7. 0.18
8. 160,000,000

6.6 Scientific Notation

Scientists use very large and very small numbers. Such numbers are often expressed in scientific notation. Here are some examples.

a. 3×10^5 **b.** 1.6×10^2 **c.** 2.3×10^{-7}

Definition

> ***Scientific notation*** is a numeral of the form $N \times 10^a$, in which a is an integer and N is a rational number such that $1 \leq N < 10$.
>
> $53{,}000. = 5.3 \times 10^4$
>
> $0.0048 = 4.8 \times 10^{-3}$

The sentence $1 \leq N < 10$ means that "$N \geq 1$ and $N < 10$."

P–1 **What is the value of N in each number below? the value of a?**

a. 5.8×10^3 **b.** 3.76×10^{-1} **c.** 1.0059×10^{-6}

The table below shows some powers of 10.

Powers of Ten	
$10^4 =$	$10{,}000$
$10^3 =$	$1{,}000$
$10^2 =$	100
$10^1 =$	10
$10^0 =$	1
$10^{-1} =$	0.1 or $\frac{1}{10}$
$10^{-2} =$	0.01 or $\frac{1}{100}$
$10^{-3} =$	0.001 or $\frac{1}{1000}$
$10^{-4} =$	0.0001 or $\frac{1}{10{,}000}$

EXAMPLE 1 Write 1.2×10^4 in decimal form.

Solution:

[1] Find the value of 10^4 from the table. ⟶ $10^4 = 10{,}000$

[2] Substitute. ⟶ $1.2 \times 10^4 = (1.2)(10{,}000)$

[3] Multiply. ⟶ $= 12{,}000.$

The result is the same as moving the decimal point four places to the right in 1.2.

EXAMPLE 2 Write 5.3×10^{-3} in decimal form.

Solution:

[1] Find the value of 10^{-3} from the table. ⟶ $10^{-3} = 0.001$

[2] Substitute. ⟶ $5.3 \times 10^{-3} = (5.3)(0.001)$

[3] Multiply. ⟶ $= 0.0053$

The result is the same as moving the decimal point three places to the left in 5.3.

P–2 **How is the exponent of 10 in scientific notation related to the location of the decimal point in decimal form?**

> To change scientific notation $N \times 10^a$ to decimal form:
> 1. move the decimal point of N to the right a places if a is positive, or
> 2. move the decimal point of N to the left a places if a is negative.

EXAMPLE 3 Write 186,000 in scientific notation.

Solution:

[1] Place a caret to locate the decimal point in N. ⟶ 1‸86000.

[2] Count the number of places from the caret to the decimal point. ⟶ 1‸86000. 5 places

[3] Write in scientific notation. ⟶ 1.86×10^5

EXAMPLE 4 Write 0.0000053 in scientific notation.

Solution:

[1] Place a caret to locate the decimal point in N. ⟶ 0.0000053‸ (caret after 5)

[2] Count the number of places from the caret to the decimal point. ⟶ 0.0000053 6 places

[3] Write in scientific notation. ⟶ 5.3×10^{-6}

CLASSROOM EXERCISES

Write the decimal form for each number. (Examples 1–2)

1. 3×10^2 **2.** 2.0×10^{-2} **3.** 7.03×10^3 **4.** 3.9×10^{-3}

5. 4.62×10^1 **6.** 8.1627×10^{-1} **7.** 5.0001×10^4 **8.** 9.99×10^{-4}

Write the scientific notation for each number. (Examples 3–4)

9. 530 **10.** 0.0926 **11.** 48.6 **12.** 0.2738

13. 76,000. **14.** 6.318 **15.** 0.000483 **16.** 0.008092

WRITTEN EXERCISES

Goal: To write numbers in scientific notation
Sample Problems: a. 418,200 **b.** 0.00531
Answers: a. 4.182×10^5 **b.** 5.31×10^{-3}

For Exercises 1–16, write each number in decimal form. (Examples 1–2)

1. 4.2×10^3 **2.** 3.7×10^2 **3.** 2.9×10^{-4} **4.** 9.1×10^{-3}

5. 8.62×10^4 **6.** 6.01×10^5 **7.** 1.00726×10^{-1} **8.** 2.053×10^{-2}

9. 7.280×10^6 **10.** 3.1709×10^7 **11.** 5.607×10^{-5} **12.** 9.090×10^{-6}

13. The diameter of the earth is about 1.28×10^7 meters.

14. The diameter of the sun is about 1.39×10^9 meters.

15. A helium atom has a diameter of 2.2×10^{-8} centimeters.

16. If one foot of copper wire is heated 1° Celsius, it expands 1.6×10^{-5} feet.

For Exercises 17–32, write each number in scientific notation. (Examples 3–4)

17. 0.0092 **18.** 0.000184 **19.** 2705.

20. 937. **21.** 0.01734 **22.** 0.80672

23. 27,000,000. **24.** 186,000. **25.** 0.0000275

26. 0.00000031801 **27.** 2 million **28.** 3.5 billion

29. The speed of light is about 297,600 kilometers per second.

30. Our galaxy has about 100,000,000,000 stars.

31. The sun is about 150,000,000 kilometers from the earth.

32. Special balances can weigh something as small as 0.00000001 gram.

Focus on Reasoning: Number Patterns

A **number sequence** is a succession of numbers that follow a fixed pattern. Each number in the sequence is related to the preceding number according to a definite plan. **Identifying the plan** is the key to writing other numbers in each sequence.

EXAMPLE Write the next two numbers in each sequence.

a. $1, \frac{2}{3}, \frac{1}{3}, 0, -\frac{1}{3}, -\frac{2}{3}, \cdots$ **b.** $1, -2, -4, 8, 16, -32, -64, \cdots$

Solution: **a. Think:** $1, \frac{2}{3}, \frac{1}{3}, 0, -\frac{1}{3}, -\frac{2}{3}$

$-\frac{1}{3}$ $-\frac{1}{3}$ $-\frac{1}{3}$ $-\frac{1}{3}$ $-\frac{1}{3}$

Rule: Subtract $\frac{1}{3}$ from the preceding number.

Next two numbers: $\mathbf{-1, -1\frac{1}{3}}$

b. Think: $1, -2, -4, 8, 16, -32, -64$

$\times(-2)$ $\times 2$ $\times(-2)$ $\times 2$ $\times(-2)$ $\times 2$

Rule: Alternate multiplying by -2 and 2.
Next two numbers: **128, 256**

EXERCISES

Write the next four numbers in each sequence.

1. $-11, -7, -3, 1, 5, 9, \cdots$

2. $-7, -4, -1, 2, 5, 8, \cdots$

3. $-\frac{2}{9}, -\frac{2}{3}, 2, 6, -18, -54, \cdots$

4. $-\frac{5}{4}, -\frac{5}{2}, 5, 10, -20, -40, \cdots$

5. $5.5, 1, -3.5, -8, -12.5, -17, \cdots$

6. $6.3, 3.2, 0.1, -3, -6.1, -9.2, \cdots$

7. $1, \frac{1}{3}, \frac{1}{5}, \frac{1}{7}, \frac{1}{9}, \frac{1}{11}, \cdots$

8. $\frac{1}{2}, \frac{1}{4}, \frac{1}{6}, \frac{1}{8}, \frac{1}{10}, \frac{1}{12}, \cdots$

9. $1, \frac{1}{4}, \frac{1}{9}, \frac{1}{16}, \frac{1}{25}, \frac{1}{36}, \cdots$

10. $\frac{1}{2}, \frac{3}{4}, \frac{5}{6}, \frac{7}{8}, \frac{9}{10}, \frac{11}{12}, \cdots$

11. $1, 9, 25, 49, 81, 121, \cdots$

12. $4, 16, 36, 64, 100, 144, \cdots$

13. $12, 10, 6, 0, -8, -18, \cdots$

14. $21, 18, 13, 6, -3, -14, \cdots$

15. $1, 4, 10, 22, 46, 94, \cdots$

16. $1, 3, 7, 15, 31, 63, \cdots$

Focus on Reasoning:
Geometric Patterns

Sometimes a sequence is a succession of geometric figures. Look for what changes and what remains the same. This will help you to identify the pattern.

EXAMPLE What is the next figure?

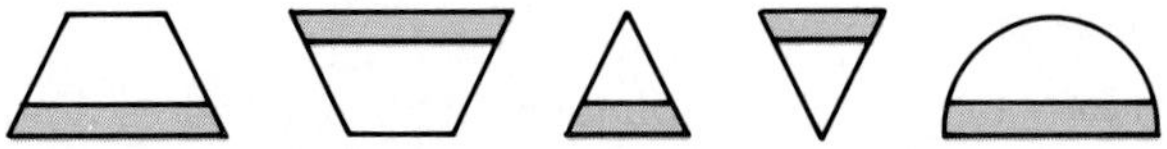

Think: The figures are in pairs.
The pairs have the same shape.
The second figure of each pair is upside down.

Next figure: 

EXERCISES

Identify the pattern. Then draw the next figure in each sequence.

1. 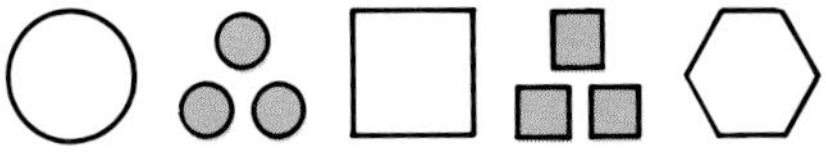

2. 

3. 

4. 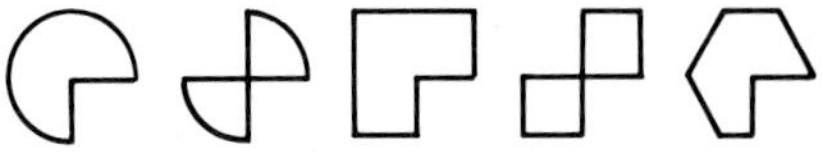

5. 

6. 

7. 

8. 

9. 

10. 

11. 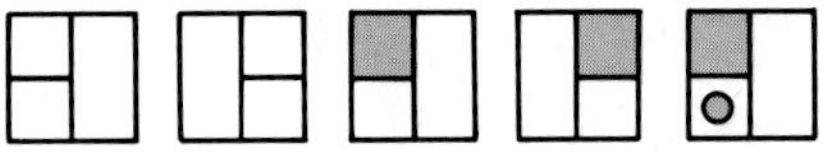

12. 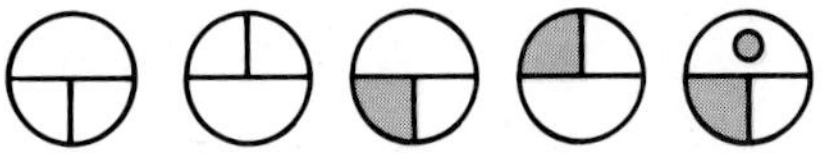

CHAPTER SUMMARY

IMPORTANT TERMS

Power of a power *(p. 159)*
Scientific notation *(p. 163)*

IMPORTANT IDEAS

1. If a is a positive integer, then x^a means that x is a factor a times.

2. *Product Property of Powers:* If a and b are integers and x is any nonzero real number, then $x^a \cdot x^b = x^{a+b}$.

3. *Quotient Property of Powers:* If a and b are integers and x is a nonzero real number, then $\frac{x^a}{x^b} = x^{a-b}$.

4. *Property of a Zero Exponent:* If x is any nonzero real number, then $x^0 = 1$.

5. *Property of Negative Exponents:* If $-a$ represents a negative integer and x is a nonzero real number, then $x^{-a} = \frac{1}{x^a}$.

6. *Property of the Power of a Product:* If x and y are nonzero real numbers and a is any integer, then $(xy)^a = x^a y^a$.

7. *Property of the Power of a Quotient:* If x and y are nonzero real numbers and a is any integer, then $\left(\frac{x}{y}\right)^a = \frac{x^a}{y^a}$.

8. *Property of the Power of a Power:* If x is a nonzero real number and a and b are integers, then $(x^a)^b = x^{ab}$.

9. To change scientific notation $N \times 10^a$ to decimal form:
a. move the decimal point of N to the right a places if a is positive, or
b. move the decimal point of N to the left a places if a is negative.

CHAPTER REVIEW

SECTION 6.1

Write each product as one power.

1. $(27)^6(27)^3$ **2.** $(-8.5)^4(-8.5)^4$ **3.** $r^2 \cdot r^5$

4. $m^3 \cdot m^{12}$ **5.** $t^3 \cdot t \cdot t^5$ **6.** $p^4 \cdot p^5 \cdot p$

SECTION 6.2

Write each quotient as one power.

7. $\frac{19^8}{19^3}$ **8.** $\frac{24^{12}}{24^7}$ **9.** $\frac{(2.7)^{15}}{(2.7)^{13}}$

10. $\frac{(-3.9)^6}{(-3.9)^3}$

11. $\frac{(\sqrt{7})^8}{(\sqrt{7})^5}$

12. $\frac{\pi^{19}}{\pi^{10}}$

Simplify.

13. $(35)^0$

14. $(242)^0$

15. $(-23.9)^0$

16. $\left(\frac{5}{4}\right)^0$

17. $(13.4)^0 - 2x^0$

18. $5n^0 + (-95)^0$

SECTION 6.3

Simplify.

19. $(t)^{-10}$

20. $(2)^{-6}$

21. $5x^{-2}$

22. $100y^{-1}$

23. $p^{-2}q$

24. $m^{-3}n$

Write each product or quotient as one power. Then simplify where necessary.

25. $k^{-4} \cdot k^2$

26. $r^{-5} \cdot r^{-3}$

27. $\frac{m^{-2}}{m^3}$

28. $\frac{t^4}{t^{-3}}$

29. $\frac{p^2 \cdot p^{-3}}{p^4}$

30. $\frac{x^{-1} \cdot x^{-3}}{x^4}$

SECTION 6.4

Simplify.

31. $(4t)^2$

32. $(2q)^4$

33. $(-2r)^5$

34. $(-3y)^3$

35. $\left(\frac{5}{p}\right)^2$

36. $\left(\frac{-3}{s}\right)^3$

37. $(-4m)^{-2}$

38. $(-3n)^{-3}$

39. $\left(\frac{-1}{-y}\right)^3$

SECTION 6.5

Simplify.

40. $(d^3)^5$

41. $(g^5)^{-2}$

42. $\left(\frac{y^4}{2}\right)^5$

43. $\left(\frac{3m^2n^3}{p}\right)^3$

44. $(-2p^{-2}q^3r^{-4})^2$

45. $(2^{-2}x^2y^{-1}z^{-4})^3$

SECTION 6.6

In Exercises 46–53, write each number in decimal form.

46. 1.83×10^{-2}

47. 3.187×10^{4}

48. 2.1692×10^{5}

49. 5.279×10^{-3}

50. 7.10036×10^{-4}

51. 8.00002×10^{3}

52. The diameter of the planet Mars is about 4.214×10^{3} miles.

53. The orbit of the Space Shuttle is about 2.224×10^{4} miles above the equator.

In Exercises 54–60, write each number in scientific notation.

54. 47,280.

55. 0.005219

56. 0.00002009

57. 512,000,000.

58. The speed of light is approximately 300,000 kilometers per second. Write 300,000 in scientific notation.

59. There are about 31,536,000 seconds in one year. Write 31,536,000 in scientific notation.

60. The average concentration of radium in seawater is about 0.00000000006 milligrams per liter. Write 0.00000000006 in scientific notation.

CUMULATIVE REVIEW: CHAPTERS 1–6

Perform the indicated operations. (Sections 1.2 through 1.5)

1. $27 + (-19)$ **2.** $-6.3 - 2.7$ **3.** $\frac{2}{3}(-27)$ **4.** $-65 \div (-5)$

Evaluate each expression. (Section 1.7)

5. $3x^2 - 8$ when $x = 4$ **6.** $m^2 - 3m$ when $m = -2$ **7.** $5x^2 - 6x$ when $x = -3$

Multiply and simplify. (Sections 2.3 through 2.5)

8. $(-7)(2)(x)(-5)$ **9.** $(3x)(12xy)(-y)$ **10.** $-3(2x - 5)$ **11.** $(2y + 11)5y$

Factor. (Section 2.5)

12. $7a + 7q$ **13.** $8c - 12$ **14.** $9x - 3xy$ **15.** $-4a^2 - 4a$

Simplify. (Section 2.6)

16. $-7w + w$ **17.** $4m - 2 - 8m - 1$ **18.** $2a^2 - 7a - 3a + 8a^2$

Solve. Check each answer. (Sections 3.1 through 3.3)

19. $y + 1 = -8$ **20.** $-\frac{3}{4}j = -30$ **21.** $-22 = 3k + 5$

22. $-3 = \frac{t}{4} - 7$ **23.** $7h - 2 + h = 14$ **24.** $11 = 2(3x - 2) - 3$

Use Condition 1 to represent the unknowns. Use Condition 2 to write an equation for the problem. Solve. (Sections 4.1 and 4.2)

25. Joyce scored 15 fewer points in her first game of bowling than in her second game. Her total number of points for the two games was 309. Find the number of points Joyce scored in the first game.

26. One special triangle has sides in the ratio 3:5:7. The perimeter of the triangle is 165 centimeters. Find the lengths of the sides.

Solve. Check each answer. (Section 4.3)

27. $7x - 2 = 2x + 13$ **28.** $2(2 + r) - 3r = r + 8$

Solve. (Section 4.4)

29. O'Leary's Clothing Store is having a sale on winter coats. The rate of discount of a particular coat is 35%. The discount is $39.90. Find the list price of the coat.

30. All of the gas grills at Big John's Hardware Store were on sale at a 30% rate of discount. Sally paid the sale price of $126 for a grill. Find the list price of the grill and the discount.

Use Conditions 1 and 2 to make a table. Use Condition 2 to write an equation for the problem. Solve. (Sections 4.4 and 4.5)

31. A football player attempted 6 more field goals this season than last season. This season he made 60% of his attempts. Last season he made 75% of his attempts. He made the same number of field goals this season as last season. How many did he attempt this season?

32. Anthony's average rate driving to work is 15 miles per hour faster than his average rate returning home. It takes him 0.6 hours to drive to work and 0.9 hours to drive home. Find Anthony's average rate driving to work and his average rate returning home.

Draw the graph of each inequality. (Section 5.1)

33. $x < 4$ **34.** $x > -2$ **35.** $x \geq -3.7$ **36.** $x \leq 1\frac{1}{2}$

Solve each inequality. (Sections 5.3 through 5.6)

37. $x + 9 < 3$ **38.** $a - 3.9 > -1.7$ **39.** $11 > 3y + 7 - 4y$

40. $3(x - 7) < 4x + 16$ **41.** $\frac{3}{5}a < -15$ **42.** $-11 < -2b$

43. $2g + 5 > 13$ **44.** $2 - k > 6k + 23$ **45.** $2(10 + t) - 5 < 5t$

Use Condition 1 to represent the unknowns. Use Condition 2 to write an equation for the problem. Solve. (Section 5.7)

46. Ryan estimates that a television costs $100 less than twice the cost of a V.C.R. He plans to spend less than $950 for both items. Find the greatest amount he can spend for the television.

47. Meghan's scores on four math tests were 73, 87, 85, and 82. She wants her mean score on five tests to be greater than 84. What is the lowest score she can get on the fifth test?

Write each product or quotient as one power. Then simplify where necessary. (Sections 6.1 through 6.3)

48. $x \cdot x^9$ **49.** $b^2 \cdot b^7 \cdot b^7$ **50.** $\frac{2^{19}}{2}$ **51.** $\frac{x^9}{x^2}$

52. $\frac{c^4}{c^7}$ **53.** $\frac{r^0}{r^{-3}}$ **54.** $x^{11} \cdot x^{-17}$ **55.** $\frac{m^2 \cdot m^{-4}}{m^{-8}}$

Simplify. (Sections 6.4 and 6.5)

56. $(2a)^3$ **57.** $\left(\frac{4}{x}\right)^2$ **58.** $(4a^{-7})^{-2}$ **59.** $\left(\frac{3x^4w}{y^5}\right)^3$

Write each number in decimal form. (Section 6.6)

60. 9.2×10^3 **61.** 4.12×10^{-4} **62.** 6.72×10^{-2} **63.** 7.3×10^5

CHAPTER 7

Roots

On December 3, 1980, *Janice Brown* made the first solar-powered flight in the *Solar Challenger.* The formula $D = 3.56\sqrt{x}$ gives the approximate distance to the horizon in kilometers from an aircraft that is x meters above the ground (altitude). How far could Janice see to the horizon at an altitude of 30 meters?

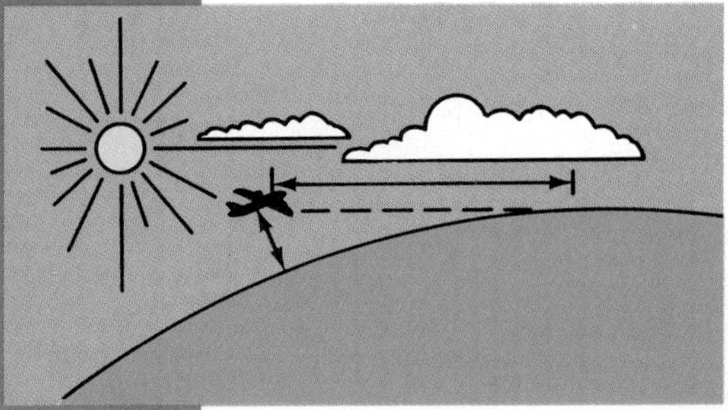

7.1 Numbers in Radical Form

P–1 **What is the simplest form of $\sqrt{36}$? of $-\sqrt{36}$?**

> $\sqrt{\ }$ *is a radical symbol.*

A ***square root*** is one of two equal factors of a number. Both 6 and −6 are square roots of 36. Also, 36 is the ***square*** of both 6 and −6.

$$(6)(6) = 36 \qquad (-6)(-6) = 36$$

P–2 **What is the value of $\sqrt{0}$?**

P–3 **Why doesn't $\sqrt{-16}$ have a real-number value?**

The square of a positive number is positive.
The square of a negative number is positive.
The square of 0 is 0.

> ***A negative number cannot have a real number as a square root.***

P–4 **What is the square of each number below?**

a. -7 **b.** $\frac{1}{2}$ **c.** -0.3 **d.** 20 **e.** $-\frac{3}{4}$

P–5 **Simplify each expression below.**

a. $-\sqrt{49}$ **b.** $\sqrt{\frac{1}{4}}$ **c.** $-\sqrt{0.09}$ **d.** $\sqrt{10{,}000}$

It may be helpful to review prime factorization at this time. The set of prime numbers less than 25 is shown below.

$$\{2, 3, 5, 7, 11, 13, 17, 19, 23\}$$

Definition

> A ***prime number*** is a counting number greater than 1 that has exactly two counting-number factors, 1 and the number itself.

P–6 **Why is 15 not a prime number?**

The set of counting-number factors of 15 is $\{1, 3, 5, 15\}$.

> ***The prime factorization of 15 is $3 \cdot 5$.***

You will recall that ***prime factorization*** is writing a number as a product of its prime-number factors. The table below shows the prime factorization of some numbers. The numbers 4, 9, 25, and 36 in the table are ***perfect squares***. Note that each prime factor of a perfect square occurs an even number of times.

Number	Prime Factorization	Number	Prime Factorization
4	$2 \cdot 2$	20	$2 \cdot 2 \cdot 5$
6	$2 \cdot 3$	25	$5 \cdot 5$
9	$3 \cdot 3$	27	$3 \cdot 3 \cdot 3$
12	$2 \cdot 2 \cdot 3$	36	$2 \cdot 2 \cdot 3 \cdot 3$

EXAMPLE 1 Write the prime factorization of 150.

Solution:

[1] Divide by 2 (the smallest prime number). Write the quotient 75.

$$\begin{array}{r|l} \mathbf{2} & 150 \\ \hline & 75 \end{array}$$

[2] Check to see whether 2 is a factor of 75. It is not.

[3] Select the next greater prime number 3. It is a factor of 75. Divide and write the quotient 25.

$$\begin{array}{r|l} 2 & 150 \\ \hline \mathbf{3} & 75 \\ \hline & 25 \end{array}$$

[4] Check again to see whether 3 is a factor of 25. It is not.

[5] Select the next greater prime number 5. Divide by 5 and write the quotient 5. The quotient is prime.

$$\begin{array}{r|l} 2 & 150 \\ \hline 3 & 75 \\ \hline \mathbf{5} & 25 \\ \hline & \mathbf{5} \end{array}$$

[6] Write the prime factorization of 150.

$$150 = 2 \cdot 3 \cdot 5 \cdot 5$$

EXAMPLE 2 Simplify $\sqrt{441}$.

Prime factorization can help in simplifying radicals.

Solution:

[1] Write the prime factorization of 441. ⟶ $\sqrt{441} = \sqrt{3 \cdot 3 \cdot 7 \cdot 7}$

[2] Write two equal factors of 441. ⟶ $= \sqrt{(3 \cdot 7)(3 \cdot 7)}$

$= \sqrt{21 \cdot 21}$

[3] Simplify. ⟶ $= 21$

$$\begin{array}{r|l} \mathbf{3} & 441 \\ \hline \mathbf{3} & 147 \\ \hline \mathbf{7} & 49 \\ \hline & \mathbf{7} \end{array}$$

CLASSROOM EXERCISES

Write the square of each number. (P-4)

1. -10 **2.** 11 **3.** -12 **4.** $\frac{1}{2}$ **5.** $-\frac{1}{3}$

6. -0.2 **7.** 0.4 **8.** -13 **9.** $\frac{5}{2}$ **10.** -1.2

Simplify. (P-1, P-5)

11. $\sqrt{9}$ **12.** $-\sqrt{1}$ **13.** $\sqrt{0}$ **14.** $-\sqrt{36}$ **15.** $\sqrt{100}$

16. $\sqrt{\frac{9}{4}}$ **17.** $-\sqrt{\frac{1}{16}}$ **18.** $\sqrt{\frac{9}{25}}$ **19.** $-\sqrt{0.01}$ **20.** $\sqrt{0.0004}$

Write the prime factorization. (Example 1)

21. 21 **22.** 12 **23.** 8 **24.** 25

25. 26 **26.** 35 **27.** 49 **28.** 77

29. 39 **30.** 18 **31.** 45 **32.** 81

WRITTEN EXERCISES

Goal: To write square roots of numbers
Sample Problems: a. $\sqrt{225}$ **b.** $-\sqrt{2^2 \cdot 3^4}$
Answers: a. 15 **b.** -18

Simplify. (Steps 2 and 3 of Example 2)

1. $\sqrt{121}$ **2.** $\sqrt{196}$ **3.** $-\sqrt{100}$ **4.** $-\sqrt{81}$ **5.** $\sqrt{\frac{4}{25}}$

6. $\sqrt{\frac{9}{49}}$ **7.** $-\sqrt{\frac{1}{9}}$ **8.** $-\sqrt{\frac{25}{36}}$ **9.** $\sqrt{0.81}$ **10.** $\sqrt{0.25}$

11. $\sqrt{(2 \cdot 5)(2 \cdot 5)}$ **12.** $\sqrt{(5 \cdot 11)(5 \cdot 11)}$ **13.** $\sqrt{2 \cdot 3 \cdot 3 \cdot 2 \cdot 2 \cdot 2}$

14. $\sqrt{3 \cdot 2 \cdot 5 \cdot 3 \cdot 2 \cdot 5}$ **15.** $\sqrt{2^2 \cdot 5^4}$ **16.** $\sqrt{3^4 \cdot 5^2 \cdot 11^2}$

Write the prime factorization. (Example 1)

17. 65 **18.** 99 **19.** 18 **20.** 400 **21.** 54

22. 60 **23.** 120 **24.** 108 **25.** 198 **26.** 264

Simplify. (Example 2)

27. $\sqrt{484}$ **28.** $\sqrt{676}$ **29.** $\sqrt{784}$ **30.** $\sqrt{1225}$ **31.** $\sqrt{1936}$

32. $\sqrt{4096}$ **33.** $\sqrt{3136}$ **34.** $\sqrt{4624}$ **35.** $\sqrt{5776}$ **36.** $\sqrt{6561}$

37. $\sqrt{5184}$ **38.** $\sqrt{7225}$ **39.** $\sqrt{9025}$ **40.** $\sqrt{4356}$ **41.** $\sqrt{3844}$

7.2 Products

Radicals such as $\sqrt{9}$, $-\sqrt{49}$, and $\sqrt{64}$ represent rational numbers.

A square root of a perfect square is rational.

Radicals such as $\sqrt{7}$, $\sqrt{15}$, $-\sqrt{34}$, and $\sqrt{50}$ represent irrational numbers.

> Any whole number that is not a perfect square has an irrational square root.

P–1 **Which numbers below are rational? Which are irrational?**

a. $\sqrt{19}$ **b.** $-\sqrt{144}$ **c.** $-\sqrt{80}$ **d.** $\sqrt{0}$

EXAMPLE 1 Multiply and simplify: $\sqrt{13} \cdot \sqrt{13}$

Solution:

$$\sqrt{13} \cdot \sqrt{13} = (\sqrt{13})^2$$
$$= 13$$

A square root is one of two equal factors of a number.

> If a is any nonnegative real number, then $\sqrt{a} \cdot \sqrt{a} = a$.
>
> $\sqrt{20} \cdot \sqrt{20} = 20$
>
> $\sqrt{3.2} \cdot \sqrt{3.2} = 3.2$

P–2 **Multiply and simplify.**

a. $\sqrt{29} \cdot \sqrt{29}$ **b.** $\sqrt{\frac{1}{4}} \cdot \sqrt{\frac{1}{4}}$ **c.** $\sqrt{1029} \cdot \sqrt{1029}$

EXAMPLE 2 Multiply and simplify: $(\sqrt{3} \cdot \sqrt{10})^2$

Solution:

1 Property of the Power of a Product ⟶ $(\sqrt{3} \cdot \sqrt{10})^2 = (\sqrt{3})^2(\sqrt{10})^2$

2 Simplify. ⟶ $= (3)(10)$ or 30

The product $\sqrt{3} \cdot \sqrt{10}$ is positive. Example 2 shows that $\sqrt{3} \cdot \sqrt{10}$ represents the positive square root of 30.

$$\sqrt{3} \cdot \sqrt{10} = \sqrt{30}$$

Product Property of Radicals

If a and b are nonnegative real numbers, then $\sqrt{a} \cdot \sqrt{b} = \sqrt{ab}$.

$\sqrt{5} \cdot \sqrt{11} = \sqrt{55}$

$\sqrt{2} \cdot \sqrt{8} = \sqrt{16}$
$= 4$

P–3 **Multiply and simplify.**

a. $\sqrt{7} \cdot \sqrt{5}$ **b.** $\sqrt{6} \cdot \sqrt{11}$ **c.** $\sqrt{3} \cdot \sqrt{12}$

EXAMPLE 3 Multiply and simplify: $-\sqrt{10} \cdot \sqrt{13}$

Solution:

1. Product of a positive and negative number → $-\sqrt{10} \cdot \sqrt{13} = -(\sqrt{10} \cdot \sqrt{13})$
2. Product Property of Radicals → $= -(\sqrt{130})$
3. Simplify. → $= -\sqrt{130}$

EXAMPLE 4 Multiply and simplify: $(-\sqrt{8})(-\sqrt{18})$

Solution:

$$\begin{aligned}(-\sqrt{8})(-\sqrt{18}) &= (\sqrt{8})(\sqrt{18}) \\ &= \sqrt{8 \cdot 18} \\ &= \sqrt{144} \\ &= 12\end{aligned}$$

EXAMPLE 5 Multiply and simplify: $\sqrt{3} \cdot \sqrt{10} \cdot \sqrt{7}$

Solution:

$$\begin{aligned}\sqrt{3} \cdot \sqrt{10} \cdot \sqrt{7} &= (\sqrt{3} \cdot \sqrt{10}) \cdot \sqrt{7} \\ &= (\sqrt{30})(\sqrt{7}) \\ &= \sqrt{210}\end{aligned}$$

You may group the factors as you please.

CLASSROOM EXERCISES

Multiply and simplify. (Examples 1–4)

1. $\sqrt{2} \cdot \sqrt{5}$
2. $\sqrt{3} \cdot \sqrt{7}$
3. $\sqrt{8} \cdot \sqrt{8}$
4. $\sqrt{8} \cdot \sqrt{2}$
5. $\sqrt{2} \cdot \sqrt{50}$
6. $\sqrt{4} \cdot \sqrt{25}$
7. $\sqrt{\frac{1}{5}} \cdot \sqrt{10}$
8. $\sqrt{\frac{1}{2}} \cdot \sqrt{10}$
9. $\sqrt{\frac{1}{3}} \cdot \sqrt{\frac{1}{3}}$
10. $\sqrt{0.5} \cdot \sqrt{0.5}$
11. $\sqrt{0} \cdot \sqrt{5}$
12. $(-\sqrt{4})(\sqrt{9})$
13. $(-\sqrt{16})(-\sqrt{25})$
14. $(-\sqrt{2})(-\sqrt{3})$
15. $(-\sqrt{2})(\sqrt{2})$
16. $(-\sqrt{3})^2$

WRITTEN EXERCISES

Goal: To simplify the products of radicals

Sample Problem: $(\sqrt{2})(\sqrt{5})(-\sqrt{7})$ **Answer:** $-\sqrt{70}$

Multiply and simplify. (Examples 1–5)

1. $\sqrt{13} \cdot \sqrt{13}$
2. $\sqrt{17} \cdot \sqrt{17}$
3. $\sqrt{12} \cdot \sqrt{12}$
4. $\sqrt{19} \cdot \sqrt{19}$
5. $\sqrt{81} \cdot \sqrt{36}$
6. $\sqrt{144} \cdot \sqrt{16}$
7. $\sqrt{2} \cdot \sqrt{18}$
8. $\sqrt{3} \cdot \sqrt{27}$
9. $\sqrt{3} \cdot \sqrt{17}$
10. $\sqrt{2} \cdot \sqrt{19}$
11. $(-\sqrt{5})(\sqrt{2})$
12. $(\sqrt{7})(-\sqrt{2})$
13. $(-\sqrt{3})(-\sqrt{10})$
14. $(-\sqrt{5})(-\sqrt{6})$
15. $\sqrt{\frac{1}{3}} \cdot \sqrt{15}$
16. $\sqrt{\frac{1}{2}} \cdot \sqrt{34}$
17. $\sqrt{\frac{1}{2}} \cdot \sqrt{\frac{1}{2}}$
18. $\sqrt{\frac{2}{3}} \cdot \sqrt{\frac{2}{3}}$
19. $\sqrt{\frac{1}{3}} \cdot \sqrt{48}$
20. $\sqrt{\frac{1}{5}} \cdot \sqrt{20}$
21. $(\sqrt{2} \cdot \sqrt{5})^2$
22. $(\sqrt{3} \cdot \sqrt{14})^2$
23. $(\sqrt{\frac{1}{2}} \cdot \sqrt{\frac{1}{3}})^2$
24. $(\sqrt{\frac{1}{4}} \cdot \sqrt{8})^2$
25. $\sqrt{2} \cdot \sqrt{3} \cdot \sqrt{6}$
26. $\sqrt{3} \cdot \sqrt{5} \cdot \sqrt{15}$
27. $\sqrt{2} \cdot \sqrt{3} \cdot \sqrt{5}$
28. $\sqrt{3} \cdot \sqrt{5} \cdot \sqrt{7}$
29. $\sqrt{2} \cdot \sqrt{5} \cdot \sqrt{10}$
30. $\sqrt{3} \cdot \sqrt{7} \cdot \sqrt{21}$

MORE CHALLENGING EXERCISES

Solve each equation.

31. $\frac{1}{\sqrt{3}}x = \sqrt{12}$
32. $\frac{1}{\sqrt{5}}x = \sqrt{20}$
33. $\frac{1}{\sqrt{7}}x = \sqrt{2}$
34. $\frac{1}{\sqrt{11}}x = \sqrt{3}$
35. $x^2 = 36$
36. $x^2 = 100$
37. $x^2 = 14$
38. $x^2 = 33$

Give an example to show that each of the following is a false statement.

39. If $\sqrt{a} \cdot \sqrt{b} = \sqrt{c}$, then $c > a$ and $c > b$.
40. If $\sqrt{a} \cdot \sqrt{\frac{1}{b}} = 1$, then $a > b$.

7.3 Simplifying Radicals

P–1 **Simplify each radical below.**

a. $\sqrt{49}$ **b.** $-\sqrt{81}$ **c.** $\sqrt{5^2}$ **d.** $-\sqrt{13^2}$

$\sqrt{49}$ is a <u>radical</u> and 49 is the <u>radicand</u>.

EXAMPLE 1 Simplify $\sqrt{12}$.

Solution:

1. Write the prime factorization of 12. ⟶ $\sqrt{12} = \sqrt{2 \cdot 2 \cdot 3}$
2. Product Property of Radicals ⟶ $= \sqrt{2 \cdot 2} \cdot \sqrt{3}$
3. Meaning of square root ⟶ $= 2\sqrt{3}$

$\sqrt{12}$ is irrational because 12 is not a perfect square.

The simplest form of $\sqrt{12}$ is $2\sqrt{3}$.

A square root radical is in <u>simplest form</u> if its radicand has no perfect square factor.

EXAMPLE 2 Simplify $\sqrt{200}$.

Solution:

1. Write the prime factorization. ⟶ $\sqrt{200} = \sqrt{2^3 \cdot 5^2}$
2. Product Property of Radicals ⟶ $= \sqrt{2^2 \cdot 5^2} \cdot \sqrt{2}$
3. Meaning of square root ⟶ $= 2 \cdot 5\sqrt{2}$
4. Simplify. ⟶ $= 10\sqrt{2}$

Even powers of factors are grouped under one radical symbol.

The simplest form of $\sqrt{200}$ is $10\sqrt{2}$.

Other roots besides square roots are possible. A ***cube root*** of a number is one of its three equal factors. A ***fourth root*** of a number is one of its four equal factors.

$2 \cdot 2 \cdot 2 = 8$ ***2 is a cube root of 8.***

$3 \cdot 3 \cdot 3 \cdot 3 = 81$ ***3 is a fourth root of 81.***

The Product Property of Radicals can be extended to other roots.

$\sqrt[3]{x} \cdot \sqrt[3]{y} = \sqrt[3]{xy}$

$\sqrt[4]{x} \cdot \sqrt[4]{y} = \sqrt[4]{xy}$

True for all nonnegative values of x and y

EXAMPLE 3 Simplify $\sqrt[3]{54}$.

Solution:

1 Write the prime factorization. ⟶ $\sqrt[3]{54} = \sqrt[3]{2 \cdot 3 \cdot 3 \cdot 3}$

2 Product Property of Radicals ⟶ $= \sqrt[3]{3 \cdot 3 \cdot 3} \cdot \sqrt[3]{2}$

3 Simplify. ⟶ $= 3\sqrt[3]{2}$

The simplest form of $\sqrt[3]{54}$ is $3\sqrt[3]{2}$.

Read "3 times the cube root of 2."

The table below shows the third and fourth powers of some counting numbers.

n	n^3	n^4	n	n^3	n^4
1	1	1	6	216	1296
2	8	16	7	343	2401
3	27	81	8	512	4096
4	64	256	9	729	6561
5	125	625	10	1000	10,000

P–2 **Simplify each radical below.**

a. $\sqrt[4]{2^4}$ **b.** $\sqrt[3]{2^3 \cdot 3^3}$ **c.** $\sqrt[3]{125}$ **d.** $\sqrt[4]{256}$

CLASSROOM EXERCISES

Simplify. (Examples 1–2)

1. $\sqrt{2^2}$ **2.** $\sqrt{3^2}$ **3.** $-\sqrt{5^2}$ **4.** $\sqrt{7^2}$ **5.** $\sqrt{2^4}$

6. $\sqrt{3^4}$ **7.** $\sqrt{5^4}$ **8.** $\sqrt{7^4}$ **9.** $\sqrt{11^2}$ **10.** $\sqrt{3^6}$

11. $\sqrt{6^4}$ **12.** $-\sqrt{10^2}$ **13.** $\sqrt{2^2 \cdot 5^2}$ **14.** $\sqrt{3^2 \cdot 5^2}$ **15.** $\sqrt{2^2 \cdot 7^2}$

16. $\sqrt{2^4 \cdot 5^2}$ **17.** $\sqrt{2^2 \cdot 3^2 \cdot 5^2}$ **18.** $\sqrt{2 \cdot 5 \cdot 2}$ **19.** $\sqrt{3 \cdot 3 \cdot 5}$ **20.** $\sqrt{2 \cdot 3 \cdot 2}$

Simplify. (Example 3)

21. $\sqrt[3]{3^3}$ 22. $\sqrt[3]{8}$ 23. $\sqrt[4]{1296}$

24. $\sqrt[3]{729}$ 25. $\sqrt[4]{2401}$ 26. $\sqrt[3]{2^3 \cdot 7}$

27. $\sqrt[3]{3 \cdot 5 \cdot 3 \cdot 5 \cdot 5}$ 28. $\sqrt[4]{5^4 \cdot 11^4}$ 29. $\sqrt[4]{2 \cdot 3 \cdot 2 \cdot 3 \cdot 5 \cdot 3 \cdot 3}$

WRITTEN EXERCISES

Goal: To simplify radicals
Sample Problem: $\sqrt{288}$ **Answer:** $12\sqrt{2}$

Simplify. (Examples 1–2)

1. $\sqrt{6^2}$ 2. $\sqrt{4^2}$ 3. $\sqrt{38^2}$ 4. $\sqrt{19^2}$

5. $\sqrt{2^2 \cdot 3^2} \cdot \sqrt{3}$ 6. $\sqrt{3^2 \cdot 5^2} \cdot \sqrt{5}$ 7. $\sqrt{5^2 \cdot 7^2} \cdot \sqrt{5 \cdot 7}$

8. $\sqrt{2^2 \cdot 11^2} \cdot \sqrt{2 \cdot 11}$ 9. $\sqrt{2^4} \cdot \sqrt{2 \cdot 3}$ 10. $\sqrt{3^4} \cdot \sqrt{3 \cdot 5}$

11. $\sqrt{2^4 \cdot 3^2} \cdot \sqrt{5 \cdot 2}$ 12. $\sqrt{3^4 \cdot 5^2} \cdot \sqrt{2 \cdot 11}$ 13. $\sqrt{2 \cdot 3 \cdot 3 \cdot 5 \cdot 2}$

14. $\sqrt{3 \cdot 5 \cdot 2 \cdot 3 \cdot 5}$ 15. $\sqrt{5 \cdot 7 \cdot 5 \cdot 3 \cdot 2 \cdot 2}$ 16. $\sqrt{3 \cdot 11 \cdot 7 \cdot 3 \cdot 7 \cdot 3}$

17. $\sqrt{8}$ 18. $\sqrt{18}$ 19. $\sqrt{27}$ 20. $\sqrt{20}$ 21. $\sqrt{24}$

22. $\sqrt{28}$ 23. $\sqrt{40}$ 24. $\sqrt{44}$ 25. $-\sqrt{48}$ 26. $-\sqrt{60}$

27. $\sqrt{120}$ 28. $\sqrt{108}$ 29. $\sqrt{96}$ 30. $\sqrt{180}$ 31. $\sqrt{270}$

(Example 3)

32. $\sqrt[3]{64}$ 33. $\sqrt[4]{16}$ 34. $\sqrt[4]{625}$ 35. $\sqrt[3]{81}$ 36. $\sqrt[3]{72}$

37. $\sqrt[4]{144}$ 38. $\sqrt[4]{243}$ 39. $\sqrt[3]{1250}$ 40. $\sqrt[3]{2160}$ 41. $\sqrt[4]{1250}$

MORE CHALLENGING EXERCISES

42. $\sqrt[3]{-512}$ 43. $\sqrt[3]{-1000}$ 44. $\sqrt[3]{-24}$ 45. $\sqrt[3]{-56}$ 46. $\sqrt[3]{-375}$

NON-ROUTINE PROBLEMS

47. Meg spent $\frac{1}{4}$ of the quarters she had saved. She gave her brother $\frac{3}{4}$ of the rest. Then she had 3 quarters left. How many quarters had she saved?

48. Two positive integers are each less than 10. The sum of their squares, added to their product, equals a perfect square. Find the integers.

7.4 Simplifying Radical Expressions

P–1 **What is $\sqrt{x}$ for each value of x below?**

a. $x = 100$ **b.** $x = \frac{1}{4}$ **c.** $x = 17$ **d.** $x = \frac{4}{9}$

P–2 **Why cannot x have a negative value in $\sqrt{x}$?**

The domain of each variable will be {nonnegative real numbers} unless stated otherwise.

EXAMPLE 1 Simplify $\sqrt{x^2y^4}$.

Solution:

1 Product Property of Radicals ⟶ $\sqrt{x^2y^4} = \sqrt{x^2} \cdot \sqrt{y^4}$

2 Meaning of square root ⟶ $= xy^2$

An even power of a variable is called a perfect square because each of its square roots can be expressed without the radical symbol.

P–3 **Simplify each radical.**

a. $\sqrt{a^6}$ **b.** $\sqrt{x^8}$ **c.** $\sqrt{b^{12}}$ **d.** $\sqrt{y^{28}}$

EXAMPLE 2 Simplify $\sqrt{x^3y^5}$.

Solution:

1 Product Property of Powers ⟶ $\sqrt{x^3y^5} = \sqrt{(x^2 \cdot x)(y^4 \cdot y)}$

x^2 and y^4 are perfect squares.

2 Commutative and Associative Properties of Multiplication ⟶ $= \sqrt{(x^2y^4)(xy)}$

Group the perfect square factors.

3 Product Property of Radicals ⟶ $= \sqrt{x^2y^4} \cdot \sqrt{xy}$

4 Meaning of square root ⟶ $= xy^2\sqrt{xy}$

The simplest form is $xy^2\sqrt{xy}$.

EXAMPLE 3 Simplify $\sqrt{12c^3d^7}$.

Solution:

$$\sqrt{12c^3d^7} = \sqrt{(2 \cdot 2 \cdot 3)(c \cdot c^2)(d \cdot d^6)}$$
$$= \sqrt{(2^2 \cdot c^2 \cdot d^6)(3cd)}$$
$$= \sqrt{2^2 \cdot c^2 \cdot d^6} \cdot \sqrt{3cd}$$
$$= 2cd^3\sqrt{3cd}$$

EXAMPLE 4 Simplify $\sqrt[3]{r^4s^5}$.

Solution:

1. Find the greatest third power factors of r^4 and s^5. ⟶ $r^4 = r^3 \cdot r$; $s^5 = s^3 \cdot s^2$
2. Product Property of Powers ⟶ $\sqrt[3]{r^4s^5} = \sqrt[3]{(r^3 \cdot r)(s^3 \cdot s^2)}$
3. Associative and Commutative Properties of Multiplication ⟶ $= \sqrt[3]{(r^3s^3)(rs^2)}$ ◀ ***Group the perfect cube factors.***
4. Product Property of Radicals ⟶ $= \sqrt[3]{r^3s^3} \cdot \sqrt[3]{rs^2}$
5. Meaning of cube root ⟶ $= rs\sqrt[3]{rs^2}$

P–4 **What is $\sqrt[3]{a^3}$? $\sqrt[3]{x^6}$? $\sqrt[3]{y^9}$?**

You know that a square root radical is in simplest form if its radicand has no perfect square factor. A cube root radical is in simplest form if its radicand has no perfect cube factor.

The following radicals are in simplest form.

a. $\sqrt{x}$ **b.** $\sqrt{10}$ **c.** $\sqrt[3]{p^2q}$ **d.** $\sqrt{2t}$

P–5 **Which of the following radicals are in simplest form?**

a. $\sqrt{28x^3}$ **b.** $\sqrt{21st}$ **c.** $\sqrt[3]{9m^2n^2}$ **d.** $\sqrt[3]{24c^5d^7}$

CLASSROOM EXERCISES

Simplify. (Examples 1–4)

1. $\sqrt{a^2}$ **2.** $\sqrt{x^4}$ **3.** $\sqrt{a^2b^2}$ **4.** $\sqrt{x^6}$ **5.** $\sqrt{x^2y^4}$

6. $\sqrt{4x^2}$ 7. $\sqrt{9y^2}$ 8. $\sqrt{16a^2}$ 9. $\sqrt{25n^2}$ 10. $\sqrt{4x^4}$

11. $\sqrt{9x^6}$ 12. $\sqrt{16x^2y^4}$ 13. $\sqrt{x^4} \cdot \sqrt{x}$ 14. $\sqrt{a^6} \cdot \sqrt{a}$

15. $\sqrt{2^2 \cdot a^2} \cdot \sqrt{2a}$ 16. $\sqrt{3^2 \cdot x^4} \cdot \sqrt{2x}$ 17. $\sqrt{x^2y^2} \cdot \sqrt{2x}$ 18. $\sqrt{3^4 \cdot x^4} \cdot \sqrt{2x}$

19. $\sqrt{6x^2}$ 20. $\sqrt{10y^4}$ 21. $\sqrt[3]{r^3t^3}$ 22. $\sqrt[3]{m^6n^3}$

23. $\sqrt[3]{x^{12}}$ 24. $\sqrt[3]{p^3q^9}$ 25. $\sqrt[3]{8k^6}$ 26. $\sqrt[3]{27r^3}$

WRITTEN EXERCISES

Goal: To simplify radicals containing variables
Sample Problem: $\sqrt[3]{16a^4b^6}$
Answer: $2ab^2\sqrt[3]{2a}$

Simplify. The domain is {nonnegative real numbers}. (Examples 1–4)

1. $\sqrt{36y^2}$ 2. $\sqrt{9a^4}$ 3. $\sqrt{y^2z^6}$ 4. $\sqrt{a^6b^8}$

5. $\sqrt{x^5}$ 6. $\sqrt{a^3}$ 7. $\sqrt{4x}$ 8. $\sqrt{9y}$

9. $\sqrt{16y^3}$ 10. $\sqrt{25y^5}$ 11. $\sqrt{8x^2}$ 12. $\sqrt{8y^4}$

13. $\sqrt{8a^5}$ 14. $\sqrt{8x^3}$ 15. $\sqrt{12ab}$ 16. $\sqrt{18xy}$

17. $\sqrt{49a^3b^3}$ 18. $\sqrt{36x^3y^3}$ 19. $\sqrt{24a^2b^3}$ 20. $\sqrt{40x^3y^2}$

21. $\sqrt{75r^5s^6}$ 22. $\sqrt{45c^4d^7}$ 23. $\sqrt{20r^2s^3t^5}$ 24. $\sqrt{27a^5b^4c^3}$

25. $\sqrt[3]{x^{12}}$ 26. $\sqrt[3]{r^6}$ 27. $\sqrt[3]{x^{13}}$ 28. $\sqrt[3]{8y^7}$

29. $\sqrt[3]{27x^3y^6}$ 30. $\sqrt[3]{64r^6s^9}$ 31. $\sqrt[3]{16a^2b^3}$ 32. $\sqrt[3]{24m^4n^6}$

33. $\sqrt[3]{54m^5n^8}$ 34. $\sqrt[3]{32a^{11}b^5}$ 35. $\sqrt[3]{32r^7s^{10}}$ 36. $\sqrt[3]{128a^4b^9}$

MORE CHALLENGING EXERCISES

Write each product as one radical. Then simplify.

EXAMPLE: $\sqrt{2x} \cdot \sqrt{2x^3}$ **SOLUTION:** $\sqrt{2x} \cdot \sqrt{2x^3} = \sqrt{4x^4}$
$= 2x^2$

37. $\sqrt{x} \cdot \sqrt{x^3}$ 38. $\sqrt{x} \cdot \sqrt{x^5}$ 39. $\sqrt{2} \cdot \sqrt{2x}$

40. $\sqrt{3} \cdot \sqrt{3y}$ 41. $\sqrt{2x} \cdot \sqrt{2x^3}$ 42. $\sqrt{3y^3} \cdot \sqrt{3y}$

43. $\sqrt[3]{4x} \cdot \sqrt[3]{3x^2}$ 44. $\sqrt[3]{2r^4} \cdot \sqrt[3]{4r^2}$ 45. $\sqrt[3]{9a} \cdot \sqrt[3]{3a^5}$

46. $\sqrt[3]{16r^2s^2} \cdot \sqrt[3]{4rs^4}$ 47. $\sqrt[3]{8b^2} \cdot \sqrt[3]{8b^7}$ 48. $\sqrt[3]{27rs^{11}} \cdot \sqrt[3]{r^8s}$

APPLICATIONS

You can use a calculator with a square root key, [√] *, to find roots other than square roots. For example, you can think of a fourth root as a "square root of a square root."*

EXAMPLE 1 $\sqrt{39}$ **SOLUTION** [3] [9] [√] 6.2449979

EXAMPLE 2 $\sqrt[4]{36}$ **SOLUTION** Think of $\sqrt[4]{36}$ as $\sqrt{\sqrt{36}}$.

[3] [6] [√] [√] 2.4494897

Evaluate each of the following.

49. $\sqrt{64}$ **50.** $\sqrt{121}$ **51.** $\sqrt{43}$ **52.** $\sqrt[4]{16}$ **53.** $\sqrt[4]{1296}$ **54.** $\sqrt[8]{6581}$

MID-CHAPTER REVIEW

Simplify. (Section 7.1)

1. $\sqrt{(13 \cdot 17)(13 \cdot 17)}$ **2.** $\sqrt{2 \cdot 5 \cdot 7 \cdot 2 \cdot 7 \cdot 5}$ **3.** $\sqrt{5^4 \cdot 7^2 \cdot 11^2}$ **4.** $\sqrt{3^2 \cdot 5^4 \cdot 13^2}$
5. $\sqrt{729}$ **6.** $\sqrt{1024}$ **7.** $\sqrt{1296}$ **8.** $\sqrt{2025}$ **9.** $\sqrt{2304}$

Multiply and simplify. (Section 7.2)

10. $\sqrt{23} \cdot \sqrt{23}$ **11.** $\sqrt{9} \cdot \sqrt{49}$ **12.** $\sqrt{12} \cdot \sqrt{3}$ **13.** $(-\sqrt{7})(\sqrt{2})$
14. $(-\sqrt{\frac{1}{2}})(-\sqrt{162})$ **15.** $\sqrt{\frac{1}{5}} \cdot \sqrt{125}$ **16.** $\sqrt{\frac{5}{6}} \cdot \sqrt{\frac{5}{6}}$ **17.** $\sqrt{5} \cdot \sqrt{7} \cdot \sqrt{6}$

Simplify. (Section 7.3)

18. $\sqrt{360}$ **19.** $\sqrt{490}$ **20.** $\sqrt{320}$ **21.** $\sqrt{700}$ **22.** $\sqrt{98}$
23. $\sqrt{405}$ **24.** $\sqrt[3]{40}$ **25.** $\sqrt[3]{108}$ **26.** $\sqrt[4]{48}$ **27.** $\sqrt[4]{162}$

Simplify. The domain is {nonnegative real numbers}. (Section 7.4)

28. $\sqrt{81a^4b^2}$ **29.** $\sqrt{49m^6n^2}$ **30.** $\sqrt{24p^3}$ **31.** $\sqrt{27r^4s^5}$
32. $\sqrt{75p^2q^3r^6}$ **33.** $\sqrt{108a^3b^5c^7}$ **34.** $\sqrt[3]{216m^3n^4}$ **35.** $\sqrt[3]{54p^7q^6}$

REVIEW CAPSULE FOR SECTION 7.5

Combine like terms. (Pages 50–51)

1. $9m + 8m$ **2.** $5.6t - 2.9t$ **3.** $76k + k$ **4.** $4.8w - 5.9w + 0.8w$

7.5 Sums and Differences

P–1 **How is each sum or difference below expressed as a product by the Distributive Property?**

a. $7x + ax$ **b.** $ty - ry$ **c.** $11m + 3m$

Find a common factor.

The Distributive Property can also be applied to sums or differences of radicals.

EXAMPLE 1 Add: $3\sqrt{5} + 4\sqrt{5}$

Solution:

1. Distributive Property ⟶ $3\sqrt{5} + 4\sqrt{5} = (3 + 4)\sqrt{5}$
2. Simplify. ⟶ $= 7\sqrt{5}$

Adding $3\sqrt{5}$ and $4\sqrt{5}$ is much like adding 3x and 4x.

P–2 **Add: $6\sqrt{2} + 5\sqrt{2}$**

Radicals such as $6\sqrt{2}$ and $5\sqrt{2}$ are called ***like radicals.*** They have equal radicands.

P–3 **Add or subtract as indicated.**

a. $4\sqrt{5} + \sqrt{5}$ **b.** $9\sqrt{3} - 2\sqrt{3}$ **c.** $\sqrt{10} - 4\sqrt{10}$

P–4 **What are the prime factors of 12?**

EXAMPLE 2 Add: $\sqrt{12} + 7\sqrt{3}$

Solution:

1. Write the prime factorization of 12. ⟶ $\sqrt{12} + 7\sqrt{3} = \sqrt{2 \cdot 2 \cdot 3} + 7\sqrt{3}$
2. Product Property of Radicals ⟶ $= \sqrt{2^2} \cdot \sqrt{3} + 7\sqrt{3}$
3. Meaning of square root ⟶ $= 2\sqrt{3} + 7\sqrt{3}$
4. Distributive Property ⟶ $= (2 + 7)\sqrt{3}$
5. Simplify. ⟶ $= 9\sqrt{3}$

Simplest form

P–5 **Simplify: a. $\sqrt{3}$ b. $\sqrt{15}$**

EXAMPLE 3 Add: $\sqrt{3} + \sqrt{15}$

Solution: Each radical is in simplest form. Therefore, $\sqrt{3} + \sqrt{15}$ is in simplest form.

P–6 **What is the prime factorization of 8? of 50?**

EXAMPLE 4 Add: $\sqrt{8} + 3\sqrt{2} + \sqrt{50}$

Solution:

$$\begin{aligned}\sqrt{8} + 3\sqrt{2} + \sqrt{50} &= \sqrt{2 \cdot 2 \cdot 2} + 3\sqrt{2} + \sqrt{2 \cdot 5 \cdot 5}\\ &= \sqrt{2^2} \cdot \sqrt{2} + 3\sqrt{2} + \sqrt{5^2} \cdot \sqrt{2}\\ &= 2\sqrt{2} + 3\sqrt{2} + 5\sqrt{2}\\ &= (2 + 3 + 5)\sqrt{2}\\ &= 10\sqrt{2}\end{aligned}$$

◀ ***Simplest form***

EXAMPLE 5 Subtract: $\sqrt{12n} - \sqrt{27n}$

Solution:

1. Write the prime factorization. ⟶ $\sqrt{12n} - \sqrt{27n} = \sqrt{2 \cdot 2 \cdot 3 \cdot n} - \sqrt{3 \cdot 3 \cdot 3 \cdot n}$
2. Product Property of Radicals ⟶ $= \sqrt{2^2} \cdot \sqrt{3n} - \sqrt{3^2} \cdot \sqrt{3n}$
3. Meaning of square root ⟶ $= 2\sqrt{3n} - 3\sqrt{3n}$
4. Distributive Property ⟶ $= (2 - 3)\sqrt{3n}$
5. Simplify. ⟶ $= -1\sqrt{3n}$ or $-\sqrt{3n}$

CLASSROOM EXERCISES

Add or subtract as indicated. (Example 1)

1. $2\sqrt{3} + 5\sqrt{3}$ **2.** $3\sqrt{5} + 2\sqrt{5}$ **3.** $3\sqrt{2} + \sqrt{2}$

4. $\sqrt{7} + \sqrt{7}$ **5.** $7\sqrt{10} + 5\sqrt{10}$ **6.** $8\sqrt{3} - 5\sqrt{3}$

Add or subtract as indicated. Simplify where necessary. (Examples 1 and 3)

7. $4\sqrt{2} - \sqrt{2}$ **8.** $5\sqrt{5} - 4\sqrt{5}$ **9.** $3\sqrt{11} - 5\sqrt{11}$ **10.** $3\sqrt{13} - 4\sqrt{13}$

11. $2\sqrt{x} + 3\sqrt{x}$ **12.** $\frac{1}{2}\sqrt{y} + 2\sqrt{y}$ **13.** $\frac{3}{4}\sqrt{a} - \frac{1}{4}\sqrt{a}$ **14.** $\sqrt{n} + \sqrt{n}$

15. $9\sqrt{m} + \sqrt{m}$ **16.** $\sqrt{16} - \sqrt{1}$ **17.** $\sqrt{4} + \sqrt{9}$ **18.** $\sqrt{2} + \sqrt{5}$

WRITTEN EXERCISES

Goal: To add and subtract radicals
Sample Problem: $3\sqrt{5} - 7\sqrt{5} + \sqrt{5}$ **Answer:** $-3\sqrt{5}$

Add or subtract as indicated. Simplify where necessary. (Examples 1–3, 5)

1. $10\sqrt{3} + 5\sqrt{3}$ **2.** $8\sqrt{5} + 3\sqrt{5}$ **3.** $7\sqrt{2} - 3\sqrt{2}$

4. $8\sqrt{7} - 3\sqrt{7}$ **5.** $2\sqrt{3} - 5\sqrt{3}$ **6.** $3\sqrt{14} - 7\sqrt{14}$

7. $5\sqrt{2} - \sqrt{2}$ **8.** $6\sqrt{6} - \sqrt{6}$ **9.** $\sqrt{11} + \sqrt{11}$

10. $\sqrt{15} + \sqrt{15}$ **11.** $\sqrt{5} + \sqrt{15}$ **12.** $2\sqrt{3} + \sqrt{7}$

13. $\sqrt{19} - \sqrt{3}$ **14.** $3\sqrt{5} - 2\sqrt{3}$ **15.** $3\sqrt{x} + 9\sqrt{x}$

16. $5\sqrt{a} + 10\sqrt{a}$ **17.** $-2\sqrt{t} - 3\sqrt{t}$ **18.** $-\sqrt{r} - 5\sqrt{r}$

19. $\sqrt{s} + \sqrt{s}$ **20.** $\sqrt{12} + 3\sqrt{3}$ **21.** $3\sqrt{8} + 5\sqrt{2}$

22. $3\sqrt{12} + 5\sqrt{3}$ **23.** $\sqrt{20} - 2\sqrt{5}$ **24.** $\sqrt{18} - 6\sqrt{2}$

25. $\sqrt{27} + \sqrt{12}$ **26.** $\sqrt{32} + \sqrt{8}$ **27.** $2\sqrt{45} + \sqrt{5}$

28. $\sqrt{24} + \sqrt{96}$ **29.** $2\sqrt{72} - \sqrt{2}$ **30.** $4\sqrt{27} - \sqrt{12}$

31. $\sqrt{28a} + 3\sqrt{7a}$ **32.** $\sqrt{44n} + 5\sqrt{11n}$ **33.** $\sqrt{2r} + \sqrt{18r}$

34. $\sqrt{16x} - \sqrt{4x}$ **35.** $\sqrt{25y} + \sqrt{49y}$ **36.** $\sqrt{2x} - 3\sqrt{32x}$

(Example 4)

37. $2\sqrt{3} + \sqrt{3} + 4\sqrt{3}$ **38.** $7\sqrt{2} + \sqrt{2} + 2\sqrt{2}$

39. $3\sqrt{7} - 4\sqrt{7} + 2\sqrt{7}$ **40.** $4\sqrt{5} - 6\sqrt{5} + \sqrt{5}$

41. $7\sqrt{y} - 2\sqrt{y} - 5\sqrt{y}$ **42.** $-3\sqrt{b} + 10\sqrt{b} - 7\sqrt{b}$

REVIEW CAPSULE FOR SECTION 7.6

Multiply and simplify. (Pages 177–179)

1. $\sqrt{3} \cdot \sqrt{12}$ **2.** $\sqrt{5x^3} \cdot \sqrt{5x}$ **3.** $\sqrt{13x} \cdot \sqrt{13x}$ **4.** $\sqrt{\frac{1}{9}x} \cdot \sqrt{x^3}$

7.6 Quotients

P–1 **Simplify: a.** $\sqrt{\frac{4}{9}}$ **b.** $\frac{\sqrt{4}}{\sqrt{9}}$

> ***Quotient Property of Radicals***
>
> If a is any nonnegative real number and b is any positive real number, then $\sqrt{\frac{\mathbf{a}}{\mathbf{b}}} = \frac{\sqrt{\mathbf{a}}}{\sqrt{\mathbf{b}}}$.
>
> $\sqrt{\frac{4}{9}} = \frac{\sqrt{4}}{\sqrt{9}}$
>
> $\sqrt{\frac{3}{7}} = \frac{\sqrt{3}}{\sqrt{7}}$

EXAMPLE 1 Simplify $\sqrt{\frac{8}{r^2}}$.

Solution:

1 Quotient Property of Radicals ⟶ $\sqrt{\frac{8}{r^2}} = \frac{\sqrt{8}}{\sqrt{r^2}}$

2 Product Property of Radicals ⟶ $= \frac{\sqrt{4} \cdot \sqrt{2}}{\sqrt{r^2}}$

3 Meaning of square root ⟶ $= \frac{2\sqrt{2}}{r}$ ◀ ***Simplest form***

P–2 **Simplify each radical.**

a. $\sqrt{\frac{n^2}{9}}$ **b.** $\sqrt{\frac{2}{x^2}}$ **c.** $\sqrt{\frac{r^3}{16}}$ **d.** $\sqrt{\frac{64}{a^2}}$

EXAMPLE 2 Simplify $\frac{\sqrt{18x^3}}{\sqrt{2x}}$.

Solution:

1 Quotient Property of Radicals ⟶ $\frac{\sqrt{18x^3}}{\sqrt{2x}} = \sqrt{\frac{18x^3}{2x}}$

2 Divide. ⟶ $= \sqrt{9x^2}$

3 Meaning of square root ⟶ $= 3x$ ◀ ***Simplest form***

Sometimes it is inconvenient to have a radical in a denominator. The procedure used in Example 3 below is called **rationalizing the denominator.**

EXAMPLE 3

Rationalize the denominator of $\frac{\sqrt{3}}{\sqrt{2}}$.

Solution:

1. Multiplication Property of One ⟶ $\frac{\sqrt{3}}{\sqrt{2}} = \frac{\sqrt{3}}{\sqrt{2}} \cdot \frac{\sqrt{2}}{\sqrt{2}}$ — ***$\frac{\sqrt{2}}{\sqrt{2}}$ is another name for 1.***
2. Product Property of Radicals ⟶ $= \frac{\sqrt{6}}{\sqrt{4}}$
3. Meaning of square root ⟶ $= \frac{\sqrt{6}}{2}$ — ***$\frac{\sqrt{6}}{2}$ has no radical in the denominator.***

P–3 **What name for 1 would be used to rationalize the denominator of each fraction below?**

a. $\frac{\sqrt{2}}{\sqrt{5}}$ **b.** $\frac{\sqrt{m}}{\sqrt{2n}}$ **c.** $\frac{\sqrt{5r}}{\sqrt{12s}}$ **d.** $\frac{\sqrt{3y}}{\sqrt{x^3}}$

Remember that $\sqrt{x^3}\sqrt{x} = x^2$.

EXAMPLE 4

Rationalize the denominator of $\frac{\sqrt{5r}}{\sqrt{12s}}$.

Then simplify.

Solution:

1. Multiplication Property of One ⟶ $\frac{\sqrt{5r}}{\sqrt{12s}} = \frac{\sqrt{5r}}{\sqrt{12s}} \cdot \frac{\sqrt{12s}}{\sqrt{12s}}$
2. Product Property of Radicals ⟶ $= \frac{\sqrt{60rs}}{12s}$
3. Product Property of Radicals ⟶ $= \frac{\sqrt{4} \cdot \sqrt{15rs}}{12s}$
4. Meaning of square root ⟶ $= \frac{2\sqrt{15rs}}{12s}$
5. Simplify. ⟶ $= \frac{\overset{1}{\cancel{2}}\sqrt{15rs}}{\underset{6}{\cancel{12}}s}$ or $\frac{\sqrt{15rs}}{6s}$

CLASSROOM EXERCISES

Simplify. (Example 1)

1. $\sqrt{\frac{1}{4}}$ 2. $\sqrt{\frac{1}{9}}$ 3. $\sqrt{\frac{9}{25}}$ 4. $\sqrt{\frac{36}{49}}$ 5. $\sqrt{\frac{64}{100}}$ 6. $\sqrt{\frac{100}{144}}$

Simplify. (Example 2)

7. $\frac{\sqrt{12}}{\sqrt{3}}$ 8. $\frac{\sqrt{8}}{\sqrt{2}}$ 9. $\frac{\sqrt{20}}{\sqrt{5}}$ 10. $\frac{\sqrt{24}}{\sqrt{6}}$ 11. $\frac{\sqrt{28}}{\sqrt{7}}$

WRITTEN EXERCISES

Goal: To simplify quotients of radicals

Sample Problem: $\frac{\sqrt{132r^3}}{\sqrt{6r}}$ **Answer:** $r\sqrt{22}$

Simplify. (Example 1)

1. $\sqrt{\frac{x^2}{16}}$ 2. $\sqrt{\frac{y^2}{25}}$ 3. $\sqrt{\frac{49}{a^2}}$ 4. $\sqrt{\frac{64}{y^2}}$ 5. $\sqrt{\frac{y^4}{100}}$

6. $\sqrt{\frac{x^4}{121}}$ 7. $\sqrt{\frac{2}{x^2}}$ 8. $\sqrt{\frac{3}{r^2}}$ 9. $\sqrt{\frac{12}{a^2}}$ 10. $\sqrt{\frac{18}{t^2}}$

Simplify. (Example 2)

11. $\frac{\sqrt{50}}{\sqrt{2}}$ 12. $\frac{\sqrt{75}}{\sqrt{3}}$ 13. $\frac{\sqrt{108}}{\sqrt{3}}$ 14. $\frac{\sqrt{72}}{\sqrt{2}}$ 15. $\frac{\sqrt{98}}{\sqrt{2}}$

16. $\frac{\sqrt{180}}{\sqrt{5}}$ 17. $\frac{\sqrt{125}}{\sqrt{5}}$ 18. $\frac{\sqrt{147}}{\sqrt{3}}$ 19. $\frac{\sqrt{x^3}}{\sqrt{x}}$ 20. $\frac{\sqrt{a}}{\sqrt{a^3}}$

21. $\frac{\sqrt{4y}}{\sqrt{y^3}}$ 22. $\frac{\sqrt{r^3}}{\sqrt{9r}}$ 23. $\frac{\sqrt{ab^2}}{\sqrt{a^3}}$ 24. $\frac{\sqrt{x^3y^3}}{\sqrt{xy}}$ 25. $\frac{\sqrt{10}}{\sqrt{2}}$

26. $\frac{\sqrt{14}}{\sqrt{7}}$ 27. $\frac{\sqrt{18}}{\sqrt{3}}$ 28. $\frac{\sqrt{22}}{\sqrt{11}}$ 29. $\frac{\sqrt{33}}{\sqrt{3}}$ 30. $\frac{\sqrt{38}}{\sqrt{2}}$

Rationalize each denominator. Then simplify. (Examples 3–4)

31. $\frac{\sqrt{3}}{\sqrt{5}}$ 32. $\frac{\sqrt{2}}{\sqrt{7}}$ 33. $\frac{\sqrt{a^2}}{\sqrt{b}}$ 34. $\frac{\sqrt{r^3}}{\sqrt{p}}$ 35. $\frac{\sqrt{3s}}{\sqrt{5t}}$ 36. $\frac{\sqrt{2x}}{\sqrt{3y}}$

REVIEW CAPSULE FOR SECTION 7.7

Write the prime factorization. (Pages 174–176)

1. 168 2. 180 3. 244 4. 288 5. 207 6. 234

7.7 Approximations

A Table of Squares and Square Roots is provided on page 194 for the positive integers from 1 to 150. Square roots in this table are approximated to three decimal places.

P–1 **What is each power below based on the table?**

a. 28^2 **b.** 76^2 **c.** 26^2

Each number in the **Number** column is the positive square root of a number in the **Square** column.

P–2 **What is the value of each square root below?**

a. $\sqrt{784}$ **b.** $-\sqrt{6084}$ **c.** $\sqrt{729}$

P–3 **What is the approximate value of each square root below?**

a. $\sqrt{26}$ **b.** $\sqrt{78}$ **c.** $-\sqrt{28}$

The process of simplifying radicals can sometimes be used to approximate the square root of numbers not in the table.

EXAMPLE 1 Approximate $\sqrt{160}$ to three decimal places.

Write the radical in simplest form. → **Use the table of square roots.**

Solution:

1 Write the prime factorization. → $\sqrt{160} = \sqrt{2 \cdot 2 \cdot 2 \cdot 2 \cdot 2 \cdot 5}$

2 Product Property of Radicals → $= \sqrt{2^4} \cdot \sqrt{2 \cdot 5}$

3 Simplify. → $= 4\sqrt{10}$

4 Use the table on page 194. → $\approx 4(3.162)$

≈ 12.648

The approximate value of $\sqrt{160}$ is 12.648.

Table of Squares and Square Roots

No.	Square	Square Root	No.	Square	Square Root	No.	Square	Square Root
1	1	1.000	**51**	2601	7.141	**101**	10,201	10.050
2	4	1.414	**52**	2704	7.211	**102**	10,404	10.100
3	9	1.732	**53**	2809	7.280	**103**	10,609	10.149
4	16	2.000	**54**	2916	7.348	**104**	10,816	10.198
5	25	2.236	**55**	3025	7.416	**105**	11,025	10.247
6	36	2.449	**56**	3136	7.483	**106**	11,236	10.296
7	49	2.646	**57**	3249	7.550	**107**	11,449	10.344
8	64	2.828	**58**	3364	7.616	**108**	11,664	10.392
9	81	3.000	**59**	3481	7.681	**109**	11,881	10.440
10	100	3.162	**60**	3600	7.746	**110**	12,100	10.488
11	121	3.317	**61**	3721	7.810	**111**	12,321	10.536
12	144	3.464	**62**	3844	7.874	**112**	12,544	10.583
13	169	3.606	**63**	3969	7.937	**113**	12,769	10.630
14	196	3.742	**64**	4096	8.000	**114**	12,996	10.677
15	225	3.873	**65**	4225	8.062	**115**	13,225	10.724
16	256	4.000	**66**	4356	8.124	**116**	13,456	10.770
17	289	4.123	**67**	4489	8.185	**117**	13,689	10.817
18	324	4.243	**68**	4624	8.246	**118**	13,924	10.863
19	361	4.359	**69**	4761	8.307	**119**	14,161	10.909
20	400	4.472	**70**	4900	8.367	**120**	14,400	10.954
21	441	4.583	**71**	5041	8.426	**121**	14,641	11.000
22	484	4.690	**72**	5184	8.485	**122**	14,884	11.045
23	529	4.796	**73**	5329	8.544	**123**	15,129	11.091
24	576	4.899	**74**	5476	8.602	**124**	15,376	11.136
25	625	5.000	**75**	5625	8.660	**125**	15,625	11.180
26	676	5.099	**76**	5776	8.718	**126**	15,876	11.225
27	729	5.196	**77**	5929	8.775	**127**	16,129	11.269
28	784	5.292	**78**	6084	8.832	**128**	16,384	11.314
29	841	5.385	**79**	6241	8.888	**129**	16,641	11.358
30	900	5.477	**80**	6400	8.944	**130**	16,900	11.402
31	961	5.568	**81**	6561	9.000	**131**	17,161	11.446
32	1024	5.657	**82**	6724	9.055	**132**	17,424	11.489
33	1089	5.745	**83**	6889	9.110	**133**	17,689	11.533
34	1156	5.831	**84**	7056	9.165	**134**	17,956	11.576
35	1225	5.916	**85**	7225	9.220	**135**	18,225	11.619
36	1296	6.000	**86**	7396	9.274	**136**	18,496	11.662
37	1369	6.083	**87**	7569	9.327	**137**	18,769	11.705
38	1444	6.164	**88**	7744	9.381	**138**	19,044	11.747
39	1521	6.245	**89**	7921	9.434	**139**	19,321	11.790
40	1600	6.325	**90**	8100	9.487	**140**	19,600	11.832
41	1681	6.403	**91**	8281	9.539	**141**	19,881	11.874
42	1764	6.481	**92**	8464	9.592	**142**	20,164	11.916
43	1849	6.557	**93**	8649	9.644	**143**	20,449	11.958
44	1936	6.633	**94**	8836	9.695	**144**	20,736	12.000
45	2025	6.708	**95**	9025	9.747	**145**	21,025	12.042
46	2116	6.782	**96**	9216	9.798	**146**	21,316	12.083
47	2209	6.856	**97**	9409	9.849	**147**	21,609	12.124
48	2304	6.928	**98**	9604	9.899	**148**	21,904	12.166
49	2401	7.000	**99**	9801	9.950	**149**	22,201	12.207
50	2500	7.071	**100**	10,000	10.000	**150**	22,500	12.247

EXAMPLE 2 Approximate $\sqrt{\frac{3}{8}}$ to three decimal places.

Rationalize the denominator. → **Use the table of square roots.** → **Divide.**

Solution:

1. Quotient Property of Radicals → $\sqrt{\frac{3}{8}} = \frac{\sqrt{3}}{\sqrt{8}}$
2. Multiplication Property of One → $= \frac{\sqrt{3}}{\sqrt{8}} \cdot \frac{\sqrt{8}}{\sqrt{8}}$
3. Product Property of Radicals → $= \frac{\sqrt{24}}{8}$
4. Simplify the numerator. → $= \frac{2\sqrt{6}}{8}$
5. Write in lowest terms. → $= \frac{\sqrt{6}}{4}$
6. Use the table on page 194. → $\approx \frac{2.449}{4}$

≈ 0.612 ◀ ***Approximate value of $\sqrt{\frac{3}{8}}$***

EXAMPLE 3 The area of a square plaque is 75 square inches. Find the length of each side of the plaque to the nearest tenth of an inch.

Solution:

1. Write the formula for the area of a square. → $A = s^2$
2. Replace A with 75. → $75 = s^2$
3. Solve for s. → $\sqrt{75} = s$
4. Simplify. → $\sqrt{25} \cdot \sqrt{3} = s$

$5\sqrt{3} = s$ ◀ ***Use the table on page 194.***

5. Approximate the value of s. → $5(1.732) \approx s$

$8.660 \approx s$

The length of each side to the nearest tenth is 8.7 inches.

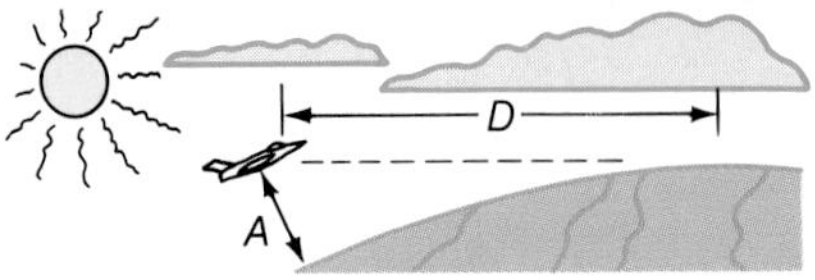

If you are sitting in a plane that is high in the sky, you can see the earth's horizon several miles away. The distance to the horizon depends on the plane's ***altitude.*** The altitude is the plane's distance above the ground.

The approximate distance to the horizon is given by the following formula, where D is the distance to the horizon in miles and A is the altitude of the plane in feet.

$$\mathbf{D = 1.22\sqrt{A}}$$

EXAMPLE 4 Find the approximate distance in miles to the horizon from a plane that is flying at an altitude of 25,000 feet. Round the distance to the nearest mile.

Solution:

1 Write the formula. ⟶ $D = 1.22\sqrt{A}$

2 Replace A with 25,000. ⟶ $D = 1.22\sqrt{25{,}000}$

3 Simplify. ⟶ $D = 1.22\sqrt{2500} \cdot \sqrt{10}$

$D = (1.22)(50)\sqrt{10}$

$D \approx 61(3.162)$ ◀ ***Use the table on page 194.***

$D \approx 192.882$

The distance to the horizon is approximately 193 miles.

CLASSROOM EXERCISES

Find each value by referring to the Table of Squares and Square Roots. (P-1, P-2, P-3)

1. 29^2
2. 56^2
3. 93^2
4. 117^2
5. $\sqrt{5184}$
6. $\sqrt{961}$
7. $-\sqrt{7744}$
8. $\sqrt{13{,}225}$
9. $\sqrt{38}$
10. $\sqrt{131}$
11. $\sqrt{78}$
12. $-\sqrt{43}$
13. $\sqrt{3481}$
14. $\sqrt{13{,}456}$
15. $-\sqrt{150}$
16. $\sqrt{109}$

WRITTEN EXERCISES

Goal: To approximate the values of square roots to two or three decimal places

Sample Problem: Approximate $\sqrt{304}$ to three decimal places.

Answer: 17.436

Approximate each square root to three decimal places. Use the table on page 194. (Examples 1–2)

1. $\sqrt{168}$ **2.** $\sqrt{180}$ **3.** $\sqrt{244}$ **4.** $\sqrt{288}$

5. $\sqrt{207}$ **6.** $\sqrt{234}$ **7.** $\sqrt{275}$ **8.** $\sqrt{425}$

9. $\sqrt{468}$ **10.** $\sqrt{684}$ **11.** $\sqrt{\frac{3}{4}}$ **12.** $\sqrt{\frac{5}{16}}$

13. $\sqrt{\frac{2}{3}}$ **14.** $\sqrt{\frac{5}{6}}$ **15.** $\sqrt{\frac{7}{8}}$ **16.** $\sqrt{\frac{5}{8}}$

17. $\sqrt{\frac{11}{12}}$ **18.** $\sqrt{\frac{7}{12}}$ **19.** $\sqrt{\frac{3}{32}}$ **20.** $\sqrt{\frac{7}{18}}$

APPLICATIONS

Solve each problem. Use the table on page 194. Round the answers to the nearest tenth. (Example 3)

21. A homesite is in the shape of a square. It has an area of 4840 square yards. Find the length of each side of the homesite in yards.

22. The Andersens have a square–shaped backyard. It has an area of 396 square meters. Find the length of each side of the yard in meters.

For Exercises 23–24, use the formula $D = 1.22\sqrt{A}$. (Example 4)

23. Find the approximate distance in miles to the horizon from a plane that is at an altitude of 5000 feet. Round the distance to the nearest mile.

24. Find the approximate distance in miles to the horizon from a plane that is at an altitude of 15,000 feet. Round the distance to the nearest mile.

For Exercises 25–26, use the formula $D = 3.56\sqrt{A}$ where A is the altitude in meters and D is the distance in kilometers.

25. Find the approximate distance in kilometers to the horizon from a plane that is at an altitude of 4500 meters. Round the distance to the nearest kilometer.

26. Find the approximate distance in kilometers to the horizon from a plane that is at an altitude of 7500 meters. Round the distance to the nearest kilometer.

PROBLEM SOLVING AND APPLICATIONS

7.8 The Rule of Pythagoras

A triangle with one angle of 90° is called a **right triangle.** The symbol ⌝ is used to indicate the 90° angle. The longest side of a right triangle is the **hypotenuse.** The two shorter sides are the **legs.** In right triangle *ABC* at the right, the hypotenuse is side *AB* and the legs are sides *AC* and *BC.*

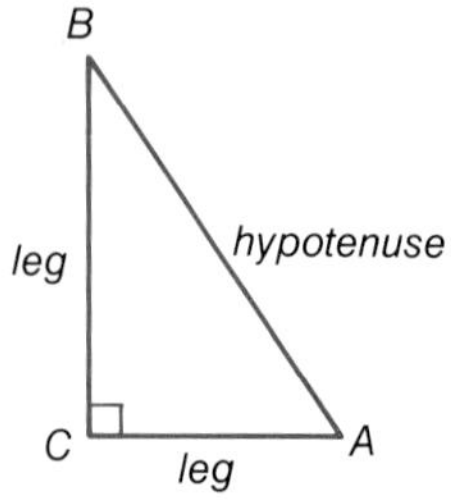

A rule about right triangles is named for Pythagoras, the Greek mathematician who developed it.

Rule of Pythagoras

In a right triangle, the square of the length of the hypotenuse, *c,* equals the sum of the squares of the lengths of the legs, *a* and *b.*

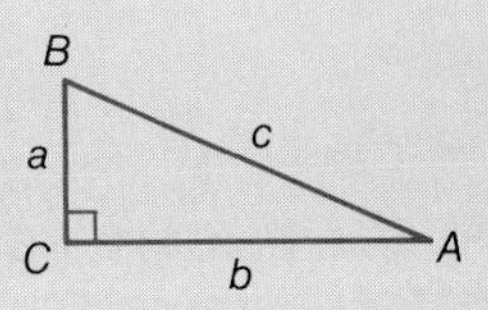

$$\mathbf{c^2 = a^2 + b^2}$$

EXAMPLE 1 Use the Rule of Pythagoras to determine whether the triangle shown at the right is a right triangle.

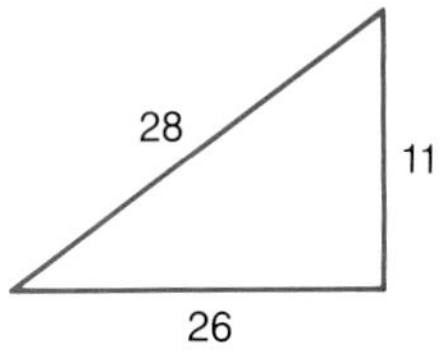

Solution: Let $a = 11$, $b = 26$, and $c = 28$.

1. Write the Rule of Pythagoras in symbols. ⟶ $c^2 = a^2 + b^2$
2. Replace the variables with their values. ⟶ $28^2 \stackrel{?}{=} 11^2 + 26^2$
3. Perform the operations. ⟶ $784 \stackrel{?}{=} 121 + 676$

$784 \neq 797$ ◀ **Not an equality**

Since the result is not an equality, the triangle is not a right triangle.

P-1 **Which of the following are the lengths of the three sides of a right triangle?**

a. 3 meters, 4 meters, 5 meters **b.** 8 feet, 12 feet, 16 feet

When the lengths of two sides of a right triangle are known, the Rule of Pythagoras can be used to find the length of the third side.

EXAMPLE 2

The legs of a right triangle have lengths of 9 centimeters and 12 centimeters. Find the length of the hypotenuse.

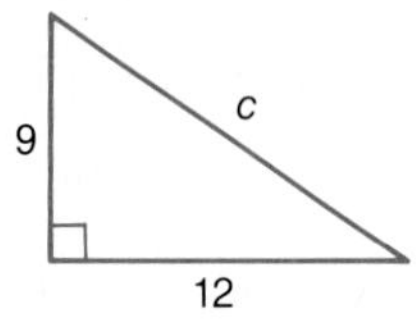

Solution: Let $a = 9$ and $b = 12$. Find c.

1. Write the Rule of Pythagoras in symbols. ⟶ $c^2 = a^2 + b^2$
2. Replace the known variables with their values. ⟶ $c^2 = 9^2 + 12^2$
3. Solve for c. ⟶ $c^2 = 81 + 144$

   $c^2 = 225$
4. Find the square root. ⟶ $c = \sqrt{225}$

   $c = 15$

Use the table on page 194.

The length of the hypotenuse is 15 centimeters.

EXAMPLE 3

Find the approximate height of a 14-inch TV screen if the width is 11 inches. Round the height to the nearest tenth. (A 14-inch screen has a diagonal length of 14 inches.)

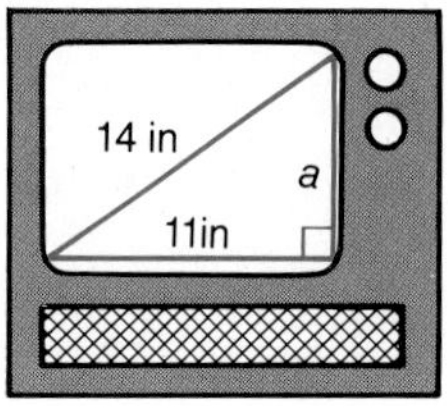

Solution: Let $b = 11$ and $c = 14$. Find a.

1. Write the Rule of Pythagoras in symbols. ⟶ $c^2 = a^2 + b^2$
2. Replace the known variables with their values. ⟶ $14^2 = a^2 + 11^2$
3. Solve for a. ⟶ $196 = a^2 + 121$

   $196 - \mathbf{121} = a^2 + 121 - \mathbf{121}$

   $75 = a^2$
4. Find the square root. ⟶ $\sqrt{75} = a$

   $8.660 \approx a$

Use the table on page 194.

The height of the screen is 8.7 inches to the nearest tenth.

CLASSROOM EXERCISES

Determine whether the given lengths are the lengths of the sides of a right triangle. (Example 1)

1. 2 feet, 3 feet, 4 feet

2. 6 meters, 8 meters, 10 meters

3. 3 inches, 5 inches, 6 inches

4. 12 yards, 5 yards, 13 yards

Use the Rule of Pythagoras to write an equation for finding the unknown length in each right triangle. (Steps 1 and 2 in Examples 2 and 3)

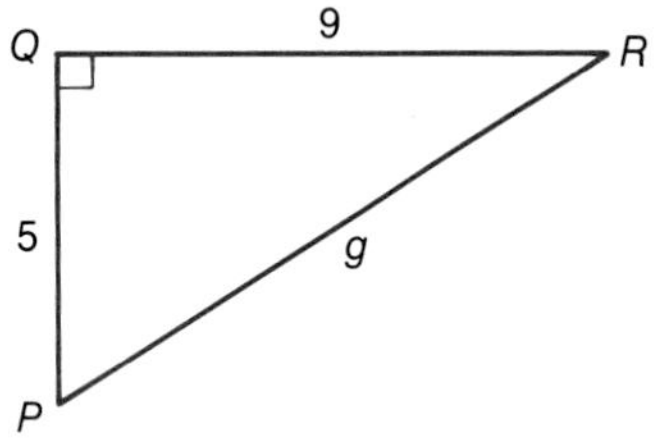

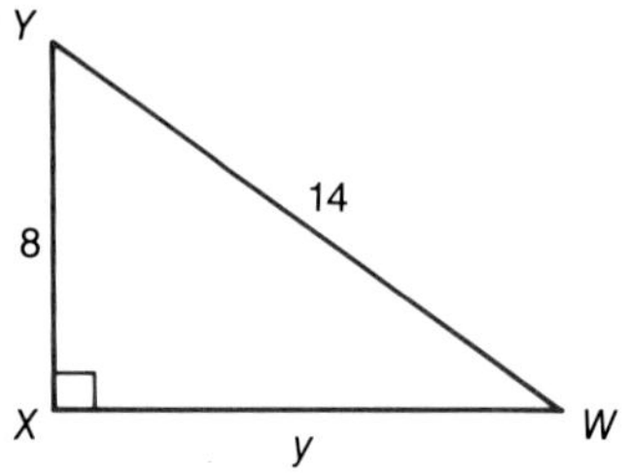

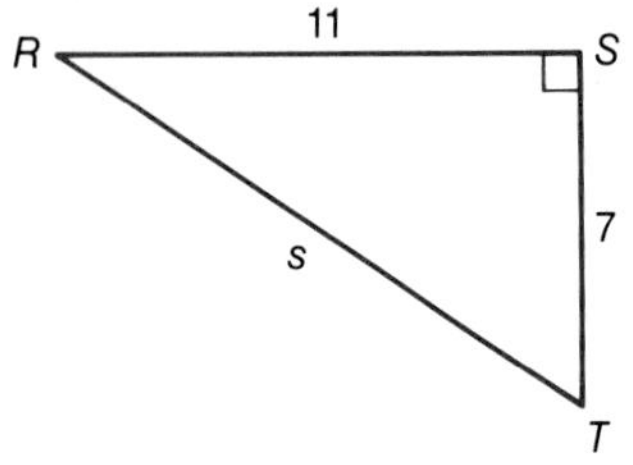

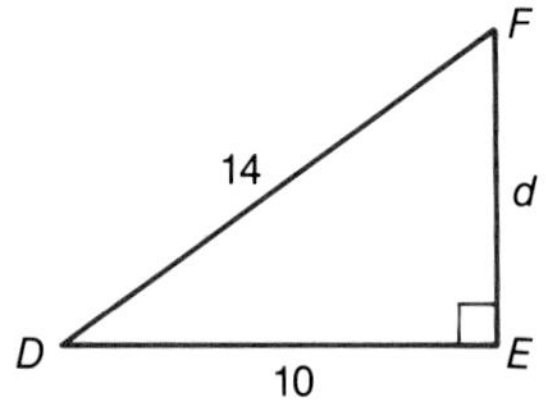

WRITTEN EXERCISES

Goal: To solve for the length of one side of a right triangle when the lengths of the other two sides are known

Sample Problem: Find the length of the hypotenuse of a right triangle when the lengths of the legs are 10 inches and 24 inches.

Answer: 26 inches

Determine whether the given lengths are the lengths of the sides of a right triangle. (Example 1)

1. 12 meters, 9 meters, 15 meters

2. 14 feet, 16 feet, 21 feet

3. 16 inches, 35 inches, 38 inches

4. 15 feet, 36 feet, 39 feet

For Exercises 5–16, refer to triangle ABC at the right. Use the Rule of Pythagoras and the table on page 194 to find the missing length. Round to the nearest tenth where necessary. (Examples 2 and 3)

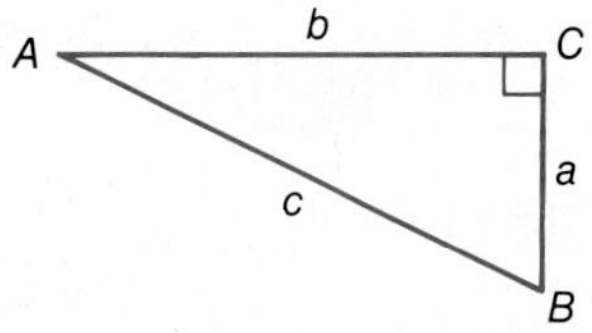

5. $a = 12$, $b = 16$

6. $a = 10$, $b = 24$

7. $a = 15$, $c = 25$

8. $a = 8$, $c = 10$

9. $b = 7$, $c = 25$

10. $b = 12$, $c = 13$

11. $c = 39$, $b = 36$

12. $c = 50$, $b = 48$

13. $c = 5$, $a = 2$

14. $c = 18$, $a = 12$

15. $a = 8$, $b = 10$

16. $a = 3$, $b = 5$

Solve each problem. Use the table on page 194. Round to the nearest tenth where necessary. (Example 3)

17. A TV with a 20-inch screen in on sale. The width of the screen is 16 inches. Find the height of the screen.

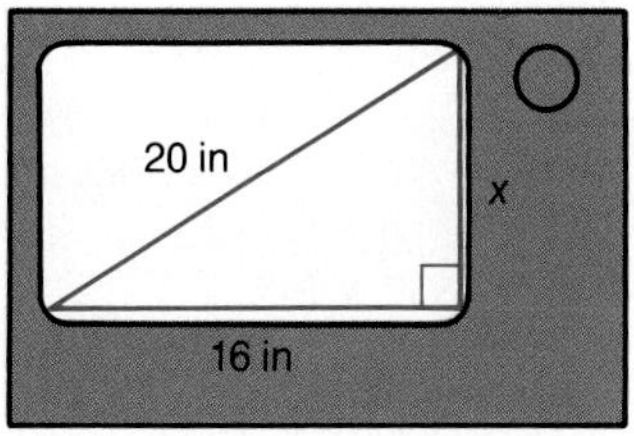

18. The distance between consecutive bases on a baseball diamond is 90 feet. Find the distance, *d*, from home plate to second base.

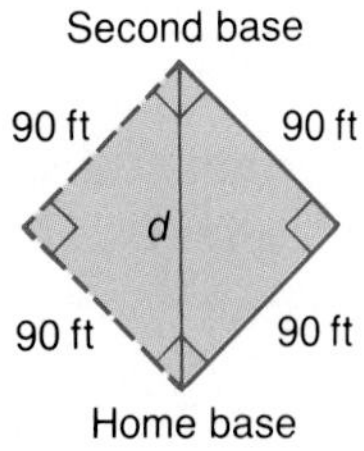

19. A utility pole has a support wire which is 36 feet long. The wire is anchored to the ground 20 feet from the base of the pole. Find the height of the pole.

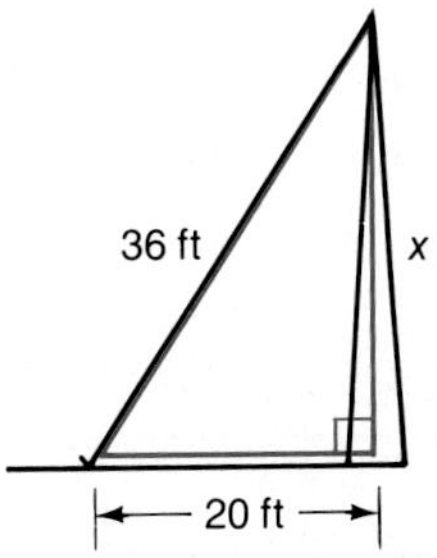

20. A boat sailed 20 kilometers due south from point *A* to point *C*. The boat then sailed 14 kilometers due west to point *B*. Find the direct sailing distance from point *A* to point *B*.

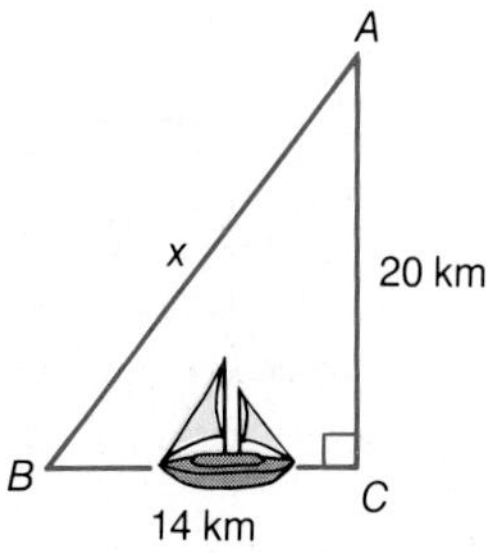

Meteorology

Meteorology is the study of the atmosphere. The best known activity of meteorologists is **weather forecasting.**

The table at the right shows the **Temperature–Humidity Index,** or **THI.** THI is a measure of how comfortable (or uncomfortable) people feel under given conditions of temperature and humidity during the warm season of the year.

THE TEMPERATURE–HUMIDITY INDEX

Temperature	Relative Humidity 10%	20%	30%	40%	50%	60%	70%	80%	90%	100%
68°	63	64	64	65	65	66	67	67	68	68
69°	64	64	65	65	66	67	67	68	68	69
70°	64	65	65	66	67	67	68	69	69	70
71°	65	65	66	67	67	68	69	70	70	71
72°	65	66	67	67	68	69	70	71	71	72
73°	66	66	67	68	69	70	70	71	72	73
74°	66	67	68	69	70	71	71	72	73	74
75°	67	67	68	69	70	71	72	73	74	75
76°	67	68	69	70	71	72	73	74	75	76
77°	68	69	70	71	72	73	74	75	76	77
78°	68	69	70	71	73	74	75	76	77	78
79°	69	70	71	72	73	74	76	77	78	79
80°	69	70	72	73	74	75	76	78	79	80
81°	70	71	72	73	75	76	77	78	80	81
82°	70	72	73	74	75	77	78	79	81	82
83°	71	72	73	75	76	78	79	80	82	83
84°	71	73	74	75	77	78	79	81	83	84
85°	72	73	75	76	78	79	80	82	84	85
86°	72	74	75	77	78	80	81	83	84	86
87°	73	74	76	77	79	81	82	84	85	87
88°	73	75	76	78	80	81	83	85	86	88
89°	74	75	77	79	81	82	84	86	87	89
90°	74	76	77	79	81	83	85	87	88	90
91°	75	76	78	80	82	84	85	87	89	91
92°	75	77	79	81	83	85	86	88	90	92
93°	76	78	80	81	83	85	87	89	91	93
94°	76	78	80	82	84	86	88	90	92	94
95°	77	79	81	83	85	87	89	91	93	95

few people feel uncomfortable ↓

about one-half of all people feel uncomfortable ↓

nearly everyone feels uncomfortable ↓

rapidly decreasing work efficiency ↓

extreme danger ⇩

EXAMPLE 1: Find the temperature–humidity index above which most people feel uncomfortable.

SOLUTION: Locate the curved band in the table for "nearly everyone feels uncomfortable." The band passes through the number 79 in each column.

Temperature–Humidity Index: **79**

In cold climates, the combined effect of high winds and cold temperatures can make a person feel colder than the actual thermometer reading. This is called **wind chill.** Weather forecasters obtain wind chill temperatures from tables such as the one in Example 2.

EXAMPLE 2: Find the wind chill temperature when the thermometer reading is −5°F and the wind speed is 10 miles per hour.

WIND CHILL TEMPERATURE TABLE

Wind Speed (mph)	Thermometer Reading (°F) −25	−20	−15	−10	−5	0	5	10	15	20
5	−31	−26	−21	−15	−10	−5	0	6	11	16
10	−52	−46	−40	−34	−27	−22	−15	−9	−3	3
15	−65	−58	−51	−45	−38	−31	−25	−18	−11	−5
20	−74	−67	−60	−53	−46	−39	−31	−24	−17	−10
25	−81	−74	−66	−59	−51	−44	−36	−29	−22	−15
30	−86	−79	−71	−64	−56	−49	−41	−33	−25	−18
35	−89	−82	−74	−67	−58	−52	−43	−35	−27	−20
40	−92	−84	−76	−69	−68	−53	−45	−37	−29	−21

SOLUTION: Locate the wind speed, 10 miles per hour, along the left side of the table. Look horizontally to the right to the column headed "−5."

Wind Chill Temperature: **−27°F**

The following formula can be used to approximate the wind chill temperature in degrees Celsius.

$$C = 33 - \frac{(10\sqrt{r} + 10.45 - r)(33 - t)}{22.1}$$

C: wind chill temperature in degrees Celsius

t: air temperature in degrees Celsius

r: wind speed in meters per second

EXAMPLE 3: On a fall day the air temperature is 5°C. The wind speed is 8 meters per second. Find the wind chill temperature.

SOLUTION: $C = 33 - \dfrac{(10\sqrt{r} + 10.45 - r)(33 - t)}{22.1}$

$C = 33 - \dfrac{(10\sqrt{8} + 10.45 - 8)(33 - 5)}{22.1}$ ◀ $r = 8$, $t = 5$

$= 33 - \dfrac{(10\sqrt{8} + 10.45 - 8)(28)}{22.1}$

8 [√] [×] 10 [+] 10.45 [−] 8 [×]

28 [÷] 22.1 [=] [+/−] [+] 33 [=]   -5.939347

The wind chill temperature is **about −6°C.**

EXERCISES

In Exercises 1–4, refer to the tables on page 202.

1. Find the temperature–humidity index above which about 50% of most people feel uncomfortable.
2. Find the temperature–humidity index when the temperature is 90°F and the relative humidity is 80%.
3. What wind speed produces a wind chill temperature of −53°F when the thermometer reading is −10°F?
4. Does the wind chill temperature increase or decrease when the wind speed increases and the thermometer reading stays the same?

Use the formula above to find the wind chill temperature to the nearest degree Celsius (°C).

5. Air temperature: −3°C
   Wind speed: 7 meters per second
6. Air temperature: −12°C
   Wind speed: 6 meters per second

CHAPTER SUMMARY

IMPORTANT TERMS

Radical symbol *(p. 174)*
Square root *(p. 174)*
Square *(p. 174)*
Prime number *(p. 174)*
Prime-factorization *(p. 175)*
Perfect square *(p. 175)*
Radical *(p. 180)*
Radicand *(p. 180)*
Cube root *(p. 180)*
Fourth root *(p. 180)*
Perfect cube *(p. 184)*
Like radicals *(p. 187)*
Rationalizing the denominator *(p. 191)*
Right Triangle *(p. 198)*
Hypotenuse *(p. 198)*
Legs *(p. 198)*

IMPORTANT IDEAS

1. A negative number cannot have a real number as a square root.
2. Each prime factor of a perfect square occurs an even number of times.
3. A square root of a perfect square counting number is rational.
4. Any whole number that is not a perfect square has an irrational square root.
5. If a is any nonnegative real number, then $\sqrt{a} \cdot \sqrt{a} = a$.
6. *Product Property of Radicals:* If a and b are nonnegative real numbers, then $\sqrt{a} \cdot \sqrt{b} = \sqrt{ab}$.
7. A square root radical is in simplest form if its radicand has no perfect square factor.
8. A cube root radical is in simplest form if its radicand has no perfect cube factor.
9. *Quotient Property of Radicals:* If a is any nonnegative real number and b is any positive real number, then $\sqrt{\frac{a}{b}} = \frac{\sqrt{a}}{\sqrt{b}}$.
10. *Rule of Pythagoras:* In a right triangle, the square of the length of the hypotenuse equals the sum of the squares of the lengths of the legs.

CHAPTER REVIEW

SECTION 7.1

Simplify.

1. $-\sqrt{49}$
2. $\sqrt{\frac{1}{16}}$
3. $\sqrt{0.36}$
4. $\sqrt{3^4 \cdot 7^2}$

SECTION 7.2

Multiply and simplify.

5. $\sqrt{23} \cdot \sqrt{23}$

6. $\sqrt{81} \cdot \sqrt{36}$

7. $(\sqrt{3})(-\sqrt{11})$

8. $(\sqrt{2})(\sqrt{7})$

9. $\left(\sqrt{\frac{1}{2}}\right)(\sqrt{72})$

10. $\left(\sqrt{\frac{1}{3}}\right)(\sqrt{27})$

SECTION 7.3

Simplify.

11. $\sqrt{2 \cdot 3 \cdot 2 \cdot 3 \cdot 3}$

12. $\sqrt{3 \cdot 5 \cdot 5 \cdot 2 \cdot 2 \cdot 2}$

13. $\sqrt{52}$

14. $\sqrt{60}$

15. $\sqrt{99}$

16. $\sqrt{63}$

17. $\sqrt[3]{24}$

18. $\sqrt[3]{56}$

SECTION 7.4

Simplify.

19. $\sqrt{r^6}$

20. $\sqrt{t^{10}}$

21. $\sqrt{9x^5}$

22. $\sqrt{16y^3}$

23. $\sqrt{28a^3b^5}$

24. $\sqrt{45r^4s^7}$

25. $\sqrt[3]{16m^3n^4}$

26. $\sqrt[3]{27p^5q^6}$

SECTION 7.5

Add or subtract as indicated. Simplify where necessary.

27. $8\sqrt{13} + 2\sqrt{13}$

28. $12\sqrt{3} - \sqrt{3}$

29. $2\sqrt{12} + 5\sqrt{3}$

30. $3\sqrt{8} + 5\sqrt{2}$

31. $\sqrt{24} - 9\sqrt{6}$

32. $12\sqrt{5} - \sqrt{45}$

SECTION 7.6

Simplify.

33. $\sqrt{\frac{3}{24}}$

34. $\sqrt{\frac{x}{81}}$

35. $\sqrt{\frac{20}{t^2}}$

36. $\sqrt{\frac{54}{y^4}}$

Rationalize each denominator. Then simplify.

37. $\frac{\sqrt{7}}{\sqrt{10}}$

38. $\frac{\sqrt{3}}{\sqrt{14}}$

39. $\frac{\sqrt{3x^3}}{\sqrt{2y}}$

40. $\frac{\sqrt{2a^5}}{\sqrt{5m}}$

SECTION 7.7

Approximate each square root to three decimal places. Use the table on page 194.

41. $\sqrt{153}$ **42.** $\sqrt{200}$ **43.** $\sqrt{\frac{11}{16}}$ **44.** $\sqrt{\frac{7}{25}}$ **45.** $\sqrt{\frac{17}{24}}$ **46.** $\sqrt{\frac{11}{20}}$

Solve each problem. Use the table on page 194. Round each answer to the nearest tenth.

47. A backyard patio is in the shape of a square. It has an area of 12 square meters. Find the length of each side of the patio in meters.

48. The Mellott family has a square-shaped kitchen floor. The floor has an area of 108 square feet. Find the length of each side of the kitchen floor in feet.

For Exercises 49–50, use the formula $D = 1.22\sqrt{A}$ where A is the altitude in feet and D is the distance in miles. Use the table on page 194. Round the answers to the nearest mile.

49. Find the approximate distance in miles to the horizon from a plane that is at an altitude of 4000 feet.

50. Find the approximate distance in miles to the horizon from a plane that is at an altitude of 9000 feet.

SECTION 7.8

Determine whether the given lengths are the lengths of the sides of a right triangle.

51. 6 meters, 8 meters, 10 meters

52. 5 inches, 6 inches, 8 inches

53. 8 feet, 9 feet, 12 feet

54. 60 yards, 11 yards, 61 yards

Solve each problem using the Rule of Pythagoras. Use the table on page 194. Round the length to the nearest tenth.

55. An appliance store is holding a sale on a TV with a 14-inch screen. The width of the screen is 10 inches. Find the height of the screen in inches.

56. A construction worker is making a rectangular form for a patio. The length of the form is 20 feet and the width is 14 feet. He measures the diagonal to see if the form makes a right angle. What should be the length of the diagonal?

CHAPTER 8

Polynomials

The photo below is the earth as seen from 400,000 kilometers in space. The formula $K = \frac{R^3}{T^2}$ says that for any planet in the solar system, the cube of its average distance from the sun divided by the square of its time for one revolution is a constant.

8.1 Monomials

The numerals and expressions below are monomials.

a. $\frac{1}{3}x$ **b.** 5 **c.** $3x^2$ **d.** $-x$

Definition

A ***monomial*** of one variable over {rational numbers} is an expression of the form ax^n in which a is any rational number and n is any nonnegative integer.

$-5x^3$

$\frac{1}{4}x^5$

$-3x^0$ or -3

P-1 **What are the values of *a* and *n* in each monomial below?**

a. $3.7x^2$ **b.** $-x$ **c.** -15

Since $-15 = -15x^0$, -15 is a monomial.

If ax^n is a monomial, n cannot be negative. The following expressions are <u>not</u> monomials.

a. $3x^{-2}$ **b.** $-\frac{1}{5}x^{-1}$ **c.** $0.7x^{-3}$

P-2 **Why is $\frac{3}{x^2}$ not a monomial?**

$\frac{3}{x^2}$ is equivalent to $3x^{-2}$.

In a monomial of the form ax^n, the number represented by a is the ***coefficient*** or ***numerical coefficient.***

P-3 **What is the coefficient of each monomial below?**

a. $\frac{1}{3}x^2$ **b.** x^3 **c.** $-x$ **d.** $-\frac{x}{5}$

Monomials are classified by their numerical coefficients. The monomials below are monomials ***over the set of integers.***

a. $2x^3$ **b.** -10 **c.** $5x$ **d.** $-x^5$

EXAMPLE Evaluate $-7x^3$ if x equals -2.

Solution:

$$-7x^3 = -7(-2)^3$$

Raise to a power first.

$$= -7(-2)(-2)(-2)$$

$$= (-7)(-8)$$

$$= 56$$

CLASSROOM EXERCISES

Write Yes or No to tell whether each numeral or expression is a monomial over {rational numbers}. (P-2)

1. $5x^3$ **2.** $5x^{-1}$ **3.** $\frac{3}{x}$ **4.** $\frac{x}{3}$ **5.** $\sqrt{4} \cdot x^3$ **6.** $-7x^0$

Write the coefficient of each monomial. (P-3)

7. $2x^2$ **8.** $-x^3$ **9.** $-7\frac{1}{2}$ **10.** $-2.8x^3$ **11.** $3\frac{2}{3}x^0$

WRITTEN EXERCISES

Goal: To evaluate monomials
Sample Problem: Evaluate $0.6x^3$ if $x = 20$. **Answer:** 4800

Write Yes or No to show whether each numeral or expression is a monomial over {rational numbers}. (P-2)

1. $13x$ **2.** $27x^2$ **3.** $-5x^3$ **4.** $-3x^5$ **5.** $2x^{-2}$

6. $-3x^{-5}$ **7.** $\frac{x^2}{10}$ **8.** $\frac{-x^3}{7}$ **9.** $\frac{3}{x^2}$ **10.** $\frac{-2}{x^2}$

11. $0.6x^0$ **12.** $5.7x^0$ **13.** $\sqrt{7} \cdot x^2$ **14.** $-\sqrt{3} \cdot x^3$ **15.** $x + 3$

Write the coefficient of each monomial. (P-3)

16. $2x^2$ **17.** $5x^4$ **18.** $\frac{1}{3}x^2$ **19.** $\frac{3}{5}x^3$ **20.** $-0.3x$ **21.** $1.8x^3$

22. x **23.** $-x^3$ **24.** $\frac{x}{10}$ **25.** $\frac{3}{5}$ **26.** $2x^0$ **27.** $\frac{3x}{5}$

Evaluate each monomial. (Example)

28. $-5x^2$ if $x = 3$ **29.** $3x^3$ if $x = -1$ **30.** $\frac{1}{4}x^4$ if $x = -2$

31. $\frac{3}{4}x^5$ if $x = 2$ **32.** $-0.5x^3$ if $x = 10$ **33.** $-8.3x^0$ if $x = -19$

REVIEW CAPSULE FOR SECTION 8.2

Write each product or quotient as one power. (Pages 144–150)

1. $m^2 \cdot m^4$ **2.** $t \cdot t^7$ **3.** $r^5 \cdot r^3 \cdot r^2$ **4.** $y^4 \cdot y^4 \cdot y^4$

5. $\frac{w^6}{w^2}$ **6.** $\frac{q^4}{q}$ **7.** $\frac{n^8}{n^4}$ **8.** $\frac{a^{10}}{a^5}$

8.2 Operations with Monomials

P–1 **Multiply.**

a. $x^3 \cdot x^4$ **b.** $y^2 \cdot y^9$ **c.** $r \cdot r^5$

Remember to add exponents.

EXAMPLE 1 Multiply: $(3x^2)(-2x^3)$

Solution:

1 Commutative and Associative Properties of Multiplication ⟶ $(3 \cdot -2)(x^2 \cdot x^3)$

2 Simplify. ⟶ $-6x^5$

Product Property of Powers

Monomials of more than one variable are shown below.

a. $2x^3y$ **b.** $-3r^2st$ **c.** $\frac{1}{2}p^3q^2$

EXAMPLE 2 Multiply: $(8st^2)(-7r^2st)$

Solution:

1 Commutative and Associative Properties of Multiplication ⟶ $(8 \cdot -7)(r^2)(s \cdot s)(t^2 \cdot t)$

2 Simplify. ⟶ $-56r^2s^2t^3$

P–2 **What is the simplest form of $4 \cdot -5 \cdot -\frac{1}{2}$?**

P–3 **What is the simplest form of $(rs)(s^2)(r)$?**

EXAMPLE 3 Multiply: $(4rs)(-5s^2)(-\frac{1}{2}r)$

Solution:

1 Commutative and Associative Properties of Multiplication ⟶ $(4 \cdot -5 \cdot -\frac{1}{2})(r \cdot r)(s \cdot s^2)$

2 Simplify. ⟶ $10r^2s^3$

Steps for multiplying two or more monomials

1. Multiply the numerical factors.
2. Add exponents of powers of the same base.

$(6x^2y)(-7xy^3) =$

$-42x^3y^4$

P–4 **Simplify each quotient.**

a. $\frac{x^5}{x^2}$ **b.** $\frac{y^4}{y}$ **c.** $\frac{r^6}{r^5}$

Remember to subtract exponents.

EXAMPLE 4 Divide: $\frac{-36n^8}{9n^5}$

Solution:

1 Product Rule of Fractions ⟶ $\frac{-36}{9} \cdot \frac{n^8}{n^5}$

2 Simplify. ⟶ $-4n^3$ ***Quotient Property of Powers***

EXAMPLE 5 Divide: $\frac{-7r^3s^2}{-4rs}$

Solution:

1 Product Rule of Fractions ⟶ $\frac{-7}{-4} \cdot \frac{r^3}{r} \cdot \frac{s^2}{s}$

2 Simplify. ⟶ $\frac{7}{4}r^2s$ or $1\frac{3}{4}r^2s$

Steps for dividing monomials

1. Divide the numerical factors.
2. Subtract exponents of powers of the same base.

$\frac{56a^4bc^3}{-7a^2b} =$

$-8a^2c^3$

CLASSROOM EXERCISES

Multiply. (Examples 1–3)

1. $(4y^2)(5y)$ **2.** $(-r^3)(2r^2)$ **3.** $(-6x^2)(7x^3)$

4. $(-8t^5)(-4t)$ **5.** $(16xy)(-\frac{1}{2}x^2y)$ **6.** $(2x)(3x)(5x)$

7. $(-x)(2x^2)(3y)$ **8.** $(r^2)(-4rs)(-s^3)$ **9.** $(2xy)(3x^2)(4xy^2)$

Divide. (Examples 4 and 5)

10. $\frac{12t^5}{-3t^2}$ 11. $\frac{-9x^4}{-x^2}$ 12. $\frac{-15x^3}{5x}$ 13. $\frac{14b^6}{-7b^2}$

14. $\frac{-63r^5s^4}{-9rs^3}$ 15. $\frac{-54s^5t^3}{6st^2}$ 16. $\frac{18r^2s^3t}{6rs}$ 17. $\frac{-48p^3qr^5}{6pr^4}$

WRITTEN EXERCISES

Goals: To multiply and to divide monomials

Examples: a. $(-12r^3s^2)(5r^2s)$ **b.** $\frac{-24x^3y}{-8xy}$ **Answers: a.** $-60r^5s^3$ **b.** $3x^2$

Multiply. (Examples 1–3)

1. $(8m^3)(9m^2)$
2. $(7a^4)(3a^3)$
3. $(-12r^3t^2)(8t^3)$
4. $(15xy^4)(-4x^2)$
5. $(-\frac{1}{3}pq^3)(-15p^3q^2)$
6. $(-20s^4t^2)(-\frac{1}{4}st^4)$
7. $(4t^2)(-3t)(2t)$
8. $(5n)(3n^2)(-3n^3)$
9. $(3mn)(-7n^2)(-m^2n)$
10. $(-4rt^2)(2rt)(-t^2)$
11. $(-3ab)(-3ab^2)(-3a^2b^3)$
12. $(-2q^2r)(-2qr^2)(-2qr^3)$
13. $(0.4rs^2t)(-3.8rst^3)$
14. $(-2.8a^2bc^3)(7ab^2c)$
15. $(-6xy^2z^3)(-1.4u^3x^2y)$

Divide. (Examples 4 and 5)

16. $\frac{-20y^5}{4y^2}$
17. $\frac{12k^4}{-3k}$
18. $\frac{-32r^2}{-8r}$
19. $\frac{-36y^3}{-9y}$
20. $\frac{9.6p^6q^3}{-6p^2q}$
21. $\frac{-4.8m^8n^5}{8m^2n^3}$
22. $\frac{-13pq^3r}{-qr}$
23. $\frac{-29a^4bc^3}{-a^3c}$
24. $\frac{-17r^5s^6t^9}{-5r^2s^2t^3}$
25. $\frac{22x^{12}y^8z^6}{-7x^4y^2z^2}$
26. $\frac{-5k^7p^{10}y^5}{12k^3py^4}$
27. $\frac{-4m^{12}n^7q^3}{-15m^3n^5q}$

NON-ROUTINE PROBLEM

28. A five-digit number contains the digits shown at the right. Use the clues below to find the number.

1 4 0 3 2

Clue 1: The number is less than 30,000.
Clue 2: The number is even.
Clue 3: The 0 is next to the 1.
Clue 4: The 3 is not next to the 2 or 0.

8.3 Polynomials

P–1 **What monomials are in the following expression?**

$$2x^2 + (-9x) + \frac{1}{4}$$

A polynomial

Definition

A ***polynomial*** is a monomial or the sum of two or more monomials.

$-1.8x^6$

$-5x^4 + 2x^2 - 3x + 1$

The monomials that form a polynomial are called ***terms***.

A ***binomial*** is a polynomial of two terms.

A ***trinomial*** is a polynomial of three terms.

"Bi" means "two."
"Tri" means "three."

P–2 **Which polynomials below are binomials? trinomials?**

a. $2x^2 + 6$ **b.** $x^2 - x + 1$

c. $\frac{1}{2} - \frac{1}{4}x^2 + \frac{3}{4}x$ **d.** $3.8x^2 - 1.3x$

The table below shows three polynomials expressed in simplest form.

Polynomials	Simplest Form
$3x^2 - 2x - x + 7$	$3x^2 - 3x + 7$
$5x - x^3 - 10 + 2x^2$	$-x^3 + 2x^2 + 5x - 10$
$1.3x^2 + 0.6 - x^4 + 5.6x^4 - x$	$4.6x^4 + 1.3x^2 - x + 0.6$

EXAMPLE 1 Simplify $2x^2 - 2x^3 + 3x - 5x - x^2 - 12 + x^4$.

Solution:

1 Combine like terms. $2x^2 - 2x^3 + 3x - 5x - x^2 - 12 + x^4 = x^2 - 2x^3 - 2x - 12 + x^4$

2 Arrange terms in order by exponents. ⟶ $= x^4 - 2x^3 + x^2 - 2x - 12$

Polynomials of more than one variable are shown below.

a. x^2y **b.** $3r^2st + 5rst^2$ **c.** $4x^3y^2 - 2x^2y + 5xy^3$

EXAMPLE 2 Simplify $1.3rs - 2.4r^2 + 3.8rs + 6.7s^2$.

Solution:

[1] Combine like terms. ⟶ $1.3rs - 2.4r^2 + 3.8rs + 6.7s^2 = 5.1rs - 2.4r^2 + 6.7s^2$

[2] Arrange in descending order of powers of r. ⟶ $= -2.4r^2 + 5.1rs + 6.7s^2$

OR

[2] Arrange in descending order of powers of s. ⟶ $= 6.7s^2 + 5.1rs - 2.4r^2$

Since monomials represent real numbers in this course, polynomials also represent real numbers.

EXAMPLE 3 Evaluate $2x^2 - 5x + 7$ if $x = -2$.

Solution:

$$\begin{aligned} 2x^2 - 5x + 7 &= 2(-2)^2 - 5(-2) + 7 \\ &= 2(4) + 10 + 7 \\ &= 8 + 10 + 7 \\ &= 25 \end{aligned}$$

CLASSROOM EXERCISES

Write Yes or No to tell whether each numeral or expression is a polynomial over {rational numbers}. (Definition)

1. $3x + 5$ **2.** $-5x^2$ **3.** x^3

4. $\frac{1}{2}$ **5.** $-\frac{2}{3}x^3 + 3x^2$ **6.** $\frac{x}{2}$

7. $\frac{2}{x}$ **8.** $5x^0$ **9.** $3x^{-1}$

10. $100x^{100}$ **11.** $\sqrt{16} \cdot x - 1$ **12.** $0.3x^2$

13. $3x^3 + \frac{1}{2}x^2$ **14.** $\sqrt{2} \cdot x^2 + x$ **15.** 0

Simplify. (Examples 1 and 2)

16. $x^2 - 3x - 7x + 1$ **17.** $3 + 5x^2 - 2x$

18. $x^2 - 3 + x - x^3 + 2x^4$ **19.** $12 - 2x^3 + 5x^3$

20. $8 - 3x + 7 - 2x^2$

21. $2x^2y + 3x^2y - 5$

22. $4x^2y^2 - 1 - 3x^2y^2 - 8$

23. $-3x^5 + 5x^8 - 3x^2 + x^{12}$

WRITTEN EXERCISES

Goals: To simplify and to evaluate polynomials
Sample Problems: a. $x^2 + y^2 + 5x^2$ **b.** Evaluate $-2x^3 + x$ if $x = -3$.
Answers: a. $6x^2 + y^2$ **b.** 51

Write Yes *or* No *to show whether each numeral or expression is a polynomial over {rational numbers}.* (Definition)

1. $2x - 5$

2. $4x + 3$

3. $x^2 + 3\pi$

4. $5x^2 - \sqrt{3}$

5. $\frac{x^2}{2} - \frac{3}{x} + 5$

6. $\frac{-x^2}{5} + \frac{x}{2} + 5$

7. $-3x^0$

8. $\frac{1}{3}x^5$

9. $1.3x^2 + 8.5x - 7.2$

10. $3x^{-2} + 4x^{-1} - 5^0$

Simplify. (Examples 1 and 2)

11. $3x^2 - x - 3x + 7$

12. $-2x^3 + x^2 - 4x^2 + 5$

13. $5x - 3x^2 + x^5 - 1 + 4x^3$

14. $12x^2 - 5x + x^3 - 2x^4 + 3$

15. $7x^2 - 9x + 8 - 3x^2 - x^2$

16. $15x^2 - x + 12 - 3x^2 - 6x$

17. $3x^3y - 5 + xy^3 + 5x^2y^2$

18. $6rs^3 - r^2s^2 + 3r^4s - 5s^4$

19. $m^2n + 5mn - 3m^2n + m^3 - 10$

20. $p^3q + 5q^3 + 3p^3q - pq^2$

21. $1.5x - 2.3x^4 + 0.9x - 1.9x^2 + 1.8x^4$

22. $0.3x^3 + 2.9x^5 - 5.2x + 6.3x - 1.8x^3$

Evaluate each polynomial. (Example 3)

23. $-x^2 + 3x - 5$ if $x = -3$

24. $3x^2 - 4x + 5$ if $x = -4$

25. $x^3 - x^2 + x - 1$ if $x = -2$

26. $-x^3 + x^2 - x + 1$ if $x = -2$

27. $2x^4 - x^3 + 3x + 2$ if $x = 2$

28. $x^5 - 2x^4 - x^2 + 5$ if $x = 1$

29. $x^2y^2 - 2xy + 1$ if $x = -2$ and $y = 3$

30. $r^3s - r^2s^2 + 2$ if $r = -1$ and $s = 2$

MORE CHALLENGING EXERCISES

Simplify.

31. $x^2 - (3x + 5)$

32. $2x^3 - (-x^2 + x)$

33. $(x^2 + 4x) + (12 - 3x)$

34. $(x^3 - 2x^2) + (5x - 4x^2)$

APPLICATIONS

Polynomials can easily be evaluated with a calculator by grouping terms with parentheses in a special way.

$$5n^3 - 4n^2 + 12n - 17 = ((5n - 4)n + 12)n - 17$$

EXAMPLE Evaluate $5x^2 - 16x + 3$ if $x = 42$.

SOLUTION [1] Group terms in a special way. ⟶ $5x^2 - 16x + 3 = (5x - 16)x + 3$

[2] Substitute and evaluate. ⟶ $= (5 \cdot 42 - 16)42 + 3$

[5] [×] [4] [2] [−] [1] [6]

[×] [4] [2] [+] [3] [=] 8151.

Note that you can:
1. work from left to right;
2. "ignore" the exponent.

Evaluate each polynomial for the given value of x.

35. $120x^2 - 75x + 29;\ x = 24$ **36.** $76x^2 + 128x - 97;\ x = 29$ **37.** $x^3 - 2x^2 + 3x - 4;\ x = 12$

MID-CHAPTER REVIEW

Evaluate each monomial. (Section 8.1)

1. $-4x^3$ if $x = -2$ **2.** $\frac{1}{2}x^2$ if $x = \frac{1}{2}$ **3.** $0.5x^4$ if $x = -2$ **4.** $-(x^5)$ if $x = -1$

Multiply or divide. (Section 8.2)

5. $(-3rt)(2r^2)$ **6.** $(-10m^2n)(mn)$ **7.** $\dfrac{-30s^2t}{6st}$ **8.** $\dfrac{-48m^3n^2}{-8mn}$

Simplify. (Section 8.3)

9. $-x + 3 - 2x^2 + 7x - x^2 + 9$

10. $-5x^3 - 12 + 3x - x^3 + 2x^2 - 8x$

11. $-1.9x^2 + 0.2x^3 - 1.3x^2 + 2.8x - 1.4$

12. $0.9x - 1.8x^4 - 1.3x + 4.3x^3 - 2.7x^4$

Evaluate for the given value of the variable. (Section 8.3)

13. $r^2 - 2r + 5;\ r = -3$ **14.** $-5x^3 - x^2 + 2x - 3;\ x = 2$ **15.** $-p^2 + 7p - 10;\ p = -10$

REVIEW CAPSULE FOR SECTION 8.4

Arrange the terms of each polynomial in descending order of powers. (Pages 213–215)

1. $2x - 3x^2$ **2.** $-4x - 3x^3$ **3.** $12 - 6x^2 + 4x$ **4.** $7 - 5x$ **5.** $3x - 5 + x^2$

8.4 Addition and Subtraction

EXAMPLE 1 Add: $(x^2 - 3x + 4) + (3x^2 + 5x - 7)$

Solution:

[1] Write the sum of all terms. ⟶ $(x^2 - 3x + 4) + (3x^2 + 5x - 7) = x^2 - 3x + 4 + 3x^2 + 5x - 7$

[2] Combine like terms. ⟶ $= 4x^2 + 2x - 3$ ◀ ***Simplest form***

P–1 **Add: a. $(x^2 - 8x) + (7x + 3)$ b. $(3x^2 - 5) + (4x - x^2)$**

Polynomials can be added by arranging them vertically in columns.

EXAMPLE 2 Add: $(3x^3 - x^2 + 4) + (3x^2 - 7x - 9) + (-x^3 + 3x - 6)$

Solution:

$$\begin{array}{rrrrr} & 3x^3 & -\ x^2 & & +\ 4 \\ & & 3x^2 & -\ 7x & -\ 9 \\ (+) & -x^3 & & +\ 3x & -\ 6 \\ \hline & 2x^3 & +\ 2x^2 & -\ 4x & -\ 11 \end{array}$$

◀ ***Leave space for the missing term, and write in the + or − between terms.***

The sum of any two polynomials is a polynomial.

EXAMPLE 3 Subtract: $(x^2 - 12x) - (3x^2 + 7x - 5)$

Solution:

[1] Meaning of subtraction ⟶ $(x^2 - 12x) - (3x^2 + 7x - 5) = (x^2 - 12x) + -(3x^2 + 7x - 5)$

[2] Property of the Opposite of a Sum ⟶ $= x^2 - 12x - 3x^2 - 7x + 5$

[3] Simplify. ⟶ $= -2x^2 - 19x + 5$

P–2 **Subtract: a. $(2x^2 + 6x) - (2x - 3)$ b. $(x^3 - 2x^2) - (-5x + 1)$**

P–3 **Subtract.**

a. $x^3 - (-3x^3)$ **b.** $0 - 2x^2$ **c.** $(-4x) - (-3x)$ **d.** $7 - (-1)$

In adding or subtracting polynomials vertically, you arrange the terms in descending order of powers.

EXAMPLE 4

Subtract: $(7 + x^3 - 4x) - (2x^2 - 1 - 3x^3 - 3x)$

Solution:

$$\begin{array}{rrrr} x^3 & & -4x & +7 \\ (-)\ -3x^3 & +2x^2 & -3x & -1 \\ \hline 4x^3 & -2x^2 & -x & +8 \end{array}$$

Leave space for the x^2-term.

Compare the monomials in the answer to Example 4 with your answers to P-3.

> The difference of any two polynomials is a polynomial.

In Example 5, the lengths of the sides of a polygon are represented by polynomials.

EXAMPLE 5

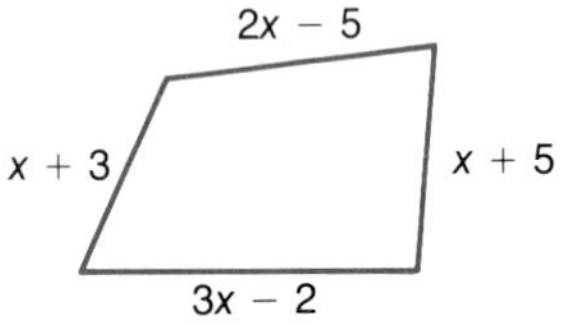

Find the perimeter of the polygon at the left. Write the perimeter as a polynomial in simplest form.

Solution: The perimeter of a polygon is the sum of the lengths of its sides.

1. Write the sum of the polynomials. ⟶ $(x + 3) + (2x - 5) + (x + 5) + (3x - 2)$
2. Combine like terms. ⟶ $(x + 2x + x + 3x) + (3 - 5 + 5 - 2)$
3. Simplify. ⟶ $7x + 1$

The polynomial $7x + 1$ represents the perimeter.

CLASSROOM EXERCISES

Add. (Example 2)

1. $\begin{array}{r} 4x - 9 \\ \underline{3x + 2} \end{array}$

2. $\begin{array}{r} -x + 15 \\ \underline{5x - 7} \end{array}$

3. $\begin{array}{rrr} 2x^2 & -x & +3 \\ -x^2 & & -4 \\ \hline -3x^2 & -4x & \end{array}$

Subtract. (Example 4)

4. $\begin{array}{r} 6x + 12 \\ (-)\ \underline{5x + 3} \end{array}$

5. $\begin{array}{r} 3x^2 + 6x + 4 \\ (-)\ \underline{x^2 + 2x + 3} \end{array}$

6. $\begin{array}{rrr} 4x^2 & & -1 \\ (-)\ 6x^2 & -x & +5 \\ \hline \end{array}$

Add or subtract as indicated. (Examples 1 and 3)

7. $(3x^2 - 2x) + (5x - 3)$

8. $(-7x^2 + 8x) - (x + 5)$

9. $(2x^2 - 3x) - (3x^2 + x)$

WRITTEN EXERCISES

Goals: To add and subtract polynomials

Sample Problem: $(-4x^2 + 7) + (9x - 12) - (x - x^2)$ **Answer:** $-3x^2 + 8x - 5$

Add. (Example 2)

1. $$\begin{array}{r} 14x^2 - 9x + 4 \\ \underline{-3x^2 + 7x - 12} \end{array}$$

2. $$\begin{array}{r} -10x^2 + 11x - 6 \\ \underline{7x^2 - 9x + 13} \end{array}$$

3. $$\begin{array}{r} 3.2x^2 + 0.5x - 1.8 \\ \underline{-1.7x^2 + 2.4x + 2.6} \end{array}$$

4. $$\begin{array}{r} -4.1x^2 + 0.9x - 3.3 \\ \underline{2.7x^2 - 2.4x - 1.6} \end{array}$$

5. $$\begin{array}{r} x^3 \qquad\quad - 6x + 10 \\ -2x^3 + x^2 + 3x - 7 \\ \underline{-x^3 - 5x^2 \qquad\quad + 6} \end{array}$$

6. $$\begin{array}{r} -23x^3 + 14x^2 \qquad\quad - 11 \\ x^3 - 9x^2 \qquad\quad + 17 \\ \underline{-8x^3 \qquad\quad + 15x - 2} \end{array}$$

Subtract. (Example 4)

7. $$\begin{array}{rr} & x^2 + 5x - 10 \\ (-) & \underline{3x^2 - x + 8} \end{array}$$

8. $$\begin{array}{rr} & -4x^2 + x - 9 \\ (-) & \underline{x^2 - 3x + 8} \end{array}$$

9. $$\begin{array}{rr} & 2x^3 \qquad\quad + 6x - 3 \\ (-) & \underline{-x^3 + 5x^2 + 9x - 5} \end{array}$$

10. $$\begin{array}{rr} & 3x^3 + 8x^2 \qquad\quad - 5 \\ (-) & \underline{4x^3 \qquad\quad + 12x - 13} \end{array}$$

MIXED PRACTICE *Add or subtract as indicated.*

11. $(5x^2 - 3x + 12) + (3 - x^2 - 4x)$

12. $(8 - 6x - 4x^2) + (5x^2 - x - 13)$

13. $(x^3 - 2x + 5) - (4x^3 - 5x^2 + x - 2)$

14. $(8x^3 - 2x^2 + 3x - 4) - (x^3 + 5x^2 - x + 7)$

15. $(4x^3 - 5x^2 - 6) - (2x^2 + 3x + 10) - (-3x^3 + x - 7)$

16. $(12x^2 - x - 10) - (x^3 + 4x^2 + 8) - (-3x^3 + 5x + 7)$

17. $(1.3x^2 - 2.7x - 0.5) + (2.5x^2 + 4.3x + 2.3) - (1.7x^2 - 2.8x - 4.9)$

18. $(0.9x^2 + 4.1x - 2.8) - (1.4x^2 - 2.7x + 5.1) + (-3.2x^2 + 4.5x - 1.8)$

APPLICATIONS *Write a polynomial to represent each perimeter.* (Example 5)

19.

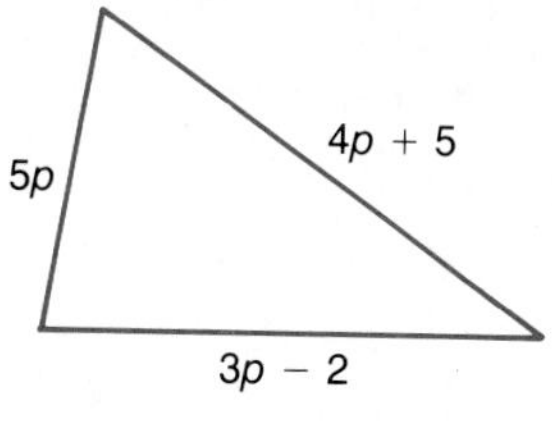

20.

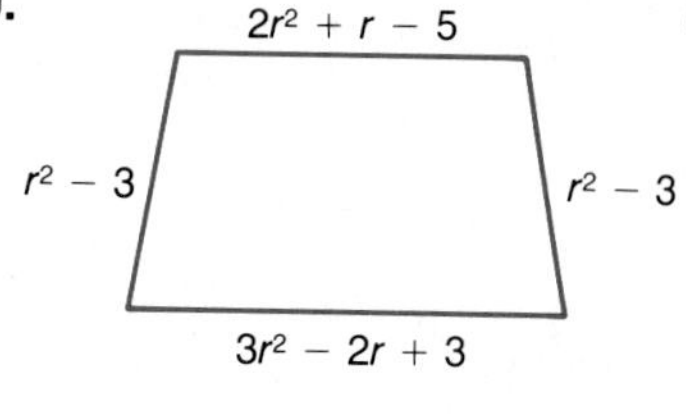

21.

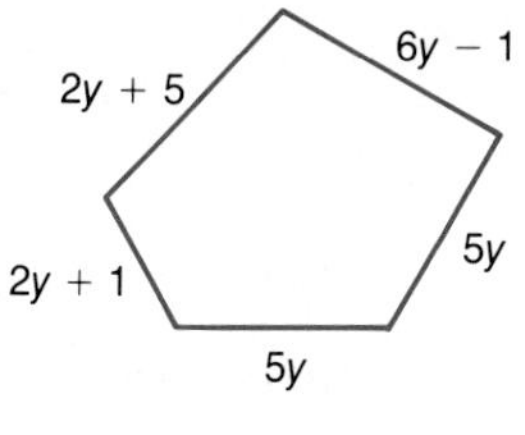

8.5 Multiplication

P–1 **Find each product.**

a. $(-2x)(x)$ **b.** $(3x)(-5)$ **c.** $(-3)(-4)$

Products of monomials

EXAMPLE 1

Multiply: $(x + 3)(3x - 4)$

Two binomials

Solution:

1. Distributive Property → $(x + 3)(3x - 4) = (x + 3)3x + (x + 3)(-4)$
2. Distributive Property → $= x(3x) + 3(3x) + x(-4) + 3(-4)$
3. Simplest forms of products → $= 3x^2 + 9x - 4x - 12$
4. Combine like terms. → $= 3x^2 + 5x - 12$

Simplest form

In a polynomial, a term with no variables is called the ***constant term.***

P–2 **What is the constant term in each polynomial below?**

a. $x^2 + 5$ **b.** $3x^2 + 5x - 12$ **c.** $x^3 - 7x + 9$

Binomials can be multiplied by a short method.

EXAMPLE 2

Multiply: $(5x - 2)(x - 3)$

Multiply the first terms. → **Multiply the outside terms.** → **Multiply the inside terms.** → **Multiply the last terms.**

Solution:

$(5x - 2)(x - 3)$	$(5x - 2)(x - 3)$	$(5x - 2)(x - 3)$	$(5x - 2)(x - 3)$
$5x^2$	$-15x$	$-2x$	6

$$(5x - 2)(x - 3) = 5x^2 - 15x - 2x + 6$$
$$= 5x^2 - 17x + 6$$

P–3 **Find each product.**

a. $(2x + 1)(x + 4)$ **b.** $(3x - 2)(2x + 3)$

c. $(4x - 1)(3x - 5)$ **d.** $(2x + 5)(x - 4)$

Try to combine like terms mentally.

Any two polynomials are multiplied by multiplying each term of one polynomial by each term of the other.

EXAMPLE 3 Multiply: $(3x + 4)(2x^2 - x + 3)$

Multiply each term of $2x^2 - x + 3$ by $3x$. → **Multiply each term of $2x^2 - x + 3$ by 4.** → **Add like terms.**

Solution: **Method 1**

$$(3x + 4)(2x^2 - x + 3) = 3x(2x^2) + 3x(-x) + 3x(3) + 4(2x^2) + 4(-x) + 4(3)$$
$$= 6x^3 - 3x^2 + 9x + 8x^2 - 4x + 12$$
$$= 6x^3 + 5x^2 + 5x + 12$$

Method 2

$$\begin{array}{r} 2x^2 - x + 3 \\ 3x + 4 \\ \hline 6x^3 - 3x^2 + 9x \\ 8x^2 - 4x + 12 \\ \hline 6x^3 + 5x^2 + 5x + 12 \end{array}$$

Multiply by 3x. *Multiply by 4.*

EXAMPLE 4 Multiply: $(x - 3)(5 + 3x^3 - x^2)$

Solution:

$$\begin{array}{rrrrr} & 3x^3 & - x^2 & & + 5 \\ & & & x & - 3 \\ \hline 3x^4 & - x^3 & & + 5x & \\ & - 9x^3 & + 3x^2 & & - 15 \\ \hline 3x^4 & - 10x^3 & + 3x^2 & + 5x & - 15 \end{array}$$

Arrange terms in order. Leave space for x term.

Columns of like terms

The product of any two polynomials is a polynomial.

EXAMPLE 5

Write a polynomial to represent the area of the tennis court.

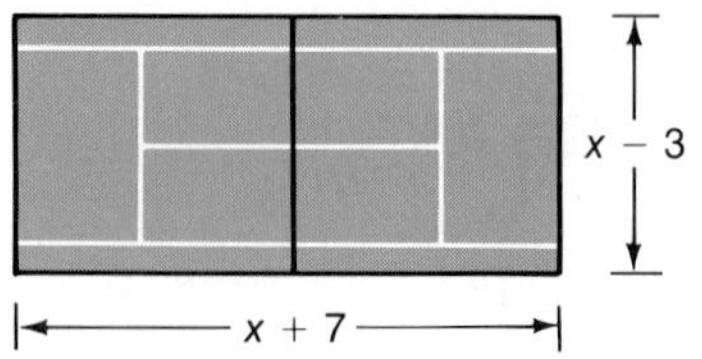

Solution: The area of a rectangle equals the product of its length and width.

[1] Represent the length and the width. ⟶ Length $= x + 7$; width $= x - 3$

[2] Multiply. ⟶ $(x + 7)(x - 3) = x^2 - 3x + 7x - 21$

$= x^2 + 4x - 21$ ◀ **Area**

CLASSROOM EXERCISES

Multiply. (Example 2)

1. $(x + 1)(x + 2)$
2. $(x + 3)(x + 4)$
3. $(x - 1)(x + 3)$
4. $(x + 2)(x - 3)$
5. $(x - 2)(x - 4)$
6. $(x - 3)(x - 7)$
7. $(2x + 1)(x + 2)$
8. $(3x - 1)(x + 2)$
9. $(4x + 1)(x - 3)$
10. $(2x - 1)(2x - 1)$

Write the missing row in each multiplication problem. (Examples 3–4)

11.
$$\begin{array}{r} 2x + 5 \\ \underline{x - 6} \\ \boxed{} \\ \underline{-12x - 30} \\ 2x^2 - 7x - 30 \end{array}$$

12.
$$\begin{array}{r} -3x - 4 \\ \underline{2x + 1} \\ -6x^2 - 8x \\ \underline{\boxed{}} \\ -6x^2 - 11x - 4 \end{array}$$

13.
$$\begin{array}{r} x^2 - 2x + 3 \\ \underline{x - 2} \\ x^3 - 2x^2 + 3x \\ \underline{-2x^2 + 4x - 6} \\ \boxed{} \end{array}$$

14.
$$\begin{array}{r} x^3 - 2x^2 \qquad + 5 \\ \underline{2x - 1} \\ 2x^4 - 4x^3 \qquad + 10x \\ \underline{-x^3 + 2x^2 \qquad - 5} \\ \boxed{} \end{array}$$

WRITTEN EXERCISES

Goal: To multiply a polynomial by a polynomial

Sample Problem: $(5x^2 - 2x + 3)(x + 2)$ **Answer:** $5x^3 + 8x^2 - x + 6$

Multiply. (Examples 1 and 2)

1. $(x + 3)(x + 5)$
2. $(x + 2)(x + 7)$
3. $(x - 4)(x + 2)$
4. $(x - 3)(x + 1)$

5. $(x-5)(x-3)$
6. $(x-2)(x-4)$
7. $(x+8)(x-5)$
8. $(x+10)(x-6)$
9. $(2x+3)(x+5)$
10. $(3x+2)(x+4)$
11. $(3x-1)(x+4)$
12. $(4x-3)(x+5)$
13. $(5x-2)(x-3)$
14. $(3x-4)(x-4)$
15. $(4x+5)(2x-3)$
16. $(2x+5)(3x-7)$

Multiply. (Examples 3 and 4)

17. $\begin{array}{r} x^2-2x+3 \\ \underline{x-4} \end{array}$

18. $\begin{array}{r} x^2+3x-2 \\ \underline{x-3} \end{array}$

19. $\begin{array}{r} 2x^2+3x-4 \\ \underline{x+5} \end{array}$

20. $\begin{array}{r} 3x^2-x+4 \\ \underline{x+4} \end{array}$

21. $\begin{array}{r} x^2 \qquad -5 \\ \underline{x+6} \end{array}$

22. $\begin{array}{r} x^2-4x \\ \underline{x-5} \end{array}$

23. $\begin{array}{r} 2x^2-3x+7 \\ \underline{3x-2} \end{array}$

24. $\begin{array}{r} 3x^2-2x-4 \\ \underline{2x+3} \end{array}$

25. $\begin{array}{r} -5x^2+2x-4 \\ \underline{4x+5} \end{array}$

26. $\begin{array}{r} 5x^3+4x^2 \qquad -6 \\ \underline{-2x+5} \end{array}$

27. $\begin{array}{r} 7x^3+6x^2 \qquad +5 \\ \underline{6x+3} \end{array}$

28. $\begin{array}{r} 2x^3 \qquad -x+2 \\ \underline{4x-5} \end{array}$

APPLICATIONS

Write a polynomial to represent each area. (Example 5)

29. A rectangular parking lot has a length of $(x-4)$ units and a width of $(x-7)$ units.

30. The length of a rectangular garden is $(2x+3)$ units. The width of the garden is $(x+4)$ units.

31. A rectangular swimming pool has a length of $(3x-1)$ units and a width of $(x+5)$ units.

32. Each side of a square piece of carpet has a length of $(x+6)$ units.

REVIEW CAPSULE FOR SECTION 8.6

Subtract. (Pages 217–219)

1. $\begin{array}{r} 4x^2-3x+2 \\ \underline{4x^2+\ \ x\ \ \ \ \ } \end{array}$

2. $\begin{array}{r} 5x^2 \qquad -4 \\ \underline{5x^2+2x\ \ \ \ } \end{array}$

3. $\begin{array}{r} -2x^2+4x-3 \\ \underline{-2x^2-4x\ \ \ \ } \end{array}$

4. $\begin{array}{r} x^4-3x^2+6 \\ \underline{x^4-5x^2-2} \end{array}$

8.6 Division

To divide a polynomial by a monomial, divide each term of the polynomial by the monomial.

P–1 **Divide:** **a.** $\frac{12x^3}{2x}$ **b.** $\frac{8x^2}{2x}$ **c.** $\frac{4x}{2x}$

EXAMPLE 1

Divide: $(12x^3 + 8x^2 - 4x) \div 2x$

2x is the divisor. $12x^3 + 8x^2 - 4x$ is the dividend.

Solution:

$$(12x^3 + 8x^2 - 4x) \div 2x = \frac{12x^3}{2x} + \frac{8x^2}{2x} - \frac{4x}{2x}$$

$$= 6x^2 + 4x - 2$$

Examples 2 and 3 show division of a trinomial by a binomial.

EXAMPLE 2

Divide: $(2x^2 + x - 15) \div (x + 3)$

Solution:

[1] Divide $2x^2$, the first term of $2x^2 + x - 15$, by x, the first term of $x + 3$. ⟶

$$\begin{array}{r} 2x \phantom{{}+x-15} \\ x+3 \overline{)\,2x^2 + x - 15} \end{array}$$

[2] Multiply $(x + 3)$ by $2x$. Subtract the product from $2x^2 + x - 15$. ⟶

$$\begin{array}{rr} & 2x \phantom{{}+x-15} \\ & x+3 \overline{)\,2x^2 + x - 15} \\ 2x(x+3) \longrightarrow & \underline{2x^2 + 6x} \phantom{{}-15} \\ & -5x - 15 \end{array}$$

[3] Divide $-5x$ by x. This is the second term of the quotient. ⟶

$$\begin{array}{r} 2x - 5 \phantom{{}-15} \\ x+3 \overline{)\,2x^2 + x - 15} \\ \underline{2x^2 + 6x} \phantom{{}-15} \\ -5x - 15 \end{array}$$

[4] Multiply $(x + 3)$ by -5. Subtract the product from $-5x - 15$. ⟶

$$\begin{array}{rr} & 2x - 5 \phantom{{}-15} \\ & x+3 \overline{)\,2x^2 + x - 15} \\ & \underline{2x^2 + 6x} \phantom{{}-15} \\ & -5x - 15 \\ -5(x+3) \longrightarrow & \underline{-5x - 15} \\ & 0 \end{array}$$

The quotient is $2x - 5$.

P–2 **Divide:** **a.** $\frac{4x^3}{2x}$ **b.** $\frac{-2x^2}{2x}$

EXAMPLE 3

Divide: $(4x^3 - 3x - 1) \div (2x + 1)$

Solution:

1 Divide $4x^3$ by $2x$ to get the first term of the quotient. ⟶

$$\begin{array}{r l} & 2x^2 \qquad\qquad\quad \\ 2x+1 \,\big)\!\!\!\! & \overline{4x^3 \qquad\quad - 3x - 1} \end{array}$$

Leave space for an x^2 term.

2 Multiply $(2x + 1)$ by $2x^2$. Subtract the product from $4x^3 - 3x - 1$. ⟶

$$\begin{array}{r l} & 2x^2 \qquad\qquad\quad \\ 2x+1 \,\big)\!\!\!\! & \overline{4x^3 \qquad\quad - 3x - 1} \\ 2x^2(2x+1) \longrightarrow & \underline{4x^3 + 2x^2 \qquad\quad} \\ & \quad -2x^2 - 3x - 1 \end{array}$$

3 Divide $-2x^2$ by $2x$ for the second term of the quotient. ⟶

$$\begin{array}{r l} & 2x^2 - x \qquad\quad \\ 2x+1 \,\big)\!\!\!\! & \overline{4x^3 \qquad\quad - 3x - 1} \\ & \underline{4x^3 + 2x^2 \qquad\quad} \\ & \quad -2x^2 - 3x - 1 \end{array}$$

4 Multiply $(2x + 1)$ by $-x$. Subtract the product from $-2x^2 - 3x - 1$. ⟶

$$\begin{array}{r l} & 2x^2 - x \qquad\quad \\ 2x+1 \,\big)\!\!\!\! & \overline{4x^3 \qquad\quad - 3x - 1} \\ & \underline{4x^3 + 2x^2 \qquad\quad} \\ & \quad -2x^2 - 3x - 1 \\ -x(2x+1) \longrightarrow & \quad \underline{-2x^2 - \;\; x \qquad} \\ & \qquad\qquad -2x - 1 \end{array}$$

5 Divide $-2x$ by $2x$ for the last term of the quotient. Multiply $(2x + 1)$ by -1 and subtract. ⟶

$$\begin{array}{r l} & 2x^2 - x - 1 \\ 2x+1 \,\big)\!\!\!\! & \overline{4x^3 \qquad\quad - 3x - 1} \\ & \underline{4x^3 + 2x^2 \qquad\quad} \\ & \quad -2x^2 - 3x - 1 \\ & \quad \underline{-2x^2 - \;\; x \qquad} \\ & \qquad\qquad -2x - 1 \\ -1(2x+1) \longrightarrow & \qquad\qquad \underline{-2x - 1} \\ & \qquad\qquad\qquad\quad 0 \end{array}$$

The quotient is $2x^2 - x - 1$.

CLASSROOM EXERCISES

Divide. (Example 1)

1. $\frac{25x^4}{-5x^2}$ **2.** $\frac{-20x^5}{4x}$ **3.** $(15x^4 - 25x^2) \div 5x^2$ **4.** $(9x^3 + 6x^2) \div 3x^2$

Complete each division problem. (Examples 2–3)

5. $$\begin{array}{r} x \qquad\qquad \\ x-2\overline{)x^2+3x-10} \\ \underline{x^2-2x \qquad\;} \end{array}$$

6. $$\begin{array}{r} x \qquad\qquad \\ x+5\overline{)x^2-7x-60} \\ \underline{x^2+5x \qquad\;} \end{array}$$

7. $$\begin{array}{r} 3x \qquad\qquad \\ 2x-3\overline{)6x^2-5x-6} \\ \underline{6x^2-9x \qquad} \end{array}$$

8. $$\begin{array}{r} 2x \qquad\qquad\quad \\ 5x-3\overline{)10x^2-31x+15} \\ \underline{10x^2-\;\;6x \qquad\;\;} \end{array}$$

WRITTEN EXERCISES

Goal: To divide a polynomial by a binomial
Sample Problem: $(2x^3 + 3x^2 - 1) \div (2x - 1)$ **Answer:** $x^2 + 2x + 1$

Divide. (Example 1)

1. $(4x^2 + 6x) \div 2x$
2. $(9y^3 - 12y) \div 3y$
3. $(18xy - 12x^2y^2) \div 6xy$
4. $(36x^3 + 12x) \div 12x$

Divide. (Examples 2 and 3)

5. $x - 3\overline{)3x^2 - 5x - 12}$
6. $x + 2\overline{)4x^2 + 5x - 6}$
7. $2x - 1\overline{)10x^2 - 17x + 6}$
8. $3x - 1\overline{)12x^2 + 2x - 2}$
9. $4x + 1\overline{)12x^2 - 13x - 4}$
10. $2x + 3\overline{)8x^2 + 10x - 3}$
11. $x - 5\overline{)x^3 - 8x^2 + 17x - 10}$
12. $x + 4\overline{)x^3 + 8x^2 + 14x - 8}$
13. $2x + 3\overline{)4x^3 \qquad\quad - 7x + 3}$
14. $3x + 1\overline{)9x^3 + 6x^2 + 4x + 1}$
15. $x^2 - 2\overline{)x^3 - 5x^2 - 2x + 10}$
16. $x^2 + 3\overline{)2x^3 - 5x^2 + 6x - 15}$
17. $x^2 - 2x\overline{)x^4 - 2x^3 + 3x^2 - 6x}$
18. $x^2 - 5\overline{)x^4 \qquad\quad - 3x^2 \qquad\quad - 10}$
19. $(10x^3 - 29x^2 + 60x - 20) \div (5x - 2)$
20. $(18x^3 + 9x^2 - 62x + 35) \div (6x - 5)$
21. $(x^3 - 8) \div (x - 2)$
22. $(x^3 + 27) \div (x + 3)$
23. $(x^4 - 81) \div (x - 3)$
24. $(x^4 - 16) \div (x - 2)$

NON-ROUTINE PROBLEM

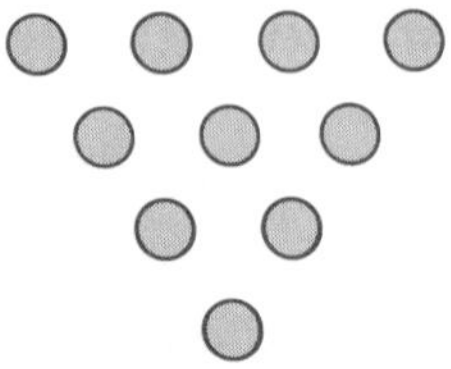

25. In a bowling alley, bowling pins are set up as shown in the figure at the right. Move exactly three of the pins to form a triangle exactly like the original triangle, but facing the opposite direction.

8.7 Division with Remainders

P–1 **Divide:** $\frac{3x^2}{x}$.

EXAMPLE 1 Divide: $x - 4\overline{)3x^2 - 10x - 6}$

Solution:

1. Divide $3x^2$ by x. Multiply $(x - 4)$ by $3x$ and subtract. ⟶

$$\begin{array}{rl} & \quad 3x \\ x - 4 & \overline{)3x^2 - 10x - 6} \\ 3x(x-4) \longrightarrow & \underline{3x^2 - 12x} \\ & \qquad\quad 2x - 6 \end{array}$$

2. Divide $2x$ by x. Multiply $(x - 4)$ by 2 and subtract. ⟶

$$\begin{array}{rl} & \quad 3x + 2 \\ x - 4 & \overline{)3x^2 - 10x - 6} \\ & \underline{3x^2 - 12x} \\ & \qquad\quad 2x - 6 \\ 2(x-4) \longrightarrow & \qquad\quad \underline{2x - 8} \\ & \qquad\qquad\quad 2 \end{array}$$

The remainder is 2. The quotient can be written as $3x + 2 + \frac{2}{x - 4}$.

Since $\frac{2}{x - 4}$ is not a polynomial, the quotient in Example 1 is not a polynomial.

> The quotient of two polynomials is not necessarily a polynomial.

P–2 **Arrange $10x^2 - 12x - 8x^3 - 19$ in descending powers of x.**

EXAMPLE 2 Divide: $(10x^2 - 12x - 8x^3 - 19) \div (2x^2 + 3)$

Solution:

1. Arrange terms in order. ⟶ $2x^2 + 3\overline{)-8x^3 + 10x^2 - 12x - 19}$

2. Divide $-8x^3$ by $2x^2$. Multiply $(2x^2 + 3)$ by $-4x$ and subtract. ⟶

$$\begin{array}{rl} & \quad -4x \\ 2x^2 + 3 & \overline{)-8x^3 + 10x^2 - 12x - 19} \\ -4x(2x^2+3) \longrightarrow & \underline{-8x^3 \qquad\qquad - 12x \qquad} \\ & \qquad\quad 10x^2 \qquad\quad - 19 \end{array}$$

3 Divide $10x^2$ by $2x^2$. Multiply $(2x^2 + 3)$ by 5 and subtract. ⟶

$$\begin{array}{r}
-4x + 5 \\
2x^2 + 3\overline{)-8x^3 + 10x^2 - 12x - 19} \\
\underline{-8x^3 \qquad\quad - 12x \qquad} \\
10x^2 \qquad\quad - 19 \\
5(2x^2+3) \longrightarrow \underline{10x^2 \qquad\quad + 15} \\
-34
\end{array}$$

The quotient is written as $-4x + 5 - \dfrac{34}{2x^2 + 3}$.

Not a polynomial

P–3 **Find the first term in the quotient.**

$$4x - 3y\overline{)8x^3 - 10x^2y - xy^2 + 2y^3}$$

EXAMPLE 3

Divide: $4x - 3y\overline{)8x^3 - 10x^2y - xy^2 + 2y^3}$

Solution:

1 Divide $8x^3$ by $4x$. Multiply $(4x - 3y)$ by $2x^2$ and subtract. ⟶

$$\begin{array}{r}
2x^2 \qquad\qquad\qquad\qquad \\
4x - 3y\overline{)8x^3 - 10x^2y - xy^2 + 2y^3} \\
2x^2(4x-3y) \longrightarrow \underline{8x^3 - \;\; 6x^2y \qquad\qquad\qquad} \\
-4x^2y - xy^2 + 2y^3
\end{array}$$

2 Divide $-4x^2y$ by $4x$. Multiply $(4x - 3y)$ by $-xy$ and subtract. ⟶

$$\begin{array}{r}
2x^2 - xy \qquad\qquad\quad \\
4x - 3y\overline{)8x^3 - 10x^2y - xy^2 + 2y^3} \\
\underline{8x^3 - \;\; 6x^2y \qquad\qquad\qquad} \\
-4x^2y - xy^2 + 2y^3 \\
-xy(4x-3y) \longrightarrow \underline{-4x^2y + 3xy^2 \qquad\quad} \\
-4xy^2 + 2y^3
\end{array}$$

3 Divide $-4xy^2$ by $4x$. Multiply $(4x - 3y)$ by $-y^2$ and subtract. ⟶

$$\begin{array}{r}
2x^2 - xy - y^2 \qquad\qquad \\
4x - 3y\overline{)8x^3 - 10x^2y - xy^2 + 2y^3} \\
\underline{8x^3 - \;\; 6x^2y \qquad\qquad\qquad} \\
-4x^2y - \;\; xy^2 + 2y^3 \\
\underline{-4x^2y + 3xy^2 \qquad\quad} \\
-4xy^2 + 2y^3 \\
-y^2(4x-3y) \longrightarrow \underline{-4xy^2 + 3y^3} \\
-y^3
\end{array}$$

The quotient is $2x^2 - xy - y^2 - \dfrac{y^3}{4x - 3y}$.

CLASSROOM EXERCISES

Write the quotient and remainder in each problem. (Examples 1 and 2)

1. $x + 2\overline{)x^2 - 3x - 7}$

$\quad x^2 + 2x$

$\quad\quad -5x - 7$

$\quad\quad \underline{-5x - 10}$

2. $x - 3\overline{)x^2 + x + 5}$

$\quad x^2 - 3x$

$\quad\quad 4x + 5$

$\quad\quad \underline{4x - 12}$

3. $3x - 1\overline{)6x^2 + 4x - 3}$

$\quad 6x^2 - 2x$

$\quad\quad 6x - 3$

$\quad\quad \underline{6x - 2}$

4. $2x + 1\overline{)4x^2 - 6x + 1}$

$\quad 4x^2 + 2x$

$\quad\quad -8x + 1$

$\quad\quad \underline{-8x - 4}$

5. $x^2 - 1\overline{)x^3 + x^2 - 3x + 1}$

$\quad x^3 \quad\quad - x$

$\quad\quad x^2 - 2x + 1$

$\quad\quad \underline{x^2 \quad\quad - 1}$

6. $x - 1\overline{)x^3 \quad\quad\quad - 6}$

$\quad x^3 - x^2$

$\quad\quad x^2 \quad - 6$

$\quad\quad x^2 - x$

$\quad\quad\quad x - 6$

$\quad\quad\quad \underline{x - 1}$

Write the polynomial missing from each row. (Example 2)

7. $\quad\quad 5x + 3$

$2x - 3\overline{)10x^2 - 9x - 11}$

$\quad \square$

$\quad\quad 6x - 11$

$\quad\quad \square$

$\quad\quad\quad -2$

8. $\quad\quad 3x - 5$

$4x + 1\overline{)12x^2 - 17x - 1}$

$\quad \square$

$\quad\quad -20x - 1$

$\quad\quad \square$

$\quad\quad\quad 4$

9. $\quad\quad x + 2$

$x^2 - 3\overline{)x^3 + 2x^2 - x + 5}$

$\quad \square$

$\quad\quad 2x^2 + 2x + 5$

$\quad\quad \square$

$\quad\quad\quad \square$

10. $\quad\quad 4x - 3$

$x^2 - 2\overline{)4x^3 - 3x^2 - 7x + 2}$

$\quad \square$

$\quad\quad -3x^2 + x + 2$

$\quad\quad \square$

$\quad\quad\quad \square$

11. $\quad\quad x^2 + x - 3$

$x^2 + 2\overline{)x^4 + x^3 - x^2 + 2x - 6}$

$\quad x^4 \quad + 2x^2$

$\quad \square$

$\quad x^3 \quad\quad + 2x$

$\quad\quad -3x^2 \quad - 6$

$\quad\quad \square$

$\quad\quad\quad 0$

12. $\quad\quad 3x^2 - x - 2$

$x^2 + 4\overline{)3x^4 - x^3 + 10x^2 - 4x - 8}$

$\quad 3x^4 \quad + 12x^2$

$\quad \square$

$\quad - x^3 \quad\quad - 4x$

$\quad\quad - 2x^2 \quad - 8$

$\quad\quad \square$

$\quad\quad\quad 0$

WRITTEN EXERCISES

Goal: To divide a polynomial by a polynomial

Sample Problem: $x + 2\overline{)x^3 + 6x^2 + 11x + 10}$ **Answer:** $x^2 + 4x + 3 + \frac{4}{x + 2}$

Divide. (Examples 1–3)

1. $x + 4\overline{)2x^2 + 3x - 17}$
2. $x - 3\overline{)4x^2 - 11x - 1}$
3. $2x - 3\overline{)4x^2 + 8x - 25}$
4. $3x + 2\overline{)12x^2 - x - 10}$
5. $5x - 6\overline{)15x^2 + 2x - 29}$
6. $6x - 1\overline{)18x^2 + 27x - 11}$
7. $2x + 5\overline{)10x^2 + 13x - 40}$
8. $4x + 1\overline{)12x^2 - 33x - 15}$
9. $2x^2 - 1\overline{)12x^3 - 10x^2 - 5x + 5}$
10. $3x^2 + 5\overline{)-12x^3 + 9x^2 - 21x + 15}$
11. $x^2 + x - 1\overline{)x^4 - 2x^3 + 2x^2 + x + 2}$
12. $2x^2 - x + 1\overline{)2x^4 - x^3 - 3x^2 + x - 1}$
13. $x^2 + 8x + 2\overline{)x^3 + 3x^2 - 38x - 16}$
14. $x^2 - x - 1\overline{)2x^4 - 2x^3 - 5x^2 + 3x + 7}$
15. $2x^2 + 9x + 4\overline{)2x^4 + 13x^3 + 16x^2 - 19x + 12}$
16. $a^2 - 4a + 4\overline{)a^3 - 6a^2 + 13a - 9}$
17. $x + y\overline{)x^3 + x^2y - xy^2 - 2y^3}$
18. $x - y\overline{)x^3 + x^2y - xy^2 - 2y^3}$

MORE CHALLENGING EXERCISES

19. $x + 2\overline{)x^3 - 8}$
20. $x - 3\overline{)x^3 + 27}$
21. $2x - 3y\overline{)2x^3 - 7x^2y + 8xy^2 - 2y^3}$
22. $3y - x\overline{)-9y^3 + 9y^2x - 5yx^2 - x^3}$
23. $2x - 1\overline{)8x^3 - 1}$
24. $5y + 4\overline{)125y^3 + 64}$
25. $a^2 - 2\overline{)a^3 + 6a^2 - 2a - 12}$
26. $3x + 2y\overline{)3x^3 + 8x^2y + 7xy^2 + 5y^3}$
27. $x + 5\overline{)x^3 + 4x - 3}$
28. $x + 2\overline{)x^3 - 6x^2 + 1}$
29. $2a - 1\overline{)5a + 6a^2 - 4}$
30. $x + 5\overline{)4 - 6x^2 + 3x}$
31. $a - 3\overline{)4a^3 - 24a - 3a^2 - 9}$
32. $2a + 3\overline{)2a^3 + 18 - 3a - 7a^2}$
33. $x + y\overline{)x^3 + y^3}$
34. $x - y\overline{)x^3 - y^3}$
35. $x^2 + xy + y^2\overline{)x^3 - y^3}$
36. $a^2 - ab + b^2\overline{)a^3 + b^3}$
37. $x - 1\overline{)x^4 - 1}$
38. $x^2 + x + 1\overline{)x^4 + x^2 + 2}$

Using Formulas

Compound Interest

Compound interest paid by a bank on deposits is computed on the total of the **principal** (amount deposited) plus the **interest** previously earned. The interest is usually paid more than once a year. This formula is used to compute compound interest.

$$A = p(1 + r)^n$$

A = amount of the new balance
r = rate of interest per period
p = principal
n = number of interest periods per year

EXAMPLE: The First City Bank pays 5% interest compounded quarterly (four times a year) on regular savings accounts. Anthony deposited $400 in a regular savings account. What will be the amount of his new balance after one year?

SOLUTION: $A = p(1 + r)^n$

$p = \$400,\ n = 4,\ r = 0.0125$ ◀ ***r = 0.05 ÷ 4, or 0.0125***

$A = (400)(1 + 0.0125)^4$, or $400(1.0125)^4$

1 . 0 1 2 5 ×

= = = × 4 0 0 = **420.37808** ◀ ***Round to the nearest cent.***

Anthony will have **$420.38** in his account after one year.

EXERCISES

For Exercises 1–3, find the new balance after one year.

1. $200; 6% interest Compounded quarterly

2. $500; 7% interest Compounded twice a year

3. $600; 6% interest Compounded monthly

4. Juan has $900 in a savings account. The bank pays 6% interest compounded twice a year. What will be the new balance after one year?

5. Cynthia has $300 in a savings account. The bank pays 5% interest compounded quarterly. What will be the new balance after one year?

CHAPTER SUMMARY

IMPORTANT TERMS

Monomial *(p. 208)*	Trinomial *(p. 213)*
Coefficient *(p. 208)*	Constant term *(p. 220)*
Polynomial *(p. 213)*	Divisor *(p. 224)*
Term *(p. 213)*	Dividend *(p. 224)*
Binomial *(p. 213)*	

IMPORTANT IDEAS

1. Monomials or polynomials are said to be over {integers} if all the coefficients are integers.
2. The sum of any two polynomials is a polynomial.
3. In adding or subtracting polynomials vertically, you arrange the terms in descending order of powers.
4. The difference of any two polynomials is a polynomial.
5. Any two polynomials are multiplied by multiplying each term of one polynomial by each term of the other.
6. The product of any two polynomials is a polynomial.
7. To divide a polynomial by a monomial, divide each term of the polynomial by the monomial.
8. The quotient of two polynomials is not necessarily a polynomial.

CHAPTER REVIEW

SECTION 8.1

Write Yes or No to show whether each numeral or expression is a monomial over {rational numbers}.

1. $\frac{1}{3}x^2$ **2.** $\frac{3}{x^2}$ **3.** $-5x^{-3}$ **4.** $\sqrt{3} \cdot x$

Evaluate each monomial.

5. $4x^3$ if $x = -2$ **6.** $-5x^2$ if $x = -3$

7. $-0.8x^4$ if $x = 2$ **8.** $1.2x^3$ if $x = -4$

SECTION 8.2

Multiply.

9. $(3t^2)(7t^4)$ **10.** $(-5xy^3)(13x^2y)$

11. $(\frac{1}{3}ab)(-6ab^2)(-a^2b^2)$ **12.** $(1.4mn^2)(7m^2n)(-2n^2)$

Divide.

13. $\dfrac{-108t^5}{9t^3}$

14. $\dfrac{1.44r^3s^6t^2}{-12rs^5t}$

15. $\dfrac{-3x^{12}y^7z^9}{-18x^3y^5z^7}$

16. $\dfrac{23a^4bc^2}{a^3c}$

17. $\dfrac{-13x^4y^5}{-4x^2y}$

18. $\dfrac{16a^5c^6}{2a^4c^5}$

SECTION 8.3

Write Yes or No to show whether each numeral or expression is a polynomial over {rational numbers}.

19. $\dfrac{2}{x^2} - \dfrac{1}{x} + \dfrac{1}{2}$

20. $x^3 - 10$

21. $3\sqrt{x} - \sqrt{3}$

22. $x^{-3} + 2x^{-2} - 5x^{-1} - 7$

Simplify.

23. $7x^2 - 2x + 5x - 16$

24. $4x^3 - 2x^2 - x^2 + 5x$

25. $12x - x^2 + 5 - 3x^3$

26. $3x^2 - 5 + 4x - 2x^3 + x^4$

SECTION 8.4

Add.

27. $\begin{array}{r} -3x^2 + 2x - 5 \\ \underline{7x^2 - 9x - 12} \end{array}$

28. $\begin{array}{r} 4.3x^2 - 1.7x + 2.5 \\ \underline{-3.9x^2 + 4.8x - 8.3} \end{array}$

29. $(5x^3 - 3x + 7) + (2x^2 - 8x - 12)$

30. $(-2x^3 + x^2 - 5) + (5x^3 - x^2 + 7x - 12)$

Subtract.

31. $\begin{array}{r} x^2 - 3x + 4 \\ \underline{-5x^2 + 4x + 9} \end{array}$

32. $\begin{array}{r} 2x^3 - 3x - 5 \\ \underline{6x^3 - x^2 + 5x - 6} \end{array}$

33. $(x^4 - 3x^2 + 7) - (5x^3 - x^2 + 2x - 5)$

34. $(3x^3 - x^2 - 4x) - (-x^3 + 4x^2 - x - 10)$

Write a polynomial to represent the perimeter of each polygon.

35.

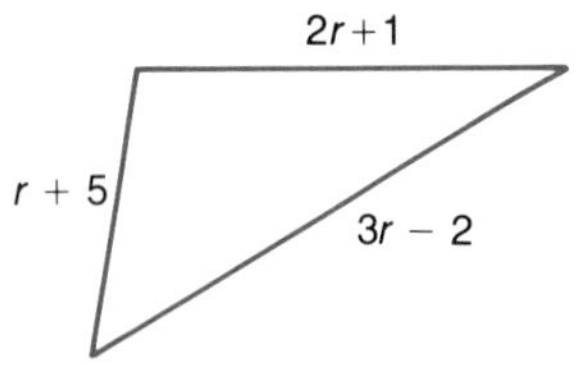

36.

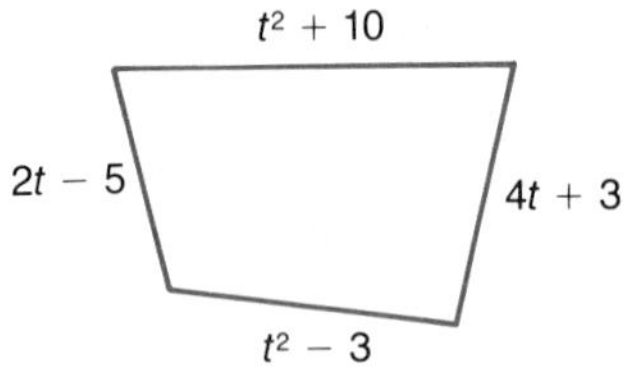

37.

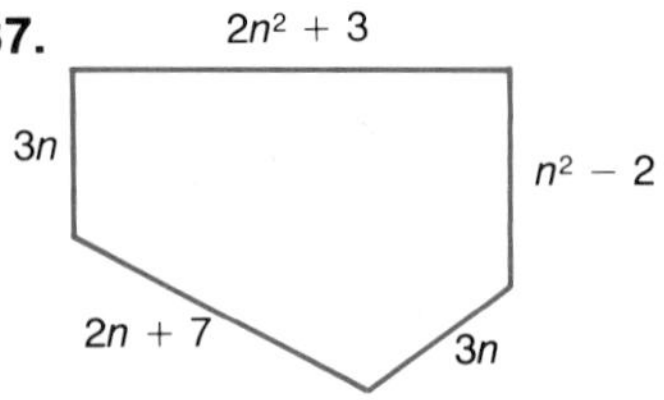

SECTION 8.5

Multiply.

38. $(2x + 7)(x - 3)$

39. $(3x - 8)(x + 4)$

40. $(5x - 2)(4x + 5)$

41. $(6x - 5)(2x - 9)$

42.
$$\begin{array}{r} 3x^3 - x^2 \qquad\qquad + 5 \\ 2x - 7 \\ \hline \end{array}$$

43.
$$\begin{array}{r} -4x^3 \qquad\qquad -5x + 3 \\ 4x - 3 \\ \hline \end{array}$$

Write a polynomial to represent each area.

44. The top of a rectangular table has a length of $(x + 3)$ units and a width of $(x - 5)$ units.

45. The length of a soccer field is $(3x - 1)$ units and the width is $(2x + 8)$ units.

SECTION 8.6

Divide.

46. $(10x^3 + 6x^2 - 8x) \div 2x$

47. $(24x^2y^2 - 16xy) \div 8xy$

48. $2x - 5\overline{)2x^2 + 9x - 35}$

49. $3x - 4\overline{)15x^2 + 4x - 32}$

50. $x^2 - 3\overline{)x^4 - 3x^3 + x^2 + 9x - 12}$

51. $x^2 + 2\overline{)-x^4 + x^3 - 7x^2 + 2x - 10}$

SECTION 8.7

Divide.

52. $4x + 11\overline{)-8x^2 - 2x + 60}$

53. $6x - 4\overline{)-18x^2 + 36x - 35}$

54. $x^2 + 3\overline{)x^4 - x^3 - 3x - 11}$

55. $2x^2 - 5\overline{)2x^4 + 3x^2 - 26}$

Factoring Polynomials

CHAPTER 9

This complicated formula is used to predict the weight of an average adult after a given number of days of dieting.

$$W = \frac{1}{a}\left[N + (aW_0 - N)\left(\frac{3500 - a}{3500}\right)^t\right]$$

Milk Group

Meat Group

Fruit and Vegetable Group

Bread and Cereal Group

9.1 Common Monomial Factors

P–1 **Express $4x^2 - 6x$ as a product.**

Here are some ways to factor $4x^2 - 6x$.

"Factor" means "write as a product of factors."

$4x^2 - 6x = 2(2x^2 - 3x)$ $\qquad$ $4x^2 - 6x = \frac{1}{2}x(8x - 12)$

$4x^2 - 6x = 4x(x - 1.5)$ $\qquad$ $4x^2 - 6x = x^{-1}(4x^3 - 6x^2)$

$$4x^2 - 6x = 2x(2x - 3)$$

The factors $2x$ and $(2x - 3)$ in the last factorization are special.

1. They are both polynomials.
2. Their coefficients are integers.
3. They are prime.

The coefficients are 2, 2, and −3.

There are only two ways to write $7x^2 - 2$ as a product of two or more polynomials that have integers as their coefficients.

$7x^2 - 2 = 1(7x^2 - 2)$

$7x^2 - 2 = -1(-7x^2 + 2)$

Either 1 times the polynomial or −1 times the opposite of the polynomial.

For this reason polynomials such as $7x^2 - 2$ are called **prime polynomials.**

EXAMPLE 1 $\quad$ Factor $6x^2 - 9x$.

Always find the prime polynomial factors.

Solution:

[1] Find the greatest number that is a common factor of 6 and 9. ⟶ 3

$6 = 3 \cdot 2$
$9 = 3 \cdot 3$

[2] Find the greatest power of x that is common to x^2 and x. ⟶ x

$x^2 = x \cdot x$
$x = x$

[3] Write the common monomial factor. ⟶ $3x(\quad)$

[4] Divide each term of $6x^2 - 9x$ by $3x$. ⟶ $3x(2x - 3)$

Check your answer by multiplying the factors.

$$\begin{array}{r} 2x - 3 \\ 3x \\ \hline 6x^2 - 9x \end{array}$$

Definition

The ***greatest common factor*** of two or more counting numbers is the greatest number that is a factor of all the numbers.	The greatest common factor of 6 and 9 is 3.

EXAMPLE 2 Factor $8x^3 + 12x^2$.

Solution:

1. Write the greatest common factor of 8 and 12. ⟶ 4
2. Write the greatest power of x common to x^3 and x^2. ⟶ x^2
3. Write the common monomial factor. ⟶ $4x^2(\quad)$
4. Divide each term of $8x^3 + 12x^2$ by $4x^2$. ⟶ $4x^2(2x + 3)$

P–2 **What is the common monomial factor of $8x^3 + 12x^2$?**

The prime factors of $8x^3 + 12x^2$ are $2 \cdot 2 \cdot x \cdot x(2x + 3)$. The simplest factored form is $4x^2(2x + 3)$. Expressing $8x^3 + 12x^2$ as $4x^2(2x + 3)$ is an example of ***common monomial factoring.***

EXAMPLE 3 Factor $10x^4 - 15x^3 + 25x^2$.

Solution:

1. Find the greatest common factor of 10, 15, and 25. ⟶ 5
2. Find the greatest power of x common to x^4, x^3, and x^2. ⟶ x^2
3. Write the common monomial factor. ⟶ $5x^2(\quad)$
4. Divide each term of $10x^4 - 15x^3 + 25x^2$ by $5x^2$. ⟶ $5x^2(2x^2 - 3x + 5)$

Check:

$$\begin{array}{r} 2x^2 - 3x + 5 \\ 5x^2 \\ \hline 10x^4 - 15x^3 + 25x^2 \end{array}$$

◀ ***Multiply the factors.***

In factoring, you should always check two conditions.

1. Any polynomial factors other than monomial factors should be prime.
2. The product of the factors must equal the given polynomial.

CLASSROOM EXERCISES

Write Yes or No to tell whether each polynomial is prime.

1. $x + 2$
2. $x - 5$
3. $2x + 4$
4. $5x + 10$
5. $x^2 + x$
6. $x^2 + 2$
7. $3x - 2$
8. $x^2 + 3x$

Factor. (Examples 1 and 2)

9. $4x + 2$
10. $3x - 3$
11. $5x + 15$
12. $x^2 - 3x$
13. $2x^2 + 10$
14. $3x^2 + 9x$
15. $-3x + 6$
16. $-x^2 + 3x$
17. $5x^3 - 10x$
18. $12x^2 + 18x$
19. $24 - 8x^2$
20. $14x^5 - 21x^2$

WRITTEN EXERCISES

Goal: To factor a polynomial by common monomial factoring
Sample Problem: Factor $9x^4 - 54x^3 + 18x^2$. **Answer:** $9x^2(x^2 - 6x + 2)$

Factor. (Examples 1–3)

1. $4x - 4$
2. $3x + 3$
3. $5x + 10$
4. $6x - 18$
5. $2x^2 + 3x$
6. $4x^2 - 5x$
7. $-3x^2 + 9x$
8. $-4x^2 + 20x$
9. $8x^3 - 12x^2$
10. $10x^3 + 25x^2$
11. $12x^3 - 18x$
12. $18x^3 - 27x$
13. $x^3 - 2x^2 + 3x$
14. $3x^3 + x^2 - 5x$
15. $2x^4 - 4x^2 - 10x$
16. $6x^4 + 8x^3 - 26x$
17. $-5x^2 - 10x$
18. $-4x^2 - 16x$
19. $8x^5 - 12x^3 + 20x^2$
20. $18x^4 + 12x^3 - 30x^2$
21. $5x^4 - 35x^2 - 10$
22. $3x^5 + 12x^3 - 9$
23. $21x^6 + 14x^5 - 10$
24. $18x^7 - 27x^4 + 9x^3$

REVIEW CAPSULE FOR SECTION 9.2

Write each trinomial in the form $x^2 + rx + sx + rs$.

EXAMPLE: $x^2 + 7x + 10$ **SOLUTION:** $x^2 + 5x + 2x + 5 \cdot 2$

1. $x^2 + 4x + 4$
2. $x^2 + 6x + 8$
3. $x^2 + 10x + 21$
4. $x^2 + 5x + 6$
5. $x^2 + 12x + 11$
6. $x^2 + 9x + 20$

9.2 Factoring Trinomials

A monomial such as $2x^5$ is said to be of the fifth degree.

> The ***degree of a monomial*** in one variable is determined by the exponent of the variable. — $-8x^3$ is of the third degree.

P–1 **What is the degree of each of these monomials?**

a. $-3x^2$ **b.** $5x$ **c.** -16 **d.** $16x^8$

Definition

> The ***degree of a polynomial*** is the highest degree of the monomials of the polynomial. — $5x^2 - 2x + 6$ is of second degree.

Second degree polynomials are called ***quadratic polynomials.***

$5x^2 - 2x + 6$ is a quadratic trinomial.

P–2 **What is the degree of each of these polynomials?**

a. $x - 5$ **b.** $x^3 - 2x^2 + 3x - 7$ **c.** $3x^2 - 2x + 2x^4 - 7$

EXAMPLE 1

Factor $x^2 + 2x + 3x + 6$.

Remember always to write the prime factors.

Solution:

1. Write as the sum of two binomials. ⟶ $x^2 + 2x + 3x + 6 = (x^2 + 2x) + (3x + 6)$
2. Factor each binomial. ⟶ $= x(x + 2) + 3(x + 2)$
3. Distributive Property ⟶ $= (x + 3)(x + 2)$

EXAMPLE 2

Factor $x^2 + rx + sx + rs$. (r and s represent integers.)

Solution:

1. Write as the sum of two binomials. ⟶ $x^2 + rx + sx + rs = (x^2 + rx) + (sx + rs)$
2. Factor each binomial. ⟶ $= x(x + r) + s(x + r)$
3. Distributive Property ⟶ $= (x + s)(x + r)$

P–3 **Factor each of the following.**

a. $x(x+3)+7(x+3)$ **b.** $x(x+10)+2(x+10)$

P–4 **What number in $x^2+5x+2x+10$ corresponds to r in $x^2+rx+sx+rs$? to s? to rs?**

EXAMPLE 3

Factor $x^2+5x+2x+10$. ◀ $x^2+rx+sx+rs$

Solution:

$$x^2+5x+2x+10=(x^2+5x)+(2x+10)$$
$$=x(x+5)+2(x+5)$$
$$=(x+2)(x+5)$$

P–5 **What two integers have a sum of 13 and a product of 36?**

EXAMPLE 4

Write $x^2+13x+36$ in the form $x^2+rx+sx+rs$. Then factor.

Solution:

$$x^2+13x+36=x^2+4x+9x+36$$ ◀ $r=4$, $s=9$
$$=(x^2+4x)+(9x+36)$$
$$=x(x+4)+9(x+4)$$
$$=(x+9)(x+4)$$

CLASSROOM EXERCISES

Write the degree of each polynomial. (P-2)

1. x^2-5
2. $3x^3-5x+6$
3. $x-3x^0$
4. $5+2x-x^2$
5. $x-x^5+x^3-x^2$
6. $-3x+7x^5-5x^{10}$

Factor. (P-3)

7. $x(x+3)+4(x+3)$
8. $x(x+2)+7(x+2)$
9. $x(x+5)+3(x+5)$
10. $x(x+2)+8(x+2)$
11. $x(x+4)+5(x+4)$
12. $x(x+7)+3(x+7)$
13. $x(x+8)+4(x+8)$
14. $x(x+9)+5(x+9)$

WRITTEN EXERCISES

Goal: To write two binomial factors of a quadratic trinomial
Sample Problem: Factor $x^2 + 16x + 63$.
Answer: $(x + 9)(x + 7)$

Factor. (Examples 1–3)

1. $x^2 + 3x + 4x + 12$
2. $x^2 + 2x + 7x + 14$
3. $x^2 + 5x + 3x + 15$
4. $x^2 + 2x + 8x + 16$
5. $x^2 + 4x + 5x + 20$
6. $x^2 + 7x + 3x + 21$
7. $x^2 + 8x + 4x + 32$
8. $x^2 + 9x + 5x + 45$
9. $x^2 + 3x + 11x + 33$
10. $x^2 + 6x + 8x + 48$

Write each trinomial in the form $x^2 + rx + sx + rs$. Then factor. (Example 4)

EXAMPLE: $x^2 + 18x + 77$ **SOLUTION:**
$$\begin{aligned} x^2 + 18x + 77 &= x^2 + (7x + 11x) + 77 \\ &= (x^2 + 7x) + (11x + 77) \\ &= x(x + 7) + 11(x + 7) \\ &= (x + 11)(x + 7) \end{aligned}$$

11. $x^2 + 7x + 10$
12. $x^2 + 9x + 14$
13. $x^2 + 5x + 6$
14. $x^2 + 8x + 15$
15. $x^2 + 10x + 21$
16. $x^2 + 13x + 22$
17. $x^2 + 15x + 26$
18. $x^2 + 14x + 33$
19. $x^2 + 13x + 30$
20. $x^2 + 11x + 18$
21. $x^2 + 11x + 28$
22. $x^2 + 11x + 10$

MORE CHALLENGING EXERCISES

Find the value of k that will make the second polynomial a factor of the first.

23. $x^2 + 7x + k;\ x + 4$
24. $x^2 + kx + 30;\ x + 5$
25. $x^2 + 11x + k;\ x + 9$

REVIEW CAPSULE FOR SECTION 9.3

Factor. (Pages 47–49)

1. $x(x + 5) - 2(x + 5)$
2. $x(x - 3) + 2(x - 3)$
3. $x(x - 6) - 4(x - 6)$
4. $x(x - 10) - 2(x - 10)$
5. $x(x - 5) - 5(x - 5)$
6. $x(x - 9) - 9(x - 9)$

9.3 Factoring $x^2 + bx + c$

The simplest polynomial form of $x^2 + rx + sx + rs$ is shown below.

$$x^2 + (r + s)x + rs$$

P–1 **In the polynomial $x^2 + 10x + 16$, what integer corresponds to $r + s$? What integer corresponds to rs?**

P–2 **What represents the value of b when $x^2 + (r + s)x + rs$ is compared with $x^2 + bx + c$?**

P–3 **What represents the constant term c?**

> To factor a trinomial of the form $x^2 + bx + c$, find two integers r and s such that
>
> $r + s = b$ and $r \cdot s = c$.

EXAMPLE 1 Factor $x^2 + 10x + 16$.

Solution: $r + s = 10$ $rs = 16$ ◀ ***Find two numbers whose sum is 10 and whose product is 16.***

1 Write all pairs of integers with product 16.

(1)(16)	(−1)(−16)
(2)(8)	(−2)(−8)
(4)(4)	(−4)(−4)

2 Select the pair with a sum of 10. ⟶ $\begin{cases} r = 8 \\ s = 2 \end{cases}$

3 Write in the form $x^2 + rx + sx + rs$. ⟶ $x^2 + 10x + 16 = x^2 + 8x + 2x + 16$

4 Write as the sum of two binomials. ⟶ $= (x^2 + 8x) + (2x + 16)$

5 Factor each binomial. ⟶ $= x(x + 8) + 2(x + 8)$

6 Distributive Property ⟶ $= (x + 2)(x + 8)$

P–4 **What integer in $x^2 - 7x + 12$ corresponds to $r + s$? to rs?**

EXAMPLE 2 Factor $x^2 - 7x + 12$.

Solution:

1. Write all pairs of integers with product 12.

(1)(12)	(−1)(−12)
(2)(6)	(−2)(−6)
(3)(4)	(−3)(−4)

2. Select the pair with a sum of −7. ⟶ $\begin{cases} r = -3 \\ s = -4 \end{cases}$

3. Write in the form $x^2 + rx + sx + rs$. ⟶ $x^2 - 7x + 12 = x^2 - 3x - 4x + 12$

4. Write as a sum of two binomials. ⟶ $= (x^2 - 3x) + (-4x + 12)$

5. Factor each binomial. ⟶ $= x(x - 3) - 4(x - 3)$

6. Distributive Property ⟶ $= (x - 4)(x - 3)$

Use $-4(x - 3)$, not $4(-x + 3)$.

P–5 **What integer in $x^2 - x - 20$ corresponds to $r + s$? to rs?**

EXAMPLE 3 Factor $x^2 - x - 20$.

Solution:

1. Write all pairs of integers with product −20.

(1)(−20)	(−1)(20)
(2)(−10)	(−2)(10)
(4)(−5)	(−4)(5)

2. Select the pair with a sum of −1. ⟶ $\begin{cases} r = 4 \\ s = -5 \end{cases}$

3. $x^2 - x - 20 = x^2 + 4x - 5x - 20$

4. $= (x^2 + 4x) + (-5x - 20)$

5. $= x(x + 4) - 5(x + 4)$

6. $= (x - 5)(x + 4)$

P–6 **What integer in $x^2 + 5x + 10$ corresponds to $r + s$? to rs?**

EXAMPLE 4 Factor $x^2 + 5x + 10$.

Solution:

[1] Write all pairs of integers with product 10.

(1)(10) (−1)(−10)
(2)(5) (−2)(−5)

Pairs of integers with product 10

[2] Select the pair with a sum of 5.

None of the pairs of integers meet this condition.

Since no two integers meet both conditions, $x^2 + 5x + 10$ is a prime polynomial.

CLASSROOM EXERCISES

Name two integers, if they exist, that have a product and a sum as given. (Steps 1 and 2, Examples 1 and 2)

1. The product is 6 and the sum is 5.
2. The product is 12 and the sum is 8.
3. The product is 18 and the sum is 9.
4. The product is 2 and the sum is 3.
5. The product is 3 and the sum is 2.
6. The product is 8 and the sum is −2.
7. The product is −8 and the sum is 2.
8. The product is −8 and the sum is −2.
9. The product is 8 and the sum is 6.
10. The product is 8 and the sum is −6.
11. The product is 10 and the sum is 3.
12. The product is 12 and the sum is 6.

Express each trinomial in the form $x^2 + rx + sx + rs$. (Examples 1–3)

13. $x^2 + 7x + 12$
14. $x^2 + 6x + 9$
15. $x^2 + 8x + 15$
16. $x^2 + 11x + 10$
17. $x^2 - 5x + 6$
18. $x^2 - 10x + 16$
19. $x^2 + 4x - 5$
20. $x^2 - 7x - 8$
21. $x^2 + 5x - 6$
22. $x^2 - 5x - 6$
23. $x^2 + 6x - 7$
24. $x^2 - 6x - 7$

WRITTEN EXERCISES

Goal: To factor quadratic trinomials of the form $x^2 + bx + c$
Sample Problems: a. $x^2 + 11x + 30$ **b.** $x^2 - 8x - 20$
Answers: a. $(x + 6)(x + 5)$ **b.** $(x - 10)(x + 2)$

For each trinomial, write the values of $r + s$ and $r \cdot s$. (P-1, P-4, P-5, P-6)

1. $x^2 + 7x + 12$ **2.** $x^2 + 14x + 15$ **3.** $x^2 + 14x + 33$
4. $x^2 + 9x + 20$ **5.** $x^2 + 4x - 32$ **6.** $x^2 + x - 30$
7. $x^2 - 8x - 48$ **8.** $x^2 - 10x - 64$ **9.** $x^2 - 17x + 72$
10. $x^2 + 5x - 36$ **11.** $x^2 - 3x + 2$ **12.** $x^2 + 7x - 30$

Write each trinomial in the form $x^2 + rx + sx + rs$. (Examples 1–3)

13. $x^2 + 13x + 12$ **14.** $x^2 + 16x + 15$ **15.** $x^2 + 11x + 30$
16. $x^2 + 13x + 42$ **17.** $x^2 - 8x + 15$ **18.** $x^2 - 9x + 14$
19. $x^2 - 11x + 24$ **20.** $x^2 - 14x + 45$ **21.** $x^2 - 3x - 40$
22. $x^2 - 2x - 35$ **23.** $x^2 + 4x - 21$ **24.** $x^2 + 4x - 32$
25. $x^2 - 9x - 36$ **26.** $x^2 - 13x + 36$ **27.** $x^2 - 7x + 6$
28. $x^2 - 7x - 8$ **29.** $x^2 - 11x - 42$ **30.** $x^2 - 3x + 2$

MIXED PRACTICE

Factor. When a polynomial cannot be factored, write Prime. (Examples 1–4)

31. $x^2 + 2x + 1$ **32.** $x^2 + 4x + 4$ **33.** $x^2 + 12x + 35$
34. $x^2 + 10x + 21$ **35.** $x^2 - 7x + 12$ **36.** $x^2 - 9x + 20$
37. $x^2 - 11x + 30$ **38.** $x^2 - 10x + 24$ **39.** $x^2 - x - 12$
40. $x^2 - x - 20$ **41.** $x^2 - x - 30$ **42.** $x^2 - 2x - 24$
43. $x^2 + x - 30$ **44.** $x^2 + x - 24$ **45.** $x^2 - 9x + 14$
46. $x^2 - 11x + 24$ **47.** $x^2 + 5x - 14$ **48.** $x^2 + 5x - 24$
49. $x^2 - 3x + 36$ **50.** $x^2 - 20x + 64$ **51.** $x^2 - 2x - 15$
52. $x^2 - 10x - 24$ **53.** $x^2 + 19x + 34$ **54.** $x^2 - 11x + 28$
55. $x^2 - 22x - 48$ **56.** $x^2 - 26x + 48$ **57.** $x^2 + 3x + 2$
58. $x^2 + 11x + 16$ **59.** $x^2 - x - 56$ **60.** $x^2 - 10x - 56$

9.4 Factoring $ax^2 + bx + c$

In the last section you factored quadratic trinomials. Each of these trinomials had a leading coefficient of 1.

Definition

The coefficient of the term of highest degree in a polynomial is called the ***leading coefficient.***	The leading coefficient of $x - 2x^2 - 5$ is -2.

P–1 **What is the leading coefficient in each polynomial below?**

a. $5x^3 - 2x^2 + 3x - 7$ **b.** $x - 5$ **c.** $2x^3 - x^4 + 3$

Example 1 reviews multiplying two binomials.

EXAMPLE 1 Multiply: $(2x + 1)(x + 3)$

Solution:

[1] Distributive Property ——→ $(2x + 1)(x + 3) = (2x + 1)x + (2x + 1)3$

[2] Distributive Property ——→ $= 2x^2 + x + 6x + 3$

[3] Combine like terms. ——→ $= 2x^2 + 7x + 3$

Quadratic trinomials are factored by reversing steps [1], [2], and [3] above. The key step is in writing $2x^2 + 7x + 3$ as $2x^2 + x + 6x + 3$. You must find values for r and s in the polynomial form

$$2x^2 + rx + sx + 3$$

or

$$2x^2 + (r + s)x + 3.$$

◀ $2x^2 + x + 6x + 3$

To factor a quadratic trinomial of the form $ax^2 + bx + c$, you must find two integers, r and s, that meet these conditions.

1. The sum of r and s equals the coefficient of the first-degree term. $(r + s = b)$
2. The product of r and s equals the product of the leading coefficient and the constant term. $(r \cdot s = a \cdot c)$

EXAMPLE 2 Factor $6x^2 + 11x - 10$.

Solution: $r + s = 11 \qquad rs = -60$

Find two numbers whose sum is 11 and whose product is −60.

1. Write all pairs of integers with product −60.

$(-1)(60)$	$(1)(-60)$	$(-2)(30)$	$(2)(-30)$	$(-3)(20)$	$(3)(-20)$
$(-4)(15)$	$(4)(-15)$	$(-5)(12)$	$(5)(-12)$	$(-6)(10)$	$(6)(-10)$

2. Select the pair with a sum of 11. ⟶ $r = -4$ and $s = 15$
3. Write in the form $6x^2 + rx + sx - 10$. ⟶ $6x^2 + 11x - 10 = (6x^2 - 4x) + (15x - 10)$
4. Factor each binomial. ⟶ $= 2x(3x - 2) + 5(3x - 2)$
5. Distributive Property ⟶ $= (2x + 5)(3x - 2)$

EXAMPLE 3 Factor $2x^2 + 7x + 3$.

Solution: $r + s = 7 \qquad rs = 6$

$2x^2 + 7x + 3 = 2x^2 + (r + s)x + 3$

1. Write all pairs of integers with product 6.

$(1)(6) \qquad (2)(3) \qquad (-1)(-6) \qquad (-2)(-3)$

2. Select the pair with a sum of 7. ⟶ $r = 1$ and $s = 6$
3. $2x^2 + 7x + 3 = (2x^2 + x) + (6x + 3)$
4. $= x(2x + 1) + 3(2x + 1)$
5. $= (x + 3)(2x + 1)$

EXAMPLE 4 Factor $12x^2 - 11x + 2$.

Solution: $r + s = -11 \qquad rs = 24$

1. $(1)(24) \qquad (2)(12) \qquad (3)(8) \qquad (4)(6)$
   $(-1)(-24) \qquad (-2)(-12) \qquad (-3)(-8) \qquad (-4)(-6)$

   Pairs of integers with product 24
2. $r = -3$ and $s = -8$
3. $12x^2 - 11x + 2 = (12x^2 - 3x) + (-8x + 2)$
4. $= 3x(4x - 1) - 2(4x - 1)$
5. $= (3x - 2)(4x - 1)$

CLASSROOM EXERCISES

Write the value of $r + s$. Then write the value of rs. (Examples 2–4)

1. $2x^2 + 3x + 1$
2. $2x^2 + 5x + 2$
3. $3x^2 + 10x + 3$
4. $3x^2 + 7x + 2$
5. $3x^2 + 11x + 6$
6. $3x^2 + 14x + 15$
7. $6x^2 + 11x + 3$
8. $6x^2 + 7x + 2$
9. $6x^2 + 17x + 5$
10. $6x^2 + 13x + 6$
11. $3x^2 - x - 2$
12. $3x^2 - 5x - 2$
13. $2x^2 - 5x - 25$
14. $2x^2 - 3x - 9$
15. $15x^2 - 7x - 2$
16. $15x^2 + 4x - 4$
17. $6x^2 - 17x + 12$
18. $10x^2 - 29x + 10$

WRITTEN EXERCISES

Goal: To factor trinomials of the form $ax^2 + bx + c$
Sample Problem: $10x^2 - 11x - 6$
Answer: $(5x + 2)(2x - 3)$

Multiply. (Example 1)

1. $(2x - 1)(x - 1)$
2. $(2x - 1)(x - 2)$
3. $(3x - 2)(x + 1)$
4. $(3x - 1)(x + 2)$

Factor. (Examples 2–4)

5. $2x^2 + 3x + 1$
6. $2x^2 + 5x + 2$
7. $3x^2 + 10x + 3$
8. $3x^2 + 7x + 2$
9. $3x^2 + 11x + 6$
10. $3x^2 + 14x + 15$
11. $6x^2 + 11x + 3$
12. $6x^2 + 7x + 2$
13. $6x^2 + 17x + 5$
14. $6x^2 + 13x + 6$
15. $3x^2 - x - 2$
16. $3x^2 - 5x - 2$
17. $2x^2 - 5x - 25$
18. $2x^2 - 3x - 9$
19. $15x^2 - 7x - 2$
20. $15x^2 + 4x - 4$
21. $2x^2 - 7x + 6$
22. $3x^2 - 10x + 3$
23. $8x^2 - 14x + 3$
24. $15x^2 - 13x + 2$
25. $2x^2 + 5x - 3$
26. $3x^2 - 7x - 6$
27. $15x^2 + 14x - 8$
28. $12x^2 - 7x - 5$
29. $18x^2 + 23x + 7$
30. $15x^2 + 11x - 12$
31. $12x^2 - 33x - 9$
32. $4x^2 - 29x + 7$
33. $8x^2 + 35x + 12$
34. $12x^2 + 25x + 12$

NON-ROUTINE PROBLEMS

For Exercises 35–38, look for a pattern in the products in the box at the right. Then use the pattern to evaluate each expression without using pencil and paper.

$999 \times 2 = 1998$
$999 \times 3 = 2997$
$999 \times 4 = 3996$
$999 \times 5 = 4995$

35. 999×6 **36.** 999×8 **37.** $6993 \div 999$ **38.** $8991 \div 999$

39. In the figure below, a belt runs over six wheels in the direction of the arrows. How many wheels are turning clockwise?

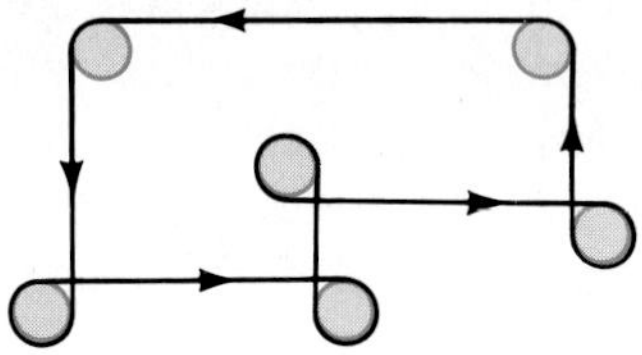

40. There are six water glasses in a row. Three of the glasses are empty and three are full. The arrangement of the glasses is empty, empty, empty, full, full, full. By moving one glass, change the arrangement to empty, full, empty, full, empty, full.

MID-CHAPTER REVIEW

Factor. (Section 9.1)

1. $12x^2 - 4x$ **2.** $-5x^3 - 10x^2$ **3.** $3x^3 + 9x^2 - 15x$ **4.** $4x^4 - 12x^2 - 2x$

5. $2x^2y + 6xy$ **6.** $3xy^2 - 9x^2y$ **7.** $5x^3y - 10x^2y + 25x^2y^2$ **8.** $12rs^2t - 6r^2st^2$

Factor. (Section 9.2)

9. $x^2 + 12x + 35$ **10.** $x^2 + 15x + 54$ **11.** $x^2 + 15x + 50$ **12.** $x^2 + 16x + 48$

Factor. (Section 9.3)

13. $x^2 + 3x - 40$ **14.** $x^2 - 16x + 60$ **15.** $x^2 - 3x - 88$ **16.** $x^2 - 2x - 63$

17. $x^2 + 6x - 72$ **18.** $x^2 + x - 72$ **19.** $x^2 - 20x + 99$ **20.** $x^2 - x - 99$

Factor. (Section 9.4)

21. $2x^2 + x - 6$ **22.** $8x^2 - 6x - 5$ **23.** $20x^2 - 3x - 2$ **24.** $6x^2 - 19x - 20$

REVIEW CAPSULE FOR SECTION 9.5

Factor. (Pages 239–241)

1. $(x^2 + 12x) - (12x + 144)$ **2.** $(x^2 + 2xy) - (2xy + 4y^2)$ **3.** $4x^2 - 6x + 6x - 9$

Using Probability

Making a Model

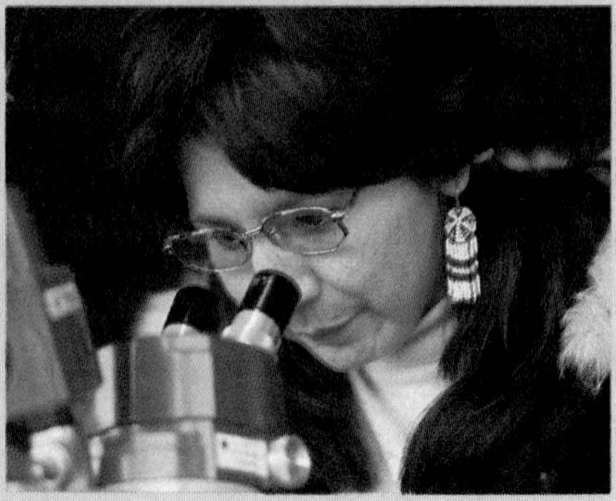

Life scientists are specialists in living things. Since biology is the science of life, many life scientists call themselves **biologists.** However, most are named by the type of living thing that they study or by the nature of the work that they perform. Some life scientist careers are agronomist, biochemist, botanist, horticulturist, microbiologist, nutritionist, and zoologist.

In the study of **genetics,** life scientists learn that the traits of parents are transmitted to their offspring by tiny particles called **genes.** Each living thing has two genes for each trait—one gene from each parent.

The genes for tall and short pea plants are represented below.

TT	Pure tall
tt	Pure short
Tt	Hybrid tall

T = gene for tallness
t = gene for shortness

When a pure tall pea plant and a pure short pea plant are crossed, all offspring are hybrid tall.

Suppose that two hybrid tall pea plants are crossed. This problem is related to the algebra problem of squaring a binomial.

Let $(\frac{1}{2}T + \frac{1}{2}t)$ represent the two genes for tallness and shortness in each hybrid tall pea plant. Then

$$(\tfrac{1}{2}T + \tfrac{1}{2}t)(\tfrac{1}{2}T + \tfrac{1}{2}t) = \tfrac{1}{4}TT + \tfrac{1}{2}Tt + \tfrac{1}{4}tt.$$

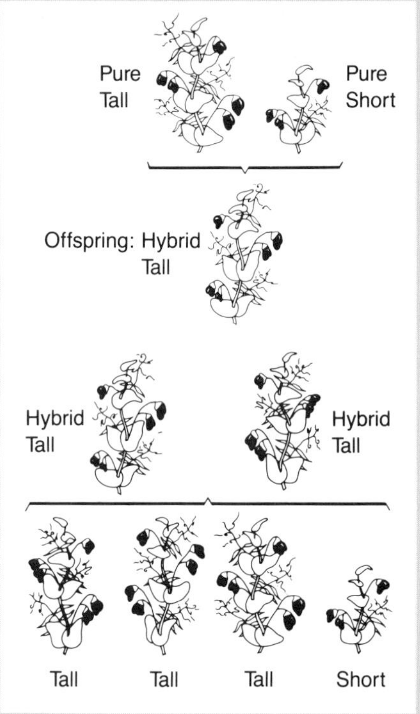

This result indicates the following probabilities.

The probability (chances) that an offspring will be pure tall (TT) is 1 out of 4. This means that, over a long enough period of time, you can expect about one fourth of all such offspring to be pure tall.

The probability (chances) that an offspring will be hybrid tall (Tt) is 1 out of 2. That is, over a long period of time about one half of all such offspring can be expected to be hybrid tall.

The probability (chances) that an offspring will be pure short (tt) is 1 out of 4.

EXAMPLE 1: There are 24 offspring of two hybrid tall parent pea plants. Find the probable number of each type.

SOLUTION: Refer to the trinomial on page 250.

$$\frac{1}{4}TT + \frac{1}{2}Tt + \frac{1}{4}tt \quad \begin{cases} \frac{1}{4} \times 24 = 6 \text{ pure tall} \\ \frac{1}{2} \times 24 = 12 \text{ hybrid tall} \\ \frac{1}{4} \times 24 = 6 \text{ pure short} \end{cases}$$

Since this is merely the *probable* distribution, you would not always expect to obtain exactly these numbers of the various types.

In guinea pigs, a rough coat gene R is **dominant** to a smooth coat gene r. This means that guinea pigs will have coats as shown below.

RR: Pure rough coat Rr: Hybrid rough coat rr: Pure smooth coat

EXAMPLE 2: A guinea pig with a pure rough coat (RR) is mated with a guinea pig which has a hybrid rough coat (Rr). There are 36 offspring. Find the probable number of each type.

SOLUTION:
$$\left(\tfrac{1}{2}R + \tfrac{1}{2}R\right)\left(\tfrac{1}{2}R + \tfrac{1}{2}r\right) = \tfrac{1}{4}RR + \tfrac{1}{4}Rr + \tfrac{1}{4}RR + \tfrac{1}{4}Rr$$
$$= \tfrac{1}{2}RR + \tfrac{1}{2}Rr$$

$\frac{1}{2} \times 36 = 18$ pure rough coat (RR) $\frac{1}{2} \times 36 = 18$ hybrid rough coat (Rr)

All of the offspring will have rough coats.

EXERCISES

There are 128 offspring of two pea plants. Find the probable number of pure tall (TT), pure short (tt), and hybrid tall (Tt) for each combination of parent plants.

1. Hybrid tall and pure short
2. Pure short and pure short
3. Pure tall and hybrid tall
4. Pure tall and pure short

There are 56 offspring of two guinea pigs. Find the probable number of offspring with pure rough coats (RR), hybrid rough coats (Rr), and pure smooth coats (rr) for each combination of parents.

5. Hybrid rough coat; hybrid rough coat
6. Pure rough coat; pure smooth coat
7. Pure rough coat; hybrid rough coat

9.5 Difference of Two Squares

A binomial such as $x^2 - 16$ is called a ***difference of two squares.***

You can write $x^2 - 16$ as $(x)^2 - (4)^2$.

The first term and the absolute value of the constant term are perfect squares.

P–1 **Which polynomials below represent the difference of two squares?**

a. $x^2 - 4$ **b.** $x^3 - 9$ **c.** $x^4 - 1$ **d.** $x^2 + 25$

The product of two special binomials suggests how to factor a difference of two squares.

EXAMPLE 1 Multiply: $(x + a)(x - a)$

Solution:

$$(x + a)(x - a) = x^2 + ax - ax - a^2$$
$$= x^2 - a^2$$

Difference of two squares

P–2 **What are the binomial factors of $x^2 - a^2$?**

The prime factorization of a binomial such as $x^2 - a^2$ is $(x + a)(x - a)$.

$$x^2 - 49 = (x + 7)(x - 7)$$

P–3 **What is the positive square root of 121?**

EXAMPLE 2 Factor $x^2 - 121$.

Solution:

$$x^2 - 121 = (x + 11)(x - 11)$$

(Sum of square roots) times (Difference of square roots)

EXAMPLE 3 Factor $x^4 - 9$.

Solution: Both x^4 and 9 are perfect squares. You can write x^4 as $(x^2)^2$ and 9 as 3^2.

$$x^4 - 9 = (x^2 + 3)(x^2 - 3)$$

EXAMPLE 4 Factor $4x^2 - 9y^2$.

Solution: You can write $4x^2$ as $(2x)^2$ and $9y^2$ as $(3y)^2$.

$$4x^2 - 9y^2 = (2x + 3y)(2x - 3y)$$

P-4 Simplify.

a. $\sqrt{x^6}$ b. $\sqrt{r^8}$ c. $\sqrt{t^{12}}$

Any power with an even exponent is a perfect square.

EXAMPLE 5 Factor $x^{10} - y^8$.

Write one factor as a sum of square roots. → **Write one factor as a difference of square roots.**

Solution: $x^{10} - y^8 = (x^5 + y^4)(x^5 - y^4)$

To factor the difference of squares, write one factor as the sum of square roots and the other factor as the difference of square roots.

CLASSROOM EXERCISES

Factor. (Examples 2–4)

1. $a^2 - 1$
2. $y^2 - 4$
3. $p^2 - 9$
4. $r^2 - 16$
5. $x^2 - 25$
6. $a^2 - 36$
7. $t^2 - 49$
8. $x^2 - 64$
9. $b^2 - 81$
10. $x^2 - 100$
11. $p^2 - q^2$
12. $c^2 - d^2$
13. $4x^2 - y^2$
14. $x^2 - 9y^2$
15. $x^4 - 25$

WRITTEN EXERCISES

Goal: To factor the difference of two squares
Sample Problem: $25a^2 - 9b^2$
Answer: $(5a + 3b)(5a - 3b)$

Multiply. (Example 1)

1. $(x - 8)(x + 8)$
2. $(x - 9)(x + 9)$
3. $(x + 11)(x - 11)$
4. $(x + 12)(x - 12)$
5. $(2x - y)(2x + y)$
6. $(3a - b)(3a + b)$
7. $(xy - 3)(xy + 3)$
8. $(rs - 5)(rs + 5)$
9. $(5x + 3)(5x - 3)$
10. $(4x + 3)(4x - 3)$
11. $(x^2 + 2)(x^2 - 2)$
12. $(y^2 + 3)(y^2 - 3)$

Factor. (Examples 2–5)

13. $a^2 - 25$
14. $y^2 - 36$
15. $4x^2 - 1$
16. $9y^2 - 1$
17. $9a^2 - 16$
18. $4b^2 - 25$
19. $x^4 - 4$
20. $y^4 - 9$
21. $4x^4 - 9$
22. $9a^4 - 1$
23. $4x^2 - y^2$
24. $16a^2 - 9b^2$
25. $a^2b^2 - 1$
26. $x^2y^2 - 4$
27. $16t^2 - 81$
28. $25r^2 - 16$
29. $x^2y^2 - r^2$
30. $a^2b^2 - c^2$
31. $16x^2y^4 - 9$
32. $9a^2b^4 - 25$
33. $x^6 - 4$
34. $a^6 - 9$
35. $x^{10} - y^8$
36. $r^8 - s^{14}$

MORE CHALLENGING EXERCISES

Multiply by the method shown in the example below. Find the answer without using pencil and paper.

EXAMPLE: $(19)(21) = (20 - 1)(20 + 1)$
$= 400 - 1 = 399$

37. (14)(16)
38. (15)(17)
39. (11)(15)
40. (14)(18)
41. (22)(18)
42. (23)(17)
43. (26)(34)
44. (36)(44)
45. (43)(37)
46. (54)(46)
47. (59)(61)
48. (89)(91)

REVIEW CAPSULE FOR SECTION 9.6

Multiply. (Pages 220–223)

1. $(x + 6)(x + 6)$
2. $(x - 3)(x - 3)$
3. $(2x - 5)(2x - 5)$
4. $(5x + 1)(5x + 1)$
5. $(2x + 7)(2x + 7)$
6. $(x - y)(x - y)$
7. $(3x + 2)(3x + 2)$
8. $(5x - 1)^2$

9.6 Trinomial Squares

A trinomial such as $x^2 + 6x + 9$ is called a ***trinomial square*** because it has two equal binomial factors.

$$x^2 + 6x + 9 = (x + 3)(x + 3)$$

> A trinomial of the form $ax^2 + bx + c$ is a trinomial square if both of the following conditions are true.
> 1. ax^2 and c are perfect squares.
> 2. $(\frac{1}{2}b)^2 = ac$

In $x^2 + 6x + 9$, both x^2 and 9 are perfect squares.

P-1 **What is the value of $\frac{1}{2}b$? of $(\frac{1}{2}b)^2$? of ac?**

You can see that $x^2 + 6x + 9$ meets both conditions.

EXAMPLE 1 Write Yes or No to show whether $4x^2 - 20x + 25$ is a trinomial square.

Solution: [1] $4x^2$ and 25 are both perfect squares.

$4x^2 = (2x)(2x)$ $\qquad$ $25 = (5)(5)$

[2] "$(\frac{1}{2}b)^2 = ac$" is true.

$$(\tfrac{1}{2}b)^2 = (\tfrac{1}{2} \cdot -20)^2 \qquad ac = 4 \cdot 25$$
$$= (-10)^2 \qquad = 100$$
$$= 100$$

Yes, $4x^2 - 20x + 25$ is a trinomial square.

A trinomial square is also called a "perfect square trinomial."

EXAMPLE 2 Factor $4x^2 - 20x + 25$.

This is a trinomial square.

Solution:

[1] Write the first term of each binomial as a square root of $4x^2$. → $4x^2 - 20x + 25 = (2x \quad)(2x \quad)$

[2] Write the second term as a square root of 25. → $= (2x \quad 5)(2x \quad 5)$

[3] Write each binomial as a difference. → $= (2x - 5)(2x - 5)$

P–2 **Why is each factor of Example 2 a difference?**

P–3 **Multiply.**

a. $(2x + 5)(2x + 5)$ **b.** $(x - 7)(x - 7)$ **c.** $(3x - 1)(3x - 1)$

Trinomials in more than one variable can be trinomial squares.

EXAMPLE 3 Factor $x^2y^2 + 8xy + 16$.

Solution: [1] x^2y^2 and 16 are perfect squares.

[2] "$(\frac{1}{2} \cdot 8)^2 = 1 \cdot 16$" is true.

$x^2y^2 + 8xy + 16$ is a trinomial square.

$$x^2y^2 + 8xy + 16 = (xy + 4)(xy + 4)$$

To check, multiply the factors.

P–4 **Factor.**

a. $x^2 + 2x + 1$ **b.** $x^2 - 12x + 36$ **c.** $9x^2 - 24x + 16$

CLASSROOM EXERCISES

Write Yes or No to tell whether each trinomial is a trinomial square. (Example 1)

1. $16x^2 + 8x + 1$ **2.** $x^2 + 9x - 16$ **3.** $x^2 + 10x + 25$

4. $x^2 + 4x + 3$ **5.** $x^2 - 2x + 1$ **6.** $100 - 20x + x^2$

7. $x^2 - 49x + 14$ **8.** $x^2 + 20x + 100$ **9.** $2x^2 + 8x - 8$

10. $2x^2 - 8x + 8$ **11.** $81x^2 + 50x + 3$ **12.** $64x^2 + 48x + 9$

Factor each trinomial square. (Example 2)

13. $x^2 + 4x + 4$ **14.** $x^2 - 2x + 1$

15. $x^2 - 6x + 9$ **16.** $x^2 - 10x + 25$

17. $x^2 + 16x + 64$ **18.** $x^2 + 18x + 81$

19. $x^2 - 24x + 144$ **20.** $x^2 - 14x + 49$

21. $4x^2 + 4x + 1$ **22.** $25x^2 - 20x + 4$

WRITTEN EXERCISES

Goal: To factor trinomial squares
Sample Problem: $4x^2 - 20x + 25$ **Answer:** $(2x - 5)(2x - 5)$

Write Yes *or* No *to show whether each trinomial is a trinomial square.* (Example 1)

1. $x^2 - 8x + 25$
2. $x^2 + 10x + 36$
3. $x^2 - 22x + 121$
4. $x^2 - 14x + 49$
5. $x^2 + 14x - 49$
6. $x^2 + 16x - 64$
7. $2x^2 - 12x + 25$
8. $3x^2 - 22x + 36$
9. $9x^2 + 18x + 1$
10. $16x^2 + 32x + 1$
11. $4x^2 - 4x + 1$
12. $5x^2 - 10x + 25$

Multiply. (P-3)

13. $(x - 5)(x - 5)$
14. $(x - 3)(x - 3)$
15. $(x + 7)(x + 7)$
16. $(x + 9)(x + 9)$
17. $(2x + 3)(2x + 3)$
18. $(3x + 2)(3x + 2)$
19. $(5x - 1)(5x - 1)$
20. $(4x - 3)(4x - 3)$
21. $(4x + 5)^2$
22. $(6x + 5)^2$
23. $(2x - 1)^2$
24. $(3x - 7)^2$

Factor. (Examples 2–3)

25. $36x^2 - 12x + 1$
26. $25x^2 - 10x + 1$
27. $25x^2 + 20x + 4$
28. $16x^2 + 24x + 9$
29. $9x^2 - 24x + 16$
30. $4x^2 - 20x + 25$
31. $49x^2 + 140x + 100$
32. $36x^2 + 84x + 49$
33. $49x^2 - 14x + 1$
34. $81x^2 + 18x + 1$
35. $x^2 + 4xy + 4y^2$
36. $x^2 + 6xy + 9y^2$
37. $4x^2 + 12xy + 9y^2$
38. $16x^2 + 40xy + 25y^2$
39. $x^2y^2 - 16xy + 64$
40. $4x^2y^2 - 20xy + 25$
41. $9x^2y^2 + 42xy + 49$
42. $x^2y^2 - 2xy + 1$

REVIEW CAPSULE FOR SECTION 9.7

Match each polynomial with its prime factors. (Pages 236–254)

1. $4x^2 - 8x + 3$
2. $4x^2 - 10x - 6$
3. $4x^2 + 4x - 3$
4. $4x^2 - 9$

A. $(2x + 3)(2x - 1)$
B. $(2x - 3)(2x - 1)$
C. $(2x - 3)(2x + 3)$
D. $2(x - 3)(2x + 1)$

9.7 Combined Types of Factoring

You have learned the following ways of factoring polynomials.

1. Common monomial factoring (Pages 236–238)
2. Factoring the difference of two squares (Pages 252–254)
3. Factoring quadratic trinomials (Pages 239–241)
4. Factoring trinomial squares (Pages 255–257)

When you factor a polynomial always write it as a product of prime polynomial factors.

In factoring a polynomial, always look for a common monomial factor as the first step.

P–1 **What is a common monomial factor of $2x^2 - 32$?**

EXAMPLE 1 Factor $2x^2 - 32$.

Solution:

1. Write the common monomial factor. ⟶ $2x^2 - 32 = 2(\quad\quad)$
2. Divide each term of $2x^2 - 32$ by 2. ⟶ $= 2(x^2 - 16)$
3. Factor $x^2 - 16$ as a difference of two squares. ⟶ $= 2(x + 4)(x - 4)$

Prime polynomial factors

When you write a polynomial as the product of prime polynomial factors, you are factoring completely. To "factor" always means to "factor completely."

P–2 **What is a common monomial factor of $3x^2 - 6x + 3$?**

EXAMPLE 2 Factor $3x^2 - 6x + 3$.

Solution:

1. Write the common factor. ⟶ $3x^2 - 6x + 3 = 3(\quad\quad)$
2. Divide each term of $3x^2 - 6x + 3$ by 3. ⟶ $= 3(x^2 - 2x + 1)$
3. Factor $x^2 - 2x + 1$ as a trinomial square. ⟶ $= 3(x - 1)(x - 1)$

EXAMPLE 3 Factor $30x^2 - 5x - 10$.

Solution:

[1] Use common monomial factoring. ⟶ $30x^2 - 5x - 10 = 5(6x^2 - x - 2)$

[2] Determine whether $6x^2 - x - 2$ is a trinomial square. ⟶ ***$6x^2 - x - 2$ is not a trinomial square.***

[3] Write $-x$ as a sum. ⟶ $6x^2 - x - 2 = 6x^2 + rx + sx - 2$

[4] Write all pairs of integers with product -12.

$(-1)(12)$	$(-2)(6)$	$(-3)(4)$
$(1)(-12)$	$(2)(-6)$	$(3)(-4)$

[5] Select the pair with a sum of -1. ⟶ $\begin{cases} r = 3 \\ s = -4 \end{cases}$

[6] $6x^2 - x - 2 = 6x^2 + 3x - 4x - 2$

[7] $= 3x(2x + 1) - 2(2x + 1)$

[8] $= (3x - 2)(2x + 1)$

Remember: 5 is also a factor of $30x^2 - 5x - 10$.

Thus, $30x^2 - 5x - 10 = 5(6x^2 - x - 2)$

$= 5(3x - 2)(2x + 1)$

EXAMPLE 4 Factor $4x^2 - 16y^2$.

Solution: You can see that $4x^2 - 16y^2$ is a difference of two squares.

$4x^2 - 16y^2 = 4(x^2 - 4y^2)$ ***Common monomial factor***

$= 4(x + 2y)(x - 2y)$

CLASSROOM EXERCISES

Write the common factor. Then divide each term by the common factor. (Steps 1 and 2 of Examples 1 and 2)

1. $4x^2 - 12$

2. $3x^2 + 6x + 6$

3. $4x^2 - 20x + 28$

4. $2x^2 - 8x + 12$

5. $15x^2 - 5x - 25$

6. $16x^2 - 64$

Write the common factor. Then divide each term by the common factor. (Steps 1 and 2 of Examples 1 and 2)

7. $x^2y^2 - 3xy + 5y$

8. $ax^2 - 3ax + 12a$

9. $-4x^2 - 20$

10. $12ax^2 + 3a$

11. $3a^2b^2 - 9ab + 6b$

12. $24 - 36x^2$

13. $3x^2y - 30xy + 75y$

14. $-12x^2y - 14xy + 6y$

WRITTEN EXERCISES

Goal: To factor polynomials using more than one type of factoring
Sample Problem: $2a^2b^2 + 14a^2b + 24a^2$
Answer: $2a^2(b + 4)(b + 3)$

Factor. Remember factors should be prime polynomials. (Examples 1–4)

1. $2y^2 - 18$

2. $2a^2 - 50$

3. $2x^2 + 8x + 8$

4. $2x^2 + 4x + 2$

5. $3x^2 + 30x + 63$

6. $3x^2 + 36x + 105$

7. $4x^2 - 40x + 96$

8. $5x^2 - 55x + 150$

9. $5x^2 - 5x - 100$

10. $3x^2 - 3x - 36$

11. $4y^2 - 9$

12. $x^2y^2 - 16$

13. $ax^2 + 4ax + 4a$

14. $x^2y + 6xy + 9y$

15. $6x^2 - 10x - 4$

16. $6x^2 - 2x - 4$

17. $12x^2 - 40x + 12$

18. $20x^2 - 70x + 60$

19. $-3x^2 - 12$

20. $-5a^2 - 125$

21. $4ax^2 - 9a$

22. $8ab^2 - 18a$

23. $x^2y - 10xy + 25y$

24. $a^2b^2 - 6a^2b + 9a^2$

25. $3x^2 - 75$

26. $27y^2 - 3$

27. $32 - 16d + 2d^2$

28. $-36 + 15a^2 - 12a$

29. $3x^2 - 6x - 24$

30. $10y^2 - 26y + 12$

31. $2 - 128x^2$

32. $x^3 - x$

33. $ab^2 - ab - 20a$

34. $pq^2 + p^2q^2 - p$

35. $24a^2 - 30a + 9$

36. $30a^2 - 5a - 10$

MORE CHALLENGING EXERCISES

37. If $x - y = 6$ and $x + y = 1$, find $x^2 - y^2$.

38. If $x \neq 0$ and $x^2 - x = x$, what is the value of $x - 1$?

39. If $s^2 - t^2 = 6$ and $s - t = 3$, find $s + t$.

40. If $m^2 - 4n^2 = 32$, and $m + 2n = 8$, find $m - 2n$.

Focus on Reasoning: Mental Computation

Logical reasoning can help you to solve problems more efficiently. Sometimes it will enable you to solve problems using **mental computation** only. The ability to identify efficient strategies to solve problems is an important test-taking skill.

EXAMPLE 1 If $12 + y \cdot z = 9$, evaluate $(12 + y \cdot z)^2 - 20$.

Solution: Since $12 + y \cdot z = 9$, $(12 + y \cdot z)^2 - 20 = (9)^2 - 20$
$= 81 - 20$
$= \mathbf{61}$

EXAMPLE 2 If $r + 7 - w = 36$, evaluate $15 + \sqrt{r + 7 - w}$.

Solution: Since $r + 7 - w = 36$, $15 + \sqrt{r + 7 - w} = 15 + \sqrt{36}$
$= 15 + 6$
$= \mathbf{21}$

EXERCISES

1. If $3x - 2y = 4$, evaluate $(3x - 2y)^2$.
2. If $14p - 3k = -9$, evaluate $(14p - 3k)^2$.
3. If $9 + d \cdot f = -2$, evaluate $(9 + d \cdot f)^3 + 3$.
4. If $11 - x \cdot y = -3$, evaluate $(11 - x \cdot y)^3 + 14$.
5. If $r - w + 3 = 49$, evaluate $\sqrt{r - w + 3}$.
6. If $k + n + 2 = 64$, evaluate $\sqrt{k + n + 2}$.
7. If $g \cdot m - 6 = 81$, evaluate $\frac{1}{3}\sqrt{g \cdot m - 6}$.
8. If $q \cdot z - 7 = 144$, evaluate $\frac{1}{6}\sqrt{q \cdot z - 7}$.
9. If $6p + v = 27$, evaluate $\sqrt[3]{6p + v}$.
10. If $8f - u = 8$, evaluate $\sqrt[3]{8f - u}$.
11. If $a \cdot b - 5 = 7$, evaluate $(a \cdot b - 6)^2 - 10$.
12. If $t \cdot w + 4 = 9$, evaluate $(t \cdot w + 5)^2 - 18$.
13. If $d \div f + 11 = 35$, evaluate $2 + \sqrt{d \div f + 12}$.
14. If $h \div k - 9 = 17$, evaluate $5 + \sqrt{h \div k - 10}$.

Examples 3 and 4 apply logical reasoning to problems involving **factoring.** The key is in identifying a binomial product or the factors of a binomial product.

EXAMPLE 3 If $x + y = 6$ and $x - y = -3$, find the value of $x^2 - y^2$.

Think: $x^2 - y^2 = (x + y)(x - y)$

Solution: Since $x + y = 6$ and $x - y = -3$, $x^2 - y^2 = 6 \cdot (-3)$, or $\mathbf{-18}$.

EXAMPLE 4 If $a^2 - b^2 = 10$ and $a + b = 5$, find the value of $a - b$.

Think: $a^2 - b^2 = (a + b)(a - b)$

Solution: Since $a^2 - b^2 = 10$ and $a + b = 5$, $5(a - b) = 10$.

So
$$\frac{\overset{1}{\cancel{5}}(a - b)}{\underset{1}{\cancel{5}}} = \frac{\overset{2}{\cancel{10}}}{\underset{1}{\cancel{5}}}$$
$$a - b = \mathbf{2}$$

EXERCISES

15. If $x + y = -2$ and $x - y = 3$, find the value of $x^2 - y^2$.

16. If $r + t = 4$ and $r - t = -3$, find the value of $r^2 - t^2$.

17. If $c + 2d = -6$ and $c - 2d = -5$, find the value of $c^2 - 4d^2$.

18. If $n + 3p = -4$ and $n - 3p = -7$, find the value of $n^2 - 9p^2$.

19. If $3h - f = 2.4$ and $3h + f = 3.1$, find the value of $9h^2 - f^2$.

20. If $5k - m = 2.3$ and $5k + m = 1.8$, find the value of $25k^2 - m^2$.

21. If $2w - 5z = \frac{3}{5}$ and $2w + 5z = \frac{5}{12}$, find the value of $4w^2 - 25z^2$.

22. If $3x - 4y = \frac{2}{3}$ and $3x + 4y = \frac{3}{4}$, find the value of $9x^2 - 16y^2$.

23. If $x^2 - y^2 = 12$ and $x + y = 3$, find the value of $x - y$.

24. If $w^2 - z^2 = 24$ and $w + z = 8$, find the value of $w - z$.

25. If $4n^2 - p^2 = 1.8$ and $2n - p = 0.9$, find the value of $2n + p$.

26. If $9x^2 - y^2 = 3.6$ and $3x - y = 1.2$, find the value of $3x + y$.

27. If $3d + 5w = \frac{1}{6}$ and $9d^2 - 25w^2 = 4$, find the value of $3d - 5w$.

28. If $6r + 5t = \frac{1}{4}$ and $36r^2 - 25t^2 = 10$, find the value of $6r - 5t$.

29. If $c^2 - d^2 = 30$ and $c - d = -5$, find the value of $(c + d) - 11$.

30. If $f^2 - h^2 = 40$ and $f - h = -8$, find the value of $(f + h) + 12$.

CHAPTER SUMMARY

IMPORTANT TERMS

Prime polynomial *(p. 236)*
Common monomial factor *(p. 236)*
Greatest common factor *(p. 237)*
Degree of a monomial *(p. 239)*
Degree of a polynomial *(p. 239)*
Quadratic polynomial *(p. 239)*
Leading coefficient *(p. 246)*
Difference of two squares *(p. 252)*
Trinomial square *(p. 255)*

IMPORTANT IDEAS

1. In a prime polynomial the only polynomial factors with coefficients that are integers are
   a. 1 and the polynomial, and
   b. −1 and the opposite of the polynomial.

2. To factor a quadratic trinomial you must find two integers r and s meeting the following conditions.
   a. The sum of r and s equals the coefficient of the first-degree term.
   b. The product of r and s equals the product of the leading coefficient and the constant term.

3. The prime factorization of $x^2 - a^2$ is $(x + a)(x - a)$.

4. For a quadratic trinomial such as $ax^2 + bx + c$ to be a trinomial square,
   a. ax^2 and c must be perfect squares, and
   b. "$(\frac{1}{2}b)^2 = ac$" must be true.

5. In factoring a polynomial, always look for a common monomial factor as the first step.

CHAPTER REVIEW

SECTION 9.1

Factor.

1. $4x^2 - 10x$

2. $3a^3 + 15a^2$

3. $18a^2b + 30a^2b^2 - 42ab$

4. $24ax + 48ay - 96at$

5. $a^2bc^3 - ab^2c + a^2b^2c^2$

6. $2x^5 - 4x^4 + 6x^3 - 8x^2 + 10x$

SECTION 9.2

Factor.

7. $x^2 + 7x + 3x + 21$

8. $x^2 + 8x + 5x + 40$

9. $x^2 + 5x + 10x + 50$

10. $x^2 + 9x + 6x + 54$

11. $x^2 + 10x + 21$

12. $x^2 + 13x + 22$

SECTION 9.3

Factor.

13. $x^2 + 3x - 18$
14. $x^2 - x - 20$
15. $x^2 - 9x + 18$
16. $x^2 - 14x + 33$
17. $x^2 - 5x - 81$
18. $x^2 - 5x - 36$

SECTION 9.4

Factor.

19. $2x^2 - x - 3$
20. $3x^2 - 5x - 2$
21. $4x^2 + 7x - 2$
22. $5x^2 + 4x - 1$
23. $3x^2 - 2x - 5$
24. $4x^2 - 4x - 3$

SECTION 9.5

Factor.

25. $9a^2 - 16b^2$
26. $4m^2 - 81n^2$
27. $r^2s^2 - x^2y^2$
28. $4t^2 - 121x^2$
29. $x^4 - 169$
30. $25a^2b^4 - 144c^2$

SECTION 9.6

Write Yes or No to show whether each trinomial is a trinomial square.

31. $x^2 - 12x + 36$
32. $x^2 + 3x + 9$
33. $4x^2 + 10x + 13$
34. $9x^2y^2 - 16xy + 25$
35. $8x^2 - 12x + 9$
36. $4x^2 - 20x + 25$

Factor.

37. $x^2 - 8x + 16$
38. $x^2 + 12x + 36$
39. $4x^2 - 12x + 9$
40. $9x^2 - 6x + 1$
41. $x^2 - 4xy + 4y^2$
42. $16x^2 - 24x + 9$

SECTION 9.7

Factor.

43. $5x^2 - 20$
44. $28x^2 - 7$
45. $3x^2 - 12x + 12$
46. $2x^2 + 4x - 70$
47. $3x^2 - 12x + 96$
48. $x^2y - 2xy - 24y$

CUMULATIVE REVIEW: CHAPTERS 1–9

Perform the indicated operations. (Sections 1.2 through 1.5)

1. $-31 + 18$ **2.** $-5.7 - (-8.3)$ **3.** $(-4\frac{2}{3})(-6)$ **4.** $64 \div (-4)$

Multiply and simplify. (Section 2.5)

5. $(3x - 7)8$ **6.** $-4(6t - 2)$ **7.** $-3n(4n + 2)$ **8.** $(2k + 5)6k$

Solve. Check your answer. (Section 3.5)

9. $x + 35 - 4x = -1$ **10.** $2a - 3(a + 1) = 5$

Choose a variable to represent the unknowns. Write an equation for each problem. Solve. (Section 3.7)

11. A particular house is worth \$4000 more than the lot on which it is located. The house is worth \$20,000. Find the worth of the lot.

12. The number of hours that Jeff worked was 2 less than the number of hours that Dan worked. Jeff worked 14 hours. How many hours did Dan work?

Use Condition 1 to represent the unknowns. Use Condition 2 to write an equation for the problem. Solve. (Section 4.1 and 4.2)

13. There are 80 more single-family homes in Fairfield Subdivision than multi-family homes. There are a total of 188 homes. Find the number of single-family homes.

14. The length of a transparency is 3 inches less than twice the width. The perimeter is 42 inches. Find the length and width.

Solve. Check your answer. (Section 4.3)

15. $5x + 3 = 3x - 5$ **16.** $3k - 2(5 - k) = 10k$

Draw the graph of each inequality. (Section 5.1)

17. $x \leq 2$ **18.** $x \geq -3$ **19.** $x > -1\frac{1}{2}$ **20.** $x < 5.7$

Solve each inequality. (Sections 5.3 through 5.6)

21. $8x - 1 > 7x - 3$ **22.** $12 < -3x$ **23.** $2(2 + x) > 3x + 8 + x$

Simplify. (Sections 6.4 and 6.5)

24. $(-2hk)^3$ **25.** $(xy)^{-4}$ **26.** $(3a^2b^6)^4$ **27.** $\left(\frac{5cd^2}{x^3}\right)^2$

Simplify. (Sections 7.1 through 7.4)

28. $\sqrt{81}$ **29.** $-\sqrt{\frac{9}{16}}$ **30.** $\sqrt{2} \cdot \sqrt{7} \cdot \sqrt{14}$ **31.** $\sqrt{12} \cdot \sqrt{27}$

32. $\sqrt{500}$ **33.** $\sqrt[3]{24}$ **34.** $\sqrt{a^4b^7}$ **35.** $\sqrt{8x^7y^9}$

Add or subtract. Simplify where necessary. (Section 7.5)

36. $5\sqrt{2} + 8\sqrt{2}$ **37.** $2\sqrt{3} - \sqrt{27}$ **38.** $\sqrt{5} + \sqrt{10}$ **39.** $\sqrt{50p} - \sqrt{18p}$

Simplify. Rationalize the denominator where necessary. (Section 7.6)

40. $\sqrt{\frac{32}{w^2}}$ **41.** $\frac{\sqrt{a^5}}{\sqrt{16a^3}}$ **42.** $\frac{\sqrt{5}}{\sqrt{6}}$ **43.** $\frac{\sqrt{2m}}{\sqrt{7p}}$

Approximate each square root to three decimal places. Use the table on page 194. (Section 7.7)

44. $\sqrt{80}$ **45.** $\sqrt{150}$ **46.** $\sqrt{\frac{3}{5}}$ **47.** $\sqrt{\frac{2}{27}}$

Solve each problem. (Section 7.8)

48. The legs of a right triangle have lengths of 15 inches and 36 inches. Find the length of the hypotenuse.

49. Find the approximate width of a 20-inch TV screen if the height is 12 inches.

Evaluate each expression. (Sections 8.1 and 8.3)

50. $-8x^3$ when $x = -2$ **51.** $y^2 - 3y$ when $y = 2$ **52.** $2a^2 + 5a - 9$ when $a = -3$

Multiply or divide as indicated. (Section 8.2)

53. $(12a^3)(-0.6a^2)$ **54.** $(5x^2y)(-9x^2y^4w)$ **55.** $\frac{-48c^7}{16c^2}$ **56.** $\frac{-24h^2k^5m}{-40hk^3m}$

Perform the indicated operation. (Sections 8.4 through 8.7)

57. $(a^2 - 3a + 7) + (3a^2 - a - 19)$ **58.** $(y^2 + y - 11) - (7y^2 + 5)$

59. $(4w - 1)(3w + 5)$ **60.** $(g - 3)(2g^2 + 3g - 1)$

61. $3x + 2 \overline{)\, 12x^2 - x - 6}$ **62.** $n - 4 \overline{)\, n^3 - 19n + 4}$

Factor. (Sections 9.1 through 9.7)

63. $10y^3 - 15y^2$ **64.** $6x^4 + 21x^3 - 3x^2$ **65.** $x^2 + 7x + 2x + 14$

66. $h^2 + 19h + 18$ **67.** $g^2 - 10g - 24$ **68.** $3m^2 - 7m + 2$

69. $6n^2 + 7n - 3$ **70.** $c^2 - 9$ **71.** $4y^2 - 25d^2$

72. $9x^2 - 6x + 1$ **73.** $9x^2 - 36$ **74.** $24x^2 + 30x - 9$

Products and Quotients of Rational Expressions

CHAPTER 10

Glide ability is related to the ratio of the wing spread and the wing area. Electrical resistance in a wire is related to the ratio of its length and diameter. The velocity of sound waves is related to an increase or decrease in temperature.

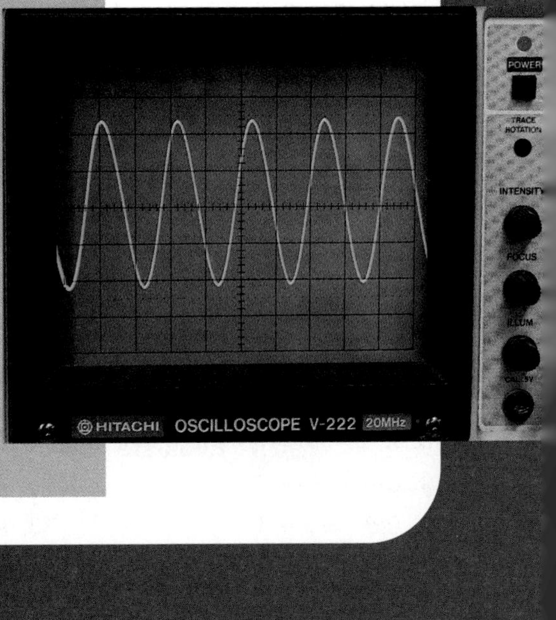

10.1 Rational Expressions

Definition

A ***rational expression*** is one that can be written as the quotient of two polynomials.

$$\frac{x^2+x-3}{x^2-3x+5};\ \frac{2}{x};$$

$$\frac{4}{3};\ \frac{x^2-5}{1}$$

P–1 **Write $x^2 + x - 1$ as a quotient.**

Any polynomial is also a rational expression since it can be expressed as a quotient with 1 as divisor.

P–2 **Which of the expressions below are rational expressions?**

a. $\frac{2x}{5}$ b. $\frac{3x+7}{x^2-3x+10}$ c. $\frac{|x^2+1|}{5}$

d. $\frac{-7}{10}$ e. $\frac{\sqrt{x-3}}{\sqrt{x+3}}$ f. 75

Remember that both dividend and divisor must be polynomials.

EXAMPLE 1 Evaluate the following rational expression if x equals -2.

$$\frac{2x^2-3x+1}{-x^2+x-3}$$

Solution:

1 Replace x by -2. ⟶ $\frac{2x^2-3x+1}{-x^2+x-3} = \frac{2(-2)^2-3(-2)+1}{-(-2)^2+(-2)-3}$

2 Simplify. ⟶ $= \frac{8+6+1}{-4-2-3}$

$$= \frac{15}{-9} \text{ or } -\frac{15}{9}$$

P–3 **Evaluate $\frac{-x^2+5x-1}{2x^2-x+4}$ if x equals -3.**

Any number that will result in a divisor of 0 must be restricted from the domain of the variable in a rational expression.

$$\frac{x+5}{x^2-1}$$

$x \neq 1$; $x \neq -1$

EXAMPLE 2 Write the number or numbers that cannot be in the domain of x in $\frac{x-2}{3x+4}$.

Solution: You can see that $3x + 4$ must not equal 0.

Solve for the number that makes 3x + 4 equal 0. → **This number cannot be in the domain.**

$$3x + 4 = 0$$

$$3x + 4 - 4 = 0 - 4$$

$$3x = -4$$

$$\frac{1}{3}(3x) = \frac{1}{3}(-4)$$

$$x = -\frac{4}{3}$$ The number $-\frac{4}{3}$ cannot be in the domain.

CLASSROOM EXERCISES

Write Yes or No to tell whether each expression is a rational expression. (P-2)

1. $x^2 - 2x + 1$
2. $\frac{x^2+1}{5}$
3. $\sqrt{x+1}$
4. 15
5. $\frac{2}{5}$
6. $\frac{2}{\sqrt{x}}$
7. $\frac{x-3}{x+1}$
8. $\frac{x(x+3)}{x^2+2x+1}$

Write the number or numbers that cannot be in the domain of each variable. (Example 2)

9. $\frac{5}{x}$
10. $\frac{3}{x+1}$
11. $\frac{x+3}{x-3}$
12. $\frac{x+10}{-5x}$
13. $\frac{x}{2x-1}$
14. $\frac{3x+1}{2x+1}$
15. $\frac{5}{x^2-1}$
16. $\frac{x}{(x-3)(x+2)}$

WRITTEN EXERCISES

Goal: To evaluate rational expressions

Sample Problem: Evaluate $\frac{x^2 - 4}{x + 3}$ if x equals 3.

Answer: $\frac{5}{6}$

Write Yes or No to show whether each expression is a rational expression. (P-2)

1. $x^2 + 10$
2. $x^2 - 3x + 5$
3. $\frac{3}{4}$
4. $-\frac{5}{6}$
5. $\frac{x - 2}{x + 2}$
6. $\frac{x^2 - 5}{x^2 + 10}$
7. $\frac{\sqrt{x} - 1}{\sqrt{x} + 1}$
8. $\frac{x^2 + \sqrt{x}}{x^2 - \sqrt{x}}$
9. $\frac{|x|}{-x}$
10. $\frac{x - 3}{|x + 2|}$
11. $\frac{x^2 - 2x + 3}{3x^2 - 5x + 2}$
12. $\frac{5 - x^2 + x}{x - 6 + 2x^2}$

Evaluate each rational expression if x equals 1. (Example 1)

13. $\frac{5}{-2x}$
14. $\frac{-7}{5x}$
15. $\frac{x + 3}{x + 2}$
16. $\frac{x - 2}{x + 2}$
17. $\frac{x^2 - 2x + 5}{x^2 + 3x - 1}$
18. $\frac{x^2 - 1}{x^2 + 1}$
19. $\frac{2x^2 - 3x + 1}{x^2 - 5x - 3}$
20. $\frac{x^2 - x - 1}{x^2 - 7x + 3}$

Write the number or numbers that cannot be in the domain of each variable. (Example 2)

21. $\frac{x^2 + 2}{x}$
22. $\frac{x - 5}{-x}$
23. $\frac{2x}{x - 10}$
24. $\frac{3x}{x + 15}$
25. $\frac{x^2 + 3}{5x}$
26. $\frac{x^2 - 4}{-3x}$
27. $\frac{2}{x^2 - 4}$
28. $\frac{5}{x^2 - 9}$
29. $\frac{2}{3x - 1}$
30. $\frac{5}{4x + 3}$
31. $\frac{3}{(x - 1)(x + 7)}$
32. $\frac{7}{(x + 4)(x - 9)}$

REVIEW CAPSULE FOR SECTION 10.2

Write the prime factorization of each polynomial. (Pages 236–238)

1. $18x^3$
2. $12ab^2$
3. $x^2 - 25$
4. $c^2 - 81$
5. $x^2 + 7x$
6. $a^2 - b^2$
7. $x^2 - 6x + 9$
8. $3ab + 3b^2$
9. $3a^2 + 6ab$
10. $5a^2 - ab$
11. $x^2 + 4x + 4$
12. $x^2 - y^2$

10.2 Simplifying Rational Expressions

The Product Rule for Fractions can be applied to rational expressions.

Product Rule for Rational Expressions

If $\frac{a}{b}$ and $\frac{c}{d}$ are rational expressions,

then $\frac{\mathbf{a}}{\mathbf{b}} \cdot \frac{\mathbf{c}}{\mathbf{d}} = \frac{\mathbf{ac}}{\mathbf{bd}}$. (*b* and *d* are not zero.)

$$\frac{x}{x+1} \cdot \frac{x+3}{x-2} = \frac{x^2+3x}{x^2-x-2}$$

You can simplify a rational expression by applying the Product Rule for Rational Expressions.

In working with rational expressions, it will be assumed that no divisor equals 0.

EXAMPLE 1 Simplify $\frac{10x}{15x^2}$.

Solution:

1 Write the prime factorizations. → $\frac{10x}{15x^2} = \frac{2 \cdot 5 \cdot x}{3 \cdot 5 \cdot x \cdot x}$

2 Product Rule for Rational Expressions → $= \frac{5x}{5x} \cdot \frac{2}{3x}$ ◀ ***$\frac{5x}{5x}$ is a name for 1.***

3 Multiplication Property of One → $= \frac{2}{3x}$ ◀ ***Simplest form***

EXAMPLE 2 Simplify $\frac{x^2+x}{x^2+2x+1}$.

Solution:

1 $\frac{x^2+x}{x^2+2x+1} = \frac{x(x+1)}{(x+1)(x+1)}$

2 $= \frac{x+1}{x+1} \cdot \frac{x}{x+1}$

3 $= \frac{x}{x+1}$ ◀ ***x is not a factor of x + 1.***

P–1 **What are the prime factors of $x^2 - 4$? of $x^2 + x - 6$?**

EXAMPLE 3

Simplify $\frac{x^2 - 4}{x^2 + x - 6}$.

Solution:

1. $\frac{x^2 - 4}{x^2 + x - 6} = \frac{(x + 2)(x - 2)}{(x - 2)(x + 3)}$

2. $= \frac{x - 2}{x - 2} \cdot \frac{x + 2}{x + 3}$

3. $= \frac{x + 2}{x + 3}$

EXAMPLE 4

Simplify $\frac{6x^2y - 3xy^2}{4x^2 - y^2}$.

Solution: The method shown below for identifying a name for 1 is often used to save work.

$$\frac{6x^2y - 3xy^2}{4x^2 - y^2} = \frac{3xy(2x - y)}{(2x + y)(2x - y)}$$ ◀ ***Prime factorization***

$$= \frac{3xy\cancel{(2x - y)}^{1}}{(2x + y)\cancel{(2x - y)}_{1}}$$ ◀ ***Note how a name for 1 is shown.***

$$= \frac{3xy}{2x + y}$$ ◀ ***Simplest form***

Steps for simplifying rational expressions:

1. Write the prime factorizations of both polynomials.
2. Form a name for 1 from all factors common to the dividend and divisor.
3. Use this name for 1 and write an indicated product equivalent to the given rational expression.
4. Apply the Multiplication Property of One and write the result in the simplest form.

CLASSROOM EXERCISES

Simplify. No divisor equals 0. (Examples 1–3)

1. $\frac{2}{6}$
2. $\frac{2x}{3x}$
3. $\frac{x}{x^2}$
4. $\frac{3x}{3(x-2)}$
5. $\frac{3(x+1)}{x(x+1)}$
6. $\frac{5(x^2+2)}{x^2+2}$
7. $\frac{x+2}{x+3}$
8. $\frac{(x+1)(x-2)}{(x+1)}$
9. $\frac{x(x+2)}{x^2(x+2)}$
10. $\frac{x-1}{x^2-1}$

WRITTEN EXERCISES

Goal: To simplify rational expressions

Sample Problem: Simplify $\frac{9x^2+18x}{x^2+4x+4}$. **Answer:** $\frac{9x}{x+2}$

Simplify. No divisor equals 0. (Examples 1–4)

1. $\frac{5x^2}{10x}$
2. $\frac{18x}{10x^2}$
3. $\frac{12x^3}{28x^4}$
4. $\frac{20x^3}{50x^2}$
5. $\frac{x(x-3)}{5(x-3)}$
6. $\frac{3(x+10)}{x(x+10)}$
7. $\frac{2x(x-5)}{3x(x-5)}$
8. $\frac{5x(x+7)}{4x(x+7)}$
9. $\frac{(x+2)(x-3)}{(x-3)(x+5)}$
10. $\frac{(x-4)(x+2)}{(x+2)(x-1)}$
11. $\frac{4x-4}{3x+9}$
12. $\frac{3x+3}{5x+5}$
13. $\frac{3x-6}{5x-10}$
14. $\frac{4x+12}{3x+9}$
15. $\frac{x^2+3x}{x^2+5x}$
16. $\frac{x^2-5x}{x^2+7x}$
17. $\frac{x+3}{x^2-9}$
18. $\frac{x-5}{x^2-25}$
19. $\frac{x^2-16}{x(x-4)}$
20. $\frac{x^2-36}{x(x+6)}$
21. $\frac{x^2+2x}{x^2-4}$
22. $\frac{x^2-49}{x^2+7x}$
23. $\frac{x(x-2)}{x^2-4x+4}$
24. $\frac{x^2+6x+9}{x(x+3)}$
25. $\frac{4x^2y}{10xy^2}$
26. $\frac{16a^2b}{12ab^2}$
27. $\frac{2a+2b}{2c}$
28. $\frac{3t}{3x-3y}$
29. $\frac{a^2-b^2}{3a+3b}$
30. $\frac{5x+5y}{x^2-y^2}$
31. $\frac{5a^2-ab}{3a^2+6ab}$
32. $\frac{3a^2+3ab}{3ab+3b^2}$

REVIEW CAPSULE FOR SECTION 10.3

Write the prime factors of each expression. (Pages 236–238)

1. $12x^2y^3$
2. $20r^3st^2$
3. $18a^2b^4c^3$
4. $36p^5q^2r^3$

10.3 Multiplying Rational Expressions

When multiplying fractions, you always simplify the product. This also applies to multiplying rational expressions.

P–1 **Write each product in simplest form.**

a. $\frac{10}{3} \cdot \frac{12}{5}$ **b.** $\frac{2}{6} \cdot \frac{12}{5}$ **c.** $\frac{5}{6} \cdot \frac{21}{10}$ **d.** $\frac{2}{1} \cdot \frac{3}{10} \cdot \frac{1}{12}$

EXAMPLE 1 Multiply and simplify: $\frac{10x}{3} \cdot \frac{12}{5x^2}$

Solution:

1 Product Rule for Rational Expressions ⟶ $\frac{10x}{3} \cdot \frac{12}{5x^2} = \frac{(10x)(12)}{(3)(5x^2)}$

2 Factor. ⟶ $= \frac{(2 \cdot 5 \cdot x)(2 \cdot 2 \cdot 3)}{(3)(5 \cdot x \cdot x)}$

3 Product Rule for Rational Expressions ⟶ $= \frac{3 \cdot 5 \cdot x}{3 \cdot 5 \cdot x} \cdot \frac{2 \cdot 2 \cdot 2}{x}$

4 Multiplication Property of One ⟶ $= \frac{2 \cdot 2 \cdot 2}{x}$

5 Simplify. ⟶ $= \frac{8}{x}$

Steps can often be saved by identifying common factors first.

EXAMPLE 2 Multiply and simplify: $\frac{2x}{6x^2} \cdot \frac{12x}{5x^2}$

Solution: $\frac{2x}{6x^2} \cdot \frac{12x}{5x^2} = \frac{\overset{1}{2\cancel{x}} \cdot \overset{2 \cdot 1}{\cancel{12}\cancel{x}}}{\underset{1 \cdot x}{\cancel{6}\cancel{x^2}} \cdot \underset{x}{5\cancel{x^2}}}$ ◀ ***x*, 6, and *x* are common factors.**

$= \frac{4}{5x^2}$ ◀ ***Simplest form***

EXAMPLE 3

Multiply and simplify: $\frac{5xy}{6a} \cdot \frac{21ab}{10y^2}$

More than one variable

Solution:

$$\frac{5xy}{6a} \cdot \frac{21ab}{10y^2} = \frac{\overset{1 \cdot 1}{\cancel{5}x\cancel{y}}}{\underset{2 \cdot 1}{\cancel{6}\cancel{a}}} \cdot \frac{\overset{7 \cdot 1}{\cancel{21}\cancel{a}b}}{\underset{2 \cdot y}{\cancel{10}\cancel{y^2}}}$$

$$= \frac{1 \cdot x \cdot 1}{2 \cdot 1} \cdot \frac{7 \cdot 1 \cdot b}{2 \cdot y}$$

$$= \frac{x}{2} \cdot \frac{7b}{2y} = \frac{7bx}{4y}$$

EXAMPLE 4

Multiply and simplify: $\frac{2}{x} \cdot \frac{3x}{10} \cdot \frac{x^2}{12}$

Solution:

$$\frac{2}{x} \cdot \frac{3x}{10} \cdot \frac{x^2}{12} = \frac{\overset{1}{\cancel{2}}}{\underset{1}{\cancel{x}}} \cdot \frac{\overset{1 \cdot 1}{\cancel{3}\cancel{x}}}{\underset{5}{\cancel{10}}} \cdot \frac{x^2}{\underset{4}{\cancel{12}}}$$

$$= \frac{x^2}{20}$$

P–2 **How do you know that $\frac{x^2}{20}$ is in simplest form?**

CLASSROOM EXERCISES

Multiply. Simplify where necessary. (Examples 1 and 2)

1. $\frac{2}{x} \cdot \frac{x}{5}$ **2.** $\frac{y}{3} \cdot \frac{2}{y}$ **3.** $\frac{x}{5} \cdot \frac{5}{x}$ **4.** $\frac{3a}{5} \cdot \frac{7}{3a}$

5. $\frac{2}{3} \cdot \frac{x}{y}$ **6.** $\frac{7}{x} \cdot \frac{5}{x}$ **7.** $\frac{a}{3} \cdot \frac{5}{2a}$ **8.** $\frac{5x}{7} \cdot \frac{6}{x}$

9. $\frac{3}{a^2} \cdot \frac{a}{5}$ **10.** $-\frac{x}{7} \cdot \frac{5}{x^2}$ **11.** $\frac{2}{x} \cdot \frac{x}{5}$ **12.** $\frac{x}{3} \cdot \frac{x^2}{5}$

Multiply. Simplify where necessary. (Examples 1–4)

13. $\frac{2x}{3} \cdot \frac{1}{10x}$ **14.** $\frac{1}{5x} \cdot 15x$ **15.** $\frac{3a}{b} \cdot 2b^2$ **16.** $\frac{3x^2}{2} \cdot \frac{2}{x}$

17. $\left(-\frac{1}{3}\right)\left(-\frac{2}{y}\right)\left(-\frac{x}{2}\right)$ **18.** $\left(-\frac{1}{3}\right)\left(\frac{3}{x}\right)\left(-\frac{x}{5}\right)$ **19.** $\left(\frac{2a}{b}\right)\left(\frac{b}{3}\right)\left(\frac{3}{2a}\right)$ **20.** $\frac{x}{y^2} \cdot \frac{y}{x^2}$

WRITTEN EXERCISES

Goal: To multiply rational expressions

Sample Problem: $\frac{3x}{4a} \cdot \frac{5a^2}{9} \cdot \frac{6}{x^2}$ **Answer:** $\frac{5a}{2x}$

Multiply. Simplify where necessary. (Examples 1–4)

1. $\frac{3}{8} \cdot \frac{5}{9}$ **2.** $\frac{4}{5} \cdot \frac{3}{7}$ **3.** $\frac{5}{x} \cdot \frac{y}{3}$

4. $\frac{a}{7} \cdot \frac{4}{b}$ **5.** $\frac{2x}{5} \cdot \frac{3}{2x}$ **6.** $\frac{6}{7x} \cdot \frac{7x}{11}$

7. $\frac{4x}{3} \cdot \frac{7}{10x}$ **8.** $\frac{5}{12x} \cdot \frac{4x}{7}$ **9.** $\left(-\frac{x^2}{y}\right)\left(-\frac{2y}{x}\right)$

10. $\left(-\frac{a}{b^2}\right)\left(-\frac{2b}{a^2}\right)$ **11.** $\frac{8xy}{3} \cdot \frac{15}{2x^2}$ **12.** $\frac{3}{10ab} \cdot \frac{14b^2}{15}$

13. $-\frac{21b^2}{a^2} \cdot \frac{3ab}{15b}$ **14.** $\left(\frac{5x^2y}{10x}\right)\left(-\frac{6y^2}{21xy}\right)$ **15.** $\frac{5y}{12x}(15x^2)$

16. $(21a^2)\frac{5x}{6ab}$ **17.** $\frac{36ab^2}{5c} \cdot \frac{20ac^2}{24b}$ **18.** $\frac{4x}{18yz^2} \cdot \frac{12xyz}{30x^2}$

19. $\frac{a}{b} \cdot \frac{b}{c} \cdot \frac{c}{a}$ **20.** $\frac{x}{y} \cdot \frac{z}{x} \cdot \frac{y}{z}$ **21.** $\frac{2a^2}{5b} \cdot \frac{21b^2}{14c} \cdot \frac{11c^2}{10b}$

22. $\frac{4xy}{15z^2} \cdot \frac{3y}{10y^2} \cdot \frac{25yz}{7x^2}$ **23.** $\frac{-5}{a} \cdot \frac{3a^2}{-b} \cdot \frac{-b^2}{10}$ **24.** $\frac{-x}{-y^2} \cdot \frac{-3y}{4x} \cdot \frac{-8xy}{-z}$

NON-ROUTINE PROBLEMS

25. The time now is between 10:00 A.M. and 11:00 A.M. Fourteen minutes from now, it will be as many minutes before 12:00 noon as it is past 10:00 A.M. now. What time is it right now?

26. Sally's grandfather has lived a sixth of his life in Ohio, a fourth of his life in Iowa, a third of his life in Illinois, and the other 21 years of his life in Wisconsin. How old is Sally's grandfather?

10.4 Binomial Numerators and Denominators

Binomials such as $x - 2$ and $x - 1$ are prime polynomials. The variable x is not a factor of $x - 2$ and $x - 1$. It is a term of each binomial.

EXAMPLE 1 Multiply and simplify: $\frac{x-2}{x+3} \cdot \frac{x+3}{x-1}$

Solution:

1 Product Property → $\frac{x-2}{x+3} \cdot \frac{x+3}{x-1} = \frac{(x-2)(x+3)}{(x+3)(x-1)}$

2 Commutative Property of Multiplication → $= \frac{(x+3)(x-2)}{(x+3)(x-1)}$

3 Product Property → $= \frac{x+3}{x+3} \cdot \frac{x-2}{x-1}$

4 Multiplication Property of 1 → $= \frac{x-2}{x-1}$ ◀ ***Simplest form***

EXAMPLE 2 Multiply and simplify: $\frac{x(x+3)}{x-5} \cdot \frac{x-5}{2x}$

Solution:

$$\frac{x(x+3)}{x-5} \cdot \frac{x-5}{2x} = \frac{\overset{1}{\cancel{x}}(x+3)\overset{1}{\cancel{(x-5)}}}{\underset{1}{\cancel{(x-5)}}2\underset{1}{\cancel{x}}}$$

◀ ***A name for 1 is*** $\frac{x(x-5)}{(x-5)x}$.

$$= \frac{x+3}{2}$$

◀ ***Simplest form***

P–1 **Write $-(a - b)$ without parentheses.**

You can use the Property of the Opposite of a Sum to write $-(a - b)$ without parentheses.

$$\begin{aligned} -(a-b) &= -(a + -b) \\ &= (-a) + -(-b) \\ &= (-a) + b \\ &= b + (-a) \\ &= b - a \end{aligned}$$

◀ ***Thus, a − b and b − a are opposites.***

Property of the Opposite of a Difference

If a and b are any real numbers, then $\mathbf{-(a - b) = b - a}$.

$-(y - 10) = 10 - y$

$-(p - q) = q - p$

P–2 **Write the opposite of each expression.**

a. $x - 2$ **b.** $5 - t$ **c.** $r - m$

P–3 **What binomial is the opposite of $x - 2$?**

EXAMPLE 3

Multiply and simplify: $\frac{x+3}{x-2} \cdot \frac{2-x}{x-5}$

Solution:

1. Property of the Opposite of a Difference → $\frac{x+3}{x-2} \cdot \frac{2-x}{x-5} = \frac{(x+3) \cdot -(x-2)}{(x-2)(x-5)}$
2. Multiplication Property of -1 → $= \frac{(x+3) \cdot -1(x-2)}{(x-2)(x-5)}$
3. Product Rule for Rational Expressions → $= \frac{x-2}{x-2} \cdot \frac{-1(x+3)}{x-5}$
4. Multiplication Property of 1 → $= \frac{-1(x+3)}{x-5}$
5. Simplest form → $= -\frac{x+3}{x-5}$

EXAMPLE 4

Multiply: $\frac{x+2}{x-3} \cdot \frac{x+1}{3}$

Solution:

$$\frac{x+2}{x-3} \cdot \frac{x+1}{3} = \frac{(x+2)(x+1)}{(x-3)3}$$

$$= \frac{x^2 + 3x + 2}{3x - 9}$$

P–4 **What is the polynomial form of $(x + 2)(x + 1)$? of $(x - 3)3$?**

CLASSROOM EXERCISES

Multiply. Simplify where necessary. (Examples 1–4)

1. $\frac{2}{x-2}\cdot\frac{x-2}{3}$
2. $\frac{x+5}{5x}\cdot\frac{3x}{x+5}$
3. $\frac{x+1}{x+2}\cdot\frac{x+3}{x+4}$
4. $\frac{2(x-3)}{x+2}\cdot\frac{x+2}{6}$
5. $\frac{x+2}{x-1}\cdot\frac{x-1}{x+2}$
6. $\frac{x+5}{x-2}(x-2)$
7. $\frac{1}{x-3}\cdot\frac{x-3}{x+2}$
8. $\frac{1}{x+2}(x+2)$
9. $\frac{5}{x-5}\cdot\frac{5-x}{6}$
10. $\frac{2}{-a-2}\cdot\frac{a+2}{5}$
11. $\frac{x-y}{b-a}\cdot\frac{a-b}{y-x}$
12. $\frac{2}{a-b}\cdot\frac{a+b}{3}$
13. $\frac{3(x+1)}{x-3}\cdot\frac{x-3}{3x}$
14. $\frac{y(x+y)}{x-y}\cdot\frac{2}{y(x+y)}$
15. $\frac{a-2}{3}\cdot\frac{2}{2-a}$

WRITTEN EXERCISES

Goal: To multiply rational expressions that involve binomials

Sample Problem: $\frac{x-2}{2}\cdot\frac{x+5}{2-x}$ **Answer:** $-\frac{x+5}{2}$

Multiply. Simplify where necessary. (Examples 1–4)

1. $\frac{11}{x-2}\cdot\frac{x-2}{12}$
2. $\frac{x+7}{9}\cdot\frac{5}{x+7}$
3. $-\frac{7}{x+y}\cdot\frac{2(x+y)}{15}$
4. $-\frac{2(a+b)}{11}\cdot\frac{5}{a+b}$
5. $\frac{x-3}{x+2}\cdot\frac{x+2}{x-1}$
6. $\frac{a+7}{a-3}\cdot\frac{a-1}{a+7}$
7. $\frac{x(x-3)}{y(x+5)}\cdot\frac{x+5}{2x}$
8. $\frac{a+6}{3(a+2)}\cdot\frac{12a(a+2)}{4b}$
9. $\frac{x-3}{x+2}(x+2)$
10. $\frac{a+7}{a-5}(a-5)$
11. $\frac{1}{x-3}\cdot\frac{2(x-3)}{x+2}$
12. $\frac{x(x-y)}{x+y}\cdot\frac{1}{x}$
13. $\frac{x-3}{x+2}\cdot\frac{x-2}{x+3}$
14. $\frac{2a+b}{a-b}\cdot\frac{a+b}{a+2b}$
15. $\frac{x-5}{x}\cdot\frac{5}{5-x}$
16. $\frac{a}{a-7}\cdot\frac{7-a}{5}$
17. $\frac{a-3}{a+4}\cdot\frac{-a-4}{3-a}$
18. $\frac{-x-2}{x-3}\cdot\frac{3-x}{x+2}$
19. $\frac{2(x-y)}{2x-3y}\cdot\frac{2x-3y}{6(y-x)}$
20. $\frac{-2a-3b}{15(a-b)}\cdot\frac{6(b-a)}{2a+3b}$
21. $\frac{x+2}{(x-1)(x+3)}\cdot\frac{x(x-1)}{x+2}$
22. $\frac{(x-5)(x-2)}{x+3}\cdot\frac{x+3}{2(x-2)}$

APPLICATIONS

A fraction such as the one shown at the right is called a ***continued fraction.*** *You can use a calculator to approximate its value.*

$$\cfrac{1}{1+\cfrac{1}{1+\cfrac{1}{1+\frac{1}{5}}}}$$

EXAMPLE Find a decimal approximation of the above continued fraction.

SOLUTION Start by finding the reciprocal of 5.

[5] [÷] [=] [+] [1] [=] [÷] [=] [+] [1]

[=] [÷] [=] [+] [1] [=] [÷] [=]

On some calculators press "[÷] [=] [=]" to get the reciprocal.

0.6470588

Find a decimal approximation.

23. $\cfrac{1}{1+\cfrac{1}{1+\cfrac{1}{1+\frac{1}{7}}}}$

24. $\cfrac{1}{1+\cfrac{1}{1+\cfrac{1}{1+\frac{1}{8}}}}$

25. $\cfrac{1}{1+\cfrac{1}{1+\cfrac{1}{1+\cfrac{1}{1+\frac{1}{11}}}}}$

MID-CHAPTER REVIEW

Evaluate each rational expression if x equals -2. (Section 10.1)

1. $\frac{x-5}{x+7}$ **2.** $\frac{x^2+3}{x^2-2}$ **3.** $\frac{x^2-x+3}{2x^2+x-1}$ **4.** $\frac{-x^2+2x-5}{3x^2-x+2}$

Simplify. No divisor equals 0. (Section 10.2)

5. $\frac{3x(x+5)}{6y(x+5)}$ **6.** $\frac{5x-15}{10x+20}$ **7.** $\frac{x^2-36}{2x^2-12x}$ **8.** $\frac{x^2-2x-8}{3x^2-12}$

Multiply. Simplify where necessary. (Section 10.3)

9. $\frac{6x}{35}\cdot\frac{14}{9x}$ **10.** $\frac{21a}{8bc}\cdot\frac{6c^2}{7a^2b}$ **11.** $-\frac{5r^2}{4st^2}\cdot\frac{10s^2}{3r}$ **12.** $\frac{mn^2}{6}\cdot\frac{4}{-m^2}\cdot\frac{-3}{10n}$

Multiply. (Section 10.4)

13. $\frac{3(x-5)}{2x}\cdot\frac{4x}{5(x-5)}$ **14.** $\frac{n+5}{n-2}\cdot\frac{n+3}{n+5}$ **15.** $\frac{2r-3}{3r+5}\cdot(3r+5)$ **16.** $\frac{m-3}{5m}\cdot\frac{10}{3-m}$

REVIEW CAPSULE FOR SECTION 10.5

Factor. (Pages 246–248)

1. x^2-2x-3 **2.** x^2-5x+6 **3.** $2x^2+5x-3$ **4.** $4x^2+4x-3$

10.5 Trinomial Numerators and Denominators

Note, in Example 1, that numerators and denominators are factored <u>before</u> multiplying the rational expressions.

EXAMPLE 1 Multiply and simplify: $\frac{x^2 - x - 6}{x^2 + 2x + 1} \cdot \frac{x + 1}{x + 2}$

Solution:

[1] Factor. ⟶ $\frac{x^2 - x - 6}{x^2 + 2x + 1} \cdot \frac{x + 1}{x + 2} = \frac{(x - 3)(x + 2)}{(x + 1)(x + 1)} \cdot \frac{x + 1}{x + 2}$ ◀ ***Factor <u>first</u>.***

[2] Product Rule for Rational Expressions ⟶ $= \frac{(x - 3)(x + 2)(x + 1)}{(x + 1)(x + 1)(x + 2)}$

[3] Identify common factors. ⟶ $= \frac{(x - 3)\overset{1}{\cancel{(x + 2)}}\overset{1}{\cancel{(x + 1)}}}{(x + 1)\underset{1}{\cancel{(x + 1)}}\underset{1}{\cancel{(x + 2)}}}$

[4] Simplify. ⟶ $= \frac{x - 3}{x + 1}$

P–1 **What are the prime factors of $x^2 + 2x$? of $x^2 - 4$?**

EXAMPLE 2 Multiply and simplify: $\frac{x^2 + 2x}{x^2 - 4} \cdot \frac{x - 2}{x + 2}$

Solution: $\frac{x^2 + 2x}{x^2 - 4} \cdot \frac{x - 2}{x + 2} = \frac{x(x + 2)}{(x + 2)(x - 2)} \cdot \frac{x - 2}{x + 2}$ ◀ ***Prime factorization***

$= \frac{x\overset{1}{\cancel{(x + 2)}}}{(x + 2)\underset{1}{\cancel{(x - 2)}}} \cdot \frac{\overset{1}{\cancel{x - 2}}}{\underset{1}{\cancel{x + 2}}}$ ◀ ***A name for 1 is identified.***

$= \frac{x}{x + 2}$ ◀ ***Simplest form***

P–2 **Write the prime factors of each polynomial below.**

a. $2x^2 + x - 1$ **b.** $x^2 - 36$ **c.** $6x^3 - 4x^2$

It may be helpful to review each of these types of factoring.

$$2x^2 + x - 1 = 2x^2 + rx + sx - 1$$

$r + s = 1$; $r \cdot s = -2$; Let $r = 2$ and $s = -1$.

$$= 2x^2 + 2x - x - 1$$

$$= 2x(x + 1) - 1(x + 1)$$

$$= (2x - 1)(x + 1)$$

$$x^2 - 36 = (x + 6)(x - 6)$$

Difference of two squares

$$6x^3 - 4x^2 = 2x^2(3x - 2)$$

Common monomial factor

P–3 **What are the prime factors of $x^2 - 3x - 10$? of $x^2 - 9$? of $x^2 + 4x + 4$?**

EXAMPLE 3

Multiply and simplify: $\dfrac{x^2 - 3x - 10}{x^2 - 9} \cdot \dfrac{x + 3}{x^2 + 4x + 4}$

Solution:

$$\frac{x^2 - 3x - 10}{x^2 - 9} \cdot \frac{x + 3}{x^2 + 4x + 4} = \frac{(x - 5)(x + 2)}{(x + 3)(x - 3)} \cdot \frac{x + 3}{(x + 2)(x + 2)}$$

$$= \frac{(x - 5)\overset{1}{\cancel{(x + 2)}}\overset{1}{\cancel{(x + 3)}}}{\underset{1}{\cancel{(x + 3)}}(x - 3)\underset{1}{\cancel{(x + 2)}}(x + 2)}$$

$$= \frac{x - 5}{(x - 3)(x + 2)}$$

Steps for multiplying rational expressions:

1. Factor all numerators and denominators.
2. Apply the Product Rule for Rational Expressions.
3. Identify a name for 1.
4. Use the Multiplication Property of One and simplify the result.

CLASSROOM EXERCISES

Simplify. (Examples 1–3)

1. $\frac{(x-2)(x+3)}{(x+3)(x+4)}$
2. $\frac{(x+5)(x+2)}{(x+5)(x+5)(x+2)}$
3. $\frac{(x+5)(x-2)}{x^2-4}$
4. $\frac{x-3}{x+7}\cdot\frac{x+7}{x+3}$
5. $\frac{x^2-2x+1}{2x}\cdot\frac{x}{x-1}$
6. $\frac{x^2+2x}{x+3}\cdot\frac{x+3}{x(x-1)}$
7. $\frac{5x}{x^2-9}\cdot\frac{x(x-3)}{x^2-4x+3}$
8. $\frac{x^2-x}{2x+4}\cdot\frac{x^2-4}{x-1}$
9. $\frac{x^2-x-2}{x-3}\cdot\frac{3x-9}{3x^2-6x}$

WRITTEN EXERCISES

Goal: To multiply rational expressions that involve trinomials

Sample Problem: $\frac{x^2-7x+12}{x^2-16}\cdot\frac{x+4}{3x-9}$ **Answer:** $\frac{1}{3}$

Multiply and simplify. (Examples 1–3)

1. $\frac{3x+6}{x^2+x}\cdot\frac{4x+4}{12}$
2. $\frac{x^2-2x}{15}\cdot\frac{3x-9}{2x-4}$
3. $\frac{x^2-4}{x+2}\cdot\frac{5x+10}{3x-6}$
4. $\frac{2x-6}{25}\cdot\frac{5x+15}{x^2-9}$
5. $\frac{x^3-2x^2}{x^2-25}\cdot\frac{5x+10}{x^2-2x}$
6. $\frac{9y^2+9}{y^2-1}\cdot\frac{2y^3+2y^2}{3y}$
7. $\frac{x^2-x-2}{10}\cdot\frac{2x+4}{x^2-4}$
8. $\frac{x^2-2x-3}{3x+3}\cdot\frac{15}{x^2-9}$
9. $\frac{x^2+8x+16}{12}\cdot\frac{3x+3}{x^2+5x+4}$
10. $\frac{x^2-5x+6}{5x-10}\cdot\frac{15}{x^2-6x+9}$
11. $\frac{x^2-16}{2x}\cdot\frac{3x^2-6x}{16-x^2}$
12. $\frac{5a}{a^2-9}\cdot\frac{9-a^2}{5a^2-a}$

MORE CHALLENGING EXERCISES

13. $\frac{2x^2-5x-3}{4x^2-1}\cdot\frac{3}{x^2+2x-15}$
14. $\frac{3x^2-14x-5}{9x^2-1}\cdot\frac{1}{x^2-6x+5}$
15. $\frac{2x^2-4x+8}{6x^2+7x-3}(2x^2+x-3)$
16. $\frac{3x^2+15x-3}{3x^2-27}(3x^2-7x-6)$
17. If $\frac{x}{6}\cdot\frac{3}{y}=1$, how does the value of x compare to the value of y?
18. If $\frac{ax^2+ax}{6x}\cdot\frac{2x-6}{x^2-2x-3}=3$, what is the value of a?
19. If $\frac{a}{b}\cdot\frac{3}{5}=1$, can you conclude that the value of a is 5 and the value of b is 3? Explain.
20. Write a rational expression that, when multiplied by $\frac{x^2+6x+5}{x+1}$, gives a product of 3.

10.6 Dividing Rational Expressions

The Quotient Rule for Fractions can be applied to rational expressions.

Quotient Rule for Rational Expressions

If $\frac{a}{b}$ and $\frac{c}{d}$ are rational expressions,

then $\frac{\mathbf{a}}{\mathbf{b}} \div \frac{\mathbf{c}}{\mathbf{d}} = \frac{\mathbf{a}}{\mathbf{b}} \cdot \frac{\mathbf{d}}{\mathbf{c}}$. ($b$, c, and d are not zero.)

Dividing by a rational expression is the same as multiplying by its reciprocal.

$$\frac{x}{x+2} \div \frac{x-1}{x}$$
$$\frac{x}{x+2} \cdot \frac{x}{x-1}$$
$$\frac{x^2}{(x+2)(x-1)}$$

Expressions such as $\frac{x}{x+1}$ and $\frac{x+1}{x}$ are reciprocals. The product of a rational expression and its reciprocal is 1.

P–1 **Write each quotient below as a product.**

a. $\frac{x-3}{x+2} \div \frac{x}{x-2}$ **b.** $\frac{3x}{5} \div \frac{4}{9x}$ **c.** $\frac{2x^2}{5} \div (x+1)$

EXAMPLE 1 Divide and simplify: $\frac{2x}{3} \div \frac{5x}{6}$

Solution:

$$\frac{2x}{3} \div \frac{5x}{6} = \frac{2x}{3} \cdot \frac{6}{5x}$$

$$= \frac{\overset{1}{\cancel{2x}}}{\underset{1}{\cancel{3}}} \cdot \frac{\overset{2}{\cancel{6}}}{\underset{1}{5\cancel{x}}}$$

$$= \frac{4}{5}$$ **_Simplest form_**

P–2 **Write $\frac{x-2}{x} \div \frac{x-2}{5}$ as a product.**

EXAMPLE 2 Divide and simplify: $\frac{x-2}{x} \div \frac{x-2}{5}$

Solution:

$$\frac{x-2}{x} \div \frac{x-2}{5} = \frac{x-2}{x} \cdot \frac{5}{x-2}$$

$$= \frac{5\overset{1}{\cancel{(x-2)}}}{x\underset{1}{\cancel{(x-2)}}}$$

$$= \frac{5}{x}$$

P–3 **What is the reciprocal of $\frac{9pq}{4(r-3)}$?**

EXAMPLE 3 Divide and simplify: $\frac{3p^2q}{2r(r-3)} \div \frac{9pq}{4(r-3)}$

Solution:

$$\frac{3p^2q}{2r(r-3)} \div \frac{9pq}{4(r-3)} = \frac{3p^2q}{2r(r-3)} \cdot \frac{4(r-3)}{9pq}$$

$$= \frac{\overset{1 \cdot p \cdot 1}{\cancel{3}\cancel{p^2}\cancel{q}} \cdot \overset{2}{\cancel{4}}\overset{1}{\cancel{(r-3)}}}{\underset{1}{2r}\underset{1}{\cancel{(r-3)}} \cdot \underset{3 \cdot 1 \cdot 1}{\cancel{9}\cancel{p}\cancel{q}}}$$

$$= \frac{2p}{3r}$$

CLASSROOM EXERCISES

Write each quotient as an equivalent product. (P-1, P-2)

1. $\frac{2}{3} \div \frac{5}{6}$ **2.** $\frac{x}{5} \div \frac{3}{x}$ **3.** $\frac{2x}{6} \div \frac{x}{3}$ **4.** $\frac{x}{3y} \div \frac{3x}{y}$

5. $\frac{2}{x-2} \div \frac{x}{5}$

6. $\frac{x+1}{2} \div \frac{3}{5}$

7. $\frac{x-7}{5x} \div \frac{x+1}{4}$

8. $\frac{x+3}{x} \div \frac{x-5}{2x}$

9. $\frac{1}{x+2} \div (x-3)$

10. $(x+3) \div \frac{1}{x+5}$

11. $\frac{x+1}{x} \div \frac{x-10}{2x}$

12. $\frac{x+3}{x-4} \div \frac{x+7}{x-5}$

13. $\frac{x+8}{x-5} \div \frac{x+2}{x-10}$

WRITTEN EXERCISES

Goal: To divide rational expressions

Sample Problem: $\frac{5a^2b^2}{a(1-b)} \div \frac{25ab^2}{b(1-b)}$ **Answer:** $\frac{b}{5}$

Divide and simplify. (Examples 1–3)

1. $\frac{2}{5} \div \frac{3}{5}$

2. $\frac{5}{8} \div \frac{3}{16}$

3. $\frac{3}{x} \div \frac{1}{2x}$

4. $\frac{x}{5} \div \frac{3x}{10}$

5. $\frac{a^2}{3} \div \frac{5a}{9}$

6. $\frac{2}{b^2} \div \frac{4}{3b}$

7. $\frac{3}{x-5} \div \frac{6}{x-5}$

8. $\frac{x+2}{5} \div \frac{x+2}{3}$

9. $\frac{x^2y}{7} \div \frac{3x}{14}$

10. $\frac{5}{2a^2b} \div \frac{15}{ab}$

11. $\frac{2}{x^2y^2} \div \frac{4}{5xy}$

12. $\frac{5xy}{3z} \div \frac{10x^2}{9z^2}$

13. $\frac{x-5}{x+2} \div \frac{5}{x+2}$

14. $\frac{3}{a+b} \div \frac{a+3}{a+b}$

15. $\frac{x^2+2}{5} \div \frac{x^2}{10}$

16. $\frac{3}{x+6} \div \frac{9}{x}$

17. $\frac{x-6}{x+5} \div 2(x-6)$

18. $\frac{x+3}{x-10} \div 5(x+3)$

19. $x(x^2+3) \div \frac{3x}{x^2+3}$

20. $3a(a^2+2) \div \frac{a^2}{a-2}$

MORE CHALLENGING EXERCISES

21. $\frac{5}{x} \div \frac{3}{5x} \div \frac{10}{3x}$

22. $\frac{a}{3} \div \frac{a^2}{2} \div \frac{7}{3a}$

23. $\frac{x+1}{x+3} \div \frac{x-1}{2(x+3)} \div \frac{3(x-1)}{x+1}$

NON-ROUTINE PROBLEM

24. The figure at the right shows three different views of a cube.
    a. What shape is shown on that side of the cube which is opposite the side showing a circle?
    b. What shape is shown on that side of the cube which is opposite the side showing a triangle?

10.7 Quotients and Complex Fractions

P–1 **What is the reciprocal of $(x^2 - 25)$?**

EXAMPLE 1 Divide and simplify: $\dfrac{x^2+4x-5}{3x-3} \div (x^2-25)$

Solution:

1 Factor. ⟶ $\dfrac{x^2+4x-5}{3x-3} \div (x^2-25) = \dfrac{(x+5)(x-1)}{3(x-1)} \div (x+5)(x-5)$

2 Quotient Property of Rational Expressions ⟶ $= \dfrac{(x+5)(x-1)}{3(x-1)} \cdot \dfrac{1}{(x+5)(x-5)}$

3 Identify common factors. ⟶ $= \dfrac{\overset{1}{\cancel{(x+5)}}\overset{1}{\cancel{(x-1)}}}{3\underset{1}{\cancel{(x-1)}}} \cdot \dfrac{1}{\underset{1}{\cancel{(x+5)}}(x-5)}$

4 Simplify. ⟶ $= \dfrac{1}{3(x-5)}$

EXAMPLE 2 Divide and simplify: $\dfrac{x^2-3x-4}{6} \div \dfrac{x^2-1}{2x-2}$

Solution:

$$\frac{x^2-3x-4}{6} \div \frac{x^2-1}{2x-2} = \frac{(x-4)(x+1)}{6} \div \frac{(x+1)(x-1)}{2(x-1)}$$

$$= \frac{(x-4)\overset{1}{\cancel{(x+1)}}}{\underset{3}{\cancel{6}}} \cdot \frac{\overset{1\cdot 1}{\cancel{2}\cancel{(x-1)}}}{\underset{1}{\cancel{(x+1)}}\underset{1}{\cancel{(x-1)}}}$$

$$= \frac{x-4}{3}$$

A quotient such as $\left(\dfrac{x^2}{2} + \dfrac{x}{3}\right) \div \dfrac{x}{6}$ is often written as a fraction.

$$\left(\frac{x^2}{2} + \frac{x}{3}\right) \div \frac{x}{6} = \frac{\dfrac{x^2}{2} + \dfrac{x}{3}}{\dfrac{x}{6}}$$

Such fractions are called complex fractions.

EXAMPLE 3

Simplify $\dfrac{\dfrac{x^2}{2} + \dfrac{x}{3}}{\dfrac{x}{6}}$

First find a common multiple.

Solution:

1. Multiplication Property of One → $\dfrac{\dfrac{x^2}{2} + \dfrac{x}{3}}{\dfrac{x}{6}} = \dfrac{6}{6} \cdot \dfrac{\dfrac{x^2}{2} + \dfrac{x}{3}}{\dfrac{x}{6}}$

   6 is a common multiple of 2, 3, and 6.

2. Product Property of Rational Expressions → $= \dfrac{6\left(\dfrac{x^2}{2} + \dfrac{x}{3}\right)}{6\left(\dfrac{x}{6}\right)}$

3. Distributive Property → $= \dfrac{6\left(\dfrac{x^2}{2}\right) + 6\left(\dfrac{x}{3}\right)}{6\left(\dfrac{x}{6}\right)}$

4. Multiply. → $= \dfrac{3x^2 + 2x}{x}$

5. Factor. → $= \dfrac{x(3x + 2)}{x}$

6. Simplify. → $= 3x + 2$

Steps for simplifying a complex fraction:

1. Determine m, a common multiple of all the denominators.
2. Multiply the complex fraction by $\dfrac{m}{m}$.
3. Simplify the result.

CLASSROOM EXERCISES

Write the prime factors of each polynomial. (Step 1 of Examples 1 and 2)

1. $2x - 6$ **2.** $x^2 - 9$ **3.** $3x^2 - x$ **4.** $5x + 10$ **5.** $x^2 - 4$

6. $x^2 - 81$ **7.** $4x - 16$ **8.** $x^2 + 4x$ **9.** $3x - 21$ **10.** $x^2 - 49$

11. $x^2 - 7x$

12. $x^2 + x$

13. $2x^2 - 10x$

14. $x^2 - 10x + 25$

15. $x^2 - 16$

16. $5x - 20$

17. $3x^2 - 15x$

18. $x^2 + 2x - 15$

19. $5x^2 + 15x$

WRITTEN EXERCISES

Goal: To simplify complex fractions

Sample Problem: $\dfrac{\frac{a}{2} - \frac{a}{4}}{\frac{a}{6}}$ **Answer:** $\dfrac{3}{2}$

Divide and simplify. (Examples 1 and 2)

1. $\dfrac{2x - 6}{5x^2} \div \dfrac{x^2 - 9}{x}$

2. $\dfrac{3x^2 - x}{5x + 10} \div \dfrac{3x}{x^2 - 4}$

3. $\dfrac{x^2 - 16}{4x - 16} \div \dfrac{x^2 + 4x}{2x}$

4. $\dfrac{3x - 21}{x^2 - 49} \div \dfrac{3x}{x^2 - 7x}$

5. $\dfrac{x^2 - 10x + 25}{x + 1} \div \dfrac{2x^2 - 10x}{x^2 + x}$

6. $\dfrac{x^2 - 16}{6x} \div \dfrac{5x - 20}{3x^2 - 15x}$

7. $\dfrac{x^2 + 2x - 15}{x + 3} \div \dfrac{x^2 + 7x + 10}{x - 2}$

8. $\dfrac{x - 5}{x^2 + 3x - 10} \div \dfrac{9x^2}{3x^2 - 6x}$

9. $\dfrac{x^2 + 10x + 24}{3x^2 - 12x} \div (x^2 - 3x - 18)$

10. $\dfrac{x^2 - 2x - 15}{5x^2 + 15x} \div (x^2 - 6x + 5)$

11. $\dfrac{x^2 - y^2}{6x} \div \dfrac{x^2y + xy^2}{3x^2y^2}$

12. $\dfrac{4x^2 - 25y^2}{2x^2y + 5xy^2} \div \dfrac{6x^2 - 15xy}{9x^2y^2}$

Simplify. (Example 3)

13. $\dfrac{\frac{x}{5}}{\frac{x}{2} + \frac{x}{5}}$

14. $\dfrac{\frac{y}{3} + \frac{y}{5}}{\frac{y}{5}}$

15. $\dfrac{\frac{a}{6} + \frac{3}{5}}{\frac{2}{5}}$

16. $\dfrac{\frac{b}{8} + \frac{1}{3}}{\frac{5}{3}}$

17. $\dfrac{\frac{a}{10} + \frac{b}{5}}{\frac{a}{3} + \frac{b}{6}}$

18. $\dfrac{\frac{x}{6} + \frac{y}{8}}{\frac{x}{2} + \frac{y}{4}}$

MORE CHALLENGING EXERCISES

Divide and simplify.

19. $\dfrac{6x^2 - 7x - 3}{10x - 15} \div \dfrac{3x^2 - 14x - 5}{5x - 25}$

20. $\dfrac{10x^2 - 13x - 3}{15x + 3} \div \dfrac{2x^2 + 7x - 15}{3x^2 + 15x}$

21. $\dfrac{10x + 15}{2x^2 - 3x - 9} \div \dfrac{x^2 + 6x + 9}{9 - x^2}$

22. $\dfrac{3x^2 - 15x}{2x^2 + x} \div \dfrac{25 - x^2}{2x^2 + 7x + 5}$

Using Proportions

Electricity

Electricians keep electric equipment such as **transformers** in working order. The main parts of a transformer are the core, the primary coil to which the input power is supplied, and the secondary coil from which output power is delivered. The following **direct proportion** relates the input (V_P) and output (V_S) voltages and the number of turns on the primary (N_P) and the secondary (N_S) coils.

$$\frac{V_S}{V_P} = \frac{N_S}{N_P}$$

The ratio $\frac{N_S}{N_P}$ is called the turns ratio.

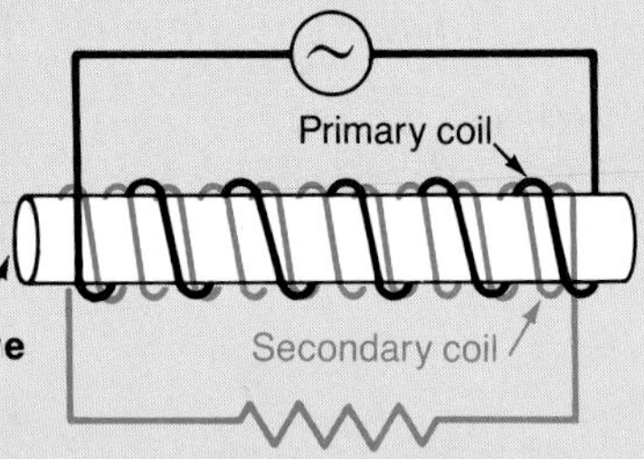

The following **indirect proportion** relates the input (I_P) and output (I_S) amperages and the number of turns on the coils.

$$\frac{I_S}{I_P} = \frac{N_P}{N_S}$$

Note that the turns ratio is inverted.

EXAMPLE: The number of turns on the primary coil of a transformer is 200. The number of turns on the secondary coil is 800. The input voltage is 110 volts. The input current is 1.2 amperes. Find **a.** the output voltage in volts and **b.** the output current in amperes.

SOLUTION: **a.** $\frac{V_S}{V_P} = \frac{N_S}{N_P}$ $\quad \frac{V_S}{110} = \frac{800}{200}$

$$V_S = \frac{(110)(800)}{200}$$

$$V_S = \mathbf{440\ volts}$$

b. $\frac{I_S}{I_P} = \frac{N_P}{N_S}$ $\quad \frac{I_S}{1.2} = \frac{200}{800}$

$$I_S = \frac{(1.2)(200)}{800}$$

$$I_S = \mathbf{0.3\ amperes}$$

EXERCISES

Find the output voltage (V_S) in volts of each transformer.

1. Turns Ratio: $\frac{20}{1}$
Input voltage: 115 volts

2. Turns Ratio: $\frac{1}{3}$
Input voltage: 690 volts

Find the output current (I_S) in amperes for each transformer.

3. Turns Ratio: $\frac{4}{5}$
Input current: 1.3 amperes

4. Turns Ratio: $\frac{25}{3}$
Input current: 12 amperes

CHAPTER SUMMARY

IMPORTANT TERMS

Rational expression *(p. 268)*
Complex fraction *(p. 287)*

IMPORTANT IDEAS

1. Any polynomial is also a rational expression.

2. Any number that will result in a divisor of 0 must be restricted from the domain of the variable in a rational expression.

3. *Product Rule of Rational Expressions:* If $\frac{a}{b}$ and $\frac{c}{d}$ are rational expressions, then $\frac{a}{b} \cdot \frac{c}{d} = \frac{ac}{bd}$. ($b$ and d are not zero.)

4. *Property of the Opposite of a Difference:* If a and b are any real numbers, then $-(a - b) = b - a$.

5. *Quotient Rule for Rational Expressions:* If $\frac{a}{b}$ and $\frac{c}{d}$ are rational expressions, then $\frac{a}{b} \div \frac{c}{d} = \frac{a}{b} \cdot \frac{d}{c}$. ($b$, c, and d are not zero.)

CHAPTER REVIEW

SECTION 10.1

Write Yes or No to show whether each expression is a rational expression.

1. $\frac{2x^2 - 5x + 3}{x^2 + 7}$ **2.** $-5x^3 - 2x$ **3.** $3x^{-2} + 4x^{-1} + 10$ **4.** $\frac{\sqrt{x + 1}}{|x - 3|}$

Write the number or numbers that cannot be in the domain of each variable.

5. $\frac{x + 1}{x + 3}$ **6.** $\frac{x - 3}{x - 2}$ **7.** $\frac{x + 3}{2x - 1}$ **8.** $\frac{x + 6}{3x + 2}$

SECTION 10.2

Simplify. No divisor equals 0.

9. $\frac{10x^2y}{12xy}$ **10.** $\frac{8ab^2c}{20a^2bc}$ **11.** $\frac{x^2 - 100}{3x^2 - 300}$ **12.** $\frac{-3x^2 - 36x}{x^2 - 144}$

SECTION 10.3

Multiply and simplify.

13. $\dfrac{12x}{5} \cdot \dfrac{15}{18x^2}$

14. $\dfrac{9}{14x} \cdot \dfrac{8x^3}{30}$

15. $\dfrac{3xy^2}{5a^2b} \cdot \dfrac{10a}{9by}$

16. $\dfrac{5ab}{18xy} \cdot \dfrac{8x^2y}{15a^2b^2}$

SECTION 10.4

Multiply and simplify.

17. $\dfrac{ax}{x^2 - a^2} \cdot \dfrac{x^2 + xa}{a^2x^2}$

18. $\dfrac{5x}{x^2 - 25} \cdot \dfrac{x^2 + 5x}{2x^2y}$

19. $\dfrac{x^2 - 4}{14} \cdot \dfrac{6x^2}{3x^2 + 6x}$

20. $\dfrac{5r^2 - 15r}{22} \cdot \dfrac{2r + 6}{r^2 - 9}$

21. $\dfrac{3 - x}{6} \cdot \dfrac{15}{3x - 9}$

22. $\dfrac{2y - 4}{21} \cdot \dfrac{7y}{6 - 3y}$

SECTION 10.5

Multiply and simplify.

23. $\dfrac{x^2 + 2x - 15}{x^2 - 6x + 9} \cdot \dfrac{x^2 - 2x - 3}{x^2 + 7x + 10}$

24. $\dfrac{x^2 + 6x + 8}{x^2 + 8x + 16} \cdot \dfrac{x^2 - x - 20}{x^2 + 4x + 4}$

25. $\dfrac{x^2 + 4x + 4}{x^2 - x - 6} \cdot \dfrac{x^2 - 5x + 6}{x^2 - 4}$

26. $\dfrac{x^2 - 1}{x^2 - 6x + 5} \cdot \dfrac{x^2 - 10x + 25}{x^2 - 4x - 5}$

SECTION 10.6

Divide and simplify.

27. $\dfrac{-10abc}{21xy} \div \dfrac{-2a^2bc}{-14x^2y}$

28. $\dfrac{49xy^2}{24c^2d^2} \div \dfrac{21x^2y}{-8cd}$

29. $\dfrac{x - 2}{x(x + 3)} \div \dfrac{x + 2}{2(x + 3)}$

30. $\dfrac{2x + 1}{x + 7} \div \dfrac{5(2x - 1)}{x(x + 7)}$

SECTION 10.7

Divide and simplify.

31. $\dfrac{2x^3 - 6x^2 + 4x}{x - 2} \div \dfrac{3x^3 - 3x^2}{x^2 + 4}$

32. $\dfrac{x^3 + 2x^2 - 3x}{x^2 - 5} \div \dfrac{x^2 - 4x - 12}{2x^2}$

Simplify.

33. $\dfrac{\frac{a}{10} + \frac{a}{3}}{\frac{5}{6}}$

34. $\dfrac{\frac{x}{12}}{\frac{3}{4} - \frac{x}{8}}$

35. $\dfrac{\frac{a}{6} - \frac{b}{5}}{\frac{a}{12} + \frac{b}{20}}$

36. $\dfrac{\frac{x}{7} + \frac{x}{5}}{\frac{y}{2} - \frac{y}{3}}$

Sums and Differences of Rational Expressions

CHAPTER 11

A silicon chip smaller than the size of a human fingernail can contain as many as 256,000 separate circuits. Each circuit can be as small as $\frac{1}{2000}$ millimeter, or $\frac{1}{500,000}$ inch.

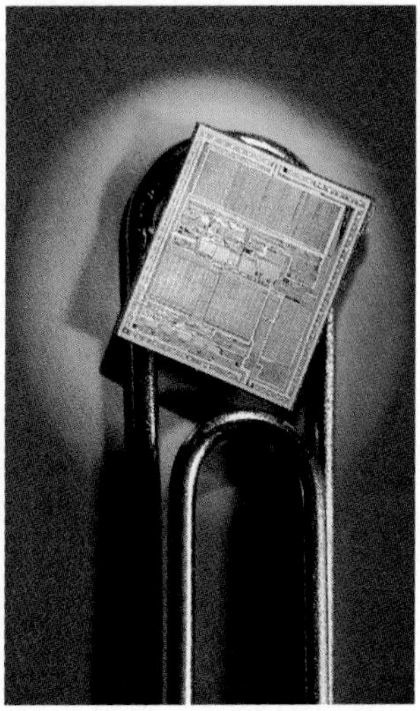

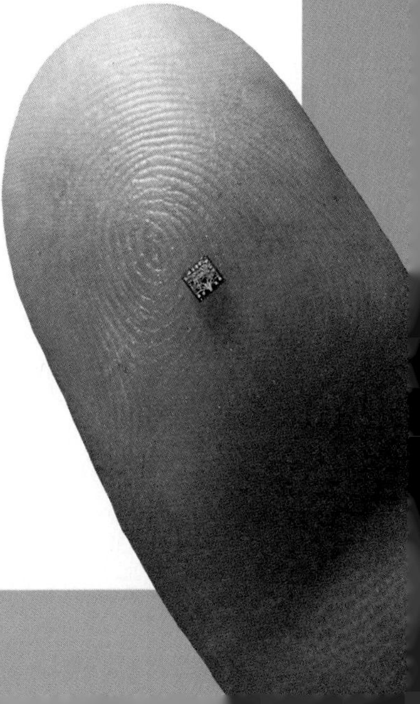

11.1 Sums/Differences: Like Denominators

P–1 **Add or subtract: a.** $\frac{3}{7}+\frac{2}{7}$ **b.** $\frac{5}{12}-\frac{11}{12}$

The Sum and Difference Rules for Fractions also apply to rational expressions.

Sum and Difference Rules for Rational Expressions

If $\frac{a}{c}$ and $\frac{b}{c}$ are rational expressions, then

1. $\frac{a}{c}+\frac{b}{c}=\frac{a+b}{c}$ 2. $\frac{a}{c}-\frac{b}{c}=\frac{a-b}{c}$.

(c is not zero.)

$$\frac{3}{x+1}+\frac{2}{x+1}=\frac{5}{x+1}$$

$$\frac{3x+2}{11x}-\frac{4}{11x}=\frac{3x-2}{11x}$$

P–2 **Add or subtract.**

a. $\frac{4}{x}+\frac{5}{x}$ **b.** $\frac{10}{a}-\frac{3}{a}$ **c.** $\frac{x}{3y}+\frac{2x}{3y}$

P–3 **In $\frac{2}{x-3}$, what number must be restricted from the domain of x?**

EXAMPLE 1 Add: $\frac{2}{x-3}+\frac{12}{x-3}$

Solution:

[1] Sum Rule for Rational Expressions ⟶ $\frac{2}{x-3}+\frac{12}{x-3}=\frac{2+12}{x-3}$

[2] Simplify. ⟶ $=\frac{14}{x-3}$

EXAMPLE 2 Subtract: $\frac{x}{x+y}-\frac{y}{x+y}$

Solution: $\frac{x}{x+y}-\frac{y}{x+y}=\frac{x-y}{x+y}$ ◀ *Simplest form*

EXAMPLE 3 Subtract: $\frac{3x}{x+5} - \frac{10x}{x+5}$

Solution:

[1] Difference Rule for Rational Expressions ⟶ $\frac{3x}{x+5} - \frac{10x}{x+5} = \frac{3x - 10x}{x+5}$

[2] Combine like terms. ⟶ $= \frac{-7x}{x+5}$

[3] Simplify. ⟶ $= -\frac{7x}{x+5}$

EXAMPLE 4 Subtract: $\frac{3x}{5y} - \frac{x}{5y} - \frac{4x}{5y}$

Solution: Remember that addition and subtraction in an expression are performed from left to right.

$$\frac{3x}{5y} - \frac{x}{5y} - \frac{4x}{5y} = \frac{3x - x - 4x}{5y}$$ ◀ ***Difference rule***

$$= \frac{-2x}{5y}$$ ◀ ***Like terms are combined.***

$$= -\frac{2x}{5y}$$

CLASSROOM EXERCISES

Add or subtract. Simplify where necessary. (Examples 1–3)

1. $\frac{1}{5} + \frac{2}{5}$ **2.** $\frac{2}{9} + \frac{5}{9}$ **3.** $\frac{5}{12} - \frac{3}{12}$ **4.** $\frac{8}{15} - \frac{11}{15}$

5. $\frac{2}{x} + \frac{1}{x}$ **6.** $\frac{8}{a} - \frac{5}{a}$ **7.** $\frac{5}{3x} + \frac{2}{3x}$ **8.** $\frac{7}{5a} - \frac{4}{5a}$

9. $\frac{5}{x+1} + \frac{6}{x+1}$ **10.** $\frac{10}{x-2} - \frac{3}{x-2}$ **11.** $\frac{x}{x+3} + \frac{2x}{x+3}$ **12.** $\frac{3y}{y-2} - \frac{y}{y-2}$

13. $\frac{a}{a+1} - \frac{5a}{a+1}$ **14.** $\frac{3x}{x-5} - \frac{3x}{x-5}$ **15.** $\frac{2x}{x+y} + \frac{3y}{x+y}$ **16.** $\frac{2a}{2a+3b} - \frac{3b}{2a+3b}$

Add or subtract. (Example 4)

17. $\frac{x}{11} - \frac{3x}{11} + \frac{5x}{11}$

18. $\frac{4}{a} + \frac{9}{a} - \frac{6}{a}$

19. $\frac{3x}{y+1} + \frac{5}{y+1} - \frac{x}{y+1}$

WRITTEN EXERCISES

Goal: To add or subtract rational expressions with equal denominators

Sample Problems: a. $\frac{6}{x-4} + \frac{x}{x-4}$ **b.** $\frac{6}{x-4} - \frac{x}{x-4}$

Answers: a. $\frac{6+x}{x-4}$ **b.** $\frac{6-x}{x-4}$

Add or subtract. (Examples 1–4)

1. $\frac{5}{13} + \frac{6}{13}$

2. $\frac{2}{11} + \frac{8}{11}$

3. $\frac{3}{17} - \frac{8}{17}$

4. $\frac{3}{7} - \frac{5}{7}$

5. $\frac{5}{a} + \frac{11}{a}$

6. $\frac{2}{b} + \frac{13}{b}$

7. $\frac{8}{y} - \frac{13}{y}$

8. $\frac{12}{x} - \frac{27}{x}$

9. $\frac{5}{x+3} + \frac{7}{x+3}$

10. $\frac{3}{y+2} + \frac{7}{y+2}$

11. $\frac{3}{a-3} - \frac{4}{a-3}$

12. $\frac{10}{b+7} - \frac{11}{b+7}$

13. $\frac{2x}{x+2} + \frac{5x}{x+2}$

14. $\frac{y}{y-1} + \frac{4y}{y-1}$

15. $\frac{a}{a+1} - \frac{5a}{a+1}$

16. $\frac{2r}{r-6} - \frac{3r}{r-6}$

17. $\frac{a}{2a+b} + \frac{b}{2a+b}$

18. $\frac{x}{x-3y} + \frac{y}{x-3y}$

19. $\frac{3r}{r-2s} - \frac{5s}{r-2s}$

20. $\frac{c}{c-d} - \frac{2d}{c-d}$

21. $\frac{3y}{19} - \frac{8y}{19} + \frac{y}{19}$

22. $\frac{t}{8} + \frac{4t}{8} - \frac{10t}{8}$

23. $\frac{4x}{3y} - \frac{x}{3y} - \frac{5x}{3y}$

24. $\frac{3p}{10q} - \frac{2p}{10q} - \frac{8p}{10q}$

25. $\frac{5}{9t} - \frac{12}{9t} + \frac{3}{9t}$

26. $\frac{4}{7y} - \frac{9}{7y} + \frac{1}{7y}$

REVIEW CAPSULE FOR SECTION 11.2

Subtract. Then combine like terms. (Pages 217–219)

1. $(4x - 2) - (3x + 1)$

2. $(5t + 1) - (6t - 5)$

3. $(3m - 6) - (-m + 10)$

4. $(r^2 - r) - (3r + 2)$

5. $(3n^2 + n) - (12 - n)$

6. $(s^2 - 3s) - (3s^2 - 5s + 3)$

11.2 More Sums and Differences

In many of the examples and exercises of this section, it will be necessary to simplify the result after applying the Sum or Difference Rule.

P–1 **In $\frac{a}{a-b}$, what values cannot be assigned to a and b?**

EXAMPLE 1 Subtract and simplify: $\frac{a}{a-b} - \frac{b}{a-b}$

Solution:

[1] Difference Rule for Rational Expressions ⟶ $\frac{a}{a-b} - \frac{b}{a-b} = \frac{a-b}{a-b}$

[2] Simplify. ⟶ $= 1$

EXAMPLE 2 Add: $\frac{2x+3}{x} + \frac{x-5}{x}$

Solution:

[1] Sum Rule for Rational Expressions ⟶ $\frac{2x+3}{x} + \frac{x-5}{x} = \frac{(2x+3)+(x-5)}{x}$

[2] Add binomials in the numerator. ⟶ $= \frac{3x-2}{x}$

P–2 **How do you know that $\frac{3x-2}{x}$ in Example 2 cannot be simplified?**

Recall how to subtract binomials.

Vertical Method	**Horizontal Method**
$\begin{array}{rr} & 4x-3 \\ (-) & x+1 \\ \hline & 3x-4 \end{array}$	$\begin{aligned}(4x-3)-(x+1) &= (4x-3)+-(x+1) \\ &= (4x-3)+(-x-1) \\ &= 3x-4\end{aligned}$

The binomial $x + 1$ is subtracted by adding its opposite $(-x - 1)$.

EXAMPLE 3 Subtract: $\frac{5x-2}{x-3} - \frac{2x+3}{x-3}$

Solution:

[1] Difference Rule → $\frac{5x-2}{x-3} - \frac{2x+3}{x-3} = \frac{(5x-2)-(2x+3)}{x-3}$

[2] Meaning of subtraction → $= \frac{(5x-2)+(-2x-3)}{x-3}$

2x + 3 is subtracted by adding its opposite.

[3] Combine like terms. → $= \frac{3x-5}{x-3}$

EXAMPLE 4 Subtract and simplify: $\frac{3x+2}{x+2} - \frac{x-2}{x+2}$

Solution:

[1] Difference Rule → $\frac{3x+2}{x+2} - \frac{x-2}{x+2} = \frac{(3x+2)-(x-2)}{x+2}$

[2] Meaning of subtraction → $= \frac{(3x+2)+(-x+2)}{x+2}$

[3] Combine like terms. → $= \frac{2x+4}{x+2}$

[4] Factor the numerator. → $= \frac{2(x+2)}{x+2}$

[5] Identify a name for 1. → $= \frac{(x+2)}{(x+2)} \cdot \frac{2}{1}$

[6] Simplify. → $= 2$

Simplifying a fraction is also called reducing the fraction.

CLASSROOM EXERCISES

Add or subtract. Simplify where necessary. (Examples 1–4)

1. $\frac{x+3}{x+2} + \frac{2}{x+2}$
2. $\frac{3a}{a-5} + \frac{2a+3}{a-5}$
3. $\frac{3y+5}{y+3} - \frac{2}{y+3}$
4. $\frac{2x-3}{x+1} - \frac{1}{x+1}$
5. $\frac{3b+1}{b+2} + \frac{b}{b+2}$
6. $\frac{5r-2}{r-4} - \frac{2r}{r-4}$
7. $\frac{2a+b}{a+b} - \frac{a}{a+b}$
8. $\frac{x+y}{2x+y} + \frac{x}{2x+y}$
9. $\frac{x+2}{x-3} + \frac{x+3}{x-3}$
10. $\frac{x+5}{x+1} + \frac{3-x}{x+1}$
11. $\frac{3x+3}{x+y} - \frac{x+1}{x+y}$
12. $\frac{3x+1}{x+4} - \frac{x+4}{x+4}$

WRITTEN EXERCISES

Goal: To add or subtract rational expressions

Sample Problem: $\frac{4x + 13}{x + 4} - \frac{x + 1}{x + 4}$

Answer: 3

Add or subtract. Simplify where necessary. (Examples 1–4)

1. $\frac{2x + 1}{x + y} + \frac{4}{x + y}$
2. $\frac{2}{a - b} + \frac{3a + 5}{a - b}$
3. $\frac{6r + 3}{r + s} - \frac{4}{r + s}$
4. $\frac{5d + 1}{c + d} - \frac{3}{c + d}$
5. $\frac{3x - 5}{x + 2} + \frac{x - 2}{x + 2}$
6. $\frac{5y - 1}{y - 3} + \frac{2y - 7}{y - 3}$
7. $\frac{5a - 2}{a + b} - \frac{a + 3}{a + b}$
8. $\frac{7c - 1}{c - d} - \frac{2c + 3}{c - d}$
9. $\frac{5x - 7}{2x - 3} + \frac{4 - 3x}{2x - 3}$
10. $\frac{10 - 3y}{5y + 1} + \frac{8y - 9}{5y + 1}$
11. $\frac{5a + 1}{a + 2} - \frac{3a - 3}{a + 2}$
12. $\frac{4b - 9}{b - 1} - \frac{2b - 7}{b - 1}$
13. $\frac{x^2 + 3x}{4x + 3} + \frac{3x^2}{4x + 3}$
14. $\frac{4y^2}{y - 1} + \frac{y^2 - 5y}{y - 1}$
15. $\frac{9a - 2a^2}{a + 2} - \frac{3a - 5a^2}{a + 2}$
16. $\frac{b^2 + 5b}{b + 2} - \frac{-5b - 4b^2}{b + 2}$
17. $\frac{x^2 + 3x}{x + 4} - \frac{16 + 3x}{x + 4}$
18. $\frac{2a^2 + 2x}{a + 3} - \frac{a^2 + 2x + 9}{a + 3}$
19. $\frac{y^2 - 3y}{y - 4} + \frac{16 - 5y}{y - 4}$
20. $\frac{25 - 8b}{b - 5} + \frac{b^2 - 2b}{b - 5}$

MORE CHALLENGING EXERCISES

21. $\frac{3x + 2}{x - 3} - \frac{x + 7}{x - 3} + \frac{2x - 7}{x - 3}$
22. $\frac{x + 3}{2x - 1} + \frac{3x - 5}{2x - 1} - \frac{1 - 2x}{2x - 1}$
23. $\frac{x - 5}{x + 5} - \frac{2x}{x + 5} - \frac{1 - x}{x + 5}$
24. $\frac{a^2 + a}{a + 4} + \frac{a}{a + 4} - \frac{a^2 + 2a}{a + 4}$

REVIEW CAPSULE FOR SECTION 11.3

Write the prime factors of each polynomial. (Pages 258–260)

1. $24x^3y^2$
2. $56r^2s^4t^3$
3. $4m^2 - 36$
4. $x^2 - 6x + 5$
5. $6x^2 + 6x - 72$
6. $(a^2 - 4)(4a^2 - 4a - 8)$

11.3 Least Common Multiple of Polynomials

Definition

The ***least common multiple*** (*LCM*) of two or more counting numbers is the smallest counting number that is divisible by the given numbers.

The LCM of 4 and 6 is 12.

EXAMPLE 1 Find the *LCM* of 12, 30, and 45.

Solution:

[1] Write the prime factorization of each number. ⟶ $12 = 2^2 \cdot 3$

$30 = 2 \cdot 3 \cdot 5$

$45 = 3^2 \cdot 5$

[2] Write a product using each prime factor only once. ⟶ $2 \cdot 3 \cdot 5$

[3] For each factor, write the greatest exponent used in any of the prime factorizations. ⟶ $2^2 \cdot 3^2 \cdot 5$

[4] Multiply. ⟶ $4 \cdot 9 \cdot 5 = 180$

The method of Example 1 can also be used with polynomials.

EXAMPLE 2 Find the *LCM* of $10x^2$, $4xy$, and $15x^2y$.

Solution: [1] $10x^2 = 2 \cdot 5 \cdot x^2$

$4xy = 2^2 \cdot x \cdot y$

$15x^2y = 3 \cdot 5 \cdot x^2 \cdot y$

[2] $2 \cdot 3 \cdot 5 \cdot x \cdot y$

[3] $2^2 \cdot 3 \cdot 5 \cdot x^2 \cdot y$

[4] $60x^2y$

You can write the *LCM* in factored form if one or more of the factors is a binomial.

EXAMPLE 3 Find the *LCM* of $(2x - 4)$ and $(x^2 - 4)$.

Solution:

[1] Factor. ⟶ $2x - 4 = 2(x - 2)$

$x^2 - 4 = (x + 2)(x - 2)$

[2] Write a product using each factor only once. ⟶ $2(x + 2)(x - 2)$

[3] Write the greatest exponent for each factor. ⟶ $2(x + 2)(x - 2)$ ◀ ***Factored form***

EXAMPLE 4 Find the *LCM* of $(x - 3)$ and $(x + 5)$.

Solution: Both of the binomials are prime.

$LCM = (x - 3)(x + 5)$ ◀ ***Since the factors are prime, the LCM is their product.***

The least common multiple of prime polynomials is their product.

P–1 **What is the *LCM* of the polynomials below?**

$(x - y)$ $(x + y)$ $3xy$

EXAMPLE 5 Find the *LCM* of $x^2 - 9$, $5x^2 + 5x - 30$, and $x^2 - 4x + 4$.

Solution:

[1] Factor. ⟶

$$\begin{cases} x^2 - 9 = (x + 3)(x - 3) \\ 5x^2 + 5x - 30 = 5(x^2 + x - 6) \\ \qquad = 5(x + 3)(x - 2) \\ x^2 - 4x + 4 = (x - 2)(x - 2) \end{cases}$$

◀ ***Prime factorizations***

[2] Write a product using each factor only once. ⟶ $5(x + 3)(x - 3)(x - 2)$

[3] Write the greatest exponent for each factor. ⟶ $5(x + 3)(x - 3)(x - 2)^2$

Steps in finding the LCM of two or more polynomials:

1. Write the prime factorization of each polynomial.
2. Write a product using each prime factor only once.
3. For each factor, write the greatest exponent used in any prime factorization.
4. Multiply where necessary.

CLASSROOM EXERCISES

Write the factored form of each LCM. (Examples 1–4)

1. $3;\ 7$
2. $2x;\ y$
3. $3x^2;\ 2y$
4. $2;\ 3;\ 5$
5. $2a;\ 3b;\ 5c$
6. $5x;\ 7y;\ 2z$
7. $(x-3);\ (x-5)$
8. $(2x+3);\ (3x-5)$
9. $(a+b);\ (a-b)$
10. $12x^2y;\ 3xy$
11. $3(a-b);\ (a-b)$
12. $2x;\ 3x$
13. $(2x-y);\ (x+2y)$
14. $(5x-3);\ (x+2);\ (x-4)$
15. $5xy;\ 3xy$
16. $2(x+y);\ 3(x+y)$
17. $5(a+b);\ 3(a-b)$

WRITTEN EXERCISES

Goal: To find the least common multiple of two or more polynomials
Sample Problem: $2x+2y;\ x^2-y^2;\ x^2+xy$ **Answer:** $2x(x+y)(x-y)$

Find each LCM. (Examples 1–5)

1. $10;\ 6$
2. $12;\ 30$
3. $4x^2;\ 6xy$
4. $2ab^2;\ 8a^2b$
5. $12a^2b;\ 3ab;\ 10a^2b$
6. $4xy;\ 6xy^2;\ 15x^2$
7. $(x^2-9);\ (x+3)$
8. $(a-5);\ (a^2-25)$
9. $x^2-y^2;\ 3x-3y$
10. $5a+5b;\ a^2-b^2$
11. $3b^2+15b;\ 2b^2-12b$
12. $5r^2-15r;\ 2r^2+6r$
13. $x^2-5x;\ x^2-25;\ x^3+5x^2$
14. $a^2-36;\ 2a^2+12a;\ a^3-6a^2$
15. $x^2+4x+4;\ 5x+10$
16. $x^2-6x+9;\ 4x-12$
17. $x^2-8x+16;\ x^2-2x-8$
18. $x^2+12x+36;\ x^2+3x-18$
19. $x^2-5x+6;\ x^2+x-6$
20. $x^2-2x-8;\ x^2-6x+8$

MORE CHALLENGING EXERCISES

21. The least common multiple of two monomials is $24cd$. One of the monomials is $8d$. Must the other monomial be $3c$? Explain.
22. If the least common multiple of the polynomials $2x^2-32$ and $ax-4a$ is $6(x-4)(x+4)$, find two possible values for a.

11.4 Unlike Monomial Denominators

EXAMPLE 1 Add: $\frac{2}{a} + \frac{3}{b}$

Solution:

1 Find the *LCM* of *a* and *b*. ⟶ ab

2 Multiplication Property of One ⟶ $\frac{2}{a} = \frac{2}{a}\left(\frac{b}{b}\right)$ and $\frac{3}{b} = \frac{3}{b}\left(\frac{a}{a}\right)$

You must get ab as the common denominator.

3 Substitute. ⟶ $\frac{2}{a} + \frac{3}{b} = \frac{2}{a} \cdot \frac{b}{b} + \frac{3}{b} \cdot \frac{a}{a}$

4 Product Rule ⟶ $= \frac{2b}{ab} + \frac{3a}{ab}$

5 Sum Rule ⟶ $= \frac{2b + 3a}{ab}$

Simplest form

EXAMPLE 2 Subtract: $\frac{5}{2x} - \frac{2}{3x}$

Solution:

1 Find the *LCM* of $2x$ and $3x$. ⟶ $6x$

2 Multiplication Property of One ⟶ $\frac{5}{2x} = \frac{5}{2x}\left(\frac{3}{3}\right)$ and $\frac{2}{3x} = \frac{2}{3x}\left(\frac{2}{2}\right)$

3 Substitute. ⟶ $\frac{5}{2x} - \frac{2}{3x} = \frac{5}{2x} \cdot \frac{3}{3} - \frac{2}{3x} \cdot \frac{2}{2}$

4 Product Rule ⟶ $= \frac{15}{6x} - \frac{4}{6x}$

5 Difference Rule ⟶ $= \frac{15 - 4}{6x}$

6 Simplify. ⟶ $= \frac{11}{6x}$

P–1 In Step 2 of Example 2, how were $\frac{3}{3}$ and $\frac{2}{2}$ chosen as names for 1?

P–2 **What is the *LCM* of $10x^2y$ and $8xy$?**

EXAMPLE 3

Add: $\frac{3}{10x^2y} + \frac{5}{8xy}$

Solution:

$$\frac{3}{10x^2y} + \frac{5}{8xy} = \frac{3}{10x^2y}\left(\frac{4}{4}\right) + \frac{5}{8xy}\left(\frac{5x}{5x}\right)$$

◀ *$10x^2y = 2 \cdot 5 \cdot x^2 \cdot y$*
$8xy = 2^3 \cdot x \cdot y$
$LCM = 2^3 \cdot 5 \cdot x^2 \cdot y$ or $40x^2y$

$$= \frac{12}{40x^2y} + \frac{25x}{40x^2y}$$

$$= \frac{12 + 25x}{40x^2y}$$

P–3 **How do you know that $\frac{12 + 25x}{40x^2y}$ in Example 3 is in simplest form?**

P–4 **How can $5r$ be written as a fraction?**

EXAMPLE 4

Subtract: $5r - \frac{3r}{8t}$

Solution:

$$\frac{5r}{1} - \frac{3r}{8t} = \frac{5r}{1}\left(\frac{8t}{8t}\right) - \frac{3r}{8t}$$

$$= \frac{40rt}{8t} - \frac{3r}{8t}$$

$$= \frac{40rt - 3r}{8t}$$

CLASSROOM EXERCISES

Write the missing name for 1.

1. $\frac{2}{x}\left(\frac{?}{?}\right) = \frac{2x}{x^2}$
2. $\frac{3}{ab}\left(\frac{?}{?}\right) = \frac{15}{5ab}$
3. $\frac{2}{5y}\left(\frac{?}{?}\right) = \frac{4}{10y}$
4. $\frac{3}{a^2b}\left(\frac{?}{?}\right) = \frac{3b}{a^2b^2}$
5. $\frac{5}{6xy}\left(\frac{?}{?}\right) = \frac{10x}{12x^2y}$
6. $\frac{3a}{2b}\left(\frac{?}{?}\right) = \frac{18a^2b}{12ab^2}$
7. $\frac{2}{5r}\left(\frac{?}{?}\right) = \frac{6s}{15rs}$
8. $\frac{5}{12xy}\left(\frac{?}{?}\right) = \frac{15xy}{36x^2y^2}$

Write the missing polynomials in each exercise. (Step 2 of Examples 1 and 2)

9. $\frac{2}{3ab}\left(\frac{?}{?}\right) = \frac{?}{9a^2b^2}$

10. $\frac{9s}{5r^2}\left(\frac{?}{?}\right) = \frac{?}{15r^2s}$

11. $\frac{?}{5y}\left(\frac{2x}{2x}\right) = \frac{6ax}{?}$

12. $\frac{3b}{?}\left(\frac{10b}{10b}\right) = \frac{?}{40a^2b}$

13. $\frac{15x}{22y}\left(\frac{?}{?}\right) = \frac{?}{66xy}$

14. $\frac{7s}{5r}\left(\frac{3rs}{3rs}\right) = \frac{?}{?}$

15. $\frac{2}{3x} + \frac{5}{2y} = \frac{2}{3x}\left(\frac{?}{?}\right) + \frac{5}{2y}\left(\frac{?}{?}\right)$
$= \frac{?}{6xy} + \frac{15x}{6xy}$
$= \frac{?}{6xy}$

16. $\frac{3}{a^2b} - \frac{5}{2ab} = \frac{3}{a^2b}\left(\frac{2}{2}\right) - \frac{5}{2ab}\left(\frac{?}{?}\right)$
$= \frac{?}{?} - \frac{5a}{2a^2b}$
$= \frac{?}{2a^2b}$

WRITTEN EXERCISES

Goal: To add or subtract rational expressions that have unlike monomial denominators

Sample Problem: $\frac{7}{4x^2} + \frac{5}{6xy}$ **Answer:** $\frac{21y + 10x}{12x^2y}$

Add or subtract. (Examples 1–4)

1. $\frac{3}{7} + \frac{2}{5}$
2. $\frac{2}{3} + \frac{5}{11}$
3. $\frac{5}{a} + \frac{7}{b}$
4. $\frac{1}{x} + \frac{4}{y}$
5. $\frac{3}{2x} - \frac{5}{x}$
6. $\frac{3}{y} - \frac{4}{3y}$
7. $\frac{2}{5x} + \frac{3}{10x}$
8. $\frac{5}{3a} + \frac{2}{15a}$
9. $\frac{5}{4a} - \frac{5}{12b}$
10. $\frac{5}{6x} - \frac{3}{24y}$
11. $\frac{2}{a^2b} + \frac{3}{ab^2}$
12. $\frac{5}{xy} + \frac{3}{xy^2}$
13. $\frac{5}{2a^2} - \frac{1}{6b^2}$
14. $\frac{2}{5xy} - \frac{7}{10y^2}$
15. $\frac{2a}{bc} + \frac{3b}{ac}$
16. $\frac{t}{2rs} + \frac{3s}{rt}$
17. $\frac{2}{3a^2b} + 5a$
18. $\frac{5}{7xy} + 4y$
19. $5cd - \frac{3d}{8c}$
20. $10rt - \frac{4r}{9t}$

MORE CHALLENGING EXERCISES

21. If $\frac{1}{3c} - \frac{1}{6d} = 0$, how does the value of c compare with the value of d?

22. If $x^a - \frac{1}{x} = \frac{x^2 - 1}{x}$, what is the value of a?

23. Write a rational expression that, when subtracted from $\frac{3x - 7}{2x + 3}$, gives a difference of $\frac{x - 5}{2x + 3}$.

24. Write a rational expression that, when added to $\frac{3}{x + 4}$, gives a sum of $\frac{7x + 5}{x^2 + 4x}$.

NON-ROUTINE PROBLEMS

25. How is it possible to put nine pigs in four pens so that there is an odd number of pigs in each pen?

26. What is the least number of straight lines you can draw to make exactly 14 squares?

27. Jim and Todd each ran two laps around a track. When Jim completed the first lap, he noted that he was 50 meters behind Todd. Jim then began to run a distance of eight meters for every seven meters that Todd ran. The two boys were tied after completing the second lap. How long was the track?

28. In a shop class of ten students, exactly two students are given the responsibility of cleaning the shop each day after class. A schedule is made pairing each student with every other student in the class exactly once. How many pairs of students are listed on the schedule?

MID-CHAPTER REVIEW

Add or subtract. (Section 11.1)

1. $\frac{5}{t}-\frac{9}{t}$ **2.** $\frac{3}{5t}+\frac{1}{5t}$ **3.** $\frac{3m}{m+8}+\frac{5m}{m+8}$ **4.** $\frac{6r}{r-3}-\frac{r}{r-3}$

Add or subtract. Simplify where necessary. (Section 11.2)

5. $\frac{3a-1}{a+1}-\frac{2}{a+1}$ **6.** $\frac{4r+8}{r+5}+\frac{r+7}{r+5}$ **7.** $\frac{m^2-3}{m+4}-\frac{5-2m}{m+4}$

Find each LCM. (Section 11.3)

8. $10a^2b$; $12abc$ **9.** $2t^2-18$; $3t^2-9t$ **10.** x^2+x-2; x^2-x-6

Add or subtract. (Section 11.4)

11. $\frac{2}{5a}-\frac{3}{b}$ **12.** $\frac{4}{3t}+\frac{7}{12t}$ **13.** $\frac{c}{6d}+\frac{d}{3c}$ **14.** $\frac{2m}{6np}-\frac{5n}{8mp}$

REVIEW CAPSULE FOR SECTION 11.5

Simplify. (Pages 271–273)

1. $\frac{x^2+3x-10}{(x+5)(x-1)}$ **2.** $\frac{5x^2-15x}{(x-3)(x-3)}$ **3.** $\frac{2x^2-2x-4}{(x-4)(x-2)}$ **4.** $\frac{4x^3-2x^2}{3x(2x-1)}$

11.5 Sums/Differences: Unlike Denominators

EXAMPLE 1 Add: $\frac{3}{x} + \frac{2}{x-2}$

Solution:

[1] Find the *LCM* of x and $x - 2$. ⟶ $x(x-2)$

[2] Multiplication Property of One ⟶ $\frac{3}{x} = \frac{3}{x}\left(\frac{x-2}{x-2}\right)$ and $\frac{2}{x-2} = \frac{2}{x-2}\left(\frac{x}{x}\right)$

[3] Substitute. ⟶ $\frac{3}{x} + \frac{2}{x-2} = \frac{3}{x}\left(\frac{x-2}{x-2}\right) + \frac{2}{x-2}\left(\frac{x}{x}\right)$

[4] Product Rule ⟶ $= \frac{3(x-2)}{x(x-2)} + \frac{2x}{x(x-2)}$

[5] Sum Rule ⟶ $= \frac{3(x-2)+2x}{x(x-2)}$

[6] Distributive Property ⟶ $= \frac{3x-6+2x}{x(x-2)}$

[7] Combine like terms. ⟶ $= \frac{5x-6}{x(x-2)}$

EXAMPLE 2 Subtract: $\frac{5}{x+3} - \frac{4}{x-1}$

Solution:

[1] Find the *LCM* of $x + 3$ and $x - 1$. ⟶ $(x+3)(x-1)$

[2] Multiplication Property of One ⟶ $\frac{5}{x+3} = \frac{5}{x+3}\left(\frac{x-1}{x-1}\right)$ and $\frac{4}{x-1} = \frac{4}{x-1}\left(\frac{x+3}{x+3}\right)$

[3] Substitute. ⟶ $\frac{5}{x+3} - \frac{4}{x-1} = \frac{5}{x+3}\left(\frac{x-1}{x-1}\right) - \frac{4}{x-1}\left(\frac{x+3}{x+3}\right)$

[4] Product Rule ⟶ $= \frac{5(x-1)}{(x+3)(x-1)} - \frac{4(x+3)}{(x+3)(x-1)}$

[5] Difference Rule ⟶ $= \frac{5(x-1)-4(x+3)}{(x+3)(x-1)}$

[6] Distributive Property ⟶ $= \frac{5x-5-4x-12}{(x+3)(x+1)}$

[7] Combine like terms. ⟶ $= \frac{x-17}{(x+3)(x+1)}$

P–1 **What names for 1 are needed as suggested below?**

$$\frac{x+1}{x-1}\left(\frac{?}{?}\right)=\frac{?}{(x-1)(x+4)} \qquad \frac{x+3}{x+4}\left(\frac{?}{?}\right)=\frac{?}{(x-1)(x+4)}$$

EXAMPLE 3

Add: $\frac{x+1}{x-1}+\frac{x+3}{x+4}$

Solution:

[1] Multiplication Property of 1 ——→ $\frac{x+1}{x-1}+\frac{x+3}{x+4}=\frac{x+1}{x-1}\left(\frac{x+4}{x+4}\right)+\frac{x+3}{x+4}\left(\frac{x-1}{x-1}\right)$

[2] Product Rule ——→ $=\frac{(x+1)(x+4)}{(x-1)(x+4)}+\frac{(x+3)(x-1)}{(x+4)(x-1)}$

[3] Sum Rule ——→ $=\frac{(x+1)(x+4)+(x+3)(x-1)}{(x-1)(x+4)}$

[4] Distributive Property ——→ $=\frac{x^2+5x+4+x^2+2x-3}{(x-1)(x+4)}$

[5] Add like terms. ——→ $=\frac{2x^2+7x+1}{(x-1)(x+4)}$

Test to see whether the result in Step [5] is in simplest form.
Try to factor $2x^2+7x+1$.

$$2x^2+rx+sx+1$$

$$r+s=7 \qquad r\cdot s=2$$

There are no integers for r and s that meet both of these conditions.
The polynomial $2x^2+7x+1$ is prime.

$$\frac{2x^2+7x+1}{(x-1)(x+4)}$$ ◀ ***Simplest form***

CLASSROOM EXERCISES

Write the LCM. (Step 1 of Examples 1 and 2)

1. x and $(x+1)$ **2.** b and $(b-1)$ **3.** $(x+5)$ and $(x-3)$

4. $(y-2)$ and $(y+2)$ **5.** r and $(r+4)$ **6.** $(a+b)$ and $(a-b)$

7. $(2x-1)$ and $(x+3)$ **8.** $(3x+2)$ and $(2x+3)$ **9.** $x(x-1)$ and $(x-1)$

Write the missing name for 1. Then write the numerator of the expression on the right. (Step 2 of Examples 1 and 2)

10. $\frac{5}{x-2}\left(\frac{?}{?}\right) = \frac{?}{x(x-2)}$

11. $\frac{3}{a}\left(\frac{?}{?}\right) = \frac{?}{a(a+3)}$

12. $\frac{x}{x+5}\left(\frac{?}{?}\right) = \frac{?}{x(x+5)}$

13. $\frac{2}{y+2}\left(\frac{?}{?}\right) = \frac{?}{(y+2)(y-2)}$

14. $\frac{a}{a+b}\left(\frac{?}{?}\right) = \frac{?}{(a+b)(a-b)}$

15. $\frac{x-1}{x+3}\left(\frac{?}{?}\right) = \frac{?}{(x-2)(x+3)}$

16. $\frac{b+5}{b-7}\left(\frac{?}{?}\right) = \frac{?}{(b-7)(b+1)}$

17. $\frac{x+5}{2(x+3)}\left(\frac{?}{?}\right) = \frac{?}{2(x+3)(x-2)}$

WRITTEN EXERCISES

Goal: To add or subtract rational expressions that have binomial denominators

Sample Problem: $\frac{7}{x-4} - \frac{3}{x+1}$ **Answer:** $\frac{4x+19}{(x-4)(x+1)}$

Add or subtract. (Examples 1–3)

1. $\frac{1}{x+1} + \frac{2}{x}$
2. $\frac{3}{x-1} + \frac{5}{x}$
3. $\frac{4}{a} - \frac{2}{a+2}$
4. $\frac{2}{b} - \frac{3}{b-1}$
5. $\frac{x}{x+5} + \frac{5}{x-3}$
6. $\frac{3}{y-2} + \frac{y}{y+2}$
7. $\frac{a}{a-5} - \frac{2}{a}$
8. $\frac{r}{r+4} - \frac{1}{r}$
9. $\frac{a}{a+b} + \frac{b}{a-b}$
10. $\frac{r}{r-s} - \frac{s}{r+s}$
11. $\frac{a+1}{a-3} + \frac{3}{a+2}$
12. $\frac{y-2}{y+1} + \frac{5}{y+3}$
13. $\frac{3}{2x-1} - \frac{5}{x+3}$
14. $\frac{2}{3x+2} - \frac{1}{2x+3}$
15. $\frac{x+3}{x-1} + \frac{x+4}{x-2}$
16. $\frac{y+5}{y+2} + \frac{y-3}{y-1}$

MORE CHALLENGING EXERCISES

17. $\frac{b+2}{b-1} - \frac{b+5}{b-2}$
18. $\frac{t+2}{t-4} - \frac{t+1}{t+2}$
19. $\frac{x+1}{x(x-1)} + \frac{5}{x-1}$

REVIEW CAPSULE FOR SECTION 11.6

Find the LCM. (Pages 300–302)

1. $4x - 2$ and $6x - 3$
2. $5x + 15$ and $x^2 - 9$
3. $x^2 - 2x$ and $x^2 + 3x - 10$

11.6 More Sums and Differences

P–1 **What are the factors of $2x + 4$? of $3x + 6$?**

EXAMPLE 1 Add: $\dfrac{3}{2x+4} + \dfrac{2}{3x+6}$

Solution:

1 Factor the denominators. ⟶ $2x + 4 = 2(x+2)$ and $3x + 6 = 3(x+2)$

2 Use the factors to write the *LCM*. ⟶ $2 \cdot 3(x+2)$ or $6(x+2)$

3 Multiplication Property of 1 ⟶ $\dfrac{3}{2(x+2)} + \dfrac{2}{3(x+2)} = \dfrac{3}{2(x+2)}\left(\dfrac{3}{3}\right) + \dfrac{2}{3(x+2)}\left(\dfrac{2}{2}\right)$

4 Product Rule for Rational Expressions ⟶ $= \dfrac{9}{6(x+2)} + \dfrac{4}{6(x+2)}$

5 Sum Rule for Rational Expressions ⟶ $= \dfrac{13}{6(x+2)}$ ◀ ***Simplest form***

EXAMPLE 2 Subtract: $\dfrac{3}{5x-10} - \dfrac{2}{x^2-4}$

Solution:

1 Factor the denominators. ⟶ $5x - 10 = 5(x-2)$ and $x^2 - 4 = (x-2)(x+2)$

2 Use the factors to write the *LCM*. ⟶ $5 \cdot (x-2)(x+2)$

3 Multiplication Property of 1 ⟶ $\dfrac{3}{5x-10} - \dfrac{2}{x^2-4} = \dfrac{3}{5(x-2)}\left(\dfrac{x+2}{x+2}\right) - \dfrac{2}{(x-2)(x+2)}\left(\dfrac{5}{5}\right)$

4 Product Rule ⟶ $= \dfrac{3(x+2)}{5(x-2)(x+2)} - \dfrac{2(5)}{5(x-2)(x+2)}$

5 Distributive Property and the Difference Rule ⟶ $= \dfrac{3x+6-10}{5(x-2)(x+2)}$

6 Combine like terms. ⟶ $= \dfrac{3x-4}{5(x-2)(x+2)}$

P–2 **How do you know that $\frac{3x-4}{5(x-2)(x+2)}$ in Example 2 is in simplest form?**

P–3 **What are the prime factors of $x^2 + x$?**

P–4 **What are the prime factors of $x^2 + 3x + 2$?**

P–5 **What is the *LCM* of $x(x + 1)$ and $(x + 1)(x + 2)$?**

EXAMPLE 3

Add: $\frac{4}{x^2+x} + \frac{1}{x^2+3x+2}$

Solution:

1. Factor the denominators. → $x^2 + x = x(x+1)$; $(x^2 + 3x + 2) = (x+2)(x+1)$
2. Use the factors to write the *LCM*. → $x(x+1)(x+2)$
3. Multiplication Property of 1 → $\frac{4}{x^2+x} + \frac{1}{x^2+3x+2} = \frac{4}{x(x+1)}\left(\frac{x+2}{x+2}\right) + \frac{1}{(x+1)(x+2)}\left(\frac{x}{x}\right)$
4. Product Rule → $= \frac{4(x+2)}{x(x+1)(x+2)} + \frac{x}{(x+1)(x+2)x}$
5. Sum Rule → $= \frac{4(x+2)+x}{x(x+1)(x+2)}$
6. Distributive Property → $= \frac{4x+8+x}{x(x+1)(x+2)}$
7. Combine like terms. → $= \frac{5x+8}{x(x+1)(x+2)}$

The steps of Examples 1–3 are summarized.

Steps in adding or subtracting rational expressions:

1. Factor the denominators.
2. Determine the *LCM*.
3. Change the rational expressions to equivalent expressions with the *LCM* as the denominator of each.
4. Add or subtract.
5. Simplify.

CLASSROOM EXERCISES

Write the LCM. (Step 2 of Examples 1–3)

1. $x(x-1);\ (x-1)(x+1)$

2. $2(x+2);\ 3(x+2)$

3. $(x+3)(x+3);\ (x-2)(x+3)$

4. $a(a-1);\ a(a+1)$

5. $(y+2)(y-3);\ (y+2)(y+1)$

6. $10(x-5);\ 2(x-2)$

7. $2(x+1)(x-2);\ (x-2)(x-2)$

8. $3(a+2);\ (a-2)(a-2)$

Write the prime factors. (P-3, P-4)

9. $2x-6$

10. $6x+6$

11. $4x+12$

12. $8x-8$

13. x^2+x-2

14. x^2-x-6

15. $a^2+3a-10$

16. $a^2+2a-15$

WRITTEN EXERCISES

Goal: To add or subtract rational expressions that have binomial or trinomial denominators

Sample Problem: $\dfrac{5}{4x-12}+\dfrac{1}{x^2-6x+9}$ **Answer:** $\dfrac{5x-11}{4(x-3)(x-3)}$

Copy each exercise. Replace each question mark with the correct polynomial. (Examples 1 and 2)

1. $\dfrac{3}{2x+2}+\dfrac{1}{x^2-1}=\dfrac{3}{2(?)}+\dfrac{1}{(x-1)(?)}$

$$=\frac{3}{2(?)}\left(\frac{?}{?}\right)+\frac{1}{(x-1)(?)}\left(\frac{?}{?}\right)$$

$$=\frac{3(?)}{2(x+1)(x-1)}+\frac{1(?)}{2(x+1)(x-1)}=\frac{?}{2(x+1)(x-1)}$$

2. $\dfrac{2}{x^2-25}+\dfrac{4}{3x-15}=\dfrac{2}{(x-5)(?)}+\dfrac{4}{3(?)}$

$$=\frac{2}{(x-5)(?)}\left(\frac{?}{?}\right)+\frac{4}{3(?)}\left(\frac{?}{?}\right)$$

$$=\frac{2(?)}{3(x-5)(x+5)}+\frac{4(?)}{3(x-5)(x+5)}=\frac{?}{3(x-5)(x+5)}$$

Copy each exercise. Replace each question mark with the correct polynomial. (Example 3)

3. $\frac{1}{x^2+x-6} - \frac{2}{x^2-2x} = \frac{1}{(x-2)(?)} - \frac{2}{x(?)}$

$$= \frac{1}{(x-2)(?)}\left(\frac{?}{?}\right) - \frac{2}{x(?)}\left(\frac{?}{?}\right)$$

$$= \frac{1(?)}{x(x+3)(x-2)} - \frac{2(?)}{x(x+3)(x-2)} = \frac{?}{x(x+3)(x-2)}$$

4. $\frac{2}{x^2+4x} - \frac{3}{x^2+3x-4} = \frac{2}{x(?)} - \frac{3}{(x-1)(?)}$

$$= \frac{2}{x(?)}\left(\frac{?}{?}\right) - \frac{3}{(x-1)(?)}\left(\frac{?}{?}\right)$$

$$= \frac{2(?)}{x(x+4)(x-1)} - \frac{3(?)}{x(x+4)(x-1)} = \frac{?}{x(x+4)(x-1)}$$

MIXED PRACTICE *Add or subtract.*

5. $\frac{1}{2x-6} + \frac{1}{6x-6}$

6. $\frac{1}{4x+12} + \frac{1}{8x-8}$

7. $\frac{3}{x^2+x-2} - \frac{2}{x^2-x-6}$

8. $\frac{2}{x^2-9} + \frac{1}{2x+6}$

9. $\frac{7}{10x-50} - \frac{3}{x^2-25}$

10. $\frac{4}{a^2+3a-10} - \frac{5}{a^2+2a-15}$

11. $\frac{x}{2x-4} + \frac{x-1}{x^2-4}$

12. $\frac{5a}{a^2+2a-15} + \frac{a+3}{a^2+6a+5}$

13. $\frac{a+7}{a^2-a-6} - \frac{a}{a^2-10a+21}$

14. $\frac{2x+1}{x^2-81} - \frac{x}{x^2-10x+9}$

15. $\frac{x-2}{x^2-1} + \frac{x}{x^2+2x-3}$

16. $\frac{y}{y^2+2y-8} + \frac{y+3}{y^2-16}$

17. $\frac{2a-1}{2a^2-5a-3} - \frac{3a+1}{2a^2-a-1}$

18. $\frac{2x+1}{3x^2+8x-3} - \frac{x-5}{3x^2+2x-1}$

REVIEW CAPSULE FOR SECTION 11.7

Multiply. (Pages 274–279)

1. $6r^2s \cdot \frac{3}{2rs}$

2. $(t+1)(t-1) \cdot \frac{t}{t+1}$

3. $3q(q+2)\left(\frac{3}{q} - \frac{5}{3q}\right)$

11.7 Equations with Rational Expressions

P-1 **In $\frac{3}{x-3}$, what real number must be excluded from the domain of x?**

EXAMPLE 1 Solve $\frac{3}{x-3} = 5$.

Solution:

1. Multiply each side by $x - 3$. → $\frac{3}{x-3}\mathbf{(x-3)} = 5\mathbf{(x-3)}$
2. Simplify the left side. → $3 = 5(x-3)$
3. Distributive Property → $3 = 5x - 15$
4. Addition Property of Equality → $18 = 5x$
5. Division Property of Equality → $\frac{18}{5} = x$

To check the solution set of Example 1, replace x in $\frac{3}{x-3} = 5$ with $\frac{18}{5}$.

EXAMPLE 2 Solve $\frac{2}{x-1} = \frac{3}{x+1}$.

Solution:

1. Multiply each side by $(x-1)(x+1)$. → $\frac{2}{x-1}\mathbf{(x-1)(x+1)} = \frac{3}{x+1}\mathbf{(x-1)(x+1)}$
2. Simplify each side. → $\frac{2}{\cancel{x-1}_1} \cdot \frac{\overset{1}{\cancel{(x-1)}}(x+1)}{1} = \frac{3}{\cancel{x+1}_1} \cdot \frac{(x-1)\overset{1}{\cancel{(x+1)}}}{1}$

   $2(x+1) = 3(x-1)$
3. Distributive Property → $2x + 2 = 3x - 3$
4. Subtract $2x$ from each side. → $2 = x - 3$
5. Add 3 to each side. → $5 = x$

Check:

$\frac{2}{x-1}$	$= \frac{3}{x+1}$
$\frac{2}{5-1}$	$\frac{3}{5+1}$
$\frac{2}{4}$	$\frac{3}{6}$
$\frac{1}{2}$	$\frac{1}{2}$

P–2 **What is the *LCM* of x, 3x, and x + 2?**

EXAMPLE 3

Solve $\frac{1}{x} + \frac{1}{3x} = \frac{2}{x+2}$.

Solution:

1. Multiply each side by $3x(x + 2)$. → $\mathbf{3x(x+2)\left(\frac{1}{x} + \frac{1}{3x}\right) = 3x(x+2) \cdot \frac{2}{x+2}}$
2. Distributive Property → $3x(x+2) \cdot \frac{1}{x} + 3x(x+2) \cdot \frac{1}{3x} = 3x(x+2) \cdot \frac{2}{x+2}$
3. Simplify each side. → $3x + 6 + x + 2 = 6x$
4. Combine like terms. → $4x + 8 = 6x$
5. Subtract $4x$ from each side. → $8 = 2x$
6. Divide each side by 2. → $4 = x$

Check:

$\frac{1}{x} + \frac{1}{3x}$	$= \frac{2}{x+2}$
$\frac{1}{4} + \frac{1}{12}$	$\frac{2}{4+2}$
$\frac{3}{12} + \frac{1}{12}$	$\frac{2}{6}$
$\frac{4}{12}$	$\frac{1}{3}$
$\frac{1}{3}$	

The steps of Examples 1–3 are summarized.

Steps in solving equations with rational expressions:

1. Exclude numbers that will make zero divisors.
2. Multiply each side by the *LCM* of the denominators.
3. Solve the resulting equation.

CLASSROOM EXERCISES

Which real number(s) must be excluded from each domain? (P-1)

1. $\frac{2}{x}=\frac{5}{x-3}$
2. $\frac{3}{x+2}=\frac{1}{x-5}$
3. $\frac{4}{3x}=\frac{2}{x(x+1)}$
4. $\frac{x}{x+7}=12$
5. $\frac{3x}{2x+1}=5$
6. $\frac{5}{3x+1}=\frac{4}{2x-1}$

Write the LCM of the denominators. (P-1)

7. $\frac{x}{x-3}=\frac{3}{4}$
8. $\frac{6}{x}=\frac{8}{x-5}$
9. $\frac{3}{2x}=\frac{6}{5x-1}$
10. $\frac{5}{3}=\frac{2}{6(x-1)}+3$
11. $\frac{5}{2x}=\frac{1}{x+3}$
12. $\frac{10}{7x}=\frac{3}{7(2x+1)}$

WRITTEN EXERCISES

Goal: To solve equations that have rational expressions

Sample Problem: $\frac{1}{2x}+\frac{1}{x-1}=\frac{1}{x}$

Answers: $x=-1$

Solve and check each equation. (Examples 1–3)

1. $\frac{12}{x}=3$
2. $\frac{18}{x}=6$
3. $\frac{6}{x+2}=3$
4. $\frac{4}{x+5}=2$
5. $\frac{3}{4}=\frac{12}{x}$
6. $\frac{9}{x}=\frac{3}{8}$
7. $\frac{2}{x-1}=\frac{3}{x-1}$
8. $\frac{6}{x+3}=\frac{5}{x+3}$
9. $\frac{5}{x}=\frac{4}{x-3}$
10. $\frac{2}{x-7}=\frac{3}{x}$
11. $\frac{7}{x+1}=\frac{3}{x+5}$
12. $\frac{2}{x+4}=\frac{3}{x-3}$
13. $\frac{2}{3}+\frac{3}{x}=\frac{5}{2x}$
14. $\frac{1}{3x}+\frac{2}{5}=\frac{5}{x}$
15. $\frac{1}{x}+\frac{2}{x-2}=\frac{1}{2x}$
16. $\frac{1}{x-4}+\frac{1}{3x}=\frac{1}{x}$

NON-ROUTINE PROBLEM

17. A large block of cheese is 16 centimeters long, 8 centimeters wide, and $7\frac{1}{2}$ centimeters high. What is the greatest number of blocks 5 centimeters long, 3 centimeters wide, and $2\frac{1}{2}$ centimeters high that can be cut from the large block?

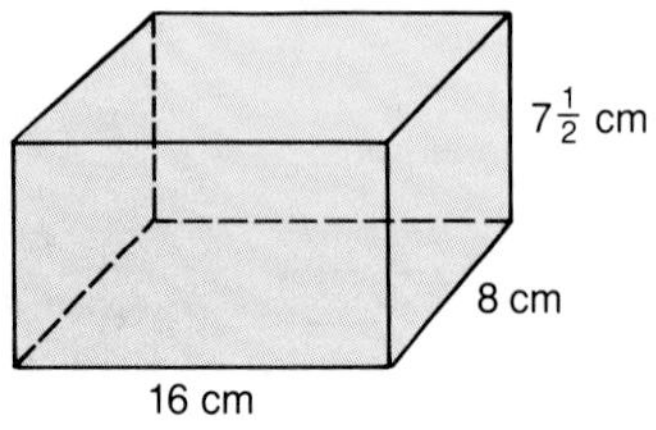

Using Statistics Misleading Graphs

Statistical data is often presented in bar graphs. **Graphs** are an efficient, easily-read way to display information. However, graphs can be **misleading** in the way in which they display data. Special attention should be paid to the horizontal and vertical scales used in the graphs.

In the graph at the right, are the profits really going up as indicated by the rising line?

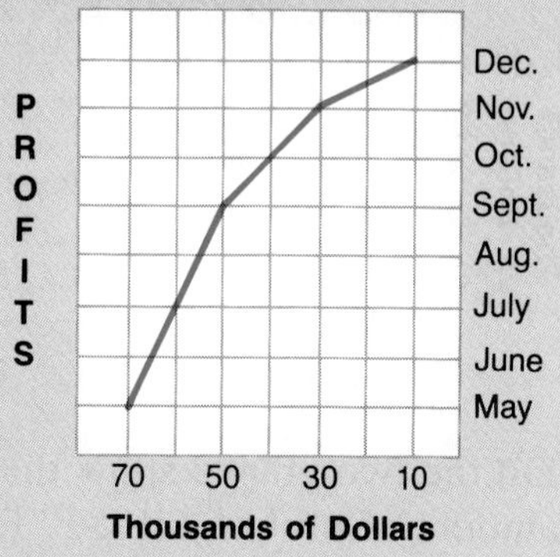

EXERCISES

The two bar graphs below show the number of stereos sold by two electronics stores for each of the past five years.

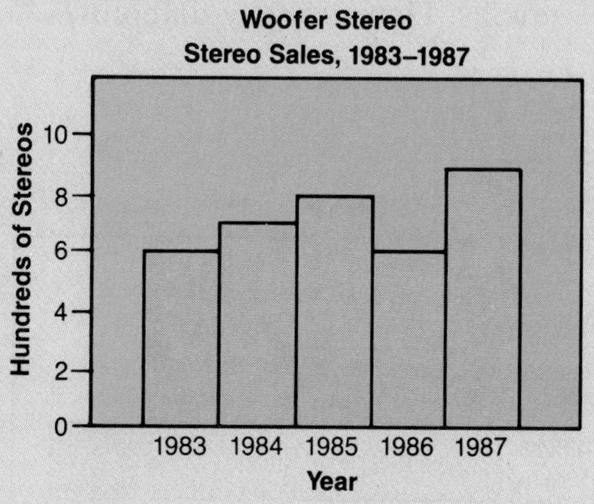

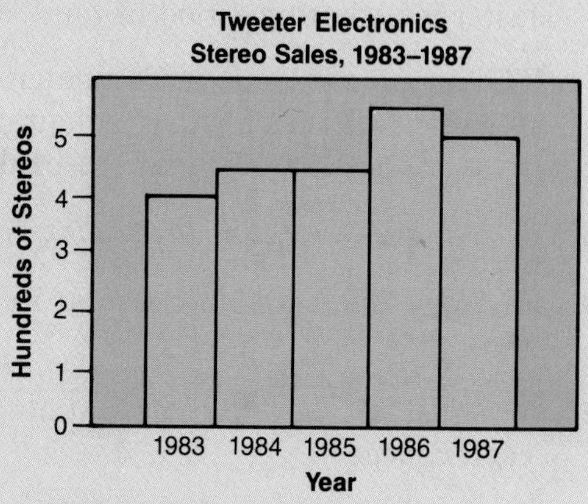

1. Which company appears to have sold the greater number of stereos each year?

2. What is the value of each vertical unit on the graph for Woofer Stereo? for Tweeter Electronics?

3. How many stereos did Woofer Stereo sell in 1983? in 1984? in 1987?

4. How many stereos did Tweeter Electronics sell in 1983? in 1984? in 1987?

5. Did Tweeter Electronics sell more stereos than Woofer Stereo in <u>any</u> year from 1983 to 1987?

6. The graphs are the same size and are placed side by side. Do you think this encourages the consumer to compare the height of the bars without examining the scales? Explain.

7. Suppose the graphs were <u>not</u> side by side. Do you think Tweeter Electronics would still appear to have sold more stereos? Explain.

The owner of Hacker's Computer Outlet drew the two line graphs below.

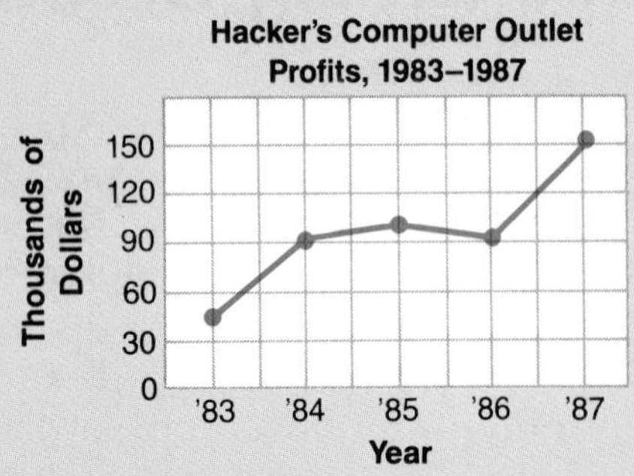

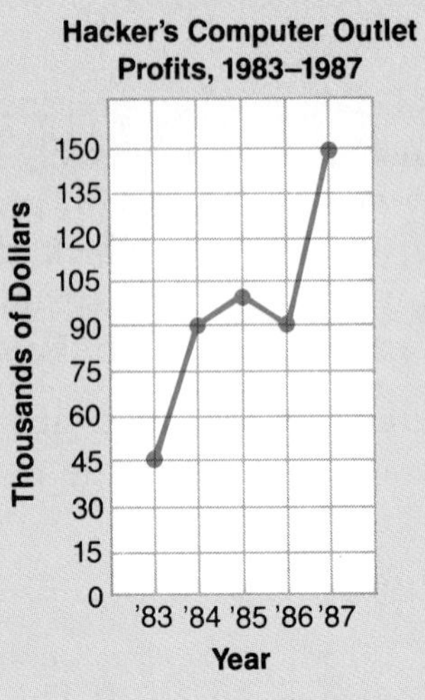

8. Do the two graphs show the same amount of profit for 1983? for 1986? for 1987?

9. In which graph does the amount of profit seem to be increasing faster in a shorter period of time?

10. Compare the scales of the two graphs. How are they different?

11. The owner of Hacker's Computer Outlet wants to borrow some money to expand the business. Which graph do you think the owner should show to the loan officer at the bank? Explain.

The graph at the right is misleading.

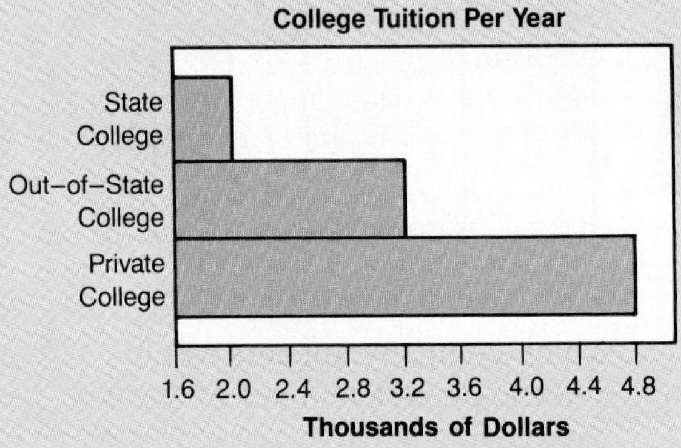

12. Measure the length of each bar in the graph.

13. What is the yearly tuition cost at each college?

14. How many times as long as the bar for State College is the bar for Out-of-State College?

15. Is the tuition at Out-of-State College actually 4 times the tuition at State College?

16. How many times as long as the bar for State College is the bar for Private College?

17. Is the tuition at Private College actually 8 times the tuition at State College?

18. Redraw the graph using a scale that begins with 0.

19. On your graph, is the bar for Out-of-State College 4 times as long as the bar for State College?

20. On your graph, is the bar for Private College 8 times as long as the bar for State College?

21. Is your graph misleading? Explain.

CHAPTER SUMMARY

IMPORTANT TERMS

Least common multiple of counting numbers *(p. 300)*
Least common multiple of polynomials *(p. 301)*

IMPORTANT IDEAS

1. Sum and Difference Rules for Rational Expressions:
If $\frac{a}{c}$ and $\frac{b}{c}$ are rational expressions, then
$\frac{a}{c}+\frac{b}{c}=\frac{a+b}{c}$ and $\frac{a}{c}-\frac{b}{c}=\frac{a-b}{c}$. $(c \neq 0)$

2. The least common multiple of prime polynomials is their product.

3. Steps in finding the *LCM* of two or more polynomials:
- ***a.*** Write the prime factorization of each polynomial.
- ***b.*** Write a product using each prime factorization only once.
- ***c.*** For each factor, write the greatest exponent used in any prime factorization.
- ***d.*** Multiply.

4. Steps in adding or subtracting rational expressions:
- ***a.*** Factor the denominators.
- ***b.*** Determine the *LCM*.
- ***c.*** Change the rational expressions to equivalent expressions with the *LCM* as the denominator of each.
- ***d.*** Add or subtract.
- ***e.*** Simplify.

5. Steps in solving an equation with rational expressions:
- ***a.*** Exclude numbers from the domain that will make zero divisors.
- ***b.*** Multiply each side by the *LCM* of the denominators.
- ***c.*** Solve the resulting equation.

CHAPTER REVIEW

SECTION 11.1

Add or subtract.

1. $\frac{13}{3x}-\frac{5}{3x}$ **2.** $\frac{3}{5a}+\frac{1}{5a}$ **3.** $\frac{2a}{a+5}-\frac{a}{a+5}$ **4.** $\frac{b}{b+2}+\frac{-3b}{b+2}$

SECTION 11.2

Add or subtract. Simplify where necessary.

5. $\frac{x+1}{2x+3}+\frac{x+2}{2x+3}$ **6.** $\frac{3x+1}{x-1}-\frac{x+3}{x-1}$ **7.** $\frac{3a+1}{5a-1}-\frac{a-3}{5a-1}$

8. $\frac{a-5}{3a+2}+\frac{5a+2}{3a+2}$

9. $\frac{5x-4}{2x-3}-\frac{x+2}{2x-3}$

10. $\frac{8x+3}{5x-1}-\frac{-2x+5}{5x-1}$

SECTION 11.3

Find each LCM.

11. $2x$; $3x$; $x-2$

12. $5a$; $10a^2$; $a+5$

13. $y-2$; $y-3$; y^2

14. $b+2$; $b+5$; $3b$

15. y^2-4; $y+2$; 2

16. a^2-9; $a-3$; $3a$

SECTION 11.4

Add or subtract.

17. $\frac{1}{2a}+\frac{1}{3a}$

18. $\frac{5}{4x}+\frac{3}{10x}$

19. $\frac{3}{4x}-\frac{5}{6y}$

20. $\frac{5}{8a}-\frac{7}{12b}$

SECTION 11.5

Add or subtract.

21. $\frac{1}{a+b}+\frac{a}{a-b}$

22. $\frac{1}{2x+3}+\frac{3}{x-1}$

23. $\frac{5a-1}{a+7}-\frac{a}{a-3}$

24. $\frac{y+1}{y-1}-\frac{1}{y+3}$

25. $\frac{x-2}{x+1}+\frac{x-3}{x+2}$

26. $\frac{x+4}{x+2}+\frac{x-5}{x-4}$

SECTION 11.6

Add or subtract.

27. $\frac{1}{3x+6}+\frac{x-1}{x+2}$

28. $\frac{2}{x-3}+\frac{x+1}{x^2-3x}$

29. $\frac{x+1}{4x^2-9}-\frac{5}{2x+3}$

30. $\frac{3}{x+y}-\frac{2x+y}{x^2-y^2}$

31. $\frac{2}{x^2-2x+1}+\frac{3}{x^2+x}$

32. $\frac{4}{x^2-x-2}+\frac{1}{x^2-2x}$

SECTION 11.7

Solve and check each equation.

33. $\frac{2}{x-3}=3$

34. $\frac{10}{2x-1}=2$

35. $\frac{3}{x}=\frac{1}{x+5}$

36. $\frac{4}{3x}=\frac{2}{x+1}$

37. $\frac{5}{x+2}=\frac{6}{x+2}$

38. $\frac{1}{x-5}=\frac{2}{x-5}$

CHAPTER 12

Relations and Functions

Locating points in space requires a method different from that used to locate points on earth. To locate points on earth, you use *latitude* and *longitude.* Early Polynesian sailors constructed a navigational chart out of shells, which represent islands, and sticks.

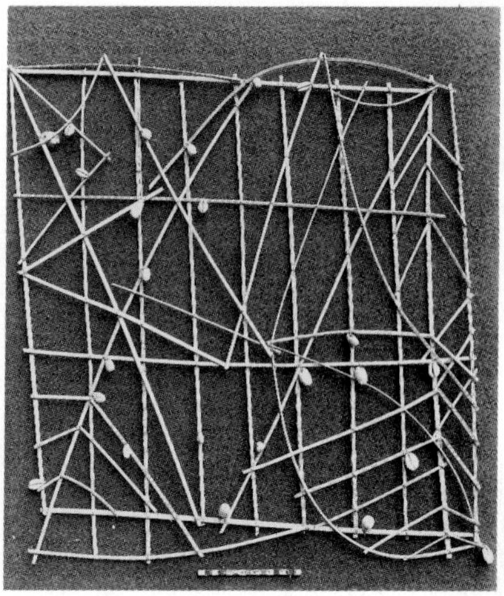

12.1 Graphs of Ordered Pairs

Figure 1 shows two intersecting real-number lines and two points, *P* and *Q*. The horizontal line is the ***x axis,*** and the vertical line is the ***y axis.*** The point where the two lines meet is called the ***origin.***

P–1 **On the x axis, what direction is positive? negative?**

P–2 **On the y axis, what direction is positive? negative?**

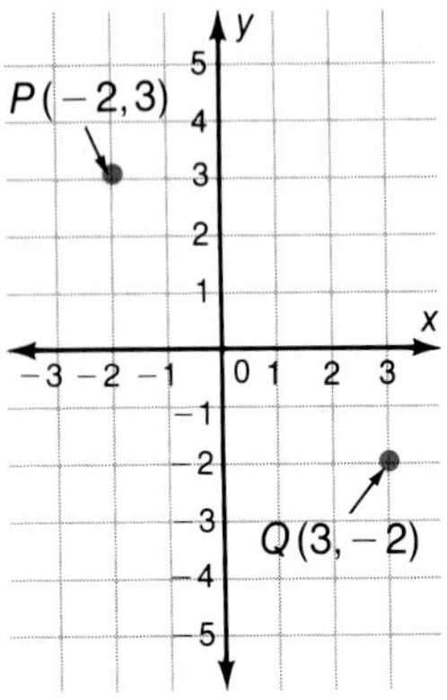

Figure 1

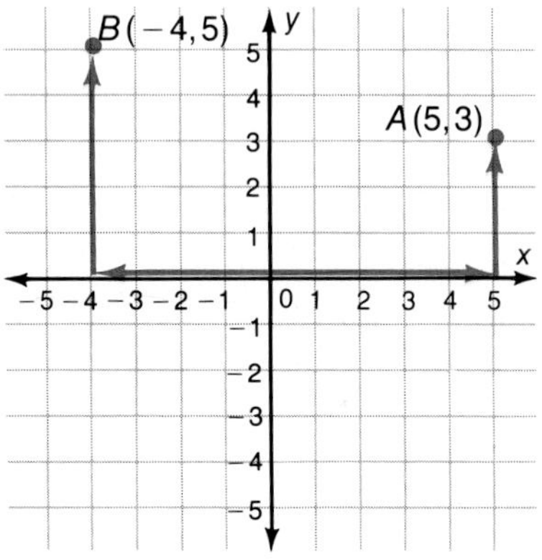

Figure 2

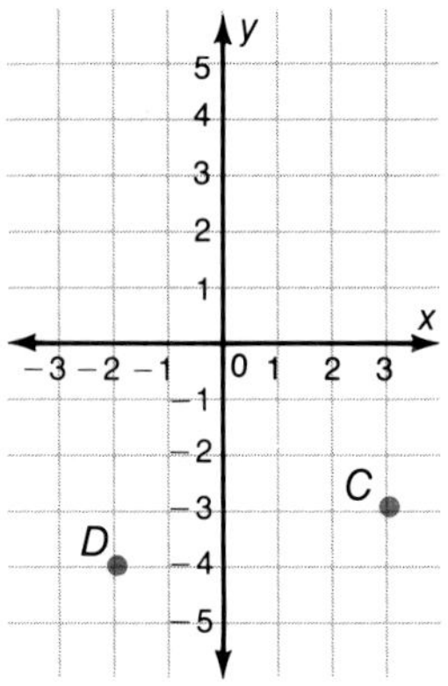

Figure 3

The pair of real numbers (−2, 3) are the ***coordinates*** of point *P* in Figure 1. Their order is important. For that reason they are called an ***ordered pair.***

P–3 **What is the first coordinate of point Q in Figure 1? the second coordinate of point Q?**

P–4 **How does Figure 1 show that (−2, 3) and (3, −2) are not the same?**

EXAMPLE 1 In Figure 2, locate *A*(5, 3).

A(5, 3) is the Point A which corresponds to the ordered pair (5, 3).

Solution:

1. Start at 0, and move five units to the right along the *x* axis.
2. Next, move up three units in a direction parallel to the *y* axis.

P–5 **Why is the move of five units made along the x axis?**

P–6 **Why do you move five units to the right?**

P–7 **Why is the second move of three units up and not down?**

EXAMPLE 2

Locate $B(-4, 5)$ in Figure 2.

Solution: [1] Start at 0, and move four units to the left along the x axis.

Left because the first coordinate is negative.

[2] Next, move five units up in a direction parallel to the y axis.

EXAMPLE 3

Write the ordered pair that corresponds to point C in Figure 3. The ordered pair is $(3, -3)$.

Solution:

[1] Start at the origin.

[2] Move to the right, directly over point C.

Right means a positive coordinate.

[3] Write the first coordinate, 3.

[4] Move down, counting the units.

Down means a negative coordinate.

[5] Write the second coordinate, -3.

The ordered pair is $(3, -3)$.

P–8 **To name the ordered pair that corresponds to point *D*, what move do you make first? What is the first coordinate?**

P–9 **What is the second move? the second coordinate?**

P–10 **What is the ordered pair for point *D*?**

CLASSROOM EXERCISES

Write the two moves needed to locate the point corresponding to each ordered pair. (Examples 1 and 2)

1. $(3, -2)$ **2.** $(-10, -3)$ **3.** $(-3, 7)$ **4.** $(5, 5)$ **5.** $(7, -1)$

6. $(10, -10)$ **7.** $(8, 3)$ **8.** $(1, -1)$ **9.** $(-3, -3)$ **10.** $(-5, 5)$

Write the ordered pair that corresponds to each point. (Example 3, P-8, P-9)

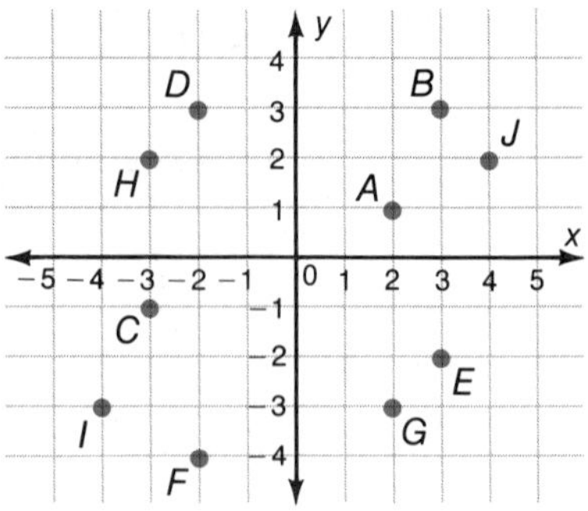

11. *A* **12.** *B*

13. *C* **14.** *D*

15. *E* **16.** *F*

17. *G* **18.** *H*

19. *I* **20.** *J*

WRITTEN EXERCISES

Goals: To locate the point that corresponds to a given ordered pair of integers, and to name the ordered pair that corresponds to a given point

Sample Problems: a. Locate $G(4, -1)$ **b.** Name the ordered pair for Z in Figure 4 below.

Answers: a. Move four units right, one unit down, and draw the point. Label it $G(4, -1)$. **b.** $(-5, 4)$

Draw an x axis and y axis on graph paper. Locate the point corresponding to each of the following. Name it with the capital letter. (Examples 1 and 2)

1. $A(1, 4)$ **2.** $B(5, 2)$ **3.** $C(-5, 3)$ **4.** $D(-2, 4)$

5. $E(6, -1)$ **6.** $F(3, -5)$ **7.** $G(-4, -1)$ **8.** $H(-2, -5)$

9. $I(-4, -4)$ **10.** $J(-5, 5)$ **11.** $K(3, 3)$ **12.** $L(2, -2)$

Write the ordered pair that corresponds to each point. (Example 3)

13. *P* **14.** *Q* **15.** *R* **16.** *S* **17.** *T* **18.** *U*

19. *V* **20.** *W* **21.** *M* **22.** *N* **23.** *K* **24.** *L*

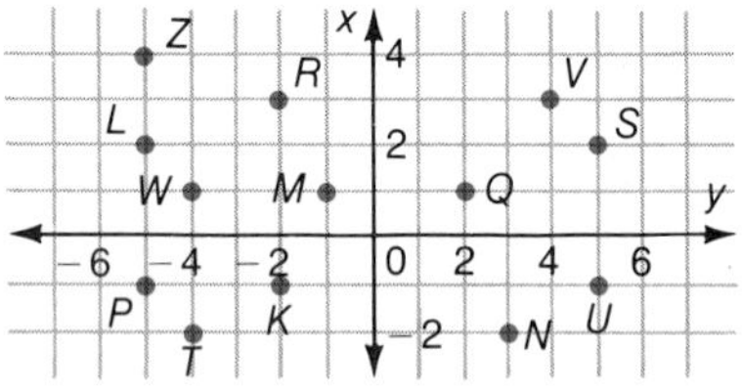

Figure 4

REVIEW CAPSULE FOR SECTION 12.2

Name the two consecutive integers that each number lies between.

EXAMPLE: -1.6 **SOLUTION:** $-2 < -1.6$; $-1.6 < -1$; The integers are -2 and -1.

1. $-3\frac{1}{4}$ **2.** $5\frac{7}{8}$ **3.** -0.9 **4.** -3.08 **5.** $-\pi$ **6.** $-\sqrt{3}$

12.2 Points and Ordered Pairs

The plane of Figure 1 is called a ***coordinate plane***. Any point of the coordinate plane may be represented as an ordered pair of real numbers (x, y).

The four regions in Figure 1 are called ***quadrants***. Both x and y are positive real numbers for each point in quadrant I. In quadrant II, x is negative, and y is positive.

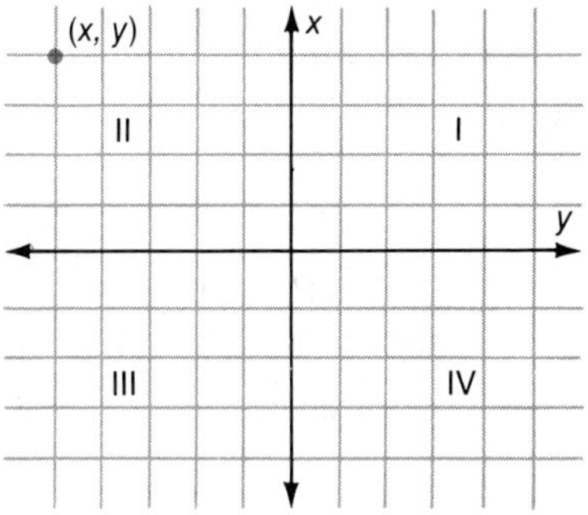

Figure 1

P–1 **How can you describe the ordered pair of coordinates corresponding to each point in quadrant III? in quadrant IV?**

EXAMPLE 1 Graph $A(3\frac{1}{2}, 5)$.

Solution: 1 Start at the origin, and move three and one-half units to the right on the x axis.

2 Move up five units parallel to the y axis. *See Figure 2.*

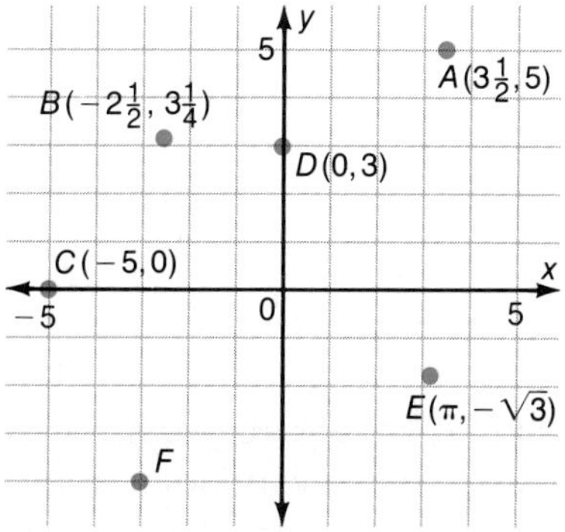

Figure 2

P–2 **To locate $B(-2\frac{1}{2}, 3\frac{1}{4})$ in Figure 2 what move is made first?**

P–3 **What move is made to represent the y value?**

EXAMPLE 2

Graph $C(-5, 0)$.

A coordinate of 0 indicates no motion.

Solution: 1 Start at 0 and move five units to the left.

2 Do not move up or down.

Point C is on the x axis.

C is between quadrants II and III. (Figure 2, p. 325)

EXAMPLE 3

Graph $D(0, 3)$.

Solution: 1 Start at 0 and do not move to the left or right.

2 Move up three units.

Point D is on the y axis.

D is between quadrants I and II. (Figure 2, p. 325)

EXAMPLE 4

Graph $E(\pi, -\sqrt{3})$.
Use approximate values of π and $-\sqrt{3}$.

$$\pi \approx 3.1 \qquad -\sqrt{3} \approx -1.7$$

The symbol "≈" means "is approximately equal to."

Solution: 1 Start at 0 and move right 3.1 units.

2 Move down 1.7 units.

See Figure 2.

P–4 **What are the coordinates of the origin? of point *F* on page 325?**

Every ordered pair of real numbers has a corresponding point in the coordinate plane. Also, every point in the coordinate plane has an ordered pair of real numbers corresponding to it.

> There is a one-to-one correspondence between points of a coordinate plane and ordered pairs of real numbers.

CLASSROOM EXERCISES

Write the quadrant in which each point lies. (Figure 1)

1. $A(-2, 7)$ **2.** $B(-5, -6)$ **3.** $C(1.7, 2.9)$ **4.** $D(3\frac{1}{2}, -5)$

Locate each point. (Examples 2 and 3)

5. $E(0, -5)$ **6.** $F(10, 0)$ **7.** $G(-5\frac{1}{2}, 0)$ **8.** $H(0, 2.7)$

WRITTEN EXERCISES

Goal: To graph points with real-number coordinates

Sample Problem: Graph $P(2\frac{1}{4}, -1\frac{1}{2})$. **Answer:**

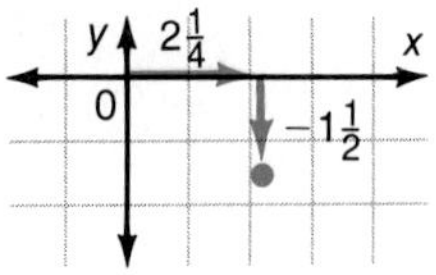

Represent a coordinate plane on graph paper. Graph the point that corresponds to each ordered pair. Label each point. (Examples 1–4)

1. $N(4\frac{1}{2}, 3)$ **2.** $Z(2\frac{1}{2}, 5)$ **3.** $P(-2, -4\frac{1}{2})$ **4.** $W(4, -3\frac{1}{2})$

5. $K(0, 4)$ **6.** $U(5, 0)$ **7.** $S(-3.5, 2.2)$ **8.** $T(1.7, -2.5)$

9. $Q(-3\frac{1}{2}, 0)$ **10.** $R(0, -5\frac{1}{2})$ **11.** $L(-3, \sqrt{5})$ **12.** $M(4.5, \sqrt{3})$

Graph these points in the same coordinate plane. (Examples 1–4)

13. $A(-6, -2)$ **14.** $E(9, 3)$ **15.** $B(0, -2)$ **16.** $F(3, 3)$

17. $C(-1, 2)$ **18.** $G(3, -3)$ **19.** $D(-7, 2)$ **20.** $H(9, -3)$

21. Connect points A and B, B and C, C and D, and D and A with line segments. What name is given to this geometric figure?

22. Connect points E and F, F and G, G and H, and H and E with line segments. What kind of figure do you get?

REVIEW CAPSULE FOR SECTION 12.3

Evaluate each expression. (Pages 21–23)

1. $-x + \frac{3}{4}$ when $x = -\frac{1}{2}$ **2.** $x^2 - x + 3$ when $x = \frac{1}{2}$

3. $-x^2 - 5$ when $x = -5$ **4.** $|x - 1.2|$ when $x = -0.9$

12.3 Relations

P–1 **What are the coordinates of the six points graphed at the right?**

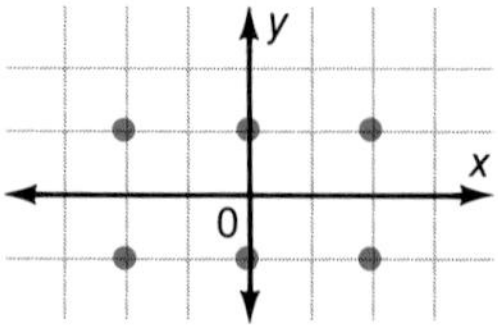

Coordinates of these points are elements of set A below.

A = {(2, 1), (2, −1), (0, 1), (0, −1), (−2, 1), (−2, −1)}

Elements of set A are ordered pairs.

Set A is a <u>relation</u>.

Definition A ***relation*** is a set of ordered pairs.

The equation $y = 2x - 1$ describes a relation.

Here are some ways to describe a relation.

Table

Kind of Description	**Description**
1. Equation	**1.** $y = 2x - 1$ when the x values are −2, −1, 0, 1, 2, 3, and 4.
2. Table of ordered pairs	**2.** (see table below)
3. Graph	**3.** (see graph below)

2.

x	−2	−1	0	1	2	3	4
y	−5	−3	−1	1	3	5	7

3.

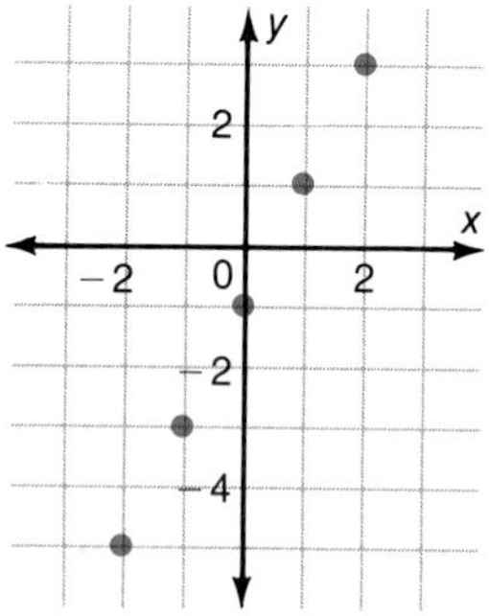

Some relations such as set A in P-1 cannot be described by an equation.

EXAMPLE 1

Write the relation T described by the equation $y = x^2 + 1$ as a set of ordered pairs when the x values are −2, −1, 0, 1, and 2.

Solution:

x	y
−2	?
−1	?
0	?
1	?
2	?

$y = x^2 + 1$

$y = (-2)^2 + 1 = 5$

$y = (-1)^2 + 1 = 2$

$y = (0)^2 + 1 = 1$

$y = (1)^2 + 1 = 2$

$y = (2)^2 + 1 = 5$

x	y
−2	5
−1	2
0	1
1	2
2	5

$T = \{(-2, 5), (-1, 2), (0, 1), (1, 2), (2, 5)\}$

EXAMPLE 2 Draw a graph of relation K described by the equation $y = |x| + 1$ when the x values are −2, −1, 0, 1, and 2.

Solution:

x	y
−2	?
−1	?
0	?
1	?
2	?

$y = |x| + 1$

$y = |-2| + 1 = 3$

$y = |-1| + 1 = 2$

$y = |0| + 1 = 1$

$y = |1| + 1 = 2$

$y = |2| + 1 = 3$

x	y
−2	3
−1	2
0	1
1	2
2	3

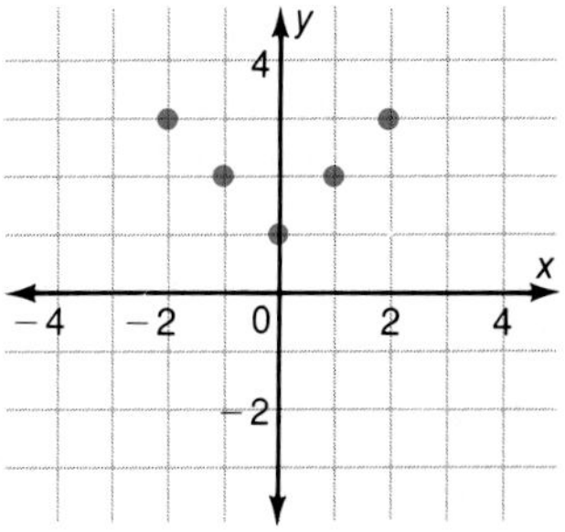

CLASSROOM EXERCISES

Name three ordered pairs of each relation below. (P-1)

1.

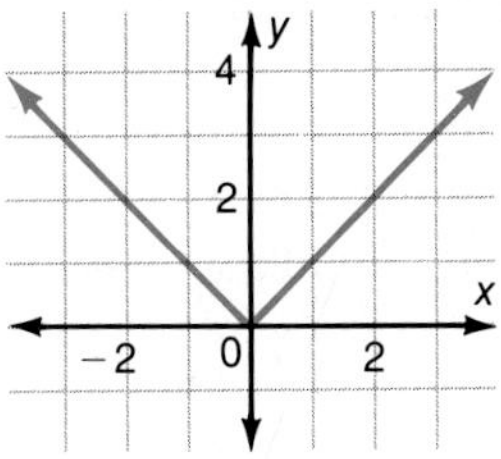

2.

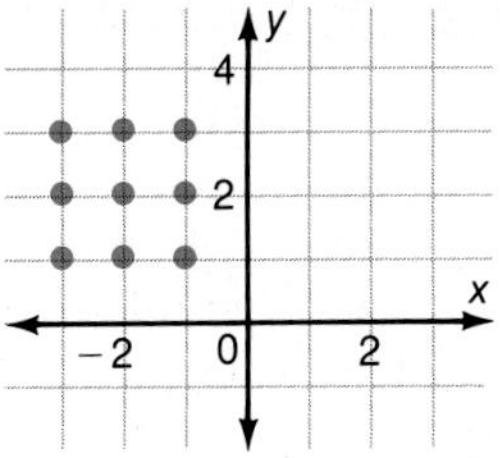

3.

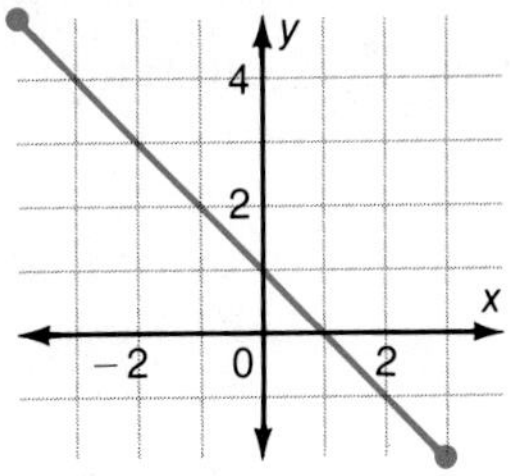

4.

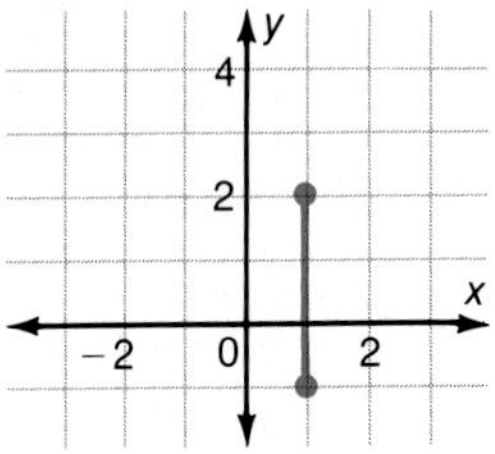

Write a table of ordered pairs for each of the following relations if the x values are −2, −1, 0, 1, and 2. (Table, Examples 1 and 2)

5. $y = x$ **6.** $y = -x$ **7.** $y = x^2$

8. $y = x - 1$ **9.** $y = x^2 + 2x + 1$ **10.** $y = |x|$

Write two ordered pairs of each relation if the x values are 1 and −1. (Example 1)

11. $y = x$ **12.** $y = -x$ **13.** $y = x^2$

14. $y = x - 1$ **15.** $y = x^2 + 2x + 1$ **16.** $y = |x|$

WRITTEN EXERCISES

Goals: To write a relation as a set of ordered pairs and to graph a relation

Sample Problem: Draw a graph of the relation described by the equation $y = x$ if the x values are 0, 1, and 2.

Set: $\{(0, 0), (1, 1), (2, 2)\}$

Graph:

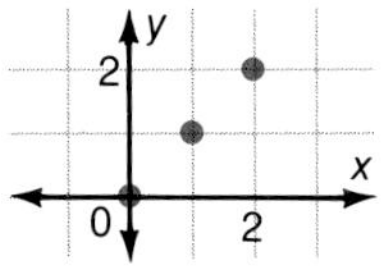

Write each relation as a set of ordered pairs if the x values are 1, 2, 3, and 4. (Example 1)

1. $y = x - 1$ **2.** $y = x - 3$

3. $y = x^2$ **4.** $y = x^2 - 1$

5. $y = |x|$ **6.** $y = \sqrt{x^2}$

7. $y = \sqrt{x}$ **8.** $y = -\sqrt{x}$

9. $y = 0 \cdot x + 1$ **10.** $y = 0 \cdot x - 2$

11. $y = x^2 - x + 1$ **12.** $y = x^2 + x - 1$

Draw a graph of each relation in a separate coordinate plane. (Example 2)

13. $\{(1, 2), (1, 3), (-3, 5), (-3, -4\frac{1}{2})\}$ **14.** $\{(-3, 4), (-3, -2), (4, 1), (4, 6\frac{1}{2})\}$

15. $\{(0, 0), (0, \pi), (0, -\pi), (\pi, 0)\}$ **16.** $\{(\pi, 3), (\pi, -3), (\pi, 5\frac{1}{2}), (\pi, 0)\}$

Draw a graph of each relation in a separate coordinate plane. The x values are −1, 0, 1 and 2. (Example 2)

17. $y = x + 1$ **18.** $y = x - 3$ **19.** $y = x^2 - 7$

20. $y = x^2 - 10$ **21.** $y = |x|$ **22.** $y = \sqrt{x^2}$

23. $y = 0 \cdot x + 1$ **24.** $y = 0 \cdot x - 2$ **25.** $y = -\sqrt{x^2}$

26. $y = x^2 - x + 1$ **27.** $y = x^2 + x - 1$ **28.** $y = 2x + 1$

NON-ROUTINE PROBLEMS

29. The owner of the Centerville Bike Shop sold two different bicycles for \$99 each. The owner made a 10% profit on the sale of one bike, but took a 10% loss on the sale of the other bike. On the two sales, did the owner make a profit, take a loss, or break even? Explain.

31. Arrange eight 8's in an addition problem so that the sum is 1000.

30. Copy the figure shown below. Then arrange the digits 2 through 9 in the empty boxes so that the sum of the number formed by the first row and the number formed by the second row equals the number formed by the third row.

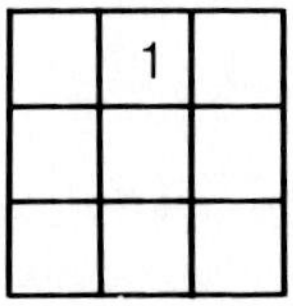

MID-CHAPTER REVIEW

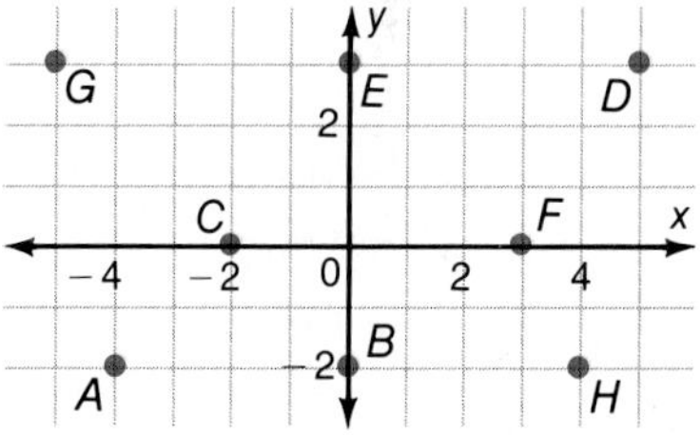

Write the ordered pair that corresponds to each point of the graph. (Section 12.1)

1. *A* **2.** *B* **3.** *C* **4.** *D*

5. *E* **6.** *F* **7.** *G* **8.** *H*

Graph these points in the same coordinate plane. (Section 12.2)

9. $A(2, -1)$ **10.** $B(-4, -3)$ **11.** $C(-2.5, -3.5)$ **12.** $D(4, 3.5)$ **13.** $E(4, -3\frac{1}{4})$

Write each relation as a set of ordered pairs and then draw its graph. The x values are $-2, -1, 0, 1,$ *and* 2. (Section 12.3)

14. $y = -2x + 3$ **15.** $y = |-x| + 2$ **16.** $y = -x^2 - 1$

REVIEW CAPSULE FOR SECTION 12.4

Write the ordered pairs for the points A, B, C, and D. (Pages 322–324)

1.

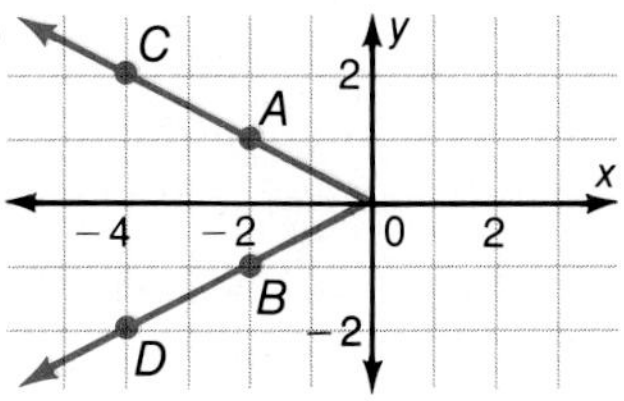

2.

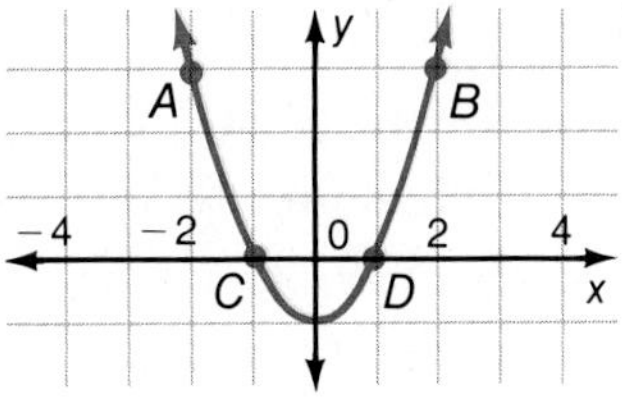

3.

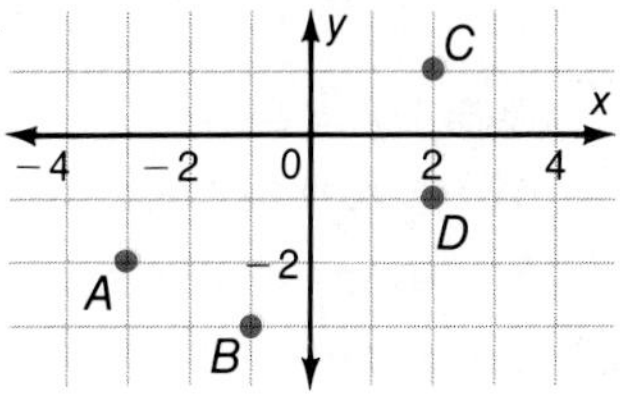

Using Graphs

Air Conditioning

One skill of an **air conditioning technician** is the ability to determine whether an air conditioner is of the right cooling capacity for a particular room. The **cooling capacity** of an air conditioner is measured in **British thermal units (Btu's).**

Monitoring a large air conditioning unit

The graph shown at the right can be used to estimate how much cooling capacity is needed for a given room of a house or apartment. (The graph can be used for any room that is about eight feet high, provided that it is not a kitchen and not directly below an attic floor.) The exposure of the room's exterior wall determines which part of the shaded band you should use.

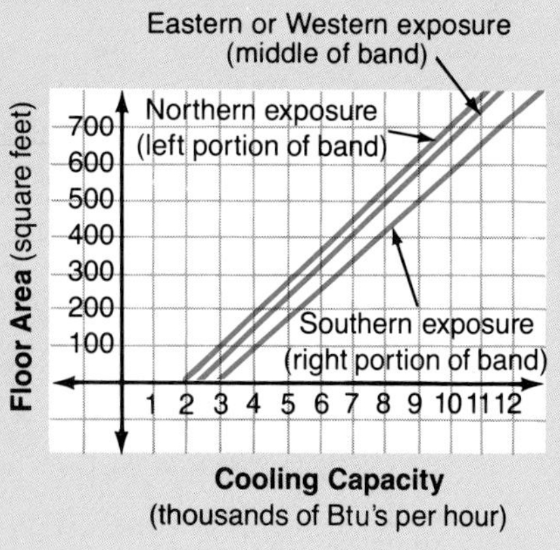

EXAMPLE: A room of a house has a northern exposure. Its floor area is 300 square feet. What must be the size (cooling capacity) of the air conditioner for this room?

SOLUTION:
1. On the vertical scale, find the room's floor area, 300 square feet.
2. Since the room has a northern exposure, find the point on the left portion of the band directly to the right of the 300 reading.
3. From this point on the band read directly down on the horizontal scale to find the correct cooling capacity, **5500 Btu's per hour.**

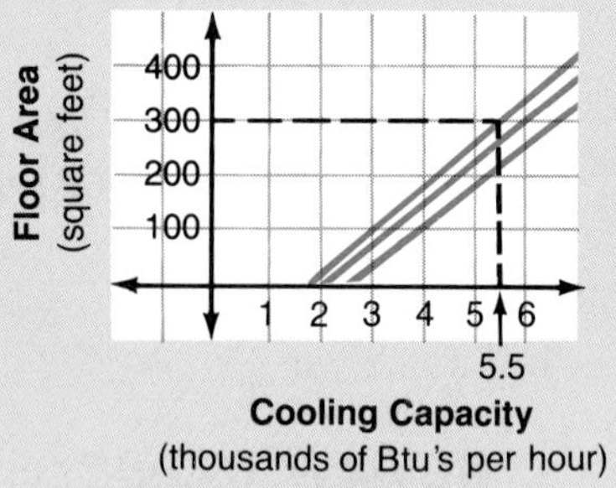

EXERCISES

Estimate the size of an air conditioner needed to cool each room described. Round your answer to the nearest 500 Btu's.

1. Southern exposure; 400 square feet
2. Northern exposure; 500 square feet
3. Eastern exposure; 300 square feet
4. Southern exposure; 250 square feet
5. Western exposure; 100 square feet
6. Northern exposure; 700 square feet

12.4 Functions

P–1 **What are the x values of relation T below? the y values?**

$$T = \{(-1, 3), (0, -2), (4, 0), (-3, -2)\}$$

Set D below is the domain of relation T. Set R is the range.

$$D = \{-1, 0, 4, -3\} \qquad R = \{3, -2, 0\}$$

Definitions

> The ***domain*** of a relation is the set of x values of its ordered pairs. The ***range*** of a relation is the set of y values of its ordered pairs.

EXAMPLE 1 Write the domain D and range R of relation W.

$$W = \{(-\tfrac{1}{2}, 0), (-2, 3), (-\tfrac{1}{4}, 2), (-\tfrac{1}{2}, -1), (2, -1)\}$$

Solution: $D = \{-\frac{1}{2}, -2, -\frac{1}{4}, 2\}$ $\qquad R = \{0, 3, 2, -1\}$

P–2 **What is different about the domains of relations A and B?**

$$A = \{(-2, 5), (4, 8), (-2, 3)\}$$
$$B = \{(-2, 5), (4, 8), (9, 3)\}$$

In A, two x values are equal. In B, no two x values are equal.

Relation B is an example of a special relation called a function.

Definition

> A ***function*** is a relation in which no two ordered pairs have the same x value. In a function, each different x value has only one y value.

P–3 **How can you show that relation T following P-1 is a function?**

P–4 **Which of the following relations are also functions?**

$$P = \{(-1, 3), (2, 7), (0, 0)\}$$
$$Q = \{(1.3, 2), (-0.5, 1), (1.3, 0)\}$$
$$R = \{(\tfrac{1}{2}, \tfrac{1}{4}), (-\tfrac{1}{4}, 0), (\tfrac{2}{3}, 5), (0.5, 0.7)\}$$
$$S = \{(\pi, -2), (3, \sqrt{5}), (-1, -\sqrt{5})\}$$

The graph of a certain function is shown in Figure 1. The "hollow dot" means that (−1, −2) is not in the function. The "heavy dot" emphasizes that (3, 1) is in the function.

The graph of this function consists of all points on the portion of the line as shown.

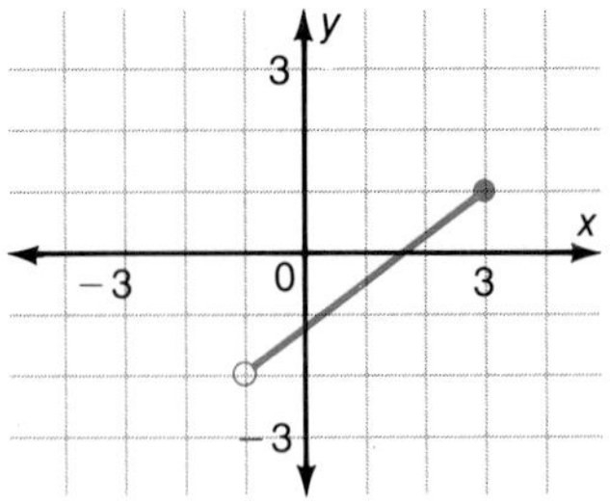

Figure 1

There is a simple test to tell whether or not a graph represents a function.

> If any vertical line has more than one point in common with a graph, then the graph does not represent a function. Otherwise, it is the graph of a function.

P–5 **Does the graph in Figure 1 represent a function? Why?**

EXAMPLE 2

Write <u>Yes</u> or <u>No</u> to show whether the graph in Figure 2 represents a function.

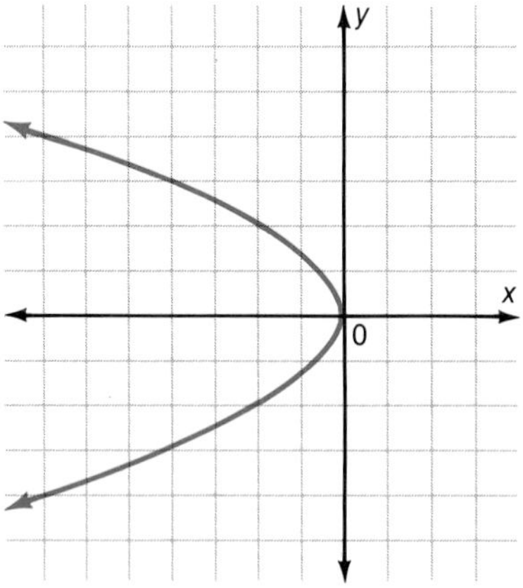

Figure 2

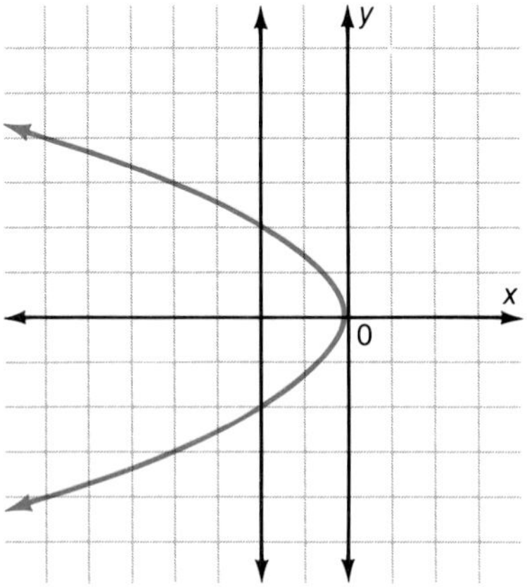

Figure 3

Solution: A vertical line is shown in Figure 3. It has two points in common with the graph. <u>No</u>, the graph of Figure 2 does not represent a function.

CLASSROOM EXERCISES

Write the domain and range of each relation. (Example 1)

1. $\{(3, 7), (-5, 2), (4, -1), (0, 0)\}$
2. $\{(\frac{1}{2}, 3), (\frac{1}{4}, 5), (\frac{1}{5}, 7), (0, 0)\}$
3. $\{(3, 8), (\sqrt{2}, -1), (\pi, -2), (\frac{1}{3}, -1)\}$
4. $\{(10, 1), (11, 2), (12, 3), (10, \frac{1}{2})\}$

5.

x	2	3	4	5
y	3	4	5	6

6.

x	$\frac{1}{2}$	$\frac{1}{3}$	$\frac{1}{4}$
y	$-\frac{1}{2}$	$-\frac{1}{3}$	$-\frac{1}{4}$

7.

x	1	−1	0	$\frac{1}{2}$	$-\frac{1}{2}$
y	1	1	0	$\frac{1}{2}$	$\frac{1}{2}$

8.

x	1	1	4	4	9
y	1	−1	2	−2	3

Write Yes or No to indicate if each set is a function. (P-2, P-3, P-4)

9. $\{(-1, 2), (2, -1)\}$
10. $\{(3, -1), (3, -2), (3, -3)\}$
11. $\{(\frac{1}{2}, 1), (\frac{1}{3}, 1), (\frac{1}{4}, 1)\}$
12. $\{(1.6, 2), (0.5, -3), (\frac{1}{2}, 0.7)\}$
13. $\{(1, 2), (3, 4), (5, 6)\}$
14. $\{(1, 1), (1, 2), (1, 3), (1, 4)\}$

15.

x	0	−3	2	−1
y	2	5	7	2

16.

x	$\frac{1}{2}$	$-\frac{1}{2}$	$\frac{3}{4}$	$-\frac{3}{4}$
y	5	6	7	8

WRITTEN EXERCISES

Goals: To determine whether a relation is a function by examining its set of ordered pairs or by applying the vertical line test to its graph

Sample Problems: a. $\{(0, 1), (1, 1)\}$ **b.** See Example 2.

Answers: a. Yes. **b.** No.

Write the domain, D, and range, R, of each relation. (Example 1)

1. $\{(5, 10), (-3, 2), (-1, 0), (1, -3)\}$
2. $\{(6, 5), (2, 3), (0, 0), (-3, -1)\}$
3. $\{(\frac{1}{2}, 2), (-1, \frac{1}{2}), (-\frac{1}{2}, -1), (2, -1)\}$
4. $\{(\frac{2}{3}, -1), (0, \frac{1}{2}), (-\frac{1}{2}, 0), (\frac{1}{4}, \frac{1}{4})\}$
5. $\{(-1, \sqrt{2}), (-2, \sqrt{3}), (\pi, -1), (0, -\pi)\}$
6. $\{(\sqrt{3}, -1), (0, \sqrt{2}), (0, \pi), (-1, \sqrt{2})\}$

7.

x	0	1	1.5	−1	$-\pi$
y	0	1	1	−1	−1

8.

x	0	1	2	1.5	−1	−1.5
y	0	1	2	2	0	−1

9.

x	0	$\frac{1}{2}$	1	$1\frac{1}{2}$	$-\frac{1}{2}$	−1
y	0	0	1	1	−1	−1

10.

x	0	1	2	$\frac{1}{2}$	−1	$-\frac{1}{2}$
y	0	1	4	$\frac{1}{4}$	1	$\frac{1}{4}$

Write Yes or No to show whether each relation is a function. (P-2, P-3, P-4)

11. $\{(5, 7), (9, 11)\}$

12. $\{(2, 4), (6, 8)\}$

13. $\{(-2, 3), (-3, 4), (-5, 3)\}$

14. $\{(-\frac{1}{2}, 1), (-\frac{3}{2}, 2), (\frac{5}{2}, 1)\}$

15. $\{(0, 0), (2, 10), (-1, 3), (0, 5)\}$

16. $\{(-2, -1), (0, -3), (-2, 0), (-3, 0)\}$

17. $\{(\pi, 3), (\frac{1}{2}, \frac{1}{3}), (0.5, 5)\}$

18. $\{(\sqrt{2}, 3), (-\pi, 2), (-\sqrt{2}, 1)\}$

Write Yes or No to show whether each graph represents a function. Explain your answer. (Example 2)

19.

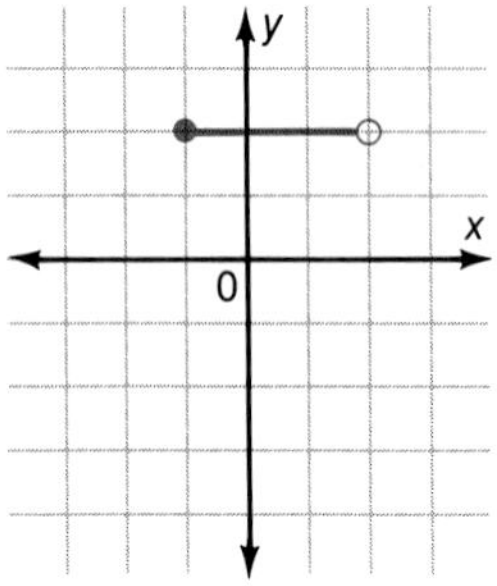

20.

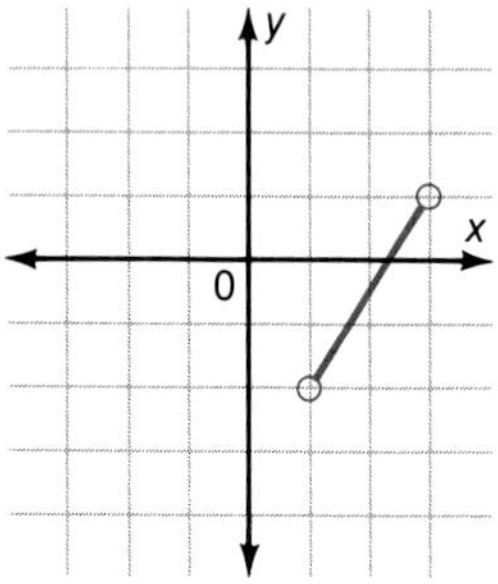

21.

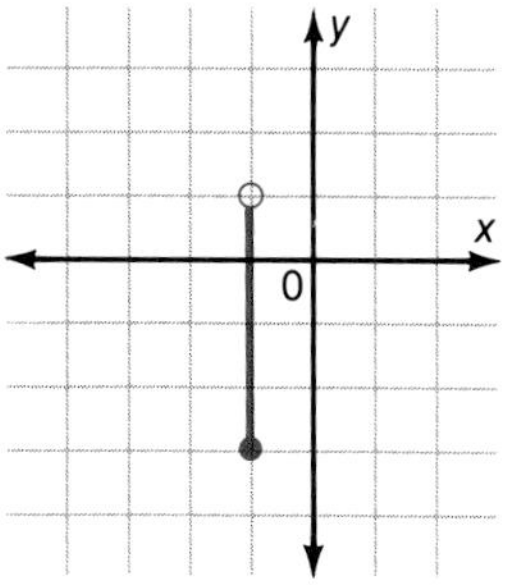

22.

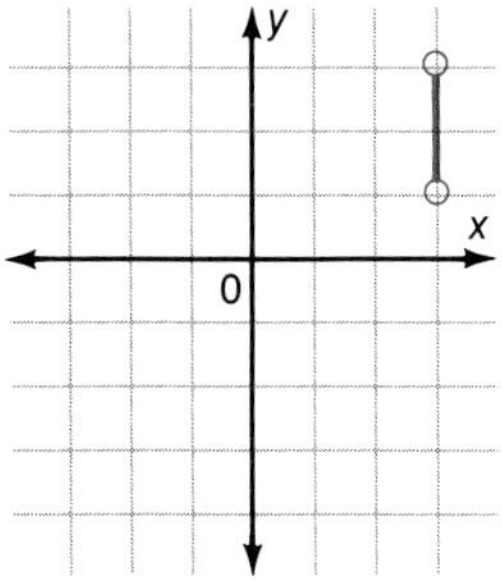

23.

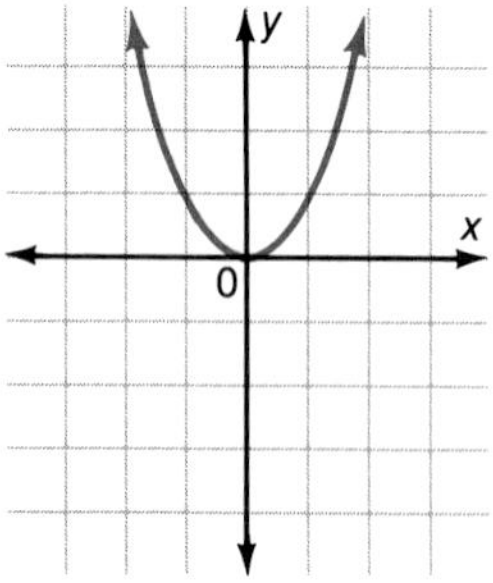

24.

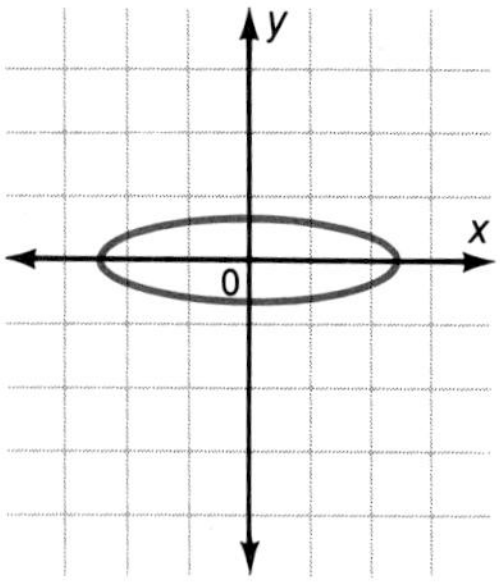

25.

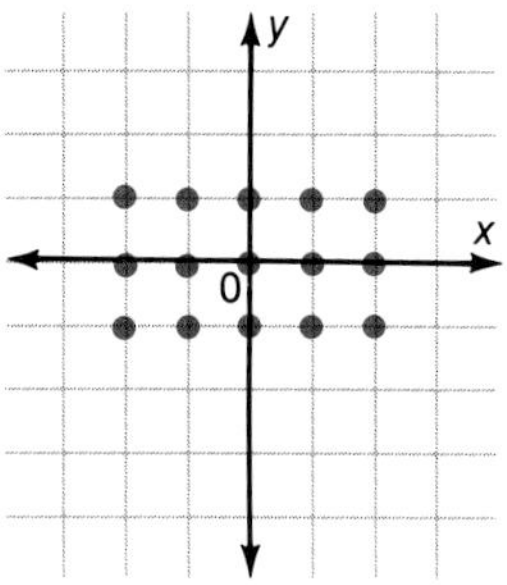

26.

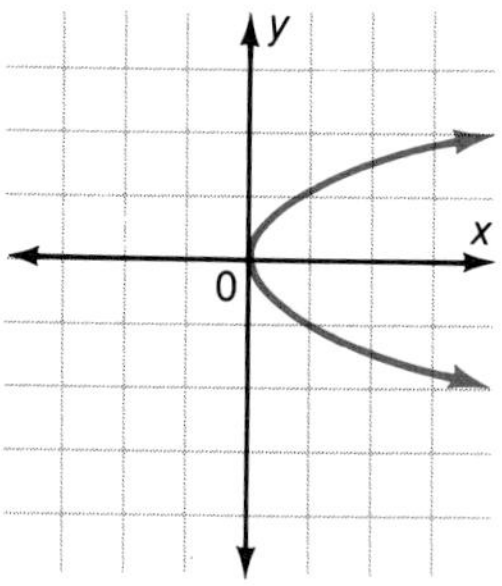

REVIEW CAPSULE FOR SECTION 12.5

Find which of these ordered pairs are in each relation. (Pages 328–330)

$(-3, 3)$ $(-2, 2)$ $(1, 4)$ $(-2, -2)$ $(1, 1)$ $(0, 0)$ $(0, 2)$ $(7, -1)$ $(-4, -2)$

1. $y = -\frac{1}{3}x + \frac{4}{3}$

2. $y = -x^2 + 2$

3. $y = |x|$

4. $y = -\sqrt{x^2}$

5. $y = -|x|$

6. $y = x^2 - x$

12.5 Graphs of Functions

EXAMPLE 1 Using {real numbers} as the domain, draw a graph of $y = 2x + 1$.

Solution: Choose a few x values that are small in absolute value. The corresponding y values are obtained from the formula to complete the table.

Values Chosen for x

x	y
-1	?
0	?
1	?
2	?

Values for y Obtained from the Formula $y = 2x + 1$

For $x = -1$, $y = 2(-1) + 1 = -1$.
For $x = 0$, $y = 2(0) + 1 = 1$.
For $x = 1$, $y = 2(1) + 1 = 3$.
For $x = 2$, $y = 2(2) + 1 = 5$.

x	y
-1	-1
0	1
1	3
2	5

These ordered pairs are graphed in Figure 1. You can see that the points lie on a line.

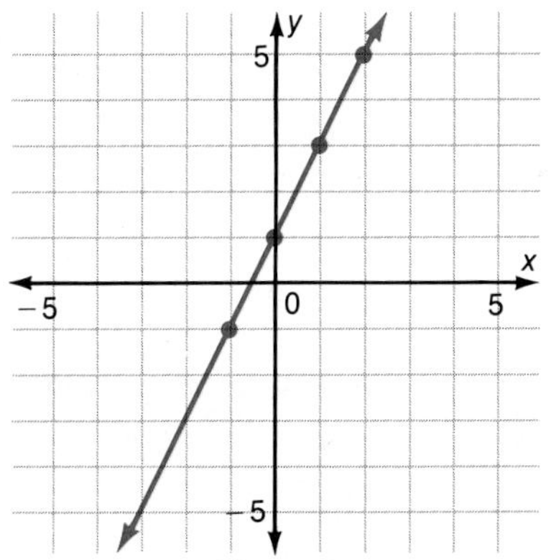

Figure 1

P–1 **If $x = -2$, what is the corresponding value for y from the formula $y = 2x + 1$?**

The point $(-2, -3)$ is on the graph of Figure 1.

EXAMPLE 2

Using {real numbers} as the domain, draw a graph of $y = x^2 - 1$.

Solution: Choose values for x and compute the corresponding values for y.

x	y
−2	?
−1	?
0	?
1	?
2	?

For $x = -2$, $y = (-2)^2 - 1 = 3$.
For $x = -1$, $y = (-1)^2 - 1 = 0$.
For $x = 0$, $y = (0)^2 - 1 = -1$.
For $x = 1$, $y = (1)^2 - 1 = 0$.
For $x = 2$, $y = (2)^2 - 1 = 3$.

x	y
−2	3
−1	0
0	−1
1	0
2	3

These ordered pairs are graphed in Figure 2, and a curved dashed line is drawn through the points. The complete graph is shown in Figure 3.

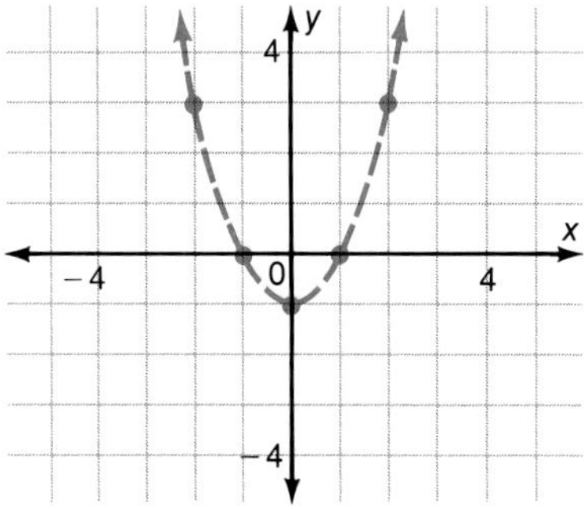

Figure 2

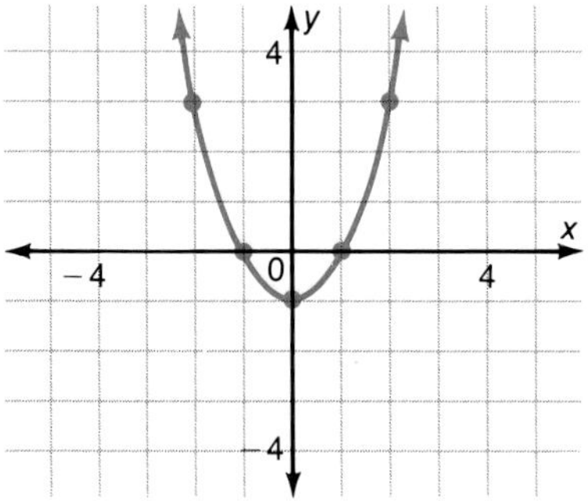

Figure 3

P–2 **What is the range of the function shown in Figure 3?**

P–3 **If $x = \frac{3}{2}$, what is the corresponding y value from the formula $y = x^2 - 1$?**

The point $(\frac{3}{2}, \frac{5}{4})$ is on the graph of Figure 3.

CLASSROOM EXERCISES

Write the y values that correspond to the given x values. (Examples 1 and 2)

1. $y = 2x$

x	0	−1	1	−2	2
y	?	?	?	?	?

2. $y = 3x + 1$

x	−2	−1	0	1	2
y	?	?	?	?	?

Refer to Figure 4 for Exercises 3–7. (P-1, P-2, P-3)

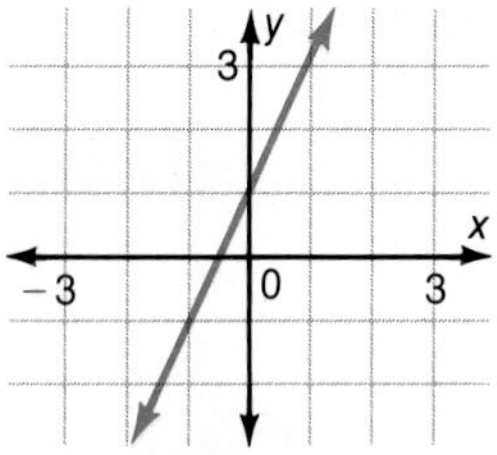

Figure 4

3. Why does the graph represent a function?
4. What is the domain?
5. What is the range?
6. What y value corresponds to $x = -1$?
7. What x value corresponds to $y = 1$?

Refer to Figure 5 for Exercises 8–12. (P-1, P-2, P-3)

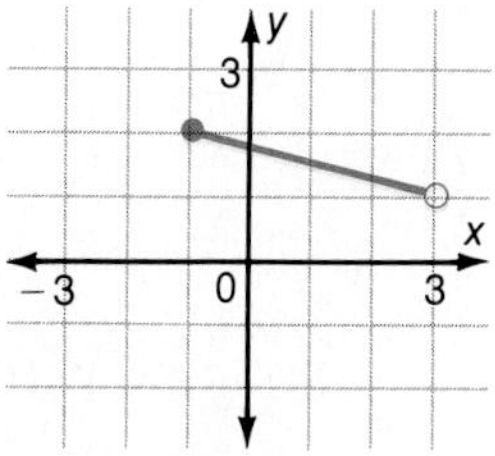

Figure 5

8. Why does the graph represent a function?
9. What is the domain?
10. What is the range?
11. What y value corresponds to $x = 1$?
12. What x value corresponds to $y = 1\frac{3}{4}$?

WRITTEN EXERCISES

Goal: To draw the graph of a function given the set description and domain

Sample Problem: Using {real numbers} as the domain, draw a graph of $y = |x|$.

Answer:

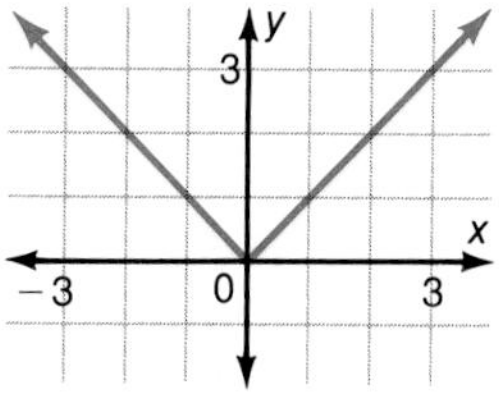

Draw a graph of each function in a separate coordinate plane. First graph five ordered pairs with x values −2, −1, 0, 1, and 2. The domain is {real numbers}. (Examples 1 and 2)

1. $y = -2x$
2. $y = -3x$
3. $y = -x + 1$
4. $y = -2x + 1$
5. $y = -|x|$
6. $y = -\sqrt{x^2}$
7. $y = |x + 1|$
8. $y = |x - 1|$
9. $y = 0 \cdot x - 3$
10. $y = 0 \cdot x + 4$
11. $y = -x^2 + 2$
12. $y = x^2 - 3$

MORE CHALLENGING EXERCISES

13. $y = \frac{1}{2}x^2 - x$
14. $y = x^2 - x + 1$
15. $y = -x^2 + x$

NON-ROUTINE PROBLEM

16. Copy the figure at the right. Then place 8 dots on the figure so that two dots are on each of the three lines and two dots are on each of the four circles.

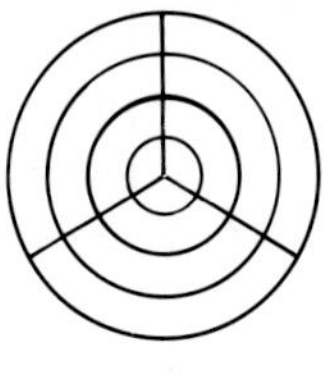

12.6 Using Graphs: Prediction

Roger is learning to parachute. His instructor told him that he will be falling faster and faster each second after leaving the airplane until he opens his parachute. The distance that he falls is a function of time.

EXAMPLE 1 The graph below describes the relationship between the distance in meters that an object falls and the time in seconds.

Find the approximate distance, in meters, for each of the following times.

a. 1 second

b. 3 seconds

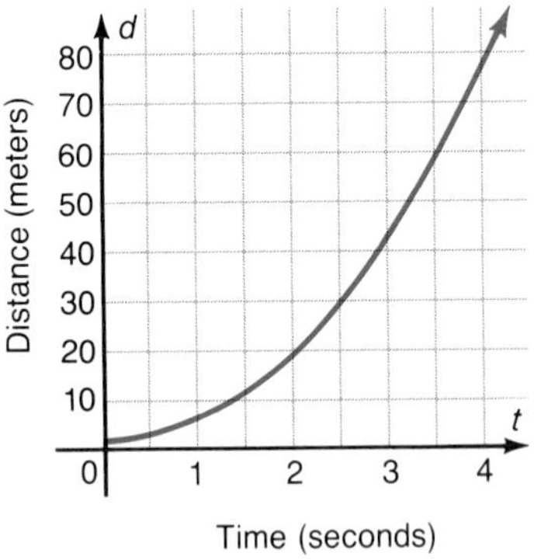

Solutions: **a.** Locate the point on the graph that corresponds to $t = 1$ on the t axis. Read the approximate corresponding distance in meters on the d axis.

When $t = 1$, d is approximately 5 meters.

b. Locate the point on the graph that corresponds to $t = 3$ on the t axis. Read the approximate corresponding distance in meters on the d axis.

When $t = 3$, d is approximately 45 meters.

By examining the change in values at regular intervals on a graph, predictions can be made for points not shown on the graph. This procedure is illustrated in Example 2. When making predictions based on a graph, you assume the graph extends without limit according to the pattern shown.

EXAMPLE 2

The graph below describes the relationship between the distance, d, that sound travels in water and the traveling time, t.

a. Use the graph to find d for $t = 2$, 4, 6, and 8.

b. By how much does d increase when t increases by 2 seconds?

c. Use your answer to **b** to predict the approximate distance traveled by sound in 10 seconds.

Solutions: **a.**

t	2	4	6	8
d	3	6	9	12

b. When t increases by 2 seconds, d increases by 3 kilometers.

c. Thus, sound will travel about 15 kilometers in 10 seconds.

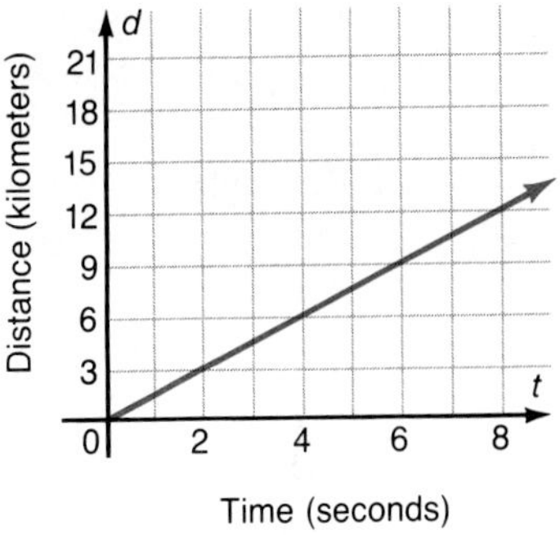

CLASSROOM EXERCISES

Use the graph in Example 1 to find the approximate distance that an object falls for each of the following times. (Example 1)

1. 2 seconds **2.** $3\frac{1}{2}$ seconds **3.** 4 seconds

4. How far would an object fall between the first and second seconds?

5. Does an object fall twice as far in the first 4 seconds as it does in the first 2 seconds?

6. How does the distance fallen between the second and third seconds compare with the distance fallen between the third and fourth seconds?

Galileo is reported to have dropped two iron balls having different weights from the Leaning Tower of Pisa. Both hit the ground at the same time, supporting Galileo's theory that gravity pulls all bodies to earth with the same acceleration, regardless of their weight.

Use the graph in Example 2 to predict the distance that sound travels in water for each of the following times.
(Example 2)

7. 12 seconds **8.** 16 seconds **9.** 20 seconds **10.** 14 seconds

WRITTEN EXERCISES

Goal: To use the graph of a function to solve word problems
Sample Problem: From the graph in Example 1, find the approximate distance that an object falls in $2\frac{1}{2}$ seconds.
Answer: 31 meters

The graph at the right shows the average stopping distance for a car traveling at various speeds. Write the stopping distance for each of the following speeds.
(Example 1)

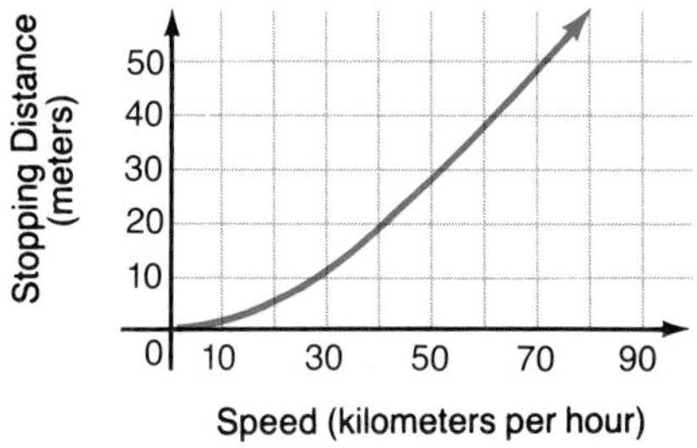

1. 20 km/hr **2.** 40 km/hr

3. 30 km/hr **4.** 50 km/hr

5. 60 km/hr **6.** 70 km/hr

Assume that the graph below represents the cost of sending first-class mail. The domain of the function is the set of weights in ounces. The range is the corresponding set of costs in cents. Use the graph to predict the approximate cost of sending first-class mail having the following weights.
(Example 2)

7. $5\frac{1}{2}$ ounces

8. $6\frac{3}{4}$ ounces

9. 8 ounces

10. $9\frac{1}{4}$ ounces

11. $7\frac{1}{2}$ ounces

12. 10 ounces

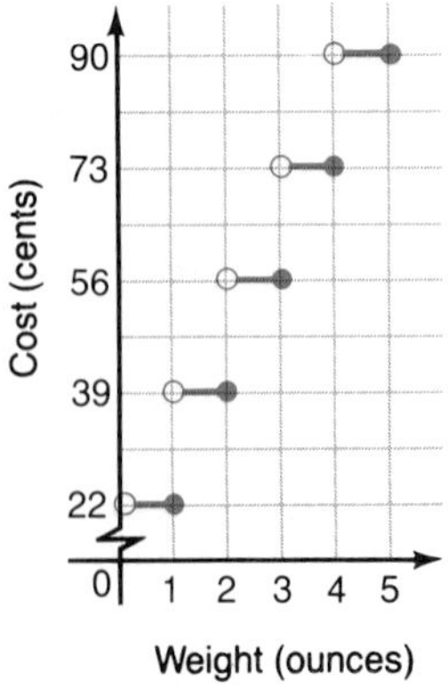

Focus on Reasoning:
Too Little/Too Much Information

When solving a problem, it is important to identify what information is useful and whether enough information is given.

If enough information is given to solve the problem, find the solution. Then state whether you used Clue 1 only, Clue 2 only, or both Clue 1 and Clue 2. If not enough information is given, write "Can't tell."

EXAMPLE 1 **CLUE 1** $5 = 2 \cdot x\%$ **CLUE 2** $x > 100$ What is the value of x?

Solution: Since $5 = 2 \cdot x\%$, $x\% = \frac{5}{2}$, or 2.5.

Since $x\% = 2.5$, $\frac{x}{100} = 2.5$. ◀ $x\% = \frac{x}{100}$

$x = 2.5(100)$

$x = \mathbf{250}$ **Clue 1 only** was used.

EXAMPLE 2 **CLUE 1** Ariel has 13 nickels and dimes. **CLUE 2** The coins have a total value of 95¢.

How many dimes does Ariel have?

Solution: Use guess and check.

Think: Since the total value is 95¢, Ariel has an odd number of nickels.

Guess 1: 5 nickels and 8 dimes **Check 1:** 25¢ + 80¢ = $1.05 ← ***Too high***
Guess 2: 11 nickels and 2 dimes **Check 2:** 55¢ + 20¢ = 75¢ ← ***Too low***
Guess 3: 7 nickels and 6 dimes **Check 3:** 35¢ + 60¢ = 95¢ ✓

Ariel has **6 dimes.** **Both Clue 1 and Clue 2** were used.

EXAMPLE 3 **CLUE 1** One number is 3 more than a second number.
CLUE 2 Both numbers are integers. Numbers: ? and ?

Solution: Use a table to make a list of possible pairs of numbers.

First integer, *n*	−7	−4	0	3	8	13 ···
Second integer, *n* + 3	−4	−1	3	6	11	16 ···

You **can't tell** what the two numbers are because not enough information is given.

EXERCISES

1. **CLUE 1** Janet and her roommate divide their rent evenly.
   CLUE 2 Janet's rent is 20% more than the \$200 she paid last month.

   How much is Janet's rent this month?

2. **CLUE 1** Video tape sales were 30% of Monday's total sales.
   CLUE 2 Video tapes were on sale for \$4.99 each.

   What was the total of Monday's video tape sales?

3. **CLUE 1** The ratio of two numbers is 2:3.
   CLUE 2 The smaller number is $\frac{2}{3}$ of the larger number.

   What is the smaller number?

4. **CLUE 1** The square of an integer is between 40 and 50.
   CLUE 2 The integer is odd.

   What is the integer?

5. **CLUE 1** A trophy weighs 20 ounces.
   CLUE 2 The trophy is made of pewter. Pewter is 60% tin and 20% lead.

   How much tin was used to make the trophy?

6. **CLUE 1** David's first four algebra test scores were 72, 72, 73, and 83.
   CLUE 2 His average after the fifth test is 77.

   What was David's score on his fifth algebra test?

7. **CLUE 1** In a 10-mile race, Beth ran the first 3 miles in 25 minutes.
   CLUE 2 She ran the final 7 miles in 1 hour 35 minutes.

   What was Beth's average speed for the race, in miles per hour?

8. **CLUE 1** The ages of two sisters are both prime numbers.
   CLUE 2 Both sisters are between 20 and 30 years old.

   What are the ages of the two sisters?

9. **CLUE 1** $s + t = 12$ and $s + w = 5$
   CLUE 2 $r + s = 15$ and $t + w = 5$

   What is the value of $r + s + t + w$?

10. **CLUE 1** $10^{3k} \cdot 10^2 = 10^8$
    CLUE 2 k is an integer.

    What is the value of k?

CHAPTER SUMMARY

IMPORTANT TERMS	x axis *(p. 322)* y axis *(p. 322)* Origin *(p. 322)* Coordinates *(p. 322)* Ordered pair *(p. 322)* Coordinate plane *(p. 325)*	Quadrants *(p. 325)* Relation *(p. 328)* Domain *(p. 333)* Range *(p. 333)* Function *(p. 333)*

IMPORTANT IDEAS

1. There is a one-to-one correspondence between points of a coordinate plane and ordered pairs of real numbers.
2. A relation can be described by an equation, by a table of ordered pairs, by a graph in a coordinate plane, or by a set of ordered pairs.
3. If any vertical line has more than one point in common with a graph, then the graph does not represent a function.

CHAPTER REVIEW

SECTION 12.1

Draw an x axis and y axis. Then locate these points.

1. $A(-3, 2)$ **2.** $B(4, -3)$ **3.** $C(-3, -4)$ **4.** $D(2, 5)$

5. $E(3, -3)$ **6.** $F(-4, -2)$ **7.** $G(3, 1)$ **8.** $H(0, 0)$

SECTION 12.2

Graph these points in the same coordinate plane.

9. $P(2, -3\frac{1}{2})$ **10.** $Q(-3, 2\frac{1}{2})$ **11.** $R(2.25, 1.75)$ **12.** $S(-4.5, -2.5)$

13. $T(0, 3)$ **14.** $W(-3, 0)$ **15.** $U(-2\frac{1}{2}, -3\frac{1}{2})$ **16.** $Z(-3\frac{1}{2}, -2\frac{1}{2})$

SECTION 12.3

Write each relation as a set of ordered pairs if the x values are −1, 0, 2, and 3. Then draw its graph.

17. $y = 3x - 5$ **18.** $y = 4 - x$

19. $y = x^2 - 1$ **20.** $y = 3 - x^2$

21. $y = |x - 1|$ **22.** $y = 2 - |x|$

SECTION 12.4

Write the domain D and the range R of each relation.

23. $\{(-\frac{1}{2}, 0), (2, 6), (4, -3)\}$

24. $\{(1.7, -2), (-2.3, -0.9), (0, -6)\}$

25. $\{(-3, \frac{1}{2}), (0, -5), (-3, 0), (-2, -4)\}$

26. $\{(3, 2), (-1.6, 3), (3.5, -2.7), (-1, 2)\}$

Write Yes or No to show whether each relation is a function.

27. $\{(12, -10), (-25, 5), (-4, -10), (17, 5)\}$

28. $\{(1, 6), (0, 5), (-4, -2.7), (1.8, -5.3)\}$

29.

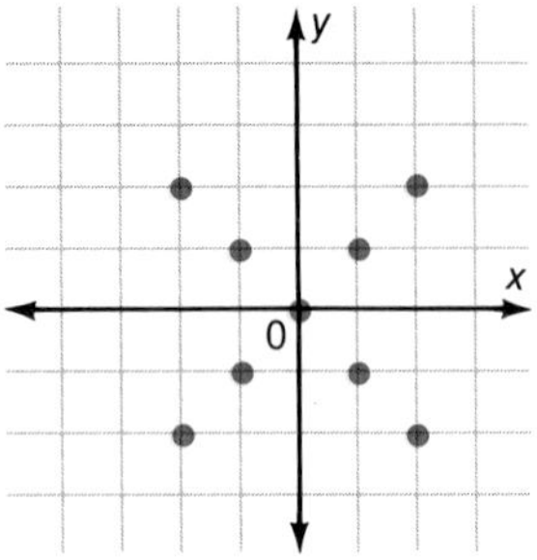

30.

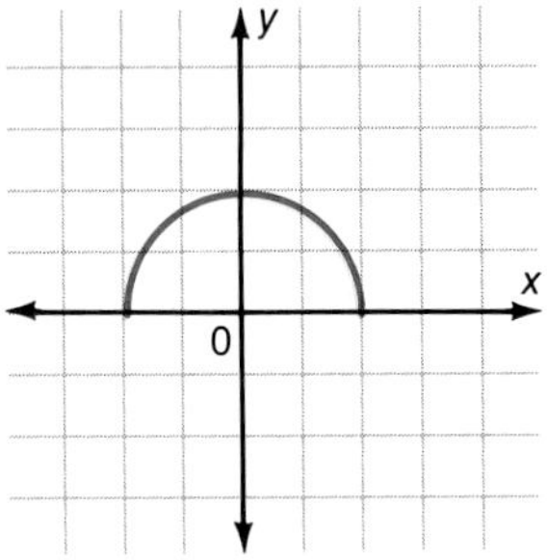

SECTION 12.5

Draw a graph of each function in a separate coordinate plane. The domain is {real numbers}.

31. $y = x - 3$

32. $y = -2x + 1$

33. $y = |x| - 2$

34. $y = 3 - \sqrt{x^2}$

SECTION 12.6

The graph at the right describes the relationship between the distance, d, a car travels in one second and the speed, s, at which it is traveling. Use the graph to determine each of the following.

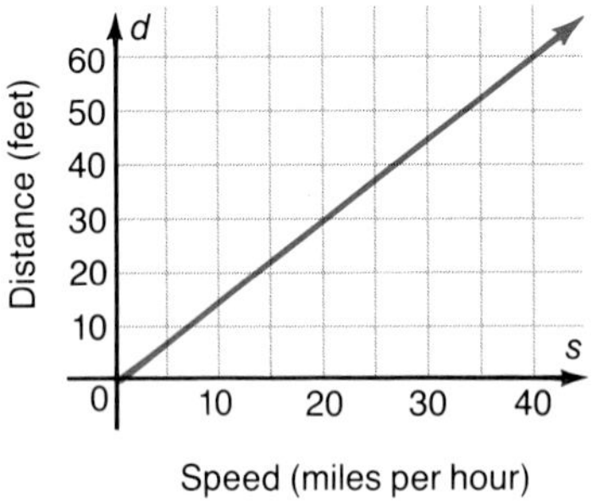

35. The approximate distance traveled in one second by a car traveling 20 mi/hr

36. The approximate distance traveled in one second by a car traveling 30 mi/hr

37. The approximate distance traveled in one second by a car traveling 40 mi/hr

38. Predict the approximate distance traveled in one second by a car traveling 50 mi/hr.

39. Predict the approximate distance traveled in one second by a car traveling 55 mi/hr.

CUMULATIVE REVIEW: CHAPTERS 1–12

Use Condition 1 to represent the unknowns. Use Condition 2 to write an equation for the problem. Solve. (Sections 4.1 and 4.2)

1. The parts used to repair a car at a local body shop cost \$500 less than the labor. The total bill was \$2100. Find the cost of the parts.

2. A sailboat race is along a triangular route. The ratio of the lengths of the three legs of the race is 7:8:10. The race has a length of 75 miles. Find the length of each leg.

Solve. Check your answer. (Section 4.3)

3. $5 - 3y = -16 + 2y$

4. $2x - 3(x - 9) = 2(x + 3)$

Solve each inequality. (Sections 5.3 through 5.6)

5. $2d - 5 < d + 11$

6. $\frac{a}{-6} > -1.5$

7. $-3(y + 8) < 2y - 1$

Simplify. (Sections 6.4 and 6.5)

8. $(7ad)^2$

9. $(-5m)^{-3}$

10. $(3a^2c^3)^2$

11. $\left(\frac{-2x^3}{y^2}\right)^4$

Simplify. (Sections 7.1 through 7.4)

12. $-\sqrt{\frac{25}{36}}$

13. $\sqrt{2} \cdot \sqrt{32}$

14. $\sqrt{150}$

15. $\sqrt{12a^6n^9}$

Add or subtract. Simplify where necessary. (Section 7.5)

16. $7\sqrt{5} + 9\sqrt{5}$

17. $4\sqrt{6} - \sqrt{24}$

18. $\sqrt{12} + \sqrt{3} - \sqrt{75}$

19. $\sqrt{72g} + \sqrt{32g}$

Evaluate each expression. (Sections 8.1 and 8.3)

20. $-6x^3$ when $x = 2$

21. $-x^2 + 7x + 9$ when $x = -5$

Multiply. (Sections 8.2 and 8.5)

22. $(-4c^6)(-3c^2)$

23. $(-6c^2d^3e)(\frac{1}{2}cd^4)$

24. $(6x + 5)(2x - 3)$

Divide. (Section 8.6)

25. $x - 3\overline{)5x^2 - 17x + 6}$

26. $x + 2\overline{)4x^3 + 6x^2 + 9}$

Factor. (Sections 9.1 through 9.7)

27. $3x^2 - 12x$

28. $n^2 + 6n + 8$

29. $g^2 + 4g - 45$

30. $5m^2 - 18m - 8$

31. $6x^2 - 17x + 5$

32. $9y^2 - 1$

33. $4h^2 - 20h + 25$

34. $48y^2 - 27$

35. $2n^2 + 6n - 36$

Write the number or numbers that cannot be in the domain of each variable. (Section 10.1)

36. $\frac{6}{7x}$ **37.** $\frac{x-6}{2x+1}$ **38.** $\frac{9}{x^2-16}$ **39.** $\frac{8x+1}{(x-2)(x+5)}$

Simplify. No divisor equals 0. (Section 10.2)

40. $\frac{18y}{24y^3}$ **41.** $\frac{3x-9}{x^2-5x+6}$ **42.** $\frac{x^2-10x+25}{x^2-25}$ **43.** $\frac{12b^2-6}{4ab^3-2ab}$

Multiply and simplify. (Sections 10.3 through 10.5)

44. $\frac{-10ab}{2}\cdot\frac{12a}{9a^2b}$ **45.** $\frac{3x^2}{8}\cdot\frac{4x}{6y^2}\cdot\frac{16y^3}{3x}$ **46.** $\frac{4x}{x+1}\cdot\frac{x+1}{2x(x-3)}$

47. $\frac{x-9}{x+9}\cdot\frac{x-1}{9-x}$ **48.** $\frac{a^2-4}{a^2-1}\cdot\frac{3a-3}{a-2}$ **49.** $\frac{x^2-6x-7}{x^2+x}\cdot\frac{x^2-x}{3x-21}$

Divide and simplify. (Sections 10.6 and 10.7)

50. $\frac{x+3}{5x}\div\frac{2(x+3)}{5}$ **51.** $\frac{x^2-x-6}{x^2-4}\div(x-3)$ **52.** $\frac{h^2-9}{h^2-6h+9}\div\frac{3h+9}{7h-21}$

Add or subtract. Simplify where necessary. (Sections 11.1, 11.2, and 11.4 through 11.6)

53. $\frac{9}{x+2}+\frac{8}{x+2}$ **54.** $\frac{3x-1}{x-4}-\frac{x+7}{x-4}$ **55.** $\frac{7}{5y}-\frac{3}{2y}$

56. $\frac{9}{x}+\frac{1}{x-1}$ **57.** $\frac{3}{2x-6}-\frac{1}{3x-9}$ **58.** $\frac{7x}{x^2-1}+\frac{8}{3x-3}$

Solve and check each equation. (Section 11.7)

59. $\frac{15}{2x-1}=-3$ **60.** $\frac{7}{x-6}=\frac{9}{x+2}$ **61.** $\frac{1}{x-2}=\frac{2}{x}+\frac{3}{x-2}$

Graph these points in the same coordinate plane. (Sections 12.1 and 12.2)

62. $P(-3, 4)$ **63.** $W(5, -6)$ **64.** $R(4, 0)$ **65.** $T(-1\frac{1}{2}, -2)$

Write Yes or No to show whether each relation is a function. (Section 12.4)

66. $\{(1,2), (3,6), (1,5)\}$ **67.** $\{(6,1), (4,-3), (-2,1)\}$

Draw a graph of each function. The domain is {real numbers}. (Section 12.5)

68. $y=3x-2$ **69.** $y=\frac{1}{2}|x|$ **70.** $y=x^2-3$

CHAPTER 13

Linear Functions

The two towers of the Verrazano-Narrows Bridge in New York City are perpendicular to the earth but are not parallel to each other. They are about four centimeters farther apart at the top than at their bases.

13.1 Graphs of Linear Functions

The steps for graphing a function are reviewed.

Select values for x. → **Compute the y values.** → **Graph the ordered pairs.** → **Draw a line containing the points.**

EXAMPLE 1 Draw a graph of $y = 2x - 1$. The domain is {real numbers}.

Solution: $y = 2x - 1$ ◀ ***Rule for the function***

x	−2	−1	0	1	2
y	−5	−3	−1	1	3

If $x = -2$, $y = 2(-2) - 1$
$= -5$.

If $x = -1$, $y = 2(-1) - 1$
$= -3$.

If $x = 0$, $y = 2(0) - 1$
$= -1$.

If $x = 1$, $y = 2(1) - 1$
$= 1$.

If $x = 2$, $y = 2(2) - 1$
$= 3$.

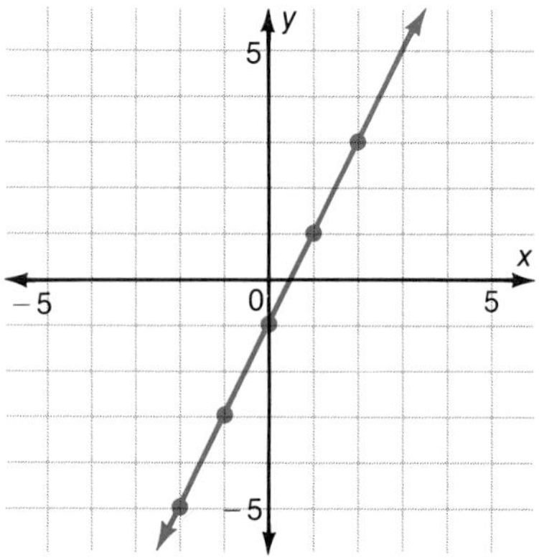

Figure 1

These ordered pairs are graphed in Figure 1. The points lie on a line.

P–1 **What are the coordinates of some other points on the line?**

P–2 **How do you know whether (3, 5) belongs to this function?**

[1] Replace x with 3. ⟶ $y = 2(3) - 1$

[2] Solve for y. ⟶ $y = 5$

Thus, (3, 5) belongs to the function. The point (3, 5) lies on the graph of Figure 1.

EXAMPLE 2 Draw a graph of $y = -\frac{1}{2}x + 2$.

Solution: When the domain is not given, assume that it is {real numbers}.

$$y = -\tfrac{1}{2}x + 2$$

$$y = -\tfrac{1}{2}(-4) + 2 = 4$$
$$y = -\tfrac{1}{2}(-2) + 2 = 3$$
$$y = -\tfrac{1}{2}(0) + 2 = 2$$
$$y = -\tfrac{1}{2}(2) + 2 = 1$$
$$y = -\tfrac{1}{2}(4) + 2 = 0$$

x	−4	−2	0	2	4
y	4	3	2	1	0

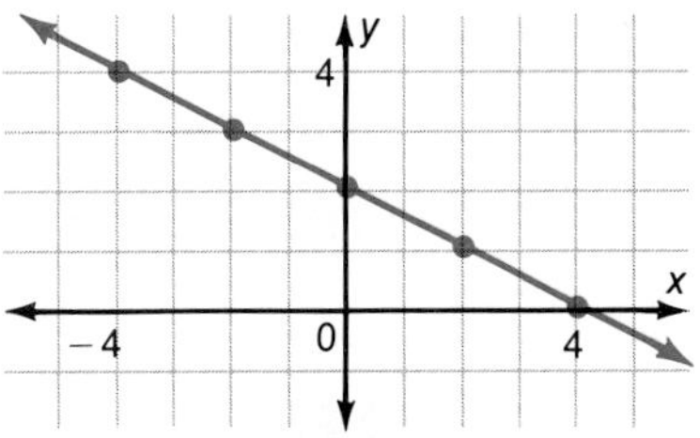

Figure 2

The graph of the function is shown in Figure 2.

Definition A function that has the points of its graph lying in a straight line is a ***linear function.***

The rule for a linear function is
$$\boldsymbol{y = mx + b.}$$
(m and b are real numbers.)

$y = 3x + 5$

$y = -\frac{5}{2}x + \sqrt{3}$

P–3 **What are the values of m and b in the rule $y = -x + \frac{1}{2}$? in the rule $y = \frac{2}{3}x - \frac{1}{4}$? in the rule $y = 10 - 3x$?**

P–4 **Which functions below are linear?**

a. $y = 2x + 5$ **b.** $y = x - 3$ **c.** $y = x^2 + 1$ **d.** $y = \sqrt{x}$

CLASSROOM EXERCISES

Write the y values that are missing from each table. (Examples 1 and 2)

1. $y = 3x$

x	−2	−1	0	2
y	?	?	?	?

2. $y = x + 3$

x	−5	−2	0	4
y	?	?	?	?

3. $y = 10x + 1$

x	−5	−2	0	3
y	?	?	?	?

Compare each rule with $y = mx + b$. Then write the values of m and b for each function. (P-3)

4. $y = -10x + 1$
5. $y = -3x + 7$
6. $y = 5 - x$
7. $y = -x - 3$
8. $y = 10x$
9. $y = -7x$
10. $y = -\frac{3}{5}$
11. $y = -0.5x + 3.8$
12. $y = -\frac{2}{3}x - \frac{1}{3}$

WRITTEN EXERCISES

Goals: To identify and to graph linear functions
Sample Problems: **a.** $y = 3x^2$ **b.** $y = -x$
Answers: **a.** Not linear **b.** Linear; the graph is at the right.

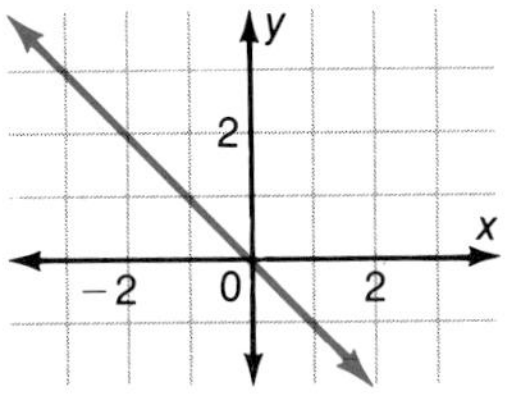

Graph at least four ordered pairs of each function. Then draw the graph. The domain is {real numbers}. (Examples 1 and 2)

1. $y = -x + 3$
2. $y = -x - 2$
3. $y = 3x - 1$
4. $y = 2x - 3$
5. $y = 2x$
6. $y = -2x$
7. $y = 0 \cdot x - 2$
8. $y = 0 \cdot x + 3$
9. $y = -4x + 3$
10. $y = \frac{1}{2}x - \frac{3}{2}$
11. $y = -\frac{1}{2}x + \frac{1}{2}$
12. $y = -2x + 1$

Write Yes or No to show whether each function is linear. (P-4)

13. $y = -4x + 10$
14. $y = x^2 + 1$
15. $y = -\frac{2}{3}x - 5$
16. $y = x(x - 1)$
17. $y = 0 \cdot x + 0$
18. $y = 0 \cdot x + \frac{1}{3}$
19. $y = -2x + 0$
20. $y = 6x + 0$
21. $y = (x + 1)(x - 1)$
22. $y = x^2 - 2x + 5$
23. $y = -4$
24. $y = 3$

NON-ROUTINE PROBLEM

25. The figure at the right shows three squares inside one another. The small square has sides which are 7 centimeters in length, the middle square has sides which are 9 centimeters in length, and the large square has sides which are 12 centimeters in length. Find the area of the shaded portion of the figure.

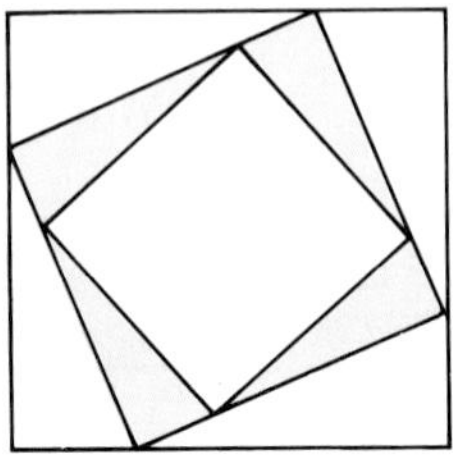

13.2 Y Intercept

P–1 **How are the equations of these functions alike?**

a. $y = \frac{1}{2}x + 2$ **b.** $y = x + 2$ **c.** $y = 2x + 2$

d. $y = -2x + 2$ **e.** $y = -x + 2$ **f.** $y = -\frac{1}{2}x + 2$

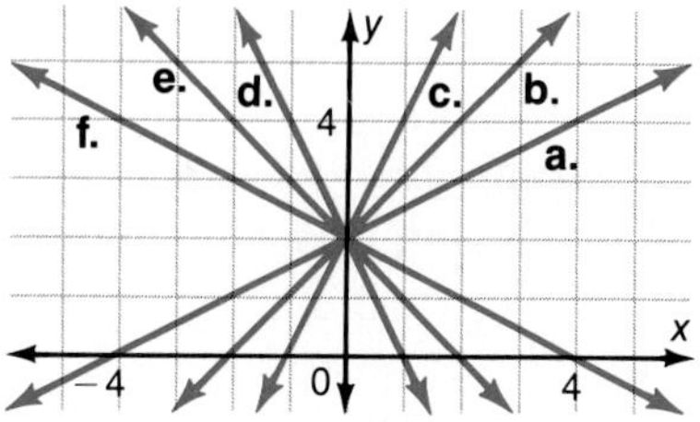

Figure 1

P–2 **What are the coordinates of the common point of the graphs in Figure 1?**

Graphs of the following functions are shown in Figure 2.

1. $y = 3x + 1$
2. $y = 2x - 3$
3. $y = -\frac{1}{2}x + 3$
4. $y = -x - 1$

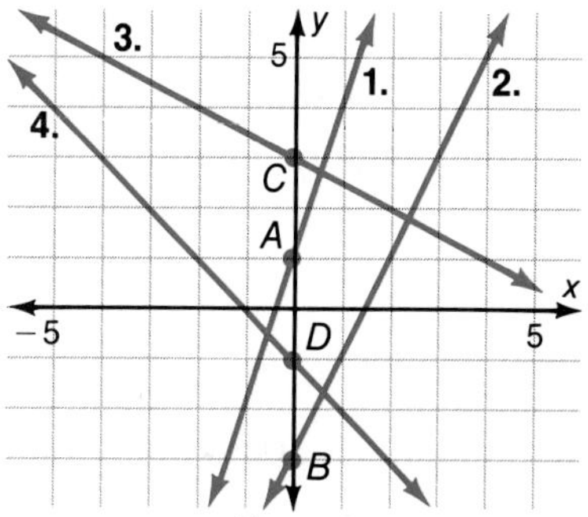

Figure 2

P–3 **What is the x value of each of the points *A*, *B*, *C*, and *D*? the y value of each point?**

Definition The *y* value of the point that a line has in common with the *y* axis is the ***y intercept.***

EXAMPLE 1 Write the *y* intercept of the graph of $y = 3x - 2$.

Solution: $y = 3x - 2$

1 Replace *x* with 0. ⟶ $y = 3(0) - 2$ — ***For any point on the y axis, x is zero.***

2 Solve for *y*. ⟶ $y = -2$ — ***y intercept***

In the rule for the linear function $y = mx + b$, b represents the y intercept of its graph.

$y = -x + \frac{1}{2}$

y intercept $= \frac{1}{2}$

P–4 **What is the y intercept of the graph of each function below?**

a. $y = \frac{3}{2}x + \frac{5}{2}$ **b.** $y = 4 - x$

EXAMPLE 2

Graph $y = -2x - 3$. Graph at least three ordered pairs, including one with the y intercept.

Solution: $y = -2x - 3$

$y = -2(0) - 3 = -3$

$y = -2(-3) - 3 = 3$

$y = -2(2) - 3 = -7$

x	y
0	−3
−3	3
2	−7

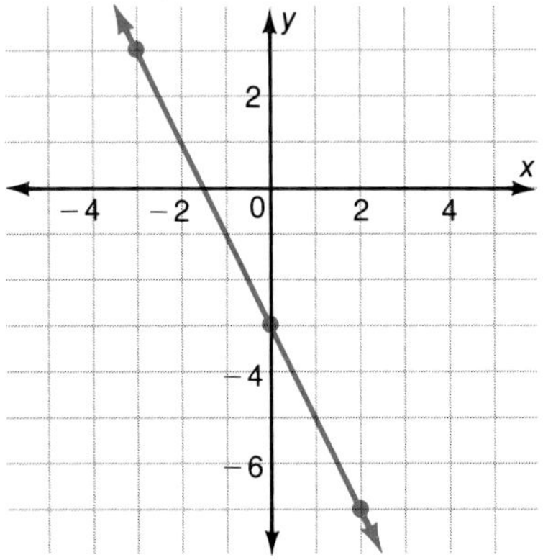

CLASSROOM EXERCISES

Estimate the y intercept of each graph. (P-3, second part)

1.
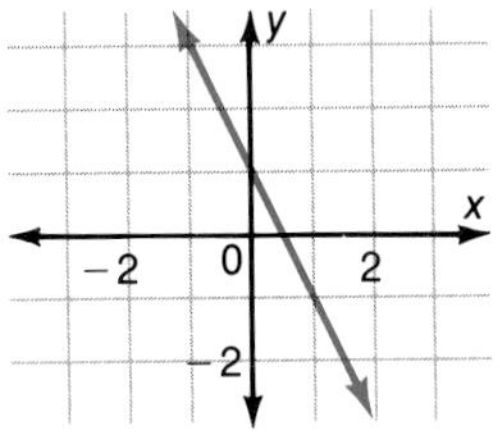

2.
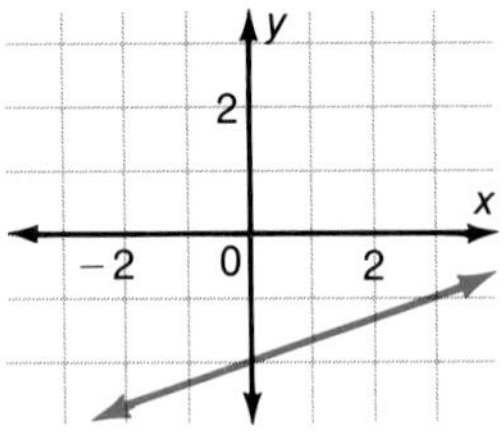

3.
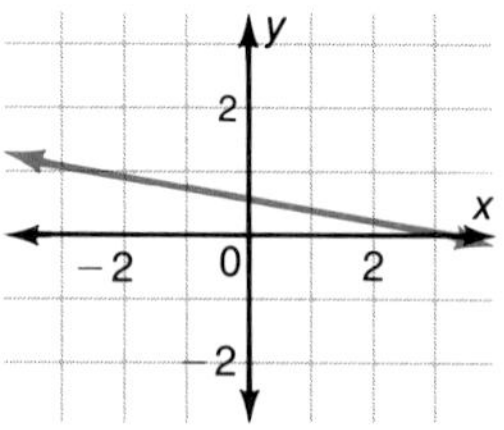

4.
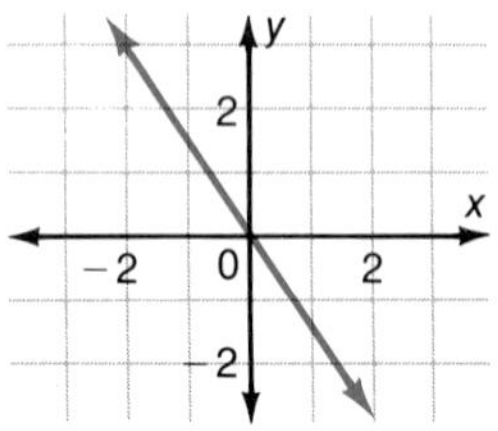

5.
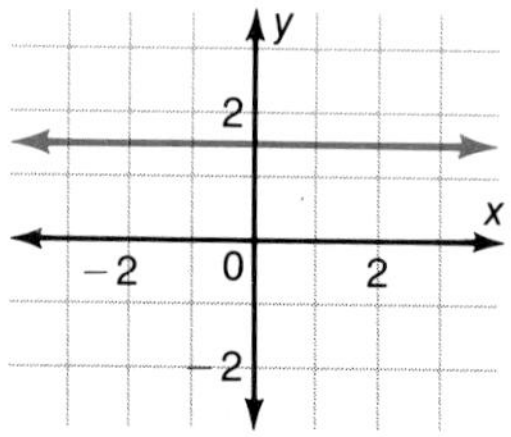

6.
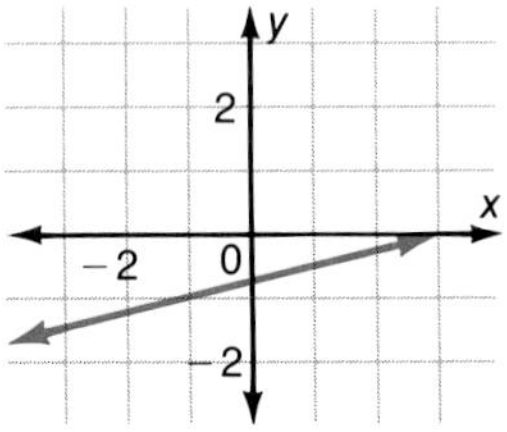

Write each y intercept. (P-4)

7. $y = 3x + \frac{1}{2}$ **8.** $y = -2x + 5$ **9.** $y = \frac{2}{3}x - \frac{1}{3}$

10. $y = -3x$ **11.** $y = -x - 1$ **12.** $y = 3x + \frac{15}{2}$

13. $y = x$ **14.** $y = 5 - 3x$ **15.** $y = 2(x + 3)$

WRITTEN EXERCISES

Goal: To identify the y intercept of a linear function and to graph the function

Sample Problem: $y = 3(2 - \frac{1}{3}x)$

Answer: 6 The graph is at the right.

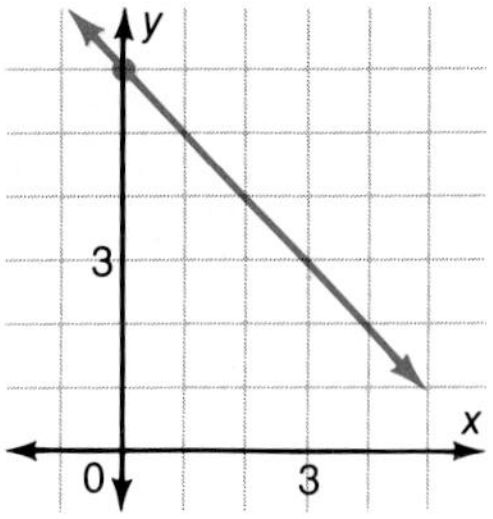

Write each y intercept. (Example 1)

1. $y = -3x + 7$ **2.** $y = 5x - 2$

3. $y = -\frac{1}{2}x + \frac{7}{2}$ **4.** $y = 1.73x + 3.85$

5. $y = \frac{2}{3}x - \frac{5}{3}$ **6.** $y = -1.17x + 3.14$

7. $y = 0.3 - 5x$ **8.** $y = 14 - \frac{1}{2}x$ **9.** $y = 3(2 - x)$

10. $y = -5(1 - 2x)$ **11.** $y = -x$ **12.** $y = \frac{7}{5} + 2x$

Graph each function. Graph at least three ordered pairs including one with the y intercept. (Example 2)

13. $y = x + 1$ **14.** $y = -x + 2$

15. $y = -2x - 3$ **16.** $y = 3x - 2$

17. $y = \frac{1}{2}x + 5$ **18.** $y = -\frac{1}{2}x + 4$

19. $y = -3x$ **20.** $y = 2x$

21. $y = \frac{1}{2}x + \frac{3}{2}$ **22.** $y = -2x - \frac{3}{2}$

REVIEW CAPSULE FOR SECTION 13.3

Graph these functions in the same coordinate plane. (Pages 350–352)

1. $y = \frac{1}{2}x + 5$ **2.** $y = \frac{1}{2}x + 3$

3. $y = \frac{1}{2}x + 1$ **4.** $y = \frac{1}{2}x$

5. $y = \frac{1}{2}x - 3$ **6.** $y = \frac{1}{2}x - 5$

13.3 Slope of a Line

P–1 **What is the *y* intercept for each graph in Figure 1?**

The rules for these linear functions are listed below.

1. $y = \frac{1}{2}x + 2$ **2.** $y = x + 2$

3. $y = \frac{3}{2}x + 2$ **4.** $y = 2x + 2$

The lines get "steeper" with respect to the *x* axis in the order from **1** to **4.**

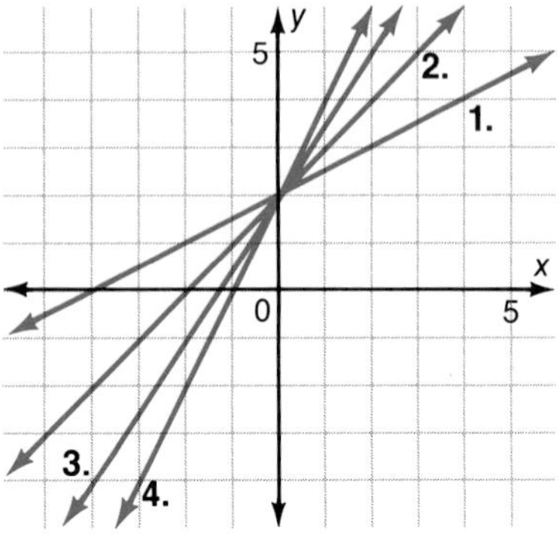

Figure 1

P–2 **What number in the rule $y = mx + b$ affects the "steepness" of a line?**

The graphs of four more linear functions are shown in Figure 2.

1. $y = -\frac{1}{2}x - 1$ **2.** $y = -x - 1$

3. $y = -\frac{3}{2}x - 1$ **4.** $y = -4x - 1$

P–3 **In what special way are the rules for these functions related?**

The "steepness" of a line is described by a number called ***slope***.

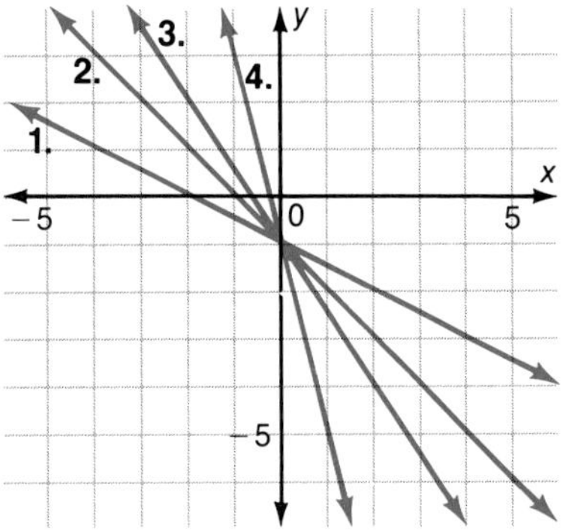

Figure 2

> In the rule for the linear function $y = mx + b$, m represents the ***slope*** of its graph.
>
> $y = -2x + 5$
>
> Slope $= -2$

The rule for a linear function, $y = mx + b$, is called the ***slope-intercept form.***

P–4 **What is the slope of the graph of the function having each rule below? the *y* intercept?**

a. $y = -\frac{3}{4}x - 2$ **b.** $y = 5x - \frac{1}{2}$ **c.** $y = \frac{5}{2} - \frac{3}{2}x$

EXAMPLE 1

Write the slope-intercept form of the rule for the line in the figure. The slope is -2.

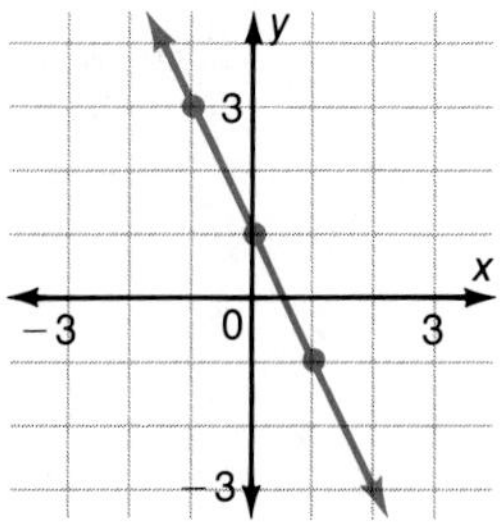

Figure 3

Solution:

$$y = mx + b$$

$$y = -2x + 1$$

$m = -2$
$b = 1$

P–5 **In what quadrants does the graph of Figure 3 lie?**

EXAMPLE 2

Graph these functions.

1. $y = 2x$ **2.** $y = -\frac{1}{3}x$

Solutions:

$y = 2x$

x	y
−2	−4
0	0
2	4

$y = -\frac{1}{3}x$

x	y
−3	1
0	0
3	−1

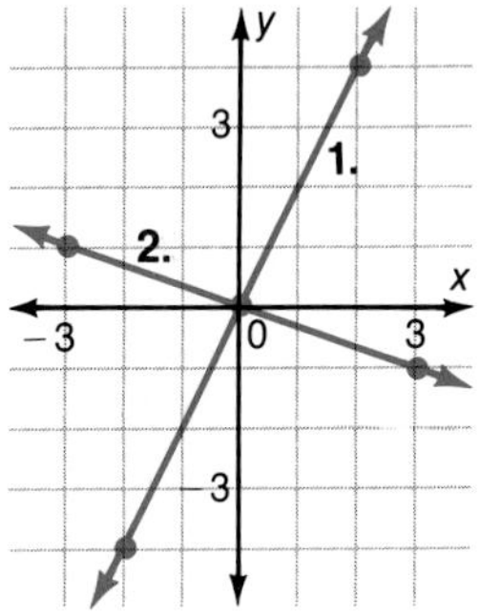

P–6 **In what quadrants does the graph of $y = 2x$ lie? the graph of $y = -\frac{1}{3}x$?**

A line such as the graph of $y = 2x$ has a positive slope. It "slopes" upward and to the right. The graph of $y = -\frac{1}{3}x$ has a negative slope. It "slopes" upward and to the left.

CLASSROOM EXERCISES

Find the slope and the y intercept of the graph of each function. (P-4)

1. $y = 3x + 2$
2. $y = -5x + 1$
3. $y = x + 5$
4. $y = -x - 1$
5. $y = \frac{1}{3}x - 6$
6. $y = -\frac{1}{2}x + 1$

Write the slope-intercept form of the rule for each function. (Example 1)

7. The slope is 5 and the y intercept is 3.
8. The slope is $\frac{1}{4}$ and the y intercept is $\frac{1}{2}$.

Write the slope-intercept form of the rule for each function. (Example 1)

9. The slope is -1 and the y intercept is 0.

10. The slope is $\frac{-2}{3}$ and the y intercept is $\frac{3}{5}$.

11. The line crosses the y axis at (0, 10) and has slope 3.

12. The line crosses the y axis at (0, -6) and has slope -1.

WRITTEN EXERCISES

Goal: To write the slope-intercept form of the rule for a linear function
Sample Problem: Slope is $\frac{2}{3}$. See graph below. **Answer:** $y = \frac{2}{3}x + 2$

Write the slope of the graph of each function. (P-4)

1. $y = x - 5$
2. $y = -x + 2$
3. $y = -\frac{1}{2}x + 7$
4. $y = \frac{1}{2}x - 6$
5. $y = 2(x - 3)$
6. $y = 3(x + 1)$
7. $y = 2 - 3x$
8. $y = 5 - 7x$
9. $y = \frac{1}{2}(3 - x)$
10. $y = -\frac{1}{3}(5 - x)$
11. $y = -\frac{1}{4}(8 + 16x)$
12. $y = \frac{1}{5}(10 - 10x)$

Write the slope-intercept form of the rule for each function graphed at the right. The slopes are given below. (Example 1)

13. Slope is -1.
14. Slope is $\frac{1}{6}$.
15. Slope is $-\frac{1}{3}$.
16. Slope is $-\frac{4}{3}$.
17. Slope is -1.
18. Slope is 1.

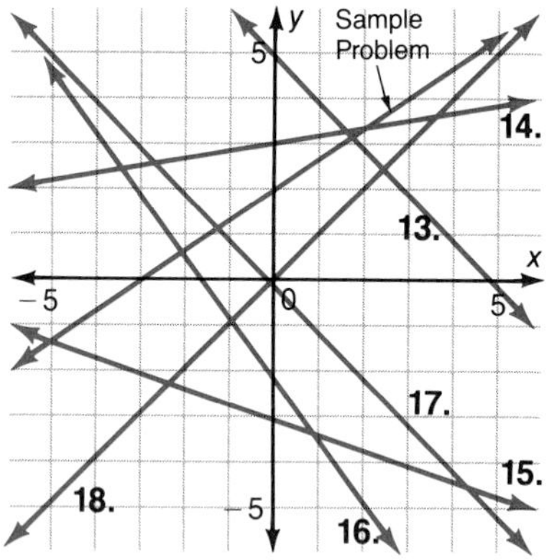

Name the quadrants in which each graph lies. (Example 2, P-5, P-6)

19. $y = 5x$
20. $y = 10x$
21. $y = -2x$
22. $y = -6x$
23. $y = -\frac{1}{2}x$
24. $y = \frac{2}{3}x$
25. $y = \frac{5}{6}x$
26. $y = -\frac{3}{4}x$
27. $y = -0.7x$
28. $y = 0.56x$
29. $y = x$
30. $y = -x$

REVIEW CAPSULE FOR SECTION 13.4

Simplify each fraction. (Pages 8–10 and 16–18)

1. $\frac{2 - 5}{7 - 2}$
2. $\frac{0 - (-3)}{2 - 4}$
3. $\frac{(-2) - 3}{4 - 5}$
4. $\frac{7 - 2}{5 - (-3)}$
5. $\frac{5 - (-3)}{7 - (-5)}$

13.4 Slope Formula

P–1 **What is the slope of the graph of $y = \frac{3}{2}x + 2$ shown in the figure at the right?**

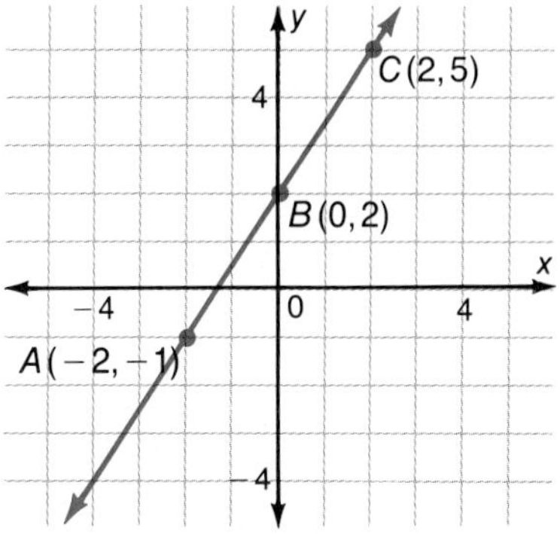

Figure 1

Suppose that you move from *B* to *C* in Figure 1 by one vertical and one horizontal motion.

P–2 **How many units and in what direction is the vertical motion? the horizontal motion?**

$$\frac{5-2}{2-0} = \frac{3}{2}$$ ◀ *Vertical motion: 3* / *Horizontal motion: 2*

Now suppose that you move from *A* to *C* by one vertical motion and one horizontal motion.

P–3 **How many units and in what direction is the vertical motion? the horizontal motion?**

Difference in y coordinates* / *Difference in x coordinates ▶ $$\frac{5-(-1)}{2-(-2)} = \frac{5+1}{2+2}$$

$$= \frac{6}{4} \text{ or } \frac{3}{2}$$ ◀ ***Vertical motion* / *Horizontal motion***

A quotient of two numbers such as $\frac{3}{2}$ is called a ratio. The slope of a line is the ratio of the vertical motion to the horizontal motion from one point to another on the line.

Slope Formula

If the coordinates of two points on a nonvertical line are (x_1, y_1) and (x_2, y_2), then m, the ***slope*** of the line, is given by

$$m = \frac{y_2 - y_1}{x_2 - x_1}.$$

◀ ***1 and 2 are subscripts. Read x sub 1, x sub 2, etc.***

(2, −3), (−5, 1)

$$m = \frac{1-(-3)}{-5-2}$$

$$m = -\frac{4}{7}$$

EXAMPLE 1

Write the slope of the line in Figure 2 by using the following ratio:

$$\frac{\text{vertical motion from } A \text{ to } B}{\text{horizontal motion from } A \text{ to } B}$$

Solution:

$$\frac{\text{vertical motion from } A \text{ to } B}{\text{horizontal motion from } A \text{ to } B} = \frac{-5-1}{1-(-3)}$$

$$\text{slope} = \frac{-6}{4} \quad \text{or} \quad -\frac{3}{2}$$

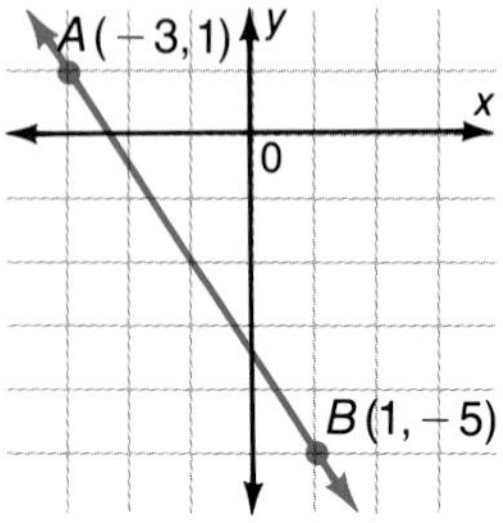

Figure 2

EXAMPLE 2

Write the slope of the line that contains the points (−2, 5) and (3, −1).

Solution:

Method 1

$x_1 = -2$, $x_2 = 3$; $y_1 = 5$, $y_2 = -1$

$$m = \frac{y_2 - y_1}{x_2 - x_1}$$

$$= \frac{-1-5}{3-(-2)}$$

$$= -\frac{6}{5}$$

Slope of the line

Method 2

$x_1 = 3$, $x_2 = -2$; $y_1 = -1$, $y_2 = 5$

$$m = \frac{y_2 - y_1}{x_2 - x_1}$$

$$= \frac{5-(-1)}{-2-3}$$

$$= -\frac{6}{5}$$

Slope of the line

EXAMPLE 3

Write the slope-intercept form of the line that contains the points (0, 3) and (5, 2).

Solution:

1 Slope formula ⟶ $\frac{y_2 - y_1}{x_2 - x_1} = \frac{2-3}{5-0}$

$= -\frac{1}{5}$

2 y intercept ⟶ $b = 3$ ***If $x = 0$, the point is on the y axis.***

3 Slope-intercept form ⟶ $y = mx + b$

4 Substitute. ⟶ $y = -\frac{1}{5}x + 3$ ***$m = -\frac{1}{5}$, $b = 3$***

CLASSROOM EXERCISES

Find the value of each ratio. (P-2, P-3)

1. $\frac{8-5}{7-2}$
2. $\frac{3-7}{9-4}$
3. $\frac{-1-2}{7-9}$
4. $\frac{5-(-3)}{2-3}$
5. $\frac{0-(-1)}{-1-2}$
6. $\frac{-3-(-2)}{7-(-1)}$
7. $\frac{-3-0}{0-(-3)}$
8. $\frac{(-3)-(-3)}{2-7}$
9. $\frac{1-2}{4-6}$

Write the slope of each line. (Example 1)

10.

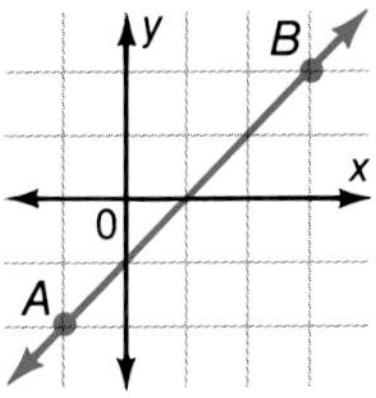

11.

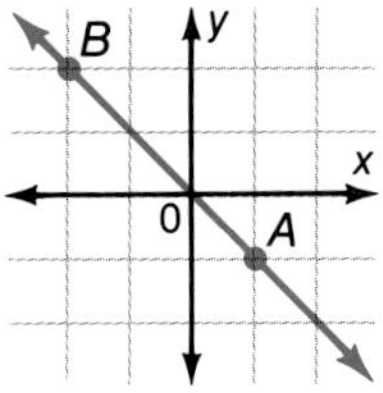

12.

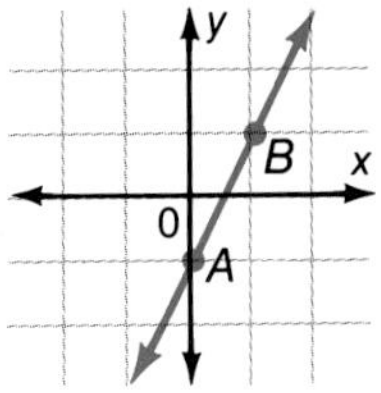

13.

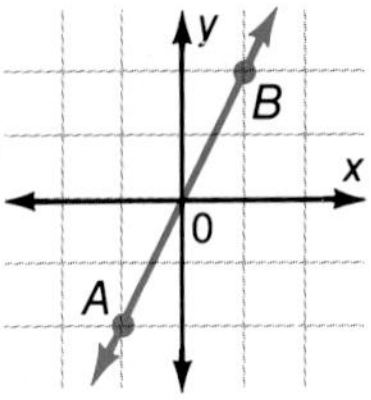

14.

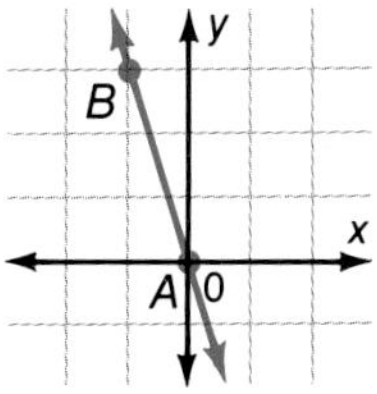

15.

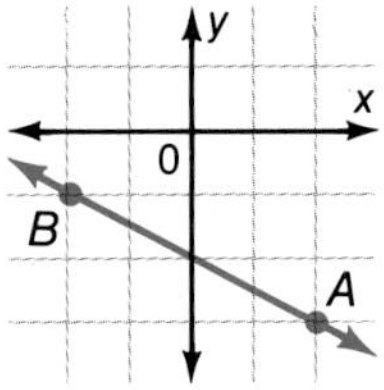

16.

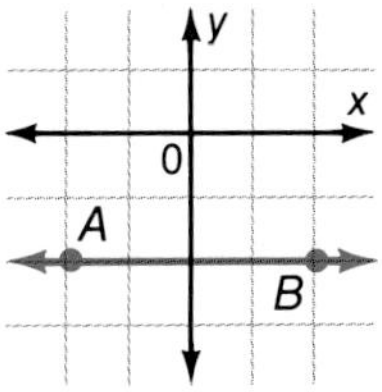

17.

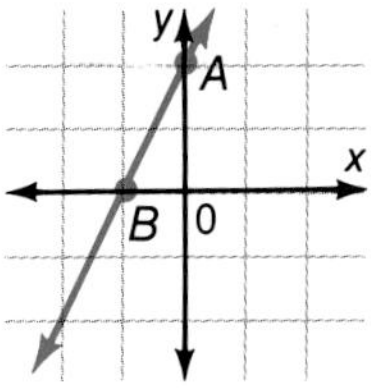

WRITTEN EXERCISES

Goal: To write the slope of a line given coordinates of two points on the line
Sample Problem: (4, 5) and (−2, −1) **Answer:** $m = 1$

Write the slope of each line. (Example 1)

1.

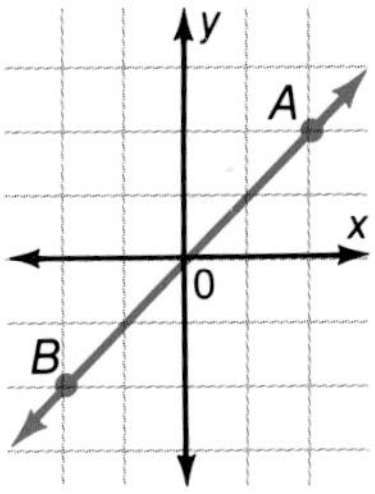

2.

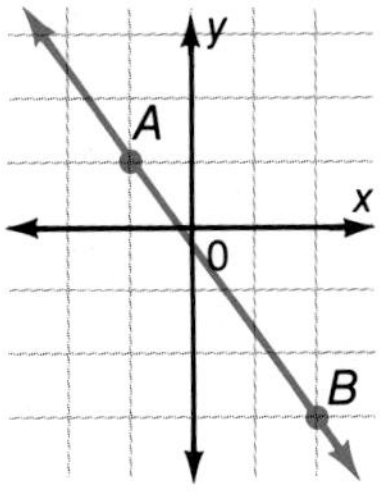

3.

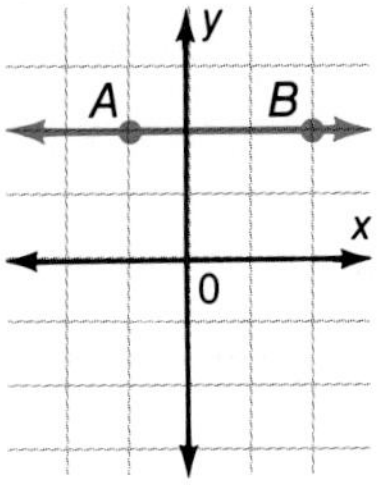

4.

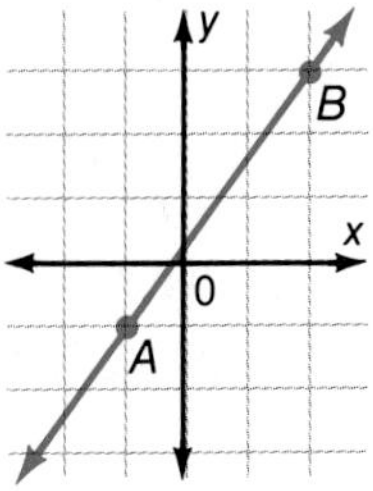

Write the slope of the line that contains the two points. (Example 2)

5. $(-1, 5); (3, 2)$ **6.** $(4, 6); (-2, 3)$ **7.** $(5, -3); (-1, 4)$

8. $(2, 3); (-1, -2)$ **9.** $(-3, -4); (-2, -5)$ **10.** $(-1, -5); (-2, -7)$

11. $(3, -5); (-1, -5)$ **12.** $(4, -3); (-1, -3)$ **13.** $(10, -5); (6, -2)$

14. $(12, 8); (-4, -10)$ **15.** $(3\frac{1}{2}, -5\frac{3}{4}); (-\frac{1}{2}, \frac{1}{4})$ **16.** $(-\frac{3}{2}, -\frac{5}{2}); (\frac{1}{2}, \frac{1}{2})$

Write the slope-intercept form of the line that contains the two points. (Example 3)

17. $(0, 5); (-3, 2)$ **18.** $(0, 1); (5, -3)$ **19.** $(0, -3); (4, 6)$

20. $(0, -5); (-1, -1)$ **21.** $(0, 0); (5, 8)$ **22.** $(0, 0); (-4, 7)$

23. $(0, -6); (2, -6)$ **24.** $(0, 10); (-1, 10)$ **25.** $(0, 5); (5, 0)$

26. $(0, -2); (-2, 0)$ **27.** $(0, 6); (3, 8)$ **28.** $(0, 2); (4, 10)$

MORE CHALLENGING EXERCISES

Draw the line with the given slope and that contains the given point.

EXAMPLE: $m = -\frac{1}{2}$; $(5, -3)$

SOLUTION: Locate the point as shown in the figure. Call it P.

Use the slope as $\dfrac{\text{vertical motion from } P \text{ to } Q}{\text{horizontal motion from } P \text{ to } Q}$

to locate a second point Q.

$$-\frac{1}{2} = \frac{1 \text{ unit upward } (+)}{2 \text{ units left } (-)}$$

Draw the line containing P and Q.

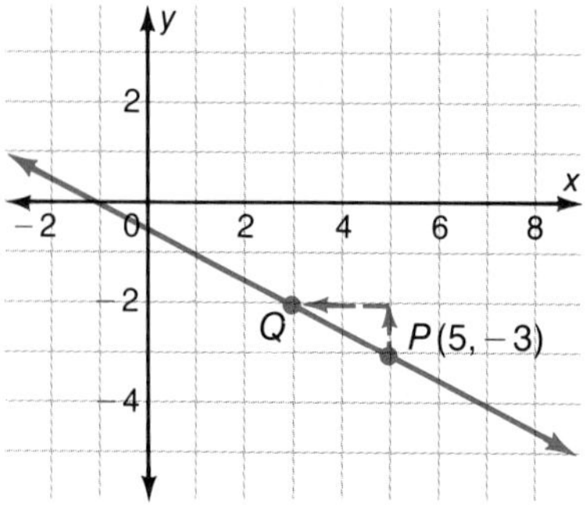

29. $m = 3$; $(0, 0)$ **30.** $m = \frac{1}{2}$; $(1, -1)$ **31.** $m = -\frac{3}{4}$; $(2, -3)$

32. $m = -5$; $(-2, -3)$ **33.** $m = 0$; $(-5, 3)$ **34.** $m = \frac{5}{3}$; $(1, -5)$

APPLICATIONS

Suppose a sheet of paper has a thickness of 0.0015 inch. What would be the thickness of the paper if it could be folded in half several times? A calculator can be used to find the answer.

EXAMPLE Find the thickness of the folded paper when the number of folds is 15.

SOLUTION

Number of Folds	Number of Thicknesses	Thickness in Inches
1	$2^1 = 2$	$2 \times 0.0015 = 0.0030$ in
2	$2^2 = 4$	$4 \times 0.0015 = 0.0060$ in
3	$2^3 = 8$	$8 \times 0.0015 = 0.012$ in
.	.	. .
.	.	. .
.	.	. .
15	$2^{15} = \underline{\ ?\ }$	$\underline{\ ?\ } \times 0.0015 = \underline{\ ?\ }$ in

[2] [×] [=] [×] [.] [0] [0] [1] [5] [=]   49.152

Press " [=] " 14 times.   *About 49 inches thick*

Find the following thicknesses for the sheet of paper in the Example.

35. 16 folds, in inches **36.** 25 folds, in feet **37.** 50 folds, in miles (HINT: $2^{50} = 2^{25} \times 2^{25}$)

MID-CHAPTER REVIEW

Graph at least four ordered pairs of each function. Then draw the graph. The domain is {real numbers}. (Section 13.1)

1. $y = x - 3$ **2.** $y = -x + 2$ **3.** $y = -2x + 1$

4. $y = 3x + 2$ **5.** $y = \frac{1}{2}x - 2$ **6.** $y = -\frac{1}{2}x + 3$

Use the y intercept and two ordered pairs to graph each function. (Section 13.2)

7. $y = -x + 3$ **8.** $y = 2x - 1$ **9.** $y = -\frac{1}{2}x + \frac{5}{2}$

Write the slope-intercept form of the rule for each function. (Section 13.3)

10. Slope: -2; y intercept: 5 **11.** Slope: $-\frac{1}{4}$; y intercept: -1 **12.** Slope: -1; y intercept: $\frac{1}{2}$

Write the slope of the line that contains the two points. (Section 13.4)

13. (2, 3);(4, 1) **14.** (−4, 0);(3, −2) **15.** (−2, −1);(3, −1) **16.** (−7, −4);(−5, −1)

13.5 Parallel Lines

EXAMPLE 1 Graph in the same coordinate plane.

1. $y = 2x - 3$ **2.** $y = 2x + 1$

Solutions:

$y = 2x - 3$

x	y
−1	−5
0	−3
2	1

$y = 2x + 1$

x	y
−2	−3
0	1
2	5

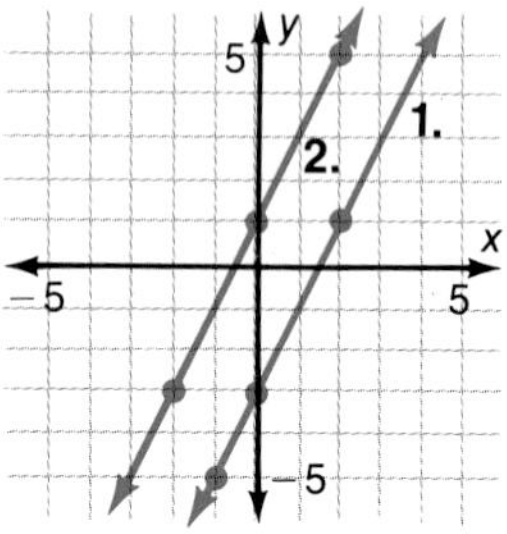

Figure 1

P–1 **What appears to be true about the two lines of Figure 1?**

Lines **1** and **2** of Figure 1 are parallel.

Definition

> ***Parallel lines*** are two lines in the same plane that do not have a common point.

Lines **1** and **2** also have the same slope.

> If two lines have the same slope, they are parallel.

P–2 **Which functions described below have parallel graphs?**

a. $y = -\frac{1}{2}x - 5$ **b.** $y = 2x - 5$ **c.** $y = 4 - \frac{1}{2}x$

EXAMPLE 2 Write the slope-intercept form of the function that has its graph parallel to the graph of $y = 2x - 5$ and that contains (0, 3).

Solution:

$m = 2$
$b = 3$ ▶ $y = mx + b$

$y = 2x + 3$ ◀ ***Slope-intercept form***

EXAMPLE 3

Graph the three functions. Use three points for the first line. Graph the other two lines using y intercepts and the meaning of slope.

1. $y = -\frac{1}{2}x + 1$ **2.** $y = -\frac{1}{2}x - 3$ **3.** $y = -\frac{1}{2}x$

Solution:

1. Draw the graph of $y = -\frac{1}{2}x + 1$.

x	−4	0	4
y	3	1	−1

2. Locate (0, −3). Move 1 unit up and 2 units left to locate (−2, −2). Draw the graph of $y = -\frac{1}{2}x - 3$.

3. Locate (0, 0). Move 1 unit up and 2 units left to locate (−2, 1). Draw the graph of $y = -\frac{1}{2}x$.

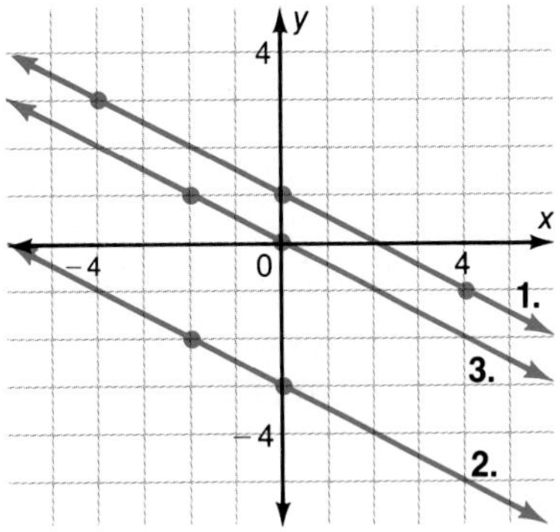

P–3 **What is the value of $0 \cdot x$ for any value of x?**

EXAMPLE 4

Graph the three functions.

1. y = 4
2. y = 2
3. y = −1

1. $y = 0 \cdot x + 4$
2. $y = 0 \cdot x + 2$
3. $y = 0 \cdot x - 1$

Solution:

All the rules are of the form $y = 0 \cdot x + b$ or simply $y = b$. The three graphs are parallel to the x axis.

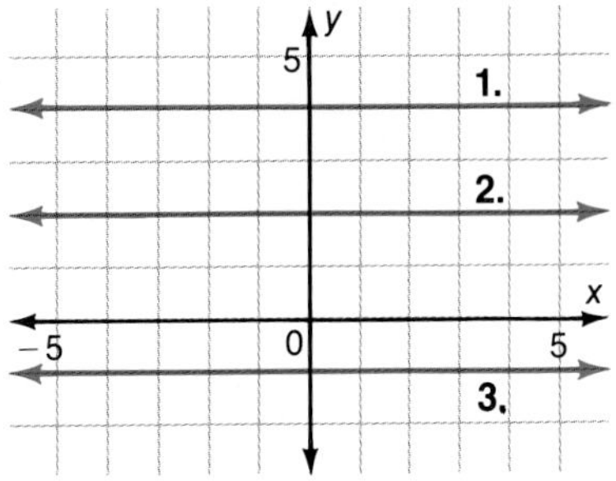

CLASSROOM EXERCISES

Write Yes *or* No *to tell whether the graphs of each pair of functions are parallel.* (P-1, P-2)

1. $y = \frac{1}{2}x + 2$
$y = 0.5x + 3$

2. $y = 5 - 2x$
$y = 5x - 2$

3. $y = 1$
$y = 3$

4. $y = \frac{3}{4}x$
$y = -\frac{3}{4}x$

Write the slope-intercept form of each function. (Example 2)

5. Its graph is parallel to the graph of $y = 4x - 3$ and contains (0, 3).

6. Its graph is parallel to the graph of $y = -5x + 2$ and contains (0, −3).

7. Its graph is parallel to the graph of $y = \frac{3}{2}x - 1$ and contains (0, 10).

8. Its graph is parallel to the graph of $y = -\frac{1}{4}x + \frac{1}{2}$ and contains $(0, \frac{5}{2})$.

9. Its graph is parallel to the graph of $y = -5.3x + 2.5$ and contains (0, 0.6).

WRITTEN EXERCISES

Goal: To use slopes to identify and graph parallel lines
Sample Problem: $y = -x$, $y = \frac{1}{2} - x$, $y = -x + 7$
Answer: Each slope: −1; lines are parallel. Graphs:

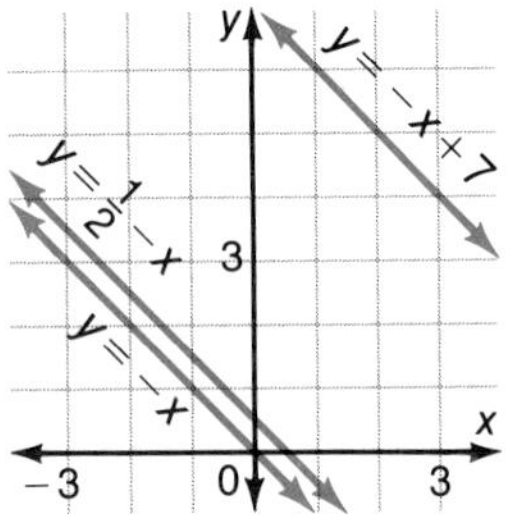

Graph each pair of functions in the same coordinate plane (Example 1)

1. $y = -2x + 3$
$y = -2x - 1$

2. $y = x + 4$
$y = x - 3$

3. $y = \frac{3}{2}x - 2$
$y = \frac{3}{2}x + 1$

4. $y = -\frac{1}{2}x$
$y = -\frac{1}{2}x - 4$

5. $y = \frac{5}{4}x + 3$
$y = \frac{5}{4}x$

6. $y = -\frac{3}{2}x + 2$
$y = -\frac{3}{2}x - 5$

Graph the three functions in the same coordinate plane. Use three points for the first line. Graph the other two lines using the y intercepts and the meaning of slope. (Example 3)

7. $y = 2x - 3$
$y = 2x + 3$
$y = 2x$

8. $y = 5x + 1$
$y = 5x - 5$
$y = 5x$

9. $y = x + 3$
$y = x - 3$
$y = x$

10. $y = -x + 2$
$y = -x - 3$
$y = -x$

Graph the functions in the same coordinate plane. (Example 4)

11. $y = 5$

12. $y = -2$

13. $y = \frac{3}{2}$

14. $y = -\frac{3}{2}$

MORE CHALLENGING EXERCISES

15. If $a = b$, are the graphs of the functions described by $ax + by = c$ and $bx + ay = d$ parallel? Explain.

16. What value of k would make the graph of $6x + 2y = 5$ and the graph of $4x + ky = 3$ parallel?

REVIEW CAPSULE FOR SECTION 13.6

Solve for x. (Pages 64–66)

1. $0 = 3x + 6$

2. $0 = \frac{1}{2}x - 1$

3. $0 = -\frac{3}{4}x + 9$

4. $0 = -3x + \frac{9}{4}$

13.6 X Intercept

P–1 **What are the coordinates of the point common to the graph of Figure 1 and the x axis?**

Any point on the x axis has a y value equal to 0.

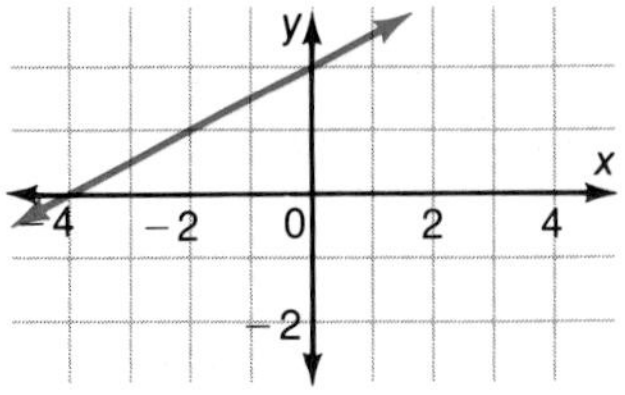

Figure 1

Definition The x value of a point common to the x axis and a straight line is called the **x intercept** of the line.

In Figure 1, the x intercept $= -4$.

EXAMPLE 1 Write the x intercept of the graph of $y = -\frac{1}{2}x + 1$.

Solution:

$$y = -\tfrac{1}{2}x + 1$$

Replace y with 0.

$$0 = -\tfrac{1}{2}x + 1$$

$$-1 = -\tfrac{1}{2}x$$

Solve for x.

$$-2(-1) = -2(-\tfrac{1}{2}x)$$

$$2 = x$$

The x intercept is 2.

P–2 **What is the x intercept of the graph of each function below?**

a. $y = x - 5$ **b.** $y = x + \frac{1}{2}$ **c.** $y = 2x + 8$

EXAMPLE 2 Graph $y = \frac{3}{2}x - 3$ by using the x and y intercepts.

Solution:

$$y = \tfrac{3}{2}x - 3$$

Let $y = 0$.

$$0 = \tfrac{3}{2}x - 3$$

$$3 = \tfrac{3}{2}x$$

$$\tfrac{2}{3}(3) = \tfrac{2}{3}(\tfrac{3}{2}x)$$

$$2 = x \text{ if } y = 0.$$

The y intercept is -3.

The line is drawn through $(2, 0)$ and $(0, -3)$.

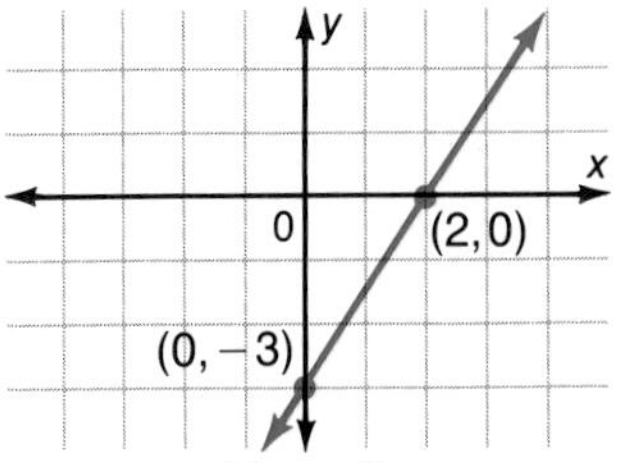

Figure 2

Figure 3 shows the graph of a linear relation that is <u>not</u> a function.

P–3 **What is the domain of the relation?**

P–4 **Why is the relation not a function?**

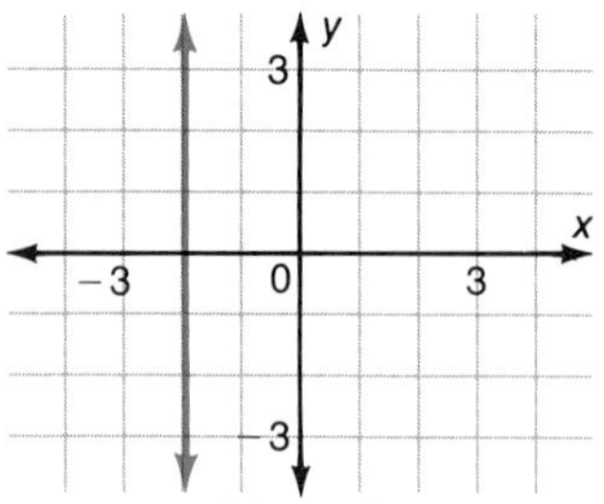

Figure 3

Let (x, y) represent the coordinates of any point on the vertical line of Figure 3. Then x equals −2 for <u>any</u> value of y.

$x = -2$ ◀ ***Rule of the relation***

> The rule of a relation having a vertical line as its graph is $x = k$ in which k is the x intercept of the line.

EXAMPLE 3 Graph $x = 2.7$.

Solution: The rule is of the form $x = k$. The graph is a vertical line 2.7 units to the right of the origin.

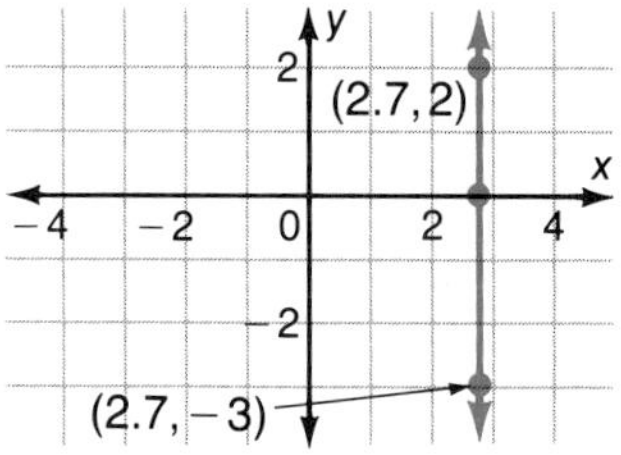

Figure 4

The ordered pairs (2.7, 2) and (2.7, −3) are coordinates of two points on the vertical line in Figure 4. Use these points to determine the slope of the line.

$$\frac{y_2 - y_1}{x_2 - x_1} = \frac{-3 - 2}{2.7 - 2.7}$$

$$= \frac{-5}{0}$$

Since you cannot divide by 0, $\frac{-5}{0}$ is not a number.

> A vertical line does not have a slope.

CLASSROOM EXERCISES

Write each x intercept. (Example 1)

1. $y = x + 3$ 2. $y = x - 1$ 3. $y = 2x - 2$ 4. $y = 3x - 5$

5. $y = 2x + 4$ 6. $y = x - \frac{1}{2}$ 7. $y = 2x - 5$ 8. $y = 8x$

9–14. *Write the x intercept of each line in Figure 5.* (P-1)

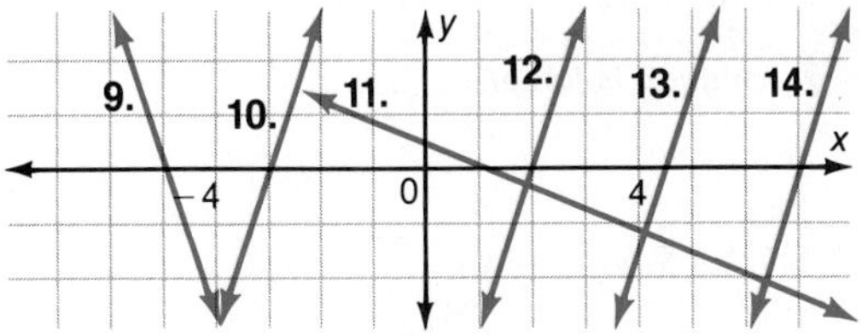

Figure 5

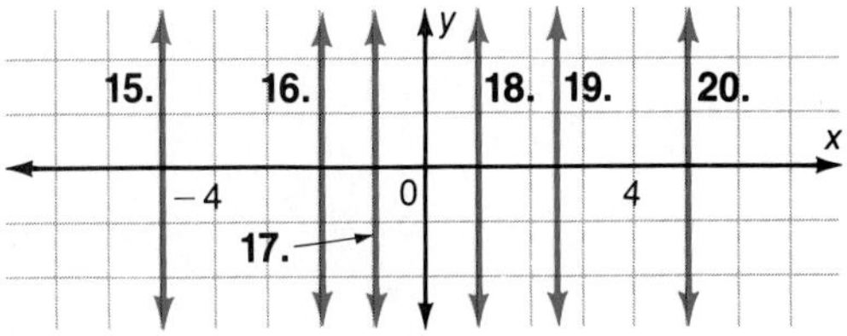

Figure 6

15–20. *Give a rule for each relation graphed in Figure 6.* (Example 3)

WRITTEN EXERCISES

Goal: To graph a linear function using *x* and *y* intercepts
Sample Problem: $y = \frac{3}{2}x - 3$ **Answer:** See Example 2.

Write each x intercept. (Example 1)

1. $y = x + 6$ 2. $y = x + 10$ 3. $y = x - 3$ 4. $y = x - 7$

5. $y = 3x + 6$ 6. $y = 2x + 10$ 7. $y = 4x - 12$ 8. $y = 3x - 15$

9. $y = \frac{1}{2}x - 2$ 10. $y = \frac{1}{2}x - 5$ 11. $y = \frac{1}{4}x + 1$ 12. $y = \frac{1}{4}x + 2$

Graph each function by using only the x and y intercepts. (Example 2)

13. $y = 2x - 4$ 14. $y = 3x + 6$ 15. $y = \frac{1}{2}x - 3$ 16. $y = \frac{1}{3}x - 2$

17. $y = 2x + 6$ 18. $y = 3x - 3$ 19. $y = 2x - 5$ 20. $y = 2x + 5$

Graph each linear relation. (Example 3)

21. $x = 5$ 22. $x = -4$ 23. $x = 2\frac{1}{2}$ 24. $x = -3\frac{1}{2}$

REVIEW CAPSULE FOR SECTION 13.7

For each formula, write the slope and y intercept. (Pages 356–358)

1. $y = 2x + 3$ 2. $F = 1.8C + 32$ 3. $c = \pi d$ 4. $K = C + 273$ 5. $s = 3 + 2.5t$

13.7 Using Graphs: Estimation

Temperature in Fahrenheit degrees is a function of temperature in Celsius degrees.

$$F = 1.8C + 32$$

Compare $F = 1.8C + 32$ with $y = 1.8x + 32$. The function is linear.

EXAMPLE 1

a. Graph $F = 1.8C + 32$.

b. Use the graph to estimate the value of F when $C = 10$.

Solutions: **a.** The ordered pairs of this function are represented by (C, F).

C	F
−20	−4
0	32
20	68

b. From the graph, the value of F is approximately 50 when $C = 10$.

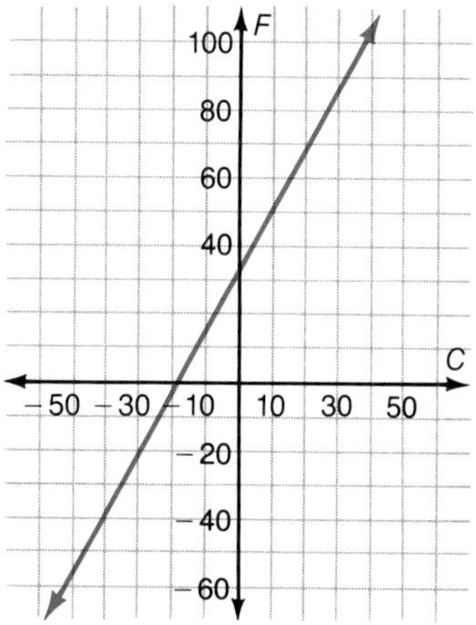

Figure 1

The distance, d, in meters that a car traveling at r kilometers per hour will cover in a certain fraction of a second is given by the function rule $d = 0.28r$.

EXAMPLE 2

a. Graph $d = 0.28r$.

Compare with $y = mx + b$. m is 0.28; b is 0.

b. Use the graph to estimate the value of d when $r = 60$.

Solutions: **a.** The ordered pairs of this function are represented by (r, d).

r	d
10	2.8
30	8.4
50	14

b. From the graph, the value of d is approximately 17 when $r = 60$.

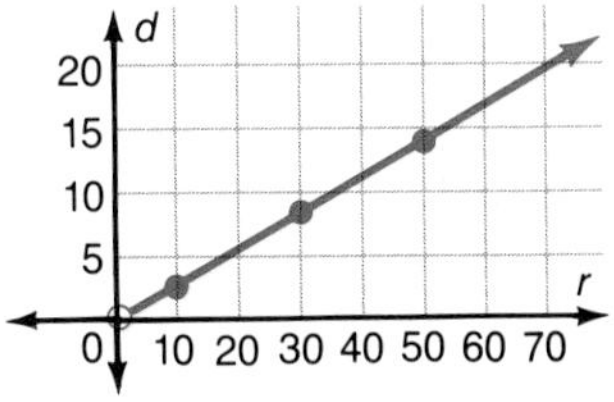

Figure 2

The circumference of a circle is a function of the length of a diameter.

$c = \pi d$ ◀ ***Compare with $y = mx + b$.***

EXAMPLE 3

a. Graph $c = \pi d$. The domain is {positive real numbers}. ($\pi \approx 3.14$)

b. Use the graph to estimate the value of c when $d = 0.5$.

Solutions: **a.** The ordered pairs of this function are represented by (d, c).

d	**c**
1	3.14
2	6.28
3	9.42

b. From the graph, the value of c is approximately 1.6 when $d = 0.5$.

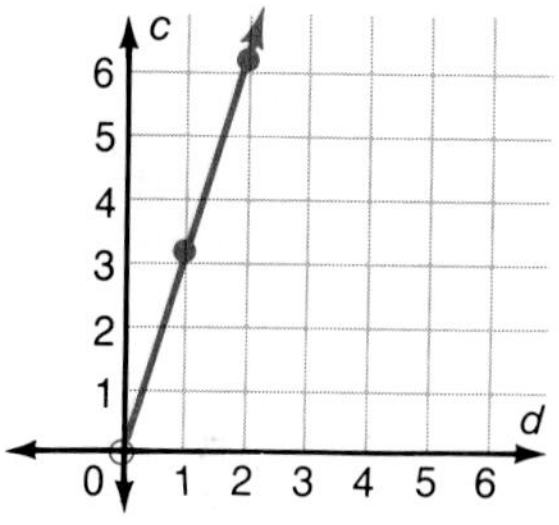

Figure 3

The simple interest on money in savings for one year at 5% is a function of the principal.

Rule: $i = 0.05p$ ◀ ***i = interest*** ***p = principal***

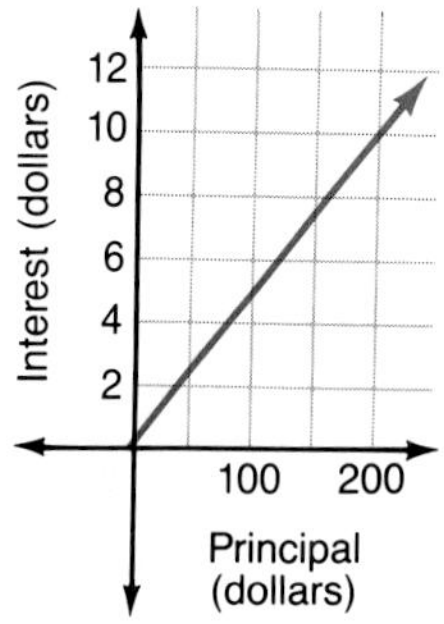

P–1 **Use the graph to estimate the simple interest for one year on a principal of $175.**

CLASSROOM EXERCISES

Refer to the graph of Figure 1. Estimate the Celsius temperature for each Fahrenheit temperature. (Example 1)

1. 100°F **2.** 60°F **3.** 20°F **4.** 0°F **5.** −20°F **6.** −60°F

Refer to the graph of Figure 2. Estimate the distance traveled at each speed. (Example 2)

7. 20 km/hr **8.** 35 km/hr **9.** 25 km/hr **10.** 40 km/hr

WRITTEN EXERCISES

Goal: To graph linear functions that represent practical applications and to estimate values from the graphs

Sample Problem: Use the graph of Figure 2 to estimate the distance traveled at a speed of 55 km/hr.

Answer: 15 meters

Graph each function as described. Unless otherwise stated the domain is {nonnegative real numbers}. (Examples 1–3)

1. The distance in millimeters that a coil spring stretches when w grams are attached is a function with this rule.
Rule: $d = 2w$

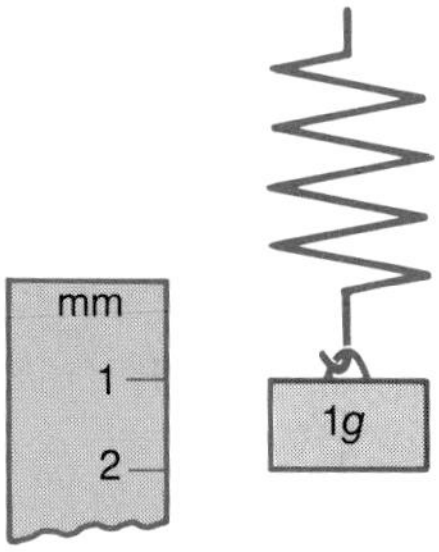

2. The simple interest on money in savings for ten years at 5% is a function of the principal.
Rule: $i = 10(0.05)p$ or $i = 0.5p$

3. The temperature in Kelvin degrees is a function of temperature in Celsius degrees.
Rule: $K = C + 273$
The domain is {real numbers greater than -273}.

4. In an auto race one racer has a 3 kilometer head start. His speed is 2.5 kilometers per minute. His total distance is a function of the number of minutes he drives.
Rule: $s = 3 + 2.5t$

Refer to the graph of Figure 1. Estimate the Fahrenheit temperature for each Celsius temperature. (Example 1)

5. 20°C **6.** 30°C **7.** 40°C **8.** -10°C **9.** -30°C **10.** -50°C

Refer to the graph of Figure 2. Estimate the speed for each distance traveled. (Example 2)

11. 5 m **12.** 10 m **13.** 15 m **14.** 20 m

Refer to the graph of Figure 3. Estimate the diameter corresponding to each circumference. (Example 3)

15. 2 cm **16.** 3 cm **17.** 4 cm **18.** 5 cm

Using Formulas

Gravity

Gravity is the force that causes an object to fall when it is released above the ground.

The speed of a falling object starting from rest increases as the object falls. The table at the right shows the object's speed in meters per second after each of the first five seconds. This increase in speed, or acceleration, is due to gravity. The **acceleration due to gravity,** represented by "g," is constant on and near the earth. It has a value of 9.8 meters per second per second.

Time (seconds)	*Speed (meters/second)*
1	9.8
2	19.6
3	29.4
4	39.2
5	49.0

The following formula computes the distance s that a falling object, starting from rest, travels in a length of time t, with acceleration g.

$$s = \frac{1}{2}gt^2$$

EXAMPLE: Find the approximate distance in meters a parachutist would fall after jumping from a plane and free-falling for 12 seconds before opening the chute.

SOLUTION: $s = \frac{1}{2}gt^2$

$s = \frac{1}{2}(9.8)(12)^2$ ◀ $g = 9.8$, $t = 12$

$= 4.9(144)$

$= 705.6$ or about **706 meters**

Skydivers fall freely at speeds of more than 160 kilometers per hour.

EXERCISES

1. A plane releases its auxiliary fuel tanks at a low altitude. The tanks take 15 seconds to reach the ground. At what altitude were they released?
2. A helicopter drops bales of hay to feed snowbound cattle. Each bale takes 5 seconds to reach the ground. At what height were the bales of hay released?
3. A rock dropped from the Royal Gorge Bridge in Colorado takes about 8 seconds to reach the river below. Find the approximate height of the bridge above the river to the nearest meter.
4. A ball dropped from the observation level of the Washington Monument reaches the ground in about $5\frac{1}{2}$ seconds. Find the height to the nearest meter.

CHAPTER SUMMARY

IMPORTANT TERMS	Linear function *(p. 351)* *y* intercept *(p. 353)* Slope *(p. 356)* Slope-intercept form *(p. 356)* Ratio *(p. 359)* Parallel lines *(p. 364)* *x* intercept *(p. 367)*
IMPORTANT IDEAS	*1.* The rule for a linear function is $y = mx + b$. **2.** In the rule for the linear function $y = mx + b$, b represents the y intercept of the graph. **3.** In the rule for the linear function $y = mx + b$, m represents the slope of its graph. **4.** The slope of a line is the ratio of the vertical motion to the horizontal motion from one point to another on the line. **5.** *Slope Formula:* If (x_1, y_1) and (x_2, y_2) are coordinates of two points on a nonvertical line, then the slope m equals $\frac{y_2 - y_1}{x_2 - x_1}$. **6.** If two lines have the same slope, they are parallel. **7.** The rule of a relation having a vertical line as its graph is $x = k$ in which k is the x intercept of the line. **8.** A vertical line does not have a slope.

CHAPTER REVIEW

SECTION 13.1

Write Yes or No to show whether each function is linear.

1. $y = x^2 + 2x$

2. $y = 5 - 7x$

3. $y = \frac{1}{2}x + \frac{3}{4}$

4. $y = |3x| - 2$

Graph at least four ordered pairs of each function. Then draw the graph with {real numbers} as the domain.

5. $y = -2x + 1$

6. $y = \frac{3}{2}x - 3$

SECTION 13.2

Write each y intercept.

7. $y = 3x - 10$

8. $y = -4x + 7$

9. $y = -0.7x + 3.8$

10. $y = 2.4x - 1.3$

SECTION 13.3

Write the slope-intercept form of the rule for each function.

11. The slope is 4 and the y intercept is -1.

12. The slope is -2 and the y intercept is 5.

13. The line crosses the y axis at (0, 2.9) and has slope -2.8.

14. The line crosses the y axis at (0, -5.3) and has slope 4.9.

Name the quadrants in which each graph lies.

15. $y = -7x$

16. $y = 4x$

17. $y = \frac{5}{2}x$

18. $y = -0.3x$

Write the slope of the graph of each function.

19. $y = -\frac{1}{2}x + 2$

20. $y = \frac{5}{4}x - 3$

21. $y = 0.8x - 1.2$

22. $y = -0.3x$

SECTION 13.4

Write the slope of the line that contains the two points.

23. $(4, -3)$; $(-1, 7)$

24. $(-2, 4)$; $(-1, -2)$

25. $(-\frac{1}{2}, \frac{5}{2})$; $(\frac{3}{2}, -\frac{7}{2})$

26. $(-\frac{3}{4}, -\frac{1}{4})$; $(\frac{5}{4}, -\frac{3}{4})$

Write the slope-intercept form of the line that contains the two points.

27. $(0, -2)$; $(3, 4)$

28. $(0, 3)$; $(-4, -2)$

29. $(0, -\frac{5}{2})$; $(4, \frac{1}{2})$

30. $(0, \frac{3}{2})$; $(-2, \frac{1}{2})$

SECTION 13.5

Graph the three functions in the same coordinate plane.

31. $y = \frac{1}{2}x - 2$
$y = \frac{1}{2}x + 3$
$y = \frac{1}{2}x$

32. $y = x - 5$
$y = x + 2$
$y = x$

Graph the functions in the same coordinate plane.

33. $y = -3$ **34.** $y = 4$ **35.** $y = 3.5$ **36.** $y = -\frac{5}{2}$

SECTION 13.6

Write each x intercept.

37. $y = -2x - 5$ **38.** $y = 3x + 8$

39. $y = -\frac{1}{2}x + 4$ **40.** $y = \frac{1}{4}x - 3$

Graph each function by using only the x and y intercepts.

41. $y = x + 1$ **42.** $y = -x + 1$

43. $y = 3x - 6$ **44.** $y = -2x + 4$

Graph each linear relation.

45. $x = -1$ **46.** $x = 2$ **47.** $x = \frac{7}{2}$ **48.** $x = -\frac{3}{2}$

SECTION 13.7

Use the graph of Figure 2 on page 370 to estimate the distance in meters that a car will travel at each given speed. Compare your estimate with the answer you obtain using the function rule $d = 0.28r$.

49. 70 km/hr **50.** 15 km/hr **51.** 45 km/hr **52.** 65 km/hr

Systems of Sentences

CHAPTER 14

The use of systems of sentences is a valuable problem-solving tool for solving a variety of problems that deal with such things as money, mixture, distance/rate/time, and digit problems.

14.1 Graphs

An equation such as $y = -2x + 3$ is called a ***linear equation*** in two variables. When you graph a linear equation, you graph its solution set. The ***solution set of a linear equation*** is the set of ordered pairs (x, y) that make the equation true.

EXAMPLE 1 Graph $x - y = -2$

"Graph" means "draw the graph of."

Determine at least three ordered pairs. → ***Graph the ordered pairs.*** → ***Draw the line.***

Solution: It is usually easy to find the value of y when x equals 0. Similarly, it is usually easy to find the value of x when y equals 0.

Let $x = 0$.	Let $y = 0$.	Let $x = 2$.
$0 - y = -2$	$x - 0 = -2$	$2 - y = -2$
$y = 2$	$x = -2$	$y = 4$

x	y
0	2
−2	0
2	4

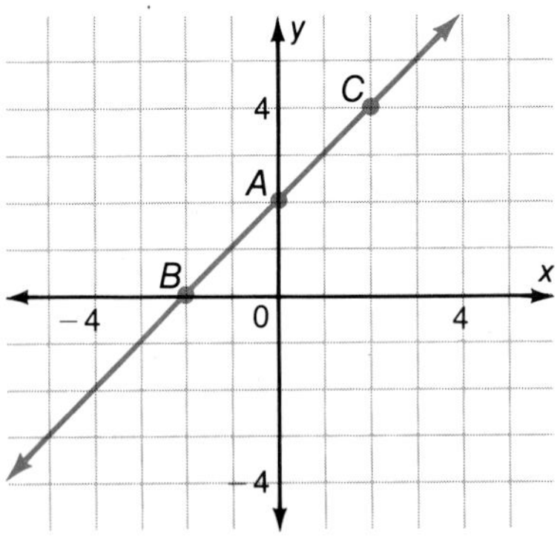

P–1 **What is the *y* value of point *A* called? the *x* value of point *B*?**

The intercepts are often used to locate two points on a graph. Other x or y values with small absolute values are then selected.

EXAMPLE 2 Graph $2x + 2y = 5$.

Solution:

$2(0) + 2y = 5$ ◀ ***Let $x = 0$.***

$2y = 5$

$y = 2\frac{1}{2}$

$2x + 2(0) = 5$ ◀ ***Let $y = 0$.***

$2x = 5$

$x = 2\frac{1}{2}$

$2(3) + 2y = 5$ ◀ ***Let $x = 3$.***

$6 + 2y = 5$

$2y = -1$

$y = -\frac{1}{2}$

x	y
0	$2\frac{1}{2}$
$2\frac{1}{2}$	0
3	$-\frac{1}{2}$

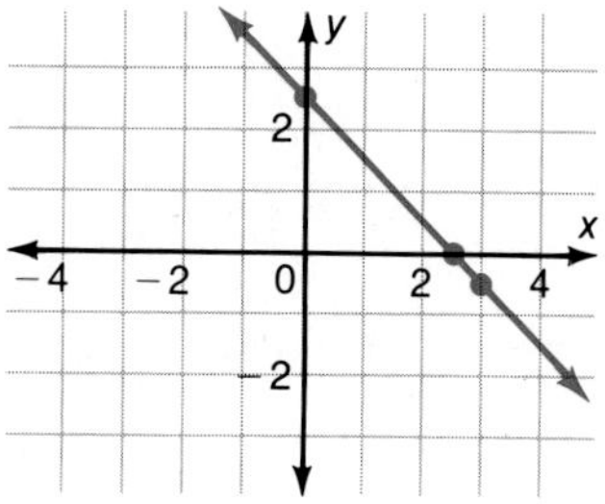

P–2 **Refer to the graph to determine what values of x will result in negative values for y.**
What values of x will give positive values for y?

EXAMPLE 3 Graph $x - 2y = 3$. Use the slope-intercept form.

Solution:

$x - 2y = 3$

$x = 3 + 2y$

$x - 3 = 2y$

$\frac{1}{2}(x - 3) = y$

$\frac{1}{2}x - \frac{3}{2} = y$

Slope-intercept form ▶ $y = \frac{1}{2}x - \frac{3}{2}$

Let $x = 0$. ▶ $y = \frac{1}{2}(0) - \frac{3}{2}$

$y = -\frac{3}{2}$

Let $y = 0$. ▶ $0 = \frac{1}{2}x - \frac{3}{2}$

$3 = x$

Let $x = -3$. ▶ $y = \frac{1}{2}(-3) - \frac{3}{2}$

$y = -3$

x	y
0	$-\frac{3}{2}$
3	0
−3	−3

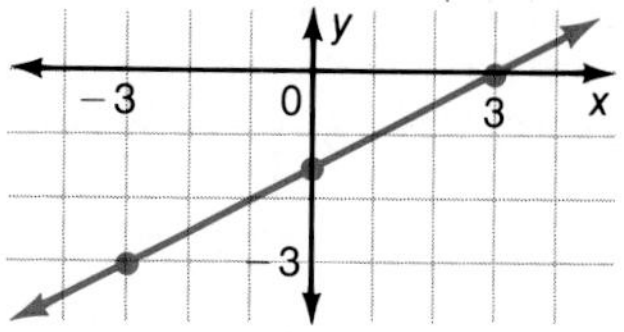

CLASSROOM EXERCISES

Write the unknown values in each table. (Examples 1–3)

1. $x + y = 2$

x	0	?	4
y	?	0	?

2. $x - y = -1$

x	0	?	?
y	?	0	3

3. $x - 2y = 0$

x	−2	?	?
y	?	0	1

4. $2x + y = 3$

x	0	?	−1
y	?	0	?

5. $x - y = 5$

x	0	?	2
y	?	0	?

6. $2x + 3y = 6$

x	0	?	?
y	?	0	−2

7. $x - 3y = 5$

x	0	?	−4
y	?	0	?

8. $y - 3 = 2x$

x	0	?	−4
y	?	0	?

WRITTEN EXERCISES

Goal: To graph the solution sets of linear equations
Sample Problem: $x + 3y = -3$
Answer: See the graph at the right.

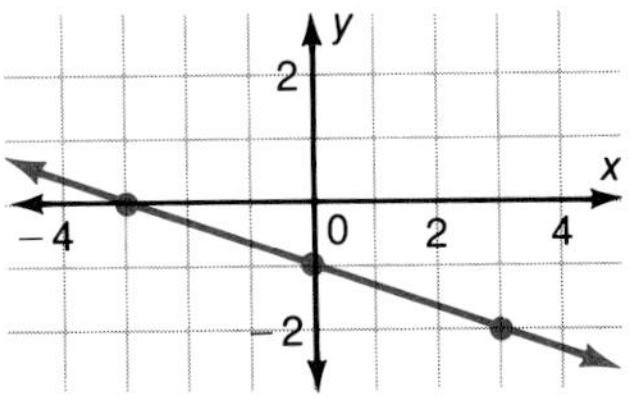

Graph each of the following. Use the tables in the Classroom Exercises. (Examples 1–3)

1. $x + y = 2$ **2.** $x - y = -1$ **3.** $x - 2y = 0$ **4.** $2x + y = 3$

5. $x - y = 5$ **6.** $2x + 3y = 6$ **7.** $x - 3y = 5$ **8.** $y - 3 = 2x$

Graph each of the following. (Examples 1–3)

9. $3x - 2y = 5$ **10.** $4x - 2y = -3$ **11.** $4y - x = -10$

12. $5y - 2x = 3$ **13.** $4x - 5y = 0$ **14.** $3y - 5x = 0$

15. $5x + 5y = -5$ **16.** $3x - 3y = 3$ **17.** $y - 3x = -6$

REVIEW CAPSULE FOR SECTION 14.2

Substitute to find one of the following ordered pairs that is in the solution set of both equations. (Pages 337–339)

(−3, 14) (−3, −2) (−1, 2) (3, −10) (−1, 1)

1. $\begin{cases} 2x - y = -4 \\ x + y = 1 \end{cases}$ **2.** $\begin{cases} x + 2y = 1 \\ x - 2y = -3 \end{cases}$ **3.** $\begin{cases} 4x + y = 2 \\ -3x - y = -5 \end{cases}$

14.2 Solving Systems by Graphing

EXAMPLE 1 Graph both linear equations in the same coordinate plane.

$$\begin{cases} \textbf{1.}\ x - 3y = 3 \\ \textbf{2.}\ 2x - y = -4 \end{cases}$$

Solution: Make a table of ordered pairs for each equation.

1.

x	y
0	−1
3	0
6	1

2.

x	y
0	4
−2	0
−1	2

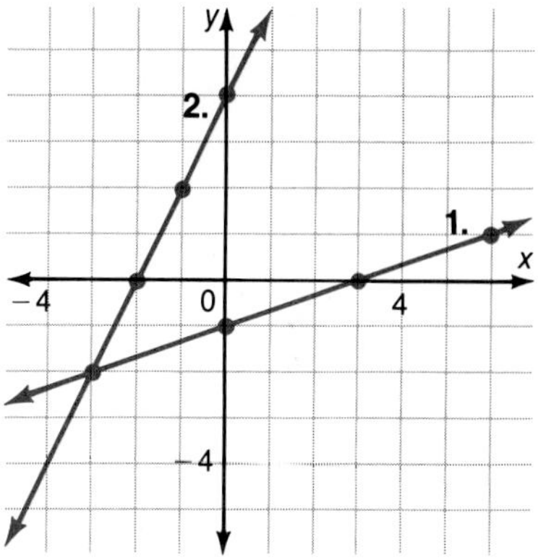

P–1 **What are the coordinates of the common point of the two graphs?**

Two equations such as **1** and **2** in Example 1 form a ***system of equations.***

The solution set of the system of Example 1 is $\{(-3, -2)\}$. This is checked below.

Check: 1.

$$x - 3y = 3$$
$$-3 - 3(-2)$$
$$-3 + 6$$
$$3$$

2.

$$2x - y = -4$$
$$2(-3) - (-2)$$
$$-6 + 2$$
$$-4$$

(−3, −2) makes both equations true.

Definition

The ***solution set of a system of two equations*** is the set of ordered pairs that make both equations true.

EXAMPLE 2 Solve the system below by graphing. Check.

$$\begin{cases} \textbf{1.}\ x - 2y = 0 \\ \textbf{2.}\ x + y = 3 \end{cases}$$

Solution: Make a table of ordered pairs for each equation.

1.

x	y
0	0
2	1
−4	−2

2.

x	y
0	3
3	0
−1	4

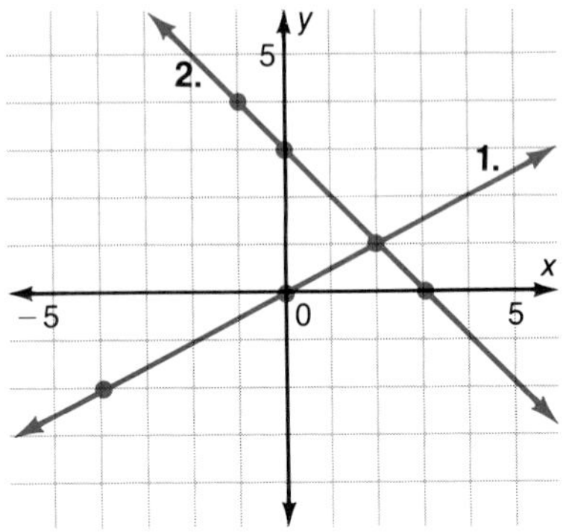

Check the point of intersection, (2, 1).

1. $x - 2y = 0$
$2 - 2(1)$
0

2. $x + y = 3$
$2 + 1$
3

Solution set: $\{(2, 1)\}$

EXAMPLE 3 Solve the system below by graphing.

$$\begin{cases} \textbf{1.}\ x - 2y = -2 \\ \textbf{2.}\ x - 2y = 4 \end{cases}$$

Solution:

1.

x	y
0	1
−2	0
2	2

2.

x	y
0	−2
4	0
2	−1

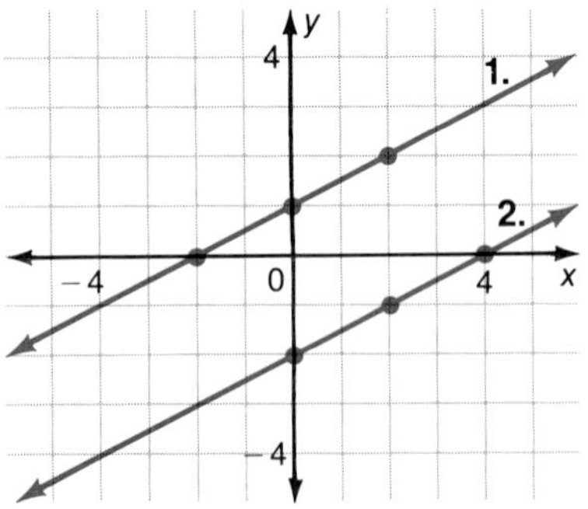

The graphs appear to be parallel. If this is true, the solution set of the system is the empty set, ϕ. To check this, write the equations in slope-intercept form.

1. $x - 2y = -2$

$x = 2y - 2$

$x + 2 = 2y$

$\frac{1}{2}(x + 2) = y$

$\frac{1}{2}x + 1 = y$

$y = \frac{1}{2}x + 1$

2. $x - 2y = 4$

$x = 4 + 2y$

$x - 4 = 2y$

$\frac{1}{2}(x - 4) = y$

$\frac{1}{2}x - 2 = y$

$y = \frac{1}{2}x - 2$

The slope of each line is $\frac{1}{2}$.

Since the slopes of the lines are equal, the lines are parallel. The solution set of the system is the empty set, ϕ.

CLASSROOM EXERCISES

Write Yes or No to tell whether each given set is the solution set of the system following it. (Checks of Examples 1 and 2)

1. $\{(1, 1)\}$; $\begin{cases} x + y = 2 \\ x - y = 3 \end{cases}$

2. $\{(3, -2)\}$; $\begin{cases} x + y = 1 \\ x - y = -1 \end{cases}$

3. $\{(0, 0)\}$; $\begin{cases} x + 2y = 0 \\ 2x - y = 0 \end{cases}$

4. $\{(0, 1)\}$; $\begin{cases} 2x - y = -1 \\ x + y = 1 \end{cases}$

5. $\{(0, 1)\}$; $\begin{cases} x - y = 2 \\ 3x - y = -1 \end{cases}$

6. $\{(2, -1)\}$; $\begin{cases} x - 2 = 0 \\ y + 1 = 0 \end{cases}$

7. ϕ; $\begin{cases} y = 2x - 1 \\ y = 2x + 3 \end{cases}$

8. $\{(-1, 5)\}$; $\begin{cases} x + 1 = 0 \\ y = 5 \end{cases}$

9. $\{(0, \frac{2}{3})\}$; $\begin{cases} 2x + 3y = 2 \\ 2x - 3y = -2 \end{cases}$

10. $\{(50, 50)\}$; $\begin{cases} x + y = 100 \\ x - y = 0 \end{cases}$

Write the solution set of each system from its graph. (P-1)

11.

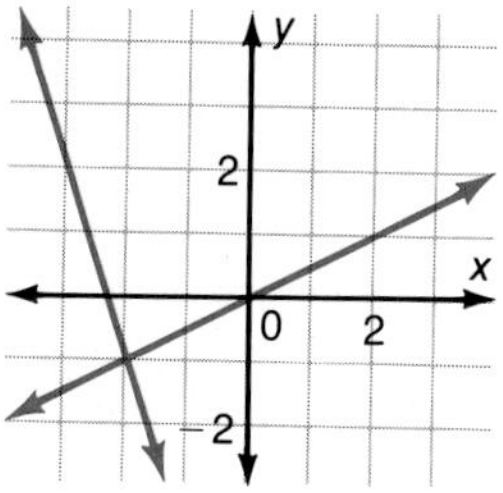

12.

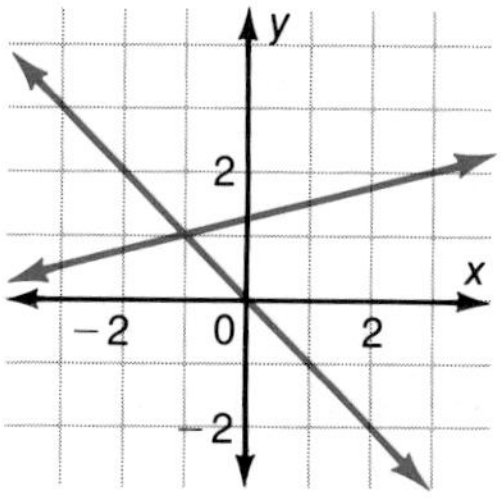

13.

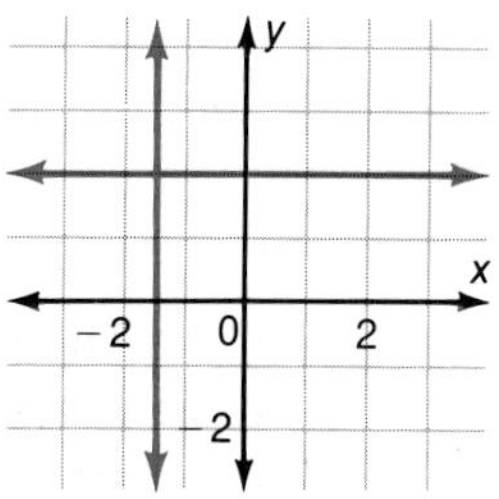

WRITTEN EXERCISES

Goal: To solve systems of linear equations by graphing

Sample Problem: Solve by graphing and check: $\begin{cases} \textbf{1.}\ 2x - y = 7 \\ \textbf{2.}\ 5x + 2y = 4 \end{cases}$

Answer: $\{(2, -3)\}$; see the graph at the right.

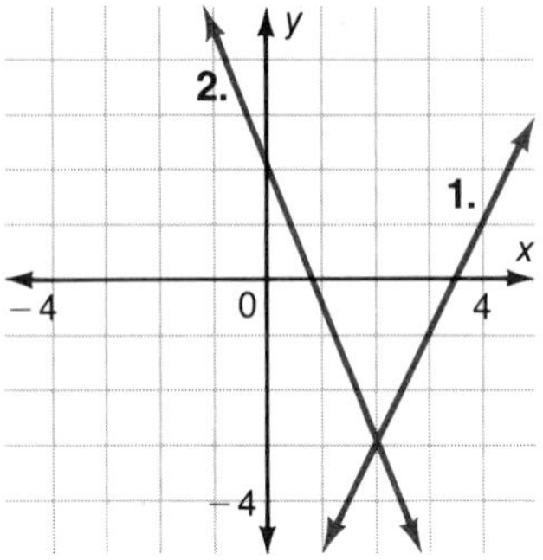

Solve each system by graphing. Check. (Examples 1–3)

1. $\begin{cases} x + y = 4 \\ x - y = -2 \end{cases}$

2. $\begin{cases} x + y = -3 \\ x - y = 1 \end{cases}$

3. $\begin{cases} 2x + y = 2 \\ x - y = -5 \end{cases}$

4. $\begin{cases} x - 2y = -3 \\ 2x + 2y = 6 \end{cases}$

5. $\begin{cases} 2x + y = -2 \\ 2x + 3y = 6 \end{cases}$

6. $\begin{cases} 2x + y = 1 \\ 4x - 2y = 6 \end{cases}$

7. $\begin{cases} x - 2y = -3 \\ 2x - 4y = 1 \end{cases}$

8. $\begin{cases} 2x + y = 1 \\ 4x + 2y = -1 \end{cases}$

9. $\begin{cases} x = 5 \\ y = -2 \end{cases}$

10. $\begin{cases} x = -3 \\ y = 4 \end{cases}$

11. $\begin{cases} 2x - 3y = 0 \\ 5x + 2y = 0 \end{cases}$

12. $\begin{cases} y - 3x = 0 \\ -2x - 3y = 0 \end{cases}$

13. $\begin{cases} x - y = 0 \\ x + y = 0 \end{cases}$

14. $\begin{cases} 2y - 3x = 0 \\ 3y + 2x = 0 \end{cases}$

15. $\begin{cases} 2x - y = 5 \\ x - 3y = 10 \end{cases}$

16. $\begin{cases} 3x - y = 5 \\ x + y = -1 \end{cases}$

17. $\begin{cases} y = -x + 6 \\ y = 5 \end{cases}$

18. $\begin{cases} 2x - y = 1 \\ x = 4 \end{cases}$

MORE CHALLENGING EXERCISES

Find the value for k that will make the lines in each system parallel.

19. $\begin{cases} 2x + y = 3 \\ kx + y = -1 \end{cases}$

20. $\begin{cases} 2x - y = -1 \\ kx - 2y = 6 \end{cases}$

21. $\begin{cases} 4y + 2x = -4 \\ kx - 2y = -2 \end{cases}$

NON-ROUTINE PROBLEM

22. The cost of a watch and a bracelet is \$77. The cost of the watch and a necklace is \$95. The cost of the necklace and the bracelet is \$108. Find the cost of each item.

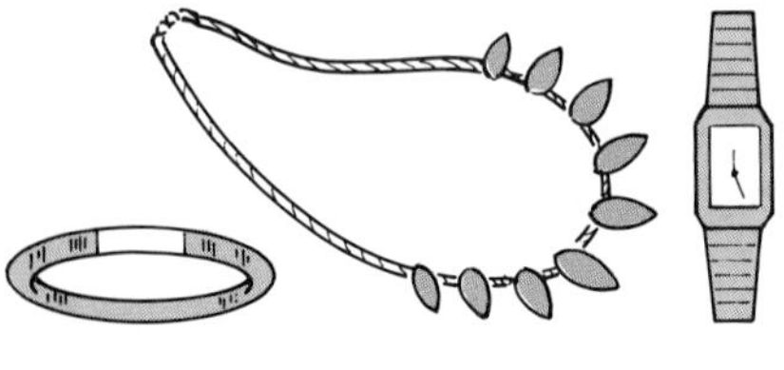

14.3 Addition Method

Sometimes you can solve a system of equations by adding the corresponding sides of the two equations to obtain a new equation with only one variable. This method is called the **addition method.**

EXAMPLE 1 Solve: $\begin{cases} \textbf{1. } x - y = -1 \\ \textbf{2. } x + y = 5 \end{cases}$

Solution:

1. Add the corresponding sides of both equations to eliminate y. Then solve for x.

$$\begin{aligned} x - y &= -1 \\ x + y &= 5 \\ \hline 2x + 0 &= 4 \\ x &= 2 \end{aligned}$$

Note: The y terms, y and −y, are opposites. $-y + y = 0$

2. Substitute 2 for x in either equation **1** or **2.**

$$\begin{aligned} x + y &= 5 \\ 2 + y &= 5 \\ y &= 3 \end{aligned}$$

Equation 2 is used.

3. **Check:** Substitute $x = 2$ and $y = 3$ in the original equations.

$x - y = -1$; $2 - 3$; -1

$x + y = 5$; $2 + 3$; 5

The solution set is $\{(2,3)\}$.

P–1 **Add the corresponding sides of both equations.**

a. $x + y = 3$
$x - y = 6$

b. $-x + y = 5$
$x + 2y = 7$

c. $3x - y = 1$
$2x + y = 9$

EXAMPLE 2 Solve: $\begin{cases} \textbf{1. } -2x - y = 2 \\ \textbf{2. } 2x + 3y = 6 \end{cases}$

Solution:

1. Add the corresponding sides of both equations to eliminate the x terms. Then solve for y.

$$\begin{aligned} -2x - y &= 2 \\ 2x + 3y &= 6 \\ \hline 2y &= 8 \\ y &= 4 \end{aligned}$$

2. Substitute 4 for y in either equation **1** or **2.**

$$\begin{aligned} -2x - y &= 2 \\ -2x - 4 &= 2 \\ -2x &= 6 \\ x &= -3 \end{aligned}$$

Equation 1 is used.

3. The check is left for you.

The solution set is $\{(-3,4)\}$.

P–2 **Add the corresponding sides of both equations.**

a. $2x + y = 0$
$-2x + y = 8$

b. $3x + 2y = 7$
$x - 2y = 5$

c. $x + 3y = -6$
$x - 3y = 0$

Steps for the Addition Method

1. Add to eliminate one of the variables. Solve the resulting equation.
2. Substitute the known value of one variable in one of the original equations of the system. Solve for the other variable.
3. Check the solution in both equations of the system.

CLASSROOM EXERCISES

Write the equation that results if you add corresponding sides of the two equations in each system. (P-1 and P-2)

1. $\begin{cases} x - y = 4 \\ x + y = 2 \end{cases}$

2. $\begin{cases} -x + y = 6 \\ x + y = 2 \end{cases}$

3. $\begin{cases} x + 3y = 9 \\ -x + 2y = 6 \end{cases}$

4. $\begin{cases} 4x + y = 8 \\ -4x + y = 0 \end{cases}$

5. $\begin{cases} x - 3y = 7 \\ 2x + 3y = 5 \end{cases}$

6. $\begin{cases} x + 2y = 1 \\ x - 2y = -5 \end{cases}$

Add the corresponding sides of the equations. Then solve for x or y. (Step 1 of Examples 1 and 2)

7. $\begin{cases} x - y = 10 \\ x + y = -2 \end{cases}$

8. $\begin{cases} -x + 2y = 5 \\ x + y = 4 \end{cases}$

9. $\begin{cases} 4x - 5y = -2 \\ 2x + 5y = 8 \end{cases}$

WRITTEN EXERCISES

Goal: To solve a system of equations by the addition method

Sample Problem: Solve: $\begin{cases} 5x - 2y = 12 \\ 4x + 2y = 6 \end{cases}$ **Answer:** $\{(2, -1)\}$

For Exercises 1–18, solve each system by the addition method. Check. (Examples 1 and 2)

1. $\begin{cases} x + y = 6 \\ x - y = -2 \end{cases}$

2. $\begin{cases} -x + y = 5 \\ x + y = 9 \end{cases}$

3. $\begin{cases} -x + 2y = 8 \\ x + 3y = 2 \end{cases}$

4. $\begin{cases} 3x - y = 5 \\ x + y = 7 \end{cases}$

5. $\begin{cases} x + 3y = 1 \\ x - 3y = -5 \end{cases}$

6. $\begin{cases} x - 5y = -2 \\ 2x + 5y = -4 \end{cases}$

7. $\begin{cases} -x - 4y = -10 \\ -3x + 4y = -6 \end{cases}$

8. $\begin{cases} 3x + y = 11 \\ -3x + 5y = -17 \end{cases}$

9. $\begin{cases} x - 6y = 6 \\ -x + 2y = -6 \end{cases}$

10. $\begin{cases} -3x + 7y = -1 \\ 2x - 7y = 3 \end{cases}$

11. $\begin{cases} 7x + 2y = -4 \\ 2x - 2y = -5 \end{cases}$

12. $\begin{cases} x + y = -8 \\ x - y = -3 \end{cases}$

13. $\begin{cases} x + 3y = 1 \\ x - 3y = -5 \end{cases}$

14. $\begin{cases} -6x + 3y = 12 \\ 6x - 5y = 4 \end{cases}$

15. $\begin{cases} -3x + y = 1 \\ 5x - y = 3 \end{cases}$

16. $\begin{cases} 2x - 3y = -2 \\ -x + 3y = -1 \end{cases}$

17. $\begin{cases} -4x - y = 2 \\ 4x - y = 3 \end{cases}$

18. $\begin{cases} 7x + 3y = 19 \\ 4x - 3y = 3 \end{cases}$

MID-CHAPTER REVIEW

Graph each of the following equations. (Section 14.1)

1. $x + y = 5$
2. $x - 2y = 3$
3. $4x + y = 1$
4. $3x - y = 4$
5. $3x - 4y = -8$
6. $y - 4 = 2x$

Solve each system by graphing. Check. (Section 14.2)

7. $\begin{cases} 5x - 6y = 8 \\ 2x - y = -1 \end{cases}$

8. $\begin{cases} x - 3y = -6 \\ 4x + 3y = -9 \end{cases}$

9. $\begin{cases} x + 2y = -2 \\ 5x + 2y = 6 \end{cases}$

Solve each system by the addition method. Check. (Section 14.3)

10. $\begin{cases} x - 2y = 13 \\ -x - 3y = 2 \end{cases}$

11. $\begin{cases} 3x + 3y = 7 \\ 4x - 3y = 7 \end{cases}$

12. $\begin{cases} 2x - y = 12 \\ -2x + 3y = 8 \end{cases}$

REVIEW CAPSULE FOR SECTION 14.4

Write the coefficient of each monomial. (Pages 208–209)

1. $3x$
2. $\frac{1}{2}x^2$
3. $-4y$
4. xy^2
5. $-y^2$

Multiply. (Pages 47–49)

6. $2(x - y)$
7. $3(x + y)$
8. $3(x + 2y)$
9. $4(2x - 3y)$
10. $3(-3x + 5y)$
11. $-2(x - y)$
12. $-2(2x - y - 3)$
13. $3(-5x + y + 5)$

14.4 Multiplication/Addition Method

When you add the corresponding sides of the equations of a system, you may get an equation that has two variables rather than only one. This occurs because the coefficients of the variables are not opposites.

To solve such a system of equations, you can multiply each side of one or both equations by a number or numbers that will make opposite coefficients for one variable. Then you add the corresponding sides of the two equations. This procedure is called the ***multiplication/addition method.***

EXAMPLE 1 Solve: $\begin{cases} \mathbf{1.}\ x - y = -1 \\ \mathbf{2.}\ 2x - 3y = -1 \end{cases}$

Solution:

1 Multiply each side of equation **1** by -2. ⟶ $-2x + 2y = 2$

2 Add equation **2**. Solve. ⟶ $\underline{2x - 3y = -1}$

$-y = 1$

$y = -1$

The coefficients of x are now opposites.

3 Substitute -1 for y in either equation **1** or **2**. Solve for x. ⟶

$x - y = -1$

$x - (-1) = -1$

$x = -2$

Equation 1 is used.

4 **Check:** Substitute $x = -2$ and $y = -1$ in $x - y = -1$ and in $2x - 3y = -1$.

The solution set is $\{(-2, -1)\}$.

P–1 **Multiply each side of equation 1 by a number that will form coefficients of x which are opposites.**

a. $\begin{cases} \mathbf{1.}\ x + y = 7 \\ \mathbf{2.}\ -3x + 2y = 4 \end{cases}$ **b.** $\begin{cases} \mathbf{1.}\ x - 6y = -5 \\ \mathbf{2.}\ x + 2y = 11 \end{cases}$ **c.** $\begin{cases} \mathbf{1.}\ x + 2y = -2 \\ \mathbf{2.}\ 7x - 5y = 43 \end{cases}$

In Example 1, it was necessary to multiply the corresponding sides of only one of the equations to eliminate a variable. It is sometimes necessary to multiply corresponding sides of <u>both</u> equations by numbers that will make opposite coefficients for one variable.

EXAMPLE 2

Solve: $\begin{cases} \textbf{1.}\ 5x - 3y = -1 \\ \textbf{2.}\ -3x + 2y = 2 \end{cases}$

Solution:

[1] Multiply each side of **1** by 2. → $10x - 6y = -2$

Multiply each side of **2** by 3. → $\underline{-9x + 6y = 6}$

The coefficients of y are now opposites.

[2] Add. $x \quad = 4$

[3] Substitute 4 for x in either **1** or **2.** → $5x - 3y = -1$

Equation 1 is used.

$5(4) - 3y = -1$

$20 - 3y = -1$

$-3y = -21$

$y = 7$

[4] The check is left for you. The solution set is $\{(4, 7)\}$.

P–2 **Multiply each side of both equations by numbers that will form coefficients of *y* which are opposites.**

a. $\begin{cases} \textbf{1.}\ 3x + 8y = -16 \\ \textbf{2.}\ 4x - 5y = 10 \end{cases}$ b. $\begin{cases} \textbf{1.}\ 2x - 3y = -11 \\ \textbf{2.}\ 5x + 4y = 30 \end{cases}$ c. $\begin{cases} \textbf{1.}\ 7x + 2y = -36 \\ \textbf{2.}\ -8x + 3y = 57 \end{cases}$

Steps for the Multiplication/Addition Method

1. Check for opposite coefficients of one variable. If necessary, multiply each side of either or both equations by numbers that will make opposite coefficients for one variable.
2. Add to eliminate one of the variables. Solve the resulting equation.
3. Substitute the known value of one variable in one of the original equations of the system. Solve for the other variable.
4. Check the solution in both equations of the system.

CLASSROOM EXERCISES

Multiply each side of one equation by a number to eliminate x or y. (Steps 1 and 2 of Example 1)

1. $\begin{cases} x - 3y = -1 \\ -2x + y = -3 \end{cases}$

2. $\begin{cases} 3x + 5y = 1 \\ 2x - y = -2 \end{cases}$

3. $\begin{cases} -2x + 4y = 1 \\ 3x + 2y = -4 \end{cases}$

Multiply each side of both equations by numbers to eliminate x or y. (Steps 1 and 2 of Example 2)

4. $\begin{cases} 2x - 4y = -1 \\ 3x - 3y = 2 \end{cases}$ **5.** $\begin{cases} 4x + 5y = 2 \\ -3x + 2y = 1 \end{cases}$ **6.** $\begin{cases} -6x - 2y = -3 \\ 4x + 3y = 1 \end{cases}$

WRITTEN EXERCISES

Goal: To solve systems of equations by the multiplication/addition method

Sample Problem: $\begin{cases} 5x - 4y = 2 \\ -x + y = 1 \end{cases}$ **Answer:** $\{(-2, -3)\}$

Solve each system by the multiplication/addition method. (Examples 1–3)

1. $\begin{cases} 2x + y = 10 \\ x - 2y = 0 \end{cases}$ **2.** $\begin{cases} 2x - y = 5 \\ x + 2y = 25 \end{cases}$ **3.** $\begin{cases} 4x + 3y = 6 \\ 2x - y = -2 \end{cases}$

4. $\begin{cases} 5x + y = 4 \\ x - 2y = 3 \end{cases}$ **5.** $\begin{cases} 2x - y = -3 \\ 4x - y = -2 \end{cases}$ **6.** $\begin{cases} 4x - y = 1 \\ 2x - y = 3 \end{cases}$

7. $\begin{cases} 2x - 3y = 13 \\ 3x + y = 3 \end{cases}$ **8.** $\begin{cases} 3x + 4y = 10 \\ 2x - y = -8 \end{cases}$ **9.** $\begin{cases} x + 3y = -3 \\ 2x - y = 4 \end{cases}$

10. $\begin{cases} 2x - 5y = -5 \\ x + 2y = -4 \end{cases}$ **11.** $\begin{cases} -5x + 3y = 0 \\ x + 2y = 0 \end{cases}$ **12.** $\begin{cases} 2x - y = 3 \\ -3x - y = -5 \end{cases}$

13. $\begin{cases} 2x + 5y = 0 \\ 3x - y = 0 \end{cases}$ **14.** $\begin{cases} 3x - y = -2 \\ -5x - y = 3 \end{cases}$ **15.** $\begin{cases} 3x + 2y = 2 \\ 2x + 3y = -2 \end{cases}$

16. $\begin{cases} 5x - 2y = 11 \\ 3x + 5y = 19 \end{cases}$ **17.** $\begin{cases} 4x - 3y = 5 \\ 2x + 9y = -1 \end{cases}$ **18.** $\begin{cases} 6x - 2y = -3 \\ 2x + 4y = 5 \end{cases}$

APPLICATIONS

You can use a calculator to check whether a given ordered pair is a solution of a system of equations.

EXAMPLE Check the solution set, $\{(4, 7)\}$, of Example 2 on page 389.

SOLUTION **1.** $5x - 3y = -1$

Store "5 · 4" in memory before adding it to "−3 · 7."

[5] [×] [4] [=] [M+] [3] [+/−] [×] [7]

[+] [MR] [=]

−1.

2. $-3x + 2y = z$ This equation is left for you to check.

19–24. Use a calculator to do Exercises 1–6 on page 383.

REVIEW CAPSULE FOR SECTION 14.5

Suppose x is the smaller and y is the greater of two unknown numbers. Write an equation for each sentence.

1. The sum of the two numbers is −12.
2. The difference of the two numbers is 7.
3. The smaller of the two numbers is 6 less than the greater number.
4. The greater of the two numbers is 13 more than the smaller number.

PROBLEM SOLVING AND APPLICATIONS

14.5 Using Systems of Equations

The addition method and multiplication/addition method of solving a system of equations can be used to solve word problems. First, you use Condition 1 and two variables to represent the unknowns. You also use Condition 1 to write one equation of the system. You use Condition 2 to write the second equation for the system.

EXAMPLE 1 The sum of two numbers is −16 (Condition 1). The difference of the two numbers is 30 (Condition 2). Find the two numbers.

Solution:

[1] Use Condition 1 to represent the unknowns. → Let x = the greater number. Let y = the smaller number.

[2] Write the equations.

Think: **The sum of two numbers is −16.** ◀ ***Condition 1***

Translate: $x + y = -16$

Think: **The difference of the two numbers is 30.** ◀ ***Condition 2***

Translate: $x - y = 30$

[3] Solve. Use the addition method. →

$$x + y = -16$$
$$x - y = 30$$
$$2x = 14$$
$$x = 7$$

◀ ***Don't forget to find y.***

$$\mathbf{x} + y = -16$$
$$\mathbf{7} + y = -16$$
$$y = -23$$

[4] **Check:** Condition 1 Is the sum of the two numbers -16?
Does $7 + (-23) = -16$? Yes ✓

Condition 2 Is the difference of the two numbers 30?
Does $7 - (-23) = 30$? Yes ✓

The two numbers are 7 and -23.

EXAMPLE 2 The width of a rectangle is 5 centimeters less than the length (Condition 1). The perimeter is 38 centimeters (Condition 2). Find the length and width.

Solution:

[1] Let $x =$ the length and let $y =$ the width.
Then $2x + 2y =$ the perimeter.

[2] Write the equations. ⟶ $\begin{cases} \textbf{1. } y = x - 5, \text{ or } -x + y = -5 \\ \textbf{2. } 2x + 2y = 38 \end{cases}$

1. Condition 1
2. Condition 2

[3] To solve the system, multiply equation **1** by 2. ⟶

$$\begin{aligned} -2x + 2y &= -10 \quad \textbf{(Equation 1)} \\ 2x + 2y &= 38 \quad \textbf{(Equation 2)} \\ \hline 4y &= 28 \\ y &= 7 \\ y &= x - 5 \\ 7 &= x - 5 \\ 12 &= x \end{aligned}$$

Don't forget to find x.

The length is 12 centimeters.
The width is 7 centimeters.

The check is left for you.

CLASSROOM EXERCISES

Write an equation in two variables that represents each sentence. Write what each variable represents. (Steps 1 and 2 of Examples 1 and 2)

1. The sum of two numbers is 25.

2. The difference of two numbers is -7.

3. The sum of the ages of Jerry and Mary is 25 years.

4. One number is 12 more than another number.

5. Half the perimeter of a rectangle is 51.

6. The perimeter of a rectangle is 93 millimeters.

7. Twice one number increased by three times another number is 19.

8. One number exceeds another number by 24.

WRITTEN EXERCISES

Goal: To solve number problems by systems of linear equations

Sample Problem: The greater of two numbers is 1 more than 3 times the smaller (Condition 1). If 8 times the smaller is decreased by 2 times the greater, the result is 10 (Condition 2). Find the numbers.

Answer: Greater number: 19 Smaller number: 6

For some problems, Condition 1 is underscored once. Condition 2 is underscored twice.

a. *Use Condition 1 and two variables to represent the unknowns.*

b. *Use Conditions 1 and 2 to write a system of equations for the problem.*

c. *Solve.* (Examples 1 and 2)

1. The sum of two numbers is 28. The difference of the two numbers is 3. Find the two numbers.

2. The sum of two numbers is 73. The difference of the two numbers is 11. Find the two numbers.

3. The sum of two numbers is -12. One number exceeds the other by 42. Find the two numbers.

4. The sum of two numbers is -65. One number exceeds the other by 27. Find the two numbers.

5. Twice the smaller of two numbers is increased by 2. The result equals the greater number. Three times the smaller number is increased by twice the greater number. The result equals 39. Find the two numbers.

6. The greater of two numbers is 4 less than twice the smaller number. From twice the greater number, three times the smaller is subtracted. The result equals 7. Find the two numbers.

7. The length of a rectangle is 8 meters more than the width. The perimeter is 34 meters. Find the length and width.

8. The width of a rectangle is 9 decimeters less than the length. The perimeter is 43 decimeters. Find the length and width.

9. John is seven years younger than his sister Ann. The sum of their ages is 45. Find their ages.

10. Kathy is 31 years younger than her father. The sum of their ages is 47. Find their ages.

11. Sam bought 6 fewer crates of corn than Tina. Twice the number of crates Tina bought is 5 more than 3 times the number of crates Sam bought. How many crates did each buy?

12. Brian played 4 fewer innings in a softball game than Denise. Twice the number of innings Brian played plus 2 equals the number of innings Denise played. How many innings did each play?

Using Tables

Chemistry

Pharmacists use their knowledge of **chemistry** to prepare various solutions. The table below shows the **molecular weights** of some chemicals.

Chemical	***Molecular Weight***
Boric acid	157.30
Calcium chloride	147.02
Magnesium sulfate	246.47
Potassium chloride	74.56

A **mole** in chemistry is the mass of a substance that equals its molecular weight. A **one molar solution** of boric acid and water would contain 157.3 grams of boric acid in 1 liter of solution.

EXAMPLE: Find the number of grams of boric acid to make 1500 milliliters of a solution with 5% molarity.

SOLUTION: **a.** Find the number of grams of boric acid to make 1 liter (1000 milliliters) of a 5% molarity solution.

Find 5% of 157.3.

1 5 7 . 3 × 5 %   7.865

One liter of a 5% molarity solution requires about 7.9 grams.

b. Find the number of grams of boric acid for 1500 milliliters. Solve the following proportion.

$$\frac{7.9}{1000} = \frac{n}{1500}$$

$$n = \frac{(7.9)(1500)}{1000}$$

$n = 11.85$ or about **11.9 grams** of boric acid

EXERCISES

Find the number of grams needed to make a solution of the given number of milliliters and molarity. Round the answer to the nearest tenth of a gram.

1. Magnesium sulfate
1200 milliliters
10% molarity

2. Potassium chloride
750 milliliters
0.3% molarity

3. Calcium chloride
1750 milliliters
8.5% molarity

CHAPTER SUMMARY

IMPORTANT TERMS

Linear equation *(p. 378)*
Solution set of a linear equation *(p. 378)*
System of equations *(p. 381)*
Solution set of a system of two equations *(p. 381)*
Addition method *(p. 385)*
Multiplication/addition method *(p. 388)*

IMPORTANT IDEAS

1. To solve a system of equations by the *multiplication/addition method:*
   a. Check for opposite coefficients of one variable. If necessary, multiply each side of either or both equations by numbers that will make opposite coefficients for one variable.
   b. Add to eliminate one of the variables. Solve the resulting equation.
   c. Substitute the known value of one variable in one of the original equations of the system. Solve for the other variable.
   d. Check the solution in both equations of the system.

2. To solve word problems using a system of equations in two variables:
   a. Use Condition 1 and two variables to represent the unknowns.
   b. Use Condition 1 to write one equation of the system.
   Use Condition 2 to write the second equation for the system.
   c. Solve the system of equations.
   d. Check your solution in the original conditions of the problem.
   Answer the question.

CHAPTER REVIEW

SECTION 14.1

Graph each of the following.

1. $2x + y = 3$ **2.** $-3x + y = -2$ **3.** $x - 5y = -1$

4. $-x - 3y = 1$ **5.** $-2x + 3y = -6$ **6.** $4x - 2y = -8$

SECTION 14.2

Solve each system by graphing. Check.

7. $\begin{cases} x + 3y = 3 \\ 2x + y = -4 \end{cases}$ **8.** $\begin{cases} x - 2y = 2 \\ x + y = 2 \end{cases}$

9. $\begin{cases} 2x - y = 0 \\ 2x - 3y = -4 \end{cases}$ **10.** $\begin{cases} x - 2y = 4 \\ x + 3y = -1 \end{cases}$

SECTION 14.3

Solve each system by the addition method. Check.

11. $\begin{cases} x - y = 3 \\ x + y = 1 \end{cases}$

12. $\begin{cases} 2x + y = 1 \\ 2x - y = 7 \end{cases}$

13. $\begin{cases} 3x + 2y = 6 \\ -3x - y = 3 \end{cases}$

SECTION 14.4

Solve each system by the multiplication/addition method. Check.

14. $\begin{cases} x - y = 1 \\ 2x + 3y = 7 \end{cases}$

15. $\begin{cases} 2x + 3y = 6 \\ x + 2y = 1 \end{cases}$

16. $\begin{cases} -2x + y = 1 \\ 4x + 3y = 23 \end{cases}$

17. $\begin{cases} 3x - 4y = 5 \\ 6x - 8y = 3 \end{cases}$

SECTION 14.5

For Exercises 18–21, Condition 1 is underscored once. Condition 2 is underscored twice.

Use Condition 1 to represent the unknowns. Use Condition 1 and Condition 2 to write a system of equations for the problem. Solve.

18. The sum of two numbers is 17. The smaller number is 33 less than the greater number. Find the two numbers.

19. The sum of two numbers is 8. Four times the greater number is two more than four times the smaller number. Find the two numbers.

20. The length of a rectangle is three meters less than twice the width. The perimeter is 24 meters. Find the length and width.

21. Three times the width of a rectangle increased by twice the length equals 40. The perimeter is 34 units. Find the number of units in the length and width.

More on Systems of Sentences

CHAPTER 15

In mathematics, the formula $d = rt$ is used to find distance. Various instruments such as the telescope, the transit, and sonar equipment are used to measure distance in space, on the earth's surface, and in the ocean, respectively.

15.1 Substitution Method

You have often replaced a variable by a number or expression. This process is called substitution. You can solve systems of linear equations by the ***substitution method.***

EXAMPLE 1 Solve the system below by the substitution method.

$$\begin{cases} \textbf{1.}\ y = 2x - 3 \\ \textbf{2.}\ 2x + 3y = 7 \end{cases}$$

Solution: $2x + 3\mathbf{y} = 7$

1. Substitute $(2x - 3)$ for y in Equation **2.** ⟶ $2x + 3\mathbf{(2x - 3)} = 7$
2. Distributive Property ⟶ $2x + 6x - 9 = 7$
3. Combine like terms. ⟶ $8x - 9 = 7$
4. Addition Property of Equality ⟶ $8x = 16$
5. Division Property of Equality ⟶ $x = 2$

Substitute 2 for x in either equation. Then solve for y.

$$\textbf{1.}\ y = 2\mathbf{x} - 3$$
$$y = 2\mathbf{(2)} - 3$$
$$y = 1$$

Check:

1. $\mathbf{y} = 2\mathbf{x} - 3$	
1	$2\mathbf{(2)} - 3$
	1

2. $2\mathbf{x} + 3\mathbf{y} = 7$	
$2\mathbf{(2)} + 3\mathbf{(1)}$	7
$4 + 3$	
7	

Solution set: $\{(2, 1)\}$

EXAMPLE 2 Solve the system by the substitution method.

$$\begin{cases} \textbf{1.}\ 2x + 3y = 1 \\ \textbf{2.}\ x - 2y = -3 \end{cases}$$

Solution: $x - 2y = -3$

1. Solve Equation **2** for x. ⟶ $x = 2y - 3$

2 Substitute $(2y - 3)$ for x in Equation **1.** Solve for y. ⟶

$$2\mathbf{x} + 3y = 1$$
$$2(\mathbf{2y - 3}) + 3y = 1$$
$$4y - 6 + 3y = 1$$
$$7y - 6 = 1$$
$$7y = 7$$
$$y = 1$$

3 Substitute 1 for y in either equation. Solve for x. ⟶

$$x - 2\mathbf{y} = -3$$
$$x - 2(\mathbf{1}) = -3$$
$$x - 2 = -3$$
$$x = -1$$

4 **Check: 1.**

$$2\mathbf{x} + 3\mathbf{y} = 1$$
$$2(\mathbf{-1}) + 3(\mathbf{1})$$
$$-2 + 3$$
$$1$$

2.

$$\mathbf{x} - 2\mathbf{y} = -3$$
$$\mathbf{-1} - 2(\mathbf{1})$$
$$-1 - 2$$
$$-3$$

Solution set: $\{(-1, 1)\}$

P–1 **What expression from $y = 2x - 3$ could be substituted for y in $x = \frac{1}{2}y + 2$?**

P–2 **What expression from $x = \frac{1}{2}y + 2$ could be substituted for x in $2x - y = 3$?**

EXAMPLE 3

Solve the system by the substitution method.

$$\begin{cases} \mathbf{1.}\ 2x - y = 3 \\ \mathbf{2.}\ x = \frac{1}{2}y + 2 \end{cases}$$

Solution:

1 Substitute $(\frac{1}{2}y + 2)$ for x in Equation **1.** ⟶

2 Solve for y. ⟶

$$2\mathbf{x} - y = 3$$
$$2(\mathbf{\tfrac{1}{2}y + 2}) - y = 3$$
$$y + 4 - y = 3$$
$$4 = 3$$

False sentence

The resulting sentence is false. Therefore, the solution set of the system is ϕ.

Steps of the Substitution Method

1. Solve one equation for one of the variables.
2. Substitute the resulting expression in the other equation. Solve the equation.
3. Substitute the value of the variable from Step **2** in either equation. Solve the resulting equation.
4. Check by substituting both values in both equations.

CLASSROOM EXERCISES

Solve the equation for one of the variables. (Step 1 of Example 2)

EXAMPLE: $2x + y = 2$ **ANSWER:** $y = 2 - 2x$ or $y = -2x + 2$

1. $x - 2y = 3$
2. $3y + x = -5$
3. $x + 3y = 1$
4. $y + 3 = 5x$
5. $x - 3 = -3y$
6. $y - x = 1$

Write the equation that results after making a substitution for one of the variables. (Step 1 of Example 1, Step 2 of Example 2)

7. $\begin{cases} y = 2x - 3 \\ x + y = 5 \end{cases}$
8. $\begin{cases} x = -3y + 4 \\ x - 2y = -1 \end{cases}$
9. $\begin{cases} -2x + 5y = 2 \\ 3y - x = -1 \end{cases}$

WRITTEN EXERCISES

Goal: To solve systems of linear equations by the substitution method

Sample Problem: $\begin{cases} x + y = 5 \\ 2x - y = -2 \end{cases}$ **Answer:** $\{(1, 4)\}$

Solve each system by the substitution method. Check. (Examples 1–3)

1. $\begin{cases} y = x - 4 \\ 2x - 5y = 2 \end{cases}$
2. $\begin{cases} y = -x + 2 \\ 2x - y = 1 \end{cases}$
3. $\begin{cases} x - y = 4 \\ 2x - 3y = -2 \end{cases}$
4. $\begin{cases} x - y = 3 \\ 5x + y = -15 \end{cases}$
5. $\begin{cases} x - 2y = -2 \\ 2x - 3y = 2 \end{cases}$
6. $\begin{cases} 2x + y = 5 \\ 8x - y = 45 \end{cases}$
7. $\begin{cases} y = -3x + 4 \\ x = -\frac{1}{3}y + 2 \end{cases}$
8. $\begin{cases} x = 2y - 1 \\ y = \frac{1}{2}x - 3 \end{cases}$
9. $\begin{cases} 4x - 8 = -y \\ 5x - 3 = -3y \end{cases}$
10. $\begin{cases} 3x - 16 = -2y \\ 7x - 19 = -y \end{cases}$
11. $\begin{cases} y = \frac{5}{2}x - 2 \\ 11x - 4y = 8 \end{cases}$
12. $\begin{cases} x = -\frac{3}{2}y + 2 \\ 2x + 4y = 4 \end{cases}$

15.2 Digit Problems

The ten numerals below are *digits*.

0, 1, 2, 3, 4, 5, 6, 7, 8, 9

Each place that a digit occupies in a whole number has a name. In this section you will work with two-digit numbers.

TEN'S DIGIT → 78 ← ONE'S DIGIT

Any two-digit number can be represented as shown below.

Let x = the ten's digit of a two-digit number.
Let y = the one's digit.

Two-digit number: $10x + y$

P–1 **Why is it incorrect to represent such a two-digit number as *xy*?**

EXAMPLE 1

The one's digit of a two-digit number is 4 more than the ten's digit (Condition 1). The sum of the digits is 14 (Condition 2). Find the one's and the ten's digits. Write the number.

Solution:

1. Use Condition 1 to represent the unknowns. → Let x = the ten's digit.
   Let y = the one's digit.
   Then $10x + y$ = the number.

2. Use Condition 2 and Condition 2 to write the equations.

Think: **The one's digit is 4 more than the ten's digit.**

Translate: $y = x + 4$ ◀ ***Condition 1***

Think: **The sum of the digits is 14.**

Translate: $x + y = 14$ ◀ ***Condition 2***

$$\begin{cases} \textbf{1.}\ y = x + 4 \\ \textbf{2.}\ x + y = 14 \end{cases}$$ ◀ ***System of linear equations***

The substitution method is used.

[3] Substitute $(x+4)$ for y in Equation **2.** Solve for x. ⟶

$$x + \mathbf{y} = 14$$
$$x + \mathbf{(x+4)} = 14$$
$$2x + 4 = 14$$
$$2x = 10$$
$$x = 5$$

◀ ***Ten's digit***

[4] Substitute 5 for x in Equation **1.** Solve for y. ⟶

$$y = \mathbf{x} + 4$$
$$y = \mathbf{5} + 4$$
$$y = 9$$

◀ ***One's digit***

[5] Write the two-digit number. ⟶

$$10\mathbf{x} + \mathbf{y} = 10\,\mathbf{(5)} + \mathbf{9}$$
$$= 50 + 9$$
$$= 59$$

Check: Does $5 + 9 = 14$? Yes ✓ Does $9 - 5 = 4$? Yes ✓

The two-digit number is 59.

EXAMPLE 2

In a two-digit number, the one's digit is 4 more than the ten's digit (Condition 1). The two-digit number is 2 more than five times the one's digit (Condition 2). Find the ten's and one's digits. Write the number.

Solution:

[1] Use Condition 1 to represent the unknowns. ⟶ Let x = the ten's digit. Let y = the one's digit. Then $10x + y$ = the number.

[2] Use Condition 1 and Condition 2 to write the equations.

Think: **The one's digit** **is** **4 more than the ten's digit.**

Translate: y $=$ $x + 4$

◀ ***Condition 1***

Think: **The number** **is** **2 more than five times the one's digit.**

Translate: $10x + y$ $=$ $5y + 2$

◀ ***Condition 2***

$$\begin{cases} \mathbf{1.}\ y = x + 4 \\ \mathbf{2.}\ 10x + y = 5y + 2 \end{cases}$$

The substitution method is used.

[3] Substitute $(x + 4)$ for y in Equation **2.** Solve for x. ⟶

$$10x + \mathbf{y} = 5\mathbf{y} + 2$$
$$10x + \mathbf{(x + 4)} = 5\mathbf{(x + 4)} + 2$$
$$11x + 4 = 5x + 20 + 2$$
$$6x + 4 = 22$$
$$6x = 18$$
$$x = 3 \quad \text{Ten's digit}$$

[4] Replace x by 3 in Equation **1.** Solve for y. ⟶

$$y = \mathbf{x} + 4$$
$$y = \mathbf{3} + 4$$
$$y = 7 \quad \text{One's digit}$$

Check: Does $7 = 3 + 4$? Yes ✓ Does $37 = 5(7) + 2$? Yes ✓

The number is 37.

CLASSROOM EXERCISES

Write an equation for each sentence. Represent each ten's digit by x and each one's digit by y. (Steps 1 and 2 of Examples 1 and 2)

1. The sum of the digits of a two-digit number is 12.

2. The ten's digit is 3 greater than the one's digit.

3. The one's digit is 5 less than the ten's digit.

4. The one's digit equals the ten's digit less 5.

5. Five times the ten's digit equals seven times the one's digit.

6. A two-digit number is 9 more than the sum of its digits.

7. The sum of the digits of a two-digit number is 1 less than twice the ten's digit.

8. A two-digit number is five times its one's digit.

9. The one's digit of a two-digit number diminished by the ten's digit equals -8.

10. Twice the one's digit of a two-digit number exceeds the ten's digit by 3.

WRITTEN EXERCISES

Goal: To solve digit problems by systems of linear equations

Sample Problem: The ten's digit of a two-digit number is 1 less than 5 times the unit's digit (Condition 1). The sum of the digits is 11 (Condition 2). Write the two-digit number.

Answer: The number is 92.

For Exercises 1–10, use Condition 1 and two variables to represent the unknowns. Use Condition 1 (underscored once) and Condition 2 (underscored twice) to write a system of equations for the problem. Solve. (Examples 1 and 2)

1. The one's digit of a two-digit number is 2 less than the ten's digit. The sum of the digits is 12. Write the two-digit number.
2. The one's digit of a two-digit number is 7 more than the ten's digit. The sum of the digits is 11. Write the two-digit number.
3. The ten's digit of a two-digit number is twice the one's digit. The ten's digit exceeds the one's digit by 4. Write the two-digit number.
4. The one's digit of a two-digit number is three times the ten's digit. The one's digit exceeds the ten's digit by 6. Write the number.
5. The sum of the digits of a two-digit number is 15. Twice the ten's digit is 5 less than three times the one's digit. Write the two-digit number.
6. The ten's digit of a two-digit number is 2 less than the one's digit. The one's digit is 4 less than twice the ten's digit. Write the number.
7. The one's digit of a two-digit number exceeds the ten's digit by 1. The number is two more than 8 times the one's digit. Write the two-digit number.
8. Three times the one's digit of a two-digit number is 2 more than the ten's digit. The number is 3 more than 10 times the ten's digit. Write the two-digit number.
9. The ten's digit of a two-digit number decreased by the one's digit equals -4. The number is 3 more than twice the sum of its digits. Write the two-digit number.
10. The one's digit of a two-digit number decreased by the ten's digit equals -5. The number is 8 times the sum of its digits. Write the two-digit number.

REVIEW CAPSULE FOR SECTION 15.3

Solve and check. (Pages 71–73 and 100–102)

1. $0.15x - 3.8 = 3.4$
2. $0.08x + 0.12(200 - x) = 0.10(200)$
3. $0.18x + 0.15(450 - x) = 0.16(450)$
4. $0.06(x - 300) + 0.015(300) = 0.03x$

15.3 Mixture Problems

Per cents are often used to express the strength of a solution or the relative amount of a specific ingredient in a solution or mixture.

a. 18 grams (g) of salt in 100 grams of a solution of salt and water is called an 18% solution $\left(\frac{18}{100} = 18\%\right)$.

b. The amount of salt in 320 grams of a 15% solution of salt and water is 15% of 320 = 0.15 (320), or 48 grams.

P–1 **How much sugar is in 80 grams of a 25% solution of sugar and water?**

Making a table will help you to organize the information and to identify Condition 1 and Condition 2.

EXAMPLE

A 10% salt solution and an 18% salt solution are mixed to obtain 320 grams of a 15% salt solution (Condition 1). How many grams of the 10% solution and of the 18% solution are needed (Condition 2)?

Solution:

1 Make a table.

Solution	Grams of Solution (Condition 1)	Grams of Salt (Condition 2)
10%	x	$0.10x$
18%	y	$0.18y$
15%	320	0.15(320)

2 Use Condition 1 and Condition 2 to write the equations.

Think: **Grams of 10% solution plus grams of 18% solution equal 320 grams.**

Translate: $x + y = 320$ ◀ **Condition 1**

Think: **Salt (g) in 10% solution plus Salt (g) in 18% solution equals Salt (g) in 15% solution**

Translate: $0.10x + 0.18y = 0.15(320)$ ◀ **Condition 2**

$$\begin{cases} \textbf{1.}\ x + y = 320 \\ \textbf{2.}\ 0.10x + 0.18y = 0.15(320) \end{cases}$$

◀ ***System of linear equations***

The substitution method is used.

[3] Solve for y in Equation **1.** ⟶ $y = 320 - x$

[4] Substitute $(320 - x)$ for y in Equation **2.** Solve for x. ⟶

$$0.10x + 0.18\mathbf{y} = 0.15(320)$$
$$0.10x + 0.18\mathbf{(320 - x)} = 0.15(320)$$
$$0.10x + 57.6 - 0.18x = 48$$
$$57.6 - 0.08x = 48$$
$$-0.08x = -9.6$$
$$\frac{-0.08x}{-0.08} = \frac{-9.6}{-0.08}$$
$$x = 120$$

◀ ***Number of grams of 10% solution***

[5] Substitute 120 for x in Equation **1.** Solve for y. ⟶

$$\mathbf{x} + y = 320$$
$$\mathbf{120} + y = 320$$
$$y = 200$$

◀ ***Number of grams of 18% solution***

Check: Does $120 + 200 = 320$? Yes ✓
Does $0.10(120) + 0.18(200) = 0.15(320)$? Yes ✓

Thus, 120 grams of the 10% solution and 200 grams of the 18% solution are needed.

CLASSROOM EXERCISES

Write the numeral missing from each row of the table for salt solutions. (Example)

	Per Cent Solution	Amount of Mixture	Amount of Salt
1.	8%	100 grams	?
2.	15%	40 grams	?
3.	4%	350 milligrams	?
4.	12%	x milligrams	?
5.	6%	y milligrams	?
6.	18%	$(x + 20)$ decigrams	?
7.	10%	?	12 grams
8.	20%	?	8 grams

Write the numeral or numerals missing from each row of the table for acid solutions. (Example)

	Per Cent Solution	Amount of Mixture	Amount of Acid	Amount of Water
9.	3%	500 grams	?	?
10.	10%	?	80 milligrams	?
11.	25%	?	10 grams	?
12.	20%	*x* decigrams	?	?

WRITTEN EXERCISES

Goal: To solve mixture problems by systems of linear equations

Sample Problem: A 30% salt solution and a 60% salt solution are mixed to obtain 60 grams of a 50% solution (Condition 1). How many grams of the 30% solution and of the 60% solution are needed (Condition 2)?

Answer: 20 grams of the 30% solution and 40 grams of the 60% solution

In Exercises 1–8, Condition 1 is underscored once. Condition 2 is underscored twice. (Example)

Use Condition 1 and Condition 2 to complete each table. Use Condition 1 and Condition 2 to write a system of equations for each problem. Solve.

1. A 4% acid solution and an 8% acid solution are mixed to obtain 400 grams of a 5% acid solution. How many grams of the 4% solution and of the 8% solution are needed?

Solution	Amount of Solution (Condition 1)	Amount of Pure Acid (Condition 2)
4%	x	$0.04x$
8%	y	?
5%	400	?

2. A 12% salt solution and a 20% salt solution are mixed to obtain 24 kilograms of a 15% salt solution. How many kilograms of the 12% solution and of the 20% solution are needed? (See the table at the top of page 408.)

Solution	Amount of Solution (Condition 1)	Amount of Salt (Condition 2)
12%	x	$0.12x$
20%	y	?
15%	24	?

3. A 12% salt solution is mixed with 48 grams of a 20% salt solution to obtain a 15% salt solution. Find the number of grams of the 12% solution and of the 15% solution.

Solution	Amount of Solution (Condition 1)	Amount of Salt (Condition 2)
12%	x	?
20%	48	?
?	y	?

4. An 8% acid solution is mixed with 200 grams of a 4% acid solution to obtain a $5\frac{1}{2}$% acid solution. Find the number of grams of the 8% solution and of the $5\frac{1}{2}$% solution.

Solution	Amount of Solution (Condition 1)	Amount of Pure Acid (Condition 2)
8%	x	?
4%	200	?
?	y	?

Use Condition 1, Condition 2, and two variables to make a table for each problem. Use Condition 1 and Condition 2 to write a system of equations for each problem. Solve.

5. A 15% vinegar solution is mixed with a 30% vinegar solution to obtain 150 grams of an 18% vinegar solution. How many grams of the 15% solution and of the 30% solution are needed?

6. A 4% iodine solution is mixed with a 12% iodine solution to obtain 300 grams of a 6% iodine solution. How many grams of the 4% solution and of the 12% solution are needed?

7. A $2\frac{1}{2}$% acid solution is mixed with 250 grams of a 10% acid solution to obtain a 4% acid solution. Find the number of grams of the $2\frac{1}{2}$% solution and of the 4% solution.

8. An 18% antifreeze solution is mixed with 200 grams of a 30% antifreeze solution to obtain a 22% antifreeze solution. Find the number of grams of the 18% solution and of the 22% solution.

NON-ROUTINE PROBLEMS

9. A woman was born in the first half of the nineteenth century. She was x years old in the year x^2. In what year was the woman born?

10. In a group of cows and chickens, the number of legs was 14 more than twice the number of heads. Find the number of cows.

11. Four friends, Luis, Maria, Carol, and Glenn, are sitting in a row. Neither Maria nor Carol are sitting next to Luis. Maria is not sitting next to Glenn. Carol is sitting just to the right of Maria. Write the seating arrangement from left to right.

12. Look at the figure below. Name two different things that you see.

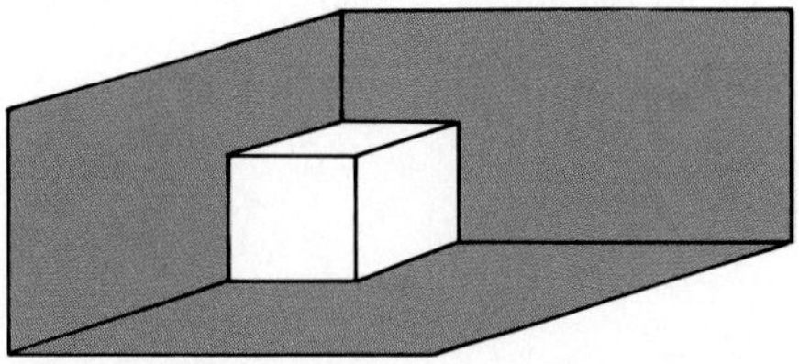

MID-CHAPTER REVIEW

Solve each system by the substitution method. (Section 15.1)

1. $\begin{cases} y = 2x - 3 \\ x - 2y = 9 \end{cases}$

2. $\begin{cases} x = -3y + 1 \\ 2x + 3y = 5 \end{cases}$

3. $\begin{cases} 4x - y = 3 \\ 2x - y = -3 \end{cases}$

For Exercises 4–7, Condition 1 is underscored once. Condition 2 is underscored twice.

Use Condition 1 and two variables to represent the unknowns for each problem. Use Condition 1 and Condition 2 to write a system of equations for each problem. Solve. (Section 15.2)

4. A two-digit number is 7 less than 4 times its one's digit. The one's digit is 1 more than 4 times the ten's digit. Write the two-digit number.

5. The ten's digit of a two-digit number is 2 more than twice the one's digit. The two-digit number is 11 more than 9 times its ten's digit. Write the two-digit number.

Use Condition 1, Condition 2, and two variables to set up a table for each problem. Use Condition 1 and Condition 2 to write a system of equations for each problem. Solve. (Section 15.3)

6. A 7% salt solution is mixed with a 15% salt solution to obtain 400 grams of a 12% salt solution. How many grams of the 7% solution and of the 15% solution are needed?

7. A 2% acid solution is mixed with a 3% acid solution to obtain 800 grams of a $2\frac{1}{2}$% acid solution. How many grams of the 2% solution and of the 3% solution are needed?

Making Tables

Rate of Work

Rate of work on a job is the fraction of the job that can be completed in one unit of time.

$\frac{1}{5}$ is the rate of work on a job requiring 5 hours.

$\frac{1}{5}$ means "one-fifth of the job per hour."

In two hours, $2(\frac{1}{5})$ or $\frac{2}{5}$ of the job can be completed.

In three hours, $3(\frac{1}{5})$ or $\frac{3}{5}$ of the job can be completed.

$\frac{2}{5} + \frac{3}{5} = 1$ ◀ ***The entire job is represented by 1.***

EXAMPLE 1: Two architects are working on the plans for a building. Jane estimates she could prepare the plans by herself in 100 hours. Melanie estimates she would need 120 hours for the entire job working alone. How many hours would the two need to prepare the plans working together? Round to the nearest whole hour.

SOLUTION:

Architect	*Number of Hours Together on Job*	*Rate of Work*	*Part of Job Done*
Jane	x	$\frac{1}{100}$	$x(\frac{1}{100})$, or $\frac{x}{100}$
Melanie	x	$\frac{1}{120}$	$x(\frac{1}{120})$, or $\frac{x}{120}$

Write an equation. ⟶ $\frac{x}{100} + \frac{x}{120} = 1$ ◀ ***The entire job is represented by 1.***

Solve. ⟶ $600(\frac{x}{100} + \frac{x}{120}) = 600(1)$

$$6x + 5x = 600$$

$$11x = 600$$

$x = 54\frac{6}{11}$, or about **55 hours**

EXAMPLE 2: A diesel–powered pump can drain a lake in 24 hours. After operating for 10 hours, the pump broke down. A smaller electric–powered pump, capable of doing the whole job in 36 hours, was substituted. How long did it take the smaller pump to complete the job?

SOLUTION:

Pump	Number of Hours on Job	Rate of Work	Part of Job Done
Diesel	10	$\frac{1}{24}$	$10(\frac{1}{24})$, or $\frac{10}{24}$
Electric	x	$\frac{1}{36}$	$x(\frac{1}{36})$, or $\frac{x}{36}$

Write an equation. ⟶ $\frac{10}{24} + \frac{x}{36} = 1$

Solve. ⟶ $72(\frac{10}{24} + \frac{x}{36}) = 72(1)$

$$30 + 2x = 72$$

$$2x = 42$$

$$x = \textbf{21 hours}$$

EXERCISES

1. An experienced roofer can roof a building alone in 20 hours. An apprentice roofer would need 30 hours working alone. How long would it take the two working together to do the job?

2. Two printers are used to print mailing labels. The faster printer could do the job alone in 15 hours. The slower printer would require 35 hours working alone. How long would it take for both printers together to do the job?

3. It takes 1000 hours for a worker to enter names and addresses onto a computer disk. After 450 hours this worker was replaced by a second worker who could do the entire job in 800 hours. How long did the second worker take to finish the job?

4. A gravity drain can drain the city pool in 5 hours. A pump could drain the pool in 3 hours. Both methods were used together for one hour until the pump broke. How many hours were needed for the gravity drain to complete the job?

15.4 2 × 2 Determinants

Another method for solving systems of equations uses determinants.

Definition

A ***determinant*** is a square array of the form $\begin{vmatrix} a & b \\ c & d \end{vmatrix}$ in which a, b, c, and d are real numbers.

$$\begin{vmatrix} -3 & \frac{1}{2} \\ 5 & \sqrt{2} \end{vmatrix}$$

Only 2 × 2 (two-by-two) determinants are used in this book. They have two rows and two columns.

ROWS → $\begin{vmatrix} a & b \\ c & d \end{vmatrix}$

COLUMNS

Each real number is an element of the determinant.

P–1 **What are the elements of the first row of the determinant at the right? the elements of the second column?**

$$\begin{vmatrix} \pi & -2 \\ 0.6 & -\sqrt{3} \end{vmatrix}$$

Each 2 × 2 determinant represents a real number.

$$\begin{vmatrix} a & b \\ c & d \end{vmatrix} = ad - bc$$

$$\begin{vmatrix} 2 & 4 \\ 1 & 5 \end{vmatrix} = 2 \cdot 5 - 4 \cdot 1$$
$$= 6$$

EXAMPLE 1 Compute the value of $\begin{vmatrix} 2 & 4 \\ -3 & 1 \end{vmatrix}$.

Solution:

$$\begin{vmatrix} 2 & 4 \\ -3 & 1 \end{vmatrix} = (2)(1) - (4)(-3)$$
$$= 2 - (-12)$$
$$= 2 + 12$$
$$= 14$$

P-2 **What is the value of each determinant below?**

a. $\begin{vmatrix} 2 & 5 \\ 3 & 0 \end{vmatrix}$ **b.** $\begin{vmatrix} 2 & 3 \\ 2 & 3 \end{vmatrix}$ **c.** $\begin{vmatrix} 1 & 1 \\ -1 & 1 \end{vmatrix}$

P-3 **What expression represents the value of each determinant?**

a. $\begin{vmatrix} p & q \\ 3 & 2 \end{vmatrix}$ **b.** $\begin{vmatrix} -x & x \\ 3 & 2 \end{vmatrix}$ **c.** $\begin{vmatrix} -s & t \\ -t & -s \end{vmatrix}$

The equations of the system must be written in ***standard form.***

$$\begin{cases} 2x - 3y = -1 \\ 4x + 3y = 16 \end{cases}$$ ***Standard form***

The ***determinant of a system of equations*** is formed with the coefficients of the variables as elements of the determinant.

Standard form $\begin{cases} 2x - 3y = -1 \\ 4x + 3y = 16 \end{cases}$ $\begin{vmatrix} 2 & -3 \\ 4 & 3 \end{vmatrix}$ ***Determinant of the system***

EXAMPLE 2 Write the determinant of the system. Then compute its value.

$$\begin{cases} y - 5 = -3x \\ -2x + y = -3 \end{cases}$$

Solution: Write the equations in standard form. ⟶ $\begin{cases} 3x + y = 5 \\ -2x + y = -3 \end{cases}$

$$\begin{vmatrix} 3 & 1 \\ -2 & 1 \end{vmatrix} = (3)(1) - (1)(-2)$$ ***Determinant of the system***

$$= 3 - (-2)$$

$$= 3 + 2$$

$$= 5$$ ***Value of the determinant***

CLASSROOM EXERCISES

Write the value of each determinant. (Example 1)

1. $\begin{vmatrix} 1 & 2 \\ 1 & 4 \end{vmatrix}$
2. $\begin{vmatrix} -2 & 0 \\ 1 & 3 \end{vmatrix}$
3. $\begin{vmatrix} 2 & 3 \\ -1 & 5 \end{vmatrix}$
4. $\begin{vmatrix} -1 & 1 \\ 0 & 1 \end{vmatrix}$
5. $\begin{vmatrix} 0 & 3 \\ 1 & 2 \end{vmatrix}$
6. $\begin{vmatrix} 5 & -1 \\ 2 & -3 \end{vmatrix}$
7. $\begin{vmatrix} 1 & -3 \\ 5 & 2 \end{vmatrix}$
8. $\begin{vmatrix} 1 & -3 \\ 0 & 2 \end{vmatrix}$
9. $\begin{vmatrix} 2 & -3 \\ -1 & 5 \end{vmatrix}$
10. $\begin{vmatrix} -1 & -4 \\ -3 & -2 \end{vmatrix}$
11. $\begin{vmatrix} -5 & 1 \\ 3 & 2 \end{vmatrix}$
12. $\begin{vmatrix} 3 & 6 \\ 2 & 4 \end{vmatrix}$
13. $\begin{vmatrix} -3 & -4 \\ -6 & 8 \end{vmatrix}$
14. $\begin{vmatrix} 2 & -3 \\ 2 & -3 \end{vmatrix}$
15. $\begin{vmatrix} 1 & -5 \\ -1 & 5 \end{vmatrix}$
16. $\begin{vmatrix} \frac{1}{2} & -3 \\ 2 & 4 \end{vmatrix}$
17. $\begin{vmatrix} 1 & 6 \\ -\frac{1}{2} & 3 \end{vmatrix}$
18. $\begin{vmatrix} 1 & 1 \\ -1 & 1 \end{vmatrix}$
19. $\begin{vmatrix} \frac{1}{2} & \frac{1}{2} \\ 2 & -4 \end{vmatrix}$
20. $\begin{vmatrix} -5 & -3 \\ -1 & -2 \end{vmatrix}$

WRITTEN EXERCISES

Goals: To write the determinant of a linear system and to compute the value of the determinant

Sample Problem: $\begin{cases} 2x - 3y = -6 \\ -x + 5y = 1 \end{cases}$ **Answer:** $\begin{vmatrix} 2 & -3 \\ -1 & 5 \end{vmatrix}$: 7

Compute the value of each determinant. (Example 1)

1. $\begin{vmatrix} -6 & 5 \\ 2 & 3 \end{vmatrix}$
2. $\begin{vmatrix} 5 & 6 \\ 2 & -7 \end{vmatrix}$
3. $\begin{vmatrix} 6 & -12 \\ -3 & 6 \end{vmatrix}$
4. $\begin{vmatrix} -4 & 16 \\ 2 & -8 \end{vmatrix}$
5. $\begin{vmatrix} 1 & 5 \\ 0 & -2 \end{vmatrix}$
6. $\begin{vmatrix} 2 & 0 \\ -10 & -3 \end{vmatrix}$
7. $\begin{vmatrix} \frac{1}{3} & -4 \\ \frac{1}{2} & 6 \end{vmatrix}$
8. $\begin{vmatrix} -\frac{1}{2} & -\frac{1}{3} \\ 9 & 8 \end{vmatrix}$
9. $\begin{vmatrix} -1 & 1 \\ 1 & -1 \end{vmatrix}$
10. $\begin{vmatrix} 5 & 3 \\ 9 & 3 \end{vmatrix}$
11. $\begin{vmatrix} 1 & -15 \\ -15 & 2 \end{vmatrix}$
12. $\begin{vmatrix} -4 & -3 \\ 3 & -4 \end{vmatrix}$

Write the determinant of the system. Then compute its value. (Example 2)

13. $\begin{cases} 2x - y = -3 \\ x - 3y = 5 \end{cases}$
14. $\begin{cases} x - 3y = -5 \\ 5x + 2y = 3 \end{cases}$
15. $\begin{cases} -2x + 5y = 1 \\ x - 3y = -8 \end{cases}$
16. $\begin{cases} 4x - 3y = -2 \\ -3x - y = 5 \end{cases}$
17. $\begin{cases} y = 2x - 5 \\ y = 5x - 2 \end{cases}$
18. $\begin{cases} y = -3x + 7 \\ y = x - 13 \end{cases}$
19. $\begin{cases} 5x - 2 = -3y \\ y - 3x = -2 \end{cases}$
20. $\begin{cases} -3y + 10 = 5x \\ 2x - 8 = -3y \end{cases}$
21. $\begin{cases} 5y - 7 = -4x \\ 5x + 6y = 8 \end{cases}$

15.5 Solving Systems by Determinants

Systems of equations can be solved by the method of determinants.

$$\begin{cases} 2x - y = 5 \\ x + 3y = -2 \end{cases}$$

$$x = \frac{\begin{vmatrix} 5 & -1 \\ -2 & 3 \end{vmatrix}}{\begin{vmatrix} 2 & -1 \\ 1 & 3 \end{vmatrix}}$$

First column has constant terms. Second column has y coefficients.

Determinant of the system

$$y = \frac{\begin{vmatrix} 2 & 5 \\ 1 & -2 \end{vmatrix}}{\begin{vmatrix} 2 & -1 \\ 1 & 3 \end{vmatrix}}$$

First column has x coefficients. Second column has constant terms.

Determinant of the system

P–1 **What is the value of the numerator determinant for x? the numerator determinant for y? the determinant of the system?**

The values of x and y are computed.

$$x = \frac{13}{7} \qquad y = \frac{-9}{7}$$

The solution set of the system is $\{(\frac{13}{7}, -\frac{9}{7})\}$.

EXAMPLE 1 Solve the following system by the method of determinants.

$$\begin{cases} x + 2 = 3y \\ 5y - 2x = 10 \end{cases}$$

Solution:

1. Write the equations in standard form. ⟶ $\begin{cases} x - 3y = -2 \\ -2x + 5y = 10 \end{cases}$

2. Write x and y as quotients of determinants.

$$x = \frac{\begin{vmatrix} -2 & -3 \\ 10 & 5 \end{vmatrix}}{\begin{vmatrix} 1 & -3 \\ -2 & 5 \end{vmatrix}} \qquad y = \frac{\begin{vmatrix} 1 & -2 \\ -2 & 10 \end{vmatrix}}{\begin{vmatrix} 1 & -3 \\ -2 & 5 \end{vmatrix}}$$

3 Solve. ⟶ $x = \dfrac{(-2)(5) - (-3)(10)}{(1)(5) - (-3)(-2)}$ $y = \dfrac{(1)(10) - (-2)(-2)}{(1)(5) - (-3)(-2)}$

$x = \dfrac{20}{-1}$ $y = \dfrac{6}{-1}$

$x = -20$ $y = -6$

The solution set of the system is $\{(-20, -6)\}$. If graphs of the equations are drawn, the lines intersect at the point $(-20, -6)$.

EXAMPLE 2

Solve the following system by the method of determinants.

$$\begin{cases} x - 2y = 5 \\ 2x - 4y = -3 \end{cases}$$

Standard form

Solution:

$$x = \frac{\begin{vmatrix} 5 & -2 \\ -3 & -4 \end{vmatrix}}{\begin{vmatrix} 1 & -2 \\ 2 & -4 \end{vmatrix}} \qquad y = \frac{\begin{vmatrix} 1 & 5 \\ 2 & -3 \end{vmatrix}}{\begin{vmatrix} 1 & -2 \\ 2 & -4 \end{vmatrix}}$$

$$x = \frac{(5)(-4) - (-2)(-3)}{(1)(-4) - (-2)(2)} \qquad y = \frac{(1)(-3) - (5)(2)}{(1)(-4) - (-2)(2)}$$

$$x = \frac{-26}{0} \qquad y = \frac{-13}{0}$$

Division by zero is not possible.

The two equations are changed to slope-intercept form below.

$x - 2y = 5$ $2x - 4y = -3$

$y = \frac{1}{2}x - \frac{5}{2}$ $y = \frac{2}{4}x + \frac{3}{4}$

The slopes are equal.

The graphs of the two equations are parallel lines.
The solution set of the system is ϕ.

Example 2 suggests the following.

> If only the determinant of a system of equations equals 0, then the solution set of the system is empty.

CLASSROOM EXERCISES

Write the missing columns of elements. (Step 2 of Example 1)

1. $\begin{cases} 3x - y = 2 \\ x + 2y = -1 \end{cases}$

$$x = \frac{\begin{vmatrix} ? & -1 \\ ? & 2 \end{vmatrix}}{\begin{vmatrix} 3 & -1 \\ 1 & 2 \end{vmatrix}}, \quad y = \frac{\begin{vmatrix} 3 & ? \\ 1 & ? \end{vmatrix}}{\begin{vmatrix} 3 & -1 \\ 1 & 2 \end{vmatrix}}$$

2. $\begin{cases} -x + 4y = 6 \\ 2x - 3y = -1 \end{cases}$

$$x = \frac{\begin{vmatrix} ? & 4 \\ ? & -3 \end{vmatrix}}{\begin{vmatrix} -1 & ? \\ 2 & ? \end{vmatrix}}, \quad y = \frac{\begin{vmatrix} -1 & ? \\ 2 & ? \end{vmatrix}}{\begin{vmatrix} ? & 4 \\ ? & -3 \end{vmatrix}}$$

For each system, write x and y as quotients of determinants. (Step 2 of Example 1)

3. $\begin{cases} 0.5x - 0.5y = 1.5 \\ -1.5x + 1.5y = -2.5 \end{cases}$

4. $\begin{cases} x - 2y = -1 \\ 2x - y = 2 \end{cases}$

WRITTEN EXERCISES

Goal: To solve systems of linear equations by the method of determinants

Sample Problem: $\begin{cases} 3x + 7y = -4 \\ 2x + 5y = -3 \end{cases}$ **Answer:** $\{(1, -1)\}$

Solve each system by the method of determinants. (Examples 1 and 2)

1. $\begin{cases} x + y = 6 \\ x - 2y = 3 \end{cases}$

2. $\begin{cases} 2x - y = 1 \\ x + y = -2 \end{cases}$

3. $\begin{cases} x - 3y = -5 \\ 2x + 5y = 1 \end{cases}$

4. $\begin{cases} 2x - y = 3 \\ x + 3y = 5 \end{cases}$

5. $\begin{cases} 4x + 3y = -6 \\ 5x - 2y = 3 \end{cases}$

6. $\begin{cases} 2x - 5y = 10 \\ x - 6y = 8 \end{cases}$

7. $\begin{cases} x + 5 = 3y \\ 5x = 3y + 2 \end{cases}$

8. $\begin{cases} -2y + 3x = 5 \\ 5y - x = -7 \end{cases}$

9. $\begin{cases} \frac{1}{3}x - 8y = -4 \\ \frac{3}{4}x + 6y = 3 \end{cases}$

10. $\begin{cases} x - 3y = -5 \\ -2x + 6y = 6 \end{cases}$

11. $\begin{cases} 2x - y = -3 \\ -6x + 3y = 6 \end{cases}$

12. $\begin{cases} -\frac{1}{4}x + 6y = -6 \\ \frac{1}{2}x - 4y = 12 \end{cases}$

REVIEW CAPSULE FOR SECTION 15.6

Write an algebraic expression. (Pages 74–77)

1. The value of *n* nickels in cents
2. The value of *t* dollars in cents
3. The cost of 5 records at *w* cents each
4. The cost of 12 pencils at *y* cents each
5. The cost in cents of *k* tickets at $8.50 each
6. The cost in cents of *n* cards at 75 cents each

PROBLEM SOLVING AND APPLICATIONS

15.6 Money Problems

Word problems that involve the cost of two items can often be solved by systems of two linear equations. One equation usually represents the total number of items (Condition 1). The second equation represents the total cost of the items (Condition 2).

EXAMPLE 1

Jane bought a total of 11 pencils and pens (Condition 1). The pencils cost 8 cents each and the pens cost 39 cents each. She paid a total of $1.50 for the pencils and pens (Condition 2). How many of each did she buy?

Solution: [1] Make a table to help you write the equations. Write all costs in cents.

	Number of Items (Condition 1)	**Price Per Item**	**Total Cost (Condition 2)**
Pencil	x	8¢	$8x$
Pen	y	39¢	$39x$
Total	11		150

[2] Equations: $\begin{cases} \textbf{1.}\ x + y = 11 \\ \textbf{2.}\ 8x + 39y = 150 \end{cases}$

1. ***Condition 1***
2. ***Condition 2***

[3] Solve. The method of determinants is shown. Other methods could also be used.

$$x = \frac{\begin{vmatrix} 11 & 1 \\ 150 & 39 \end{vmatrix}}{\begin{vmatrix} 1 & 1 \\ 8 & 39 \end{vmatrix}} = \frac{429 - 150}{39 - 8} = \frac{279}{31} = 9$$

$$y = \frac{\begin{vmatrix} 1 & 11 \\ 8 & 150 \end{vmatrix}}{31} = \frac{150 - 88}{31} = \frac{62}{31} = 2$$

[4] **Check:** Condition 1 Did Jane buy a total of 11 pencils and pens?
Does $9 + 2 = 11$? Yes ✓

Condition 2 Was the total cost $1.50, or 150 cents?
Does $9(8) + 2(39) = 150$? Yes ✓

Jane bought 9 pencils and 2 pens.

EXAMPLE 2

A school's Ski Club is planning a trip. A small bus holds 6 fewer passengers than a large bus (Condition 1). A large bus will cost \$18 per student if it is filled and a small bus will cost \$20 per student. The difference between the cost of the large bus and the small bus is \$30 (Condition 2). Find the number of passengers that each bus can carry.

Solution: [1]

	Number of Passengers (Condition 1)	Cost Per Passenger	Total Cost (Condition 2)
Large bus	x	\$18	$18x$
Small bus	y	\$20	$20y$

[2] Write the equations.

Think: **The small bus carries 6 fewer than the large bus.**

Translate: $y \quad = \quad x - 6$

Think: **Cost of large bus minus cost of small bus is \$30.**

Translate: $18x \quad - \quad 20y \quad = \quad 30$

$$\begin{cases} \mathbf{1.}\ y = x - 6 \\ \mathbf{2.}\ 18x - 20y = 30 \end{cases}$$

1. Condition 1
2. Condition 2

The substitution method is used.

[3] Substitute $(x - 6)$ for y in Equation **2.** Solve for x.

$$18x - 20\mathbf{y} = 30$$
$$18x - 20\mathbf{(x - 6)} = 30$$
$$18x - 20x + 120 = 30$$
$$-2x + 120 = 30$$
$$-2x = -90$$
$$x = 45$$

Capacity of large bus

[4] Substitute 45 for x in Equation **1.** Solve for y.

$$y = \mathbf{x} - 6$$
$$y = \mathbf{45} - 6$$
$$y = 39$$

Capacity of small bus

The check is left for you.

CLASSROOM EXERCISES

Complete the table. Then use the table to write two equations for the problem. (Examples 1 and 2)

1. Mike takes 60 coins consisting of x nickels and y dimes to the bank. The total deposit is $6.50.

	Number of Coins	Value of Each	Total Value
Nickels	x	5¢	?
Dimes	y	10¢	?
Total	60		?

2. The Junior Class sold x boxes of pecans at $6 each and y boxes of peanuts at $5 each. They sold a total of 82 boxes for $440.

	Number of Boxes	Cost Per Box	Total Cost
Pecans	x	$6	?
Peanuts	y	?	?
Total	?		440

3. During a special sale, an appliance store sold x color TV sets for $325 each and y black and white TV sets for $135 each. The total receipts from the sale were $11,415 for selling 55 sets.

	Number of Sets	Price Per Set	Total Receipts
Color	?	$325	?
Black and White	?	$135	?
Total	?		11,415

4. Gina bought x cans of peas for 85 cents each and y cans of corn for 78 cents each. She bought 2 more cans of peas than cans of corn. The total cost was $6.59.

	Number of Cans	Price Per Can	Total Cost
Peas	?	?	?
Corn	?	78¢	?

5. At a recent bake sale, the Booster Club sold x cookies at 15 cents each and y cupcakes at 25 cents each. There were 20 more cookies sold than cupcakes. The total receipts from the sale were $15.00.

	Number of Item	Cost Per Item	Total Cost
Cookies	?	15¢	?
Cupcakes	?	?	?

WRITTEN EXERCISES

Goal: To solve problems involving costs by systems of linear equations

Sample Problem: Allyson has a total of 20 dimes and quarters (Condition 1). The total value of the coins is $3.95 (Condition 2). How many of each coin does she have?

Answer: She has 7 dimes and 13 quarters.

For Exercises 1–8, Condition 1 is underscored once. Condition 2 is underscored twice.

a. *Use Condition 1, Condition 2, and two variables to make a table.*

b. *Use Condition 1 and Condition 2 to write a system of equations for the problem.*

c. *Solve.* (Examples 1 and 2)

1. Brad Whitefoot bought a total of 9 records and cassette tapes. The records cost $6 each, and the cassette tapes cost $7 each. He paid a total of $58. How many of each did he buy?

2. A school lunch stand sold 230 hamburgers and hot dogs in one day for total receipts of $176.50. The hamburgers sold for 80 cents each and the hot dogs for 65 cents each. How many of each were sold?

3. Marc's scout troop bought 50 cans of beans and tomatoes for a camping trip at a total cost of $20.30. The tomatoes cost 35 cents per can and the beans 45 cents per can. How many cans of each kind were purchased?

4. Michelle sold two types of calendars for her club. She sold a total of 64 calendars for $140.50. One type of calendar sold for $2.50 each, and the other type sold for $1.75 each. How many calendars of each type did she sell?

5. Raphael deposits a total of 86 five- and ten-dollar bills. The amount of the deposit is $675. How many of each kind of bill are there?

6. Betty has 53 nickels and dimes in her bank. The total value of the coins is $3.75. How many of each type of coin does she have?

7. Susie won the ticket selling contest for the school carnival. She sold 82 tickets for total sales of $32.45. The student tickets sold for 35 cents and the adult tickets for 50 cents. How many tickets of each kind did she sell?

8. The Northeast Riders 4H Club sold orange juice and cocoa at their horse show. The orange juice cost 50 cents per glass, and the cocoa cost 75 cents per cup. They sold 20 more cups of cocoa than glasses of orange juice. How much of each did they sell if the total receipts were $45.00?

15.7 Distance/Rate/Time

The basic formula $d = rt$ relates distance, rate, and time.

EXAMPLE 1

Alexander travels a distance of 360 kilometers. The speed of the first part of his trip by plane is 240 kilometers per hour. He completes the journey by car at a speed of 80 kilometers per hour (Condition 2). The total time for the trip is 3 hours (Condition 1). How long does he travel by plane and how long by car?

Solution: [1]

	Time (Condition 1)	Rate	Distance (Condition 2)
By plane	x hrs	240 km/hr	$240x$ km
By car	y hrs	80 km/hr	$80y$ km
Total	3 hrs		360 km

[2] $\begin{cases} \textbf{1.}\ x + y = 3 \\ \textbf{2.}\ 240x + 80y = 360 \end{cases}$ ◀ ***Use the substitution method to solve.***

[3] Solve for y in Equation **1.** ⟶ $y = -x + 3$

[4] Substitute $(-x + 3)$ for y in Equation **2.** Solve for x.

$$240x + 80\mathbf{y} = 360$$
$$240x + 80(\mathbf{-x + 3}) = 360$$
$$240x - 80x + 240 = 360$$
$$160x = 120$$
$$x = \frac{120}{160}$$
$$x = \frac{3}{4}$$ ◀ ***Number of hours by plane***

[5] Substitute $\frac{3}{4}$ for x in Equation **1.** Solve for y.

$$\mathbf{x} + y = 3$$
$$\tfrac{3}{4} + y = 3$$
$$y = 2\tfrac{1}{4}$$ ◀ ***Number of hours by car***

The check is left for you.

EXAMPLE 2

Hisako swims 200 meters against a current in 6 minutes (Condition 1). She returns in 3 minutes swimming with the current (Condition 2). What is the speed of the current and the swimmer's speed in still water?

Solution:

[1] Let $x =$ the speed of the swimmer in still water (meters per minute).
Let $y =$ the speed of the current.

	Time	Rate	Distance
Against current	6 minutes	$x - y$	200 meters
With current	3 minutes	$x + y$	200 meters

[2] $\begin{cases} \textbf{1.}\ 6(x - y) = 200 \\ \textbf{2.}\ 3(x + y) = 200 \end{cases}$ ◀ *1. Condition 1* *2. Condition 2*

[3] Distributive Property **1.** $6(x - y) = 200 \longrightarrow 6x - 6y = 200$
2. $3(x + y) = 200 \longrightarrow 3x + 3y = 200$

[4] Solve. The method of determinants is used.

$$x = \frac{\begin{vmatrix} 200 & -6 \\ 200 & 3 \end{vmatrix}}{\begin{vmatrix} 6 & -6 \\ 3 & 3 \end{vmatrix}} = \frac{600 - (-1200)}{18 - (-18)} = \frac{1800}{36} = 50$$

$$y = \frac{\begin{vmatrix} 6 & 200 \\ 3 & 200 \end{vmatrix}}{36} = \frac{1200 - 600}{36} = \frac{600}{36} = 16\tfrac{2}{3}$$

The speed of the swimmer in still water is 50 meters per minute. The speed of the current is approximately 16.7 meters per minute.

CLASSROOM EXERCISES

Write an algebraic expression for each word expression. (Examples 1 and 2)

1. The distance traveled in x seconds by an object at a speed of 20 centimeters per second
2. The distance traveled in 5 minutes by an object at a speed of y meters per minute
3. The distance traveled in 10 hours by a ship whose speed with a current is $(x + y)$ miles per hour
4. The distance traveled in 6 hours by a ship whose speed against a current is $(x - y)$ miles per hour

5. The distance between two cars 3 hours after they drive off in opposite directions, one at a speed of x kilometers per hour and the other at y kilometers per hour

6. The distance between two planes 2 hours after they take off in opposite directions, one at a speed x kilometers per hour and the other at y kilometers per hour

WRITTEN EXERCISES

Goal: To solve motion problems by systems of linear equations

Sample Problem: During a recent trip, Julia drove a total of 680 kilometers. She drove at a speed of 70 kilometers per hour on the first day and at a speed of 80 kilometers per hour on the second day (Condition 2). The total time for the trip was 9 hours (Condition 1). How long did she drive each day?

Answer: She drove 4 hours on the first day and 5 hours on the second day.

For Exercises 1–7, Condition 1 is underscored once. Condition 2 is underscored twice.

a. *Use Condition 1 and Condition 2 to complete the table.*

b. *Use Condition 1 and Condition 2 to write a system of equations for the problem.*

c. *Solve.* (Example 1)

1. Some scouts traveled by bus at a speed of 70 kilometers per hour and then hiked at a speed of 7 kilometers per hour. The distance traveled 224 kilometers. The total time for the trip was 5 hours. How long did they travel by bus and how long did they hike?

	Time	Rate	Distance
By bus	x hrs	70 km/hr	$70x$ km
Hiking	y hrs	?	?
Total	5 hrs		224 km

2. Two trains leave the same city in opposite directions. The eastbound train travels 10 hours and the westbound train travels 5 hours. They are 1300 kilometers apart at their destinations. The westbound train is 20 kilometers per hour faster. What is the speed of each train?

	Rate	Time	Distance
Eastbound train	x km/hr	10 hrs	?
Westbound train	y km/hr	5 hrs	?
Total			?

3. A plane and a carrier 3720 kilometers apart move toward each other and meet in 3 hours. The speed of the plane is 30 times the speed of the carrier in kilometers per hour. Find the speed of the plane and the speed of the carrier.

	Rate	Time	Distance
Plane	x km/hr	3 hrs	?
Carrier	y km/hr	?	?
Total			?

(Example 2)

4. Linda rowed her boat 24 miles with a current in 3 hours. Rowing against the current, she returned to the point where she started in 4 hours. Find Linda's speed in still water and the speed of the current.

	Time	Rate	Distance
Against current	4 hrs	$(x - y)$ m/hr	24 m
With current	?	$(x + y)$ m/hr	?

5. A cross country runner runs a distance of 4500 meters against a wind in 18 minutes. He runs the same distance with the wind in 15 minutes. What is the speed of the runner and the speed of the wind?

	Time	Rate	Distance
Against wind	18 min	$(x - y)$ m/min	?
With wind	?	$(x + y)$ m/min	?

6. A plane travels 2240 kilometers with a wind in 7 hours. It travels the same distance against the wind in 8 hours. Find the speed of the plane in still air and the speed of the wind.

	Time	Rate	Distance
Against wind	?	$(x - y)$ km/hr	?
With wind	7 hrs	$(x + y)$ km/hr	?

7. Blake paddles a canoe 48 kilometers against a current in 6 hours. He paddles the same distance with the current in 4 hours. Find Blake's speed in still water and the speed of the current.

	Time	Rate	Distance
Against current	?	$(x - y)$ km/hr	?
With current	?	$(x + y)$ km/hr	?

Focus on Reasoning: Comparisons

Many problems involving the **comparison of two quantities** can be solved by logical reasoning. Little or no computation with paper and pencil may be necessary.

Refer to these instructions for the Examples and Exercises.

Each problem consists of two quantities, one in Column I and one in Column II. Compare the two quantities.

- Write **A** if the quantity in Column I is greater.
- Write **B** if the quantity in Column II is greater.
- Write **C** if the two quantities are equal.
- Write **D** if there is not enough information to determine how the two quantities are related.

EXAMPLE 1

Column I	Column II
$1800 \times 9 \times 32$	$18 \times 32 \times 900$

Solution: Since 32 appears in both products, compare (1800×9) and (18×900).
Since $(1800 \times 9) = 18 \times 100 \times 9$ and $(18 \times 900) = 18 \times 9 \times 100$, the products are equal.

Answer: **C**

EXERCISES

	Column I	Column II
1.	$5 \times 600 \times 3$	$500 \times 7 \times 3$
2.	$0.25 \times 800 \times 9$	$9 \times 25 \times 8$
3.	6% of 1010	1% of 606
4.	$\frac{(19)(7)(6)}{(32)(4)}$	$\frac{(2)(10)(42)}{(16)(8)}$
5.	$\frac{8}{8 - 7.9}$	$\frac{8}{8 - 7.99}$
6.	$(4)(6)(8)(2^5)$	$(2)(6)(16)(5^2)$

Sometimes additional information is given about the quantities being compared.

EXAMPLE 2

	Column I		Column II
		$x^2 = 3, (x > 0)$	
	$2x\sqrt{3}$		$(2x)^2$

Solution: Since $x^2 = 3$ $(x > 0)$, $x = \sqrt{3}$.

Then $2x\sqrt{3} = 2(\sqrt{3})(\sqrt{3}) = 6$.

Also $(2x)^2 = (2\sqrt{3})^2 = 12$.

Answer: **B**

EXERCISES

	Column I		Column II
7.		The average of 2, 6, and x is 5.	
	$\frac{6}{x}$		$\frac{x}{6}$
8.		$8x = 1$	
	$\sqrt{2x}$		$\sqrt[3]{x}$
9.		$3x + 2 = 20$	
	$\frac{5}{6}x + 1$		$\frac{1}{6}x + 5$
10.		$y = 2x, (x > 0)$	
	$\frac{y - x}{2}$		$\frac{y}{2} - x$
11.		$3x = 4y, (y \neq 0)$	
	$\frac{x}{y}$		$\frac{3}{4}$
12.		$x^2 = 8, (x > 0)$	
	$x\sqrt{2}$		4
13.		$x < 1$ and $x > -\frac{1}{4}$	
	$4x^2 + x$		$4x + 1$

CHAPTER SUMMARY

IMPORTANT TERMS

Substitution method *(p. 398)*
Determinant *(p. 412)*
Standard form *(p. 413)*
Determinant of a system *(p. 413)*

IMPORTANT IDEAS

1. To solve a system of linear equations by the *substitution method:*
   a. Solve one equation for one of the variables.
   b. Substitute the resulting expression in the other equation. Solve the equation.
   c. Substitute the value of the variable from Step **2** in either equation. Solve the resulting equation.
   d. Check by substituting both values in both equations.
2. Any two-digit number can be represented by $10x + y$, in which x represents the ten's digit and y represents the one's digit.
3. The strength of a solution or the relative amount of a specific ingredient in a mixture is often expressed by per cents. Thus, ten grams of salt in 100 grams of a mixture of salt and water is called a 10% salt solution.
4. Each 2×2 determinant represents a real number.
   $$\begin{vmatrix} a & b \\ c & d \end{vmatrix} = ad - bc$$
5. The determinant of a system of equations is formed with the coefficients of the variables as elements of the determinant.
6. If only the determinant of a system of equations equals 0, then the solution set of the system is empty.
7. The basic formula $d = rt$ relating distance, rate, and time is often used in motion problems.
8. Sometimes speeds can be added. Thus, if you let x represent the speed of an object in still water and y the speed of the current, then $(x + y)$ will represent the speed of the object with the current and $(x - y)$ the speed against the current.

CHAPTER REVIEW

SECTION 15.1

Solve each system by the substitution method. Check.

1. $\begin{cases} y = -3x - 6 \\ x + 2y = 3 \end{cases}$

2. $\begin{cases} y = \frac{1}{2}x - \frac{3}{2} \\ 3x - 2y = -3 \end{cases}$

3. $\begin{cases} x - 3y = 5 \\ 2y + 3x = 4 \end{cases}$

4. $\begin{cases} 6 - x = 5y \\ -3x + 2y = 16 \end{cases}$

For Exercises 5–10, Condition 1 is underscored once. Condition 2 is underscored twice.

SECTION 15.2

For each problem, use Condition 1 and two variables to represent the unknowns. Use Condition 1 and Condition 2 to write a system of equations. Solve.

5. The ten's digit of a two-digit number is three times the one's digit. The sum of the digits is 12. Write the two-digit number.

6. The one's digit of a two-digit number exceeds the ten's digit by 5. Nine times the ten's digit increased by three times the one's digit equals 39. Write the two-digit number.

7. A two-digit number equals 9 times its one's digit. The sum of the digits is 9. Write the two-digit number.

8. A two-digit number is 2 less than 5 times its one's digit. The one's digit exceeds the ten's digit by 5. Write the two-digit number.

SECTION 15.3

For each problem, use Condition 1, Condition 2, and two variables to make a table. Use Condition 1 and Condition 2 to write a system of equations. Solve.

9. A 12% salt solution is mixed with a 20% salt solution to obtain 800 grams of a 17% salt solution. How many grams of the 12% solution and of the 20% are needed?

10. A 7% acid solution is mixed with 360 grams of a $2\frac{1}{2}$% acid solution to obtain a 5% acid solution. Find the number of grams of the 7% solution and of the 5% solution.

SECTION 15.4

Write the value of each determinant.

11. $\begin{vmatrix} 5 & -2 \\ 3 & 7 \end{vmatrix}$ **12.** $\begin{vmatrix} \frac{1}{2} & 0 \\ -6 & -4 \end{vmatrix}$ **13.** $\begin{vmatrix} 0 & 1 \\ 1 & 1 \end{vmatrix}$ **14.** $\begin{vmatrix} 0 & 1 \\ 1 & 0 \end{vmatrix}$

SECTION 15.5

Solve each system by the method of determinants.

15. $\begin{cases} 3x - y = 4 \\ x + 2y = -1 \end{cases}$ **16.** $\begin{cases} 3x + 2y = 13 \\ 2x + 3y = 12 \end{cases}$ **17.** $\begin{cases} x - 3y = 10 \\ 2x + 5y = 1 \end{cases}$ **18.** $\begin{cases} -2x + 5y = 6 \\ 3x - 4y = -2 \end{cases}$

For Exercises 19–22, Condition 1 is underscored once. Condition 2 is underscored twice.

SECTION 15.6

For each problem, use Condition 1, Condition 2, and two variables to make a table. Use Condition 1 and Condition 2 to write a system of equations. Solve.

19. The Gymnastics Club sold 204 tickets for its show. The total receipts were $215. The adult tickets sold for $1.25 and the student tickets for $0.75. Find the number of each kind of ticket that was sold.

20. Mrs. Bell purchased 96 greeting cards on sale. The total cost was $10.20. One kind cost 15 cents each and another kind cost 8 cents each. Find the number of each kind she purchased.

SECTION 15.7

For each problem, use Condition 1 and Condition 2 to complete the table. Use Condition 1 and Condition 2 to write a system of equations. Solve.

21. A fishing group traveled by plane for 3 hours and then by jeep for 2 hours. The total distance traveled was 1000 kilometers. The speed of the plane in kilometers per hour was six times the speed of the jeep. Find the speed of the plane and the speed of the jeep.

	Rate	Time	Distance
By plane	x km/hr	?	?
By jeep	y km/hr	?	?
Total			?

22. A motorboat travels 240 kilometers in 6 hours when it is traveling against a current. The same motorboat travels 240 kilometers in 4 hours when it is traveling with the current. Find the speed of the boat in still water and the speed of the current.

	Time	Rate	Distance
Against current	?	$(x - y)$ km/hr	?
With current	?	$(x + y)$ km/hr	?

CUMULATIVE REVIEW: CHAPTERS 1–15

Solve each inequality. (Sections 5.4 through 5.6)

1. $5x + 10 < 4x + 6$ **2.** $-6x < 18$ **3.** $4(x - 2) > 6x + 2$

Simplify. (Sections 6.4 and 6.5)

4. $(-6gh)^2$ **5.** $(ab)^{-5}$ **6.** $(2a^2b^3)^4$ **7.** $\left(\frac{-3xy^4}{5}\right)^3$

Simplify. (Sections 7.1 through 7.4)

8. $-\sqrt{144}$ **9.** $\sqrt{5} \cdot \sqrt{3}$ **10.** $\sqrt{50}$ **11.** $\sqrt{12x^6y^9}$

Multiply. (Sections 8.2 and 8.5)

12. $(2d^3)(-6d)$ **13.** $(-4cd^2e^3)(-2cd^4e^2)$ **14.** $(3x - 1)(x + 7)$

Factor. (Sections 9.4 through 9.7)

15. $12n^2 - 17n - 5$ **16.** $16m^2 - 1$ **17.** $3x^2 - 18x + 27$

Multiply or divide. Then simplify. (Sections 10.3 through 10.7)

18. $\frac{7ab^2}{8a^2} \cdot \frac{16a^3c}{14bc}$ **19.** $\frac{a^2 + 5a}{a - 4} \cdot \frac{a^2 - 4a}{a^2 - 25}$ **20.** $\frac{x^2 - 16}{x^2 + x - 12} \div \frac{x^2 - 6x + 8}{5x - 15}$

Add or subtract. Simplify where necessary.
(Sections 11.2 and 11.4 through 11.6)

21. $\frac{7x}{x + 2} - \frac{x - 12}{x + 2}$ **22.** $\frac{5}{4x} + \frac{2}{3x}$ **23.** $\frac{x + 3}{x^2 - 4} + \frac{x - 5}{x + 2}$

Solve and check each equation. (Section 11.7)

24. $\frac{18}{2x + 3} = 3$ **25.** $\frac{10}{x - 3} = \frac{9}{x - 5}$ **26.** $\frac{6}{x - 3} - \frac{4}{x - 3} = \frac{5}{x}$

Graph at least four ordered pairs of each function. Then draw the graph. The domain is {real numbers}. (Section 13.1)

27. $y = -2x + 3$ **28.** $y = \frac{1}{2}x - 4$ **29.** $y = 4x$

Find the slope and Y-intercept of the graph of each function.
(Sections 13.2 and 13.3)

30. $y = \frac{2}{3}x - 6$ **31.** $y = 7 - x$ **32.** $y = 3(x + 2)$

Write the slope of the line that contains the two points. (Section 13.4)

33. $(2,7)$; $(-3,5)$ **34.** $(-2,0)$; $(-5,9)$ **35.** $(3,-1)$; $(-1,8)$

Write Yes or No to tell whether the graphs of each pair of functions are parallel. (Section 13.5)

36. $y = 2x - 4;\ y = 2x + 3$

37. $y = \frac{1}{2}x - 4;\ y = 4x - \frac{1}{2}$

Solve each system by graphing. Check. (Section 14.2)

38. $\begin{cases} x - y = 9 \\ 2x + y = 6 \end{cases}$

39. $\begin{cases} 3x + y = 6 \\ x - 2y = 2 \end{cases}$

40. $\begin{cases} 2x - y = -3 \\ 4x - 2y = 2 \end{cases}$

Solve each system using the addition method or the multiplication/addition method. (Sections 14.3 and 14.4)

41. $\begin{cases} 4x - y = -5 \\ -4x + 5y = -7 \end{cases}$

42. $\begin{cases} 2x + 3y = 8 \\ 3x + y = 5 \end{cases}$

43. $\begin{cases} 3x - 2y = -6 \\ 4x - 5y = -1 \end{cases}$

Solve each system by the substitution method. Check. (Section 15.1)

44. $\begin{cases} y = 3 - x \\ 5x + 3y = -1 \end{cases}$

45. $\begin{cases} x - 3y = 0 \\ 2x - 5y = 4 \end{cases}$

46. $\begin{cases} 2x - y = 6 \\ 2x + 3y = -2 \end{cases}$

Use Condition 1 and two variables to represent the unknowns. Use Conditions 1 and 2 to write a system of equations for the problem. Solve. (Sections 14.5 and 15.2)

47. The width of a rectangle is 7 meters less than twice the length. The perimeter is 19 meters. Find the length and width.

48. The ten's digit of a two-digit number is 3 more than 5 times the one's digit. The sum of the digits is 9. Write the two-digit number.

Solve each system by the method of determinants. (Section 15.5)

49. $\begin{cases} 2x + 3y = 8 \\ 3x + y = 5 \end{cases}$

50. $\begin{cases} 6x - 2y = -4 \\ 3x + y = -10 \end{cases}$

51. $\begin{cases} 4x - 3y = 5 \\ 2x - 9y = 1 \end{cases}$

Use Condition 1, Condition 2, and two variables to make a table. Use Conditions 1 and 2 to write a system of equations for the problem. Solve. (Sections 15.3 and 15.6)

52. A 10% salt solution is mixed with 20 grams of a 15% salt solution to obtain a 12% salt solution. Find the number of grams of the 10% solution and of the 12% solution.

53. The junior class sold 520 tickets for a talent show. The total receipts were \$1500. The adult tickets sold for \$3, and the student tickets sold for \$2. Find the number of each kind of ticket that was sold.

Systems of Inequalities

CHAPTER 16

Given a choice of any geometric solid that you could use to enclose a given volume, you would choose the *sphere* because the sphere has the property of enclosing a given volume with a minimum amount of material.

16.1 Inequalities in Two Variables

These three inequalities are examples of ***linear inequalities***.

a. $y < 2x - 5$ **b.** $x - 3y + 2 > 0$ **c.** $-x + 2y + 1 \le 0$

You can see that if you replace each inequality symbol by "=", you obtain three linear equations.

> The ***solution set of a linear inequality*** is the set of ordered pairs (x, y) that make the inequality true.

A linear inequality is the rule for a relation.

A relation is a set of ordered pairs.

P–1 **What is one ordered pair in the solution set of $y < 2x - 5$?**

EXAMPLE 1 Write three ordered pairs having different x values that are in the solution set of $-2x + y - 1 \le 0$.

Solution:

1. Let $x = 0$. ⟶ $-2(0) + y - 1 \le 0$

$$y - 1 \le 0$$

$$y \le 1$$

Infinitely many y values

Choose a value for y that is less than or equal to 1.

The first ordered pair can be (0, 1), (0, 0.3), (0, −2), etc.

2. Let $x = 2.7$. ⟶ $-2(2.7) + y - 1 \le 0$

$$-5.4 + y - 1 \le 0$$

$$y - 6.4 \le 0$$

$$y \le 6.4$$

Choose a value for y that is less than or equal to 6.4.

The second ordered pair can be (2.7, 6.2), (2.7, 0), (2.7, −1), etc.

3. Let $x = -3\frac{1}{2}$. ⟶ $-2(-3\frac{1}{2}) + y - 1 \le 0$

$$7 + y - 1 \le 0$$

$$y \le -6$$

Choose a value for y that is less than or equal to -6.

The third ordered pair can be $(-3\frac{1}{2}, -6)$, $(-3\frac{1}{2}, -7)$, etc.

A linear inequality can be written in slope-intercept form.

The ***slope-intercept form*** of a linear inequality is one of the following:

$\mathbf{y < mx + b}$ $\mathbf{y > mx + b}$
$\mathbf{y \le mx + b}$ $\mathbf{y \ge mx + b}$

$y \le -2x + 7$

$y > \frac{1}{2}x - \frac{5}{2}$

EXAMPLE 2 Write the slope-intercept form of $2x + y < 5$.

Solution: $2x + y < 5$

1. Addition Property of Order ⟶ $\mathbf{2x + y + (-2x) < 5 + (-2x)}$
2. Simplify. ⟶ $y < 5 + (-2x)$
3. Commutative Property of Addition ⟶ $y < -2x + 5$ ◀ ***Slope-intercept form***

EXAMPLE 3 Write the slope-intercept form of $-3x - 5y + 2 \ge 0$.

Solution: $-3x - 5y + 2 \ge 0$

1. Add $3x$ to each side. ⟶ $\mathbf{-3x - 5y + 2 + 3x \ge 0 + 3x}$
2. Simplify. ⟶ $-5y + 2 \ge 3x$
3. Subtract 2 from each side. ⟶ $\mathbf{-5y + 2 - 2 \ge 3x - 2}$
4. Simplify. ⟶ $-5y \ge 3x - 2$
5. Multiply each side by $-\frac{1}{5}$. ⟶ $\mathbf{-\frac{1}{5}(-5y) \le -\frac{1}{5}(3x - 2)}$ ◀ ***The direction is reversed.***
6. Simplify. ⟶ $y \le -\frac{3}{5}x + \frac{2}{5}$

CLASSROOM EXERCISES

Write three ordered pairs having different x values that are in the solution set of each inequality. (Example 1)

1. $y < x + 1$ **2.** $y < 2x - 1$ **3.** $y < -x + 2$ **4.** $y < x$

5. $y < -x$ **6.** $y > 3x + 5$ **7.** $y > \frac{1}{2}x + 3$ **8.** $y \le \frac{1}{3}x + 1$

Write the slope-intercept form of each inequality. (Examples 2 and 3)

9. $-2x + y - 1 < 0$ **10.** $3x + y > 2$ **11.** $-3x + y - 2 \le 0$

12. $y + 5 \ge \frac{1}{2}x$ **13.** $-y - x > 10$ **14.** $2y < 2x + 4$

WRITTEN EXERCISES

Goals: To write ordered pairs in the solution set of a given inequality and to write the slope-intercept form of the inequality

Sample Problems: **a.** Write three ordered pairs in the solution set of $-2x + 3y + 5 > 0$.

b. Write the slope-intercept form of $-2x + 3y + 5 > 0$.

Answers: **a.** (0, 0), $(\frac{5}{2}, 3)$, (−1, −2), etc. **b.** $y > \frac{2}{3}x - \frac{5}{3}$

Write five ordered pairs having different x values that are in the solution set of each inequality. (Example 1)

1. $y < 2x$ **2.** $y < -5x$ **3.** $y \ge -3x + 1$

4. $y \ge 2x - 5$ **5.** $y < \frac{1}{2}x - 3$ **6.** $y > \frac{1}{3}x + 3$

7. $x + y \ge -2$ **8.** $x - y \le 4$ **9.** $y \ge 10x + 2$

10. $x + y \le 3$ **11.** $-2x + 3y - 5 > 0$ **12.** $4x - y + 3 < 0$

Write the slope-intercept form of each inequality. (Examples 2 and 3)

13. $-3x + y - 5 > 0$ **14.** $2x + y + 3 < 0$ **15.** $4x + y \le 6$

16. $-5x + y \ge -2$ **17.** $x - y < -3$ **18.** $3x - y > 7$

19. $\frac{1}{3}y > x - 5$ **20.** $-\frac{1}{5}y \le -x + 2$ **21.** $2x + 3y - 6 \ge 0$

22. $-5x + 2y + 3 \le 0$ **23.** $4x - 2y < 5$ **24.** $6x - 3y + 2 > 0$

25. $-7x + 14y - 3 \ge 0$ **26.** $-9x + 27y > 3$ **27.** $3y + 5x - 1 < 0$

REVIEW CAPSULE FOR SECTION 16.2

Graph each equation. (Pages 378–380)

1. $y = -x + 2$ **2.** $y = 2x - 1$ **3.** $y = \frac{3}{2}x - 3$ **4.** $y = \frac{5}{4}x + 2$

5. $y = \frac{1}{2}x + \frac{3}{2}$ **6.** $y = -3x - 1$ **7.** $y = -2$ **8.** $x = -3$

16.2 Graphs of Inequalities with < and >

Every linear inequality has a related linear equation.

Inequality: $y < 2x - 1$ **Related Equation:** $y = 2x - 1$

The graph of $y = 2x - 1$ is shown in Figure 1. The line separates the coordinate plane into three sets of points or ***regions***.

a. One region is one side of the line.
b. One region is the other side of the line.
c. One region is the line itself.

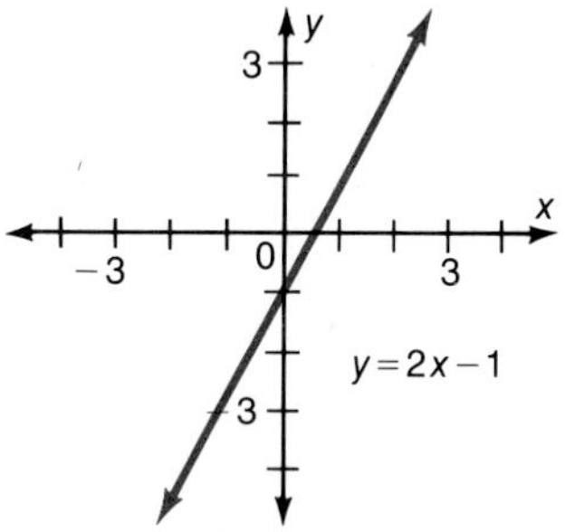

Figure 1

P–1 **In which region is the point with coordinates (0, 0)? the point with coordinates (2, 3)? the point with coordinates (5, −3)?**

(2, 3) is in the solution set of $y = 2x - 1$.
(0, 0) is in the solution set of $y > 2x - 1$.
(5, −3) is in the solution set of $y < 2x - 1$.

In Figure 1, the region above, and to the left of, the line is the graph of $y > 2x - 1$. The region below, and to the right of, the line is the graph of $y < 2x - 1$.

In graphing inequalities that contain > or <, you must remember that the line is not part of the graph. Therefore, the graph of the related equation is drawn as a dashed line.

EXAMPLE 1 Graph $y < 2x - 1$.

"To graph" means "to graph the solution set of."

Solution:

1 Graph the related equation $y = 2x - 1$. Use a dashed line.

2 Shade the region of the plane to the right of the dashed line.

The shaded region is the graph of $y < 2x - 1$.

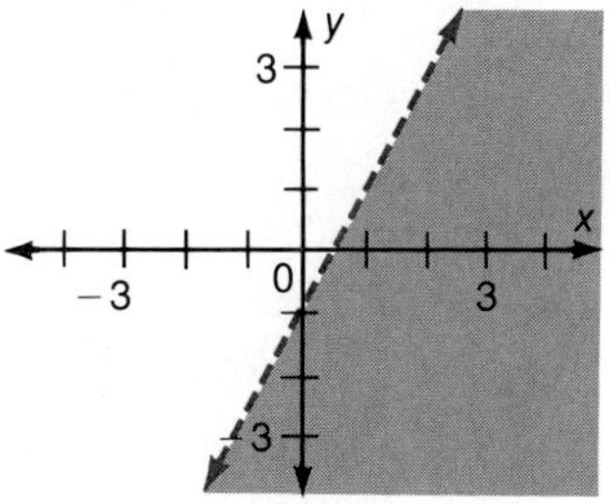

EXAMPLE 2 Graph $x + y - 1 > 0$.

Solution:

1 Write the slope-intercept form. ⟶ $y > -x + 1$

2 Graph the related equation $y = -x + 1$ as a dashed line as in Figure 2.

3 Choose a point not on the line. Check whether its coordinates make the inequality true or false.

$y > -x + 1$ ◀ ***Try the origin, (0, 0).***

$0 > -0 + 1$

$0 > 1$ ◀ ***False sentence***

The origin is not in the graph.

4 Shade the region on the side of the line not containing the origin as in Figure 3. The shaded region is the graph of $x + y - 1 > 0$.

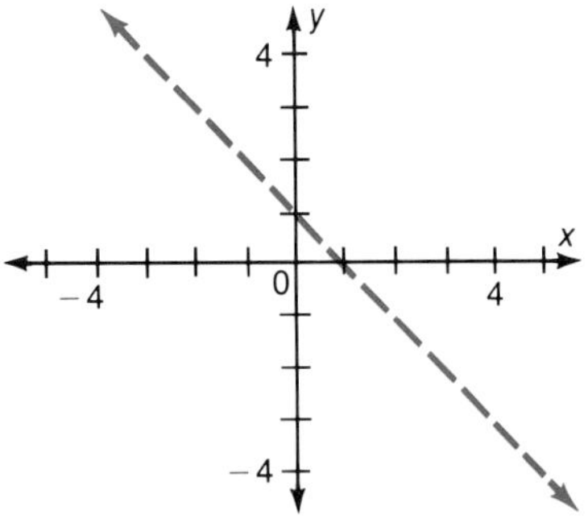

Figure 2

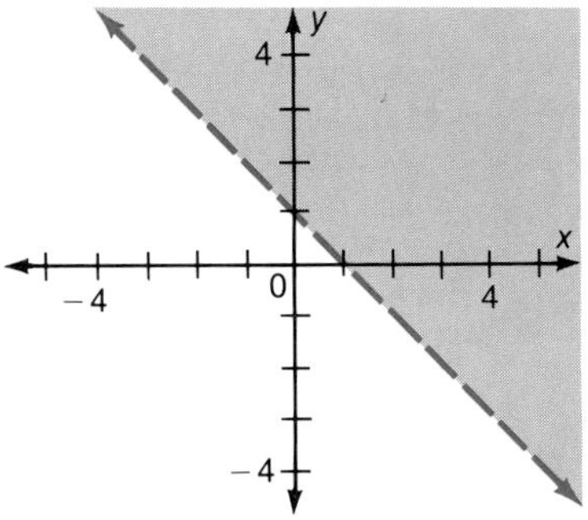

Figure 3

Steps for graphing the relations $y < mx + b$ and $y > mx + b$:

1. Draw the graph of $y = mx + b$.
2. Select a point not on the line and substitute its coordinates in the inequality.
3. Shade the region of the plane containing this point if its coordinates make the inequality true.
4. Shade the region on the other side of the line from this point if its coordinates make the inequality false.

CLASSROOM EXERCISES

Write Yes or No to tell whether (0, 0) is in each solution set. (P-1)

1. $y < 2x + 5$ **2.** $y > x - 1$ **3.** $y < -2x - 3$ **4.** $y < 3x$

5. $y = 5x$ **6.** $y > -\frac{1}{2}x - \frac{2}{3}$ **7.** $y > 3$ **8.** $x < -2$

Write an inequality for each graph. The equation of each dashed line is given.

9.

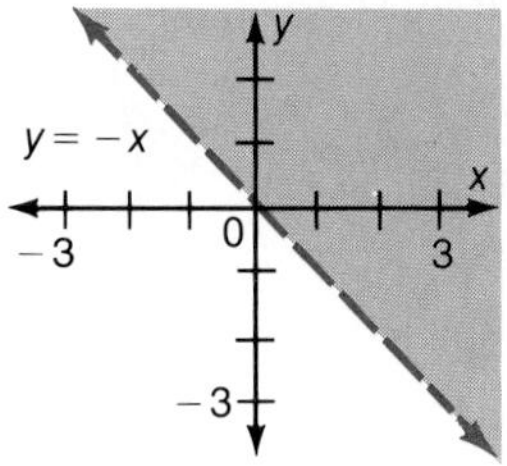

10.

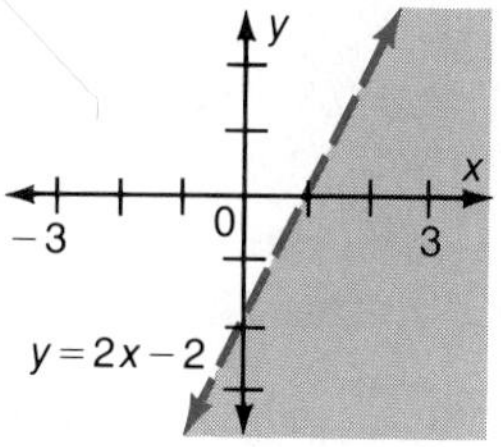

11.

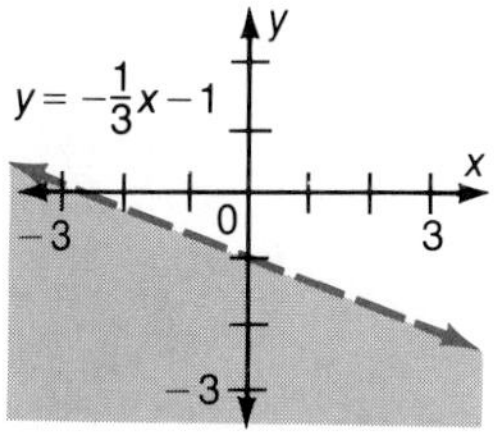

WRITTEN EXERCISES

Goal: To graph linear inequalities that use > or <.
Sample Problem: $x + y - 3 > 0$
Answer: See the graph at the right.

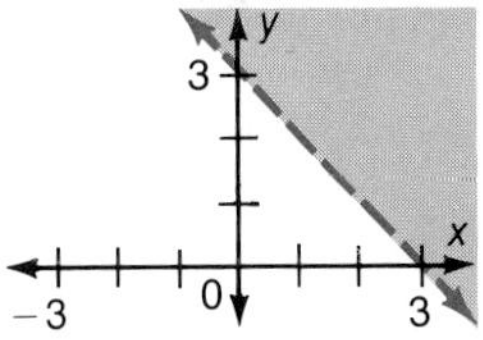

Graph each inequality. (Examples 1 and 2)

1. $y < 3x - 1$ **2.** $y < -2x + 2$ **3.** $y > -x + 3$

4. $y > x - 3$ **5.** $y < \frac{1}{2}x - 3$ **6.** $y < -\frac{1}{2}x - 1$

7. $y < x$ **8.** $y > -x$ **9.** $y > -3$

10. $y < -2$ **11.** $3x + y - 2 < 0$ **12.** $-2x + y + 3 < 0$

13. $x < -2$ **14.** $x > -3$ **15.** $-3x + 2y - 2 > 0$

16. $4x + 2y - 3 > 0$ **17.** $3x - 4y - 12 < 0$ **18.** $2x - y + 4 > 0$

REVIEW CAPSULE FOR SECTION 16.3

Write as one inequality using ≤ or ≥. (Pages 116–119)

1. $y < 2$ or $y = 2$ **2.** $y < x - 5$ or $y = x - 5$

3. $y > 2x$ or $y = 2x$ **4.** $y = \frac{1}{2} - 3x$ or $y > \frac{1}{2} - 3x$

5. $y = -x + 1$ or $y < -x + 1$ **6.** $y > 2 - x$ or $y = 2 - x$

16.3 Graphs of Inequalities with ≤ and ≥

Mathematical sentences can be classified as simple or compound. An important example of a compound sentence is the *or* sentence.

Simple Sentences	Compound *Or* Sentences
$x > 9$	$x > 1$ *or* $x = -5$
$x = -2$	$x < 4$ *or* $x = 6$
$x < 6$	$x > -3$ *or* $x = -3$

This can also be shown as $x \geq -3$.

P–1 **What are four numbers in the solution set of $x > -3$?**

P–2 **Find the only number in the solution set of $x = -3$.**

The ***union of two sets*** is the set of elements in either one of the two sets or in both of them.

$$\{1, 3, 5\} \cup \{2, 3, 4\} = \{1, 2, 3, 4, 5\}$$

$A \cup B$ means "the union of A and B."

The ***solution set of a compound or sentence*** is the union of the solution sets of its two simple sentences.

P–3 **What are four numbers in the solution set of the compound *or* sentence "$x > -3$ *or* $x = -3$"?**

When a compound *or* sentence has *two* variables its graph is drawn in a coordinate plane.

P–4 **What are four ordered pairs in the solution set of $x + y = 2$? of $x + y < 2$?**

In graphing inequalities that contain ≥ or ≤, you must remember that the line is part of the graph. Therefore, the graph of the related equation is drawn as a solid line.

EXAMPLE 1 Graph $x + y \leq 2$.

$x + y < 2$
or
$x + y = 2$

Solution:

1. Write the slope-intercept form. ⟶ $y \leq -x + 2$
2. Graph $y = -x + 2$.
3. Choose a point not on the line such as the origin, (0, 0). Check whether its coordinates make $y < -x + 2$ true or false.

$$y < -x + 2$$
$$0 < -0 + 2$$
$$0 < 2$$

True sentence

4. Shade the region on the side of the line containing the origin.

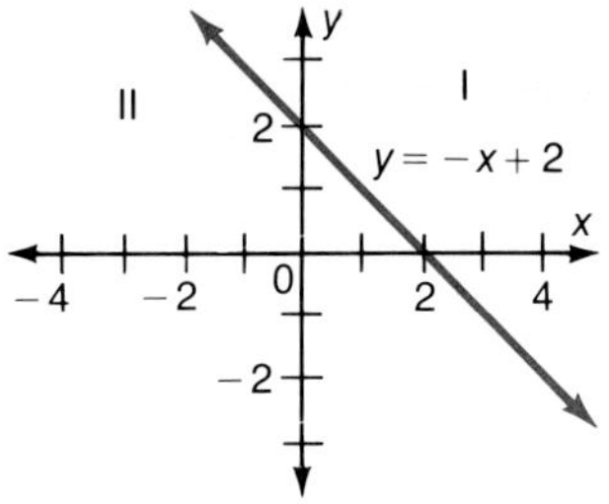

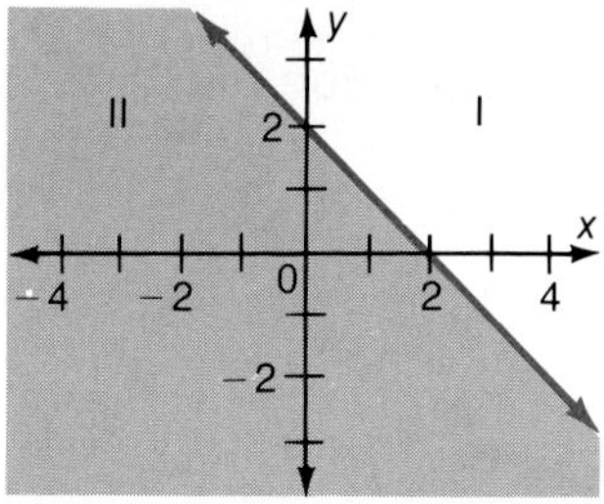

The graph is the union of the solid line and Region II.

EXAMPLE 2 Graph $x + 2y - 2 \geq 0$.

$x + 2y - 2 > 0$
or
$x + 2y - 2 = 0$

Solution:

1. Write the slope-intercept form. ⟶ $y \geq -\frac{1}{2}x + 1$
2. Graph $y = -\frac{1}{2}x + 1$.
3. Choose a point not on the line such as the origin, (0, 0). Check whether its coordinates make $y > -\frac{1}{2}x + 1$ true or false.

$$y > -\tfrac{1}{2}x + 1$$
$$0 > -\tfrac{1}{2}(0) + 1$$
$$0 > 1$$

False sentence

4. Shade Region I above the line. It does not contain the origin.

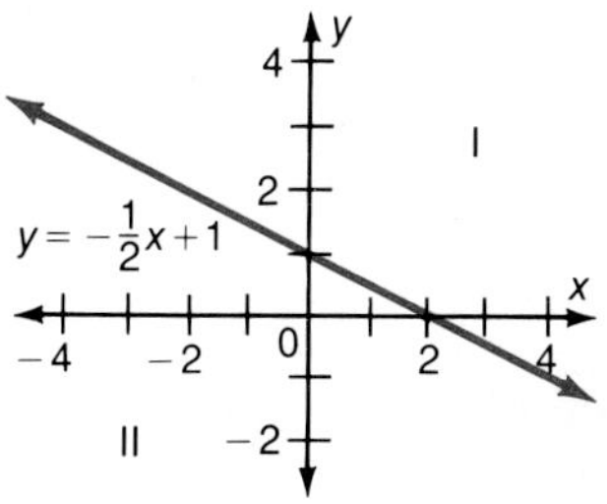

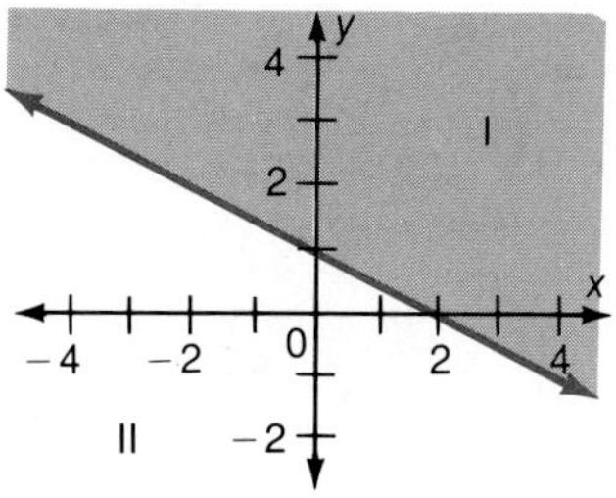

The graph is the union of the solid line and Region I.

EXAMPLE 3

Write the slope-intercept form of the inequality that is graphed below. The slope m of the line is $\frac{1}{2}$.

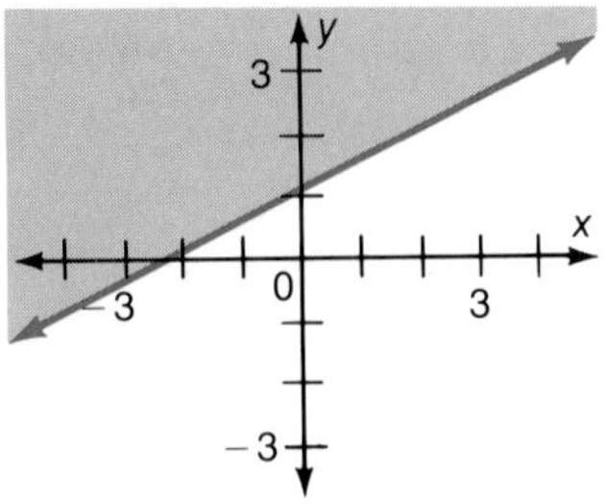

Solution: The Y intercept of the line is 1 as shown on the graph.

Equation of the line: $y = mx + b$

$y = \frac{1}{2}x + 1$

Inequality: $y \geq \frac{1}{2}x + 1$ ◀ ***The region above the line is shaded.***

Note, as a check, that the point with coordinates (0, 2) makes the inequality true.

CLASSROOM EXERCISES

Write Yes *or* No *to indicate whether (0, 0) is in each solution set.* (Step 3 of Example 1)

1. $y \leq x + 1$

2. $y \leq -x - 1$

3. $y \leq 2x + 3$

4. $y \leq -5x + 1$

5. $y \geq x$

6. $y \leq -x$

7. $x + 2y - 5 \leq 0$

8. $-2x + 3y - 1 \geq 0$

9. $x + y \geq -2$

Write the slope-intercept form for each graph. The slope, m, of each line is given. (Example 3)

10. $m = 1$

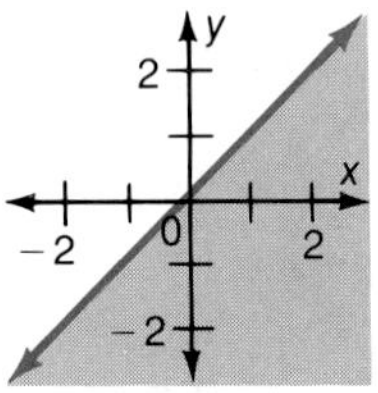

11. $m = -1$

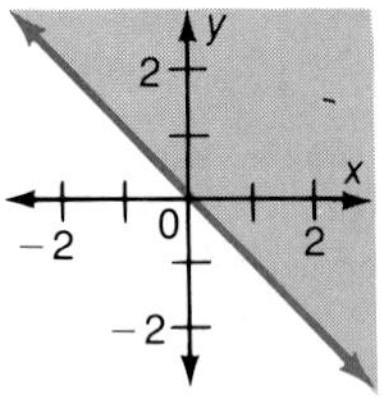

12. $m = -\frac{1}{2}$

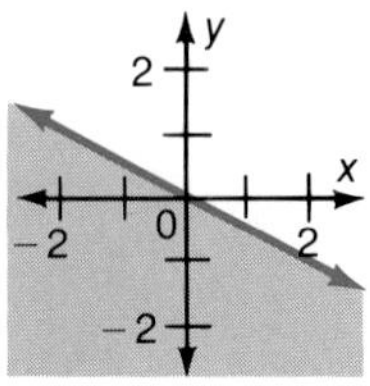

13. $m = \frac{1}{2}$

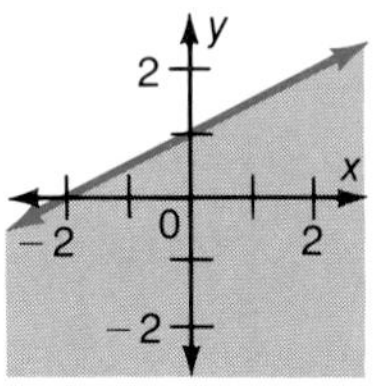

WRITTEN EXERCISES

Goal: To graph inequalities in two variables that use $\geq$ or $\leq$

Sample Problem: $x + y \leq 3$

Answer: See the graph at the right.

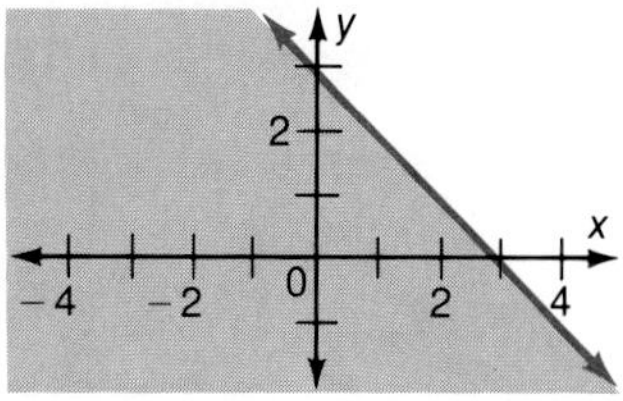

Graph each inequality. (Examples 1 and 2)

1. $y \leq x + 1$

2. $y \leq -x + 1$

3. $y \leq 2x - 3$

4. $y \leq 3x - 2$

5. $y \geq -2x + 1$

6. $y \geq -3x - 1$

7. $-\frac{1}{2}x + y - 1 \leq 0$

8. $\frac{1}{2}x + y + 2 \leq 0$

9. $2x + 2y - 3 \geq 0$

10. $3x + 3y - 6 \geq 0$

11. $-\frac{1}{3}x + y \leq 1$

12. $-\frac{1}{4}x + y \leq -2$

13. $y \geq -1$

14. $y \leq -2$

15. $x \leq 3$

16. $x \geq -3$

17. $x \geq -y + 2$

18. $x \leq -y - 3$

Write the slope-intercept form of an inequality for each graph. The slope, m, of each line is given. (Example 3)

19. $m = 1$

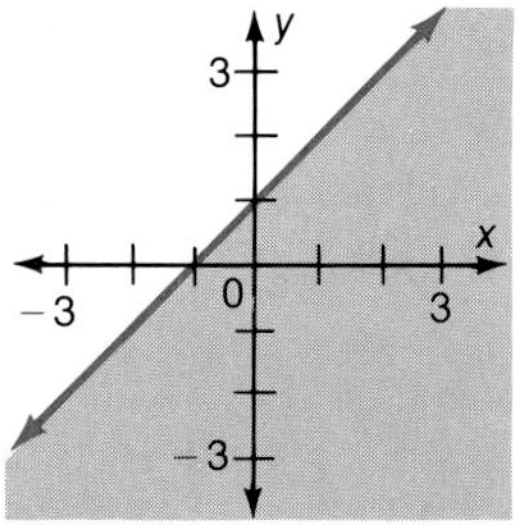

20. $m = -1$

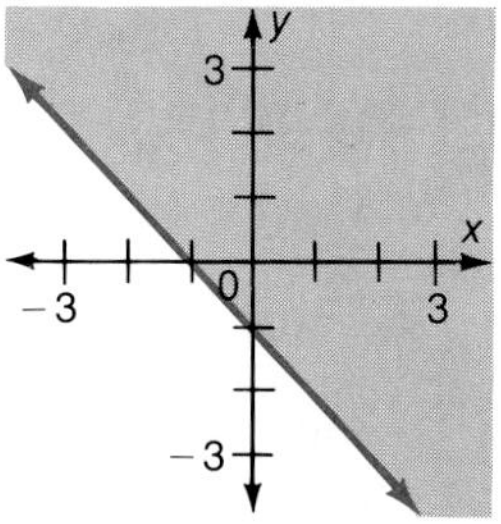

21. $m = -2$

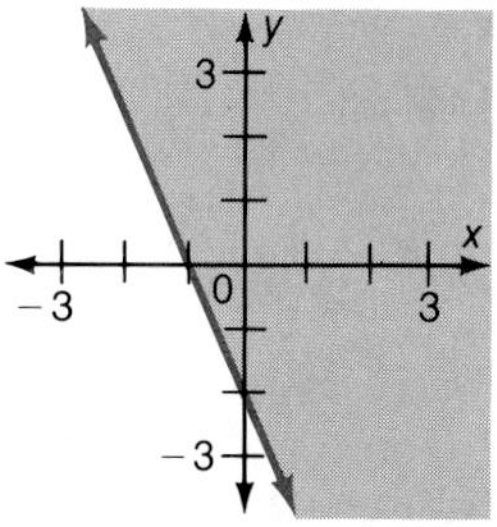

22. $m = 2$

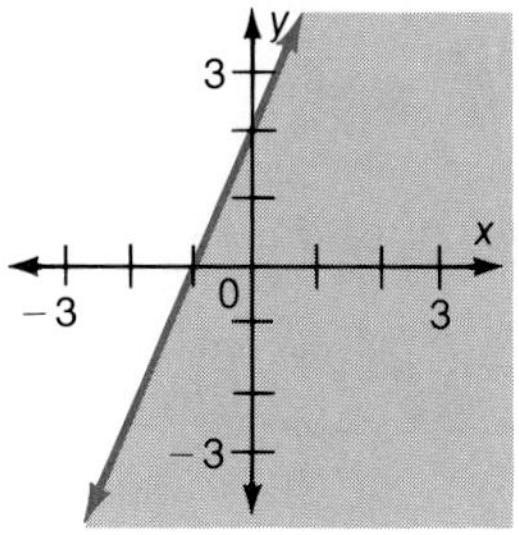

23. $m = -\frac{1}{3}$

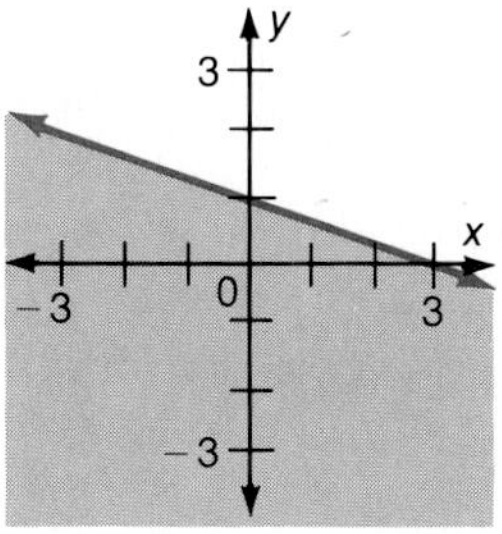

24. $m = \frac{1}{2}$

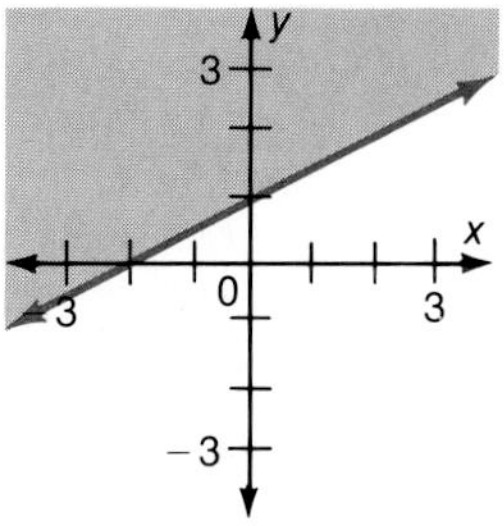

APPLICATIONS

You can use a calculator to check whether an ordered pair is in the solution set of an inequality. First, substitute the ordered pair in the inequality. Then compare the two sides of the inequality.

EXAMPLE Is $(-2, 5)$ in the solution set of $6x - y > 1 + 3y$?

SOLUTION *Left side:* [6] [×] [2] [+/−] [−] [5] [=] → -17.

Right side: [3] [×] [5] [+] [1] [=] → 16.

Note: Multiply before adding. ***See page 19.***

Since "$-17 > 16$" is <u>false</u>, $(-2, 5)$ is not in the solution set.

Check whether the given ordered pair is in the solution set of the inequality.

25. $3x - 5 < 5y + 2;(3, 2)$ **26.** $4x + y > 2x - 4;(3, -7)$ **27.** $(x + 2)5 \leq x + 2y;(-5, 6)$

MID-CHAPTER REVIEW

Write the slope-intercept form of each inequality. (Section 16.1)

1. $5x + y - 2 < 0$ **2.** $-2x + y + 6 > 0$ **3.** $3x - 2y > -5$
4. $-5x + 3y < 8$ **5.** $-3 > -8x + 6y$ **6.** $12 < 20x - 8y$

Graph each inequality. (Section 16.2)

7. $y > 2x + 3$ **8.** $y < 3x - 2$ **9.** $y < \frac{1}{2}x - 2$
10. $y > -\frac{1}{3}x + 1$ **11.** $2x - y + 5 < 0$ **12.** $-3x + y - 3 > 0$

Graph each inequality. (Section 16.3)

13. $y \geq \frac{3}{2}x - 3$ **14.** $y \leq -\frac{7}{4}x + 2$ **15.** $y \leq -x + \frac{5}{2}$
16. $y \geq -x - \frac{9}{4}$ **17.** $3x - 2y + 6 \geq 0$ **18.** $-4x + 2y - 5 \leq 0$

REVIEW CAPSULE FOR SECTION 16.4

Which of the following are in the solution sets of <u>both</u> inequalities? $\{(-2, 3), (0, -1), (3, 5), (2, -4), (-4, 3)\}$ (Pages 434–436)

1. $\begin{cases} y > 2x - 1 \\ y > \frac{1}{2}x + 3 \end{cases}$ **2.** $\begin{cases} y < 4x + 3 \\ y > x - 5 \end{cases}$ **3.** $\begin{cases} y < -x + 3 \\ y > -3x - 1 \end{cases}$ **4.** $\begin{cases} y > x + 5 \\ y < -2x - 4 \end{cases}$

16.4 Graphs of Systems with < and >

P–1 **What are three numbers in the solution set of $x < 4$?**

P–2 **What are three numbers in the solution set of $x > -2$?**

The ***intersection of two sets*** is the set of elements they have in common.

$$\{1, 5, 8\} \cap \{0, 5, 10\} = \{5\}$$

A ∩ B means "the intersection of A and B."

The ***solution set of a compound and sentence*** is the intersection of the solution sets of its simple sentences.

P–3 **What are four numbers in the solution set of the compound *and* sentence "$x < 4$ *and* $x > -2$"?**

When a compound *and* sentence has two variables, its graph is drawn in a coordinate plane. A compound *and* sentence in two variables is often written as a ***system of linear sentences*** or, simply, as a linear system.

$$\begin{cases} \textbf{1.}\ y < -x + 3 \\ \textbf{2.}\ y > 3x - 1 \end{cases}$$

$y < -x + 3$ and $y > 3x - 1$

The ***solution set of a system of linear sentences*** is the intersection of the solution sets of the sentences.

> The ***graph of the solution set of a system of sentences*** is the intersection of the graphs of the sentences.

EXAMPLE 1

Graph the following linear system.

$$\begin{cases} \textbf{1. } y < 2x + 1 \\ \textbf{2. } y < -\frac{1}{2}x - 2 \end{cases}$$

y < 2x + 1
and
$y < -\frac{1}{2}x - 2$

Solution:

[1] Graph $y = 2x + 1$ as the dashed line in Figure 1.

[2] Shade the region below the dashed line to represent the solution set of $y < 2x + 1$.

[3] Graph $y = -\frac{1}{2}x - 2$ as the dashed line in Figure 2.

[4] Shade the region below the dashed line to represent the solution set of $y < -\frac{1}{2}x - 2$.

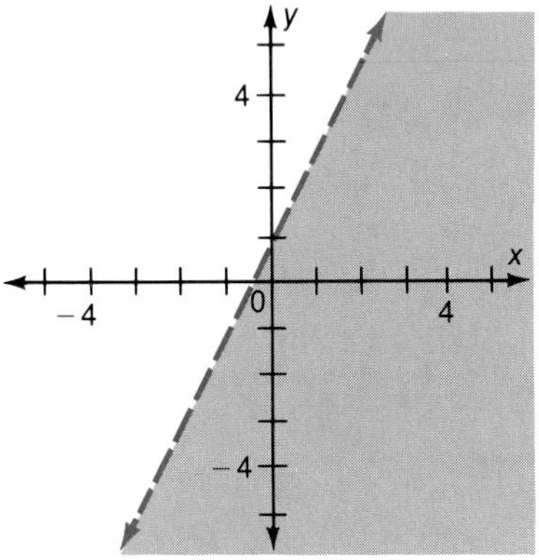

Figure 1

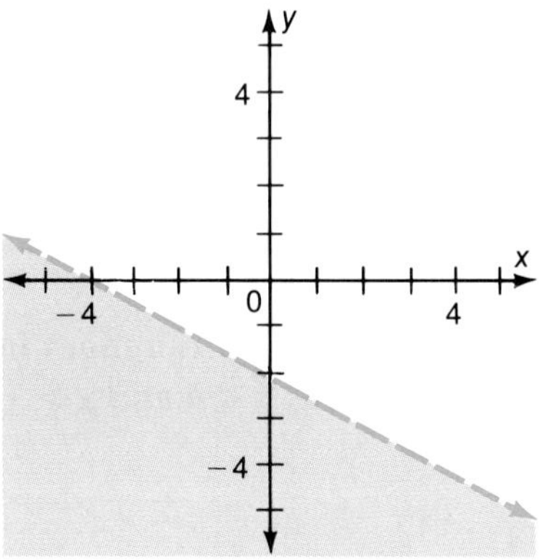

Figure 2

The graph of the solution set of the system of Example 1 is the intersection of the two graphs in Figures 1 and 2.

The intersection is shown by the darkest region in Figure 3.

The dashed lines are not included in the intersection.

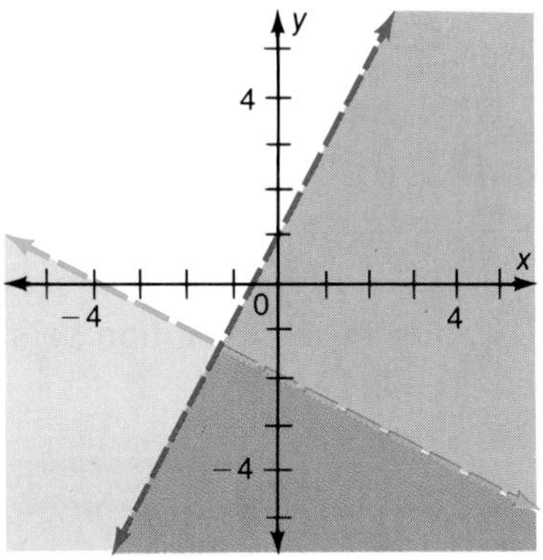

Figure 3

EXAMPLE 2 Graph the following linear system.

$$\begin{cases} \textbf{1. } y > 2x - 1 \\ \textbf{2. } y < -x + 3 \end{cases}$$

Solution:

[1] Graph $y > 2x - 1$ in the coordinate plane of Figure 4.

[2] Graph $y < -x + 3$.

The graph of the solution set of the system is the region showing the darkest shading in Figure 4.

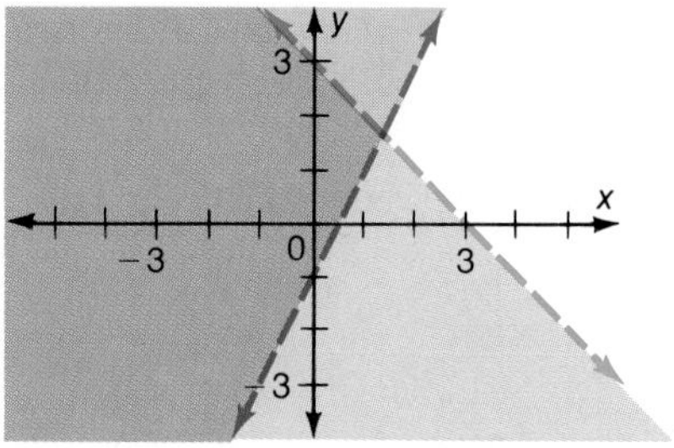

Figure 4

CLASSROOM EXERCISES

Write the region that is the graph of each system.

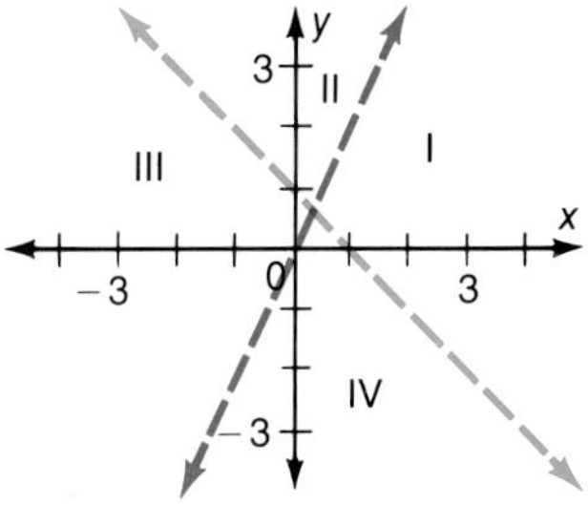

1. $\begin{cases} y < -x + 1 \\ y < 2x \end{cases}$

2. $\begin{cases} y > -x + 1 \\ y < 2x \end{cases}$

3. $\begin{cases} y > -x + 1 \\ y > 2x \end{cases}$

4. $\begin{cases} y < -x + 1 \\ y > 2x \end{cases}$

WRITTEN EXERCISES

Goal: To graph systems of linear sentences

Sample Problem: $\begin{cases} \textbf{1. } y > 2x + 5 \\ \textbf{2. } y < -x \end{cases}$

Answer: See the graph at the right.

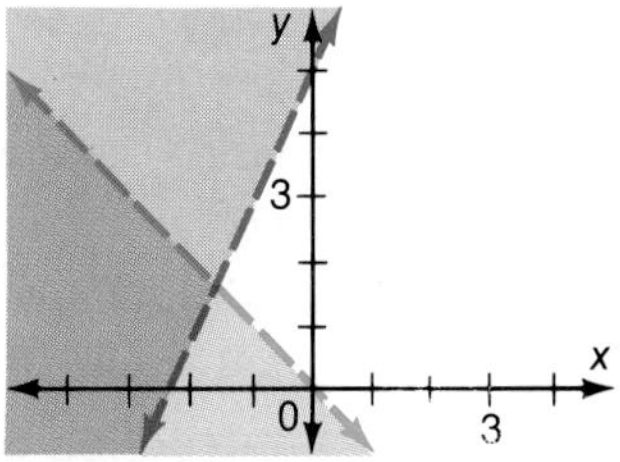

Graph each linear system. (Examples 1 and 2)

1. $\begin{cases} y < x + 1 \\ y < -x - 1 \end{cases}$

2. $\begin{cases} y > -x + 1 \\ y > x - 1 \end{cases}$

3. $\begin{cases} y > 2x + 1 \\ y < -x + 2 \end{cases}$

4. $\begin{cases} y < \frac{1}{2}x + 2 \\ y < 3x - 1 \end{cases}$

5. $\begin{cases} x + 2y < 4 \\ x - 2y < -4 \end{cases}$

6. $\begin{cases} 2x + y > -1 \\ x - 2y > 2 \end{cases}$

7. $\begin{cases} x - 3y > 3 \\ 2x + y < -1 \end{cases}$

8. $\begin{cases} 2x + y < 4 \\ x - y < 3 \end{cases}$

9. $\begin{cases} y < 2 \\ x < -1 \end{cases}$

10. $\begin{cases} y > -2 \\ x < -3 \end{cases}$

11. $\begin{cases} y < 3x \\ x > -1 \end{cases}$

12. $\begin{cases} y > -2x \\ x < 3 \end{cases}$

16.5 Graphs of Systems with $\leq$ and $\geq$

Recall that in graphing inequalities that contain $\leq$ or $\geq$, the line is part of the graph. Therefore, the graph of the related equation is drawn as a solid line.

EXAMPLE 1 Graph the following linear system.

$$\begin{cases} \textbf{1.}\ y \leq 3x - 1 \\ \textbf{2.}\ y \leq -2x + 3 \end{cases}$$

Solution:

[1] Graph $y = 3x - 1$ as the solid line in Figure 1.

[2] Shade the region below the solid line to represent the solution set of $y < 3x - 1$.

[3] Graph $y = -2x + 3$ as the solid line in Figure 2.

[4] Shade the region below the solid line to represent the solution set of $y < -2x + 3$.

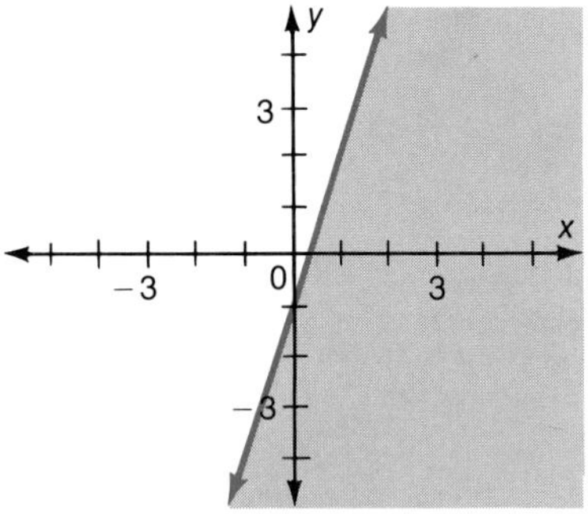

Figure 1

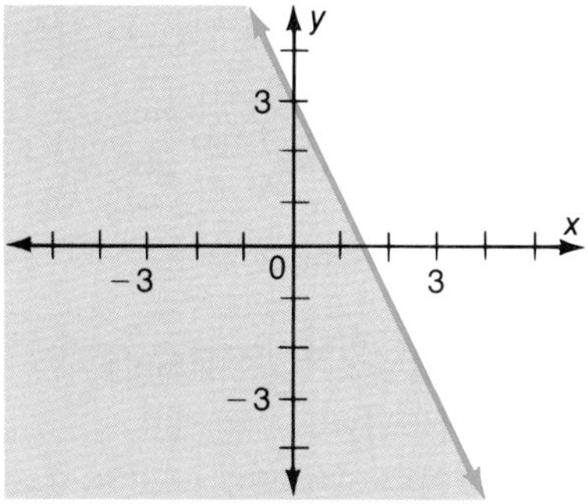

Figure 2

The graph of the solution set of the system of Example 1 is the intersection of the two graphs in Figures 1 and 2. The intersection is shown by the darkest region in Figure 3. The solid lines are included in the intersection.

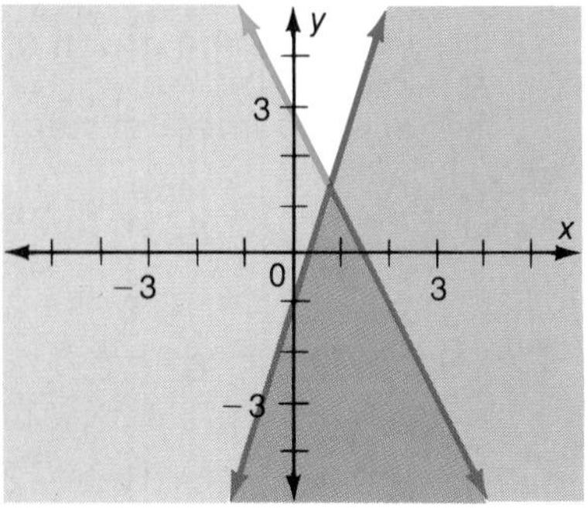

Figure 3

EXAMPLE 2

Graph the following linear system.

$$\begin{cases} \textbf{1. } y > 5x - 2 \\ \textbf{2. } y \geq \frac{1}{3}x \end{cases}$$

Solution:

[1] Graph $y > 5x - 2$ in the coordinate plane of Figure 4.

[2] Graph $y \geq \frac{1}{3}x$.

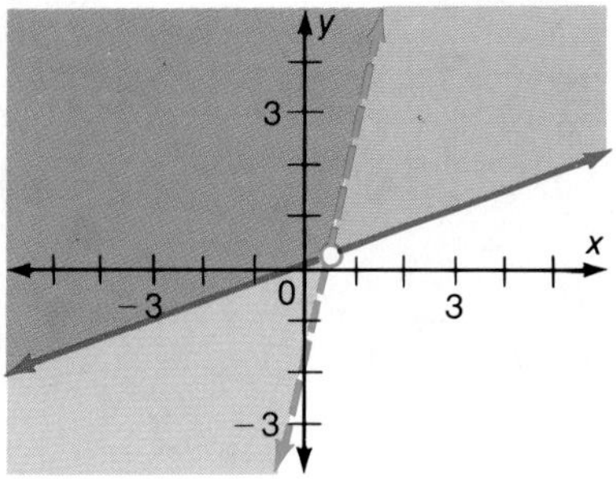

Figure 4

The graph of the solution set of the system is the region showing the darkest shading in Figure 4. A circle is drawn to emphasize that the point of intersection of the lines is not in the solution set of the system.

EXAMPLE 3

Write a system of linear sentences that corresponds to the graph of Figure 5.

Solution:

Inequality for points on or above the graph of $y = x$: $y \geq x$

Inequality for points above the graph of $y = -x$: $y > -x$

System: $\begin{cases} \textbf{1. } y \geq x \\ \textbf{2. } y > -x \end{cases}$

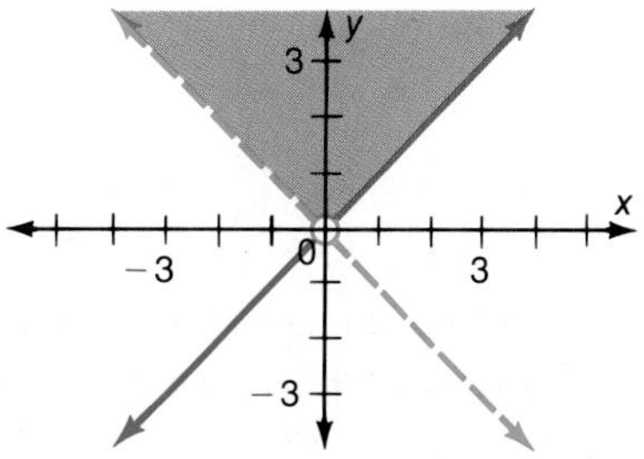

Figure 5

The circle drawn on the graph shows that the point of intersection of the lines is not in the solution set of the system.

CLASSROOM EXERCISES

Match each graph in Exercises 1–4 with one of the systems given in a–d. (Examples 1 and 2)

1.

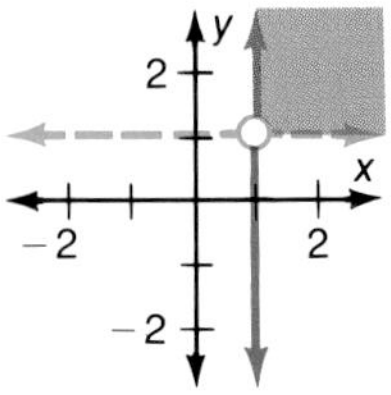

2.

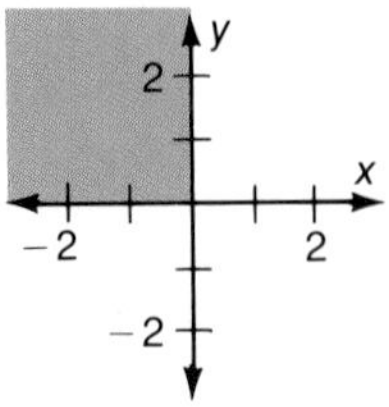

3.

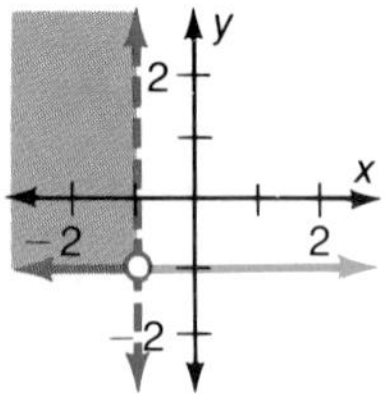

4.

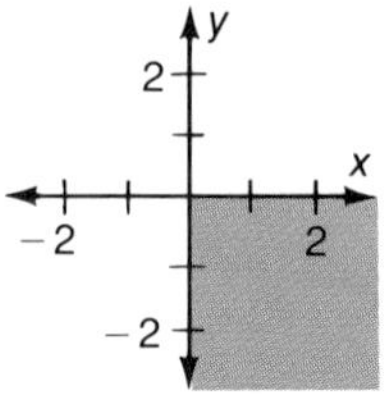

a. $\begin{cases} x \le 0 \\ y \ge 0 \end{cases}$ **b.** $\begin{cases} x \ge 0 \\ y \le 0 \end{cases}$ **c.** $\begin{cases} x \ge 1 \\ y > 1 \end{cases}$ **d.** $\begin{cases} x < -1 \\ y \ge -1 \end{cases}$

WRITTEN EXERCISES

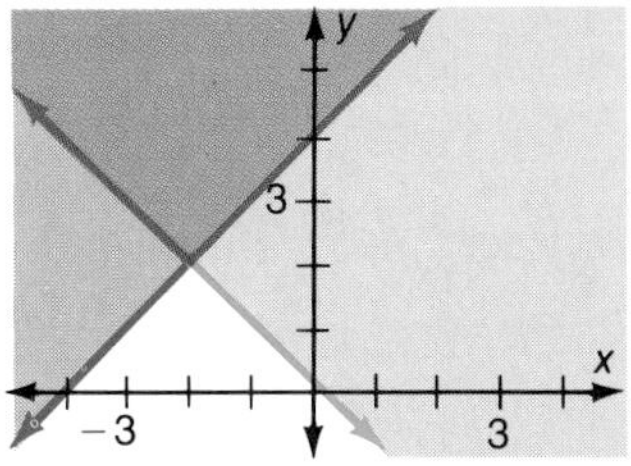

Goal: To graph systems of linear sentences

Sample Problem: $\begin{cases} \textbf{1. } y \ge x + 4 \\ \textbf{2. } y \ge -x \end{cases}$

Answer: See the graph at the right.

Graph each linear system. (Examples 1 and 2)

1. $\begin{cases} y \ge x + 1 \\ y \ge -x - 1 \end{cases}$ **2.** $\begin{cases} y \le 2x - 1 \\ y \le x \end{cases}$ **3.** $\begin{cases} y < -x + 1 \\ y \ge 2x \end{cases}$ **4.** $\begin{cases} y \ge \frac{1}{4}x + 1 \\ y > x - 1 \end{cases}$

5. $\begin{cases} x + y \le 4 \\ x - y \le -4 \end{cases}$ **6.** $\begin{cases} 2x + y \ge 1 \\ x - 2y \ge 1 \end{cases}$ **7.** $\begin{cases} x - 3y > 2 \\ 2x + y \le -2 \end{cases}$ **8.** $\begin{cases} 3x + y < -3 \\ x - y \ge 2 \end{cases}$

9. $\begin{cases} y > -x - 2 \\ y \ge x + 2 \end{cases}$ **10.** $\begin{cases} y \le -2x + 1 \\ y \le 2x - 3 \end{cases}$ **11.** $\begin{cases} y \ge 3x \\ y \le x \end{cases}$ **12.** $\begin{cases} y \ge \frac{1}{2}x - 2 \\ y \ge x - 3 \end{cases}$

13. $\begin{cases} y \le 3 \\ x < -2 \end{cases}$ **14.** $\begin{cases} y > -1 \\ x \le -2 \end{cases}$ **15.** $\begin{cases} y \le 3x \\ x > 0 \end{cases}$ **16.** $\begin{cases} y > -x \\ x \le 1 \end{cases}$

Write a system of linear sentences that corresponds to each graph. The equations of the lines are $y = x$ and $y = -x$. (Example 3)

17.

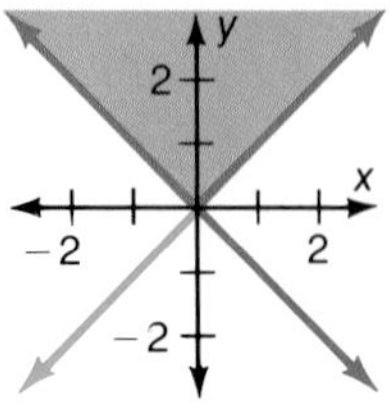

18.

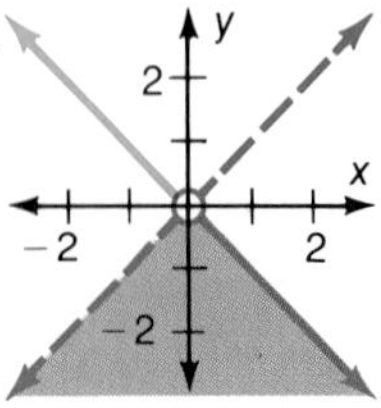

19.

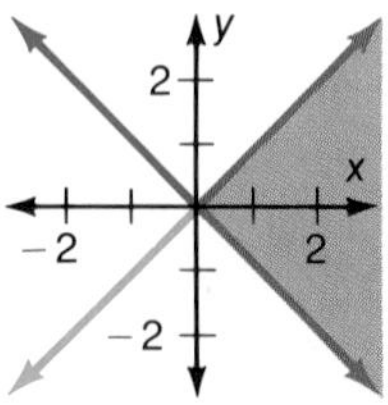

20.

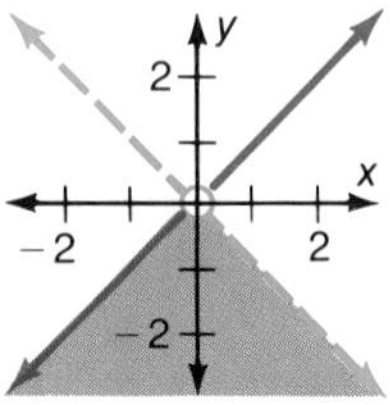

Linear Programming

In large manufacturing companies there may be **production planners** or **industrial** or **production technicians** who plan the layout of machinery and movement of materials in order to obtain the most efficient way of operating.

A method called **linear programming** has been developed for solving some of these problems. It is based on the solution of systems of linear inequalities.

EXAMPLE: A cement mixing company has two plants. There are 5 trucks at Plant I and 10 trucks at Plant II. The company must supply mixed cement for two building sites A and B. Site A can handle 8 trucks, and Site B can handle 4. The company's production planner must decide how many trucks to send from each plant to each building site in order to make transportation costs as small as possible. How many trucks should be sent from each plant to each site?

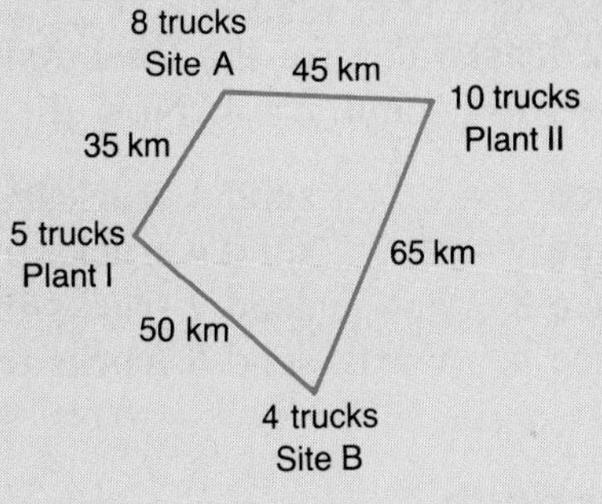

SOLUTION: Prepare a table showing the facts of the problem with the unknowns represented by two variables.

From Plant	To Site	Distance	Number of Trucks	Total Distance
I	A	35 km	x	$35x$
I	B	50 km	y	$50y$
II	A	45 km	$8-x$	$45(8-x)$
II	B	65 km	$4-y$	$65(4-y)$

Next, write a system of inequalities for the problem. Graph the system. In each case the number of trucks used must be greater than or equal to zero. Thus, six inequalities can be written.

1. $x \geq 0$
2. $y \geq 0$
3. $8 - x \geq 0$, or $x \leq 8$
4. $4 - y \geq 0$, or $y \leq 4$
5. $x + y \leq 5$
6. $(8 - x) + (4 - y) \leq 10$, or $x + y \geq 2$

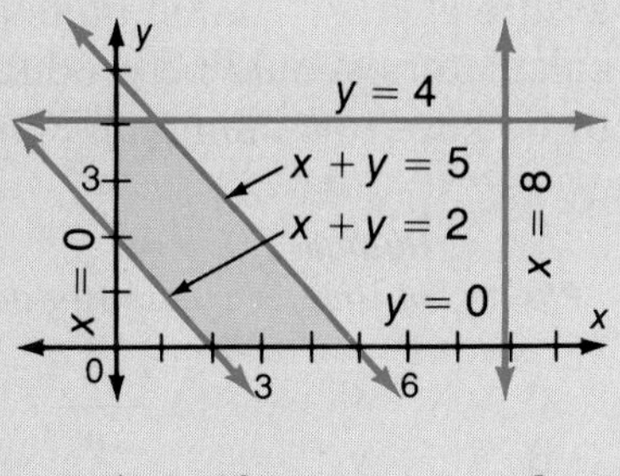

Represent the total distance by an algebraic expression.

$$\text{Total Distance} = 35x + 50y + 45(8 - x) + 65(4 - y)$$
$$= \mathbf{620 - 10x - 15y}$$

This distance must have a minimum value. Find the minimum value by substituting the x and y values of the five vertices of the polygon shown in the graph.

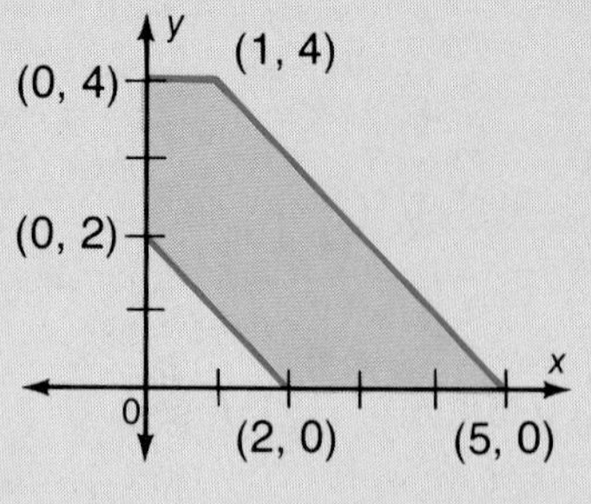

Vertex	*Total Distance*
(2, 0)	620 − 10(2) − 15(0) = 600 kilometers
(5, 0)	620 − 10(5) − 15(0) = 605 kilometers
(1, 4)	620 − 10(1) − 15(4) = 550 kilometers
(0, 4)	620 − 10(0) − 15(4) = 560 kilometers
(0, 2)	620 − 10(0) − 15(2) = 590 kilometers

The least value for the total distance is **550 kilometers.** This occurs when $x = 1$ and $y = 4$. Now the problem can be solved.

Since $x = 1$, send **1 truck from Plant I to Site A.**
Since $y = 4$, send **4 trucks from Plant I to Site B.**
Since $8 - x = 7$, send **7 trucks from Plant II to Site A.**
Since $4 - y = 0$, send **0 trucks from Plant II to Site B.**

EXERCISES

An electronics company has two plants that manufacture calculators. A special order of 10,000 calculators requires delivery within two weeks. To meet that deadline, the calculators must be produced in 5 or fewer working days (120 hours). Plant A can produce the calculators at a cost of \$1.30 each and at a rate of 60 units per hour. Plant B can produce the calculators at a cost of \$1.35 each and at a rate of 40 units per hour. On an order of 10,000 units, the quality control engineer recommends that 200 extra units be produced to assure at least 10,000 units of acceptable quality. Find how many calculators should be produced by each plant if the production cost is to be kept to a minimum. Complete the following table as the first step.

Plant	*Number of Units*	*Hours Needed*	*Production Cost*
A	x	?	?
B	y	?	?

CHAPTER SUMMARY

IMPORTANT TERMS

Linear inequality *(p. 434)*
Solution set of a linear inequality *(p. 434)*
Slope-intercept form of a linear inequality *(p. 435)*
Region *(p. 437)*
Compound sentence *(p. 440)*
Union of two sets *(p. 440)*
Solution set of a compound *or* sentence *(p. 440)*
Intersection of two sets *(p. 445)*
Solution set of a compound *and* sentence *(p. 445)*
System of linear sentences *(p. 445)*
Solution set of a system of linear sentences *(p. 445)*

IMPORTANT IDEAS

1. The graph of a linear equation separates a coordinate plane into three regions: the points on the line and the points of the plane on each side of the line.

2. Steps for drawing the graph of the relations $y < mx + b$ and $y > mx + b$:
   a. Draw the graph of $y = mx + b$.
   b. Select a point, not on the line, and substitute its coordinates in the inequality.
   c. Shade the region of the plane containing this point if its coordinates make the inequality true.
   d. Shade the region on the other side of the line from this point if its coordinates make the inequality false.

3. The graph of the solution set of a system of sentences is the intersection of the graphs of the sentences.

CHAPTER REVIEW

SECTION 16.1

Write five ordered pairs having different x values that are in the solution set of each inequality.

1. $y > -2x - 3$

2. $y < 4x - 6$

3. $y \leq \frac{5}{2}x + 1$

4. $y \geq -2.3x + 1.9$

Write the slope-intercept form of each inequality.

5. $-3x + y + 2 \geq 0$

6. $5x - y + 4 > 0$

7. $x - 3y + 5 < 0$

8. $2x + 3y - 1 \leq 0$

SECTION 16.2

Graph each inequality.

9. $y > \frac{1}{2}x + 3$

10. $y < -x + 2$

11. $y < -2x - 1$

12. $y > 2x - 3$

13. $y < -x$

14. $y > x$

SECTION 16.3

Graph each inequality.

15. $y \geq -x + 3$

16. $y \leq -\frac{1}{2}x + 1$

17. $x - 2y + 4 \leq 0$

18. $3x - 4y + 8 \geq 0$

19. $y \geq -4.5$

20. $x \leq -1.5$

SECTION 16.4

Graph each linear system.

21. $\begin{cases} y < \frac{1}{2}x + 2 \\ y < -x - 1 \end{cases}$

22. $\begin{cases} y < -2x + 3 \\ y > \frac{1}{3}x - 2 \end{cases}$

23. $\begin{cases} 2x + y < -3 \\ x + y > 2 \end{cases}$

SECTION 16.5

Graph each linear system.

24. $\begin{cases} y \geq 3x - 1 \\ y < x + 2 \end{cases}$

25. $\begin{cases} -3x + y \leq -1 \\ x + 2y \geq 4 \end{cases}$

26. $\begin{cases} y < 2 \\ x \geq -4 \end{cases}$

Quadratic Functions

CHAPTER 17

The photographs below are real-world models of quadratic functions that you may study in a later high school mathematics course: the *circle*, the *ellipse*, and the *parabola*.

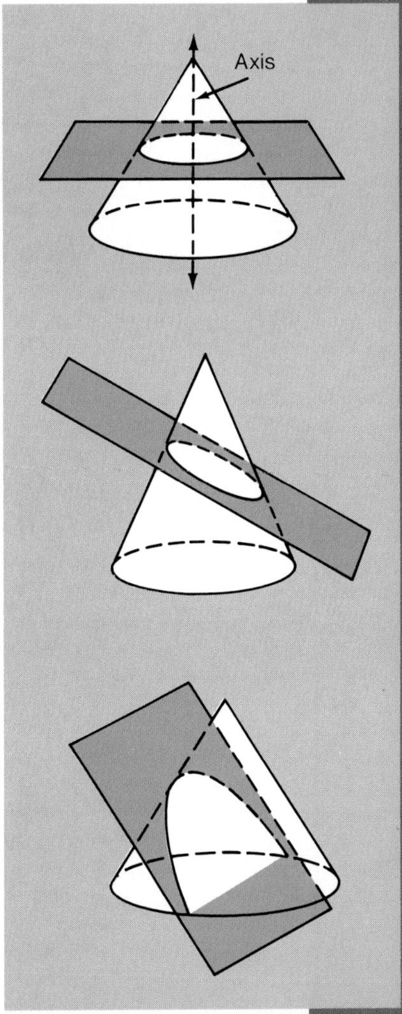

17.1 Quadratic Functions

Quadratic trinomials such as $ax^2 + bx + c$ are used to form rules for quadratic functions.

Definition

A ***quadratic function*** is a function described by the rule $y = ax^2 + bx + c$, in which a, b, and c are rational numbers with $a \neq 0$.

$$y = 2x^2 - 3x + 5$$

$$y = 0.5x^2 + 1.2x - 4.8$$

P–1 **Which functions below are quadratic?**

a. $y = 5x^2$ **b.** $y = 2x + 5$ **c.** $y = 3 - 2x^2$

EXAMPLE Write three ordered pairs of $y = -3x^2 + x - 2$.

Solution: Let the x values equal -2, 0, and 2.

$y = -3(-2)^2 + (-2) - 2$ ◀ ***Let*** $\boldsymbol{x = -2.}$

$y = -3(4) - 2 - 2$

$y = -12 - 2 - 2$

$y = -16$ The first ordered pair is $(-2, -16)$.

$y = -3(0)^2 + 0 - 2$ ◀ ***Let*** $\boldsymbol{x = 0.}$

$y = -3(0) + 0 - 2$

$y = 0 + 0 - 2$

$y = -2$ The second ordered pair is $(0, -2)$.

$y = -3(2)^2 + 2 - 2$ ◀ ***Let*** $\boldsymbol{x = 2.}$

$y = -3(4) + 2 - 2$

$y = -12 + 2 - 2$

$y = -12$ The third ordered pair is $(2, -12)$.

As with other functions, you can limit the domain of a quadratic function. Unless indicated otherwise, however, the domain will be {real numbers}.

CLASSROOM EXERCISES

Write Yes or No to indicate if each rule describes a quadratic function. (P-1)

1. $y = 5x^2 - 10x + 2$
2. $y = 3x - 10$
3. $y = x^3 + x^2 - 2x + 5$
4. $y = -3x^2$
5. $y = \frac{1}{x^2} + \frac{2}{x} + 3$
6. $y = x^2 - 3x$
7. $y = -2x + 1$
8. $y = \frac{3}{4}x^2 - \frac{1}{2}$
9. $y = \sqrt{x^2}$
10. $y = \sqrt{3}x^2 + \sqrt{2}x + \sqrt{5}$
11. $y = 2 - 3x^2 - 5x$
12. $y = (x + 1)^2$
13. $y = (x - 2)(x + 3)$
14. $y = x^{-2} + 3x^{-1} + 5$
15. $y = 3^2$

WRITTEN EXERCISES

Goal: To write ordered pairs of a quadratic equation
Sample Problem: Write three ordered pairs of $y = -x^2 + x + 1$.
Let $x = -2$, 0, and 1. **Answer:** $(-2, -5)$, $(0, 1)$, $(1, 1)$

Write Yes or No to show whether each rule describes a quadratic function. (P-1)

1. $y = 2x - 5$
2. $y = -x + 3$
3. $y = -3x^2$
4. $y = 5x^2$
5. $y = x^3 + 2x - 6$
6. $y = x^4 - 2x - 3$
7. $y = (x - 2)^2$
8. $y = (2x - 3)(x + 1)$
9. $y = \frac{1}{x^2 + x - 5}$
10. $y = \sqrt{x^2 - 2x - 3}$
11. $y = 0.5x^2 + 0.7x - 0.3$
12. $y = -1.7x^2 - 3.1x + 2.7$

Write three ordered pairs of each function. Let the x values equal −3, −1, and 2. (Example 1)

13. $y = -x^2$
14. $y = -x^2 + 1$
15. $y = x^2 + 2x + 1$
16. $y = x^2 - x + 1$
17. $y = x^2 - 3$
18. $y = x^2 - 4$
19. $y = -2x^2 - 3x$
20. $y = -3x^2 - 1$

17.2 Graphs of Quadratic Functions

You have already seen the graph of a quadratic function. It is a smooth curved line.

EXAMPLE 1 Graph $y = x^2 - 2x - 3$.

Prepare a table of ordered pairs. → **Graph the points for these pairs.** → **Draw a smooth curve containing the points.**

Solution:

$y = x^2 - 2x - 3$

$y = (-2)^2 - 2(-2) - 3 = 5$

$y = (-1)^2 - 2(-1) - 3 = 0$

$y = (0)^2 - 2(0) - 3 = -3$

$y = (1)^2 - 2(1) - 3 = -4$

$y = (2)^2 - 2(2) - 3 =$ __?__

$y = (3)^2 - 2(3) - 3 =$ __?__

$y = (4)^2 - 2(4) - 3 =$ __?__

x	y
−2	5
−1	0
0	−3
1	−4
2	?
3	?
4	?

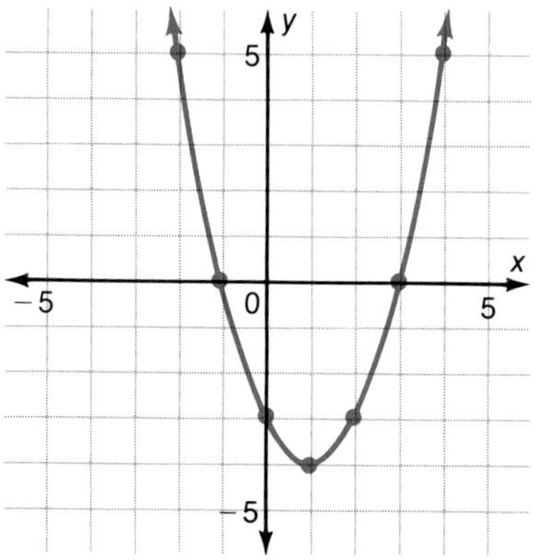

Figure 1

The points corresponding to these ordered pairs are graphed in Figure 1. A smooth curve containing the points is drawn.

The curve of Example 1 is called a ***parabola.***

P–1 **Based on the graph, what is an estimate of each unknown *y* value?**

a. $(2\frac{1}{2},$ __?__$)$ **b.** $(-1\frac{1}{2},$ __?__$)$ **c.** $(\frac{1}{2},$ __?__$)$

The two arrows on the graph of Figure 1 indicate that the parabola extends upward infinitely. The curve widens as it extends upward.

P–2 **Based on the graph, what is the range of the function?**

There is no y value less than -4. The range is {all real numbers greater than or equal to -4}.

EXAMPLE 2 Graph $y = -x^2 + 3x$.

Solution:

$y = -x^2 + 3x$

$y = -(-1)^2 + 3(-1) = -4$

$y = -(0)^2 + 3(0) = 0$

$y = -(1)^2 + 3(1) = 2$

$y = -(2)^2 + 3(2) = 2$

$y = -(3)^2 + 3(3) = 0$

$y = -(4)^2 + 3(4) = -4$

x	y
−1	−4
0	0
1	2
2	2
3	0
4	−4

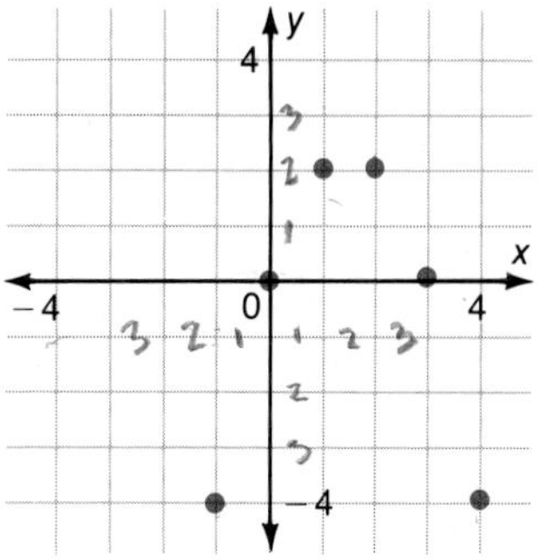

It is also helpful to locate the point at which the parabola "turns" or changes direction.

$y = -x^2 + 3x$ ◀ ***Let*** $x = \frac{3}{2}$.

$y = -\left(\frac{3}{2}\right)^2 + 3\left(\frac{3}{2}\right)$

$y = -\frac{9}{4} + \frac{9}{2}$ ◀ ***Solve for y.***

$y = -\frac{9}{4} + \frac{18}{4}$

$y = \frac{9}{4}$

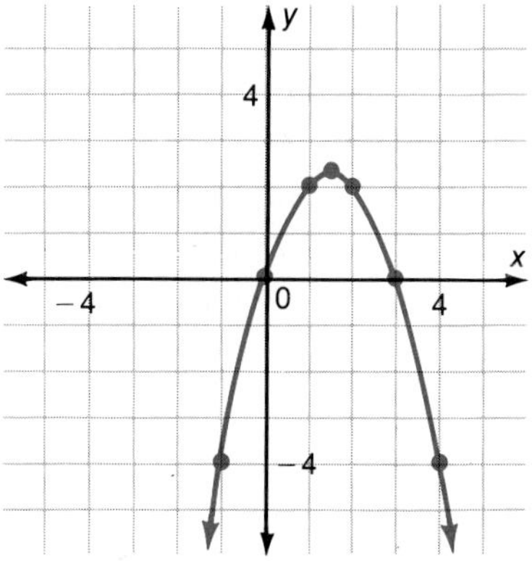

The point, $(\frac{3}{2}, \frac{9}{4})$, is located.

CLASSROOM EXERCISES

Write the unknown y values in each table. (Example 1)

1. $y = x^2$

x	y
−2	?
−1	?
0	?
1	?
2	?

2. $y = -x^2$

x	y
−2	?
−1	?
0	?
1	?
2	?

3. $y = x^2 + 2$

x	y
−2	?
−1	?
0	?
1	?
2	?

4. $y = x^2 + x + 1$

x	−2	−1	0	1	2
y	?	?	?	?	?

5. $y = 2x^2 + x$

x	−2	−1	0	1	2
y	?	?	?	?	?

6. $y = -x^2 + 2x - 3$

x	−2	−1	0	1	2
y	?	?	?	?	?

Refer to the graph at the right. Estimate one value for y and two values for x. (P-1)

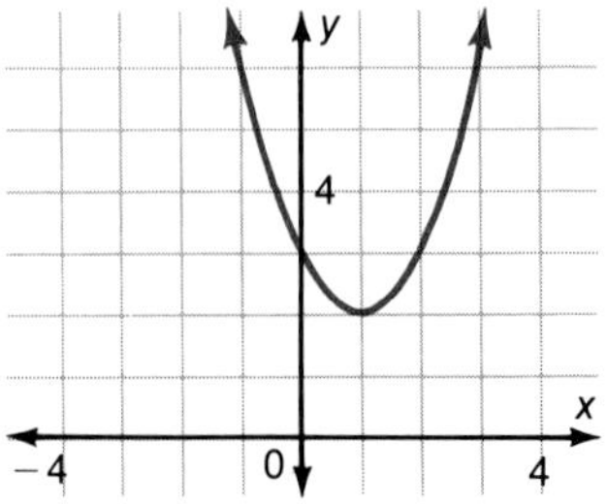

7. $(1, y)$
8. $(2, y)$
9. $(\frac{1}{2}, y)$
10. $(-\frac{1}{2}, y)$
11. $(x, 3)$
12. $(x, 6)$
13. $(x, 2\frac{1}{4})$
14. $(x, 5)$

WRITTEN EXERCISES

Goal: To graph quadratic functions
Sample Problem: Graph $y = -x^2 + 3x$. **Answer:** See Example 2.

Graph the ordered pairs of each table. Draw a parabola that contains the points. (Examples 1 and 2)

1.

x	−2	−1	0	1	2
y	5	2	1	2	5

2.

x	−2	−1	0	1	2
y	3	0	−1	0	3

3.

x	−2	−1	0	1	2	3
y	−5	0	3	4	3	0

4.

x	−3	−2	−1	0	1
y	−3	0	1	0	−3

5.

x	−1	0	1	2	3
y	6	3	2	3	6

6.

x	−2	−1	0	1	2	3
y	4	0	−2	−2	0	4

Graph each function. (Examples 1 and 2)

7. $y = x^2 - 2$
8. $y = -x^2 + 3$
9. $y = -x^2 - x$
10. $y = 2x^2 + x$
11. $y = x^2 - 6x + 10$
12. $y = -x^2 - 2x + 1$

NON-ROUTINE PROBLEM

13. Find the missing digits that make the three statements at the right true. The digits 0 through 9 can each appear exactly once in the three statements.

$7 + ■ = ■$

$■ - 6 = ■$

$■ \times ■ = 20$

17.3 Functions Defined by $y = ax^2$

You have learned that any quadratic function has a rule of the form $y = ax^2 + bx + c$.

P–1 **Find the values of a, b, and c in each rule below.**

1. $y = -\frac{5}{2}x^2$ **2.** $y = 3x^2 - 4x - 1$ **3.** $y = x^2 + 3x$

EXAMPLE 1

Graph in the same coordinate plane.

1. $y = 2x^2$ **2.** $y = x^2$ **3.** $y = \frac{1}{2}x^2$

Solutions:

1.

x	y
−2	8
−1	2
0	0
1	2
2	8

2.

x	y
−2	4
−1	1
0	0
1	1
2	4

3.

x	y
−4	8
−2	2
0	0
2	2
4	8

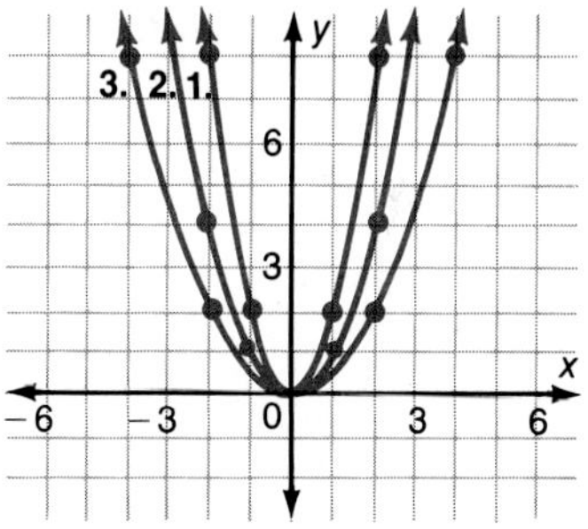

If $a > 0$, the parabola opens upward.

P–2 **Which function in Example 1 has the "widest" graph? the next widest? the least wide?**

EXAMPLE 2

Graph in the same coordinate plane.

1. $y = -2x^2$ **2.** $y = -x^2$ **3.** $y = -\frac{1}{2}x^2$

Solutions:

1.

x	y
−2	−8
−1	−2
0	0
1	−2
2	−8

2.

x	y
−2	−4
−1	−1
0	0
1	−1
2	−4

3.

x	y
−4	−8
−2	−2
0	0
2	−2
4	−8

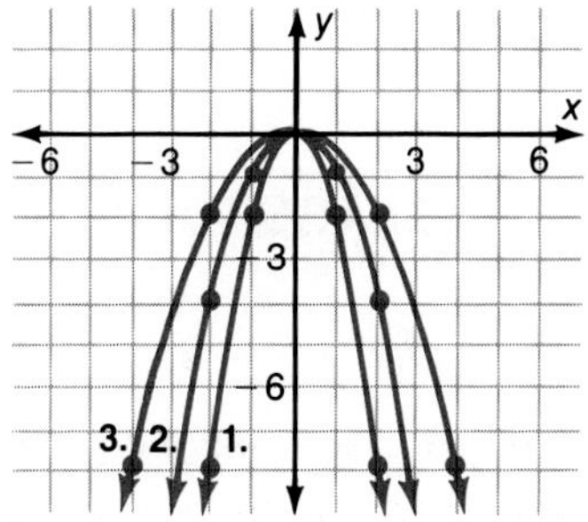

If $a < 0$, the parabola opens downward.

P–3 **Which function in Example 2 has the widest graph? the next widest? the least wide?**

Examples 1 and 2 show that as a gets smaller in absolute value, the corresponding parabola gets wider.

> ***Effect of a on the graph of $y = ax^2 + bx + c$:***
>
> 1. If a is positive, the parabola opens upward.
> 2. If a is negative, the parabola opens downward.
> 3. As $|a|$ decreases, the parabola becomes wider.

A relation defined by a rule of the form $x = ay^2$ also has a parabola for its graph.

EXAMPLE 3 Graph the parabola defined by $x = -\frac{1}{2}y^2$.

Solution: Choose values for y and compute corresponding values for x.

$x = -\frac{1}{2}y^2$
$x = -\frac{1}{2}(-2)^2 = -2$
$x = -\frac{1}{2}(-1)^2 = -\frac{1}{2}$
$x = -\frac{1}{2}(0)^2 = 0$
$x = -\frac{1}{2}(1)^2 = -\frac{1}{2}$
$x = -\frac{1}{2}(2)^2 = -2$

x	y
-2	-2
$-\frac{1}{2}$	-1
0	0
$-\frac{1}{2}$	1
-2	2

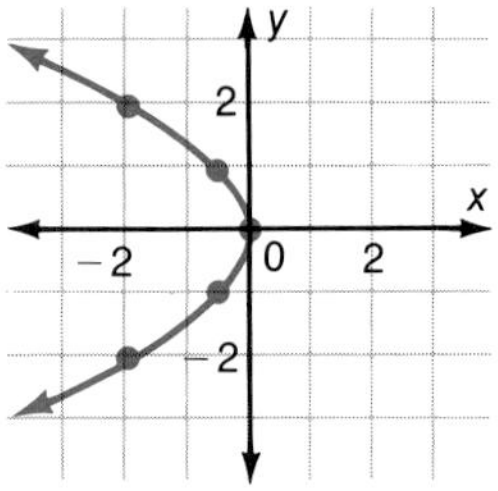

You can see that $x = -\frac{1}{2}y^2$ does not define a function. There are many ordered pairs that have the same x values.

CLASSROOM EXERCISES

Refer to the graphs of Examples 1 and 2.

1. What point is in all six graphs?

2. How do the graphs of $y = x^2$ and $y = -x^2$ compare?

3. Does the graph of $y = -2x^2$ open upward or downward?

4. Which parabola is wider, $y = -x^2$ or $y = 2x^2$?

5. What is the range of the function defined by $y = 2x^2$?

6. What is the range of the function defined by $y = -2x^2$?

Indicate if the graph of each function opens upward or downward. (Examples 1 and 2)

7. $y = -10x^2$
8. $y = 0.3x^2$
9. $y = 4 - \frac{5}{2}x^2$
10. $y = (2x - 1)^2$

Indicate which of the two functions in each exercise has the wider graph. (P-2, P-3)

11. $y = 5x^2$; $y = 3x^2$
12. $y = \frac{1}{2}x^2$; $y = \frac{3}{2}x^2$
13. $y = 0.15x^2$; $y = 0.09x^2$
14. $y = x^2 - 5x + 1$; $y = -2x^2 + x - 3$

WRITTEN EXERCISES

Goal: To use a in the rule $y = ax^2 + bx + c$ to compare the graphs of quadratic functions

Sample Problem: Graph in the same coordinate plane: $y = 2x^2$, $y = \frac{1}{2}x^2$, $y = x^2$, $y = -2x^2$, $y = -\frac{1}{2}x^2$, $y = -x^2$.

Answer: See Examples 1 and 2.

Complete each table. Graph the two functions in the same coordinate plane. (Examples 1 and 2)

1. $y = x^2$

x	y
−2	?
−1	?
0	?
1	?
2	?

$y = \frac{1}{4}x^2$

x	y
−4	?
−2	?
0	?
2	?
4	?

2. $y = -x^2$

x	y
−2	?
−1	?
0	?
1	?
2	?

$y = -\frac{1}{4}x^2$

x	y
−4	?
−2	?
0	?
2	?
4	?

3. $y = -\frac{1}{2}x^2$

x	y
−3	?
−2	?
0	?
2	?
3	?

$y = -\frac{3}{2}x^2$

x	y
−2	?
−1	?
0	?
1	?
2	?

4. $y = \frac{1}{2}x^2$

x	y
−3	?
−2	?
0	?
2	?
3	?

$y = \frac{3}{2}x^2$

x	y
−2	?
−1	?
0	?
1	?
2	?

5. $y = -2x^2$

x	y
−2	?
−1	?
0	?
1	?
2	?

$y = \frac{3}{2}x^2$

x	y
−2	?
−1	?
0	?
1	?
2	?

6. $y = 2x^2$

x	y
−2	?
−1	?
0	?
1	?
2	?

$y = -\frac{3}{2}x^2$

x	y
−2	?
−1	?
0	?
1	?
2	?

7. $y = \frac{1}{2}x^2$

x	y
−2	?
−1	?
0	?
1	?
2	?

$y = \frac{1}{2}x^2 - 2$

x	y
−2	?
−1	?
0	?
1	?
2	?

8. $y = -\frac{1}{4}x^2$

x	y
−4	?
−2	?
0	?
2	?
4	?

$y = -\frac{1}{4}x^2 + 3$

x	y
−4	?
−2	?
0	?
2	?
4	?

9. $y = -\frac{3}{4}x^2$

x	y
−4	?
−2	?
0	?
2	?
4	?

$y = -\frac{3}{4}x^2 + 1$

x	y
−4	?
−2	?
0	?
2	?
4	?

10. $y = \frac{3}{8}x^2$

x	y
−4	?
−2	?
0	?
2	?
4	?

$y = \frac{3}{8}x^2 - 2$

x	y
−4	?
−2	?
0	?
2	?
4	?

Complete each table. Then graph each parabola. (Example 3)

11. $x = y^2$

x	y
?	1
?	−1
?	0
?	2
?	−2

12. $x = -y^2$

x	y
?	1
?	−1
?	0
?	2
?	−2

13. $x = 2y^2$

x	y
?	1
?	−1
?	0
?	2
?	−2

14. $x = -2y^2$

x	y
?	1
?	−1
?	0
?	2
?	−2

REVIEW CAPSULE FOR SECTION 17.4

Multiply and simplify. (Pages 220–223)

1. $(x - 2)(x - 2)$

2. $(x + 3)(x + 3)$

3. $(x - 5)(x - 5)$

4. $(x + 6)(x + 6)$

5. $(x + 2.1)(x + 2.1)$

6. $(x - 2.7)(x - 2.7)$

17.4 Functions Defined by $y = a(x - h)^2$

The equation $y = \frac{1}{2}(x - 2)^2$ is a rule for a quadratic function.

$$\frac{1}{2}(x-2)^2 = \frac{1}{2}(x-2)(x-2)$$
$$= \frac{1}{2}(x^2 - 4x + 4)$$
$$= \frac{1}{2}x^2 - 2x + 2$$

In the form $ax^2 + bx + c$

P–1 **What are the values of *a*, *b*, and *c*?**

Replace $(x - 2)$ by A in $y = \frac{1}{2}(x-2)^2$. ⟶ $y = \frac{1}{2}A^2$

This form suggests that the graph of $y = \frac{1}{2}(x-2)^2$ is related to the graph of $y = \frac{1}{2}x^2$.

EXAMPLE 1 Graph the parabolas $y = \frac{1}{2}x^2$ and $y = \frac{1}{2}(x-2)^2$ in the same coordinate plane.

Solutions:

$y = \frac{1}{2}x^2$

x	y
−3	$4\frac{1}{2}$
−2	2
−1	$\frac{1}{2}$
0	0
1	$\frac{1}{2}$
2	2
3	$4\frac{1}{2}$

$y = \frac{1}{2}(x-2)^2$

x	y
−1	$4\frac{1}{2}$
0	2
1	$\frac{1}{2}$
2	0
3	$\frac{1}{2}$
4	2
5	$4\frac{1}{2}$

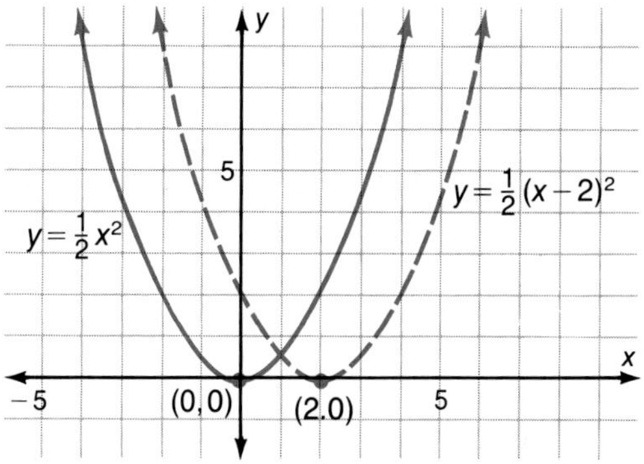

The parabolas have the same shape.

Note that the graph of $y = \frac{1}{2}x^2$ can be moved ***two*** units to the ***right*** to obtain the graph of $y = \frac{1}{2}(x-2)^2$.

EXAMPLE 2 Graph in the same coordinate plane.

1. $y = -\frac{1}{2}x^2$ **2.** $y = -\frac{1}{2}(x+3)^2$

Solutions:

$y = -\frac{1}{2}x^2$

x	y
−3	$-4\frac{1}{2}$
−2	−2
−1	$-\frac{1}{2}$
0	0
1	$-\frac{1}{2}$
2	−2
3	$-4\frac{1}{2}$

$y = -\frac{1}{2}(x + 3)^2$

x	y
−6	$-4\frac{1}{2}$
−5	−2
−4	$-\frac{1}{2}$
−3	0
−2	$-\frac{1}{2}$
−1	−2
0	$-4\frac{1}{2}$

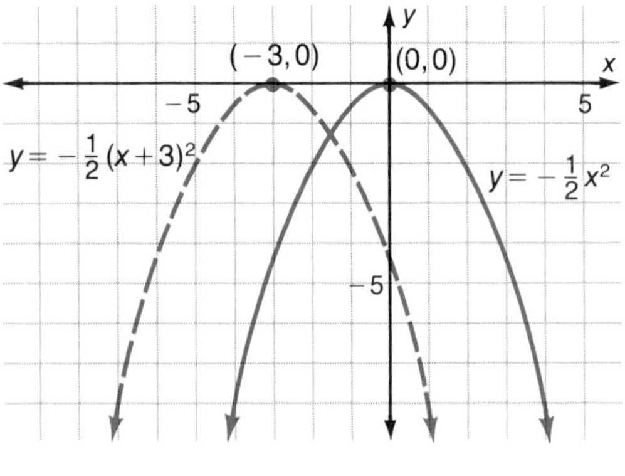

The parabolas have the same shape.

Note that the graph of $y = -\frac{1}{2}x^2$ can be moved ***three*** units to the ***left*** to obtain the graph of $y = -\frac{1}{2}(x + 3)^2$.

Graphs of $y = ax^2$ and $y = a(x - h)^2$

The graph of $y = ax^2$ can be moved horizontally $|h|$ units to obtain the graph of $y = a(x - h)^2$. If h is negative, the motion is to the left. If h is positive, the motion is to the right.

EXAMPLE 3

Graph the parabola $y = (x + 4)^2$ in the same plane as the parabola $y = x^2$ shown in Figure 1. Locate points of $y = (x + 4)^2$ that correspond to points A, B, C, D, and E.

Solution:

1 Locate points A', B', C', D', and E' in Figure 2 that are 4 units to the left of the corresponding points of Figure 1.

2 Draw the parabola.

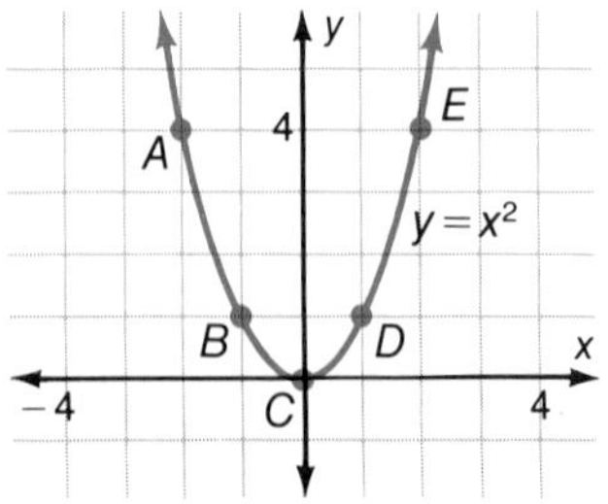

Figure 1

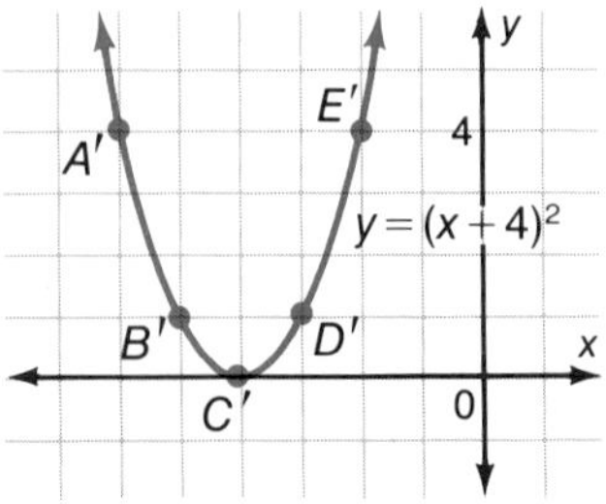

Figure 2

CLASSROOM EXERCISES

For each exercise, indicate how to move the graph of the first function to obtain the graph of the second function.

1. $y = x^2$; $y = (x - 2)^2$
2. $y = x^2$; $y = (x + 2)^2$
3. $y = 5x^2$; $y = 5(x - 1)^2$
4. $y = 5x^2$; $y = 5(x + 4)^2$
5. $y = -3x^2$; $y = -3(x + 2)^2$
6. $y = -3x^2$; $y = -3(x - 1)^2$
7. $y = \frac{1}{3}x^2$; $y = \frac{1}{3}(x + 10)^2$
8. $y = \frac{1}{3}x^2$; $y = \frac{1}{3}(x - 5)^2$
9. $y = \frac{3}{4}x^2$; $y = \frac{3}{4}(x - 7)^2$
10. $y = \frac{3}{4}x^2$; $y = \frac{3}{4}(x + 6)^2$
11. $y = 2(x - 3)^2$; $y = 2(x + 3)^2$
12. $y = 2(x - 1)^2$; $y = 2(x + 2)^2$

WRITTEN EXERCISES

Goal: To use a graph of $y = ax^2$ to draw a graph of the form $y = a(x - h)^2$
Sample Problem: Graph $y = (x - 4)^2$ and $y = (x + 4)^2$.
Answer: See Example 3.

Write the number of units and the direction the graph of the first function can be moved to obtain the graph of the second function.

EXAMPLE: $y = 5x^2$; $y = 5(x + 6)^2$ **ANSWER:** 6 units left

1. $y = -3x^2$; $y = -3(x - 5)^2$
2. $y = 4x^2$; $y = 4(x - 10)^2$
3. $y = \frac{5}{2}x^2$; $y = \frac{5}{2}(x + 8)^2$
4. $y = -\frac{7}{2}x^2$; $y = -\frac{7}{2}(x + 6)^2$
5. $y = -x^2$; $y = -(x + 3\frac{1}{2})^2$
6. $y = x^2$; $y = (x - 2\frac{1}{4})^2$
7. $y = 1.4x^2$; $y = 1.4(x - 2.7)^2$
8. $y = -3.6x^2$; $y = -3.6(x + 2.1)^2$

Graph the first parabola in each exercise using the x values −2, −1, 0, 1, and 2. Then graph the second parabola. (Examples 1–3)

9. $y = 2x^2$; $y = 2(x + 2)^2$
10. $y = -2x^2$; $y = -2(x - 2)^2$
11. $y = -\frac{1}{2}x^2$; $y = -\frac{1}{2}(x - 3)^2$
12. $y = \frac{1}{2}x^2$; $y = \frac{1}{2}(x + 3)^2$
13. $y = -x^2$; $y = -(x + \frac{3}{2})^2$
14. $y = x^2$; $y = (x - \frac{5}{2})^2$
15. $y = \frac{3}{2}x^2$; $y = \frac{3}{2}(x - 2)^2$
16. $y = -\frac{3}{2}x^2$; $y = -\frac{3}{2}(x + 2)^2$
17. $y = -3x^2$; $y = -3(x + 1)^2$
18. $y = 3x^2$; $y = 3(x - 1)^2$

APPLICATIONS

The expression $c - \frac{b^2}{4a}$ *represents the* <u>*least*</u> *value of the quadratic function* $y = ax^2 + bx + c$*, when* $a > 0$.

EXAMPLE Find the approximate least value of $y = 21x^2 + 9x + 52$.

SOLUTION $a = 21, b = 9, c = 52$ $\qquad c - \frac{b^2}{4a} = 52 - \frac{9^2}{(4)(21)}$

[9] [×] [=] [÷] [4] [÷] [2] [1]

[=] [M+] [5] [2] [−] [MR] [=] 51.035715 ← **Least value**

Find the least value.

19. $y = 34x^2 + 68x + 109$ **20.** $y = 176x^2 + 465x + 239$ **21.** $y = 1084x^2 + 1296x + 238$

MID-CHAPTER REVIEW

Write four ordered pairs for each function. Let the x values equal $-2, -1,$ *0, and 1.* (Section 17.1)

1. $y = -\frac{1}{2}x^2 - 3x$ **2.** $y = 5x^2 - 2$ **3.** $y = x^2 + x - 3$

Graph each function. (Section 17.2)

4. $y = \frac{1}{2}x^2 - 3$ **5.** $y = -\frac{1}{2}x^2 + 2$ **6.** $y = x^2 - 4x + 2$

Graph the two functions in the same coordinate plane. (Section 17.3)

7. $y = \frac{3}{4}x^2$, $y = -\frac{3}{4}x^2$ **8.** $y = \frac{5}{2}x^2$, $y = -\frac{5}{2}x^2$ **9.** $y = 2x^2$, $y = \frac{1}{2}x^2$ **10.** $y = -\frac{3}{2}x^2$, $y = -\frac{3}{4}x^2$ **11.** $y = -2x^2 - 2$, $y = \frac{1}{2}x^2 + 1$

Write the number of units and the direction the graph of the first function can be moved to obtain the graph of the second function. (Section 17.4)

12. $y = \frac{1}{4}x^2$, $y = \frac{1}{4}(x - 3)^2$ **13.** $y = -3x^2$, $y = -3(x - 5)^2$ **14.** $y = -\frac{5}{2}x^2$, $y = -\frac{5}{2}(x + \frac{3}{2})^2$ **15.** $y = 5x^2$, $y = 5(x + 4)^2$

REVIEW CAPSULE FOR SECTION 17.5

Write three ordered pairs for each function below. Let the x values equal -2*, 0, and 2.* (Pages 456–457)

1. $y = \frac{1}{2}(x - 2)^2 + 1$ **2.** $y = 2x^2 - x - 3$ **3.** $y = -2(x + 1)^2 - 3$

17.5 Functions Defined by $y = a(x - h)^2 + k$

The expression $\frac{1}{2}(x - 3)^2 + 2$ can be changed to the form $ax^2 + bx + c$.

$$\frac{1}{2}(x - 3)^2 + 2 = \frac{1}{2}(x^2 - 6x + 9) + 2$$
$$= \frac{1}{2}x^2 - 3x + \frac{9}{2} + 2$$
$$= \frac{1}{2}x^2 - 3x + \frac{13}{2}$$

In the form $ax^2 + bx + c$

P–1 **What kind of function does $y = \frac{1}{2}(x - 3)^2 + 2$ represent?**

EXAMPLE 1 Graph the following functions in the same coordinate plane.

1. $y = \frac{1}{2}x^2$ **2.** $y = \frac{1}{2}(x - 3)^2$ **3.** $y = \frac{1}{2}(x - 3)^2 + 2$

Solutions:

1.

x	y
−3	$4\frac{1}{2}$
−2	2
−1	$\frac{1}{2}$
0	0
1	$\frac{1}{2}$
2	2
3	$4\frac{1}{2}$

2.

x	y
0	$4\frac{1}{2}$
1	2
2	$\frac{1}{2}$
3	0
4	$\frac{1}{2}$
5	2
6	$4\frac{1}{2}$

3.

x	y
0	$6\frac{1}{2}$
1	4
2	$2\frac{1}{2}$
3	2
4	$2\frac{1}{2}$
5	4
6	$6\frac{1}{2}$

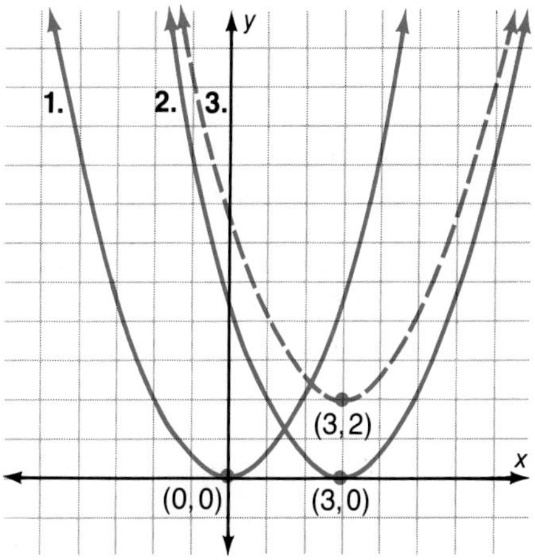

The parabolas have same shape.

P–2 **How could the graph of $y = \frac{1}{2}(x - 3)^2$ be moved to make it coincide with the graph of $y = \frac{1}{2}(x - 3)^2 + 2$?**

P–3 **How could the graph of $y = \frac{1}{2}x^2$ be moved in two steps to make it coincide with the graph of $y = \frac{1}{2}(x - 3)^2 + 2$?**

EXAMPLE 2

Graph the following functions in the same coordinate plane.

1. $y = -2x^2$ **2.** $y = -2(x + 4)^2$

3. $y = -2(x + 4)^2 - 3$

Solutions:

1.

x	y
−2	−8
−1	−2
0	0
1	−2
2	−8

2.

x	y
−6	−8
−5	−2
−4	0
−3	−2
−2	−8

3.

x	y
−6	−11
−5	−5
−4	−3
−3	−5
−2	−11

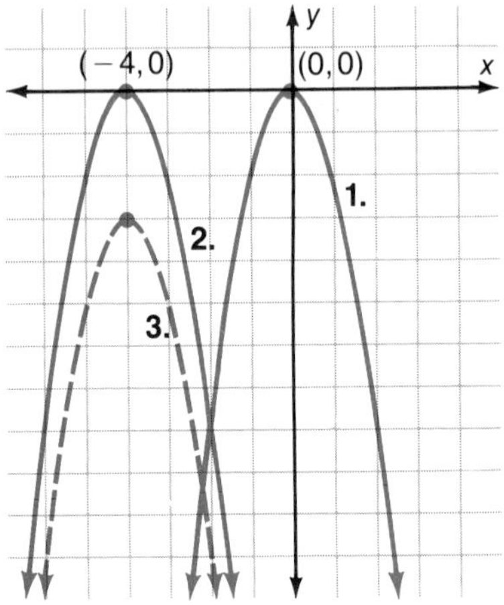

The parabolas have same shape.

P–4 **How could the graph of $y = -2(x + 4)^2$ be moved to make it coincide with the graph of $y = -2(x + 4)^2 - 3$?**

P–5 **How could the graph of $y = -2x^2$ be moved in two steps to make it coincide with the graph of $y = -2(x + 4)^2 - 3$?**

Graphs of $y = ax^2$ and $y = a(x - h)^2 + k$

1. The graph of $y = a(x - h)^2 + k$ has the same shape as the graph of $y = ax^2$.
2. The graph of $y = a(x - h)^2 + k$ is $|h|$ units to the right or to the left of the graph of $y = ax^2$.
3. The graph of $y = a(x - h)^2 + k$ is $|k|$ units above or below the graph of $y = ax^2$.

P–6 **How do you know whether the graph of $y = a(x - h)^2 + k$ is to the right or to the left of the graph of $y = ax^2$?**

P–7 **How do you know whether the graph of $y = a(x - h)^2 + k$ is above or below the graph of $y = ax^2$?**

EXAMPLE 3 Write an equation of a function for which the graph is obtained as described below.

"The graph of $y = -\frac{3}{2}x^2$ is moved 4 units to the left and 3 units upward."

Solution: Let $h = -4$, $k = 3$, $a = -\frac{3}{2}$. ◀ *The graph is 4 units to the left, 3 units upward, and has the same shape as $y = -\frac{3}{2}x^2$.*

$y = a(x - h)^2 + k$

$y = -\frac{3}{2}(x - (-4))^2 + 3$

$y = -\frac{3}{2}(x + 4)^2 + 3$ ◀ *Equation of the function*

CLASSROOM EXERCISES

Write Yes or No to indicate if the graph of each function below has the same shape as the graph of $y = \frac{1}{2}(x - 3)^2 + 5$.

1. $y = \frac{1}{2}x^2$
2. $y = 2x^2$
3. $y = 3(x - 3)^2 + 5$
4. $y = \frac{1}{2}(x + 3)^2 - 5$
5. $y = -\frac{1}{2}x^2$
6. $y = \frac{1}{2}(x - 4)^2 + 3$
7. What variable in $a(x - h)^2 + k$ represents a number that determines the shape of the parabola?
8. What variable in $a(x - h)^2 + k$ represents a number that determines the horizontal position of the parabola?
9. What variable in $a(x - h)^2 + k$ represents a number that determines the vertical position of the parabola?

Write the two motions needed to make the graph of the first function coincide with the graph of the second function. (P-3, P-5)

10. $y = x^2$; $y = (x - 2)^2 + 3$
11. $y = 3x^2$; $y = 3(x - 5)^2 - 2$
12. $y = \frac{1}{3}x^2$; $y = \frac{1}{3}(x + 3)^2 + 5$
13. $y = -\frac{1}{2}x^2$; $y = -\frac{1}{2}(x + 2)^2 - 3$
14. $y = -x^2$; $y = -(x - 2)^2 + 1$
15. $y = \frac{3}{4}x^2$; $y = \frac{3}{4}(x + 1)^2 - 1$

WRITTEN EXERCISES

Goal: To use a graph of $y = ax^2$ to graph a function of the form $y = a(x - h)^2 + k$

Sample Problem: Graph in the same coordinate plane: $y = -2x^2$, $y = -2(x + 1)^2 + 5$, $y = -2(x + 1)^2 - 5$.

Answer: See Examples 1 and 2.

Graph the first function using the x values −2, −1, 0, 1, and 2. Then graph the second function. (Examples 1 and 2)

1. $y = \frac{1}{2}x^2$; $y = \frac{1}{2}(x - 2)^2 + 3$

2. $y = -\frac{1}{2}x^2$; $y = -\frac{1}{2}(x - 5)^2 + 2$

3. $y = -x^2$; $y = -(x + 3)^2 + 2$

4. $y = x^2$; $y = (x + 2)^2 + 4$

5. $y = 2x^2$; $y = 2(x + 1)^2 - 3$

6. $y = -2x^2$; $y = -2(x - 3)^2 - 3$

7. $y = -\frac{3}{2}x^2$; $y = -\frac{3}{2}(x - 3\frac{1}{2})^2 + 2\frac{1}{2}$

8. $y = \frac{3}{2}x^2$; $y = \frac{3}{2}(x + 2\frac{1}{2})^2 - 3\frac{1}{2}$

Write an equation of a function for which the graph is obtained as described in each exercise. (Example 3)

9. The graph of $y = 2x^2$ is moved three units to the right.

10. The graph of $y = -3x^2$ is moved five units to the right.

11. The graph of $y = \frac{1}{2}x^2$ is moved two units to the left and five units upward.

12. The graph of $y = 5x^2$ is moved three units to the left and two units upward.

13. The graph of $y = \frac{1}{3}(x + 1)^2$ is moved five units downward.

14. The graph of $y = -3(x - 5)^2$ is moved four units downward.

15. The graph of $y = -x^2$ is moved three units to the right and two units downward.

16. The graph of $y = \frac{1}{2}x^2$ is moved five units to the right and six units downward.

17. The graph of $y = 3x^2$ is moved five units downward.

18. The graph of $y = -\frac{1}{2}x^2$ is moved three units downward.

REVIEW CAPSULE FOR SECTION 17.6

Find the missing term to form a trinomial square (take $\frac{1}{2}$ the coefficient of x and square it). Then factor the trinomial. (Pages 255–257)

1. $x^2 - 4x +$ ___?___

2. $x^2 + 6x +$ ___?___

3. $x^2 + x +$ ___?___

4. $x^2 - 3x +$ ___?___

5. $x^2 + 10x +$ ___?___

6. $x^2 + 18x +$ ___?___

7. $x^2 - 5x +$ ___?___

8. $x^2 - 7x +$ ___?___

17.6 Standard Form of Quadratic Polynomials

P–1 **What are the values of *a*, *h*, and *k* in each expression below?**

a. $-(x+3)^2-\frac{3}{4}$ **b.** $\frac{3}{2}(x-\frac{1}{4})^2+\frac{5}{4}$

EXAMPLE 1 Write $2(x-3)^2+5$ in the form ax^2+bx+c.

Solution:

$$2(x-3)^2+5 = 2(x^2-6x+9)+5$$
$$= 2x^2-12x+18+5$$
$$= 2x^2-12x+23$$

x^2-6x+9 is a trinomial square.

Every quadratic polynomial can be changed from the form ax^2+bx+c to the ***standard form*** $a(x-h)^2+k$.

EXAMPLE 2 Write x^2+2x+5 in standard form.

Solution:

1. Group terms to form a trinomial square. ⟶ $(x^2+2x+\underline{\ ?\ })+5+\underline{\ ?\ }$
2. Multiply the coefficient of x by $\frac{1}{2}$. Square. ⟶ $\frac{1}{2}(2)=1;\ (1)^2=1$
3. Add 1 to make a trinomial square. Then add -1. ⟶ $(x^2+2x+1)+5+(-1)$
4. Factor and simplify. ⟶ $(x+1)^2+4$
5. Write the standard form. ⟶ $(x-(-1))^2+4$

$a=1$, $h=-1$, $k=4$

EXAMPLE 3 Write x^2-6x-2 in standard form.

Solution:

1. Group terms. ⟶ $(x^2-6x+\underline{\ ?\ })-2+\underline{\ ?\ }$
2. Multiply the coefficient of x by $\frac{1}{2}$. Square. ⟶ $\frac{1}{2}(6)=3;\ 3^2=9$
3. Add 9 to make a trinomial square. Then add -9. ⟶ $(x^2-6x+9)-2+(-9)$
4. Factor and simplify. ⟶ $(x-3)^2-11$

$9+(-9)=0$

EXAMPLE 4 Write $x^2 + 5x + 1$ in standard form.

Solution:

1. Group terms. → $(x^2 + 5x + \underline{\ ?\ }) + 1 + \underline{\ ?\ }$
2. Multiply the coefficient of x by $\frac{1}{2}$. Square. → $\frac{1}{2}(5) = \frac{5}{2}$; $(\frac{5}{2})^2 = \frac{25}{4}$
3. Add $\frac{25}{4}$. Then add its opposite. → $(x^2 + 5x + \frac{25}{4}) + 1 + (-\frac{25}{4})$
4. Factor and simplify. → $(x + \frac{5}{2})^2 - \frac{21}{4}$
5. Write the standard form. → $(x - (-\frac{5}{2}))^2 - \frac{21}{4}$

$\frac{25}{4} + (-\frac{25}{4}) = 0$

CLASSROOM EXERCISES

Write each trinomial in the form $(x - h)^2$.

1. $x^2 + 2x + 1$ **2.** $x^2 + 4x + 4$ **3.** $x^2 - 6x + 9$ **4.** $x^2 - 8x + 16$

5. $x^2 + x + \frac{1}{4}$ **6.** $x^2 - x + \frac{1}{4}$ **7.** $x^2 - 5x + \frac{25}{4}$ **8.** $x^2 - 3x + \frac{9}{4}$

WRITTEN EXERCISES

Goal: To write a quadratic polynomial in standard form
Sample Problem: $x^2 - 7x + 10$ **Answer:** $(x - \frac{7}{2})^2 - \frac{9}{4}$

Write each expression in the form $ax^2 + bx + c$. (Example 1)

1. $(x - 3)^2 + 2$ **2.** $(x - 2)^2 + 5$ **3.** $2(x - 1)^2 - 7$ **4.** $-2(x + 1)^2 - 5$

5. $3(x + 2)^2 + 3$ **6.** $5(x - 5)^2 + 3$ **7.** $\frac{1}{2}(x - 2)^2 + 6$ **8.** $-\frac{1}{2}(x + 2)^2 - 3$

Write each polynomial in standard form. (Examples 2–4)

9. $x^2 + 2x + 3$ **10.** $x^2 - 2x - 5$ **11.** $x^2 - 4x + 1$ **12.** $x^2 + 4x + 3$

13. $x^2 + 6x - 5$ **14.** $x^2 - 6x + 10$ **15.** $x^2 - 10x - 1$ **16.** $x^2 + 12x + 3$

17. $x^2 - x + 5$ **18.** $x^2 + x - 2$ **19.** $x^2 + 3x - 4$ **20.** $x^2 - 5x - 3$

REVIEW CAPSULE FOR SECTION 17.7

Identify the values of a, h, and k in each function. (Pages 469–472)

1. $y = 3(x - 2)^2 + 1$ **2.** $y = 4(x - 5)^2 - 2$ **3.** $y = \frac{1}{2}(x + \frac{3}{2})^2 - 5$

17.7 Vertex and Axis of a Parabola

You have seen how the values of a, h, and k in $y = a(x - h)^2 + k$ affect the shape and position of the graph. There are other facts about the graph that the standard form can provide.

EXAMPLE 1 The graph of $y = \frac{1}{2}(x - 2)^2 + 3$ is drawn. Write the coordinates of its lowest point.

Solution: $y = \frac{1}{2}(x - 2)^2 + 3$

x	y
0	5
1	$3\frac{1}{2}$
2	3
3	$3\frac{1}{2}$
4	5

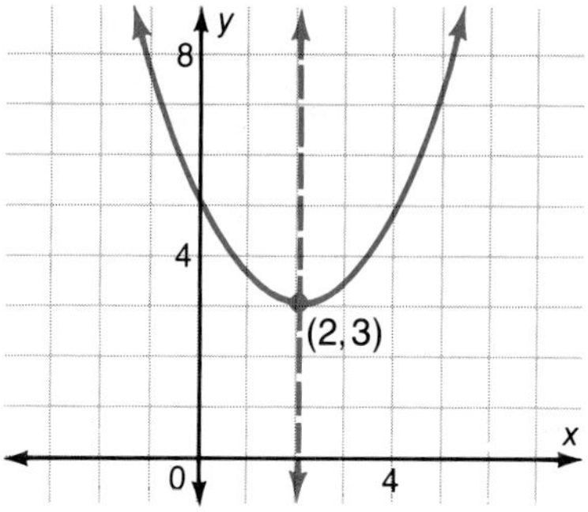

Figure 1

The coordinates of the lowest point of the graph are (2, 3).

The point with coordinates (2, 3) is called the ***vertex of the parabola.*** The vertex of a parabola that is a graph of a quadratic function is either the lowest point or the highest point of the graph.

P–1 **Is the vertex in Figure 1 the highest point or the lowest point of the graph?**

Compare $\frac{1}{2}(x - 2)^2 + 3$ with the standard form $y = a(x - h)^2 + k$.

P–2 **What is the value of a? of h? of k?**

The coordinates of the vertex of the parabola in Figure 1 are (2, 3). They correspond to (h, k).

The vertex of a parabola with a rule of the form $y = a(x - h)^2 + k$ has coordinates (h, k).

$y = -2(x - (-5))^2 - 1$

Vertex: $(-5, -1)$

P–3 **What are the coordinates of the vertex of the graph of each function below?**

a. $y = (x - \frac{1}{2})^2 + \frac{5}{2}$ **b.** $y = (x + 3)^2 - 7$

A vertical line that contains the vertex of a parabola is called the ***axis*** of the parabola. The dashed line in Figure 1 is the axis.

P–4 **What is an equation for the axis in Figure 1?**

If a parabola has the rule $y = a(x - h)^2 + k$, its axis has the equation $x = h$.	$y = 5(x - 2)^2 + 3$ Equation of axis: $x = 2$

P–5 **What is the value of h if you compare each of the following with $y = a(x - h)^2 + k$?**

a. $y = -3(x - 5)^2 + 2$ **b.** $y = -(x + 3)^2 - 2$

c. $y = \frac{1}{2}(x - \frac{1}{4})^2 + 1$ **d.** $y = -2(x + \frac{4}{3})^2 - 5$

EXAMPLE 2 Write an equation for the axis of the graph of $y = -(x + 3)^2 - 2$.

Solution: Since $h = -3$,
$x = -3$. ◀ ***Equation of the axis***

P–6 **What are the coordinates of the vertex of the parabola in Example 2?**

CLASSROOM EXERCISES

Write the coordinates of the vertex of each graph. (P-3)

1. $y = 2(x - 5)^2 + 2$

2. $y = 3(x - 1)^2 - 4$

3. $y = -\frac{1}{2}(x + 2)^2 + 6$

4. $y = (x + 1)^2 - 6$

5. $y = x^2$

6. $y = x^2 + 1$

7. $y = -(x - \frac{1}{2})^2 + \frac{3}{4}$

8. $y = 2.5(x + 0.3)^2 - 1.2$

9. $y = (x - 5)^2$

10. $y = -(x + 4)^2$

Write an equation for the axis of each graph. (Example 2)

11. $y = \frac{1}{2}(x + 1)^2 - 2$

12. $y = -2(x - 3)^2 + 4$

13. $y = -3(x - \frac{1}{2})^2 + \frac{3}{2}$

14. $y = (x + 4.5)^2 - 2.5$

15. $y = -\frac{1}{2}x^2$

16. $y = \frac{4}{3}x^2$

17. $y = 2x^2 + 3$

18. $y = -3(x + 5)^2$

WRITTEN EXERCISES

Goals: To write the coordinates of the vertex and the equation of the axis of the graph of a parabola

Sample Problem: $y = -3(x + 5)^2 - 2$

Answer: Vertex: $(-5, -2)$; Equation of axis: $x = -5$

Write the coordinates of the vertex of each graph. (Example 1, P-3)

1. $y = 3(x - 1)^2 + 10$

2. $y = \frac{1}{2}(x - 5)^2 + 2$

3. $y = -5(x - 7)^2 - 2$

4. $y = -3(x - 3)^2 - 5$

5. $y = -(x + 2)^2 + 3$

6. $y = 2(x + 5)^2 + 7$

7. $y = \frac{2}{3}(x + 1)^2 - 8$

8. $y = -\frac{1}{2}(x + 4)^2 - 3$

9. $y = \frac{3}{4}(x - \frac{1}{2})^2 - \frac{2}{3}$

10. $y = \frac{2}{3}(x - \frac{1}{3})^2 - \frac{3}{4}$

Write an equation for the axis of each graph. (Example 2)

11. $y = 3(x - 5)^2 + 2$

12. $y = -3(x - 2)^2 + 1$

13. $y = -2(x - 3)^2 + 5$

14. $y = \frac{1}{2}(x - 1)^2 + 5$

15. $y = -(x + 1)^2 - 5$

16. $y = (x + 2)^2 - 6$

17. $y = 3x^2 - 2x + 3$

18. $y = 2x^2 + 3x - 5$

19. $y = -x^2 - 7x + 2$

20. $y = -3x^2 - x + 7$

21. $y = -5x^2 - x + 3$

22. $y = -x^2 - 9x + 1$

Using Statistics Predicting

Jeanine Curren is the manager of an automobile parts supply house. She must be able to predict the arrival time for a shipment of parts sent to a customer by train.

Jeanine used records of past deliveries to construct the table below. In the table, x is the delivery distance in miles and y is the delivery time in days.

x	210	290	350	480	490	730	780	850	920	1010
y	5	7	6	11	8	11	12	8	15	12

EXAMPLE: Find the predicted delivery time for a shipment to a customer located 600 rail miles away.

SOLUTION:

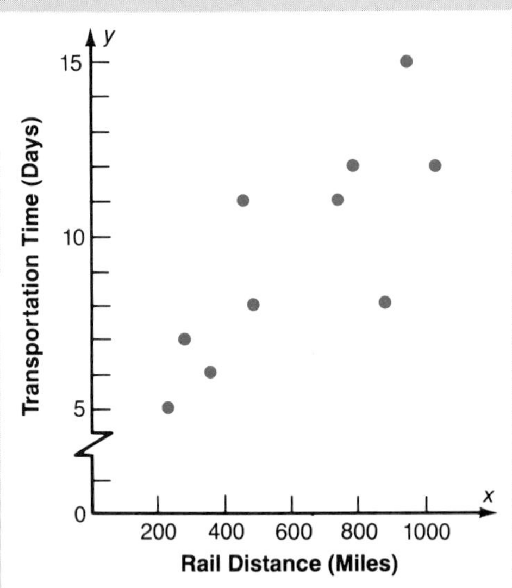

[1] Graph the ordered pairs (x, y) in the table. This is a **scatter plot.**

[2] Separate the data into three nearly equal groups with an odd number of ordered pairs in the first and third groups.

First Group

x	y
210	5
290	7
350	6

Second Group

x	y
480	11
490	8
730	11
780	12

Third Group

x	y
850	8
920	15
1010	12

[3] Find the median for the x and y values in the first and third groups. (For a review of median, see pages 52–53.) Since there is an odd number of values in each group, the **median** is the middle value.

Medians:		x	y
	First group	290	6
	Third group	920	12

4 Graph the median points (290, 6) and (920, 12). Draw the line passing through the two points. This is the **median line of fit.**

5 To predict the delivery time, locate 600 on the x axis. Draw a vertical line to the median line of fit. Then draw a horizontal line to the y axis. Read the value on the y axis.

The y value (delivery time) is about 9.

Thus, Jeanine can predict that the delivery time will be **at least 9 days.**

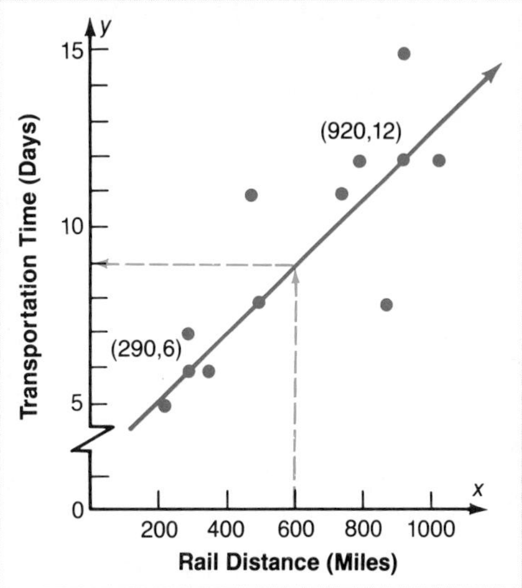

EXERCISES

For each exercise:

a. *Construct a scatter plot.*
b. *Determine the median points.*
c. *Graph the median line of fit.*
d. *Find the predicted value.*

1.

Pages	*x*	10	20	20	30	40	50	60	100	110	120
Hours to Type	*y*	6	8	12	16	18	21	32	46	60	58

Find the predicted number of hours needed to type 70 pages.

2.

Tons of Crunchola	*x*	100	150	170	180	220	240	300	350	400	500
Total Cost (thousands of dollars)	*y*	65	70	65	85	115	125	140	190	200	200

Find the predicted total cost for 200 tons of crunchola.

3.

Height in Inches	*x*	62	64	65	66	67	67	69	71	72	74
Weight in Pounds	*y*	105	115	115	135	140	150	165	170	180	180

Find the predicted weight for a person whose height is 68 inches.

4.

Number of Push-ups	*x*	15	20	22	25	27	27	32	35	40	42
Number of Chin-ups	*y*	3	5	4	6	6	9	8	9	11	10

Find the predicted number of chin-ups done by a person who can do 30 push-ups.

CHAPTER SUMMARY

IMPORTANT TERMS

Quadratic function *(p. 456)*
Parabola *(p. 458)*
Standard form of a quadratic polynomial *(p. 473)*
Vertex of a parabola *(p. 475)*
Axis of a parabola *(p. 476)*

IMPORTANT IDEAS

1. *Effect of a on the graph of $y = ax^2 + bx + c$:*
   a. If a is positive, the parabola opens upward.
   b. If a is negative, the parabola opens downward.
   c. If $|a|$ decreases, the parabola becomes wider.
2. The graph of $y = ax^2$ can be moved horizontally $|h|$ units to obtain the graph of $y = a(x - h)^2$. If h is negative, the motion is to the left. If h is positive, the motion is to the right.
3. The graph of $y = a(x - h)^2 + k$ has the same shape as the graph of $y = ax^2$.
4. The graph of $y = a(x - h)^2 + k$ is $|h|$ units to the right or to the left of the graph of $y = ax^2$.
5. The graph of $y = a(x - h)^2 + k$ is $|k|$ units above or below the graph of $y = ax^2$.
6. Every quadratic polynomial can be changed from the form $ax^2 + bx + c$ to the standard form $a(x - h)^2 + k$.
7. The vertex of a parabola with a rule of the form $y = a(x - h)^2 + k$ has coordinates (h, k).
8. If a parabola has the rule $y = a(x - h)^2 + k$, its axis has the equation $x = h$.

CHAPTER REVIEW

SECTION 17.1

Write Yes or No to show whether each rule describes a quadratic function.

1. $y = -5x + 1$

2. $y = -x^2 + 3x - 5$

Write three ordered pairs of each function below. Let the x values equal −1, 0, and 2.

3. $y = -3x^2 + x - 1$

4. $y = (2x - 1)^2$

SECTION 17.2

Graph each function.

5. $y = x^2 - 1$
6. $y = -x^2 + 2$
7. $y = -\frac{1}{2}x^2 + x$
8. $y = \frac{1}{2}x^2 - 2x$
9. $y = x^2 + 2x - 3$
10. $y = -x^2 + 4x + 5$

SECTION 17.3

Graph the two functions in the same coordinate plane.

11. $y = -\frac{1}{2}x^2;\ y = -2x^2$
12. $y = -x^2;\ y = -\frac{3}{2}x^2$
13. $y = \frac{5}{2}x^2;\ y = -\frac{5}{2}x^2$
14. $y = -\frac{1}{2}x^2;\ y = \frac{1}{2}x^2$

Graph each parabola.

15. $x = \frac{1}{2}y^2$
16. $x = \frac{3}{2}y^2$
17. $x = -y^2$
18. $x = -2y^2$

SECTION 17.4

Write the number of units and the direction the graph of the first function can be moved to obtain the graph of the second function.

19. $y = 2x^2;\ y = 2(x + 3)^2$
20. $y = -\frac{1}{2}x^2;\ y = -\frac{1}{2}(x - 5)^2$
21. $y = -x^2;\ y = -(x - \frac{5}{2})^2$
22. $y = x^2;\ y = (x + 12)^2$

SECTION 17.5

Graph the first function using the x values −2, −1, 0, 1, and 2. Then graph the second function.

23. $y = x^2$
    $y = (x - 3)^2 - 2$
24. $y = -x^2$
    $y = -(x + 2)^2 - 3$
25. $y = -\frac{1}{2}x^2$
    $y = -\frac{1}{2}(x + 1)^2 + 3$
26. $y = \frac{3}{2}x^2$
    $y = \frac{3}{2}(x - 3)^2 + 1$

Write an equation of a function for which the graph is obtained as described in each exercise.

27. The graph of $y = -10x^2$ is moved 5 units to the left.
28. The graph of $y = 4x^2$ is moved 8 units to the right.

Write an equation of a function for which the graph is obtained as described in each exercise.

29. The graph of $\frac{5}{2}x^2$ is moved 2 units to the right and 3 units downward.

30. The graph of $-3.7x^2$ is moved 4 units to the left and 2 units upward.

SECTION 17.6

Write each expression in the form $ax^2 + bx + c$.

31. $(x + 1)^2 - 5$

32. $(x - 3)^2 + 4$

33. $-3(x - 2)^2 + 1$

34. $4(x + 1)^2 - 6$

35. $\frac{1}{2}(x + 4)^2 - 8$

36. $-\frac{1}{2}(x - 2)^2 + 3$

Write each polynomial in the form $(x - h)^2 + k$.

37. $x^2 - 6x + 2$

38. $x^2 + 4x - 5$

39. $x^2 + 3x - \frac{1}{4}$

40. $x^2 - 5x + \frac{3}{4}$

41. $x^2 - 2x + \frac{1}{2}$

42. $x^2 + 7x + \frac{5}{4}$

SECTION 17.7

Write the coordinates of the vertex of each graph.

43. $y = \frac{1}{2}(x - 10)^2 + 5$

44. $y = -2(x + 8)^2 - 3$

45. $y = 3(x - 6)^2$

46. $y = -4(x + 5)^2$

Write an equation for the axis of each graph.

47. $y = \frac{3}{2}(x - 2)^2 + 5$

48. $y = -5(x + 1)^2 - 6$

49. $y = -(x + \frac{9}{2})^2 - \frac{7}{2}$

50. $y = (x - \frac{4}{3})^2 + \frac{5}{3}$

Quadratic Equations

CHAPTER 18

A golf ball being driven from the tee follows a path which is a curve of the type called a *parabola*. Quadratic functions whose graphs are parabolas have many interesting applications besides their purely mathematical uses.

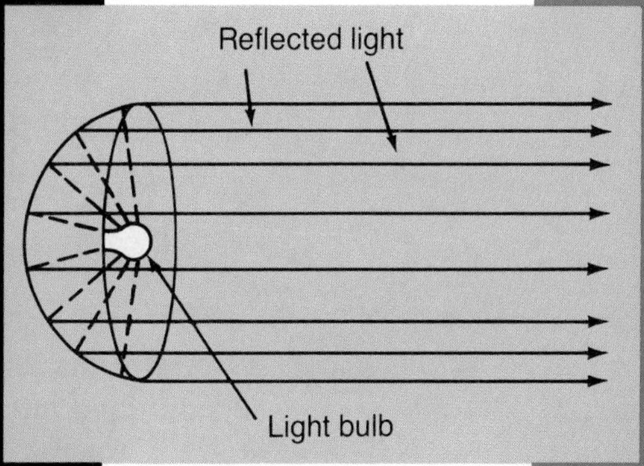

18.1 Quadratic Equations and Graphs

The equation $ax^2 + bx + c = 0$ can be obtained from the rule of the function $y = ax^2 + bx + c$ if y is replaced with 0.

Definition

A ***quadratic equation*** is an equation of the form $ax^2 + bx + c = 0$, $(a \neq 0)$. (a, b, and c represent rational numbers in this course.)

$$2x^2 - 3x + 1 = 0$$

$$-\frac{1}{2}x^2 + \frac{3}{2}x + \frac{1}{4} = 0$$

EXAMPLE 1

Change $x^2 - 3x - 5 = 2$ to the form $ax^2 + bx + c = 0$.

Solution:

1. Subtract 2 from each side. ⟶ $x^2 - 3x - 5 \mathbf{- 2} = 2 \mathbf{- 2}$
2. Simplify. ⟶ $x^2 - 3x - 7 = 0$ ◀ ***In the form*** $\boldsymbol{ax^2 + bx + c = 0}$

A graph of the function $y = x^2 - x - 2$ is shown in Figure 1.

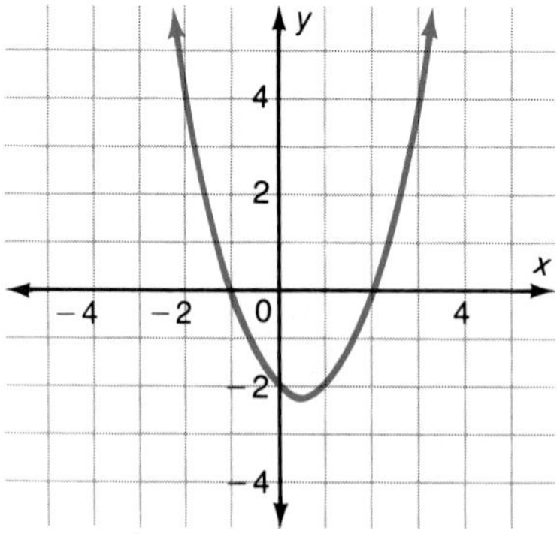

Figure 1

P–1 **What are the x values of the points which the parabola and the x axis have in common?**

The numbers -1 and 2 are ***zeros of the function*** $y = x^2 - x - 2$. They are values of x which make the value of the function 0. They are also the ***solutions*** of $x^2 - x - 2 = 0$.

EXAMPLE 2 Use the graph of $y = x^2 - x - 2$ to estimate the solutions of $x^2 - x - 2 = 4$.

Draw a graph of $y = x^2 - x - 2$. → **Draw the line with equation $y = 4$.** → **Find the x values of the points common to the two graphs.**

Solution: Notice the x values of points A and B in Figure 2.

The numbers −2 and 3 are the solutions of $x^2 - x - 2 = 4$.

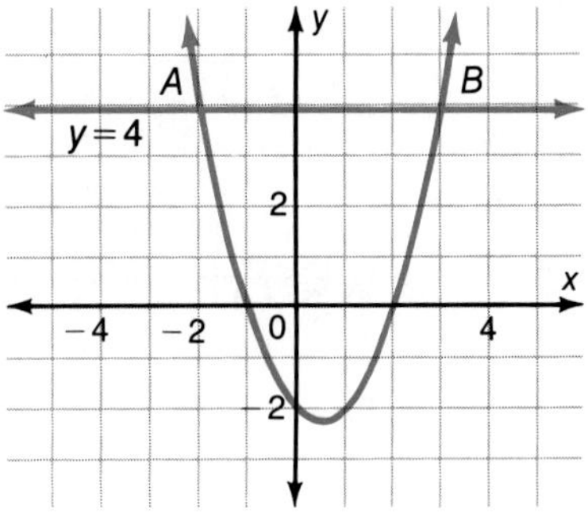

Figure 2

You can estimate the solutions of other equations from the same graph. Note the horizontal lines and their equations.

Equation	Estimated Solutions
$x^2 - x - 2 = 3$	−1.8, 2.8
$x^2 - x - 2 = -1$	−0.6, 1.6
$x^2 - x - 2 = 1$	−1.3, 2.3
$x^2 - x - 2 = -2$	0, 1

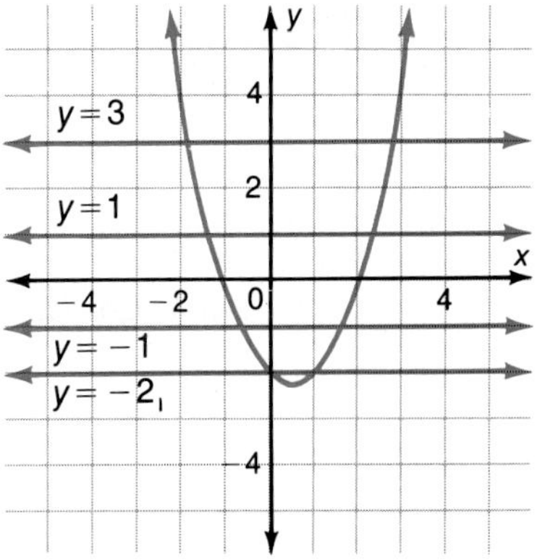

Figure 3

The line with equation $y = -3$ does not intersect the parabola. There are no real solutions for $x^2 - x - 2 = -3$.

CLASSROOM EXERCISES

Change each equation to the form $ax^2 + bx + c = 0$. (Example 1)

1. $x^2 - 2x - 1 = 2$ **2.** $2x^2 - 3x + 5 = 1$ **3.** $-3x^2 + 5x = 6$

4. $5x - x^2 + 3 = 0$ **5.** $4 + x^2 + 2x = 0$ **6.** $x^2 = 2x$

The graph of $y = x^2 - 4x + 3$ is shown at the right. Estimate the solutions of each equation from the graph. (Example 2)

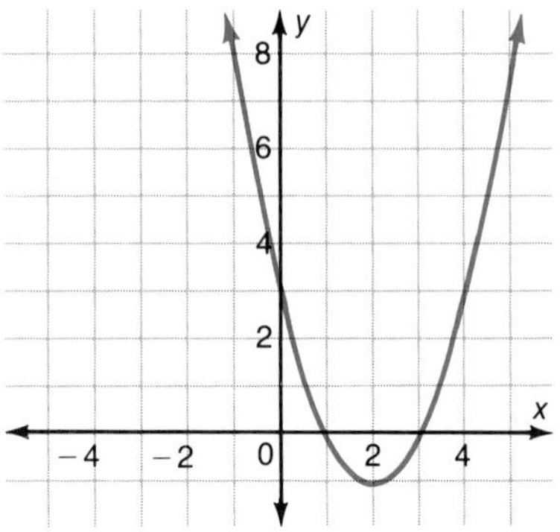

7. $x^2 - 4x + 3 = 0$
8. $x^2 - 4x + 3 = 1$
9. $x^2 - 4x + 3 = 3$
10. $x^2 - 4x + 3 = 5$
11. $x^2 - 4x + 3 = -1$
12. $x^2 - 4x + 3 = -2$

WRITTEN EXERCISES

Goal: To estimate the solutions of a quadratic equation from the graph of a related function

Sample Problem: $x^2 - x - 2 = 2$ (Refer to Figure 2.)

Answer: The solutions are approximately 2.6 and −1.6.

Change each equation to the form $ax^2 + bx + c = 0$. (Example 1)

1. $x^2 = 5x - 3$
2. $x^2 = 5 - 3x$
3. $5x - 2x^2 = 10$
4. $3x - x^2 = -5$
5. $x^2 = 9$
6. $x^2 = 3x$

Graph $y = x^2 + 2x - 3$. Refer to this graph to estimate the solutions of each equation. (Example 2)

7. $x^2 + 2x - 3 = 0$
8. $x^2 + 2x - 3 = 1$
9. $x^2 + 2x - 3 = 2$
10. $x^2 + 2x - 3 = 4$
11. $x^2 + 2x - 3 = -5$
12. $x^2 + 2x - 3 = -6$

Graph $y = -x^2 + 4x - 5$. Estimate the solutions of each equation. (Example 2)

13. $-x^2 + 4x - 5 = 0$
14. $-x^2 + 4x - 5 = 1$
15. $-x^2 + 4x - 5 = -1$
16. $-x^2 + 4x - 5 = -2$
17. $-x^2 + 4x - 5 = -3$
18. $-x^2 + 4x - 5 = -4$

REVIEW CAPSULE FOR SECTION 18.2

Factor each trinomial. (Pages 246–248)

1. $x^2 - 4x - 12$
2. $x^2 - 7x + 10$
3. $x^2 - 2x - 24$
4. $x^2 + x - 56$
5. $2x^2 + 7x - 4$
6. $3x^2 - 5x + 2$
7. $7 - x^2 - 6x$
8. $9x^2 + 14 - 65x$

18.2 Solving by Factoring

In this course, polynomials represent real numbers. Therefore, the ***Factors of Zero Property*** applies to polynomials.

> ***Factors of Zero Property***
>
> If a and b are real numbers and $ab = 0$, then $a = 0$ *or* $b = 0$.

P–1 **What is the value of x in the equation $x + 5 = 0$? in the equation $x - 2 = 0$?**

EXAMPLE 1 Solve $(x + 5)(x - 2) = 0$.

Solution: $(x + 5)(x - 2) = 0$

1 Factors of Zero Property ⟶ $x + 5 = 0$ <u>*or*</u> $x - 2 = 0$

2 Solve both equations. ⟶ $x = -5$ <u>*or*</u> $x = 2$

The solution set is $\{-5, 2\}$.

EXAMPLE 2 Solve $x^2 - 2x - 3 = 0$.

Factor the left side. ⟶ **Apply the Factors of Zero Property.** ⟶ **Solve for values of x.**

Solution: $x^2 - 2x - 3 = 0$ ◀ ***In the form*** $ax^2 + bx + c = 0$

1 Factor. ⟶ $(x - 3)(x + 1) = 0$

2 Factors of Zero Property ⟶ $x - 3 = 0$ <u>*or*</u> $x + 1 = 0$

3 Solve both equations. ⟶ $x = 3$ <u>*or*</u> $x = -1$

The solution set is $\{3, -1\}$.

EXAMPLE 3 Solve $x^2 - 6x + 9 = 0$.

Solution:

$$x^2 - 6x + 9 = 0$$

$$(x - 3)(x - 3) = 0$$

$$x - 3 = 0 \quad \underline{or} \quad x - 3 = 0$$

Both equations have the same solution, 3.
Solution set: $\{3\}$

Before solving a quadratic equation, be sure that it is in the form $ax^2 + bx + c = 0$.

EXAMPLE 4 Solve $x^2 = 5x$.

Solution: $x^2 = 5x$

1 Subtract $5x$ from each side. → $x^2 - 5x = 0$ ◀ ***In the form $ax^2 + bx + c = 0$***

2 Factor. → $x(x - 5) = 0$

3 Factors of Zero Property → $x = 0$ <u>or</u> $x - 5 = 0$

4 Solve both equations. → $x = 0$ <u>or</u> $x = 5$

Solution set: $\{0, 5\}$

EXAMPLE 5 Solve $2x^2 + x = 6$.

Solution: $2x^2 + x = 6$

1 Subtract 6 from each side. → $2x^2 + x - 6 = 0$

2 Factor. → $(2x - 3)(x + 2) = 0$

3 Factors of Zero Property → $2x - 3 = 0$ <u>or</u> $x + 2 = 0$

4 Solve both equations. → $2x - 3 + \mathbf{3} = 0 + \mathbf{3}$ <u>or</u> $x + 2 - \mathbf{2} = 0 - \mathbf{2}$

$$2x = 3 \qquad x = -2$$

$$\tfrac{1}{2}(2x) = \tfrac{1}{2}(3)$$

$$x = \tfrac{3}{2}$$

Solution set: $\{\tfrac{3}{2}, -2\}$

CLASSROOM EXERCISES

Solve. (Example 1)

1. $(x-3)(x+5)=0$
2. $x(x-1)=0$
3. $(x-6)(x+6)=0$
4. $(x-2)(x-7)=0$
5. $(x+4)(x+7)=0$
6. $(x-3)(x-3)=0$
7. $(2x-1)(x+5)=0$
8. $(x-1)(2x+1)=0$
9. $(2x+3)(x+1)=0$
10. $(4x-1)(3x+2)=0$

WRITTEN EXERCISES

Goal: To use the Factors of Zero Property to solve quadratic equations
Sample Problem: $x^2-56=-x$
Answer: $\{7, -8\}$

Solve. (Examples 2–4)

1. $x^2+3x=0$
2. $x^2-4x=0$
3. $x^2-25=0$
4. $x^2-49=0$
5. $x^2=10x$
6. $x^2=-8x$
7. $x^2=100$
8. $x^2=81$
9. $x^2+4x-5=0$
10. $x^2-5x+6=0$
11. $x^2+3x=4$
12. $x^2-3x=10$
13. $x^2+8x+16=0$
14. $x^2-12x+36=0$
15. $x^2+4=4x$
16. $x^2+9=-6x$
17. $8x-12=x^2$
18. $2x+x^2=24$
19. $x^2+24=11x$
20. $x^2+15=8x$
21. $x^2+18x=-81$
22. $x^2+100=20x$
23. $x^2=x+72$
24. $7=x^2+6x$

Solve. (Example 5)

25. $3x^2+8x-3=0$
26. $2x^2-9x+7=0$
27. $6x^2+x-2=0$
28. $9x^2-9x+2=0$
29. $4-12x^2=13x$
30. $10-15x^2=19x$
31. $8x^2-5+6x=0$
32. $6x^2-7x-5=0$
33. $2x^2-13x=45$
34. $9x^2+14=65x$
35. $4x^2+x-3=0$
36. $25x^2+9-30x=0$

18.3 Special Quadratic Equations

P–1 **Solve each equation below.**

a. $x^2 = 16$ **b.** $x^2 = 49$ **c.** $x^2 = 100$

If the left side of an equation is the square of a binomial, the equation can be solved in the same way you solved P-1.

EXAMPLE 1 Solve $(x - 3)^2 = 25$.

Solution:

[1] Write the equivalent compound *or* sentence. ⟶ $x - 3 = 5$ *or* $x - 3 = -5$

[2] Solve both equations. ⟶ $x = 8$ *or* $x = -2$

Solution set: $\{8, -2\}$

Check:

$$(x - 3)^2 = 25 \qquad (x - 3)^2 = 25$$
$$(8 - 3)^2 \qquad (-2 - 3)^2$$
$$5^2 \qquad (-5)^2$$
$$25 \qquad 25$$

Any quadratic equation of the form $(x + r)^2 = s$ in which r and s are rational numbers, $s \geq 0$, is equivalent to "$x + r = \sqrt{s}$ *or* $x + r = -\sqrt{s}$."

"$(x + 1)^2 = 5$" is equivalent to "$x + 1 = \sqrt{5}$ *or* $x + 1 = -\sqrt{5}$."

P–2 **What is the value of r if you compare $(x - 5)^2 = 81$ with $(x + r)^2 = s$?**

EXAMPLE 2 Solve $(x - 5)^2 = 81$.

Solution: $(x - 5)^2 = 81$

$x - 5 = 9$ *or* $x - 5 = -9$

$x = 14$ *or* $x = -4$

Solution set: $\{14, -4\}$

The check is left for you.

You have learned to tell whether a quadratic trinomial of the form $ax^2 + bx + c$ is a trinomial square.

1. The leading coefficient a must be a perfect square.
2. The constant term c must be a perfect square.
3. The sentence $(\frac{1}{2}b)^2 = ac$ must be true.

P–3 **What is the value of a in $x^2 - 8x + 16$? of c? of $(\frac{1}{2}b)^2$? of ac?**

EXAMPLE 3

Solve $x^2 - 8x + 16 = 36$.

Solution:

$$x^2 - 8x + 16 = 36$$

$x^2 - 8x + 16$ is a trinomial square.

$$(x - 4)^2 = 36$$

$$x - 4 = 6 \quad \underline{or} \quad x - 4 = -6$$

$$x = 10 \quad \underline{or} \quad x = -2$$

The check is left for you.

Solution set: $\{10, -2\}$

EXAMPLE 4

Solve $x^2 + 3x + \frac{9}{4} = 5$.

Solution: $x^2 + 3x + \frac{9}{4}$ is a trinomial square over {rational numbers}.

$$a = 1 \qquad b = 3 \qquad c = \frac{9}{4}$$

a and c are perfect squares.

$$(\tfrac{1}{2}b) = \frac{3}{2}$$

$$(\tfrac{1}{2}b)^2 = \frac{9}{4} \qquad ac = \frac{9}{4}$$

$(\frac{1}{2}b)^2 = ac$

$$x^2 + 3x + \frac{9}{4} = 5$$

$$(x + \tfrac{3}{2})^2 = 5$$

$$x + \frac{3}{2} = \sqrt{5} \quad \underline{or} \quad x + \frac{3}{2} = -\sqrt{5}$$

$$x = -\frac{3}{2} + \sqrt{5} \quad \underline{or} \quad x = -\frac{3}{2} - \sqrt{5}$$

Solution set: $\{-\frac{3}{2} + \sqrt{5}, -\frac{3}{2} - \sqrt{5}\}$

CLASSROOM EXERCISES

Solve. (P-1)

1. $x^2 = 1$ **2.** $x^2 = 5$ **3.** $x^2 = 121$ **4.** $x^2 = 3$

Write an equivalent compound or sentence. (Step 1 of Example 1)

5. $(x - 6)^2 = 36$
6. $(x + 3)^2 = 49$
7. $(x - \frac{1}{2})^2 = 16$

Write the values of $(\frac{1}{2}b)^2$ and ac for each trinomial. Write Yes or No to indicate if each trinomial is a trinomial square. (P-3)

8. $x^2 + 2x + 2$
9. $x^2 - 4x + 2$
10. $x^2 - 12x + 36$
11. $x^2 + 10x + 25$
12. $x^2 + x + \frac{1}{4}$
13. $x^2 - 3x + \frac{3}{4}$

Write two equal factors of each trinomial square over {rational numbers}.

14. $x^2 - 8x + 16$
15. $x^2 + 20x + 100$
16. $x^2 + x + \frac{1}{4}$
17. $x^2 - \frac{1}{2}x + \frac{1}{16}$
18. $x^2 + 5x + \frac{25}{4}$
19. $x^2 - 3x + \frac{9}{4}$

WRITTEN EXERCISES

Goal: To solve quadratic equations of the form $(x + r)^2 = s$
Sample Problem: Solve $x^2 + x + \frac{1}{4} = 9$. **Answer:** $\{\frac{5}{2}, -\frac{7}{2}\}$

Solve each equation. (P-1)

1. $x^2 = 9$
2. $x^2 = 16$
3. $x^2 = 169$
4. $x^2 = 225$
5. $x^2 = 7$
6. $x^2 = 10$
7. $x^2 = 75$
8. $x^2 = 48$

Write an equivalent compound or sentence. (Step 1 of Example 1)

9. $(x - 5)^2 = 36$
10. $(x - 4)^2 = 49$
11. $(x + 6)^2 = 25$
12. $(x - 1)^2 = 5$
13. $(x + \frac{1}{2})^2 = \frac{1}{4}$
14. $(x + \frac{2}{3})^2 = \frac{4}{9}$

Solve. (Examples 1–4)

15. $(x + 1)^2 = 121$
16. $(x + 2)^2 = 169$
17. $(x - 3)^2 = 5$
18. $(x - 5)^2 = 3$
19. $(x - \frac{1}{2})^2 = \frac{1}{4}$
20. $(x - \frac{3}{4})^2 = \frac{1}{16}$
21. $(x + \frac{2}{3})^2 = \frac{5}{9}$
22. $(x + \frac{1}{3})^2 = \frac{2}{9}$
23. $x^2 + 8x + 16 = 0$
24. $x^2 + 12x + 36 = 0$
25. $x^2 - 10x + 25 = 0$
26. $x^2 - 6x + 9 = 0$
27. $x^2 - \frac{1}{2}x + \frac{1}{16} = 0$
28. $x^2 - 3x + \frac{9}{4} = 0$
29. $x^2 - \frac{4}{5}x + \frac{4}{25} = 0$

MORE CHALLENGING EXERCISES

30. If $r = 3$ and the solution set of $(x + r)^2 = s$ is $\{-1, -5\}$, what is the value of s?

31. If $r = -2$ and the solution set of $(x + r)^2 = s$ is $\{\frac{1}{2}, \frac{7}{2}\}$, what is the value of s?

APPLICATIONS

You can use a calculator to evaluate an expression such as the following that involves several radicals.

$$\sqrt{x+\sqrt{x+\sqrt{x+\sqrt{x}}}}$$

EXAMPLE Evaluate the above expression for $x = 2$.

SOLUTION Start with 2 at the extreme right. Then work to the left.

[2] [√] [+] [2] [=] [√] [+] [2]
[=] [√] [+] [2] [=] [√]   1.9903694

Evaluate the radical expression shown above for each given value of x.

32. $x = 4$ **33.** $x = 3$ **34.** $x = 10$ **35.** $x = 16$

36. $x = 5.2$ **37.** $x = 0.67$ **38.** $x = 100$ **39.** $x = 359$

MID-CHAPTER REVIEW

Graph $y = x^2 + 3x - 4$. *Estimate the solutions in each equation.* (Section 18.1)

1. $x^2 + 3x - 4 = 0$ **2.** $x^2 + 3x - 4 = 1$ **3.** $x^2 + 3x - 4 = 2$

4. $x^2 + 3x - 4 = -1$ **5.** $x^2 + 3x - 4 = -2$ **6.** $x^2 + 3x - 4 = -3$

Solve. (Section 18.2)

7. $2x^2 + 9x - 5 = 0$ **8.** $2x^2 - 5x - 3 = 0$ **9.** $3x^2 - 2x - 1 = 0$

10. $4x^2 - 5x + 1 = 0$ **11.** $6x^2 + 5x + 1 = 0$ **12.** $4x^2 + 4x - 3 = 0$

Solve. (Section 18.3)

13. $(x - 3)^2 = 144$ **14.** $(x + 5)^2 = 81$ **15.** $x^2 - x + \frac{1}{4} = \frac{9}{4}$

16. $x^2 + x + \frac{1}{4} = \frac{9}{16}$ **17.** $x^2 + 10x + 25 = 9$ **18.** $x^2 + 14x + 49 = 121$

REVIEW CAPSULE FOR SECTION 18.4

Write each equation in the form $x^2 + bx = -c$. (Pages 58–63)

1. $x^2 - 6x + 3 = 0$ **2.** $x^2 + 5x - 4 = 0$ **3.** $x^2 = 4x - 10$

4. $x^2 - \frac{1}{2}x + \frac{3}{2} = 0$ **5.** $2x^2 - 4x + 5 = 0$ **6.** $2x^2 + 3x - 10 = 0$

18.4 Completing the Square

It is possible to write any quadratic equation in a form that makes the left side a trinomial square. This is called ***completing the square***.

EXAMPLE 1 Solve by completing the square: $x^2 - 8x - 9 = 0$

Solution:

1. Add 9 to each side. → $x^2 - 8x = 9$
2. Multiply the coefficient of x by $\frac{1}{2}$. → $\frac{1}{2}(-8) = -4$
3. Square (-4). → $(-4)^2 = 16$
4. Add 16 to each side. → $x^2 - 8x + \mathbf{16} = 9 + \mathbf{16}$
5. Factor and simplify. → $(x - 4)^2 = 25$
6. Write the equivalent compound *or* sentence. → $x - 4 = 5$ *or* $x - 4 = -5$
7. Solve both equations. → $x = 9$ *or* $x = -1$

$x^2 - 8x + 16$ is a trinomial square.

Solution set: $\{9, -1\}$

EXAMPLE 2 Solve $x^2 - x - 1 = 0$.

Solution:

1. Add 1 to each side. → $x^2 - x = 1$
2. Multiply the coefficient of x by $\frac{1}{2}$. → $\frac{1}{2}(-1) = -\frac{1}{2}$
3. Square $(-\frac{1}{2})$. → $(-\frac{1}{2})^2 = \frac{1}{4}$
4. Add $\frac{1}{4}$ to each side. → $x^2 - x + \frac{1}{4} = 1 + \frac{1}{4}$
5. Factor and simplify. → $(x - \frac{1}{2})^2 = \frac{5}{4}$
6. Write the equivalent compound *or* sentence. → $x - \frac{1}{2} = \sqrt{\frac{5}{4}}$ *or* $x - \frac{1}{2} = -\sqrt{\frac{5}{4}}$
7. Simplify the radicals. → $x - \frac{1}{2} = \frac{\sqrt{5}}{2}$ *or* $x - \frac{1}{2} = -\frac{\sqrt{5}}{2}$
8. Solve both equations. → $x = \frac{1}{2} + \frac{\sqrt{5}}{2}$ *or* $x = \frac{1}{2} - \frac{\sqrt{5}}{2}$

Solution set: $\left\{\frac{1 + \sqrt{5}}{2}, \frac{1 - \sqrt{5}}{2}\right\}$

EXAMPLE 3 Solve $2x^2 + 3x - 5 = 0$.

Solution:

1	Add 5 to each side. →	$2x^2 + 3x = 5$
2	Multiply each side by $\frac{1}{2}$. →	$x^2 + \frac{3}{2}x = \frac{5}{2}$
3	Multiply the coefficient of x by $\frac{1}{2}$. →	$\frac{1}{2}(\frac{3}{2}) = \frac{3}{4}$
4	Square $\frac{3}{4}$. →	$(\frac{3}{4})^2 = \frac{9}{16}$
5	Add $\frac{9}{16}$ to each side. →	$x^2 + \frac{3}{2}x + \frac{9}{16} = \frac{5}{2} + \frac{9}{16}$
6	Factor and simplify. →	$(x + \frac{3}{4})^2 = \frac{49}{16}$
7	Write the equivalent compound *or* sentence. →	$x + \frac{3}{4} = \sqrt{\frac{49}{16}}$ *or* $x + \frac{3}{4} = -\sqrt{\frac{49}{16}}$
8	Simplify the radicals. →	$x + \frac{3}{4} = \frac{7}{4}$ *or* $x + \frac{3}{4} = -\frac{7}{4}$
9	Solve both equations. →	$x = -\frac{3}{4} + \frac{7}{4}$ *or* $x = -\frac{3}{4} - \frac{7}{4}$
10	Simplify. →	$x = 1$ *or* $x = -2\frac{1}{2}$

Solution set: $\{1, -2\frac{1}{2}\}$

The solutions of the equations of Examples 1–3 can be checked by replacing x in the original equation with each number from the solution set.

Steps for solving $ax^2 + bx + c = 0$ by completing the square:

1. Add $-c$ to each side.
2. Multiply each side by $\frac{1}{a}$.
3. Compute $\left(\frac{1}{2} \cdot \frac{b}{a}\right)$.
4. Compute $\left(\frac{1}{2} \cdot \frac{b}{a}\right)^2$.
5. Add $\left(\frac{1}{2} \cdot \frac{b}{a}\right)^2$ to each side.
6. Factor the left side.
7. Write the equivalent compound *or* sentence.
8. Solve.

CLASSROOM EXERCISES

Write the number that you should add to each side to form a trinomial square in the left side. (Examples 1 and 2)

1. $x^2 + 4x = 1$
2. $x^2 - 6x = 5$
3. $x^2 - 2x = 3$
4. $x^2 + 10x = 12$
5. $x^2 - x = 2$
6. $x^2 + 3x = 5$
7. $x^2 + \frac{1}{3}x = 3$
8. $x^2 - \frac{2}{3}x = 4$

Solve each equation.

9. $x^2 = \frac{3}{4}$
10. $x^2 = \frac{5}{9}$
11. $x^2 = \frac{1}{2} + \frac{3}{4}$
12. $x^2 = \frac{3}{2} + \frac{5}{16}$
13. $(x - \frac{1}{2})^2 = \frac{3}{4}$
14. $(x - 3)^2 = \frac{13}{25}$

Write the factors of each trinomial square over {rational numbers}.

15. $x^2 + 18x + 81$
16. $x^2 - 14x + 49$
17. $x^2 + 3x + \frac{9}{4}$
18. $x^2 + 5x + \frac{25}{4}$
19. $x^2 - \frac{2}{5}x + \frac{1}{25}$
20. $x^2 + \frac{3}{2}x + \frac{9}{16}$

WRITTEN EXERCISES

Goal: To solve quadratic equations by completing the square
Sample Problem: Solve $2x^2 - 5x - 7 = 0$. **Answer:** $\{\frac{7}{2}, -1\}$

Solve by completing the square. (Examples 1–3)

1. $x^2 - 2x - 15 = 0$
2. $x^2 + 2x - 8 = 0$
3. $x^2 + 4x - 5 = 0$
4. $x^2 - 4x - 12 = 0$
5. $x^2 + x - 6 = 0$
6. $x^2 - x - 20 = 0$
7. $x^2 + 2x - 5 = 0$
8. $x^2 + 4x - 1 = 0$
9. $x^2 - 4x + 7 = 0$
10. $x^2 + 10x + 2 = 0$
11. $x^2 - 6x - 5 = 0$
12. $x^2 - 4x - 10 = 0$
13. $x^2 + 8x + 13 = 0$
14. $x^2 - x - 3 = 0$
15. $x^2 - 3x - 1 = 0$
16. $x^2 - 5x - 3 = 0$
17. $x^2 + 3x - 5 = 0$
18. $x^2 + \frac{2}{3}x - \frac{2}{3} = 0$
19. $x^2 + \frac{2}{5}x - \frac{1}{5} = 0$
20. $2x^2 + 5x - 3 = 0$
21. $2x^2 + x - 3 = 0$
22. $3x^2 + 8x - 3 = 0$
23. $3x^2 - 5x - 2 = 0$
24. $2x^2 - 4x - 3 = 0$
25. $3x^2 - 6x + 2 = 0$
26. $2x^2 + 4x - 5 = 0$
27. $-2x + 4x^2 - 10 = 0$

REVIEW CAPSULE FOR SECTION 18.5

Simplify. (Pages 180–182 and 187–189)

1. $\frac{3 + \sqrt{49}}{4}$; $\frac{3 - \sqrt{49}}{4}$
2. $\frac{-2 + \sqrt{36}}{2}$; $\frac{-2 - \sqrt{36}}{2}$
3. $\frac{4 + \sqrt{24}}{8}$; $\frac{4 - \sqrt{24}}{8}$

18.5 The Quadratic Formula

A special formula called the ***quadratic formula*** can be used to solve any quadratic equation.

EXAMPLE 1 Solve $ax^2 + bx + c = 0$ for x.

Solution:

1 Add $-c$ to each side. → $ax^2 + bx = -c$

2 Multiply each side by $\frac{1}{a}$. → $x^2 + \frac{b}{a}x = \frac{-c}{a}$

3 Add $\left(\frac{1}{2} \cdot \frac{b}{a}\right)^2$ or $\frac{b^2}{4a^2}$ to each side. → $x^2 + \frac{b}{a}x + \frac{b^2}{4a^2} = -\frac{c}{a} + \frac{b^2}{4a^2}$

4 Factor the left side and write the simplest name for the right side. → $\left(x + \frac{b}{2a}\right)^2 = \frac{b^2 - 4ac}{4a^2}$

5 Write the equivalent compound *or* sentence. → $x + \frac{b}{2a} = \frac{\sqrt{b^2 - 4ac}}{2a}$

or $x + \frac{b}{2a} = \frac{-\sqrt{b^2 - 4ac}}{2a}$

6 Solve both equations. → $x = \frac{-b}{2a} + \frac{\sqrt{b^2 - 4ac}}{2a}$

or $x = \frac{-b}{2a} - \frac{\sqrt{b^2 - 4ac}}{2a}$

7 Simplify. → $x = \frac{-b + \sqrt{b^2 - 4ac}}{2a}$

or $x = \frac{-b - \sqrt{b^2 - 4ac}}{2a}$

Solution set: $\left\{\frac{-b + \sqrt{b^2 - 4ac}}{2a}, \frac{-b - \sqrt{b^2 - 4ac}}{2a}\right\}$

Quadratic Formula

Any quadratic equation of the form $ax^2 + bx + c = 0$ is equivalent to the following compound sentence:

$$\mathbf{x = \frac{-b + \sqrt{b^2 - 4ac}}{2a}} \quad \textit{or} \quad \mathbf{x = \frac{-b - \sqrt{b^2 - 4ac}}{2a}}$$

EXAMPLE 2 Solve $6x^2 - x - 1 = 0$ by the quadratic formula.

Solution:

1 Write the values of a, b, and c. ⟶ $a = 6 \quad b = -1 \quad c = -1$

2 Substitute these values in the two equations of the quadratic formula. Then simplify.

$$x = \frac{-b + \sqrt{b^2 - 4ac}}{2a} \quad \text{or} \quad x = \frac{-b - \sqrt{b^2 - 4ac}}{2a}$$

$$= \frac{-(-1) + \sqrt{(-1)^2 - 4(6)(-1)}}{2(6)} \quad \text{or} \quad = \frac{1 - \sqrt{25}}{12}$$

$$= \frac{1 + \sqrt{1 + 24}}{12} \quad \text{or} \quad = \frac{1 - 5}{12}$$

$$= \frac{1 + \sqrt{25}}{12} \quad \text{or} \quad = \frac{-4}{12}$$

$$= \frac{1 + 5}{12} = \frac{1}{2} \quad \text{or} \quad = -\frac{1}{3}$$

Solution set: $\{\frac{1}{2}, -\frac{1}{3}\}$

EXAMPLE 3 Solve $2x^2 - x = 5$.

Solution:

1 Subtract 5 from each side. ⟶ $2x^2 - x - 5 = 0$ ◀ ***In the form $ax^2 + bx + c = 0$***

2 Write the values of a, b, and c. ⟶ $a = 2 \quad b = -1 \quad c = -5$

3 Substitute these values in the two equations of the quadratic formula. Then simplify.

$$x = \frac{-b + \sqrt{b^2 - 4ac}}{2a} \quad \text{or} \quad x = \frac{-b - \sqrt{b^2 - 4ac}}{2a}$$

$$= \frac{-(-1) + \sqrt{(-1)^2 - 4(2)(-5)}}{2(2)} \quad \text{or} \quad = \frac{-(-1) - \sqrt{(-1)^2 - 4(2)(-5)}}{2(2)}$$

$$= \frac{1 + \sqrt{1 + 40}}{4} \quad \text{or} \quad = \frac{1 - \sqrt{1 + 40}}{4}$$

$$= \frac{1 + \sqrt{41}}{4} \quad \text{or} \quad = \frac{1 - \sqrt{41}}{4}$$

Solution set: $\left\{\frac{1 + \sqrt{41}}{4}, \frac{1 - \sqrt{41}}{4}\right\}$

The Table of Square Roots on page 194 can be used to approximate the solutions of Example 3.

$$\frac{1}{4}+\frac{\sqrt{41}}{4}\approx 0.25+\frac{6.4}{4} \quad \underline{or} \quad \frac{1}{4}-\frac{\sqrt{41}}{4}\approx 0.25-\frac{6.4}{4}$$

$$\approx 0.25+1.6 \quad \underline{or} \quad \approx 0.25-1.6$$

$$\approx 1.9 \quad \underline{or} \quad \approx -1.35 \text{ or } -1.4$$

The solutions of $2x^2 - x = 5$ are approximately 1.9 and -1.4.

CLASSROOM EXERCISES

Compare each equation to $ax^2 + bx + c = 0$. Then write the values of a, b, and c for each equation. (Examples 2 and 3)

1. $x^2 + 3x - 2 = 0$ **2.** $-2x^2 - x + 5 = 0$ **3.** $x^2 + 5x = 0$

4. $4 - x^2 - x = 0$ **5.** $x^2 = -3x - 5$ **6.** $3x^2 - 1 = 5x$

WRITTEN EXERCISES

Goal: To solve quadratic equations by the quadratic formula
Sample Problem: Solve $3x^2 - 8x + 5$. **Answer:** $\{1, \frac{5}{3}\}$

Solve by the quadratic formula. (Examples 1–3)

1. $x^2 + 5x - 6 = 0$ **2.** $x^2 + 3x - 10 = 0$ **3.** $x^2 + 6x - 7 = 0$

4. $x^2 + 2x - 8 = 0$ **5.** $x^2 - 7x + 10 = 0$ **6.** $x^2 - 7x + 12 = 0$

7. $x^2 = 4x + 12$ **8.** $x^2 = 2x + 15$ **9.** $-x^2 + 6x - 9 = 0$

10. $-x^2 + 10x - 25 = 0$ **11.** $6x^2 + x - 2 = 0$ **12.** $4x^2 + 12x + 5 = 0$

13. $x^2 - 3 = 0$ **14.** $x^2 - 5 = 0$ **15.** $x^2 - 6x = 0$

16. $x^2 + 9x = 0$ **17.** $3x - 5x^2 + 1 = 0$ **18.** $5x^2 + 7x - 10 = 0$

Solve by the quadratic formula. Approximate to one decimal place.

19. $x^2 + 2x - 1 = 0$ **20.** $x^2 + 4x + 1 = 0$ **21.** $2x^2 - 3x - 1 = 0$

22. $-3x^2 + 6x - 2 = 0$ **23.** $2x^2 - 6x + 3 = 0$ **24.** $-5x^2 + x + 1 = 0$

REVIEW CAPSULE FOR SECTION 18.6

Find the value of $b^2 - 4ac$. (Pages 74–77)

1. $a = -2$; $b = -5$; $c = 1$ **2.** $a = 4$; $b = 3$; $c = 5$ **3.** $a = 1$; $b = 6$; $c = -2$

18.6 The Discriminant

In using the quadratic formula, you find the value of the expression $b^2 - 4ac$.

> The expression $b^2 - 4ac$ is called the ***discriminant.***

The value of the discriminant describes the kinds of numbers that belong to the solution set of a quadratic equation. It tells whether the numbers are rational, irrational, or not real.

The relationship between the solutions of quadratic equations and their discriminants is exhibited below for several cases.

Equation: $x^2 + x - 12 = 0$ **Solution Set:** $\{-4, 3\}$

$(x + 4)(x - 3) = 0$

Value of Discriminant: $b^2 - 4ac = 1^2 - 4(1)(-12) = 49$ ◀ ***Perfect square***

The solutions are rational numbers because the value of the discriminant is a perfect square.

Equation: $x^2 - 5 = 0$ **Solution Set:** $\{\sqrt{5}, -\sqrt{5}\}$

$x^2 = 5$

Value of Discriminant: $b^2 - 4ac = 0^2 - 4(1)(-5) = 20$ ◀ ***Not a perfect square***

The solutions are irrational numbers because the value of the discriminant is positive and not a perfect square.

Equation: $x^2 - 4x + 4 = 0$ **Solution Set:** $\{2\}$

$(x - 2)(x - 2) = 0$

Value of Discriminant: $b^2 - 4ac = (-4)^2 - 4(1)(4) = 0$ ◀ ***Perfect square***

The solutions are rational and equal because the value of the discriminant is 0. The solution set contains one number.

Equation: $x^2 + 4 = 0$ **Solution Set:** ϕ

$x^2 = -4$ There is no real number whose square is -4.

Value of Discriminant: $b^2 - 4ac = 0^2 - 4(1)(4) = -16$ ◀ ***The value of the discriminant is negative.***

There are no real solutions because the value of the discriminant is a negative number.

Value of the discriminant and the solution set of $ax^2 + bx + c = 0$

1. If $b^2 - 4ac$ is 0 or a perfect-square counting number, the solutions are rational.
2. If $b^2 - 4ac$ is positive and not a perfect square, the solutions are irrational.
3. If $b^2 - 4ac = 0$, the solution set contains one number. The solution is rational.
4. If $b^2 - 4ac$ is negative, no real number solutions exist.

EXAMPLE 1

Compute the value of $b^2 - 4ac$ for each equation. Write One rational number, Two rational numbers, Two irrational numbers, or No real numbers to describe the solutions.

a. $2x^2 - x + 3 = 0$ **b.** $2x^2 + 3x = 2$

Solutions:

a. $2x^2 - x + 3 = 0$

$a = 2,\ b = -1,\ c = 3$

$b^2 - 4ac = (-1)^2 - 4(2)(3)$

$= 1 - 24$

$= -23$ ◀ ***Discriminant is negative.***

No real numbers

b. $2x^2 + 3x = 2$

$2x^2 + 3x - 2 = 0$

$a = 2,\ b = 3,\ c = -2$

$b^2 - 4ac = 3^2 - 4(2)(-2)$

$= 9 + 16 = 25$ ◀ ***Discriminant is a non-zero perfect square.***

Two rational numbers

EXAMPLE 2

Compute the value of $b^2 - 4ac$ for each equation. Write One rational number, Two rational numbers, Two irrational numbers, or No real numbers to describe the solutions.

a. $x^2 + 3x = 5$ **b.** $4x^2 - 4x + 1 = 0$

Solutions:

a. $x^2 + 3x = 5$

$x^2 + 3x - 5 = 0$

$a = 1,\ b = 3,\ c = -5$

$b^2 - 4ac = 3^2 - 4(1)(-5)$

$= 9 + 20 = 29$ ◀ ***Discriminant is not a perfect square.***

Two irrational numbers

b. $4x^2 - 4x + 1 = 0$

$a = 4,\ b = -4,\ c = 1$

$b^2 - 4ac = (-4)^2 - 4(4)(1)$

$= 16 - 16$

$= 0$ ◀ ***Discriminant is zero.***

One rational number

P–1 **Use <u>One rational number</u>, <u>Two rational numbers</u>, <u>Two irrational numbers</u>, or <u>No real numbers</u> to describe the solution of each equation.**

a. $8x^2 - 2x - 1 = 0$ **b.** $x^2 - x - 1 = 0$

c. $x^2 - 2x + 4 = 0$ **d.** $x^2 - 16x + 64 = 0$

CLASSROOM EXERCISES

Write <u>One rational number</u>, <u>Two rational numbers</u>, <u>Two irrational numbers</u>, or <u>No real numbers</u> to describe the solutions for each equation having the given value of its discriminant. (Examples 1–2)

1. 25 **2.** 36 **3.** 0 **4.** 12 **5.** 17 **6.** 1

7. 100 **8.** 4 **9.** -5 **10.** 22 **11.** 9 **12.** -32

Write the value of $b^2 - 4ac$ for each equation. Then write <u>One rational number</u>, <u>Two rational numbers</u>, <u>Two irrational numbers</u>, or <u>No real numbers</u> to describe the solutions. (Examples 1–2)

13. $x^2 + x - 2 = 0$ **14.** $x^2 + 2x + 1 = 0$ **15.** $x^2 - 3x + 2 = 0$

16. $x^2 - 5x + 5 = 0$ **17.** $x^2 + 16 = 0$ **18.** $x^2 - 7 = 0$

19. $2x^2 - 3x - 1 = 0$ **20.** $5x^2 + 5x + 1 = 0$ **21.** $x^2 - x + 1 = 0$

WRITTEN EXERCISES

Goal: To describe the solutions of quadratic equations by use of the discriminant

Sample Problems: **a.** $2x^2 - 7x + 1 = 0$ **b.** $x^2 - 8x + 16 = 0$

Answers: **a.** Two irrational numbers **b.** One rational number

Compute the value of $b^2 - 4ac$ for each equation. Then write <u>One rational number</u>, <u>Two rational numbers</u>, <u>Two irrational numbers</u>, or <u>No real numbers</u> for the solutions. (Examples 1–2)

1. $x^2 - x - 6 = 0$ **2.** $x^2 + 4x - 5 = 0$ **3.** $-x^2 + 2x + 8 = 0$

4. $-x^2 + 8x - 12 = 0$ **5.** $2x^2 - x - 2 = 0$ **6.** $3x^2 + 3x - 1 = 0$

7. $-x^2 + 4x - 4 = 0$ **8.** $-x^2 + 10x - 25 = 0$ **9.** $x^2 - 7 = 0$

10. $x^2 - 13 = 0$ **11.** $x^2 + 25 = 0$ **12.** $2x^2 + x + 6 = 0$

13. $2x^2 = 3 - 5x$ **14.** $3x^2 = 9x - 5$ **15.** $x^2 = 10x$

16. $3x^2 = 12x$ **17.** $5x^2 - 2x = 0$ **18.** $3x - 2x^2 = 0$

19. $x^2 - 2x + 7 = 0$ **20.** $3x^2 - 2x + 8 = 0$ **21.** $-8x^2 - 9x + 2 = 0$

22. $9x^2 = 5$ **23.** $3x^2 - 10 = 0$ **24.** $-x^2 - x + 1 = 0$

REVIEW CAPSULE FOR SECTION 18.7

Write each equation in the form $ax^2 + bx + c = 0$. (Pages 484–486)

1. $x(x - 4) = 78$ **2.** $x(2x + 5) = 102$ **3.** $5x + 1 = x(x - 2)$

4. $3x - 4 = x(x + 5)$ **5.** $x(3x - 1) - x^2 = 128$ **6.** $2x(x + 3) - 5x = 10$

PROBLEM SOLVING AND APPLICATIONS

18.7 Using Quadratic Equations

EXAMPLE 1 The length of a rectangle is 3 centimeters more than its width (Condition 1). The area is 18 square centimeters (Condition 2). Find the length and width.

Solution:

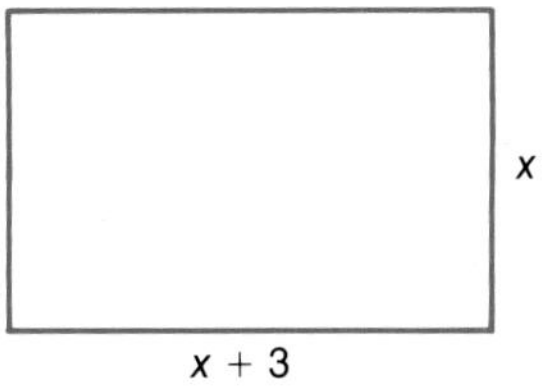

[1] Use Condition 1 to represent the unknowns. → Let x = the width. Then $x + 3$ = the length.

[2] Use Condition 2 to write an equation.

$$A = \ell \cdot w$$

Formula for the area of a rectangle

$$18 = (x + 3)\,x$$

[3] Solve the equation. → $18 = x^2 + 3x$

$$0 = x^2 + 3x - 18$$

$$0 = (x + 6)(x - 3)$$

$x + 6 = 0$ *or* $x - 3 = 0$

$x = -6$ *or* $x = 3$

The measure must be positive.

Length: $x + 3 = 6$

[1] **Check:** Condition 1 Is the length 3 centimeters more than the width? Does $6 = 3 + 3$? Yes ✓

Condition 2 Is the area 18 square centimeters? Does $(6)(3) = 18$? Yes ✓

The width is 3 centimeters, and the length is 6 centimeters.

In Examples 2 and 3, Condition 1 is underscored once and Condition 2 is underscored twice.

EXAMPLE 2

The product of two consecutive integers is 72. What are the integers?

Solution:

[1] Let $n =$ the smaller integer.
Then $n + 1 =$ the greater integer. ◀ ***Condition 1***

[2] Write an equation.

Think: The product of two consecutive integers is 72.

Translate: $n(n+1) \quad = 72$ ◀ ***Condition 2***

[3] Solve the equation. ⟶ $n^2 + n = 72$

$$n^2 + n - 72 = 0$$

$$(n+9)(n-8) = 0$$

$$n + 9 = 0 \quad \underline{or} \quad n - 8 = 0$$

$$n = -9 \quad \underline{or} \quad n = 8$$

Smaller Integer	Greater Integer
−9	−8
8	9

◀ ***Two answers are possible.***
1. −9 and −8
2. 8 and 9

[4] **Check:** Condition 1 Are the two integers consecutive?
Does $-8 - (-9) = 1$? Yes ✔ Does $9 - 8 = 1$? Yes ✔

Condition 2 Is the product 72? Does $(-9)(-8) = 72$? Yes ✔
Does $(8)(9) = 72$? Yes ✔

EXAMPLE 3

The area of a rectangle less the area of a square equals 15 square meters. The width of the rectangle equals the length of each side of the square. The length of the rectangle is 2 meters more than twice the width. Find the width and length of the rectangle.

Solution:

[1] Let $x =$ the width of the rectangle.
Let $x =$ the length of each side of the square.
Then $2x + 2 =$ the length of the rectangle.

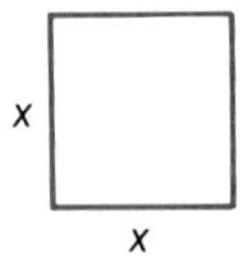

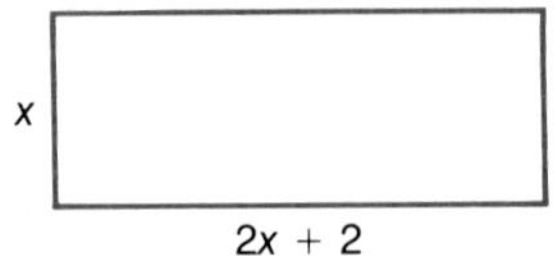

Think: The area of a rectangle less the area of a square equals 15.

[2] **Translate:** $x(2x + 2) \quad - \quad x^2 \quad = \quad 15$

[3]

$$2x^2 + 2x - x^2 = 15$$

$$x^2 + 2x - 15 = 0$$

$$(x + 5)(x - 3) = 0$$

$$x + 5 = 0 \quad \text{or} \quad x - 3 = 0$$

$$x = -5 \quad \text{or} \quad x = 3$$

◀ ***The width cannot be negative.***

Length: $2x + 2 = 8$

[4] **Check:** Condition 1 Is the length of the rectangle 2 meters more than twice the width? Does $2(3) + 2 = 8$? Yes ✓

Condition 2 Is the area of the rectangle less the area of the square 15 square meters? Does $2(3)^2 + 2(3) - 3^2 = 15$? Yes ✓

The width is 3 meters, and the length is 8 meters.

CLASSROOM EXERCISES

For Exercises 1–8, Condition 1 is underscored once. Condition 2 is underscored twice.

a. *Use Condition 1 to represent the unknowns.*

b. *Use Condition 2 to write an equation for the problem.*

(Examples 1 and 2)

1. The product of two consecutive integers is 210.
2. The product of two consecutive even integers is 48.
3. One integer is three more than another integer. Their product is 108.
4. One number is 15 less than another number. Their product is −36.
5. The length of a rectangle is 5 units greater than the width. The area is 24 square units.
6. The length of a rectangle is three units more than twice the width. The area is 65 square units.
7. One number is three times as great as another number. The square of the smaller number is 18 less than the square of the greater number.
8. One number is three more than another number. The square of the smaller number is 51 more than the square of the greater number.

WRITTEN EXERCISES

Goal: To use quadratic equations to solve problems

Sample Problem: One integer is 10 less than another integer (Condition 1). Their product is −16 (Condition 2). Find the integers.

Answer: Equation: $x(x - 10) = -16$; $x^2 - 10x + 16 = 0$
The integers are 8 and −2 or 2 and −8.

For Exercises 1–8, Condition 1 is underscored once. Condition 2 is underscored twice.

a. *Use Condition 1 to represent the unknowns.*
b. *Use Condition 2 to write an equation for the problem.*
c. *Solve.* (Examples 1–3)

1. The width of a rectangle is 2 units less than the length. If the area is 80 square units, what are the length and width?

2. The length of a rectangle is 5 units more than the width. If the area is 50 square units, what are the length and width?

3. The product of two consecutive integers is 30. Find two pairs of such integers.

4. The product of two consecutive integers is 56. Find two pairs of such integers.

5. One integer is 13 less than another integer. Their product is −40. Find all possible pairs of such integers.

6. One integer is 6 more than another integer. Their product is 55. Find all possible pairs of such integers.

7. A rectangular plate has a rectangular piece cut from one corner as shown in the figure. The area after the rectangular piece is cut out is 40 square units. Use the information given in the figure to find the length and width of the original rectangular plate.

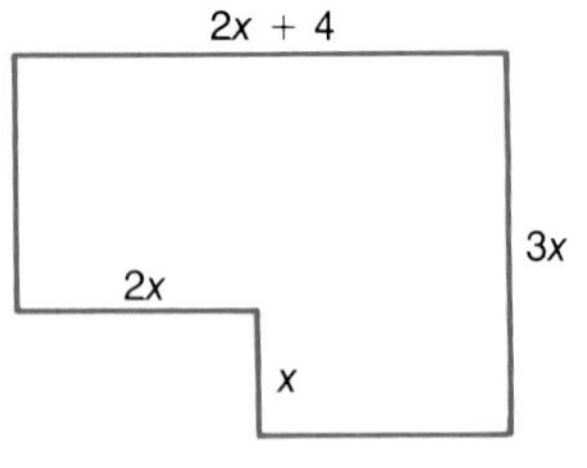

8. A square hole is cut in a rectangular plate as shown in the figure. The area of the plate after the square hole is cut out is 105 square units. The width of the rectangle is twice the length of each side of the square hole. The length of the rectangle is 13 units. Find the width of the rectangle.

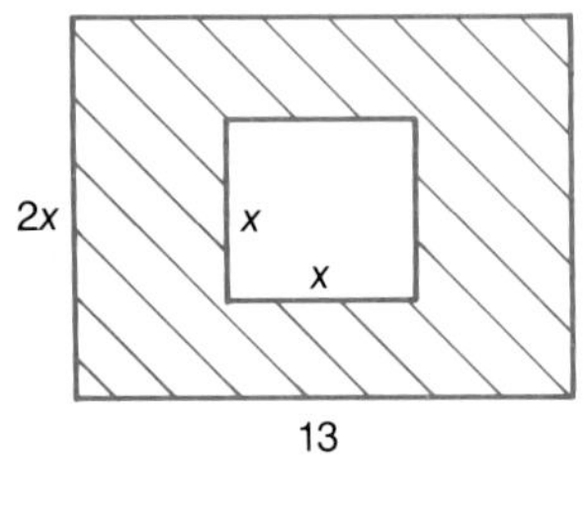

Using a Geometric Model **Factoring**

An algebraic method for solving a quadratic equation by **completing the square** was shown on pages 494–495. The following method of completing the square uses a **geometric model.** This method works for finding positive solutions only to quadratic equations of the form $ax^2 + bx + c = 0$ where $a > 0$, $b > 0$, and $c < 0$.

EXAMPLE: Solve $x^2 + 4x - 12 = 0$ by completing the square.

SOLUTION:

1 Write the equation in the form $x^2 + bx = -c$. ⟶ $x^2 + 4x = 12$

2 Represent x^2 as a square with sides of length x.

3 Represent $4x$ as two rectangles, each with an area of $\frac{1}{2}(4x)$, or $2x$.

Position the rectangles as shown.

Area of the L-shaped figure: $2x + x^2 + 2x$, or $x^2 + 4x$

From step 1, $x^2 + 4x = 12$.
Thus, the area of the L-shaped figure is 12.

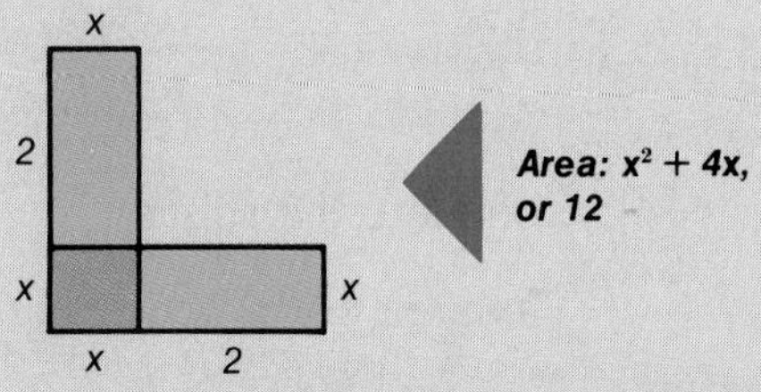

4 Complete the square by adding the red region shown.

Why must the lengths of the sides of the red region equal 2?

What is the area of the red square?

$$\text{Area of L-shaped Region} + \text{Area of Red Square} = \underline{\ ?\ }$$

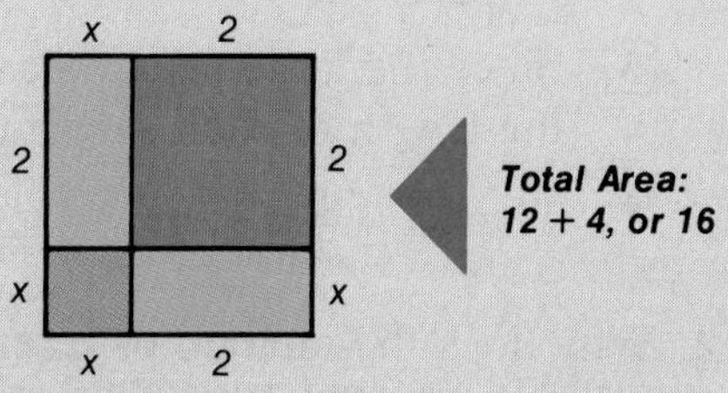

5 Since the large square has a total area of 16, the length of each side must equal 4. Why?

Thus, $x + 2 = 4$, or $x = 2$.

6 **Check:** Does $(2)^2 + 4(2) - 12 = 0$?

Does $4 + 8 - 12 = 0$? Yes ✓

EXERCISES

For Exercise 1, replace each _?_ to solve $2x^2 + 5x - 3 = 0$ by completing the square.

1. [1] Write the equation in the form $x^2 + \frac{b}{a}x = -\frac{c}{a}$ ⟶ $x^2 + \frac{5}{2}x = \underline{\ ?\ }$

[2] Represent x^2.

[3] Represent $\frac{5}{2}x$ as two rectangles, each with an area of $\frac{1}{2}(\frac{5}{2}x)$, or _?_.

Area of L-shaped figure:
$\frac{5}{4}x + x^2 + \frac{5}{4}x = \underline{\ ?\ }$

From step 1, $x^2 + \frac{5}{2}x = \frac{3}{2}$.
Thus, the area of the L-shaped figure = _?_.

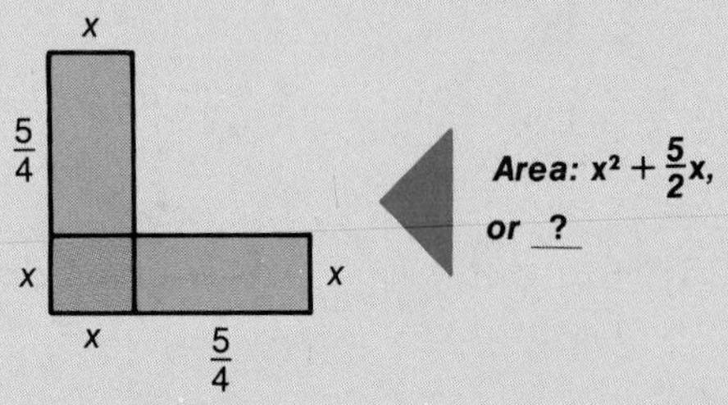

[4] Complete the square.

The area of the red square equals _?_.

$$\text{Area of L-shaped Region} + \text{Area of Red Square} = \underline{\ ?\ }$$

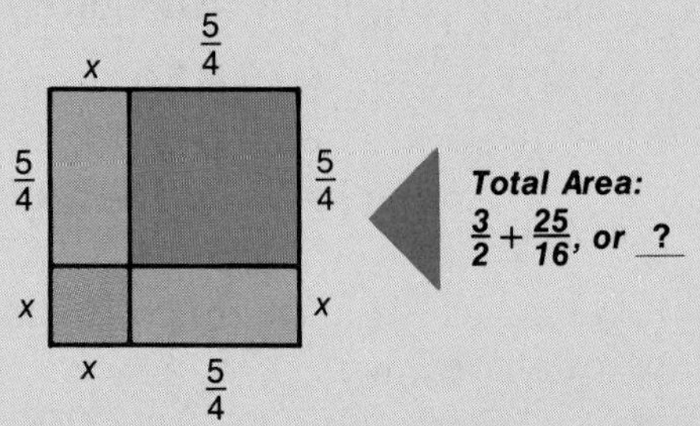

[5] Since the large square has a total area of $\frac{49}{16}$, the length of each side must equal _?_.

Thus, $x + \frac{5}{4} = \frac{7}{4}$, or $x = \underline{\ ?\ }$.

2. Why does the method of Exercise 1 work only for finding a positive solution of a quadratic equation?

Use the geometric method of completing the square to find a positive solution for each quadratic equation.

3. $x^2 + 2x - 3 = 0$
4. $x^2 + 4x - 21 = 0$
5. $x^2 + 6x - 16 = 0$
6. $x^2 + 8x - 20 = 0$
7. $2x^2 + 5x - 12 = 0$
8. $2x^2 + 3x - 9 = 0$

CHAPTER SUMMARY

IMPORTANT TERMS

Quadratic equation *(p. 484)*
Zeros of a function *(p. 484)*
Completing the square *(p. 494)*
Quadratic formula *(p. 497)*
Discriminant *(p. 500)*

IMPORTANT IDEAS

1. The solutions of a quadratic equation of the form $ax^2 + bx + c = 0$ are the same as the zeros of the function $y = ax^2 + bx + c$.
2. *Factors of Zero Property:* If a and b are real numbers and $ab = 0$, then $a = 0$ *or* $b = 0$.
3. Any quadratic equation of the form $(x + r)^2 = s$ in which r and s are rational numbers, $s \geq 0$, is equivalent to "$x + r = \sqrt{s}$ *or* $x + r = -\sqrt{s}$."
4. It is possible to write any quadratic equation in a form that makes the left side a trinomial square.
5. *Steps for solving $ax^2 + bx + c = 0$ by completing the square:*
   - **a.** Add $-c$ to each side.
   - **b.** Multiply each side by $\frac{1}{a}$.
   - **c.** Compute $\left(\frac{1}{2} \cdot \frac{b}{a}\right)$.
   - **d.** Compute $\left(\frac{1}{2} \cdot \frac{b}{a}\right)^2$.
   - **e.** Add $\left(\frac{1}{2} \cdot \frac{b}{a}\right)^2$ to each side.
   - **f.** Factor the left side.
   - **g.** Write the equivalent compound *or* sentence.
   - **h.** Solve.
6. *Quadratic Formula:* Any quadratic equation of the form $ax^2 + bx + c = 0$ is equivalent to the following compound sentence:
   $$x = \frac{-b + \sqrt{b^2 - 4ac}}{2a} \text{ or } x = \frac{-b - \sqrt{b^2 - 4ac}}{2a}$$
7. *Value of the discriminant and the solution set of $ax^2 + bx + c = 0$*
   See page 501.

CHAPTER REVIEW

SECTION 18.1

Change each equation to the form $ax^2 + bx + c = 0$.

1. $x^2 + 5x = 2$

2. $x^2 = 3 - 2x$

Graph $y = x^2 - 1$. Refer to this graph to estimate the solutions of each equation.

3. $x^2 - 1 = 2$

4. $x^2 - 1 = 0$

SECTION 18.2

Solve by using the Factors of Zero Property.

5. $x^2 - 9x - 10 = 0$

6. $x^2 = -21 + 10x$

7. $2x^2 + 10x = 0$

SECTION 18.3

Write the solutions of each equation.

8. $x^2 = 256$

9. $x^2 = 361$

10. $(x - 6)^2 = 64$

SECTION 18.4

Solve by completing the square.

11. $x^2 - 8x - 20 = 0$

12. $x^2 + 4x - 32 = 0$

13. $4x^2 - x - 3 = 0$

SECTION 18.5

Solve by the quadratic formula.

14. $8x^2 - 6x + 1 = 0$

15. $6x^2 - x - 1 = 0$

16. $x^2 + 5x + 7 = 0$

SECTION 18.6

Compute the value of $b^2 - 4ac$ for each equation. Then write One rational number, Two rational numbers, Two irrational numbers, or No real numbers to describe the solutions.

17. $x^2 - 4x + 4 = 0$

18. $x^2 + 5x - 14 = 0$

19. $-3x^2 + 4x + 5 = 0$

SECTION 18.7

For Exercises 20–21, Condition 1 is underscored once, and Condition 2 is underscored twice. Use Condition 1 to represent the unknowns. Use Condition 2 to write an equation for the problem. Solve.

20. One number is 3 greater than another number. Their product is 10. What are the two pairs of numbers that satisfy these conditions?

21. The length of a rectangle is 4 centimeters more than its width. The area is 60 square centimeters. Find the length and width.

CHAPTER 19

Probability

When all of the steel balls have fallen in the probability machine below and in the device at the right, a *bell-shaped curve* is formed. As each ball falls, it is equally likely to go to the right or to the left.

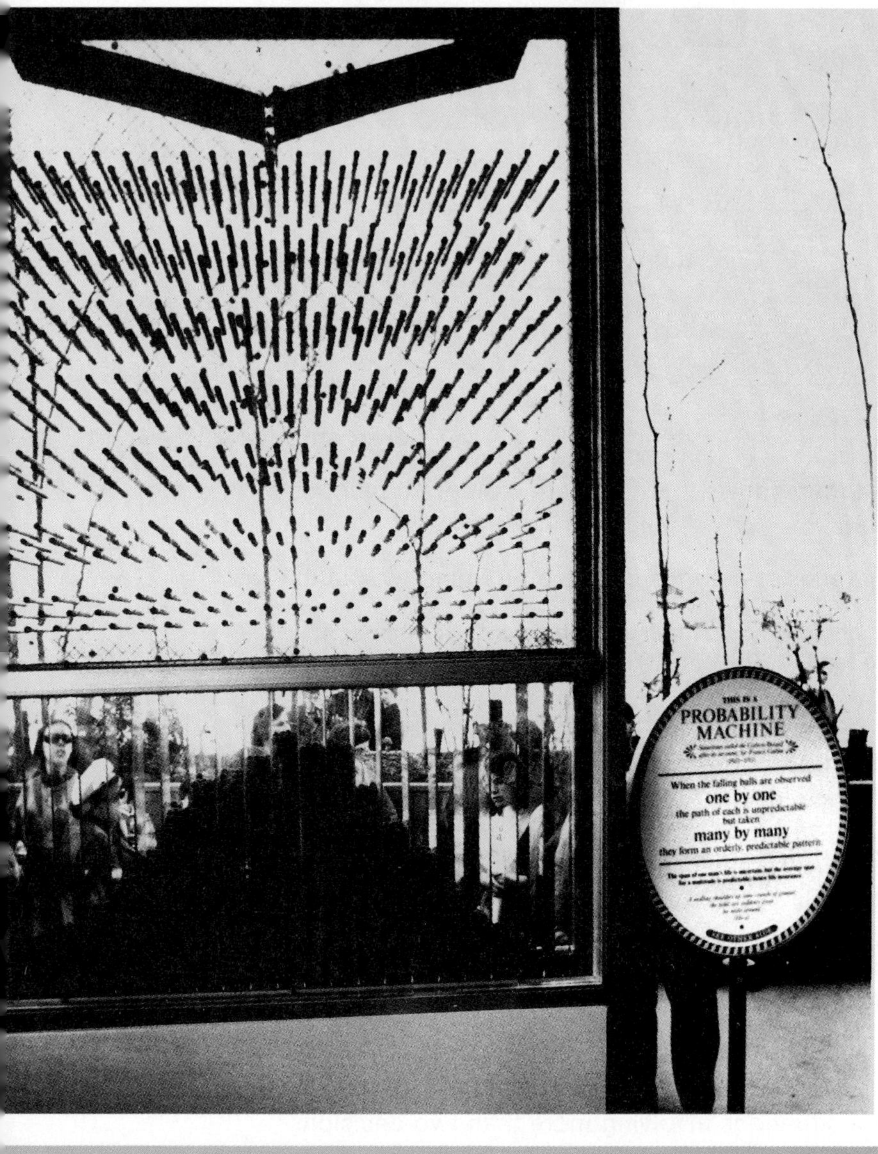

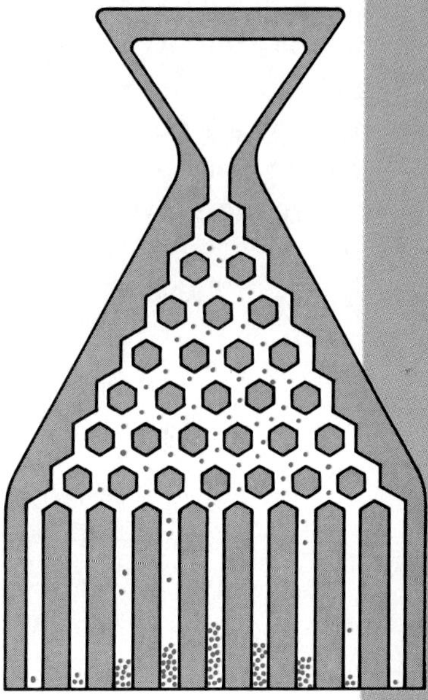

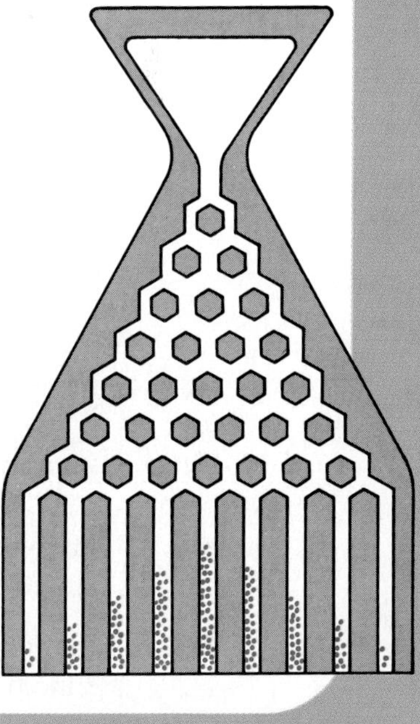

19.1 Fundamental Counting Principle

Karen and Sarah are trying out for the female lead in this year's school play. John, Mike, and Henry are auditioning for the male lead. The director asked Sarah to make a chart that shows how many pairings are possible for the two leading roles. A tree diagram can be used to show the pairings.

Table

Female Lead	Male Lead	Pairings	
Karen	John	Karen and John	1
	Mike	Karen and Mike	2
	Henry	Karen and Henry	3
Sarah	John	Sarah and John	4
	Mike	Sarah and Mike	5
	Henry	Sarah and Henry	6

The tree diagram shows that there are 6 different pairings for the two leading roles.

The total number of pairings can also be found by multiplying.

$$\left(\begin{array}{c}\textbf{Number of choices:}\\ \textbf{Female lead}\end{array}\right) \times \left(\begin{array}{c}\textbf{Number of choices:}\\ \textbf{Male lead}\end{array}\right) = \left(\begin{array}{c}\textbf{Number of}\\ \textbf{pairings}\end{array}\right)$$

$$2 \times 3 = 6$$

This method for multiplying the choices applies the Fundamental Counting Principle.

> ***Fundamental Counting Principle***
>
> If there are p choices for the first decision and q choices for the second decision, then there are $p \times q$ ways that the two decisions can be made together.

The following example shows how to use the Fundamental Counting Principle for situations involving more than two decisions.

EXAMPLE The members of the senior class are choosing class rings. They can decide on the material (white gold or yellow gold), style (traditional, contemporary, or signet), and the color of the stone (red, blue, green, or yellow). How many different choices does each student have?

Solution: Use the Fundamental Counting Principle.

$$\begin{pmatrix}\textbf{Number of}\\ \textbf{choices:}\\ \textbf{Metal type}\end{pmatrix} \times \begin{pmatrix}\textbf{Number of}\\ \textbf{choices:}\\ \textbf{Style}\end{pmatrix} \times \begin{pmatrix}\textbf{Number of}\\ \textbf{choices:}\\ \textbf{Stone color}\end{pmatrix} = \begin{pmatrix}\textbf{Total}\\ \textbf{choices}\end{pmatrix}$$

$$2 \quad \times \quad 3 \quad \times \quad 4 \quad = \quad 24$$

Each student has 24 different choices for a class ring.

CLASSROOM EXERCISES

Draw a tree diagram to illustrate each exercise. Then write the number of different choices that are possible. (Table)

1. The number of ways to choose the color of a new convertible sports car if the top can be white or black and the car body can be red, blue, brown, or black

2. The number of different outfits possible if a girl has four sweaters, three skirts, and two pairs of shoes that are all color coordinated

3. The number of ways to choose a president and vice-president for the senior class if there are six candidates for president and four for vice-president

4. The number of ways to choose a lunch in the cafeteria if there are two choices for an entree, three choices for vegetable, and two choices for dessert (only one choice is allowed per category)

5. The number of ways that a two-digit number can be written using the following digits:

1, 2, 3, 4

NOTE: The digits can be repeated.

WRITTEN EXERCISES

Goal: To use the Fundamental Counting Principle

Sample Problem: A man has two sports coats, three pairs of slacks, and five ties that are color coordinated. In how many different ways can he choose an outfit from these items?

Answer: $2 \times 3 \times 5 = 30$

Use the Fundamental Counting Principle to solve each problem. (Example)

1. A car manufacturer offers three new models in six different colors. How many cars will a car dealer need in order to display all the possible selections?

2. There are five candidates for junior class president and three candidates for vice-president. In how many possible ways can the two offices be filled?

3. An English class must choose from a list of twenty books to write a book report, and from ten topics to write a theme. In how many ways can a student choose a book and a theme topic?

4. Freight can be shipped from city A to city B either by plane or by boat. It can be shipped from city B to city C by plane, truck, or bus. In how many possible ways can freight be shipped from city A to city B to city C?

5. The identification code used by a computer operator can be any two letters. The two letters can be the same. How many different identification codes are possible under this system?

6. A dinner menu has a choice of four appetizers, nine entrees, three beverages, and five desserts. In how many ways can a dinner be ordered, choosing one item from each category?

7. A commuter going to work can walk, ride a bike, or share a ride to the station. At the station she can either take a train or a bus. After arriving in the city, she can walk, ride a shuttle bus, or take a subway. How many possible ways are there for her to get to work?

8. For a required reading assignment, a student must choose one novel from a list of ten, one book of nonfiction from a list of six, and one short story from a list of eight. In how many possible ways can the reading assignment be completed?

REVIEW CAPSULE FOR SECTION 19.2

Write each fraction in lowest terms.

1. $\frac{5}{20}$ 2. $\frac{6}{15}$ 3. $\frac{12}{28}$ 4. $\frac{10}{16}$ 5. $\frac{6}{24}$

19.2 Sample Space and Probabilities

Suppose you toss a coin and observe the way in which it lands. It can land "heads" up or "tails" up. Both results are equally likely to occur. Now you toss the coin twice. The tree diagram below shows the ways the coin can land after being tossed.

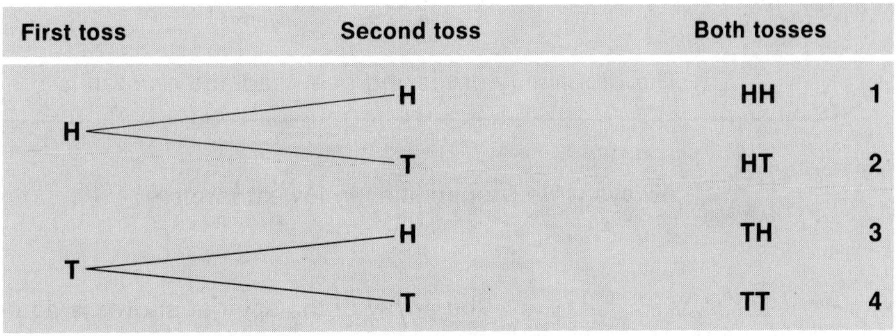

Tossing the coin twice is an example of an ***experiment.*** Each result of an experiment is called an ***outcome.*** The set of all the different possible outcomes of an experiment is called the ***sample space.*** An ***event*** is a subset of a sample space. Thus, it is a set of one or more of the possible outcomes of an experiment. The sample space for tossing a coin twice consists of the four outcomes shown in the tree diagram. The sample space is written as a set.

$$\{HH, HT, TH, TT\}$$

Each of the four outcomes in this sample space is equally likely to occur. Therefore, the probability of each outcome is 1 out of 4, or $\frac{1}{4}$.

Definition

The probability *P* of an event *E* is the ratio of the number of favorable outcomes of the event to the total number of possible outcomes in the sample space.

$$P(E) = \frac{\textbf{number of favorable outcomes}}{\textbf{total number of possible outcomes}}$$

EXAMPLE 1

A coin is tossed twice. What is the probability of tossing one head and one tail?

Solution: Possible outcomes (sample space): {HH, HT, TH, TT}

Favorable outcomes: {HT, TH}

[1] Determine the total number of possible outcomes. ⟶ 4

[2] Determine the number of favorable outcomes for the event. ⟶ 2

[3] Write the probability ratio.

$P(\text{tossing one head and one tail}) = \frac{2}{4}$ ⟵ ***Number of favorable outcomes*** / ⟵ ***Number of possible outcomes***

$= \frac{1}{2}$ ◀ ***Lowest terms***

The probability of tossing one head and one tail is $\frac{1}{2}$.

Always write probabilities in lowest terms.

EXAMPLE 2

The arrow of the spinner shown is equally likely to stop on any of the six regions. The arrow is spun once. Find each probability.

a. Spinning a number less than 2
b. Spinning a number greater than 1

Solutions: Possible outcomes: {1, 2, 3, 4, 5, 6}

	a.	**b.**
Favorable outcomes:	{1}	{2, 3, 4, 5, 6}
[1] Determine the total number of possible outcomes. ⟶	6	6
[2] Determine the number of favorable outcomes for the event. ⟶	1	5
[3] Write the probability ratio. ⟶	$P(\text{number} < 2) = \frac{1}{6}$	$P(\text{number} > 1) = \frac{5}{6}$

The probability of spinning a number less than 2 is $\frac{1}{6}$.

The probability of spinning a number greater than 1 is $\frac{5}{6}$.

In Example 2, spinning a number less than 2 and spinning a number greater than 1 are complementary events. Since two ***complementary events*** include all of the possible outcomes of an experiment, the sum of their probabilities is always 1.

EXAMPLE 3 For the spinner in Example 2, find these probabilities.

a. Spinning a whole number
b. Not spinning a whole number

Solutions: Possible outcomes: {1, 2, 3, 4, 5, 6}

	a.	**b.**
Favorable outcomes:	{1, 2, 3, 4, 5, 6}	ϕ
1 Determine the total number of possible outcomes. →	6	6
2 Determine the number of favorable outcomes for the event. →	6	0
3 Write the probability ratio. →	$P(\text{whole}) = \frac{6}{6}$, or 1	$P(\text{not whole}) = \frac{0}{6}$, or 0

All of the numbers are whole numbers.

The probability of spinning a whole number is 1.
The probability of not spinning a whole number is 0.

> The probability of an event that is certain to happen is 1.
> The probability of an event that cannot happen is 0.

CLASSROOM EXERCISES

A coin is tossed twice. Find each probability. (Examples 1 and 3)

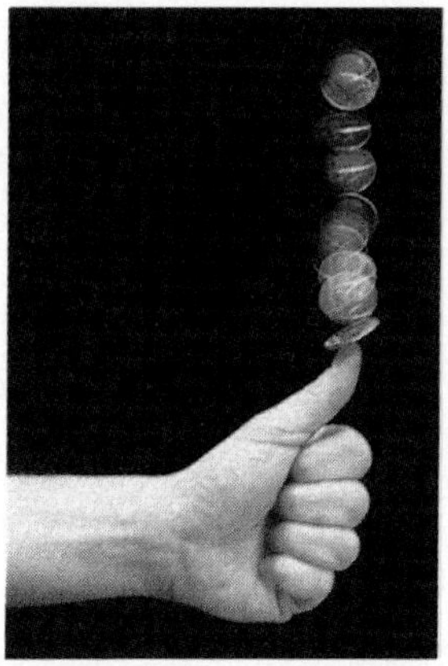

1. Tossing two heads
2. Tossing two tails
3. Not tossing a head
4. Not tossing a tail
5. Tossing more than one head
6. Tossing more than two tails

The arrow of the spinner at the right is equally likely to stop on any of the 4 regions. The arrow is spun once.

For Exercises 9–12, find each probability. (Examples 2 and 3)

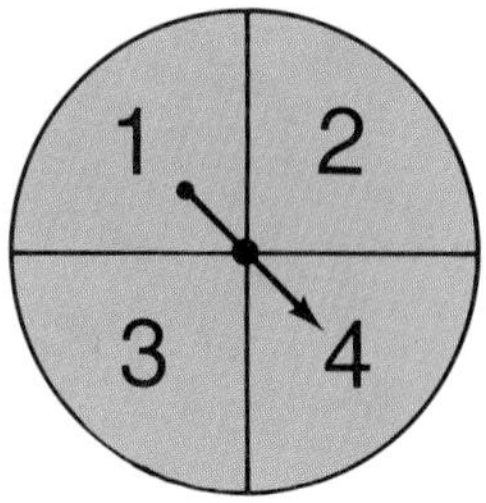

7. Stopping on an even number

8. Stopping on a prime number

9. Not stopping on 4

10. Not stopping on a whole number

11. Stopping on a whole number less than 5

12. Stopping on a whole number greater than 3

13. Are the events in Exercises 7 and 8 complementary events? Explain.

14. Are the events in Exercises 9 and 12 complementary events? Explain.

WRITTEN EXERCISES

Goal: To compute probabilities

Sample Problem: A container holds three yellow balls and four red balls. One ball is drawn without looking. What is the probability that a red ball is drawn?

Answer: $\frac{4}{7}$

A ball is drawn from a bag without looking. The bag contains 4 red balls, 3 white balls, 5 black balls, and 2 green balls. Find the probability of each event. (Example 1)

1. Drawing a green ball

2. Drawing a red ball

3. Drawing a black ball

4. Drawing a white ball

An experiment involves tossing a game marker which has four sides numbered 1, 2, 3, and 4. The marker is equally likely to land on any of the four sides. The marker is tossed once.

For Exercises 5–10, find each probability. (Examples 2 and 3)

5. Landing on 2

6. Landing on an odd number

7. Landing on a prime number

8. Landing on an even number

9. Landing on a counting number

10. Landing on a number greater than 4

11. Are the events in Exercises 5 and 7 complementary events? Explain.

12. Are the events in Exercises 6 and 8 complementary events? Explain.

The cards shown at the right are placed in a box. The cards are shuffled and one card is drawn without looking.

Find each probability.
(Examples 2 and 3)

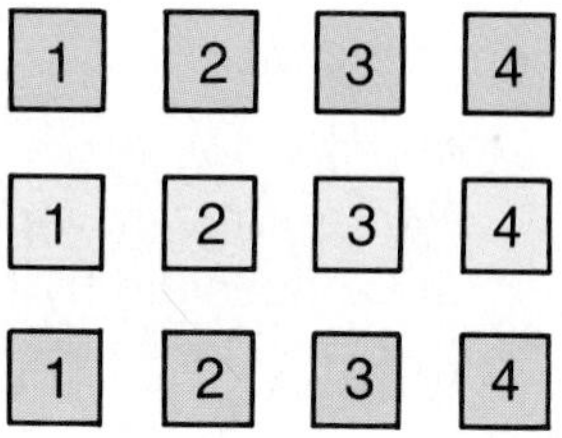

13. Drawing a green card

14. Not drawing a blue card

15. Drawing a 4

16. Drawing a blue 2

17. Drawing a number less than 3

18. Not drawing an odd number

19. Drawing a red card

20. Drawing a number less than 5

MORE CHALLENGING EXERCISES

A coin is tossed three times. Draw a tree diagram to identify the sample space for this experiment. Then find each probability.

21. Tossing two heads and one tail

22. Tossing one head and two tails

23. Tossing two or more heads

24. Tossing less than three heads

25. Tossing more tails than heads

26. Tossing no heads

NON-ROUTINE PROBLEMS

27. Doris took some coins out of her bank and arranged the coins in four rows with four coins in each row. Each of the four rows, the four columns, and the two diagonals had exactly one penny, one nickel, one dime, and one quarter. Show how Doris could have arranged the coins.

28. Roberto needs to draw a line that is six inches long, but he does not have a ruler. He does have some sheets of notebook paper that are each $8\frac{1}{2}$ inches wide and 11 inches long. Describe how Roberto can use the notebook paper to measure six inches.

19.3 Independent Events

When tossing two number cubes, the outcome of tossing the second cube is not affected by the outcome of tossing the first. The two outcomes are said to be <u>independent</u>.

Two events are ***independent*** if the occurrence of one event is not influenced by the occurrence of the other.

When two events A and B are independent, the probability that <u>both</u> events will occur together is the product of their probabilities.

$$P(A \text{ and } B) = P(A) \cdot P(B)$$

EXAMPLE 1 A red number cube and a green number cube are tossed. Find the probability of tossing a number less than 3 on the red cube <u>and</u> an even number on the green cube.

Solution: Let event A be tossing a number less than 3 on the red cube. Let event B be tossing an even number on the green cube.

1. Identify the elements of events A and B. ⟶ $A = \{1, 2\}$ $\quad B = \{2, 4, 6\}$

2. Find the probability of each event. ⟶ $P(A) = \frac{2}{6}$, or $\frac{1}{3}$ $\quad P(B) = \frac{3}{6}$, or $\frac{1}{2}$

3. Find the probability of both events. ⟶ $P(A \text{ and } B) = P(A) \cdot P(B)$

$$= \frac{1}{3} \cdot \frac{1}{2}$$

$$= \frac{1}{6}$$

The probability of tossing a number less than 3 on the red cube <u>and</u> an even number on the green cube is $\frac{1}{6}$.

P–1 **Find these probabilities for the number cube experiment in Example 1.**

a. Tossing a number greater than 4 on the red cube and a 2 on the green cube

b. Tossing a prime number on the red cube and an odd number on the green cube

Marian has 3 pennies, 5 nickels, and 4 dimes in her pocket. Suppose she draws one coin. She wants to know the probability of drawing a penny or a dime. Since no coin is both a penny and a dime, the two events are said to be ***mutually exclusive.*** Mutually exclusive events have no outcomes in common.

> The probability that one of two mutually exclusive events will occur is the sum of their probabilities.
>
> $$P(A \text{ or } B) = P(A) + P(B)$$

EXAMPLE 2

Marian needs a coin for the parking meter. Without looking, she draws one coin from the 3 pennies, 5 nickels, and 4 dimes in her pocket. What is the probability that it will be a nickel or a dime?

Solution: Let event A be drawing a nickel.
Let event B be drawing a dime.

1. Find the probability of each event. ⟶ $P(A) = \frac{5}{12}$ $P(B) = \frac{4}{12}$

2. Find the probability of drawing a nickel or a dime. ⟶ $P(A \text{ or } B) = P(A) + P(B)$

$$= \frac{5}{12} + \frac{4}{12}$$

$$= \frac{9}{12}, \text{ or } \frac{3}{4}$$ ◀ *Lowest terms*

The probability that Marian will draw a nickel or a dime is $\frac{3}{4}$.

CLASSROOM EXERCISES

Four cards are marked with the letters M, A, T, and H. They are shuffled and placed with the letters down. One card is drawn. Then the spinner is spun. Find each probability. (Example 1)

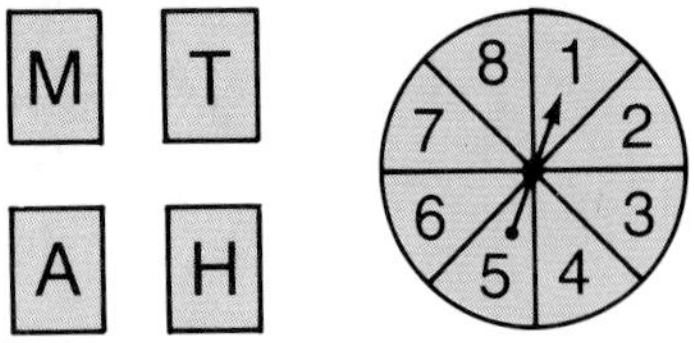

1. Drawing a consonant and spinning a number greater than 2
2. Drawing a vowel and spinning a number less than 5
3. Drawing the T and spinning a prime number
4. Drawing a consonant and spinning an even number

The seven cards shown at the right are shuffled and placed with the letters down. One card is drawn. Find each probability. (Example 2)

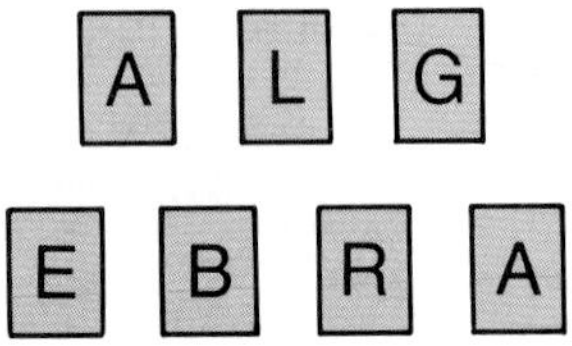

5. Drawing the G or a vowel
6. Drawing the B or an A
7. Drawing the R or an A
8. Drawing the E or a consonant

WRITTEN EXERCISES

Goal: To find the probability of independent events

Sample Problem: A coin is tossed twice. Find the probability of getting a head on the first toss and a tail on the second.

Answer: $\frac{1}{2} \cdot \frac{1}{2} = \frac{1}{4}$

A red number cube and a green number cube are tossed. Find each probability. (Example 1)

1. Tossing a 3 on the red cube and a 5 on the green cube
2. Tossing a 1 on the red cube and an even number on the green cube
3. Tossing a number less than 6 on the red cube and a number greater than 4 on the green cube
4. Tossing an odd number on the red cube and an even number on the green cube
5. Tossing a number less than 3 on the red cube and a number greater than 3 on the green cube
6. Tossing a number greater than 2 on the red cube and a number less than 4 on the green cube
7. Tossing a negative number on the red cube and a number greater than 0 on the green cube
8. Tossing a counting number on the red cube and a whole number on the green cube

A bag contains 6 white, 4 orange, 7 yellow, and 3 green balls. One ball is drawn. Find each probability. (Example 2)

9. Drawing a white ball or a yellow ball

10. Drawing an orange ball or a green ball

11. Drawing a white ball or a green ball

12. Drawing a yellow ball or a green ball

13. Drawing a white ball or an orange ball

14. Drawing an orange ball or a yellow ball

MID-CHAPTER REVIEW

Use the Fundamental Counting Principle to write the number of different choices that are possible. (Section 19.1)

1. The number of ways to choose classes for the first two periods of the school day if there are three choices for the first period and six choices for the second

2. The number of ways to choose a computer system if there are four computer models, two different disc drives, and two different printers

The arrow of the spinner is equally likely to stop on any of the eight regions. The arrow is spun once. Find each probability. (Section 19.2)

3. Stopping on a number less than 3

4. Stopping on a prime number

5. Stopping on a counting number

6. Not stopping on an integer

The arrow of the spinner is equally likely to stop on any of the four regions. The arrow is spun and then a number cube is tossed. Find each probability. (Section 19.3)

7. Spinning an even number and tossing a multiple of 3

8. Spinning a prime number and tossing a number greater than 4

A sack contains 4 red, 2 white, 6 green, and 8 blue marbles. One marble is drawn without looking. Find each probability. (Section 19.3)

9. Drawing a red marble or a blue marble

10. Drawing a white marble or a green marble

Using Statistics

Collecting Data

How do manufacturers determine which colors to use on cereal boxes in order to attract the most customers? They do it by taking a **sample.** For example, they might ask people entering supermarkets in several cities in different parts of the country which of two or three cereal boxes they prefer. The designs and colors for cereal boxes are determined on the basis of these choices.

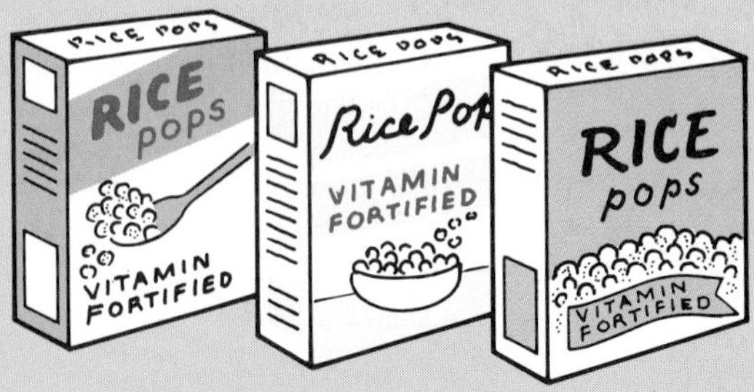

A **sample** is a group chosen to represent a larger group called the **population.** In order for the sample to represent the population, the sample must be sufficiently large and it must be chosen at random. The items in a random sample are chosen completely by chance.

EXERCISES

1. This advertisement is based on a sample of eight plumbers. Is the sample sufficiently large? Explain.

2. This advertisement is based on a sample of Senator Win's friends. Is this a random sample? Explain.

3. A book has 700 pages. To check for errors, an editor randomly chose 70 pages to be examined. Is this a good sample of the population? Explain.

4. Estimate the number of left-handed students in your school by asking 25 students. Tell how you selected the students. Explain why you think the sample does, or does not, provide a good basis for the estimate.

5. A local television station includes a poll as part of its late night newscast. A "question of the day" is flashed on the screen and viewers are asked to call in their "yes" or "no" votes. Do you think the results represent the community's opinion? Explain.

19.4 Dependent Events

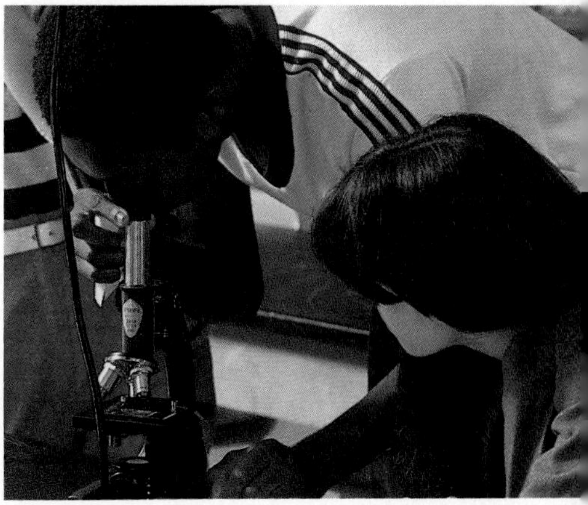

Twenty students in a biology class are assigned to one of four lab stations by drawing cards from a box. The box contains twenty cards. Five cards are marked with a 1, five are marked with a 2, five are marked with a 3, and five are marked with a 4.

Each time a student draws a card, there is one fewer card for the next student to choose from. Thus, the outcome of each drawing is dependent on the previous drawing.

> Two events are ***dependent*** if the occurrence of one event is influenced by the occurrence of the other.

As with independent events, the probability that two dependent events will both occur is the product of their probabilities. The difference is that the number of possible outcomes for the second event has been affected by the occurrence of the first event.

EXAMPLE 1 Jane is the first and Linda is the second biology student to draw a card. Find the probability that Jane draws a 1 and Linda draws a 4.

Solution: Let event A be Jane's drawing a 1.
Let event B be Linda's drawing a 4, given that Jane draws a 1.

[1] Determine the total number of possible outcomes for each event. ⟶ A: 20 B: 19 ◀ ***Jane has drawn one card.***

[2] Determine the number of favorable outcomes for each event. ⟶ A: 5 B: 5

[3] Find the probability of each event. ⟶ $P(A) = \frac{5}{20}$, or $\frac{1}{4}$ $P(B) = \frac{5}{19}$

[4] Find the probability of both events. ⟶ $P(A \text{ and } B) = P(A) \cdot P(B)$

$$= \frac{1}{4} \cdot \frac{5}{19}$$

$$= \frac{5}{76}$$

The probability that Jane draws a 1 and Linda draws a 4 is $\frac{5}{76}$.

In Example 1, the total number of possible outcomes for Linda's drawing was fewer because Jane had drawn before her. Example 2 shows a situation in which both the total number of possible outcomes and the number of favorable outcomes for Linda's drawing are fewer than Jane's.

EXAMPLE 2

As in Example 1, Jane is the first and Linda is the second student to draw a card. Find the probability that both Jane and Linda draw a 1.

Solution: Let event A be Jane's drawing a 1.
Let event B be Linda's drawing a 1, given that Jane draws a 1.

[1] Determine the total number of possible outcomes for each event. ⟶ A: 20 B: 19 ◀ ***Jane has drawn one card.***

[2] Determine the number of favorable outcomes for each event. ⟶ A: 5 B: 4 ◀ ***Jane has drawn a 1.***

[3] Find the probability of each event. ⟶ $P(A) = \frac{5}{20}$, or $\frac{1}{4}$ $P(B) = \frac{4}{19}$

[4] Find the probability of both events. ⟶ $P(A \text{ and } B) = P(A) \cdot P(B)$

$$= \frac{1}{\cancel{4}_1} \cdot \frac{\cancel{4}^1}{19}$$

$$= \frac{1}{19}$$

The probability that both Jane and Linda draw a 1 is $\frac{1}{19}$.

In Example 1, $P(B) = \frac{5}{19}$. In Example 2, $P(B) = \frac{4}{19}$. Both represent the probability that event B will occur, given that event A has already occurred. This kind of probability is called a conditional probability.

CLASSROOM EXERCISES

There are 5 seniors, 2 juniors, and 3 sophomores who are reporters for the school newspaper. Reporting assignments are handed out by drawing names from a box. If two names are chosen (without replacement), find each probability. (Examples 1 and 2).

1. Drawing a senior, and then a junior
2. Drawing a sophomore, and then a junior
3. Drawing a senior and then a sophomore
4. Drawing a junior, and then a senior
5. Drawing two seniors
6. Drawing two juniors
7. Drawing two sophomores

WRITTEN EXERCISES

Goal: To find the probability of dependent events

Sample Problem: Six cards are numbered 1 through 6 and then placed with the numbers down. A card is drawn. Without replacing the first card, a second card is drawn. What is the probability of drawing an even number first, and then a 5?

Answer: $\frac{1}{2} \cdot \frac{1}{5} = \frac{1}{10}$

The Homecoming Planning Committee consists of 4 seniors (3 girls and 1 boy), 4 juniors (2 girls and 2 boys), and 2 sophomores (both girls). The president and vice-president of the committee will be selected by drawing names from a box.

The name of the president is drawn first and then, without replacing the first name, the name of the vice-president is drawn. Find the probability of each event. (Examples 1 and 2)

1. Drawing a senior for president and a junior for vice-president
2. Drawing a girl for president and a boy for vice-president

3. Drawing a senior boy for president and a junior girl for vice-president

4. Drawing a junior girl for president and a sophomore girl for vice-president

5. Drawing a senior for president and another senior for vice-president

6. Drawing a sophomore for president and another sophomore for vice-president

7. Drawing a girl for president and another girl for vice-president

8. Drawing a boy for president and another boy for vice-president

9. Drawing a junior boy for president and another junior boy for vice-president

10. Drawing a senior girl for president and another senior girl for vice-president

MORE CHALLENGING EXERCISES

The Homecoming Planning Committee decided they need to choose a secretary/treasurer in addition to a president and a vice-president. Three names will be drawn from the box, without replacement. The first name drawn will be for president, the second will be for vice-president, and the third will be for secretary/treasurer. Find each probability.

11. Drawing the names of three girls

12. Drawing the names of three boys

13. Drawing a senior for president, a junior for vice-president, and a sophomore for secretary/treasurer

14. Drawing a girl for president, a boy for vice-president, and a boy for secretary/treasurer

15. Drawing a senior boy for president, a junior boy for vice-president, and a junior girl for secretary/treasurer

16. Drawing a junior girl for president, a junior girl for vice-president, and a sophomore girl for secretary/treasurer

NON-ROUTINE PROBLEM

17. Jerome and his sister Latasha walk to school every day. They start at the same time and walk at the same speed. Who gets to school first? Explain. Refer to the figure at the right.

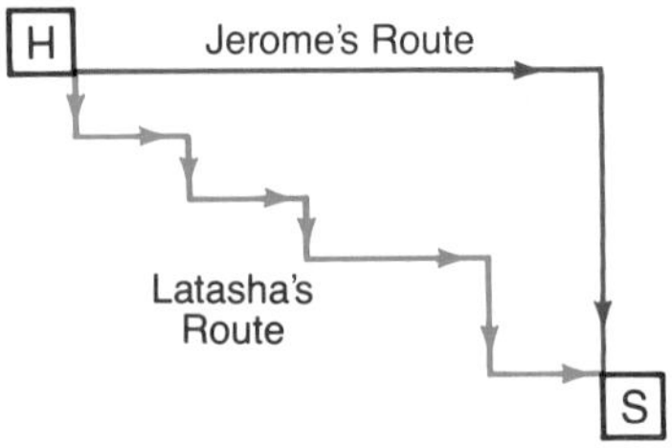

REVIEW CAPSULE FOR SECTION 19.5

Multiply.

1. $\frac{4}{5} \times 240$
2. $\frac{2}{3} \times 123$
3. $\frac{3}{8} \times 488$
4. $\frac{5}{6} \times 234$
5. $\frac{7}{12} \times 1512$
6. $\frac{3}{4} \times 1064$

19.5 Experimental Probability

In many situations, the probability of an event is determined by performing an experiment a large number of times. Probabilities determined in this way are called experimental probabilities.

If an experiment is carried out n times and a certain event A occurs a times, then the ***experimental probability*** for event A is $\frac{a}{n}$.

$$P(A) = \frac{\text{Number of times event } A \text{ occurred}}{\text{Total number of times the experiment is tried}} = \frac{a}{n}$$

EXAMPLE 1 A survey was conducted at Pleasant Mountain High School. Students were asked to recite the words to the first verse of the school song. Of the 420 students surveyed, 240 students correctly recited the words. Find the probability that a student at Pleasant Mountain High School can recite the words to the first verse of the school song.

Solution: Let event A be correctly reciting the words.

$$P(A) = \frac{\text{Number of students who correctly recited the words}}{\text{Total number of students surveyed}} = \frac{\overset{4}{\cancel{240}}}{\underset{7}{\cancel{420}}} = \frac{4}{7}$$

The probability that a student at Pleasant Mountain High School can recite the words to the first verse of the school song is $\frac{4}{7}$.

Once the probability of an event is known, that probability can be used to predict how many times the event will occur if the experiment is conducted again under the same circumstances.

If an experiment is carried out n times, the number of ***expected outcomes*** E of an event A is the product of the probability of event A and n.

$$E(A) = P(A) \cdot n$$

EXAMPLE 2

In Example 1, the 420 students surveyed were from a total of 1015 students at Pleasant Mountain High School. According to the results of the survey, how many of the remaining 595 students could be expected to correctly recite the words to the first verse of the school song?

Solution: Let event A be correctly reciting the words.

$$E(A) = P(A) \cdot n$$

$$= \frac{4}{\not{7}_{1}} \cdot \overset{85}{\not{595}}$$ ◀ ***From Example 1, $P(A) = \frac{4}{7}$.***

$$= 340$$

Of the remaining 595 students at Pleasant Mountain High School, 340 could be expected to correctly recite the words to the first verse of the school song.

CLASSROOM EXERCISES

A survey of 200 households in Pleasant Valley is conducted. The results of the survey showed that 4 households have no automobiles, 48 households have 1 automobile, 132 households have 2 automobiles, and 16 households have 3 or more automobiles. Find each probability for the households surveyed.
(Example 1)

1. Of a household having no automobiles

2. Of a household having 1 automobile

3. Of a household having 2 automobiles

4. Of a household having 3 or more automobiles

There are 650 households in Pleasant Valley. Use the probabilities found in Exercises 1–4 to find the expected number of households having the following number of automobiles.
(Example 2)

5. 0 automobiles

6. 1 automobile

7. 2 automobiles

8. 3 or more automobiles

WRITTEN EXERCISES

Goal: To find the experimental probability of an event

Sample Problem: A radio station found that 36 out of 90 telephone requests taken from 7:00 A.M. to 5:00 P.M. were for the same song. What is the probability that the next request will be for this song?

Answer: $\frac{36}{90}$, or $\frac{2}{5}$

A survey of 450 households in Riverview is conducted. The results of the survey showed that 108 households have no children, 90 households have 1 child, 180 households have 2 children, 45 households have 3 children, 18 households have 4 children, and 9 households have 5 or more children. Find each probability. (Example 1)

1. Of a household having no children

2. Of a household having 1 child

3. Of a household having 2 children

4. Of a household having 3 children

5. Of a household having 4 children

6. Of a household having 5 or more children

There are 1150 households in Riverview. Use the probabilities found in Exercises 1–6 to find the expected number of households having the following number of children. (Example 2)

7. 0 children

8. 1 child

9. 2 children

10. 3 children

11. 4 children

12. 5 or more children

Solve. (Examples 1 and 2)

13. The results of a survey of 200 students at South High School show that 35 students ride a bike to school. There are 2400 students at the school. How many of the 2400 can be expected to ride a bike to school?

14. At the Timely Company clock factory, 150 clocks are tested. The test results show that 2 of the clocks are defective. In a shipment of 1200 clocks, how many can be expected to be defective?

Focus on Reasoning: Venn Diagrams

Venn diagrams can sometimes help you to organize information so that it is easier to see relationships.

EXAMPLE In a sample of 180 students, 119 said that they had seen the movie Outer Stars 1 and 110 said that they had seen Outer Stars 2. Thirteen students had not seen either movie.

a. How many students had seen both movies?
b. How many students had seen Outer Stars 1 only?
c. How many students had seen Outer Stars 2 only?

Solutions:

The interior of the rectangle represents the students sampled.
The interior of the circle on the left represents the students who saw Outer Stars 1.
The interior of the circle on the right represents the students who saw Outer Stars 2.
The overlap of the circles represents the students who saw both movies.

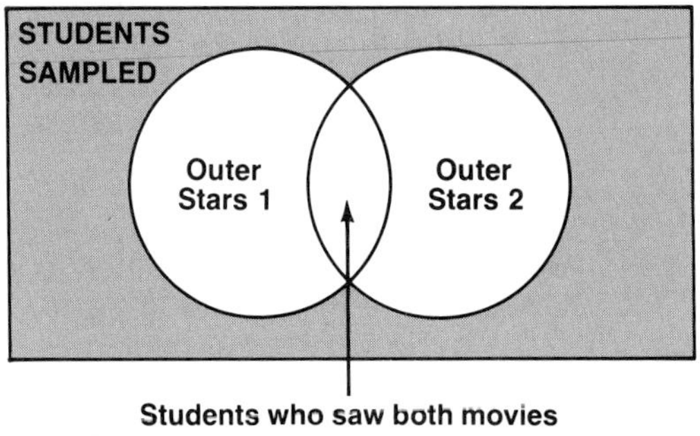

Think: Since 13 students did not see either movie, there are 180 − 13, or **167 students** who saw one or both movies.

a. Since 119 students saw Outer Stars 1 and 110 students saw Outer Stars 2, the sum of these is 229. But there are only 167 students who saw one or both movies. Therefore, the number of students who saw both movies is 229 − 167, or **62 students.**

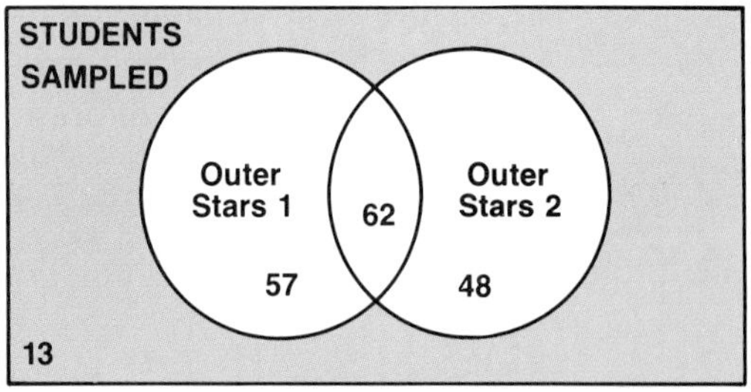

b. Since 62 students saw both movies, 119 − 62, or **57 students** saw Outer Stars 1 only.

c. Since 62 students saw both movies, 110 − 62, or **48 students** saw Outer Stars 2 only.

Check: Does 13 + 57 + 62 + 48 = 180? Yes ✓

EXERCISES

For Exercises 1–8, refer to the information below and the Venn diagram at the right.

There are 153 students taking algebra.
There are 135 students taking biology.
There are 94 students taking Spanish.
There are 55 students taking biology and Spanish.
There are 15 students taking algebra, biology, and Spanish.

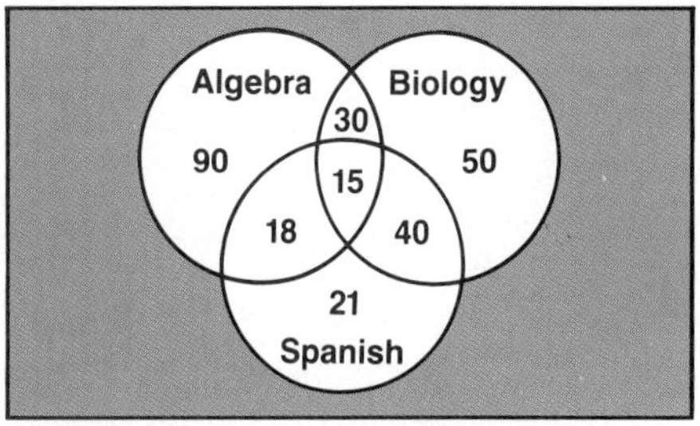

1. How many students are taking algebra and biology, but not Spanish?
2. How many students are taking algebra and Spanish, but not biology?
3. How many students are taking biology, but not algebra or Spanish?
4. How many students are taking algebra, but not biology or Spanish?
5. How many students are taking Spanish, but not algebra or biology?
6. How many students are taking exactly one of these three courses?
7. How many students are taking at least one of these three courses?
8. How many students are taking exactly two of these three courses?

For Exercises 9–10, use a Venn diagram to solve each problem.

9. In a class of 31 students, there are 17 who play softball and 22 who play tennis. There are 4 students who do not play either game. How many students play both softball and tennis?
10. At a party, 27 persons played the game "Mystery Clues" and 22 played the game "Skill Shot." Fourteen persons played both games. How many persons played one game, but not both?

CHAPTER SUMMARY

IMPORTANT TERMS

Tree diagram *(p. 512)*
Fundamental Counting Principle *(p. 512)*
Experiment *(p. 515)*
Outcome *(p. 515)*
Sample space *(p. 515)*
Event *(p. 515)*
Probability of an event *(p. 515)*
Complementary events *(p. 517)*
Independent events *(p. 520)*
Mutually exclusive events *(p. 521)*
Dependent events *(p. 525)*
Conditional probability *(p. 526)*
Experimental probability *(p. 529)*
Expected outcomes *(p. 529)*

IMPORTANT IDEAS

1. *Fundamental Counting Principle:* If there are p choices for the first decision and q choices for the second decision, then there are $p \times q$ ways that the two decisions can be made together.
2. The probability of an event that is certain to happen is 1.
3. The probability of an event that cannot happen is 0.
4. For two independent or two dependent events, A and B, $P(A \underline{\text{ and }} B) = P(A) \cdot P(B)$.
5. For two mutually exclusive events A and B, $P(A \underline{\text{ or }} B) = P(A) + P(B)$.
6. If an experiment is carried out n times, $E(A) = P(A) \cdot n$.

CHAPTER REVIEW

SECTION 19.1

Use the Fundamental Counting Principle to solve each problem.

1. A student will choose three elective courses. There are three choices for the first elective, five choices for the second, and three choices for the third. In how many ways can the electives be chosen?
2. The Freshness Bakery bakes three types of bread. Loaves of bread are made in four sizes, and sliced two different ways. How many choices are there when buying a loaf of bread at the bakery?
3. A dinner menu has a choice of three appetizers, seven entrees, four beverages, and three desserts. In how many ways can a dinner be ordered, choosing one item from each category?
4. When using the Automatic Teller Machine at the American Trust Bank, a customer must enter a four-digit security code. If the digits can be repeated, how many different security codes are possible?

SECTION 19.2

The arrow of the spinner is equally likely to stop on any of the ten regions. The arrow is spun once. Find each probability.

5. Stopping on a multiple of 3
6. Stopping on a number less than 7
7. Stopping on a number greater than 8
8. Not stopping on a prime number
9. Not stopping on a whole number
10. Stopping on a counting number

SECTION 19.3

A golfer has 10 white, 6 yellow, and 4 orange golf balls. She selects one ball without looking. Find each probability.

11. Selecting a white <u>or</u> orange ball
12. Selecting a yellow <u>or</u> orange ball

The golfer has 8 red and 12 white tees. She selects one ball and one tee without looking. Find each probability.

13. Selecting a white ball <u>and</u> a white tee
14. Selecting a yellow ball <u>and</u> a white tee
15. Selecting a yellow ball <u>and</u> a red tee
16. Selecting an orange ball <u>and</u> a red tee

SECTION 19.4

Jim has 7 nickels, 5 dimes, and 3 quarters in his pocket. He selects two coins, one at a time. Find each probability.

17. Selecting a quarter first <u>and</u> then a dime
18. Selecting a nickel first <u>and</u> then a dime
19. Selecting two nickels
20. Selecting two quarters

SECTION 19.5

Two new products are being manufactured. The first 200 samples of each product were tested for defects. Thirty-six of the Product A samples were defective and twenty-eight of the Product B samples were defective.

21. If another Product A sample is tested, what is the probability that it is defective?
22. If another Product B sample is tested, what is the probability that it is defective?
23. In a shipment of 1500, how many of Product A can be expected to be defective?
24. In a shipment of 1200, how many of Product B can be expected to be defective?

CUMULATIVE REVIEW: CHAPTERS 1–19

Solve. Check each answer. (Section 4.3)

1. $15 - 4a = 7a - 7$

2. $4x - 5 = 2(x - 1) + 6$

Multiply. (Sections 8.2 and 8.5)

3. $(5e^6)(3e^3)$

4. $(-6x^2y^4)(5x^7y)$

5. $(4x - 5)(x - 11)$

Perform the indicated operations. Simplify where necessary. (Sections 10.3, 10.5, 10.7, 11.2, 11.4, and 11.6)

6. $\frac{3ac^2}{2a^2b^3} \cdot \frac{4ab}{9c}$

7. $\frac{2x + 1}{x - 3} \cdot \frac{x^2 - 9}{4x + 2}$

8. $\frac{x^2 + 5x + 4}{x^2 - 16} \div \frac{x^2 - 1}{3x - 12}$

9. $\frac{9x}{x + 6} - \frac{7x - 12}{x + 6}$

10. $\frac{2}{3x} - \frac{3}{4x}$

11. $\frac{3}{2x - 6} + \frac{5}{x^2 - 9}$

Solve each system of equations. (Sections 14.4, 15.1, and 15.5)

12. $\begin{cases} 2x - 3y = -14 \\ 4x + 2y = -12 \end{cases}$

13. $\begin{cases} 5x + y = 7 \\ 8x + 2y = 10 \end{cases}$

14. $\begin{cases} 5x + 7y = -23 \\ 8x - y = 12 \end{cases}$

Write the slope-intercept form of each inequality. (Section 16.1)

15. $3x + y - 1 > 0$

16. $5x - y < -3$

17. $-7x + 2y \leq 8$

Graph each linear system. (Sections 16.4 and 16.5)

18. $\begin{cases} y < -x + 1 \\ y > 2x - 3 \end{cases}$

19. $\begin{cases} x \leq 3 \\ x - y \leq 3 \end{cases}$

20. $\begin{cases} x + y < 3 \\ 2x - y \geq -1 \end{cases}$

Write three ordered pairs of each function. Let the x values equal −2, 0, and 3. (Section 17.1)

21. $y = -x^2$

22. $y = x^2 - 2$

23. $y = x^2 - 2x + 1$

Graph each function. (Section 17.2)

24. $y = x^2 + 3$

25. $y = 2x^2 - x$

26. $y = -x^2 + 4x - 4$

Graph the two functions in the same coordinate plane. (Section 17.3)

27. $y = x^2;\ y = 2x^2$

28. $y = -x^2;\ y = -\frac{1}{2}x^2$

Write the motion(s) needed to make the graph of the first function coincide with the graph of the second function. (Sections 17.4 and 17.5)

29. $y = x^2$; $y = (x - 5)^2$

30. $y = \frac{1}{2}x^2$; $y = \frac{1}{2}(x - 3)^2 + 4$

Write each polynomial in the form $(x - h)^2 + k$. (Section 17.6)

31. $x^2 + 4x + 1$

32. $x^2 - 6x - 3$

33. $x^2 + 3x - 2$

Write an equation for the axis of each graph. (Section 17.7)

34. $y = 2(x - 3)^2 + 4$

35. $y = \frac{1}{4}(x - 1)^2 - 3$

36. $y = x^2 - 10x + 27$

Solve. (Sections 18.2 through 18.4)

37. $x^2 - 7x + 10 = 0$

38. $x^2 = 81$

39. $x^2 + 8x - 1 = 0$

Solve by the quadratic formula. (Section 18.5)

40. $x^2 + 7x + 2 = 0$

41. $2x^2 = 5x + 3$

42. $x^2 - 5x = 0$

Use Condition 1 to represent the unknowns. Use Condition 2 to write an equation for the problem. Solve. (Section 18.7)

43. The product of two consecutive integers is 56. Find two pairs of such integers.

44. The length of a rectangle is 3 feet more than the width. If the area is 40 square feet, what are the length and width?

A blue number cube and a red number cube are tossed. Find each probability. (Sections 19.2 and 19.3)

45. Tossing a 4 on the blue cube

46. Tossing an odd number on the red cube

47. Tossing a 5 on the blue cube and a 5 on the red cube

48. Tossing a 4 on the blue cube or a number less than 4 on the red cube

Solve. (Section 19.5)

49. The results of a survey of 350 people at an airport showed that 84 people had eaten eggs for breakfast that morning. Find the probability that a person at the airport ate eggs for breakfast that morning.

50. Professor Plank's Hardware store examined a shipment of 550 nails and found 11 to be defective. In a shipment of 1800 nails, how many can be expected to be defective?

Using Manipulatives

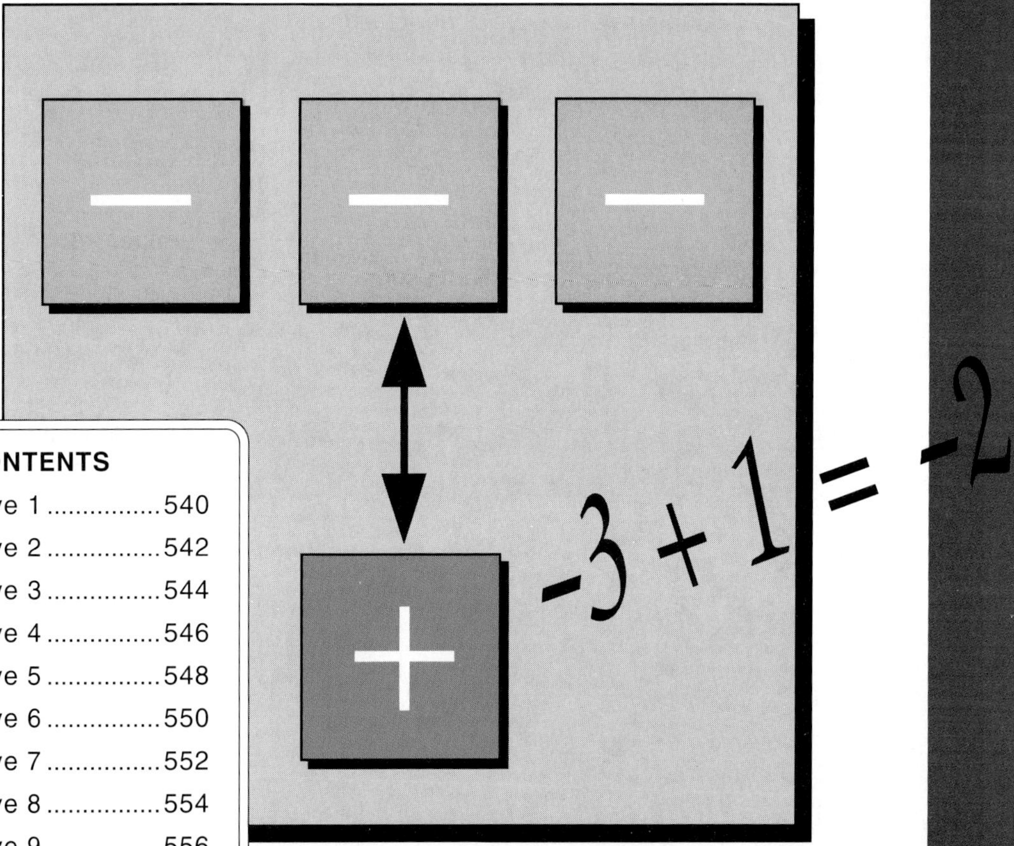

CONTENTS

Manipulative 1 *(Use with Lesson 1.2.)*

You can use [+] and [−] tiles to model addition of integers. The value of a square is determined by the number and type of tiles in the square.

Value: +2, or 2

Value: −3

Value: 0

A [+]-tile and a [−]-tile have opposite signs. A pair of tiles having opposite signs forms a **neutral pair**. A neutral pair has a value of 0.

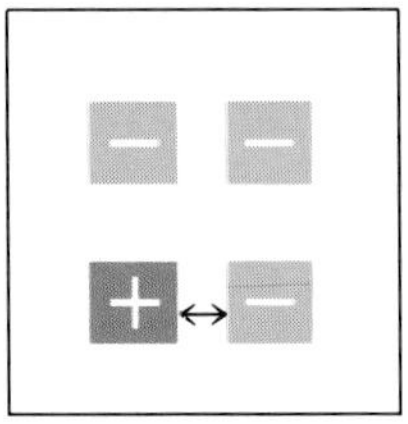

Value: −2

You can use algebra tiles to find a sum.

EXAMPLES

Use algebra tiles to find each sum.

a. $-2 + (-4) = \underline{\ ?\ }$

Solution:

A. Put in two [−]-tiles.

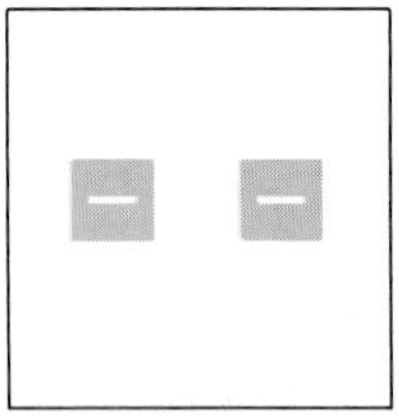

B. Add four [−]-tiles.

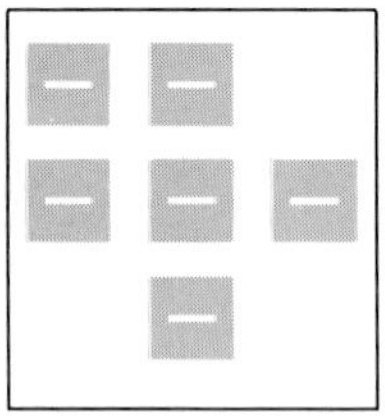

C. Count the tiles. Record the sum.

$-2 + (-4) = \mathbf{-6}$

b. $-5 + 2 = \underline{\ ?\ }$

Solution:

A. Put in five [−]-tiles.

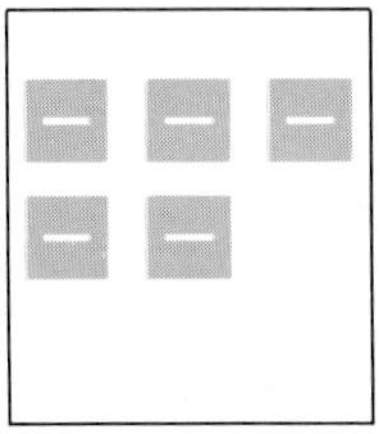

B. Add two [+]-tiles.

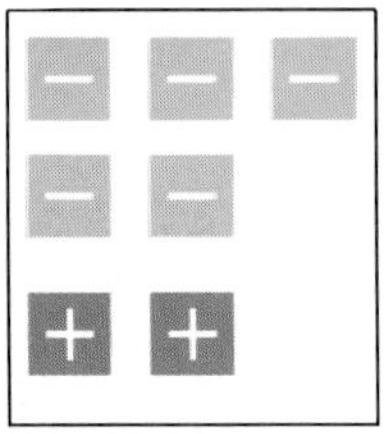

C. Remove neutral pairs. Record the sum.

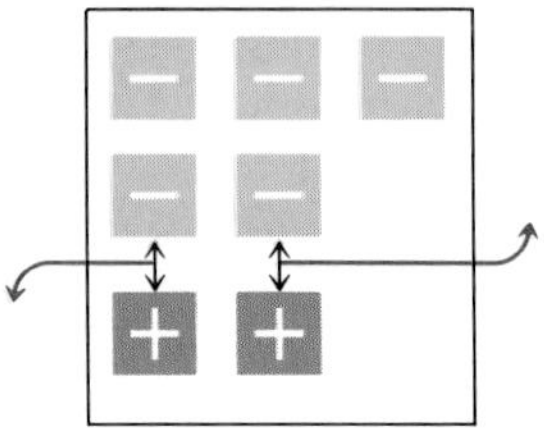

$-5 + 2 = \mathbf{-3}$

c. $6 + (-3) = \underline{\ ?\ }$

A. Put in six [+]-tiles.

B. Add three [−]-tiles.

C. Remove neutral pairs.

D. Record the sum.

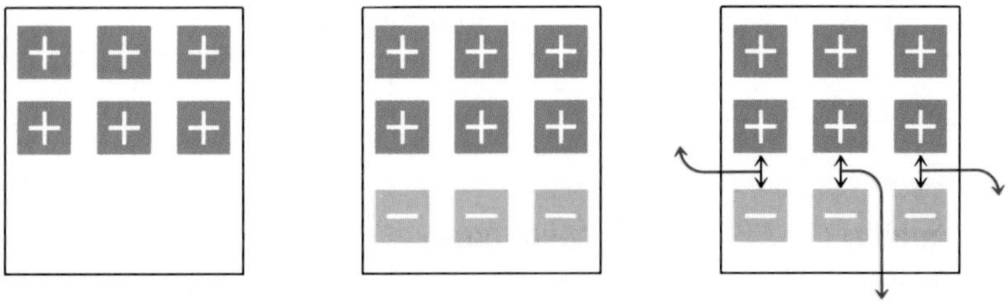

$6 + (-3) = \mathbf{3}$

EXERCISES

Goal: To add positive and negative numbers

Give the value of each square.

1.

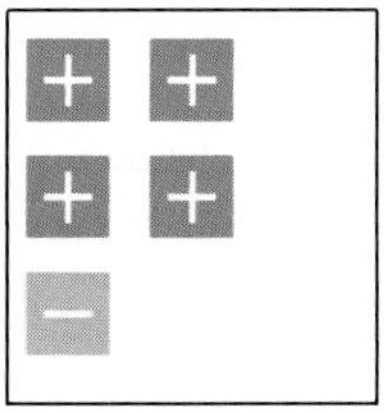

2.

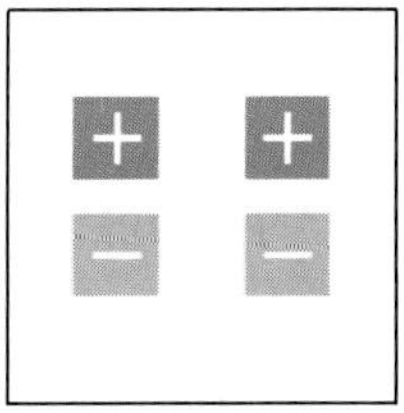

3.

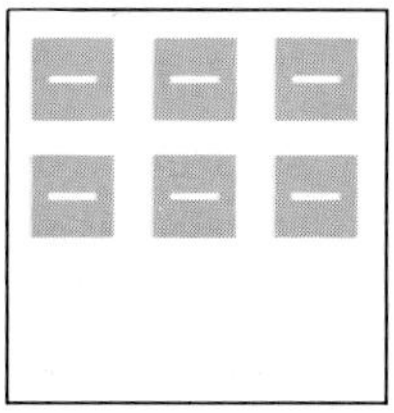

Use algebra tiles to model each sum.

4. $-3 + 6$ **5.** $-5 + (-4)$ **6.** $4 + (-7)$ **7.** $-1 + 1$

8. $7 + (-5)$ **9.** $-2 + (-2)$ **10.** $3 + (-2)$ **11.** $0 + (-3)$

12. Use algebra tiles to show that $6 + (-2) = (-2) + 6$.

Use algebra tiles to solve.

13. $8 + (-9) + 1$ **14.** $-3 + 2 + 5$ **15.** $6 + (-2) + (-7)$

16. $-2 + (-2) + (-2)$ **17.** $-4 + 9 + (-3)$ **18.** $5 + (-3) + (-7)$

19. Tom gained 4 pounds in the month of March. He lost 6 pounds in April. What is his overall loss or gain in weight?

20. At 4 A.M., the temperature was −3°C. At 10 A.M., the temperature had risen 7°. What was the temperature then?

Write true (T) or false (F) for each statement. When a statement is false, explain why.

21. The sum of two negative numbers is always a positive number.

22. The sum of a positive and a negative number is always a negative number.

Manipulative 2 *(Use with Lesson 1.3.)*

You can use algebra tiles to show that

$$4 - 3 \quad \text{means the same as} \quad 4 + (-3).$$

EXAMPLE 1

Use algebra tiles to show that $4 - 3 = 4 + (-3)$.

a. $4 - 3 =$?

Solution:

A. Put in four (+)-tiles.

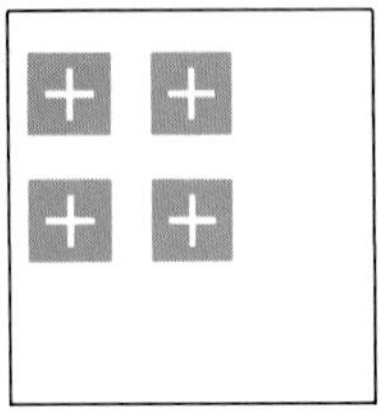

B. Remove three (+)-tiles.

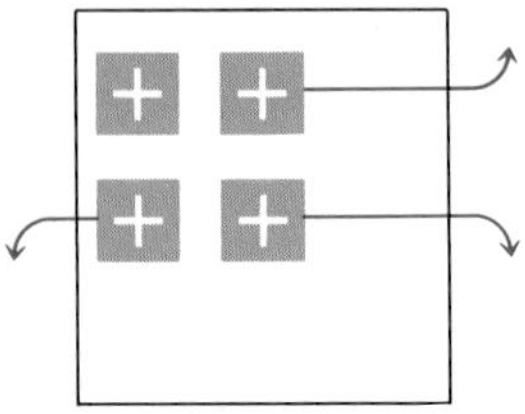

C. Record the result.

$4 - 3 = \mathbf{1}$

b. $4 + (-3) =$?

Solution:

A. Put in four (+)-tiles.

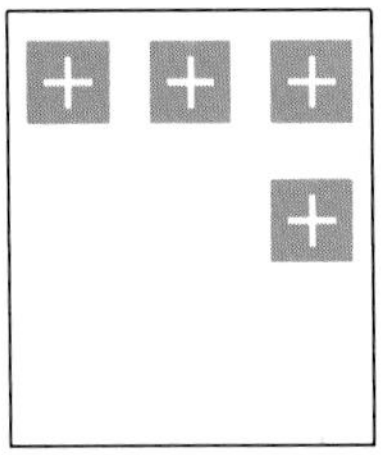

B. Add three (−)-tiles.

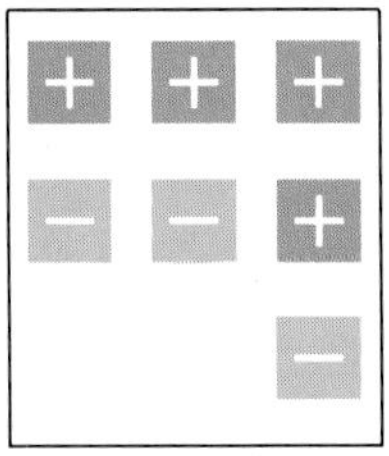

C. Remove neutral pairs. Record the result.

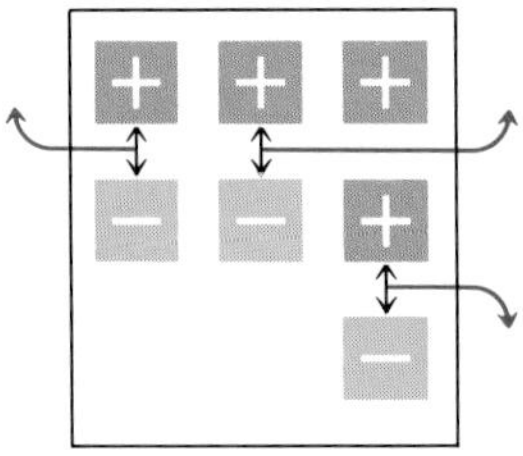

$4 + (-3) = \mathbf{1}$

Example 1 suggests the following.

To subtract an integer, add its opposite.

EXAMPLE 2

Use algebra tiles to subtract: $-1 - (-3)$

Solution: Rewrite $-1 - (-3)$ as $-1 + 3$.

A. Put in one (−)-tile.

B. Add three (+)-tiles.

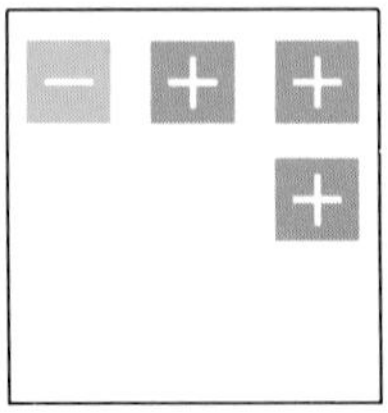

C. Remove neutral pairs. Record the result.

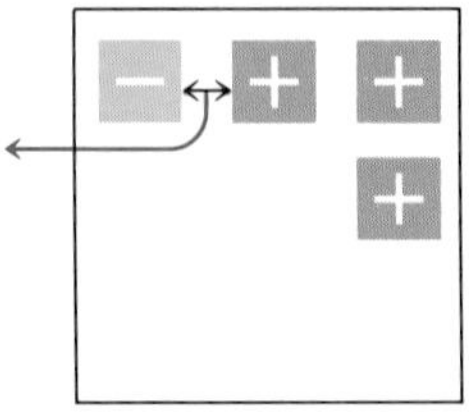

$-1 - (-3) = \mathbf{2}$

EXERCISES

Goal: To use algebra tiles to subtract integers

Write an addition problem that will give the same answer as the subtraction problem.

1. $9 - 5$
2. $-9 - 5$
3. $-9 - (-5)$
4. $9 - (-5)$
5. $-13 - (-7)$
6. $13 - (-1)$
7. $1 - 8$
8. $1 - (-10)$

Use algebra tiles to subtract.

9. $5 - 9$
10. $-4 - 6$
11. $-3 - 6$
12. $4 - (-5)$
13. $-6 - (-7)$
14. $-8 - 7$
15. $6 - (-9)$
16. $-5 - (-8)$
17. $7 - (-3)$
18. $1 - (-1)$
19. $3 - 5$
20. $-3 - 5$
21. Use algebra tiles to determine which is greater, $1 - (-8)$ or $8 - 1$.

Use algebra tiles to illustrate your answer to each question.

22. You start with 6 and subtract a negative integer having an absolute value less than 6. Will the result be a positive or a negative integer?

23. You start with 6 and subtract a positive integer greater than 6. Will the result be a positive or a negative integer?

24. You start with -6 and subtract a positive integer with a value less than $|-6|$. Will the result be a positive or a negative integer?

25. You start with -6 and subtract a positive integer with a value greater than $|-6|$. Will the result be a positive or a negative integer?

26. You start with -6 and subtract a negative integer having an absolute value greater than $|-6|$. Will the result be a positive or a negative integer?

27. You start with -6 and subtract a negative integer having an absolute value less than $|-6|$. Will the result be a positive or a negative integer?

Manipulative 3 *(Use with Lesson 2.6.)*

You can use algebra tiles like these to model algebraic expressions.

1-tile **x-tile** **x²-tile**

For each positive tile, there is a corresponding negative tile.

(−1)-tile **(−x)-tile** **(−x²)-tile**

Just as with the 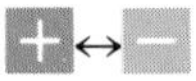 pairs you used to model a neutral pair of integers, neutral pairs can also be formed as shown below. The value of each neutral pair is 0.

Neutral Pairs

If two terms are **like terms**, the algebra tiles that model them will have the same size, but may have different signs. The pairs of tiles below show like terms.

EXAMPLE Simplify $2x^2 - x^2 + x - 1$.

Solution: Think of $2x^2 - x^2 + x - 1$ as $2x^2 + (-x^2) + x + (-1)$.

A. Model the expression.

$2x^2 + (-x^2) + x + (-1)$

B. Group like terms. Remove neutral pairs and record the result.

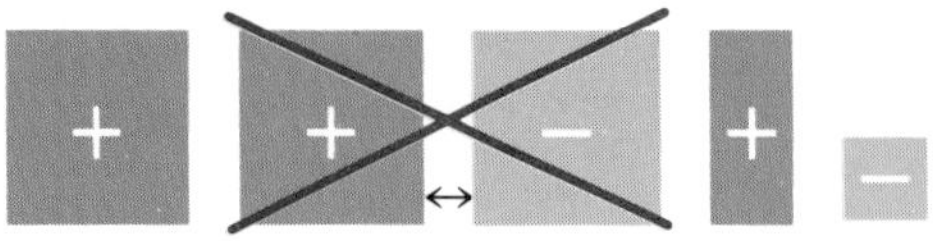

$[2x^2 + (-x^2)] + x - 1$

$x^2 + x - 1$

EXERCISES

Goal: To combine like terms

Which pairs of tiles show like terms? Answers Yes or No.

1. 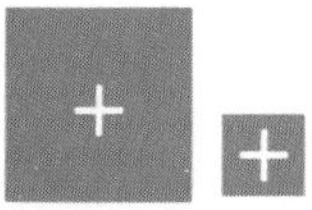

2. 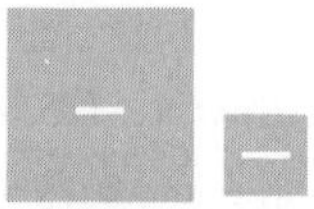

3. 

4. 

5. 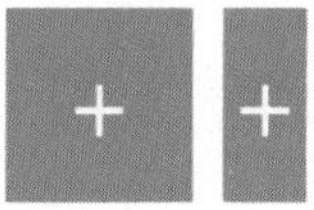

6. 

Use algebra tiles to simplify each expression.

7. $-x^2 - 5x^2$

8. $-3 + 2x - 2x$

9. $3x^2 - 5x^2$

10. $x^2 + x + 1 - x$

11. $-5 + x^2 - 6x^2$

12. $-x - 2 - 8x$

13. $x^2 - 7x - 5x - 3$

14. $3 + 4x - x + 6$

15. $x + 3 - x^2 + 2x^2$

16. $-x^2 + 3x - 7x + 4$

17. $5x - x^2 + 2x$

18. $8 - 3x^2 + 7x + 2x^2$

Use algebra tiles to determine whether each statement is true (T) or false (F).

19. $-(-x) + x = 0$

20. $x^2 - (-x^2) = 2x^2$

21. $x^2 - x \neq x - x^2$

22. Use algebra tiles to show an algebraic expression of three terms that, when simplified, can be represented as $-x$.

23. Use algebra tiles to show an algebraic expression of four terms that, when simplified, can be represented as $x^2 - 1$.

Manipulative 4 *(Use with Lesson 3.1.)*

You can use algebra tiles to solve equations involving addition and subtraction.

EXAMPLE 1

Use algebra tiles to solve $x - 5 = -1$.

Solution: **A.** Model the equation. **Think:** $x - 5 = x + (-5)$

$x + (-5) = -1$

B. Think: There are five (-1)-tiles on the left side of the equation. To get x alone, add five (1)-tiles to each side.

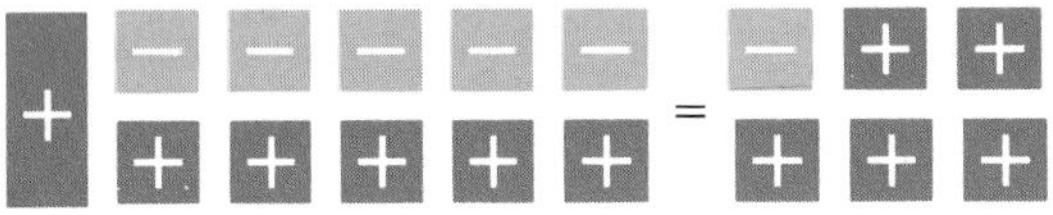

$x + (-5) + \mathbf{5} = -1 + \mathbf{5}$

C. Identify and remove any neutral pairs from each side.

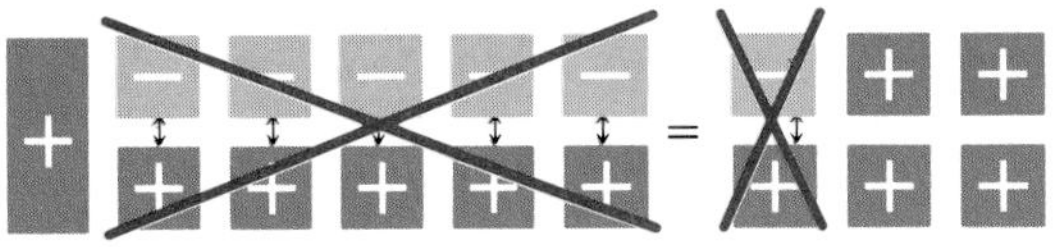

$x + 0 = 4$

D. Show the result.

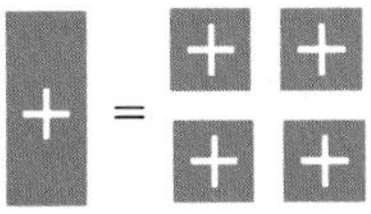

$x = 4$

EXAMPLE 2

Use algebra tiles to solve $x + 2 = 5$.

Solution: **A.** Model the equation.

$x + 2 = 5$

B. To get x alone, subtract two (1)-tiles from each side.

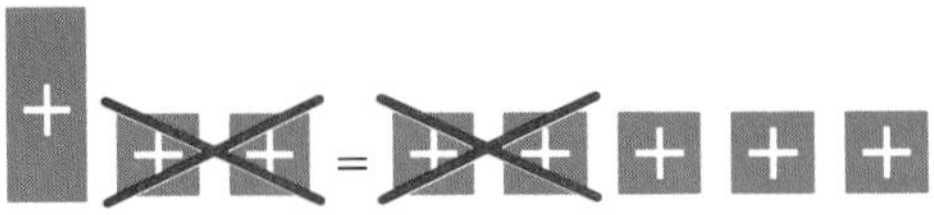

$x + 2 - \mathbf{2} = 5 - \mathbf{2}$

C. Show the result.

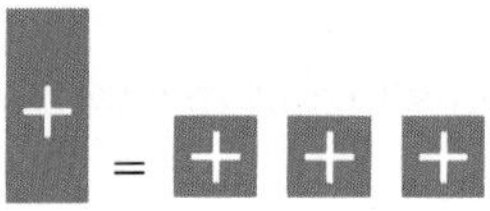

$x = 3$

EXERCISES

Goal: To use algebra tiles to solve equations of the form $x \pm a = b$

1. Use algebra tiles to show that 9 is the solution of $x - 8 = 1$.

Use algebra tiles to solve each equation.

2. $x - 3 = -4$

3. $x + 3 = 7$

4. $x - 5 = 8$

5. $x + 4 = -6$

6. $x - 7 = 1$

7. $x + 4 = 7$

8. $x - (-3) = 7$

9. $-2 = x - 4$

10. $2 = x - 7$

Use algebra tiles to model an equation having the given solution. Write the equation you have modeled.

11.

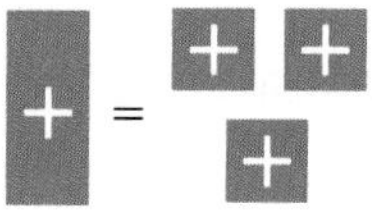

12.

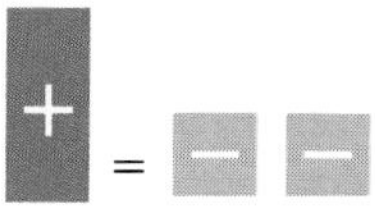

13.

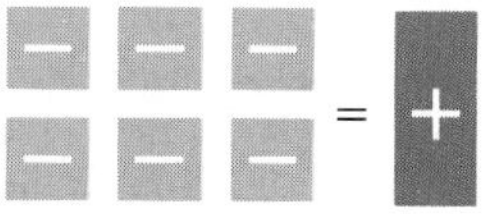

Use algebra tiles to model two equations having the given solution. Write the equations you have modeled.

14. $x = -1$

15. $x = 5$

16. $x = 0$

17. Write a paragraph that summarizes how you solve equations of the form $x \pm a = b$.

Manipulative 5 *(Use with Lesson 3.2.)*

You can also use algebra tiles to solve equations involving multiplication.

EXAMPLE 1

Use algebra tiles to solve: $3x = -12$

Solution: **A.** Model the equation.

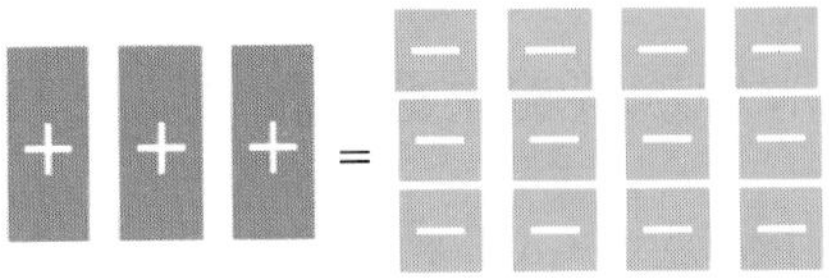

$3x = -12$

B. Think: There are three x-tiles on the left of the equation. To find the value of one x-tile, form 3 equal groups on each side.

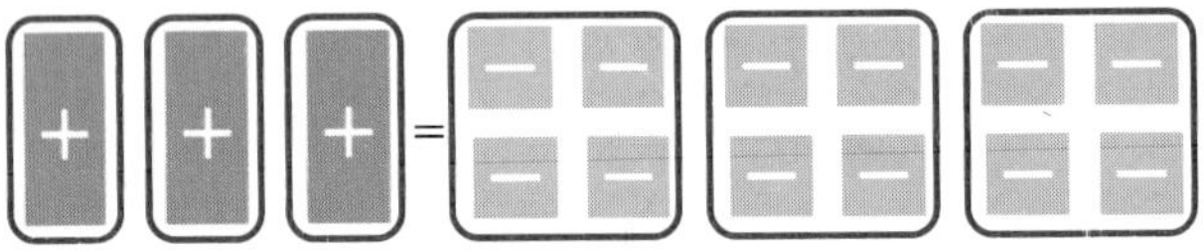

$$\frac{3x}{3} = \frac{-12}{3}$$

C. Show the value of one x-tile.

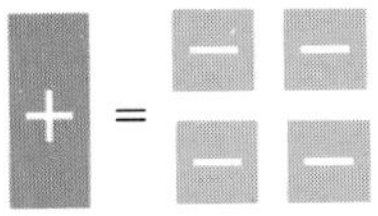

$x = -4$

Example 2 shows that you may sometimes need to add neutral pairs to solve for x.

EXAMPLE 2

Use algebra tiles to solve: $6 = -x$

Solution: **A.** Model the equation.

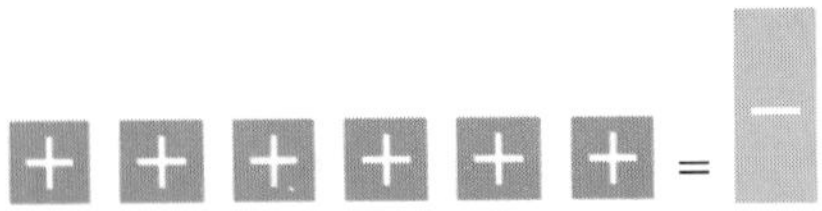

$6 = -x$

B. Think: There is a negative x-tile on the right side of the equation. Add a positive x-tile to each side.

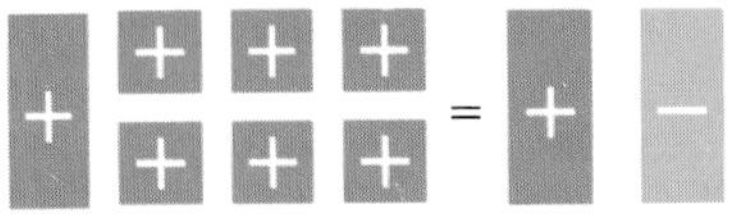

$x + 6 = x + (-x)$

C. Identify and remove neutral pairs.

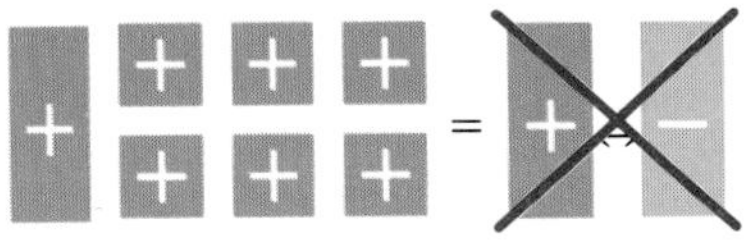

$x + 6 = 0$

D. To get x-alone, add six (-1)-tiles to each side. Remove neutral pairs.

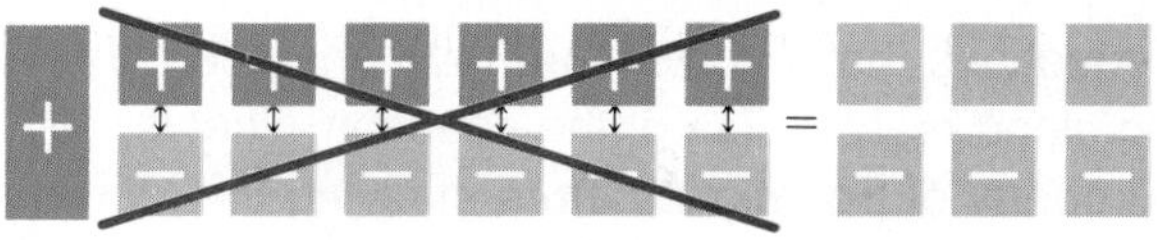

$$x + 6 + (-6) = 0 + (-6)$$
$$x + 0 = -6$$

E. Show the value of x.

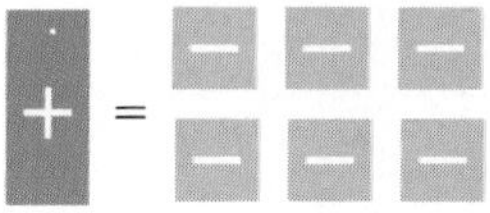

$$x = -6$$

EXERCISES

Goal: To use algebra tiles to solve equations of the form $ax = b$

Use algebra tiles to solve each equation.

1. $2x = 8$ **2.** $4x = -8$ **3.** $3x = 12$ **4.** $4x = 12$

5. $4 = -x$ **6.** $-x = 1$ **7.** $-3 = -x$ **8.** $-x = -1$

9. $4x = -4$ **10.** $5x = -15$ **11.** $-2x = -6$ **12.** $-2x = 6$

13. $9 = -3x$ **14.** $-6 = -6x$ **15.** $-5x = -10$ **16.** $-2x = 4$

Solve. Use algebra tiles if you need help.

17. If $x = -5$, what is the value of $3x$?

18. If $-x = 1$, what is the value of $-5x$?

19. If $-x = -3$, what is the value of $-4x$?

20. If $-1 = -x$, what is the value of $6x$?

21. If $-x = 2$, what is the value of $8x$?

22. List five equations of the form $ax = b$ that have a solution of $x = -2$.

Manipulative 6 *(Use with Lesson 3.3.)*

The first step in solving equations that involve more than one operation is to get the term with the variable alone on one side of the equation. That is, you apply the Addition or Subtraction Property of Equations *before* applying the Multiplication or Division Property.

EXAMPLE

Use algebra tiles to solve: $2x - 3 = 7$

Solution: Think of $2x - 3$ as $2x + (-3)$.

A. Model the equation.

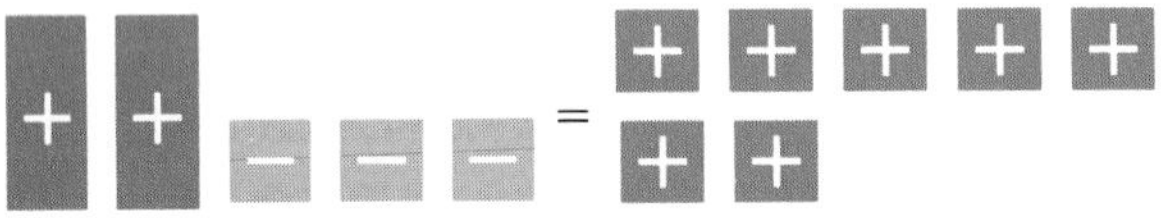

$2x+(-3)=7$

B. Think: There are three (-1)-tiles on the left side of the equation. To get the two x-tiles alone, add three (1)-tiles to each side.

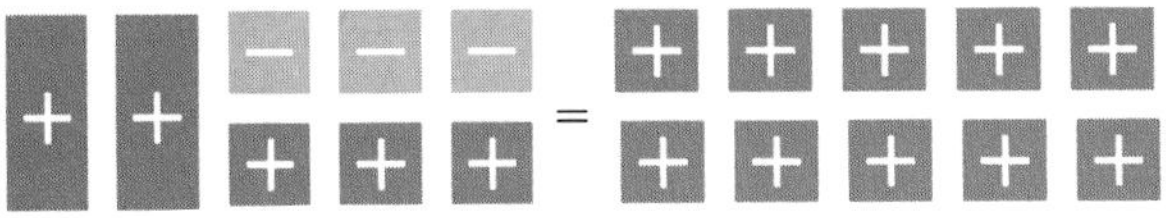

$2x+(-3)+3=7+3$

C. Identify and remove any neutral pairs.

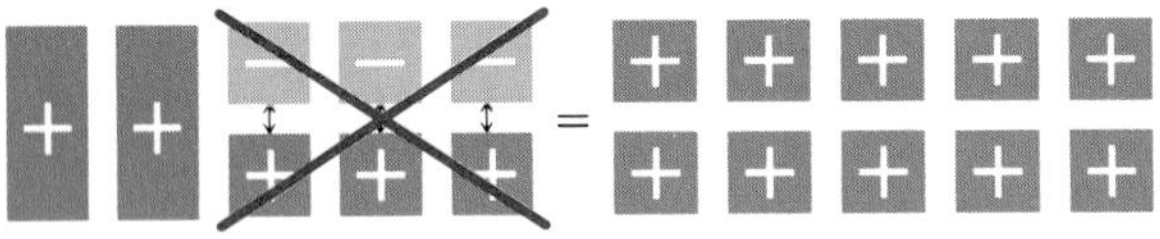

$2x=10$

D. Since there are two x-tiles on the left side of the equation, form two equal groups on each side.

$\frac{2x}{2}=\frac{10}{2}$

E. Pair one x-tile with one group of (1)-tiles.

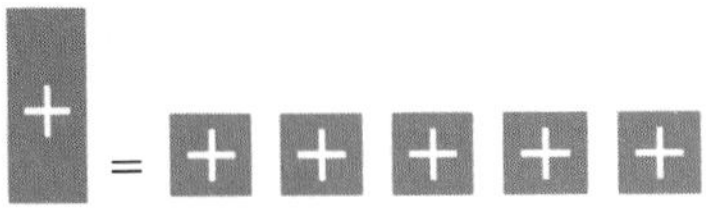

$x=5$

EXERCISES

Goal: To use algebra tiles to solve equations involving two or more operations

Write the equation shown by each model.

1. 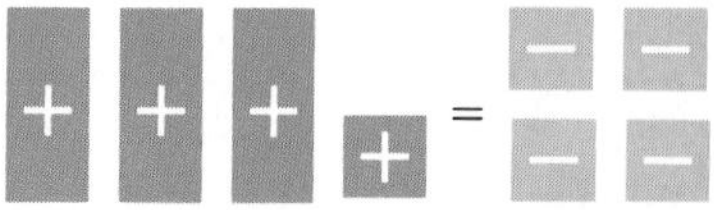

2. 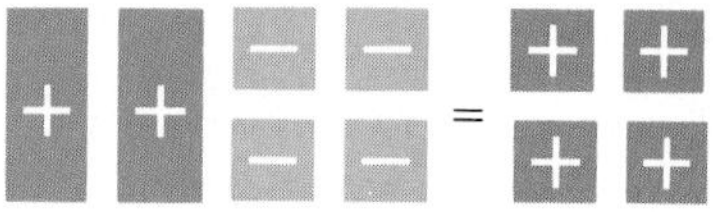

Use algebra tiles to solve each equation.

3. $2x + 1 = 5$

4. $3x - 2 = 7$

5. $2x + 4 = -6$

6. $2x - 3 = 7$

7. $9 = 2x + 5$

8. $3 + 2x = -9$

9. $7 = 2x + 3$

10. $8 = 3x - 1$

11. $-10 = 3x + 5$

Write an equation having two or more operations and the solution shown by the model.

12. 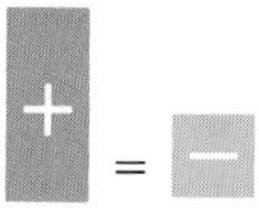

13. 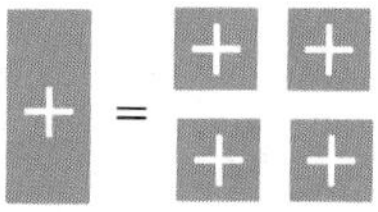

14. 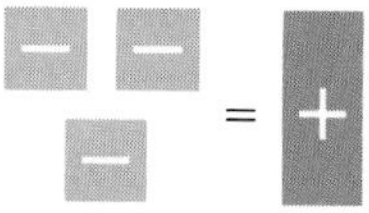

Use algebra tiles to solve each problem.

15. When 3 is subtracted from four times a number, x, the result is 1. What is the number?

16. Four more than three times a number, x, is 16. What is the number?

17. Write a word problem that can be solved by using the equation $3x + 5 = 14$.

18. Write a word problem that can be solved by using the equation $2x - 7 = 11$.

Manipulative 7 *(Use with Lesson 3.5.)*

To solve some equations, you may have to combine like terms first.

EXAMPLE Use algebra tiles to solve: $3x + 5 + x = -3$

Solution:

A. Model the equation.

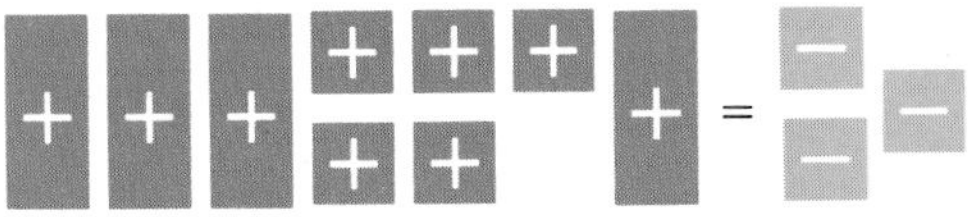

$$3x+5+x=-3$$

B. Group the like terms.

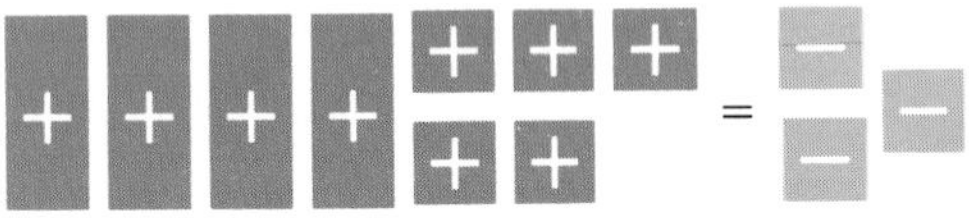

$$4x+5=-3$$

C. To get the four x-tiles alone, add five (-1)-tiles to each side of the equation.

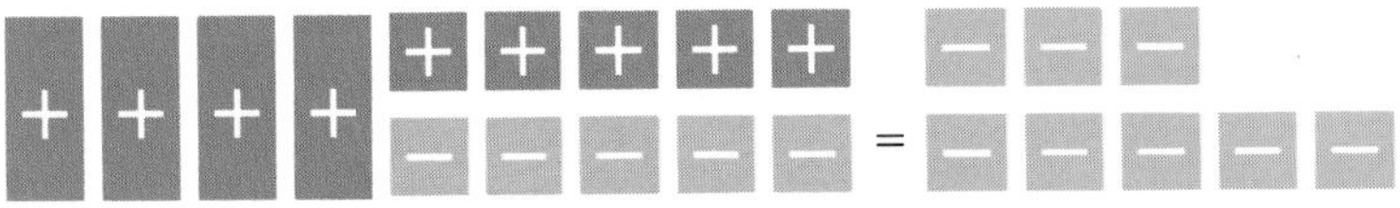

$$4x+5+(-5)=-3+(-5)$$

D. Identify and remove neutral pairs.

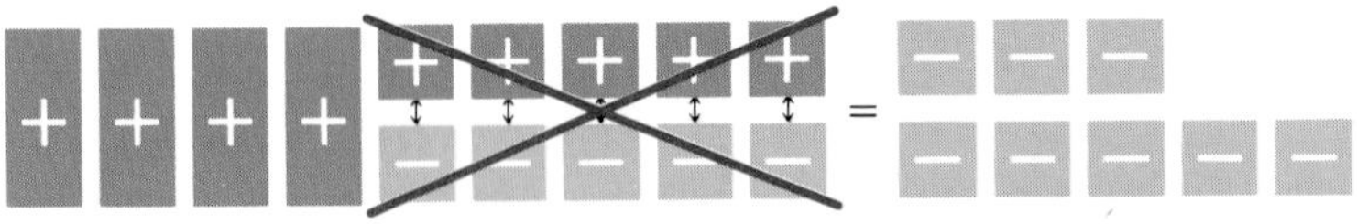

$$4x=-8$$

E. Since there are four x-tiles on the left side of the equation, form four equal groups on each side.

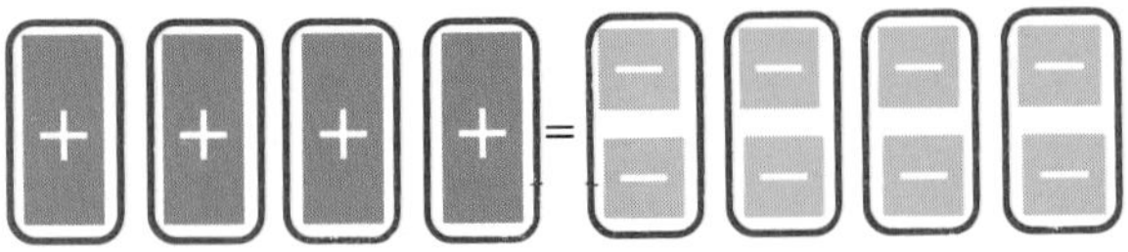

$$\frac{4x}{4}=\frac{-8}{4}$$

F. Write the value of x.

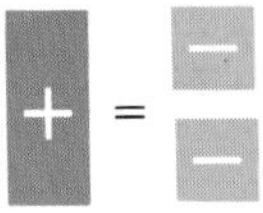

$$x=-2$$

EXERCISES

Goal: To use algebra tiles to solve equations having like terms.

1. Use algebra tiles to show that $x = -2$ is the solution of $3x + 5 + x = -3$.
2. Use algebra tiles or drawings to illustrate two pairs of tiles that represent like terms and two pairs of tiles that represent unlike terms. Check students' work.

Use algebra tiles to solve each equation.

3. $2x + 3 + x = 9$
4. $5x - 3 - x = 9$
5. $3x + 6 - x = -8$
6. $2x + 1 - 9 = 6$
7. $2x + 3x - 7 = 3$
8. $7 - 3x + x = 5$
9. $8 = 3x + 2 - x$
10. $5 = 2x + 1 - 4$
11. Use algebra tiles to show that the solution of $3(x - 1) = 6$ is $x = 3$.
12. Use algebra tiles to solve and check: $2(x - 3) = 4$.

For Exercises 13–16, write an equation that satisfies the given condition. Exchange equations with a partner and solve the equations by using algebra tiles. Check each other's solutions.

13. Write an equation involving like terms and having a solution of 0.
14. Write an equation involving like terms and having a solution of -7.
15. Write an equation having a solution of 5 and all the like terms on the right side of the equation.
16. Write an equation having a solution of -1 and having like terms of x and $(-x)$.

Manipulative 8 *(Use with Lesson 4.3.)*

You can use algebra tiles to solve equations having the variable on both sides. The first step in solving these equations is to eliminate the variable from one side.

EXAMPLE

Use algebra tiles to find the solution of $3x + 1 = x - 5$.

Solution:

A. Model the equation.

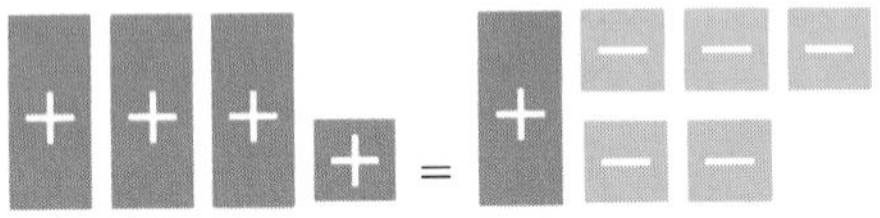

$3x + 1 = x - 5$

B. To eliminate the x-tile from the right side of the equation, add one $(-x)$-tile to each side.

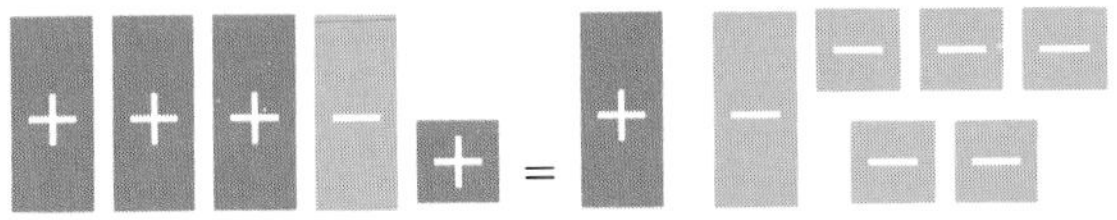

$3x + (-x) + 1 = x + (-x) - 5$

C. Identify and remove the neutral pairs.

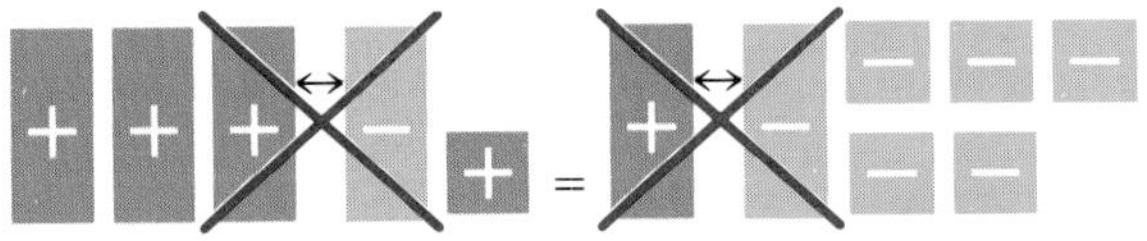

$2x + 1 = -5$

D. To eliminate the (1)-tile from the left side of the equation, add a (-1)-tile to each side.

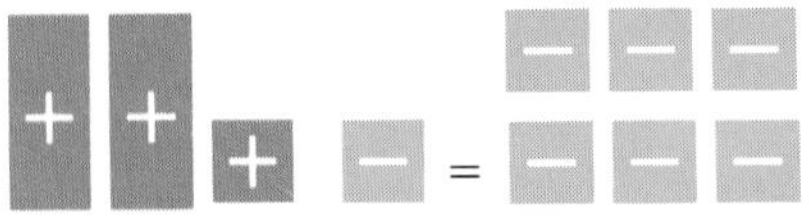

$2x + 1 + (-1) = -5 + (-1)$

E. Identify and remove any neutral pairs.

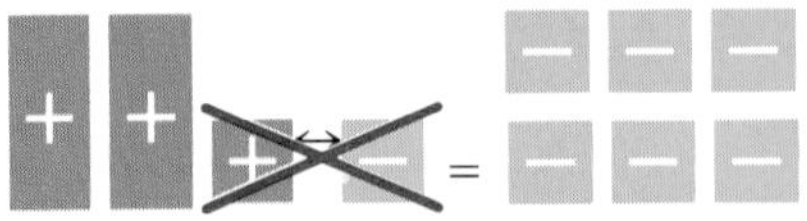

$2x = -6$

F. Since there are two x-tiles, form 2 equal groups on each side.

$\frac{2x}{2} = \frac{-6}{2}$

G. Write the value of x.

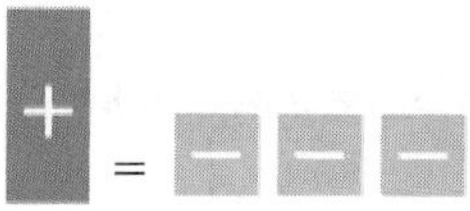

$x = -3$

EXERCISES

Goal: To use algebra tiles to solve equations that have the variable on both sides

Find the value of one x-tile.

1. 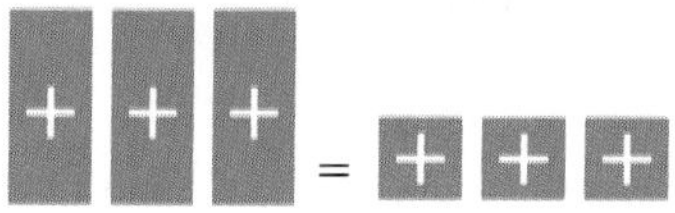

2. 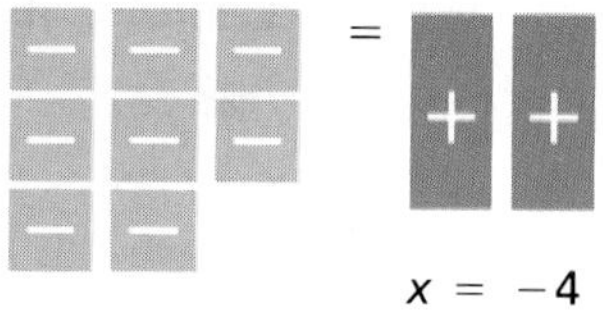

$x = -4$

3. Use algebra tiles to show that $x = -3$ is the solution of $3x + 1 = x - 5$.

Use algebra tiles to solve each equation.

4. $2x + 1 = x + 5$
5. $x + 8 = -3x + 4$
6. $3x - 4 = x + 2$
7. $4x + 1 = 6 - x$
8. $-x - 1 = -2x + 4$
9. $-2x + 6 = 3x + 1$
10. $3x + 2 = -2x - 8$
11. $3x - 4 = 8 - x$
12. $3x - 8 = 2 - 2x$

Which equation has a solution of $x = -3$? Write Yes or No.

13. 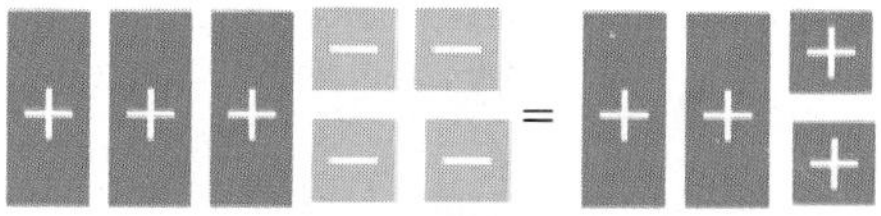

14. 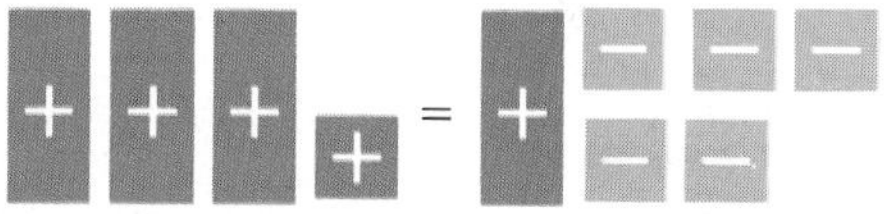

Which equation has a solution of $x = -1$? Write Yes or No.

15. 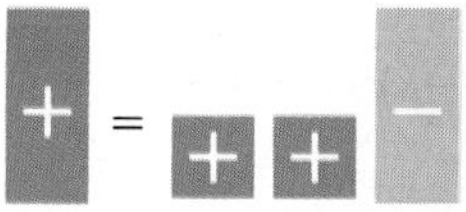

16. 

17. Write a letter to a friend explaining how to solve the equation $3 - 2x = 17 - x$ by using algebra tiles. Explain also how you would check to be sure that your solution is correct.

Manipulative 9 *(Use with Lesson 8.3.)*

To model polynomials, you will use positive and negative (1)-tiles, positive and negative x-tiles, and positive and negative x^2-tiles.

x^2-tile **$(-x^2)$-tile**

Recall that algebra tiles used to model like terms will have the same size and shape, but may have different signs.

EXAMPLE 1 Tell whether each pair of tiles represents like terms. Give a reason for your answer.

	Model	Like terms?
a.		Yes; the tiles have the same size and shape.
b.	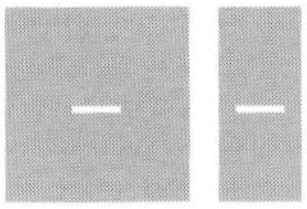	No; the x^2 – tile is wider than the x-tile.
c.		No; the x-tile is taller than the (1)-tile.

You can use algebra tiles to simplify polynomials.

EXAMPLE 2 Use algebra tiles to simplify $x^2 - x + 1 + x^2 + 2x - 3$

Solution: **A.** Model the polynomial. Think of $x^2 - x$ as $x^2 + (-x)$ and think of $2x - 3$ as $2x + (-3)$.

$x^2 + (-x) + 1 + x^2 + 2x + (-3)$

B. Group the like terms.

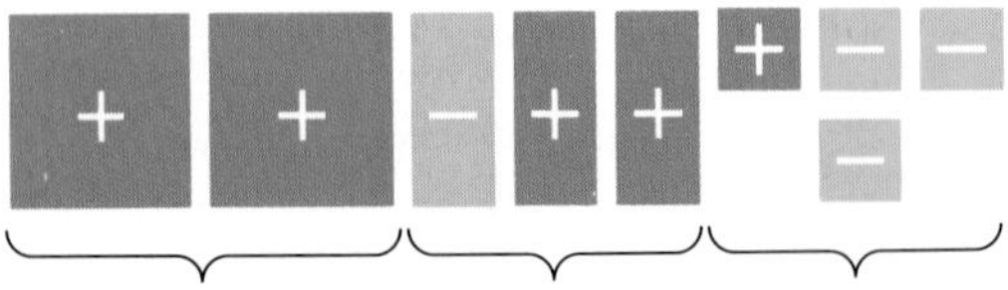

$x^2 + x^2 + (-x) + 2x + 1 + (-3)$

Like Terms **Like Terms** **Like Terms**

C. Identify and remove neutral pairs. Count the remaining tiles and record the results.

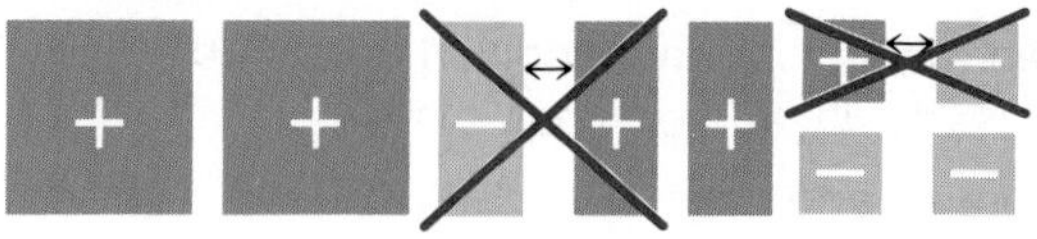

$2x^2 + x - 2$

EXERCISES

Goal: To use algebra tiles to simplify polynomials

Which pair of algebra tiles shows like terms? Answer Yes or No.

1. 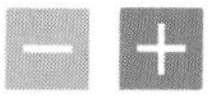

2. 

3. 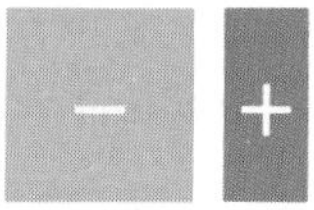

4. 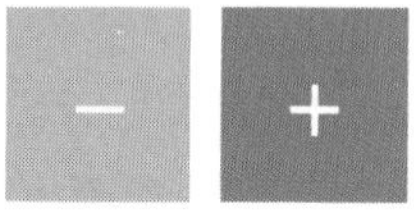

Use algebra tiles to simplify each polynomial.

5. $x^2 - 3x + 4 + x^2 + 6x - 2$

6. $x^2 - 5x - 1 - x^2 + 7x - 8$

7. $x^2 + 2 - 3x + 5 + x^2 - 4x$

8. $-x^2 + 2x + 3x^2 - 4x + 6$

9. $x^2 - 7 + 2x - x^2 - 4x + 6$

10. $3 - x - x^2 + 2x^2 - 7x + 10$

11. $2 - x^2 + 7x - 2x^2 - 3x + 4$

12. $8 - x + x^2 + 3 + x^2 - 2$

13. $7 - 8x + x^2 + 4x - 2x^2$

14. $-2x^2 + 6 - 3x + 3x^2 - 8$

Write true (T) or false (F) for each statement. When a statement is false, give an example to show why it is false.

15. The only way to model a monomial is to use one x-tile.

16. The only way to model a binomial in simplest form is to use one x^2-tile and one x-tile.

17. To model a trinomial in simplest form, you can use one x^2-tile, one x-tile, and one (1)-tile. The tiles can be positive or negative.

18. Three (-1)-tiles can be used to model a monomial.

Manipulative 10 *(Use with Lesson 8.4.)*

You can use algebra tiles to model addition and subtraction of polynomials.

EXAMPLE 1

Use algebra tiles to add:

$$\begin{array}{r} x^2 + 3x - 4 \\ \underline{-x^2 - x + 1} \end{array}$$

Solution: **A.** Model each polynomial. Align tiles representing like terms.

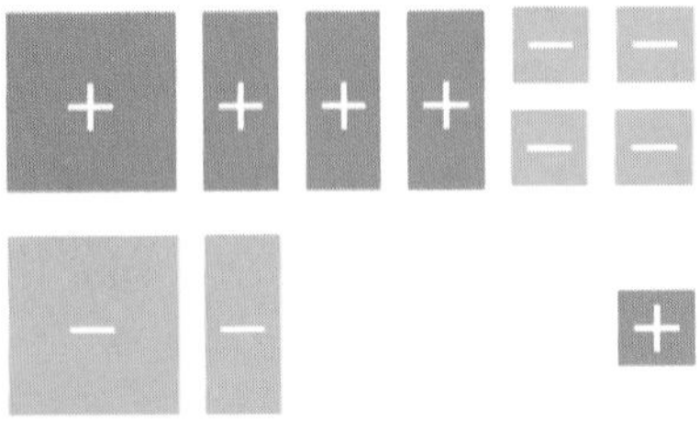

$x^2 + 3x - 4$

$x^2 - x + 1$

B. Remove neutral pairs and count the remaining tiles.

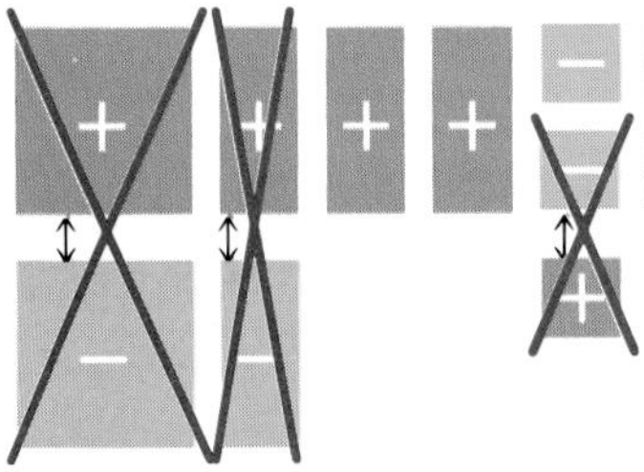

$$\begin{array}{r} x^2 + 3x - 4 \\ \underline{-x^2 - x + 1} \\ 2x - 3 \end{array}$$

Recall that subtracting a number is the same as adding its opposite. That is,
$x^2 - (3x^2 - 7x) = x^2 + [-(3x^2 - 7x)] = x^2 - 3x^2 + 7x$.

EXAMPLE 2

Subtract:

$$\begin{array}{rr} & x^2 - 3x + 2 \\ (-) & \underline{-x^2 - 2x + 1} \end{array}$$

Solution: **A.** Add the opposite of $(-x^2 - 2x + 1)$ to $x^2 - 3x + 2$.

$$\begin{array}{rr} & x^2 - 3x + 2 \\ (-) & \underline{-x^2 - 2x + 1} \end{array} \longrightarrow \begin{array}{rr} & x^2 - 3x + 2 \\ + & \underline{x^2 + 2x - 1} \end{array}$$

B. Model the polynomials. Align like terms.

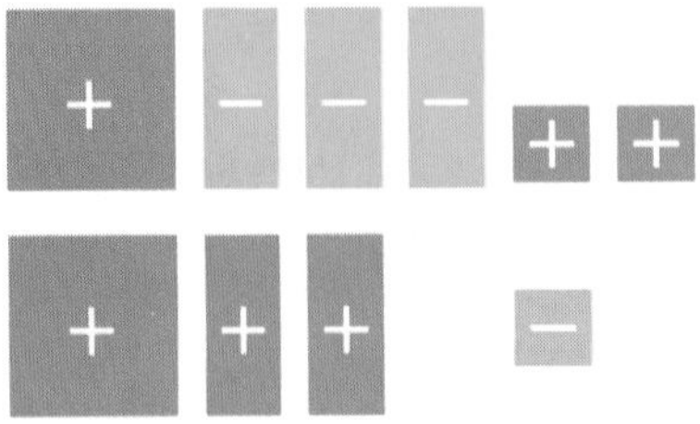

$x^2 - 3x + 2$

$x^2 + 2x - 1$

C. Remove neutral pairs and count the remaining like tiles.

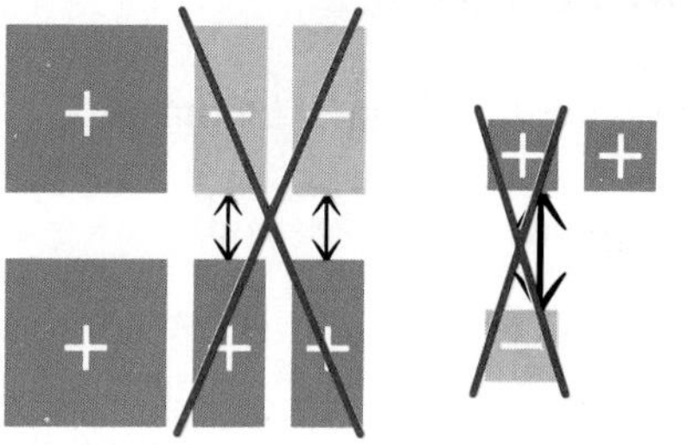

$$\begin{array}{r} x^2 - 3x + 2 \\ + \; x^2 + 2x - 1 \\ \hline 2x^2 - \;\; x + 1 \end{array}$$

EXERCISES

Goal: To use algebra tiles to add and subtract polynomials

Use algebra tiles to add.

1. $\begin{array}{r} x^2 - 2x + 1 \\ \underline{x^2 + 4x - 3} \end{array}$

2. $\begin{array}{r} 2x^2 + 7x - 4 \\ \underline{x^2 - 5x + 8} \end{array}$

3. $\begin{array}{r} -x^2 + 2x - 6 \\ \underline{2x^2 + 3x + 8} \end{array}$

Use algebra tiles to find the opposite of each polynomial.

4. $x - 7$

5. $-x^2 + x$

6. $2x^2 + 6x$

7. $2x^2 - 3x - 4$

8. $5x - x^2 + 9$

9. $5x^2 - 9x - 3$

Use algebra tiles to subtract.

10. $\begin{array}{rr} & x^2 - 5x + 6 \\ (-) & \underline{x^2 + 2x + 4} \end{array}$

11. $\begin{array}{rr} & 3x^2 + 6x - 9 \\ (-) & \underline{2x^2 - \;\; x + 4} \end{array}$

12. $\begin{array}{rr} & x^2 + 2x - 6 \\ (-) & \underline{2x^2 - 4x + 5} \end{array}$

13. $\begin{array}{rr} & -2x^2 + 5x + 4 \\ (-) & \underline{-x^2 + 7x - 5} \end{array}$

14. $\begin{array}{rr} & -x^2 + 5x - 6 \\ (-) & \underline{-x^2 + 4x - 8} \end{array}$

15. $\begin{array}{rr} & -4x^2 + \;\; x - 1 \\ (-) & \underline{x^2 - 3x + 8} \end{array}$

16. The lengths of the sides of a polygon can be represented as $5x$, $3x - 1$, and $2x + 5$.

a. Use algebra tiles to represent the perimeter of the polygon in simplest form.

b. If $x = 19$ inches, what is the perimeter of the polygon in inches?

Manipulative 11 *(Use with Lesson 8.5.)*

You can use algebra tiles to find the product of binomials.

EXAMPLE 1 Use algebra tiles to multiply: $(x + 1)(x + 2)$

Solution:

A. Model the factors $(x + 1)$ and $x + 2$.

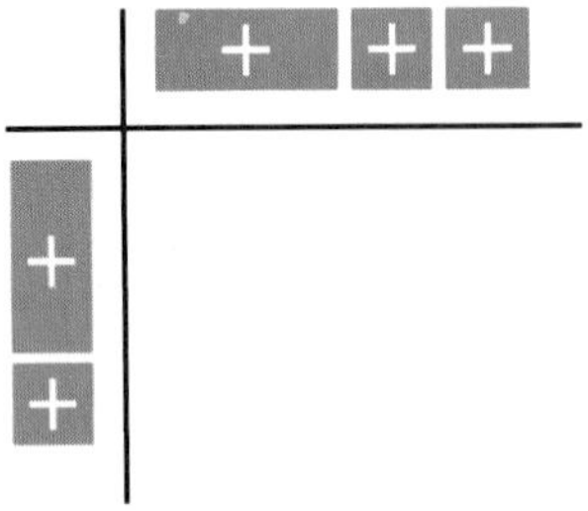

B. Build a rectangle having dimensions $(x + 1)$ and $(x + 2)$.

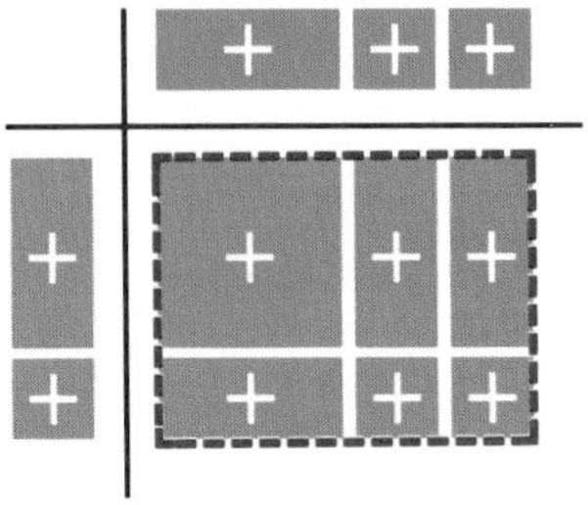

C. Count the tiles in the completed rectangle.

$$\begin{aligned}(x + 1)(x + 2) &= x^2 + 2x + x + 2 \\ &= x^2 + 3x + 2\end{aligned}$$

The product $x^2 + 3x + 2$ represents the area of the rectangle. The area is the sum of the areas of the tiles in the rectangle.

EXAMPLE 2 Use algebra tiles to multiply: $(x - 1)(x + 2)$

Solution:

A. Build a rectangle with dimensions $(x - 1)$ and $(x + 2)$. First place the x^2-tile and the x-tiles in the first row. Since the factor at the left and the factors above are positive, the tiles in the first row will be positive.

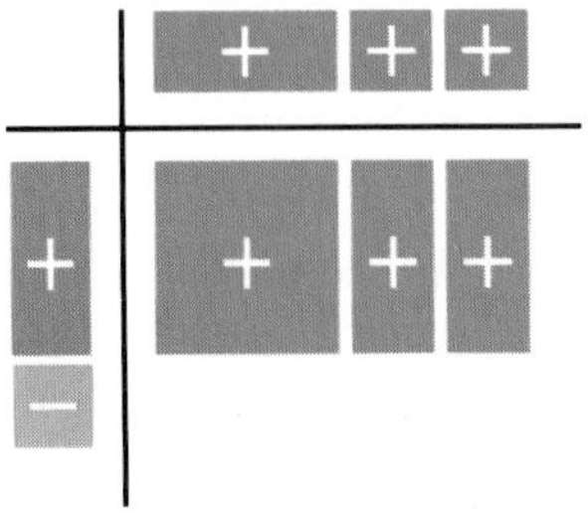

B. Complete the second row of the rectangle. Since the factor at the left is negative and the factors above are positive, the tiles in the second row will be negative.

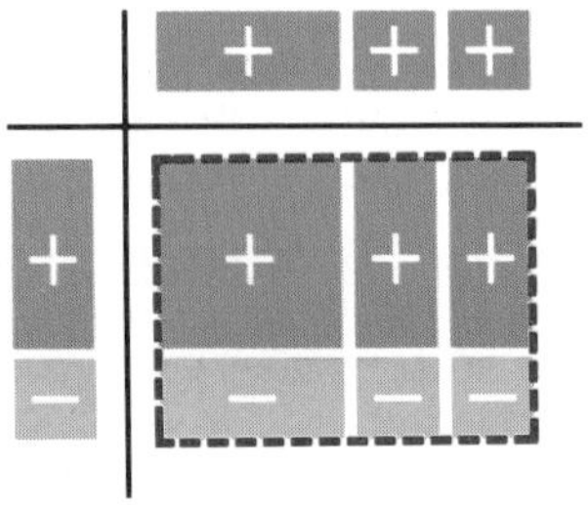

C. Count the tiles in the completed rectangle.

$$(x - 1)(x + 2) = x^2 + 2x - x - 2$$
$$= x^2 + x - 2$$

EXERCISES

Goal: To use algebra tiles to model multiplication of binomials

For each model, name the binomial factors and the product.

1.

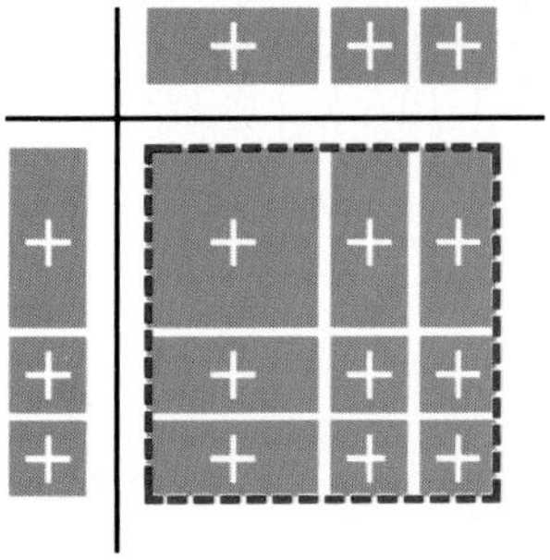

2.

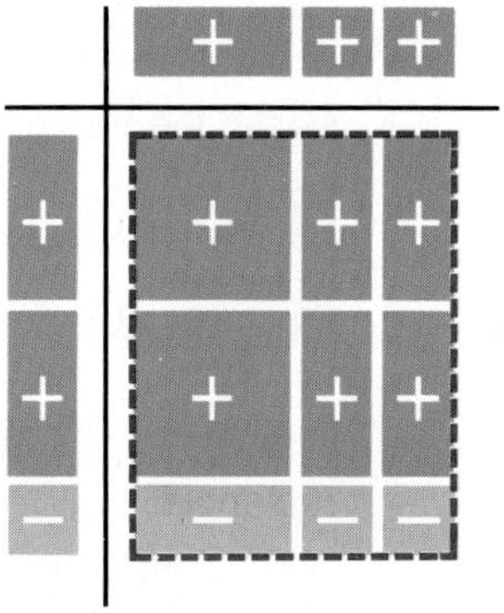

Use algebra tiles to find each product.

3. $(x + 3)(x + 2)$

4. $(x + 3)(x + 4)$

5. $(2x + 1)(x + 4)$

6. $(3x + 1)(x + 2)$

7. $(2x - 3)(x + 2)$

8. $(2x + 4)(x - 1)$

9. $(2x + 3)(x - 2)$

10. $(x - 1)(x - 1)$

11. $(3x - 2)(3x - 2)$

12. $(x + 4)(x + 4)$

13. $(x - 4)(x - 4)$

14. $(x + 4)(x - 4)$

15. $(x - 3)(x - 3)$

16. $(x + 3)(x + 3)$

17. $(x - 3)(x + 3)$

18. Explain how the product rectangle in Exercise 2 shows that $(2x - 1)(x + 2) = x(x + 2) + x(x + 2) - 1(x + 2) = 2x^2 + 3x - 2$.

Solve. Use algebra tiles if you need help.

19. The length of one side of a square piece of land is x feet. If you double the length of one pair of opposite sides and add 3 feet to the other pair of opposite sides, what will be the new dimensions of the piece of land?

20. Write two expressions for the area of the enlarged piece of land in Exercise 19.

21. If the length of one side of the original piece of land in Exercise 19 was 12 feet, what is the area of the enlarged piece of land?

Manipulative 12 *(Use with Lesson 9.3.)*

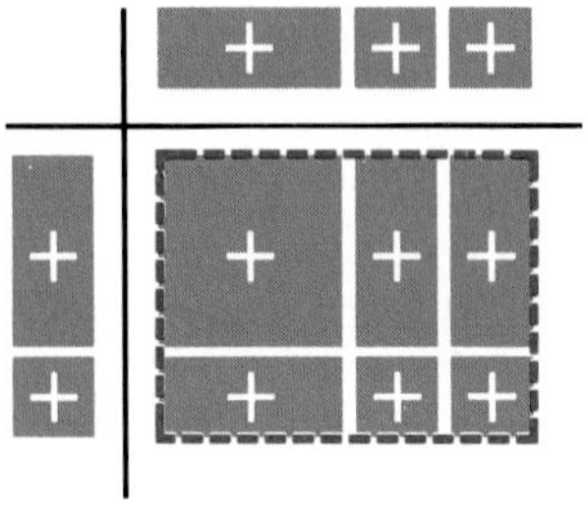

In the previous lesson, you used algebra tiles to show that

$$(x + 1)(x + 2) = x^2 + 3x + 2.$$

This means that the binomials $(x + 1)$ and $(x + 2)$ are factors of $x^2 + 3x + 2$.

Now consider the trinomial $x^2 + 4x + 3$. Is this trinomial factorable?

EXAMPLE 1

Use tiles to factor $x^2 + 4x + 3$.

Solution: **A.** Model the trinomial.

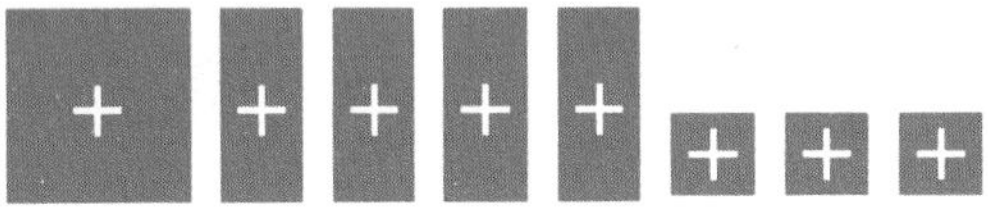

B. The trinomial is **factorable** if the tiles can be arranged to form a rectangle.

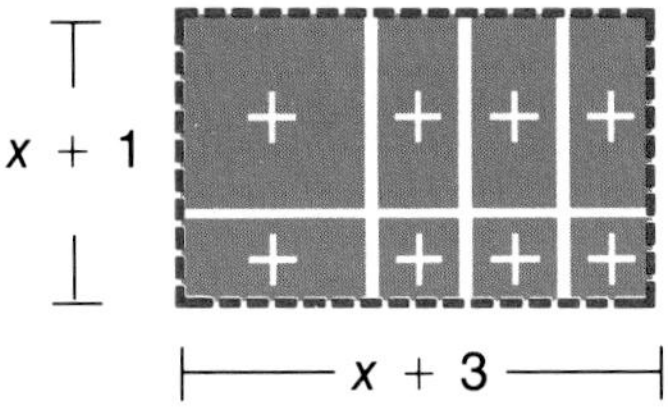

So $x^2 + 4x + 3 = (x + 1)(x + 3)$.

EXAMPLE 2

Use tiles to factor $x^2 + 7x + 10$.

Solution: **A.** Model the trinomial.

B. Build a rectangle. Position the x^2-tile and the x-tiles so that the ten (1)-tiles will fit.

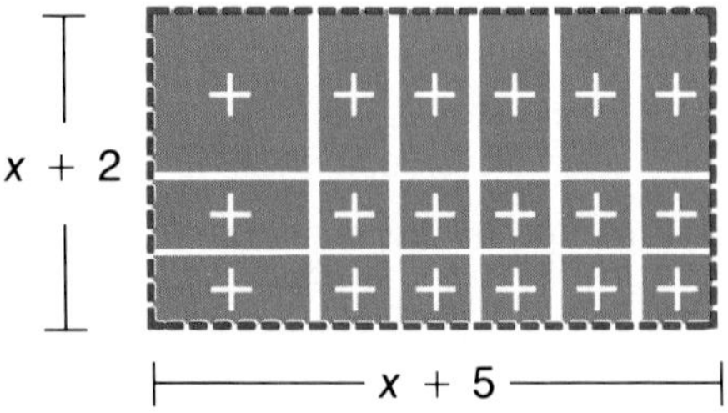

So $x^2 + 7x + 10 = (x + 2)(x + 5)$.

EXERCISES

Goal: To use algebra tiles to factor trinomials

Use algebra tiles to factor each trinomial.

1. $x^2 + 2x + 1$

2. $x^2 + 3x + 2$

3. $x^2 + 4x + 4$

4. $x^2 + 5x + 6$

5. $x^2 + 7x + 12$

6. $x^2 + 6x + 9$

7. $x^2 + 4x + 3$

8. $x^2 + 5x + 4$

9. $x^2 + 6x + 5$

10. $x^2 + 7x + 10$

11. $x^2 + 5x + 6$

12. $x^2 + 8x + 7$

Use algebra tiles to determine whether the trinomials are factorable. Answer <u>*Yes*</u> *or* <u>*No*</u>*.*

13. $x^2 + 3x + 3$

14. $x^2 + 8x + 15$

15. $x^2 + 2x + 16$

16. $x^2 + 2x + 2$

17. $x^2 + 7x + 6$

18. $x^2 + 8x + 16$

Find the missing factor. Use algebra tiles if you need help.

19. $x^2 + 14x + 13 = (x + 13)(\underline{\ ?\ })$

20. $x^2 + 12x + 11 = (\underline{\ ?\ })(x + 11)$

21. $x^2 + 8x + 12 = (\underline{\ ?\ })(x + 2)$

22. $x^2 + 10x + 21 = (x + 7)(\underline{\ ?\ })$

23. A car travels $(x^2 + 9x + 14)$ miles in $(x + 2)$ hours. Write an expression for the car's speed in miles per hour.

24. A car travels $(x^2 + 7x + 10)$ miles at an average rate of $(x + 5)$ miles per hour. Write an expression for the number of hours the car traveled.

25. Write an explanation describing how you could use algebra tiles to determine whether $x^2 + 11x + 16$ is a factorable or a prime polynomial.

Manipulative 13 *(Use with Lesson 9.5.)*

You can use algebra tiles to model the product of two binomials of the form $(x + a)(x - a)$.

EXAMPLE 1 Use algebra tiles to multiply: $(x + 3)(x - 3)$

Solution:

A. Model the factors. Then begin to build a rectangle by placing the tiles in the first row. Since the x-tile and the (1)-tiles in the factors differ in sign, the x-tiles in the first row will be negative.

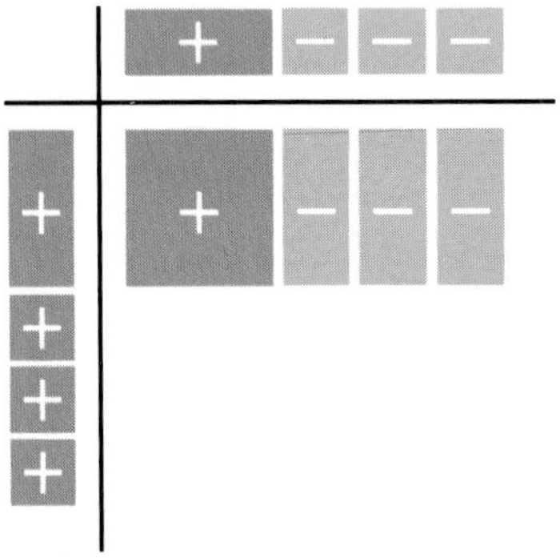

B. Complete the rectangle by placing the tiles in the remaining rows. Since the (1)-tiles in the factors differ in sign, the (1)-tiles in the product will be negative.

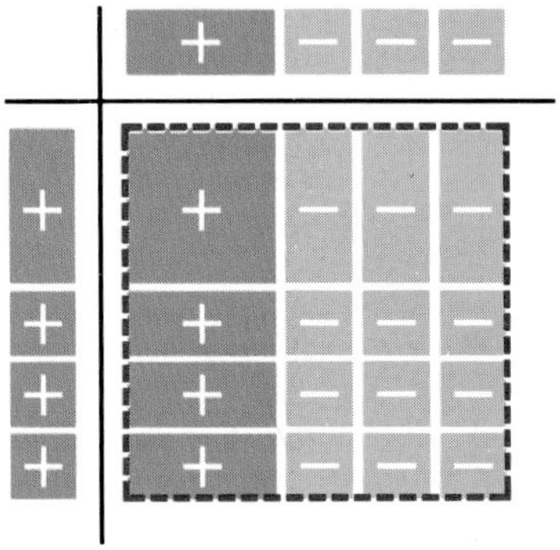

C. Identify neutral pairs. Count the remaining tiles.

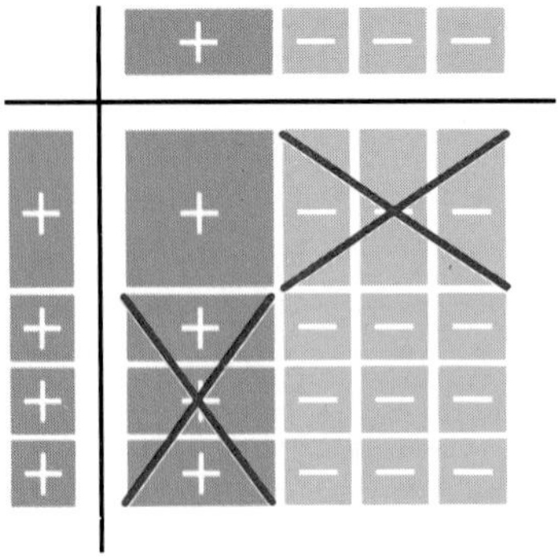

D. Write the product.

$$(x + 3)(x - 3) = x^2 - 9$$

In the expression $x^2 - 9$, x^2 and 9 are perfect squares. That is, $x^2 - 9$ can be written as $(x)^2 - (3)^2$.

This suggests that a binomial that can be expressed as the difference of perfect squares can be factored as the sum and difference of the square roots of the terms of the binomial.

EXAMPLE 2 Factor: **a.** $a^2 - 1$ **b.** $r^2 - 16$

Solutions:

a. $a^2 - 1 = (a)^2 - (1)^2$
$= (a + 1)(a - 1)$

b. $r^2 - 16 = (r)^2 - (4)^2$
$= (r + 4)(r - 4)$

EXERCISES

Goal: To use algebra tiles to find the product of binomials of the form $(x + a)(x - a)$

Use algebra tiles to find each product.

1. $(x + 2)(x - 2)$ **2.** $(x + 5)(x - 5)$ **3.** $(2x - 1)(2x + 1)$

4. $(5 - x)(5 + x)$ **5.** $(3x + 2)(3x - 2)$ **6.** $(4x + 1)(4x - 1)$

Draw a model on paper to show each product.

7. $(x + 1)(x - 1)$ **8.** $(x - 7)(x + 7)$ **9.** $(6 - x)(6 + x)$

Match each binomial in Column A with its factors in Column B.

Column A	Column B	
10. $x^2 - 64$	**a.** $(x + 9)(x - 9)$	**f.** $(5x - 8)(5x + 8)$
11. $81 - x^2$	**b.** $(x + 8)(x - 8)$	**g.** $(2x + 7)(2x - 7)$
12. $25x^2 - 64$	**c.** $(9 + x)(9 - x)$	**h.** $(10 + 3x)(10 - 3x)$
13. $4x^2 - 49$	**d.** $(5x - 64)(5x + 64)$	
14. $100 - 9x^2$	**e.** $(25x + 8)(25x - 8)$	

Factor.

15. $x^2 - 49$ **16.** $x^2 - 144$ **17.** $25 - x^2$ **18.** $x^2 - 100$

19. $(xy)^2 - 36$ **20.** $25x^2 - 4$ **21.** $-1 + 81x^2$ **22.** $-9 + x^2$

23. Write an explanation of what happens to the middle term when you multiply two binomials of the form $(x + a)$ and $(x - a)$.

Manipulative 14 *(Use with Lesson 9.6.)*

You can use algebra tiles to factor trinomials that have special patterns.

EXAMPLE 1

Use algebra tiles to factor $x^2 + 6x + 9$.

Solution: **A.** Model the polynomial.

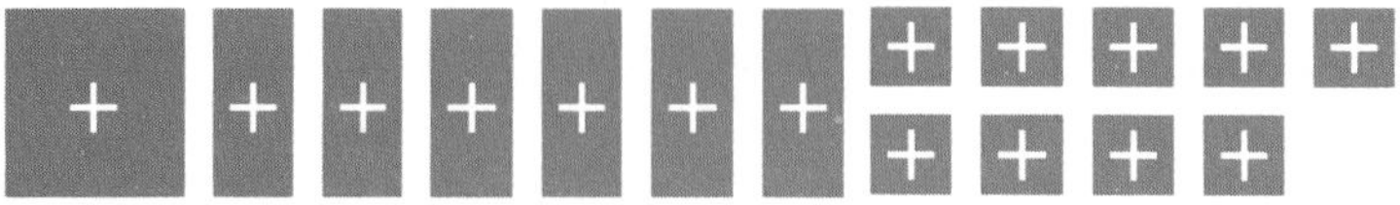

B. The polynomial is factorable if the tiles can be arranged to form a rectangle. Arrange the tiles.

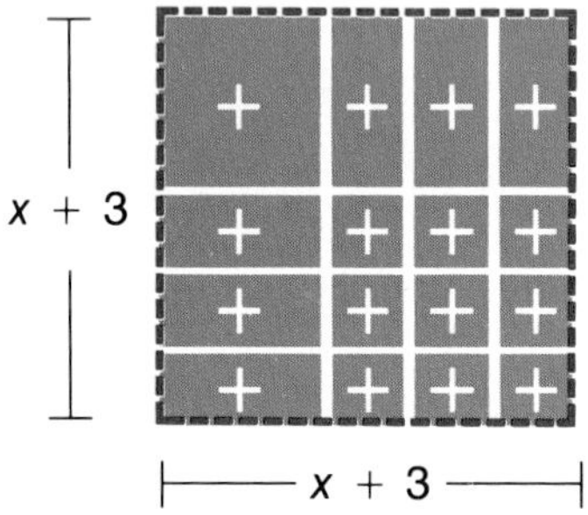

C. Write the factors.

$$x^2 + 6x + 9 = (x + 3)(x + 3)$$
$$= (x + 3)^2$$

Since the trinomial $x^2 + 6x + 9$ has two equal binomial factors, it is called a **trinomial square**. Notice that the rectangle formed by the tiles is a square.

EXAMPLE 2

Use algebra tiles to factor $4x^2 + 4x + 1$.

Solution: **A.** Model the polynomial.

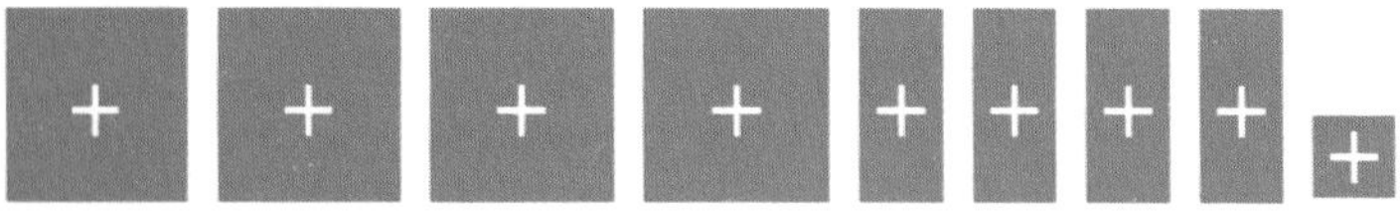

B. Arrange the tiles to form a rectangle.

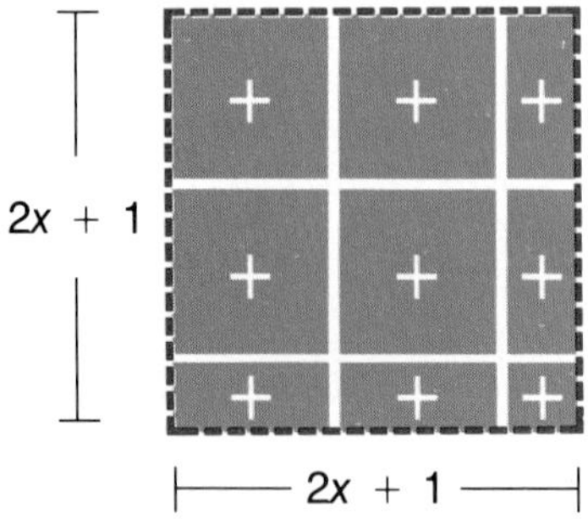

C. Write the factors.

$$4x^2 + 4x + 1 = (2x + 1)(2x + 1)$$
$$= (2x + 1)^2$$

EXERCISES

Goal: To use algebra tiles to factor trinomial squares.

Use algebra tiles to factor each polynomial.

1. x^2+2x+1 **2.** $x^2+8x+16$ **3.** x^2+4x+4

4. $9x^2+6x+1$ **5.** $4x^2+8x+4$ **6.** $25+10x+x^2$

7. Use algebra tiles to find this product: $(x - 2)(x - 2)$

8. Compare the product you found in Exercise 7 to this product: $(x + 2)(x + 2)$. Which terms are alike? Which are different?

9. What are the factors of $x^2 - 4x + 4$?

10. Is $x^2 - 4x + 4$ a trinomial square? Explain.

Use the results of Exercises 7–10 to factor each trinomial.

11. x^2-6x+9 **12.** $x^2-8x+16$ **13.** $x^2-14x+49$

14. $4x^2-4x+1$ **15.** $9x^2-6x+1$ **16.** $25x^2-10x+1$

17. Write an explanation of how you could use algebra tiles to determine whether a trinomial is a perfect square.

18. Help Yuri solve this problem.
If $x^2 + a^2x + a^2 = (x + a)^2$, what is the value of a?

Manipulative 15 *(Use with Lesson 19.5.)*

Theoretical probability is the probability predicted by mathematical reasoning. For example, we say that the probability of a fair coin landing heads up when it is tossed is $\frac{1}{2}$, because a coin has 2 sides and only one of the sides is heads.

If you toss 10 coins yourself, you might get 6 heads and 4 tails. So you would say that the **experimental (empirical) probability** of getting a head is $\frac{6}{10}$.

Since results usually vary somewhat when an experiment is repeated, experimental probabilities are not exact. One way to make an experimental probability more accurate is to repeat the experiment a large number of times. For example, as the number of times you toss a coin increases, the closer the experimental probability will get to $\frac{1}{2}$, the probability predicted by mathematical reasoning.

In the Exercises that follow, you will examine some experimental probabilities.

EXERCISES

Goal: To determine the experimental probability of events

Toss a coin 10 times. Record the outcome of each toss in order in a table like this one.

Toss	1	2	3	4	5	6	7	8	9	10
H or T										

Refer to your table to answer these questions.

1. How many of the 10 tosses were heads?
2. How many of the 10 tosses were tails?
3. How many times did you get two tails in a row?
4. How many times did you get two heads in a row?
5. What is the greatest number of successive heads or successive tails that you tossed?
6. Suppose that you toss the coin 10 more times. Would you expect that the results will show the same pattern? Explain.
7. Would you expect that the total number of heads in Exercise 6 to be close to the number of heads in Exercise 1?

Now copy and complete a table like the one below. Your table will have ten rows. Toss a coin 100 times and record the results, in order, filling the first row, then the second row, then the third row, and so on. Compute the total number of heads and tails for each row.

	1	2	3	4	5	6	7	8	9	10	Total	
											H	T
Row 1												
Row 2												

Use the results in your table to answer these questions.

8. Does each row have the same number of heads?

9. What is the greatest number of heads in any row?

10. Does each row have the same sequence (order) of heads and tails?

11. How many times did you get 4 heads or 5 heads or 6 heads in one of the rows?

12. According to this experiment, what is the probability of getting (4 heads or 5 heads or 6 heads) in one of the rows?

13. According to the experiment, what is the probability of **not** getting (4 heads or 5 heads or 6 heads) in one of the rows?

14. What is the sum of the probabilities in Exercises 12 and 13?

15. Write a summary of the results of your experiment. Include any patterns you observe, which pattern or patterns seem to occur most often, whether the results seem reasonable, and what you predict would happen if you repeated the experiment.

16. Read your summary to the class. Compare your results with those of your classmates and discuss any differences.

GLOSSARY

The following definitions and statements reflect the usage of terms in this textbook.

Absolute Value The *absolute value* of a positive number equals the number. The *absolute value* of a negative number equals the opposite of the number. The *absolute value* of 0 equals 0. (Page 2)

Algebraic Expression An *algebraic expression* is an expression with one or more variables. (Page 21)

Axis of a Parabola A vertical line that contains the vertex of a parabola is called the *axis of the parabola.* (Page 476)

Base See **Power.**

Binomial A *binomial* is a polynomial of two terms. (Page 213)

Coefficient In a monomial of the form ax^n, the number represented by a is the *coefficient* or *numerical coefficient*. (Page 208)

Completing the Square Writing a quadratic equation in a form that makes the left side a trinomial square is called *completing the square.* (Page 494)

Constant Term In a polynomial, a term with no variables is called the *constant term.* (Page 220)

Cube Root A *cube root* of a number is one of its three equal factors. (Page 180)

Degree of a Monomial The *degree of a monomial* in one variable is determined by the exponent of the variable. (Page 239)

Degree of a Polynomial The *degree of a polynomial* is the highest degree of the monomials of the polynomial. (Page 239)

Dependent Events Two events are *dependent* if the occurrence of one event is influenced by the occurrence of the other. (Page 525)

Determinant A *determinant* is a square array of the form $\begin{vmatrix} a & b \\ c & d \end{vmatrix}$ in which $a, b, c,$ and d are real numbers. (Page 412)

Difference of Two Squares A binomial such as $x^2 - 16$ is called a *difference of two squares.* The first term and the absolute value of the constant term are perfect squares. (Page 252)

Domain The set of replacements for a variable is the *domain*. (Page 116)

Domain of a Relation The *domain of a relation* is the set of x values of its ordered pairs. (Page 333)

Equation An *equation* is a sentence that contains the equality symbol "=." (Page 58)

Equivalent Equations *Equivalent equations* have the same solution. (Page 58)

Event An *event* is a subset of the sample space. It is a set of one or more of the possible outcomes of an experiment. (Page 515)

Exponent See **Power.**

Experimental Probability If an experiment is carried out n times and a certain event A occurs a times, then the *experimental probability* for event A is $\frac{a}{n}$. (Page 529)

Expression An *expression* includes at least one of the operations of addition, subtraction, multiplication, or division. (Page 19)

Fourth Root A *fourth root* of a number is one of its four equal factors. (Page 180)

Function A *function* is a relation in which no two ordered pairs have the same x value. (Page 333)

Fundamental Counting Principle If there are p choices for the first decision and q choices for the second decision, then there are $p \times q$ ways that the two decisions can be made together. (Page 512)

Greatest Common Factor The greatest common factor of two or more counting numbers is the greatest number that is a factor of all the numbers. (Page 237)

Independent Events Two events are independent if the occurrence of one event is not influenced by the occurrence of the other. (Page 520)

Integers The set of integers consists of the set of whole numbers and their opposites. (Page 30)

Irrational Number An irrational number is a number that cannot be expressed as a quotient of two integers. (Page 31)

Least Common Multiple (LCM) The least common multiple (LCM) of two or more counting numbers is the smallest counting number that is divisible by the given numbers. (Page 300)

Like Radicals Radicals such as $6\sqrt{2}$ and $5\sqrt{2}$ are called *like radicals*. They have equal radicands. (Page 187)

Like Terms *Like terms* have exactly the same variables and the same powers of these variables. (Page 50)

Linear Function A function that has the points of its graph lying in a straight line is a *linear function*. (Page 351)

Monomial of One Variable A *monomial of one variable* over $\{$rational numbers$\}$ is an expression of the form ax^n in which a is any rational number and n is any nonnegative integer. (Page 208)

Mutually Exclusive Events Events are *mutually exclusive* if they have no outcomes in common. (Page 521)

Origin The point where the x axis and the y axis meet is called the *origin*. (Page 322)

Parallel Lines *Parallel lines* are lines in the same plane that do not have a common point. (Page 364)

Perfect Square Positive numbers such as 1, 4, 9, 16, 25, and 36 are *perfect squares*, because each is the square of a counting number. (Page 31)

Perimeter The *perimeter* of a geometric figure such as a triangle, square, or rectangle is the sum of the lengths of its sides. (Page 95)

Polynomial A *polynomial* is a monomial or the sum of two or more monomials. (Page 213)

Power An expression such as $(-2)^3$ is called a *power*. The 3 is the exponent and the -2 is the base. (Page 21)

Prime Factorization The *prime factorization* of a number is the number expressed as a product of its prime-number factors. (Page 175)

Prime Number A *prime number* is a counting number greater than 1 that has exactly two counting-number factors, 1 and the number itself. (Page 174)

Probability The *probability* P of an event E is the ratio of the number of favorable outcomes of the event to the total number of possible outcomes in the sample space. (Page 515)

Quadrants The four regions of the coordinate plane are called *quadrants*. (Page 325)

Quadratic Equation A *quadratic equation* is an equation of the form $ax^2 + bx + c = 0$, $(a \neq 0)$. (Page 484)

Quadratic Function A *quadratic function* is a function described by the rule $y = ax^2 + bx + c$, in which *a, b,* and *c* are rational numbers with $a \neq 0$. (Page 456)

Quadratic Polynomials Second degree polynomials are called *quadratic polynomials*. (Page 239)

Range of a Relation The *range of a relation* is the set of *y* values of its ordered pairs. (Page 333)

Ratio A quotient of two numbers is called a *ratio*. (Page 359)

Rational Expression A *rational expression* is one that can be written as the quotient of two polynomials. (Page 268)

Rational Number A *rational number* is a number that can be expressed as a quotient of two integers, in which the denominator is not zero. (Page 30)

Real Numbers The set of *real numbers* contains all the rational numbers and all the irrational numbers. (Page 32)

Reciprocals Two numbers are *reciprocals* of each other if their product is 1. (Page 16)

Relation A *relation* is a set of ordered pairs. (Page 328)

Sample Space The set of all the different possible outcomes of an experiment is called the *sample space*. (Page 515)

Scientific Notation *Scientific notation* is a numeral of the form $N \times 10^a$, in which *a* is an integer and *N* is a rational number such that $1 \leq N < 10$. (Page 163)

Slope In the rule for the linear function $y = mx + b$, *m* represents the *slope* of its graph. (Page 356)

Square Root A *square root* is one of two equal factors of a number. (Page 174)

Solution of an Equation The *solution* or *root of an equation* is a number that makes the equation true. (Page 58)

Solution Set of a Linear Equation The *solution set of a linear equation* is the set of ordered paris (x,y) that makes the equation true. (Page 378)

Solution Set of a System of Two Equations *The solution set of a system of two equations* is the set of ordered pairs that makes both equations true. (Page 381)

Terms The monomials that form a polynomial are called its *terms*. (Page 213)

Trinomial A *trinomial* is a polynomial of three terms. (Page 213)

Trinomial Square A trinomial that has two equal binomial factors is called a *trinomial square*. (Page 255)

Variable A *variable* is a letter such as *x* that represents one or more numbers. (Page 21)

Vertex of a Parabola The *vertex of a parabola* that is a graph of a quadratic function is either the lowest point or the highest point of the graph. (Page 475)

Whole Numbers The set of *whole numbers* consists of zero and the counting numbers. (Page 30)

x intercept The *x* value of a point common to the *x* axis and a straight line is called the *x intercept* of the line. (Page 367)

y intercept The *y* value of the point that a line has in common with the *y* axis is the *y intercept*. (Page 353)

Zeros of a Function The *zeros of a function* are the values of *x* which make the value of the function 0. (Page 484)

INDEX

ANSWERS TO SELECTED EXERCISES

The answers to the odd-numbered problems in the *Classroom Exercises, Written Exercises, Mid-Chapter Reviews, Chapter Reviews, Cumulative Reviews, Special Topic,* and *Focus on Reasoning* sections are given on the following pages.

The answers are provided for all of the problems in the *Review Capsules* and *Pivotal Exercises* (P-1, P-2, etc.).

CHAPTER 1 OPERATIONS ON NUMBERS

Page 3 Classroom Exercises 1. 25 3. 0.001 5. $\frac{2}{3}$ 7. $-8\frac{7}{8}$ 9. −8.5 11. 9 13. 0 15. $\frac{4}{5}$ 17. 0.053 19. $\frac{1}{4}$ 21. −0.01 23. −2.5 25. $-2\frac{1}{2}$ 27. 12 29. −5

Page 3 Written Exercises 1. −23 3. 0.12 5. 0 7. $-\frac{3}{5}$ 9. $5\frac{6}{7}$ 11. 18 13. −227 15. $\frac{7}{8}$ 17. $5\frac{1}{8}$ 19. 0.083 21. 2.73 23. 23 25. −6.8 27. 12 29. 23 31. −13 33. 13

Pages 5-6 P-1 a. −9 b. −6 c. −4.7 **P-2** a. 2 b. $1\frac{1}{2}$ c. 2.6 **P-3** a. −3 b. $-1\frac{1}{3}$ c. −7.4

Page 6 Classroom Exercises 1. b 3. d 5. −40 7. −6 9. −0.7 11. 16 13. 3.4 15. $\frac{1}{5}$ 17. −6 19. −3.3 21. $-7\frac{1}{2}$ 23. 0 25. 0

Page 7 Written Exercises 1. −42 3. −106 5. −0.08 7. 8 9. 4 11. 0.13 13. −25 15. −32 17. −8.9 19. −53 21. 16 23. −45.5 25. 0 27. 0 29. 1 31. 5°C

Page 8 Review Capsule for Section 1.3 1. a. 4 b. 4 Both a and b are the same. 2. a. 4 b. 4 Both a and b are the same. 3. a. 7 b. 7 Both a and b are the same. 4. a. 10 b. 10 Both a and b are the same. 5. a. 21 b. 21 Both a and b are the same. 6. a. 0 b. 0 Both a and b are the same.

Page 8 P-1 10, 0.8, −37, 0 **P-2** a. 18 b. 20 c. −25 d. 4.2

Page 9 Classroom Exercises 1. 20 + (−15) 3. 0 + (−12) 5. (−12) + (−6) 7. −20 + 18 9. 5 + 23 11. −3.3 + (−6.2) 13. −4 15. −23 17. $-5\frac{1}{3}$ 19. −10 21. 16 23. −14

Pages 9-10 Written Exercises 1. 4 3. 6 5. −5 7. −9 9. −22 11. −10.8 13. −24.7 15. $-\frac{3}{4}$ 17. 31 19. 32 21. 2 23. 42 25. 8.3 27. −4 29. −13 31. −18 33. 8 35. 4.4 37. 1.3 39. $-\frac{1}{7}$ 41. 20°F 43. $38

Page 10 Review Capsule for Section 1.4 1. $\frac{5}{4}$ 2. $\frac{9}{1}$ 3. $\frac{11}{6}$ 4. $\frac{13}{1}$ 5. $\frac{27}{5}$ 6. 30 7. 5.2 8. 9 9. $\frac{1}{3}$ 10. 2.25 11. 0.56 12. 1.44 13. 0 14. 0 15. 1 16. $\frac{2}{3}$ 17. $7\frac{1}{5}$

Pages 11-12 P-1 0 **P-2** −8, −16, −24 **P-3** a. −42 b. −48 c. 0 d. −1 **P-4** 6, 12, 18 **P-5** a. 45 b. 70 c. 0.26 d. 3

Page 13 Classroom Exercises 1. −36 3. −96 5. −3 7. 0 9. −9.6 11. −2.1 13. 77 15. 150 17. 147 19. 6 21. 1 23. 1

Pages 13-14 Written Exercises 1. −70 3. 0 5. −66 7. −48 9. −10 11. $-\frac{4}{15}$ 13. −20.4 15. −8.36 17. 72 19. 1 21. 87 23. 3200 25. 0.33 27. 99 29. −432 31. 198 33. −3600 35. 0 37. 1.56 39. $-4\frac{2}{3}$ 41. $-3\frac{3}{4}$ 43. −24 45. 8.64 47. Yes, because 2 * (−4) = −16 and −4 * 2 = −16.

Pages 14-15 Mid-Chapter Review 1. −17 3. $\frac{1}{2}$ 5. 0.06 7. 0 9. 2.5 11. −17 13. 10 15. −33 17. 1 19. −26 21. 6°C 23. 11 + (−7) 25. 12.1 + 7.9 27. −7 + (−4) 29. −15 31. 9.25 33. $-6\frac{1}{2}$ 35. 25°F 37. −56 39. −3 41. 1 43. 33.6 45. 22°F

Page 15 Review Capsule for Section 1.5 1. $\frac{3}{2}$ 2. $\frac{12}{1}$ 3. $\frac{8}{5}$ 4. $\frac{7}{3}$ 5. $\frac{22}{7}$ 6. $\frac{22}{5}$ 7. $\frac{17}{3}$ 8. $\frac{13}{6}$ 9. $\frac{7}{1}$ 10. $\frac{15}{4}$ 11. 1 12. 1 13. 1 14. 1 15. 3 16. 3 17. 3 18. $\frac{3}{4}$

Pages 16-17 P-1 a. $\frac{10}{3}$ b. $-\frac{1}{3}$ c. $\frac{3}{4}$ d. 1 e. −1 **P-2** a. 64 b. $\frac{1}{9}$ c. 0 **P-3** a. $-1\frac{13}{18}$ b. $-15\frac{3}{7}$ c. −0.3

Page 18 Classroom Exercises 1. −8 × 4 3. 28 × $(-\frac{1}{7})$ 5. 175 7. 3 9. $-\frac{1}{2}$ 11. 0

Page 18 Written Exercises 1. 24 3. 3 5. $10\frac{2}{3}$ 7. $\frac{1}{6}$ 9. $-10\frac{2}{3}$ 11. −5 13. −3 15. −9 17. 0 19. −11 21. −0.9 23. −10 25. −4 27. −5 29. 52 31. 0 33. −3 35. $-\frac{1}{3}$ 37. −10 39. −41 41. −$2350

Page 19 P-1 a. Correct b. Incorrect c. Correct d. Incorrect

Page 20 Classroom Exercises 1. 16 3. 0 5. 5 7. 12 9. 43 11. 12 13. −11 15. −16

Page 20 Written Exercises 1. 9 3. −12 5. −0.5 7. 13 9. 4 11. −60 13. 9 15. −90 17. −14 19. −35 21. 20 23. 5.7 25. −15 27. −22 29. 9 31. −2 33. −5 35. −132 37. 49 years

Pages 21-22 P-1 a. 12 b. 2 c. $-7\frac{1}{2}$ **P-2** a. 4 b. −4 c. 33

Page 22 Classroom Exercises 1. 36 3. 125 5. −64 7. −4 9. 8 11. −16 13. −110 15. 11 17. −23 19. 18 21. −17 23. 30 25. 25 27. −10 29. 0

Pages 22-23 Written Exercises 1. 121 3. 1000 5. −343 7. 54 9. 37 11. −18 13. 18 15. 13 17. 60 19. 8 21. −2 23. 0 25. −84 27. 30 29. 83 31. 300 33. 190 35. 60 37. −790 39. 256 41. −216 43. 13 45. $-\frac{3}{8}$ 47. 26 49. 234 51. −132 53. −35 55. 72 57. $-12\frac{1}{2}$ 59. 15 61. −1152 63. 49 65. 625 67. 117,649

Pages 24-25 1. a. An estimate will be close enough. Reasons will vary. b. Answers will vary. c. His estimate was 17¢ higher than the actual cost. d. Yes 3. a. An exact answer should be computed. Reasons will vary. b. Answers will vary. c. Curry (62.5), Mills (62.4), Jones (52.9), Lang (48.3), Grimm (48.0) d. Answers will vary.

Pages 26-28 Chapter Review 1. 1.5 3. $\frac{3}{8}$ 5. 0 7. 2.6 9. 35 11. −3 13. 4 15. 11 17. −14 19. −13.5 21. $-3\frac{1}{4}$ points 23. 17 + (−29); −12 25. −27 + 18; 9 27. 19.2 + 4.7; 23.9 29. 19°F 31. −204 33. 450 35. −9 37. −12°F 39. −7 41. 12 43. −52 45. $\frac{2}{3}$ 47. −1.4 inches per month 49. −7 51. −16 53. 22 55. 64 57. 51 59. −28 61. −4 63. −21 65. 21 67. −1 69. −69

CHAPTER 2 REAL NUMBERS

Pages 30-32 **P-1** a. 2 b. −6 c. 0 d. −213 **P-2** a. 0.222 ··· b. 0.3636 ··· c. 0.1666 ···
P-3 a. Irrational b. Irrational c. Rational

Pages 32-33 **Classroom Exercises** 1. $\frac{7}{2}$; $\frac{14}{4}$ 3. $\frac{0}{1}$; $\frac{0}{3}$ 5. $\frac{-2}{8}$; $\frac{-4}{16}$ 7. $\frac{-5}{1}$; $\frac{-10}{2}$ 9. 0.5000 ··· 11. 0.571428571428 ··· 13. −0.375000 ··· 15. Integers 17. Counting numbers 19. Irrational 21. Rational 23. 28, −124, 17 25. 28, −1.8, $3\frac{1}{4}$, 0.1010010001 ···, −124, 17, $-\frac{22}{7}$, 5.436436436 ··· 27. 0.1010010001 ···

Pages 33-34 **Written Exercises** 1. $\frac{-14}{16}$; $\frac{-28}{32}$ 3. $\frac{35}{8}$; $\frac{70}{16}$ 5. $\frac{-183}{100}$; $\frac{-366}{200}$ 7. $\frac{20}{1}$; $\frac{40}{2}$ 9. $\frac{-3}{1}$; $\frac{-6}{2}$ 11. −24.000 ··· 13. 0.6363 ··· 15. 0.428571428571 ··· 17. Integers 19. Whole numbers 21. Rational 23. Irrational 25. Rational 27. Rational 29. Rational 31. Irrational 33. π 35. $\frac{10}{2}$, 19 37. $\frac{10}{2}$, −2, 4.7, 0, π, $-1\frac{3}{4}$, $\frac{7}{8}$, 19, $-\frac{17}{4}$, −6, −0.9 39. b, c, e 41. b, c, e 43. c, e 45. c, e 47. b, c, e 49. b, c, e

Page 34 **Review Capsule for Section 2.2** 1. 6 + (−4) 2. 7 + (−6) 3. 15 + (−8) 4. 4 + (−9) 5. 5 + (−12) 6. 23 + (−19) 7. 14 + (−18) 8. 2 + (−25) 9. 17 10. 17 11. $\frac{1}{16}$ 12. $-\frac{19}{60}$ 13. 1.1 14. 1.2 15. −9 16. −9 17. $-\frac{1}{9}$ 18. $4\frac{1}{6}$ 19. −7.1 20. −5.1 21. 2 22. 3 23. $\frac{1}{12}$ 24. $\frac{37}{60}$ 25. −1.8 26. −0.9 27. 1 28. −5 29. $\frac{11}{12}$ 30. $\frac{1}{12}$ 31. −2.6 32. 10.7

Page 36 **P-1** a. 8.2; Commutative Property for Addition b. 6; Associative Property for Addition **P-2** a. 9 b. $-\frac{7}{8}$ c. $\frac{5}{6} + y$ d. b − 0.8

Pages 36-37 **Classroom Exercises** 1. 0 3. $4\frac{1}{3}$ 5. −0.7 7. 0 9. 0 11. $3\frac{1}{2}$ 13. 0.6 15. 3 17. 0 19. 54 − a 21. 2.1 + x 23. x + y − 6 25. 5.1 − a − b

Pages 37-38 **Written Exercises** 1. −8; Addition Property of Opposites 3. 7; Commutative Property of Addition 5. 0; Addition Property of Opposites 7. 5 9. $1\frac{1}{2}$ 11. 2.5 13. 1.7 15. 2 − u 17. p + 1 19. −a − 0.3 21. f − 4.4 23. −17 + k − b 25. $d - 2\frac{1}{8} - f$ 27. a + 9 − b 29. s − t 31. 39 33. $-4\frac{1}{2}$ 35. 10.6 37. −y 39. −82 + q 41. $c - d - 11\frac{1}{2}$ 43. 119

Page 38 **Review Capsule for Section 2.3** 1. −30 2. 45 3. $-\frac{1}{4}$ 4. $\frac{6}{25}$ 5. −2.48 6. 7.02 7. −70 8. 88 9. $\frac{5}{8}$ 10. $-4\frac{4}{5}$ 11. 4.94 12. −36.04 13. $\frac{1}{5}$ 14. $\frac{5}{2}$ 15. $-\frac{1}{4}$ 16. $\frac{3}{1}$, or 3 17. $-\frac{8}{3}$ 18. $\frac{1}{10}$ 19. 1 20. $-\frac{12}{1}$, or −12 21. $-\frac{1}{13}$ 22. $\frac{1}{11}$ 23. −1 24. $\frac{5}{8}$

Page 40 **P-1** a. 8.1; Commutative Property of Multiplication b. 7; Associative Property of Multiplication c. $-\frac{1}{9}$; Multiplication Property of Reciprocals **P-2** a. −9.6 b. 48n c. 100pq

Pages 40-41 **Classroom Exercises** 1. 1 3. 62 5. 1 7. 17.8 9. 3.04 11. 16 13. 54 15. $\frac{4}{5}$ 17. $\frac{1}{2}$ 19. 570 21. −12 23. 20 25. −0.8yz 27. −120mn 29. −1.08bh

Page 41 **Written Exercises** 1. $-\frac{5}{7}$; Multiplication Property of 1 3. 8; Associative Property of Multiplication 5. $-\frac{1}{2}$; Commutative Property of Multiplication 7. 40 9. 36 11. 0.42pq 13. 288m 15. −27t 17. −5.4mp 19. 225xy 21. $-1\frac{3}{10}$rd 23. −10wz 25. p 27. n

Page 42 **Mid-Chapter Review** 1. $\frac{-10}{12}$; $\frac{-15}{18}$ 3. $\frac{14}{5}$; $\frac{28}{10}$ 5. $\frac{35}{10}$; $\frac{70}{20}$ 7. 0.444 ⋯ 9. 0.7272 ⋯ 11. 5.25000 ⋯ 13. Rational 15. Rational 17. Rational 19. Rational 21. $\sqrt{36}$ 23. $\frac{2}{3}$, $-\sqrt{5}$, $-1\frac{2}{7}$, $\sqrt{36}$, −9.3 25. −y 27. y + z − 4 29. $r - s - 4\frac{3}{8}$ 31. −15.5 − a + b 33. 4 35. −3h 37. −1600k 39. −8.82ay 41. −24pf

Page 42 **Review Capsule for Section 2.4** 1. $-12xy^2$ 2. $35a^2b^2$ 3. $3p^2qr$ 4. $-2mn^2s$ 5. x − 8 6. −t − 1 7. $6.5f^2gt^2$ 8. w − 16 9. −22 − p

Pages 43-44 **P-1** a. −x − 2 b. −4 + t c. $-\sqrt{2}$ d. mk **P-2** a. $-\frac{5k}{7}$ b. 1 c. 1 d. −1

Page 45 **Classroom Exercises** 1. −3k 3. 48y 5. −x − y 7. x − 2 − y + a 9. −4h 11. −2xy 13. $-3x^2y$ 15. $-y^2z^2$ 17. $-2a^2c^2$ 19. $-16m^2n^2$ 21. $-\frac{9}{f}$ 23. 1 25. 1 27. 1

Pages 45-46 **Written Exercises** 1. ac 3. 2t − r + 1 5. −4d 7. −m − n 9. 2a − 11 − b 11. 16cd 13. −4q 15. $-9a^2$ 17. $6x^2y^2$ 19. $\frac{4}{3}a^2b^2$ 21. $-1.56r^2s^2$ 23. $-\frac{3}{p}$ 25. 1 27. $\frac{-2-y}{x}$ 29. 1 31. 1 33. 7fg 35. −1 37. $-0.78r^2$ 39. $\frac{2}{15}x^2y$ 41. −3 − r + 3s 43. r − 2 45. −1 47. 13 triangles

Page 46 **Review Capsule for Section 2.5** 1. 14a + 16 2. 21 + 24x 3. −10 − 30b 4. −8y − 32 5. 12pq − 8p 6. $-\frac{1}{2}xy - 8x^2y$ 7. Yes 8. No 9. Yes 10. No 11. No 12. Yes

Pages 47-48 **P-1** a. −15a + 10 b. −3m − 2 c. 6rs − 15r **P-2** a. 3x b. 4c c. −7t **P-3** a. 6(y + 4) b. 2n(5m − 2) c. 3x(5x − 1) d. −3v(3 + w)

Page 48 **Classroom Exercises** 1. 10p + 10q 3. 2k − 6 5. $20x^2 + 28x$ 7. $-3m^2 + m$ 9. 8(m + n) 11. 2(3x + 5y) 13. 5(x − y) 15. t(4m − 9n) 17. −3(a + b) 19. −r(6r + 1)

Page 49 **Written Exercises** 1. 10a + 10b 3. −3x − 6 5. 7r − 56 7. 24p + 16 9. $4r^2 + 12r$ 11. $-2y^2 + 12y$ 13. $24t^2 - 20tu$ 15. $-2t^2 + 3t$ 17. 5(a + b) 19. 4(r + 5s) 21. 2k(m + 2n) 23. pq(p + q) 25. 3(r − s) 27. 8(x − 1) 29. $8x(kx - 2y^2)$ 31. $\frac{3}{4}r(x - y)$ 33. −7b(2a + 1) 35. −3x(x − 2) 37. xy(−2 + 5), or 3xy 39. −x(4x + 1 41. 5(a − 3b + 2c) 43. $3x(2x^2 + x - 6)$ 45. $6(5r^2 - 6x - 23x^2)$ 47. Yes. Let a = 2 and b = 4. (2x + 4y) = 2(x + 2y). Thus, c = 2.

Page 49 **Review Capsule for Section 2.6** 1. 7 2. −10 3. −6 4. −13 5. −6 6. 5 7. −20 8. 5

Page 50 **P-1** a. Like b. Unlike c. Unlike

Page 51 **Classroom Exercises** 1. 11y 3. 2x 5. 2ab 7. $-3y^2$ 9. $-2t^2$ 11. 8x − 1 13. −7 15. $-6x^2 - 3$ 17. $a^2 - a + 5$ 19. $2x^2 + 1$ 21. $\frac{1}{9}a^2 + 2\frac{1}{2}a + \frac{1}{2}$ 23. $2.1m^2 + 5.7m + n$ 25. $-66p^2 - p - 73$

Page 51 Written Exercises 1. 20y 3. 19t 5. −17x 7. $-2t^2$ 9. $7z^2$ 11. 6a − 6 13. $-0.2t^2 + 2t$ 15. −3g − 3.4 17. $\frac{17}{20}a - 1\frac{1}{2}b$ 19. −62b − 295bz + 200z 21. $3.3t^2 + 2.8t + 0.3$ 23. $\frac{11}{24}r^2 - \frac{8}{15}s^2 + \frac{1}{3}$ 25. $-50f^2 - 17f - 45$ 27. 0.5t 29. $ab^2 - 2a^2b + ab - a^2b^2$ 31. $8v^2 + 10w^2 - 4w$ 33. $75a^2 + 25$ 35. $-4\frac{1}{3}x^2 - \frac{1}{4}x + \frac{1}{5}$

Pages 52-53 1. Median 3. Mean: 28.7 mpg; Median: 27.9 mpg; Mode: 30.4 mpg 5. Mean: 29.2 mpg; Median: 28.2 mpg; Mode: 28.2 mpg 7. The manufacturer of Car C; Car C has the greatest mean of the three cars. 9. Median; Car A has the greatest median of the three cars. 11. Median; Reasons will vary. 13. Median; Reasons will vary.

Pages 55-56 Chapter Review 1. $\frac{-26}{32}; \frac{-39}{48}$ 3. $\frac{31}{8}; \frac{62}{16}$ 5. $\frac{-37}{1}; \frac{-74}{2}$ 7. −21.000 · · · 9. −0.333 · · · 11. 4.5000 · · · 13. Positive rational numbers 15. Rational numbers 17. $-\sqrt{5}, \pi$ 19. $-5, -\sqrt{5}, -\sqrt{9}, -0.5$ 21. 0, 13 23. $5\frac{1}{2}$; Addition Property of Zero 25. −14; Addition Property of Opposites 27. −11 29. −4.5 31. 6 − a 33. 10 − r + w 35. 400 37. −2000 39. −63 41. −36 43. −10y 45. −10d 47. −12m + 2n + p 49. w − k − 12 51. r + s − p + q 53. 12kn 55. $7t^2$ 57. $72q^3r^2$ 59. −1 61. 1 63. 1 65. −7r − 70 67. −18a − 9 69. $-18m^2 + 30m$ 71. −1.02ab − 0.78b 73. $\frac{1}{2}(a - b)$ 75. 5(3r + 2s) 77. −2y(2y − 1) 79. −22xy(y − 2) 81. $-27t^2 - 5$ 83. −1.4q − 12.1 85. $-2p^2q + pq^2$

CHAPTER 3 EQUATIONS AND PROBLEM SOLVING

Pages 58-59 **P-1** a. True b. True c. True **P-2** a. y = −1 b. x = 29 c. t = 0 d. k = 22 **P-3** a. n = −1 b. k = 1 c. x = 0

Page 59 Classroom Exercises 1. Subtract 12. 3. Subtract 8. 5. Add 13.7. 7. Add $\frac{5}{2}$. 9. Subtract 23. 11. y = 5 13. q = 26 15. b = 4 17. $h = -2\frac{5}{6}$ 19. y = 17 21. c = −17

Page 60 Written Exercises 1. x = 33 3. z = 12 5. b = −4.2 7. d = 38 9. f = 24 11. g = 1 13. j = 17 15. n = −24 17. x = −1 19. b = 4.8 21. k = 2 23. $y = -1\frac{1}{5}$ 25. x = −33 27. b = 10.1 29. d = −2.9 31. f = 8.8 33. $x = -3\frac{1}{2}$ 35. k = 12.8 37. p = 1.3 39. z = −62 41. b = 33.4

Page 60 Review Capsule for Section 3.2 1. 1 2. 1 3. 1 4. 1 5. 1 6. 1 7. 1 8. 1 9. 1 10. 1 11. 1 12. 1 13. 9 14. −7.02 15. $-7\frac{1}{2}$ 16. 117 17. $2\frac{5}{8}$ 18. −988 19. 464 20. −102.5 21. 8 22. $\frac{1}{4}$ 23. 15 24. −52 25. $-25\frac{5}{6}$ 26. 0.4 27. $-2\frac{1}{2}$ 28. $-\frac{1}{20}$

Pages 61-62 **P-1** a. True b. True c. True **P-2** a. b = −36 b. x = −100 c. n = 1 d. b = 256 **P-3** a. y = 0.1 b. x = −3 c. n = 1 d. t = −3 **P-4** a. 2 b. $-\frac{4}{3}$ c. $-\frac{3}{5}$ d. $\frac{5}{4}$

Page 63 Classroom Exercises 1. Multiply by 12. 3. Divide by 14. 5. Divide by −8.9. 7. Multiply by $-\frac{3}{2}$. 9. j = 28 11. z = 140 13. r = −4.8 15. t = 81 17. t = −6 19. a = −0.3 21. k = −16 23. k = 36

Page 63 Written Exercises 1. x = −65 3. y = −96 5. x = −116.2 7. t = 60 9. v = 184 11. x = −83.2 13. s = −33 15. r = −0.544 17. a = −7 19. d = −8 21. x = −10.2 23. x = −9

25. $x = 8\frac{3}{4}$ 27. $r = 0$ 29. $z = -88$ 31. $k = 75$ 33. $w = -30$ 35. $m = 48$ 37. $x = -35$ 39. $p = 56$
41. $x = 0.2$, or $x = -0.2$ 43. $n = 21$, or $n = -21$

Page 64 **Review Capsule for Section 3.3** 1. 5n 2. 3r 3. b 4. t 5. x 6. y

Pages 64-65 **P-1** a. $t = 7$ b. $k = 8$ c. $r = -4$ **P-2** a. $b = -7$ b. $s = 25$ c. $d = 60$

Page 65 **Classroom Exercises** 1. Subtract 5. 3. Add 5.6. 5. Subtract 3. 7. Subtract 5. 9. Add 6. 11. Subtract 6.5. 13. Yes 15. No 17. No 19. Yes 21. No

Page 66 **Written Exercises** 1. $x = 7$ 3. $b = -13$ 5. $c = -22$ 7. $e = -19$ 9. $y = -8$ 11. $y = -3\frac{3}{4}$
13. $x = -56$ 15. $x = -9.6$ 17. $a = 150$ 19. $x = -5.5$ 21. $h = -300$ 23. $g = 1\frac{1}{2}$ 25. $z = -7\frac{1}{5}$
27. $c = 15$ 29. $x = -1.55$ 31. $j = 33$ 33. $b = 50$ 35. $r = -24$ 37. $x = 25\frac{2}{3}$ 39. $x = 10\frac{1}{2}$ 41. $k = \frac{5}{6}$
43. 16 inches

Page 68 **Classroom Exercises** 1. $t = 3$ hr 3. $r = 70$ ft per sec. 5. $\ell = 45$ kilometers

Pages 68-69 **Written Exercises** 1. 300,000 km per sec. 3. 2.3 seconds 5. 0.7 m 7. Correct 9. Correct

Page 69 **Mid-Chapter Review** 1. $b = -12$ 3. $y = 4.5$ 5. $t = 150$ 7. $a = -3$ 9. $z = -3\frac{1}{3}$ 11. $x = 13$
13. $y = 7$ 15. $t = -15$ 17. $j = 12$ 19. 2.5 seconds

Page 70 1. 3560 pounds 3. decrease 5. 4786.67 pounds

Page 71 **Review Capsule for Section 3.5** 1. $3n + 3$ 2. $2.7t$ 3. $-2\frac{1}{2}k - \frac{1}{2}$ 4. $14m - 20$ 5. $30 - 5x$
6. $45 + 30n$ 7. $-8k + 48$ 8. $60 + 36x$ 9. $\frac{1}{2}y + \frac{3}{4}$

Pages 71-72 **P-1** a. $m = 3$ b. $k = -7$ **P-2** a. $y = -1$ b. $r = 5$

Pages 72-73 **Classroom Exercises** 1. $4y + 6 = 13$ 3. $-5r + 8 = 17$ 5. $-9.7 = 3.1m - 4.8$
7. $12x - 3 + 6 = 10$; $12x + 3 = 10$ 9. $\frac{1}{3} - 2q + 10 + \frac{1}{4}q = 6$; $-1\frac{3}{4}q + 10\frac{1}{3} = 6$ 11. $\frac{3}{4}x - \frac{3}{2} + 6x + \frac{1}{2} = -11$; $6\frac{3}{4}x - 1 = -11$ 13. $-6 = 12 - 8t + 5t - 20$; $-6 = -8 - 3t$ 15. 7 17. -11 19. -3

Page 73 **Written Exercises** 1. $x = 7$ 3. $z = -0.4$ 5. $x = -8$ 7. $b = 7$ 9. $x = -12$ 11. $f = 14$
13. $k = -5\frac{5}{6}$ 15. $m = -38$ 17. $n = -1$ 19. $x = -\frac{1}{9}$ 21. $h = 11$ 23. $d = -5.1$ 25. $e = 3.8$ 27. $t = 35\frac{1}{2}$
29. $y = 2\frac{1}{2}$ 31. $x = 30$

Pages 75-76 **Classroom Exercises** 1. $n + 12$ 3. $\frac{n}{8}$ 5. $1800n$ 7. $\frac{1200}{n}$ 9. $26 - n$ (For Ex. 11-20, variables may vary.) 11. Let n = low temperature; $n + 27 = 50$; $n = 23°$ 13. Let p = total number of points; $\frac{p}{20} = 18$; $p = 360$ 15. Let n = number of passengers; $104 = n + 11$; $n = 93$ 17. Let r = average amount of rainfall; $r - 9.4 = 69.3$; $r = 78.7$ cm

Pages 76-77 **Written Exercises** (For Ex. 1-18, variables may vary.) 1. $2t$ 3. $v + 25{,}000$ 5. $m - 15.8$
7. $\frac{b}{28}$ 9. $a - 512$ 11. Let c = total cost; $\frac{c}{24} = 86$; $c = \$2064$ 13. Let ℓ = length of rectangle; $14.2 = \ell -$

5.8; $\ell = 20$ m 15. Let h = high temperature for the day; $-5.3 = h - 7.9$; $h = 2.6°C$ 17. Let a = amount of sale; $0.06a = 33.60$; a = \$560 19. Let s = Tim's weekly salary; $s + 283 = 561$; s = \$278 21. ℓ 23. n 25. o 27. k

Pages 79-80 Classroom Exercises (For Ex. 1-6, variables may vary.) 1. Let m = number of miles driven on Saturday; $m - 12 = 75$ 3. Let ℓ = loss for August; $\frac{\ell}{-1.2} = 5000$ 5. Let n = number of letters mailed on Monday; $2n - 16 = 132$

Pages 80-81 Written Exercises (For Ex. 1-11, variables may vary.) 1. Let m = number of cars washed in the morning; $m + 7 = 18$; $m = 11$ 3. Let ℓ = record low temperature; $-3.5\ell = 105$; $\ell = -30°F$ 5. Let t = thermometer reading; $4t + 5 = -23$; $t = -7°F$ 7. Let n = number of people attending last year; $2n - 12 = 326$; $n = 169$ 9. Let n = number of hours needed to complete a job; $\frac{n}{6} = 7.5$; $n = 45$ 11. Let ℓ = loss for 1985; $-2.5\ell = 30{,}000$; $\ell = -\$12{,}000$

Page 83 1. Alice: drum; Nathan: saxophone; Marie: flute

3.

	Baseball	Soccer	Track	Golf
Alex	X	X	X	✓
Dee	X	✓	X	X
Rod	✓	X	X	X
Sue	X	X	✓	X

Dee is a member of the soccer team.

Pages 84-86 Chapter Review 1. $x = -7$ 3. $b = 19$ 5. $f = -42$ 7. $t = 30.3$ 9. $k = -9$ 11. $m = -112$ 13. $t = 36$ 15. $y = -8.3$ 17. $z = 11\frac{1}{4}$ 19. $m = -10$ 21. $x = 7\frac{1}{3}$ 23. $v = -57.6$ 25. 19.2 m/sec 27. 83.9 inches 29. $x = 11$ 31. $y = 5$ 33. $t = -21.5$ 35. $p = \frac{2}{7}$ 37. Let d = total distance traveled; $d - 25.6 = 278.3$; $d = 303.9$ km 39. Let s = bowler's score; $\frac{s}{5} = 168$, $s = 840$ 41. Let n = number of rainy days last year; $n + 6 = 82$; $n = 76$ 43. Let t = thermometer reading; $3t - 7 = -31$; $t = -8°F$

Pages 87-88 Cumulative Review: Chapters 1-3 1. 0.07 3. −12 5. −22 7. −100 9. −12 11. 7 13. −28 15. 170 17. 4 19. −15 21. 2 23. −1 25. 29 27. 85 29. $-5, 0, 4, \frac{18}{3}$ 31. 0 33. $k - w + 18$ 35. $-8x$ 37. $-32x^2y$ 39. $30a^2b^2$ 41. $-48a + 6$ 43. $8x - 4y$ 45. $3(4r - 3)$ 47. $-6x(x + 3)$ 49. $16a - 8$ 51. $x = -4$ 53. $h = 4$ 55. $y = -54$ 57. $x = -\frac{17}{3}$ 59. $7\frac{1}{2}$ inches 61. $n = \frac{2}{3}$ 63. $x = 3$ 65. Let n = number of apples; $\frac{n}{17} = 5$; $n = 85$ 67. Let p = programs sold last week; $p - 54 = 192$; $p = 246$

CHAPTER 4 PROBLEM SOLVING: ONE VARIABLE

Pages 91-92 Classroom Exercises 1. Let p = number of seats in first class, then $p + 186$ = number of seats in tourist class. $p + p + 186 = 234$ 3. Let p = cost of labor, then $p + 3.58$ = cost of parts. $p + p + 3.58 = 75.62$ 5. Let p = city's rainfall in May, then $2p - 4\frac{3}{5}$ = city's rainfall in June. $p + 2p - 4\frac{3}{5} = 14\frac{9}{10}$ 7. Let p = amount earned first year, then $2p - 5000$ = amount earned second year. $p + 2p - 5000 = 93{,}000$ 9. Let p = number of true/false questions, then $3p - 7$ = number of multiple choice questions. $p + 3p - 7 = 105$

Pages 92-94 Written Exercises (For Ex. 1-16, variables may vary.) 1. Let x = the number of visiting runners, then x + 448 = the number of local runners. x + x + 448 = 3640; x = 1596, x + 448 = 2044 3. Let x = number of games in first set, then x + 4 = number of games in second set. x + x + 4 = 20; x = 8, x + 4 = 12 5. Let x = amount spent on food, then 2x − 150 = amount spent on housing and transportation. x + 2x − 150 = 8500; x = \$2883.33 7. Let x = fuel capacity of compact car, then 2x − 2.3 = fuel capacity of large car. x + 2x − 2.3 = 35.2; x = 12.5 gal., 2x − 2.3 = 22.7 gal. 9. Let x = amount of cargo space, then 5x + 5 = amount of passenger space. x + 5x + 5 = 131; x = 21 ft^3, 5x + 5 = 110 ft^3 11. Let x = number of hours worked the first week, then x + $2\frac{3}{4}$ = number of hours worked in second week. x + x + $2\frac{3}{4}$ = $45\frac{3}{4}$; x = $21\frac{1}{2}$, x + $2\frac{3}{4}$ = $24\frac{1}{4}$ 13. Let x = no. of foreign coins, then 3x − 26 = no. of U.S. coins. 3x − 26 + x = 998; x = 256, 3x − 26 = 742 15. Let x = number of records entered first day, then x + 12 = number of records entered second day. x + x + 12 = 256; x = 122, x + 12 = 134 17. 9 coins 19. 61 fish

Page 94 Review Capsule for Section 4.2 1. t = 26 2. w = 61.5 3. n = 4.2 4. p = 3.25 5. x = 6.8 6. r = 10.3 7. d; 4d 8. 7d; 2d 9. 5d, 4d, 2d 10. 37d, 9d, 4d

Page 95 P-1 a. 23 in b. 24 m c. 24 cm

Page 97 Classroom Exercises 1. 18 cm 3. 4s 5. Let m = width, m + 27 = length, 2m + 2(m + 27) = 174 7. Let m = width, m + 1.1 = length. 2m + 2(m + 1.1) = 11.4 9. Let m = length of second leg, 3m = length of third leg. 25.3 + m + 3m = 67.3 11. Let 4m, 5m, and 6m = distances between cities. 4m + 5m + 6m = 184.5 13. Let 20m = length, 9m = width. 2(20m) + 2(9m) = 232 15. Let 7m = length of guy wire, 5m = height of pole. 9.8 + 7m + 5m = 33.8

Pages 98-99 Written Exercises (For Ex. 1-16, variables may vary.) 1. Let x = height, x + 6 = length. 2x + 2(x + 6) = 84; x = 18 ft, x + 6 = 24 ft 3. Let x = shorter side, x + 1.7 = longer side. 8.5 + x + x + 1.7 = 51.2; x = 20.5 m, x + 1.7 = 22.2 m 5. Let sides be 5x, 12x, and 13x. 5x + 12x + 13x = 204; x = 6.8, 5x = 34 m, 12x = 81.6 m, 13x = 88.4 m 7. Let 4x, 7x, and 9x be the lengths of the legs. 4x + 7x + 9x = 684; x = 34.2, 4x = 136.8 km, 7x = 239.4 km, 9x = 307.8 km 9. Let x = width, 2x − 4 = length. 2x + 2(2x − 4) = 148; x = 26 ft, 2x − 4 = 48 ft 11. Let x = height, x + $2\frac{1}{2}$ = width. 2x + 2(x + $2\frac{1}{2}$) = 44; x = $9\frac{3}{4}$ in, x + $2\frac{1}{2}$ = $12\frac{1}{4}$ in 13. Let 18x = cable length, 11x = tower height. 18x + 11x + 35.6 = 108.1; x = 2.5, 11x = 27.5 m 15. Let 3x, 4x, and 5x = lengths of sides of triangle, 3x = length of each side of square. 3x + 4x + 5x + 4(3x) = 96; triangle: 3x = 12 cm, 4x = 16 cm, 5x = 20 cm; square: 3x = 12 cm

Page 99 Review Capsule for Section 4.3 1. $\left.\begin{matrix}33\\15\end{matrix}\right\}$ 2. $\left.\begin{matrix}-4\\-4\end{matrix}\right\}$ 3. $\left.\begin{matrix}17.5\\17.5\end{matrix}\right\}$ 4. $\left.\begin{matrix}16\\16\end{matrix}\right\}$ 5. $\left.\begin{matrix}6\frac{3}{4}\\6\frac{3}{4}\end{matrix}\right\}$ 6. $\left.\begin{matrix}-18\\-18\end{matrix}\right\}$

Pages 100-101 P-1 a. t = −2 b. j = −4 **P-2** a. p = −14 b. f = $1\frac{1}{3}$ **P-3** a. y = $1\frac{1}{2}$ b. g = −4

Page 101 Classroom Exercises 1. x − 2 = 5; −2 = −x + 5 3. 8 + 4a = 6; 8 = −4a + 6 5. −3d − 6 = 12; −6 = 12 + 3d 7. 5p + 2 = 6 or 2 = 6 − 5p 9. −8 = 4z + 5 or −4z − 8 = 5 11. 10.8 = 9.2 + 10.5j or 10.8 − 10.5j = 9.2

Page 102 Written Exercises 1. x = 17 3. a = 13 5. c = −12 7. k = $-6\frac{2}{3}$ 9. x = 7 11. z = −5 13. p = $-21\frac{1}{4}$ 15. t = −0.8 17. x = −14 19. y = −11 21. n = 3.75 23. y = $-\frac{5}{9}$ 25. y = $-8\frac{1}{2}$ 27. q = 2 29. k = −3 31. d = −6 33. d = 6 35. p = −5 37. x = b − 2a; odd

Page 103 Mid-Chapter Review (For Ex. 1-10, variables may vary.) 1. Let x = number of boys, then x + 7 = number of girls. x + x + 7 = 45; x = 19, x + 7 = 26 3. Let x = number of cars the second hour, then x − 2900 = number of cars the first hour. x + x − 2900 = 24,500; x = 13,700, x − 2900 = 10,800 5. Let x = width, then 2x + 6.2 = length. 2(x) + 2(2x + 6.2) = 106.6; x = 15.7 m, 2x + 6.2 = 37.6 m 7. Let 2x, 4x, and 5x be the lengths of the sides. 2x + 4x + 5x = 264; x = 24, 2x = 48 in, 4x = 96 in, 5x = 120 in 9. d = −14 11. p = 4 13. x = 1

Page 103 Review Capsule for Section 4.4 1. $115.20 2. $7 3. 139 4. $11.98 5. 42 6. 42.5 7. $33.75 8. 305

Page 106 Classroom Exercises 1. $128 3. $3.80 5. $90 7. $450; $54 9. Monthly Income: x + 240, Amount Spent: 0.30x, 0.25(x + 240); 0.30x = 0.25(x + 240)

Page 107 Written Exercises 1. $575 3. $60; $12 (For Ex. 5-7, variables may vary.) 5. Let x = number of games last season, x + 4 = number of games this season. 0.55x = 0.50(x + 4); x = 40, x + 4 = 44 7. Let x = number who tried out for girls' team, x + 4 = number for boys' team. 0.75x = 0.60(x + 4); x = 16, x + 4 = 20

Page 110 Classroom Exercises 1. Rate: x − 6, Distance: 6x, 6.4(x − 6); 6x = 6.4(x − 6) 3. Fuel Tank Capacity: x − 2, Range: 10x, 12(x − 2); 10x = 12(x − 2) − 112

Page 111 Written Exercises

1.

	r	t	d
Going	x + 15	$2\frac{4}{5}$	$2\frac{4}{5}(x + 15)$
Returning	x	4	4x

$2\frac{4}{5}(x + 15) = 4x$,
x = 35,
x + 15 = 50 mi/hr

3.

	r	t	d
Going	x	1	x
Returning	x − 7	$1\frac{1}{4}$	$1\frac{1}{4}(x - 7)$

$x = 1\frac{1}{4}(x - 7)$,
x = 35 mi/hr,
x − 7 = 28 mi/hr

5.

	Fuel Capacity	Mileage	Range
Car A	x	5	5x
Car B	x − 35	8.5	8.5(x − 35)

5x = 8.5(x − 35),
x = 85 L,
x − 35 = 50 L

Page 112 1. When r = 3.7, S = 277.3. When r = 3.8, S = 277.0. When r = 3.9, S = 277.1. Thus, to the nearest tenth, 3.8 is the value of r corresponding to the smallest value of S.

Pages 113-114 Chapter Review 1. Let x = number of non-fiction books, x + 82 = number of fiction books. x + x + 82 = 372, x = 145, x + 82 = 227 3. Let x = number of miles father drove, then 2x + 100 = number of miles Alice drove. x + 2x + 100 = 1150, x = 350 mi, 2x + 100 = 800 mi 5. Let x = width, x + 11.8 = length. 2(x) + 2(x + 11.8) = 80, x = 14.1 m, x + 11.8 = 25.9 m 7. Let length = 7x, width = 3x. 2(7x) + 2(3x) = 166, x = 8.3, 7x = 58.1 mm, 3x = 24.9 mm 9. h = −7 11. $y = 9\frac{1}{2}$ 13. $t = 1\frac{1}{3}$ 15. g = 15 17. $457.50 19. Let x = Russell's monthly income, x + 200 = Brad's monthly income. 0.32x = 0.28(x + 200); x = $1400, x + 200 = $1600

21.

	r	t	d
Going	x − 5	4.25	4.25(x − 5)
Returning	x	4	4x

4.25(x − 5) = 4x,
x = 85 km/hr,
x − 5 = 80 km/hr

CHAPTER 5 INEQUALITIES

Pages 116-117 **P-1** To show that all integers less than 2 are solutions. **P-2** 2 is equal to 2. **P-3** 3 is not equal to 2. **P-4** To show that 3 is a solution of $x \le 3$. **P-5** a. {all real numbers greater than or equal to $3\frac{1}{2}$} b. {all real numbers less than $-2\frac{1}{2}$}; Assume {real numbers} as the domain.

Page 118 **Classroom Exercises** 1. {all integers greater than -2} 3. {all real numbers less than $2\frac{1}{2}$} (For Ex. 5-10, answers will vary.) 5. 0; 4; 9 7. -1.2; 0; 2.1 9. -5; $3\frac{1}{4}$; $\frac{7}{8}$

Pages 118-119 **Written Exercises** 1. $\{-2, -1, 0, 1\}$ 3. {all integers less than 1} 5. {all real numbers greater than -2} 7. {all real numbers less than or equal to -1} 9. {all real numbers greater than or equal to $-1\frac{1}{2}$} 11. On a number line, the points 2, 3, 4, 5, etc. 13. On a number line, the points -3, -4, -5, -6, etc. 15. On a number line, all points to the left of, and not including, $-\frac{3}{4}$. 17. On a number line, all points to the right of, and including, 2.7. 19. No 21. Yes 23. Yes 25. Yes

Page 119 **Review Capsule for Section 5.2** 1. 0.3, 0.5; $\frac{5}{10} > \frac{3}{10}$ 2. 0.75, 0.25; $\frac{3}{4} > \frac{1}{4}$ 3. 0.4, -0.6; $\frac{2}{5} > -\frac{3}{5}$ 4. 0.75, 0.875; $\frac{7}{8} > \frac{6}{8}$ 5. -0.4, -0.5; $-\frac{4}{10} > -\frac{5}{10}$ 6. -0.375, -0.625; $-\frac{3}{8} > -\frac{5}{8}$ 7. $0.666\cdots$, $0.8333\cdots$; $\frac{5}{6} > \frac{4}{6}$ 8. $-0.333\cdots$, $-0.666\cdots$; $-\frac{1}{3} > -\frac{2}{3}$

Pages 120-121 **P-1** $a > b$ **P-2** $a < b$ **P-3** a. $6 > -y$; $-y < 6$ b. $-t < 13$; $13 > -t$ c. $-s < r$; $r > -s$ **P-4** $r < 2$; $2 > r$ **P-5** $2 < t$; $t > 2$ **P-6** $r < t$; $t > r$ **P-7** $x < y$; $y > x$ **P-8** If $a > b$ and $b > c$, then $a > c$.

Page 121 **Classroom Exercises** For Ex. 1-7, only one of two possible inequalities is given for each pair. 1. $-5 < 2$ 3. $-12 < -9$ 5. $-16 < 6$ 7. $1.3 > 1.03$ 9. $<$ 11. $>$

Page 122 **Written Exercises** 1. $-5 < 4$ 3. $-12 < -11$ 5. $1.6 < 1.7$ 7. $-\frac{7}{8} < -\frac{3}{4}$ 9. $-3 > b$ 11. $a > 2$ 13. $a + b > -2$ 15. $b < -a$ 17. True 19. False 21. False 23. True 25. $y < x$ 27. $x < y$ 29. $y < x$ 31. $x < y$

Page 122 **Review Capsule for Section 5.3** 1. $t = 135$ 2. $x = -15$ 3. $y = 13$ 4. $w = 7.9$ 5. $t = -6.1$ 6. $r = 1\frac{5}{8}$ 7. $d = -\frac{5}{6}$ 8. $p = 3.86$ 9. $s = -0.53$

Pages 124-125 **Classroom Exercises** 1. Subtract 5. 3. Subtract 7. 5. Add 0.8. 7. {all real numbers less than 11} 9. {all real numbers less than 8} 11. {all real numbers less than 4} 13. {all real numbers less than 4} 15. {all real numbers greater than -8} 17. {all real numbers less than 8}

Pages 125-126 **Written Exercises** 1. No 3. Yes 5. No 7. Yes 9. No 11. {all real numbers greater than -5} 13. {all real numbers less than -7} 15. {all real numbers less than -1} 17. {all real numbers less than -3.1} 19. {all real numbers greater than 6} 21. {all real numbers greater than 7.9} 23. {all integers greater than -1} 25. {all integers greater than 7} 27. {all integers greater than or equal to -5} 29. $b < a$ 31. Fill the 9-quart pail. Fill the 3-quart pail from the 9-quart pail. There are 6 quarts remaining in the 9-quart pail. Fill the 2-quart pail from the 3-quart pail. Pour out the 1 quart remaining in the 3-quart pail. There are now exactly 8 quarts. 33. First bag: 27; second bag: 21; third bag: 11; fourth bag: 16; fifth bag: 25

Page 126 Mid-Chapter Review 1. On a number line, all points to the right of, and not including, −4. 3. On a number line, the points 4, 3, 2, 1, 0, −1, etc. 5. On a number line, the numbers 6, 5, 4, 3, 2, 1, 0. 7. $a < b$ 9. $b < a$ 11. $b < a$ 13. $t < 5$ 15. $x > -1.2$ 17. $r < -0.3$

Page 126 Review Capsule for Section 5.4 1. $n = 1\frac{4}{5}$ 2. $t = 2\frac{2}{3}$ 3. $w = -7$ 4. $y = -10$ 5. $p = 1\frac{2}{3}$ 6. $m = 5$

Page 128 Classroom Exercises 1. $-3 < x + 2$ 3. $a - 2 < -7$ 5. $-y + \frac{1}{2} > -12\frac{1}{4}$ 7. $3w - 5 - 2w < 2w + 12 - 2w$ 9. $3\frac{1}{2}x - 2 - 2\frac{1}{2}x > 2\frac{1}{2}x + 10 - 2\frac{1}{2}x$ 11. $0.6m - 0.7 + 0.4m < -0.4m + 0.2 + 0.4m$ 13. $-3x + 6 < -5 - 2x$ 15. $5 - x < -2x + 6 - 2$ 17. $5y - 2 - 3y < 3y - 4$ 19. $a - 5 < a + 2$ 21. $5x - 5 < 2x - 6 + 3x$

Page 128 Written Exercises 1. {all real numbers greater than −5} 3. {all real numbers less than −5} 5. {all real numbers less than $12\frac{3}{4}$} 7. {all real numbers less than 17} 9. {all real numbers greater than 12} 11. {all real numbers less than 0.9} 13. {all real numbers greater than 11} 15. {all real numbers less than −1} 17. {all real numbers greater than 2} 19. {real numbers} 21. ϕ

Page 128 Review Capsule for Section 5.5 1. $x = 36$ 2. $n = -84$ 3. $t = 60$ 4. $r = -16$ 5. $w = -5.6$ 6. $p = -5\frac{1}{2}$ 7. $q = 4\frac{1}{5}$ 8. $s = -26$

Pages 129-130 P-1 a. 2 b. −3 c. $\frac{5}{4}$ d. $-\frac{3}{2}$ **P-2** Dividing by 3: a. $x < 3$ b. $-2x > 2$ c. $4x < -1$ d. $-5x < \frac{1}{3}$ Dividing by −3: a. $-x > -3$ b. $2x < -2$ c. $-4x > 1$ d. $5x > -\frac{1}{3}$

Pages 130-131 Classroom Exercises 1. $5x > 10$ 3. $10a > 50$ 5. $-3y < -6$ 7. $r > -6$ 9. Divide by 2. 11. Multiply by −2. 13. $>$ 15. $<$

Page 131 Written Exercises 1. {all real numbers less than 5} 3. {all real numbers greater than −20} 5. {all real numbers greater than −3} 7. {all real numbers greater than −2} 9. {all real numbers less than −6} 11. {all real numbers less than 6} 13. {all real numbers less than −9} 15. {all real numbers less than $-\frac{11}{3}$} 17. {all real numbers less than −10} 19. {all real numbers greater than −3} 21. The box marked "GB."

Pages 132-133 P-1 The order of an inequality is unchanged if each side is multiplied by a positive number. **P-2** Each side is multiplied by a negative number. **P-3** {all real numbers less than 6} **P-4** {real numbers}; ϕ

Pages 133-134 Classroom Exercises 1. $2x + 1 - 1 > 3 - 1$ 3. $\frac{1}{2}x - 3 + 3 < -7 + 3$ 5. $5 - 2x - 4x > 4x - 3 - 4x$ 7. $10x - 30 < -10$ 9. $-5x + 3 > -2x - 6 - 6$ 11. {all real numbers greater than −2} 13. {all real numbers less than 1} 15. {all real numbers greater than −9} 17. {all real numbers greater than −1} 19. {all real numbers less than −3} 21. {real numbers} 23. {all real numbers greater than −4} 25. {all real numbers less than 6}

Page 134 Written Exercises 1. {all real numbers less than $\frac{8}{3}$} 3. {all real numbers less than $-\frac{13}{2}$} 5. {all real numbers greater than 8} 7. {all real numbers greater than $\frac{7}{2}$} 9. {all real numbers less than −2} 11. {all real numbers greater than $-\frac{1}{5}$} 13. {all real numbers greater than 6} 15. {all real numbers greater than 0} 17. {all real numbers less than 2} 19. ϕ 21. {real numbers}

Page 136 Classroom Exercises 1. $c < 60$ 3. $2d - 3 > 20$ 5. $\frac{32 + 45 + x}{3} < 42$ 7. $\frac{65 + 70 + 75 + x}{4} > 72$

Page 137 Written Exercises 1. Let x = cost of tennis racket, then x − 50 = cost of tennis shoes. $x + x - 50 < 120$; $x < 85$, \$84.99 3. Let x = cost of radio, then 4x + 30 = cost of television. $x + 4x + 30 > 525$; $x > 99$; Yes 5. Let x = score on fourth quiz. $\frac{85 + 72 + 81 + x}{4} > 80$; $x > 82$, 83 7. Let x = score in fourth game. $\frac{95 + 130 + 125 + x}{4} > 110$; $x > 90$, 91

Pages 138-139 1.

2	5, 7, 7, 8
3	1, 1, 2, 2, 3, 5, 5, 6
4	0, 0, 0

3. 72 5. a. Answers will vary. b. Reasons will vary. 7. Median: 188; Mode: 196 9. Theater A; Theater A

Pages 140-142 Chapter Review 1. $\{-3, -2, -1, 0\}$ 3. {all real numbers less than $1\frac{1}{2}$} 5. On a number line, the points 4, 3, 2, 1, 0. 7. On a number line, all points to the right of, and not including, −3. 9. All points to the left of, and not including, 4. 11. $-r > 12$ 13. $-a < 3$ 15. $x < 4.2$ 17. $1 < x$ 19. $c < a$; $a > c$ 21. True 23. No 25. No 27. Yes 29. {all real numbers less than 1} 31. {all real numbers less than −2} 33. {all real numbers less than −9.7} 35. {all real numbers less than −8} 37. {all real numbers greater than 11} 39. {all real numbers less than $7\frac{1}{2}$} 41. {all real numbers less than −5} 43. {all real numbers less than 4} 45. {all real numbers greater than 1.6} 47. {all real numbers greater than $10\frac{1}{3}$} 49. ϕ 51. Let x = cost of pants, then 3x + 20 = cost of coat. $x + 3x + 20 < 105$; $x < 21.25$, \$21.24 53. Let x = score on fourth test. $\frac{85 + 93 + 78 + x}{4} > 85$; $x > 84$, 85

CHAPTER 6 POWERS

Pages 144-145 **P-1** a. $2 \cdot 2 \cdot 2 \cdot 2$ b. $x \cdot x \cdot x$ c. $y \cdot y \cdot y \cdot y \cdot y$ d. $4 \cdot 4 \cdot 4$ e. $z \cdot z$ **P-2** 5 **P-3** $(-3) \cdot (-3)$; $(-3) \cdot (-3) \cdot (-3) \cdot (-3)$ **P-4** a + b times **P-5** a. $(-1.9)^5$ b. $(\frac{1}{2})^7$ c. t^9 **P-6** a. $t^6 + t^5$ b. $3k^5 - 2k^4 + 3k^3$ **P-7** x = 4

Pages 145-146 Classroom Exercises 1. 2^5 3. $(-5)^6$ 5. $(\frac{1}{2})^5$ 7. k^{22} 9. 7^6 11. 2^7 13. x = 5 15. x = 6 17. x = 4

Page 146 Written Exercises 1. 12^{10} 3. 5^4 5. $(3.2)^{12}$ 7. $(-15)^9$ 9. $(-6)^{11}$ 11. $(-\frac{2}{3})^7$ 13. a^7 15. r^9 17. a^{11} 19. 5^9 21. $y^3 + y^4$ 23. $x^9 + x^7$ 25. $2^5 - 2^7$ 27. $n^4 - n^3 + n^2$ 29. x = 11 31. x = 5 33. x = 5 35. x = 6

Page 146 Review Capsule for Section 6.2 1. True 2. True 3. False 4. False

Pages 147-148 **P-1** $x \cdot x \cdot x \cdot x \cdot x$; $x \cdot x \cdot x$ **P-2** a. 17^3 b. $(1.8)^5$ c. m^7 **P-3** 113^0 **P-4** 1 **P-5** 1 **P-6** a. 2 b. 0 c. 1

Page 149 Classroom Exercises 1. 3^3 3. $(-3)^4$ 5. a^2 7. 2^3 9. $(\frac{1}{2})^7$ 11. $(-5)^{t-4}$ 13. False 15. False 17. True 19. 1 21. 1 23. 1 25. 1

Pages 149-150 Written Exercises 1. 10^2 3. $(\frac{1}{2})^1$, or $\frac{1}{2}$ 5. x^4 7. 2^3 9. $(0.6)^1$, or 0.6 11. π^3 13. True 15. False 17. False 19. True 21. True 23. x = 5 25. x = 10 27. x = 5 29. x = 4 31. x = 1 33. y = 1 35. a = 0 37. x = 0 39. p = any nonzero real number 41. −1 43. 5 45. −1 47. Sum: 2775; 100 + 330 + 505 + 077 + 099 = 1111

Pages 151-152 **P-1** a. $\frac{1}{r^3}$ b. $\frac{1}{m^6}$ c. $\frac{1}{3^2}$ or $\frac{1}{9}$ d. $\frac{1}{2^4}$ or $\frac{1}{16}$ e. $\frac{1}{10^3}$ or $\frac{1}{1000}$ **P-2** Add the exponents. Keep the same base. **P-3** a. 17^{-1} b. $(1.2)^4$ c. r^{-3} **P-4** Subtract the exponents. Keep the same base. **P-5** a. 5^2 b. $(3.7)^{-3}$ c. y^2 d. x^5 e. r^{-4}

Page 153 **Classroom Exercises** 1. $\frac{1}{5^{10}}$ 3. $\frac{1}{(5.2)^2}$ 5. $\frac{1}{r^8}$ 7. x^7 9. 2^2

Pages 153-154 **Written Exercises** 1. $\frac{1}{r^4}$ 3. $\frac{2}{y^5}$ 5. $\frac{1}{4^2}$ or $\frac{1}{16}$ 7. $\frac{1}{(-3)^3}$ or $-\frac{1}{27}$ 9. $\frac{a}{b^3}$ 11. 2^5 or 32 13. a^{-3} or $\frac{1}{a^3}$ 15. 2 17. a^2 19. x^{-5}; $\frac{1}{x^5}$ 21. n^{-1}; $\frac{1}{n}$ 23. r^3 25. t^3 27. p^{-2}; $\frac{1}{p^2}$ 29. 10^4 31. y^{-2}; $\frac{1}{y^2}$ 33. r^0 ; 1 35. 0.0009766 37. 0.0003925 39. 6.1096817 41. 0.0000485 43. positive 45. positive

Page 154 **Mid-Chapter Review** 1. 12^{13} 3. $(-6)^{14}$ 5. $(\frac{3}{4})^{13}$ 7. $(1.4)^7$ 9. 21^2 11. $(\frac{3}{4})^3$ 13. 1 15. 0 17. $(0.6)^{-3}$; $\frac{1}{(0.6)^3}$ 19. w^{-6}; $\frac{1}{w^6}$ 21. 12^{-2}; $\frac{1}{12^2}$ 23. m^{-3}; $\frac{1}{m^3}$

Page 155 1. 490 seconds 3. 4 hours 5. c

Pages 156-157 **P-1** 2 **P-2** 8 **P-3** a. x^2y^2 b. $9a^2$ c. $8r^3$ d. $9r^2$ **P-4** a. $\frac{9}{t^2}$ b. $\frac{p^4}{q^4}$ c. $\frac{8r^3}{s^3}$ d. $\frac{1}{r^4}$

Page 158 **Classroom Exercises** 1. $4a^2$ 3. $\frac{1}{4}y^2$ 5. $25n^2$ 7. $4b^2$ 9. $0.25a^2$ 11. $-8r^3$ 13. x^2y^2 15. a^4b^4 17. $\frac{9}{x^2}$ 19. $\frac{m^5}{n^5}$ 21. $\frac{x^2}{y^2}$ 23. $-\frac{8}{x^3}$

Page 158 **Written Exercises** 1. $27r^3$ 3. $4a^2$ 5. $-27p^3$ 7. a^2b^2 9. $-r^3s^3$ 11. $4x^2y^2$ 13. $9p^2q^2$ 15. $-8a^3b^3$ 17. $\frac{8}{x^3}$ 19. $-\frac{a^3}{8}$ 21. $\frac{x^2}{y^2}$ 23. $\frac{y}{2}$ 25. $\frac{4}{pq}$ 27. $-\frac{1}{3t}$ 29. $\frac{27a^3}{b^3}$

Page 158 **Review Capsule for Section 6.5** 1. m^{12} 2. $(-3)^6$ 3. t^{-9} 4. $(2a)^{-6}$ 5. g^{-20} 6. $(-0.5)^6$

Pages 159-161 **P-1** $(x^2)(x^2)(x^2)$ **P-2** $2 \cdot 3 = 6$ **P-3** $(-4)(-3) = 12$ **P-4** a. m^{15} b. k^8 c. t^{-4} **P-5** a. 512 b. $\frac{1}{2048}$ c. $\frac{1}{1024}$ d. m^5

Page 161 **Classroom Exercises** 1. 3^6 3. $\frac{1}{a^3}$ 5. y^6 7. $(0.5)^8$ 9. $(-3)^{24}$ 11. $\frac{1}{t^6}$ 13. $\frac{q^4}{9}$ 15. $\frac{1}{x^2}$ 17. $\frac{1}{r^6s^9}$ 19. $\frac{1}{x^6y^9}$ 21. 1 23. $\frac{s^8}{m^4}$

Page 162 **Written Exercises** 1. a^8 3. $\frac{1}{y^{10}}$ 5. m^5 7. $\frac{1}{t^4}$ 9. $9r^6$ 11. $8a^6b^3$ 13. $32x^5y^{10}$ 15. $\frac{x^{15}}{32}$ 17. $\frac{32a^{10}b^{15}}{c^{105}}$ 19. $-8t^9$ 21. $\frac{1}{8x^3y^6}$ 23. $\frac{m^2}{4n^3}$ 25. $\frac{64r^2}{s^6}$ 27. $-\frac{y^5}{32x^{10}z^{15}}$ 29. $0.0000128m^{28}n^{35}$ 31. 120 miles

Page 162 **Review Capsule for Section 6.6** 1. a. 1.5678 b. 156.78 2. a. 826.605027 b. 82,660.5027 3. a. 48,278.3 b. 4,827,830 4. a. 5,280,000 b. 528,000,000 5. a. 67.5928 b. 0.675928 6. a. 502.65 b. 5.0265 7. a. 0.000018 b. 0.00000018 8. a. 16,000 b. 160

Pages 163-164 **P-1** a. N = 5.8; a = 3 b. N = 3.76; a = −1 c. N = 1.0059; a = −6 **P-2** The exponent shows how many places to "move" the decimal point.

Page 165 Classroom Exercises 1. 300 3. 7030 5. 46.2 7. 50,001 9. 5.3×10^2 11. 4.86×10^1 13. 7.6×10^4 15. 4.83×10^{-4}

Page 165 Written Exercises 1. 4200 3. 0.00029 5. 86,200 7. 0.100726 9. 7,280,000 11. 0.00005607 13. 12,800,000 15. 0.000000022 17. 9.2×10^{-3} 19. 2.705×10^3 21. 1.734×10^{-2} 23. 2.7×10^7 25. 2.75×10^{-5} 27. 2×10^6 29. 2.976×10^5 31. 1.5×10^8

Page 166 1. 13, 17, 21, 25 3. 162, 486, −1458, −4374 5. −21.5, −26, −30.5, −35 7. $\frac{1}{13}, \frac{1}{15}, \frac{1}{17}, \frac{1}{19}$ 9. $\frac{1}{49}, \frac{1}{64}, \frac{1}{81}, \frac{1}{100}$ 11. 169, 225, 289, 361 13. −30, −44, −60, −78 15. 190, 382, 766, 1534

Page 167 1. 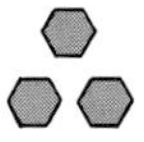 3.  5.  7. 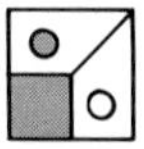 9.  11. 

Pages 168-170 Chapter Review 1. 27^9 3. r^7 5. t^9 7. 19^5 9. $(2.7)^2$ 11. $(\sqrt{7})^3$ 13. 1 15. 1 17. −1 19. $\frac{1}{t^{10}}$ 21. $\frac{5}{x^2}$ 23. $\frac{q}{p^2}$ 25. $k^{-2}; \frac{1}{k^2}$ 27. $m^{-5}; \frac{1}{m^5}$ 29. $p^{-5}; \frac{1}{p^5}$ 31. $16t^2$ 33. $-32r^5$ 35. $\frac{25}{p^2}$ 37. $\frac{1}{16m^2}$ 39. $\frac{1}{y^3}$ 41. $\frac{1}{g^{10}}$ 43. $\frac{27m^6n^9}{p^3}$ 45. $\frac{x^6}{64y^3z^{12}}$ 47. 31,870 49. 0.005279 51. 8000.02 53. 22,240 55. 5.219×10^{-3} 57. 5.12×10^8 59. 3.1536×10^7

Pages 171-172 Cumulative Review: Chapters 1-6 1. 8 3. −18 5. 40 7. 63 9. $-36x^2y^2$ 11. $10y^2 + 55y$ 13. $4(2c - 3)$ 15. $-4a(a + 1)$ 17. $-4m - 3$ 19. $y = -9$ 21. $k = -9$ 23. $h = 2$ 25. Let x = points scored in second game, then x − 15 = points scored in first game. $x + x - 15 = 309$; $x = 162$, $x - 15 = 147$ 27. $x = 3$ 29. \$114

31.

	field goals attempted	per cent field goals made	field goals made
this season	x + 6	60%	0.60(x + 6)
last season	x	75%	0.75x

$0.60(x + 6) = 0.75x$; $x = 24$, $x + 6 = 30$ 33. On a number line, the points to the left of, and not including, 4. 35. On a number line, the points to the right of, and including, −3.7. 37. {all real numbers less than −6} 39. {all real numbers greater than −4} 41. {all real numbers less than −25} 43. {all real numbers greater than 4} 45. {all real numbers greater than 5} 47. Let x = score on fifth test. $\frac{73 + 87 + 85 + 82 + x}{5} > 84$; $x > 93$, 94 49. b^{16} 51. x^7 53. r^3 55. m^6 57. $\frac{16}{x^2}$ 59. $\frac{27x^{12}w^3}{y^{15}}$ 61. 0.000412 63. 730,000

CHAPTER 7 ROOTS

Page 174 P-1 6; −6 **P-2** 0 **P-3** There are no two equal factors of −16. **P-4** a. 49 b. $\frac{1}{4}$ c. 0.09 d. 400 e. $\frac{9}{16}$ **P-5** a. −7 b. $\frac{1}{2}$ c. −0.3 d. 100 **P-6** It has more than two counting-number factors.

Page 176 Classroom Exercises 1. 100 3. 144 5. $\frac{1}{9}$ 7. 0.16 9. $\frac{25}{4}$ 11. 3 13. 0 15. 10 17. $-\frac{1}{4}$ 19. −0.1 21. $3 \cdot 7$ 23. $2 \cdot 2 \cdot 2$ 25. $2 \cdot 13$ 27. $7 \cdot 7$ 29. $3 \cdot 13$ 31. $3 \cdot 3 \cdot 5$

Page 176 Written Exercises 1. 11 3. −10 5. $\frac{2}{5}$ 7. $-\frac{1}{3}$ 9. 0.9 11. 10 13. 12 15. 50 17. $5 \cdot 13$ 19. $2 \cdot 3 \cdot 3$ 21. $2 \cdot 3 \cdot 3 \cdot 3$ 23. $2 \cdot 2 \cdot 2 \cdot 3 \cdot 5$ 25. $2 \cdot 3 \cdot 3 \cdot 11$ 27. 22 29. 28 31. 44 33. 56 35. 76 37. 72 39. 95 41. 62

Pages 177-178 P-1 a. Irrational b. Rational c. Irrational d. Rational **P-2** a. 29 b. $\frac{1}{4}$ c. 1029 **P-3** a. $\sqrt{35}$ b. $\sqrt{66}$ c. 6

Page 179 Classroom Exercises 1. $\sqrt{10}$ 3. 8 5. 10 7. $\sqrt{2}$ 9. $\frac{1}{3}$ 11. 0 13. 20 15. −2

Page 179 Written Exercises 1. 13 3. 12 5. 54 7. 6 9. $\sqrt{51}$ 11. $-\sqrt{10}$ 13. $\sqrt{30}$ 15. $\sqrt{5}$ 17. $\frac{1}{2}$ 19. 4 21. 10 23. $\frac{1}{6}$ 25. 6 27. $\sqrt{30}$ 29. 10 31. $x = 6$ 33. $x = \sqrt{14}$ 35. $x = 6$ or $x = -6$ 37. $x = \sqrt{14}$ or $x = -\sqrt{14}$ 39. $a = b = c = 1$

Pages 180-181 P-1 a. 7 b. −9 c. 5 d. −13 **P-2** a. 2 b. 6 c. 5 d. 4

Pages 181-182 Classroom Exercises 1. 2 3. −5 5. 4 7. 25 9. 11 11. 36 13. 10 15. 14 17. 30 19. $3\sqrt{5}$ 21. 3 23. 6 25. 7 27. $5\sqrt[3]{9}$ 29. $3\sqrt[4]{20}$

Page 182 Written Exercises 1. 6 3. 38 5. $6\sqrt{3}$ 7. $35\sqrt{35}$ 9. $4\sqrt{6}$ 11. $12\sqrt{10}$ 13. $6\sqrt{5}$ 15. $10\sqrt{21}$ 17. $2\sqrt{2}$ 19. $3\sqrt{3}$ 21. $2\sqrt{6}$ 23. $2\sqrt{10}$ 25. $-4\sqrt{3}$ 27. $2\sqrt{30}$ 29. $4\sqrt{6}$ 31. $3\sqrt{30}$ 33. 2 35. $3\sqrt[3]{3}$ 37. $2\sqrt[4]{9}$ 39. $5\sqrt[3]{10}$ 41. $5\sqrt[4]{2}$ 43. −10 45. $-2\sqrt[3]{7}$ 47. 16

Pages 183-184 P-1 a. 10 b. $\frac{1}{2}$ c. $\sqrt{17}$ d. $\frac{2}{3}$ **P-2** A negative number cannot have a real number as a square root. **P-3** a. a^3 b. x^4 c. b^6 d. y^{14} **P-4** a; x^2; y^3 **P-5** a. No b. Yes c. Yes d. No

Pages 184-185 Classroom Exercises 1. a 3. ab 5. xy^2 7. 3y 9. 5n 11. $3x^3$ 13. $x^2\sqrt{x}$ 15. $2a\sqrt{2a}$ 17. $xy\sqrt{2x}$ 19. $x\sqrt{6}$ 21. rt 23. x^4 25. $2k^2$

Pages 185-186 Written Exercises 1. 6y 3. yz^3 5. $x^2\sqrt{x}$ 7. $2\sqrt{x}$ 9. $4y\sqrt{y}$ 11. $2x\sqrt{2}$ 13. $2a^2\sqrt{2a}$ 15. $2\sqrt{3ab}$ 17. $7ab\sqrt{ab}$ 19. $2ab\sqrt{6b}$ 21. $5r^2s^3\sqrt{3r}$ 23. $2rst^2\sqrt{5st}$ 25. x^4 27. $x^4\sqrt[3]{x}$ 29. $3xy^2$ 31. $2b\sqrt[3]{2a^2}$ 33. $3mn^2\sqrt[3]{2m^2n^2}$ 35. $2r^2s^3\sqrt[3]{4rs}$ 37. x^2 39. $2\sqrt{x}$ 41. $2x^2$ 43. $x\sqrt[3]{12}$ 45. $3a^2$ 47. $4b^3$ 49. 8 51. 6.5574385 53. 6

Page 186 Mid-Chapter Review 1. 221 3. 1925 5. 27 7. 36 9. 48 11. 21 13. $-\sqrt{14}$ 15. 5 17. $\sqrt{210}$ 19. $7\sqrt{10}$ 21. $10\sqrt{7}$ 23. $9\sqrt{5}$ 25. $3\sqrt[3]{4}$ 27. $3\sqrt[4]{2}$ 29. $7m^3n$ 31. $3r^2s^2\sqrt{3s}$ 33. $6ab^2c^3\sqrt{3abc}$ 35. $3p^2q^2\sqrt[3]{2p}$

Page 186 Review Capsule for Section 7.5 1. 17m 2. 2.7t 3. 77k 4. −0.3w

Pages 187-188 P-1 a. $(7 + a)x$ b. $(t - r)y$ c. $(11 + 3)m$ **P-2** $11\sqrt{2}$ **P-3** a. $5\sqrt{5}$ b. $7\sqrt{3}$ c. $-3\sqrt{10}$ **P-4** 2, 2, and 3 **P-5** a. $\sqrt{3}$ b. $\sqrt{15}$ **P-6** $2 \cdot 2 \cdot 2$; $2 \cdot 5 \cdot 5$

Pages 188-189 Classroom Exercises 1. $7\sqrt{3}$ 3. $4\sqrt{2}$ 5. $12\sqrt{10}$ 7. $3\sqrt{2}$ 9. $-2\sqrt{11}$ 11. $5\sqrt{x}$ 13. $\frac{1}{2}\sqrt{a}$ 15. $10\sqrt{m}$ 17. 5

Page 189 Written Exercises 1. $15\sqrt{3}$ 3. $4\sqrt{2}$ 5. $-3\sqrt{3}$ 7. $4\sqrt{2}$ 9. $2\sqrt{11}$ 11. $\sqrt{5} + \sqrt{15}$ 13. $\sqrt{19} - \sqrt{3}$ 15. $12\sqrt{x}$ 17. $-5\sqrt{t}$ 19. $2\sqrt{s}$ 21. $11\sqrt{2}$ 23. 0 25. $5\sqrt{3}$ 27. $7\sqrt{5}$ 29. $11\sqrt{2}$ 31. $5\sqrt{7a}$ 33. $4\sqrt{2r}$ 35. $12\sqrt{y}$ 37. $7\sqrt{3}$ 39. $\sqrt{7}$ 41. 0

Page 189 Review Capsule for Section 7.6 1. 6 2. $5x^2$ 3. 13x 4. $\frac{1}{3}x^2$

Pages 190-191 P-1 a. $\frac{2}{3}$ b. $\frac{2}{3}$ **P-2** a. $\frac{n}{3}$ b. $\frac{\sqrt{2}}{x}$ c. $\frac{r\sqrt{r}}{4}$ d. $\frac{8}{a}$ **P-3** a. $\frac{\sqrt{5}}{\sqrt{5}}$ b. $\frac{\sqrt{2n}}{\sqrt{2n}}$ c. $\frac{\sqrt{12s}}{\sqrt{12s}}$ or $\frac{\sqrt{3s}}{\sqrt{3s}}$ d. $\frac{\sqrt{x}}{\sqrt{x}}$

Page 192 **Classroom Exercises** 1. $\frac{1}{2}$ 3. $\frac{3}{5}$ 5. $\frac{4}{5}$ 7. 2 9. 2 11. 2

Page 192 **Written Exercises** 1. $\frac{x}{4}$ 3. $\frac{7}{a}$ 5. $\frac{y^2}{10}$ 7. $\frac{\sqrt{2}}{x}$ 9. $\frac{2\sqrt{3}}{a}$ 11. 5 13. 6 15. 7 17. 5 19. x 21. $\frac{2}{y}$ 23. $\frac{b}{a}$ 25. $\sqrt{5}$ 27. $\sqrt{6}$ 29. $\sqrt{11}$ 31. $\frac{\sqrt{15}}{5}$ 33. $\frac{a\sqrt{b}}{b}$ 35. $\frac{\sqrt{15st}}{5t}$

Page 192 **Review Capsule for Section 7.7** 1. $2\cdot 2\cdot 2\cdot 3\cdot 7$ 2. $2\cdot 2\cdot 3\cdot 3\cdot 5$ 3. $2\cdot 2\cdot 61$ 4. $2\cdot 2\cdot 2\cdot 2\cdot 2\cdot 3\cdot 3$ 5. $3\cdot 3\cdot 23$ 6. $2\cdot 3\cdot 3\cdot 13$

Page 193 **P-1** a. 784 b. 5776 c. 676 **P-2** a. 28 b. −78 c. 27 **P-3** a. 5.099 b. 8.832 c. −5.292

Page 196 **Classroom Exercises** 1. 841 3. 8649 5. 72 7. −88 9. 6.164 11. 8.832 13. 59 15. −12.247

Page 197 **Written Exercises** 1. 12.962 3. 15.620 5. 14.388 7. 16.585 9. 21.636 11. 0.866 13. 0.816 15. 0.936 17. 0.958 19. 0.306 21. 69.6 yards 23. 86 miles 25. 239 kilometers

Page 198 **P-1** a

Page 200 **Classroom Exercises** 1. No 3. No 5. $g^2 = 5^2 + 9^2$ 7. $s^2 = 7^2 + 11^2$

Pages 200-201 **Written Exercises** 1. Yes 3. No 5. c = 20 7. b = 20 9. a = 24 11. a = 15 13. b = 4.6 15. c = 12.8 17. 12 inches 19. 29.9 feet

Page 203 1. 75 3. 20 miles per hour 5. −16°C

Pages 204-206 **Chapter Review** 1. −7 3. 0.6 5. 23 7. $-\sqrt{33}$ 9. 6 11. $6\sqrt{3}$ 13. $2\sqrt{13}$ 15. $3\sqrt{11}$ 17. $2\sqrt[3]{3}$ 19. r^3 21. $3x^2\sqrt{x}$ 23. $2ab^2\sqrt{7ab}$ 25. $2mn\sqrt[3]{2n}$ 27. $10\sqrt{13}$ 29. $9\sqrt{3}$ 31. $-7\sqrt{6}$ 33. $\frac{\sqrt{2}}{4}$ 35. $\frac{2\sqrt{5}}{t}$ 37. $\frac{\sqrt{70}}{10}$ 39. $\frac{x\sqrt{6xy}}{2y}$ 41. 12.369 43. 0.829 45. 0.842 47. 3.5 meters 49. 77 miles 51. Yes 53. No 55. 9.8 inches

CHAPTER 8 POLYNOMIALS

Page 208 **P-1** a. a = 3.7; n = 2 b. a = −1; n = 1 c. a = −15; n = 0 **P-2** n = −2, which is a negative number. **P-3** a. $\frac{1}{3}$ b. 1 c. −1 d. $-\frac{1}{5}$

Page 209 **Classroom Exercises** 1. Yes 3. No 5. Yes 7. 2 9. $-7\frac{1}{2}$ 11. $3\frac{2}{3}$

Page 209 **Written Exercises** 1. Yes 3. Yes 5. No 7. Yes 9. No 11. Yes 13. No 15. No 17. 5 19. $\frac{3}{5}$ 21. 1.8 23. −1 25. $\frac{3}{5}$ 27. $\frac{3}{5}$ 29. −3 31. 24 33. −8.3

Page 209 **Review Capsule for Section 8.2** 1. m^6 2. t^8 3. r^{10} 4. y^{12} 5. w^4 6. q^3 7. n^4 8. a^5

Pages 210-211 **P-1** a. x^7 b. y^{11} c. r^6 **P-2** 10 **P-3** r^2s^3 **P-4** a. x^3 b. y^3 c. r

Pages 211-212 **Classroom Exercises** 1. $20y^3$ 3. $-42x^5$ 5. $-8x^3y^2$ 7. $-6x^3y$ 9. $24x^4y^3$ 11. $9x^2$ 13. $-2b^4$ 15. $-9s^4t$ 17. $-8p^2qr$

Page 212 Written Exercises 1. $72m^5$ 3. $-96r^3t^5$ 5. $5p^4q^5$ 7. $-24t^4$ 9. $21m^3n^4$ 11. $-27a^4b^6$ 13. $-1.52r^2s^3t^4$ 15. $8.4u^3x^3y^3z^3$ 17. $-4k^3$ 19. $4y^2$ 21. $-0.6m^6n^2$ 23. $29abc^2$ 25. $-\frac{22}{7}x^8y^6z^4$ or $-3\frac{1}{7}x^8y^6z^4$ 27. $\frac{4}{15}m^9n^2q^2$

Page 213 P-1 $2x^2$, $-9x$, and $\frac{1}{4}$ **P-2** a and d; b and c

Pages 214-215 Classroom Exercises 1. Yes 3. Yes 5. Yes 7. No 9. No 11. Yes 13. Yes 15. Yes 17. $5x^2 - 2x + 3$ 19. $3x^3 + 12$ 21. $5x^2y - 5$ 23. $x^{12} + 5x^8 - 3x^5 - 3x^2$

Pages 215-216 Written Exercises 1. Yes 3. No 5. No 7. Yes 9. Yes 11. $3x^2 - 4x + 7$ 13. $x^5 + 4x^3 - 3x^2 + 5x - 1$ 15. $3x^2 - 9x + 8$ 17. $3x^3y + 5x^2y^2 + xy^3 - 5$ 19. $m^3 - 2m^2n + 5mn - 10$ 21. $-0.5x^4 - 1.9x^2 + 2.4x$ 23. -23 25. -15 27. 32 29. 49 31. $x^2 - 3x - 5$ 33. $x^2 + x + 12$ 35. 67,349 37. 1472

Page 216 Mid-Chapter Review 1. 32 3. 8 5. $-6r^3t$ 7. $-5s$ 9. $-3x^2 + 6x + 12$ 11. $0.2x^3 - 3.2x^2 + 2.8x - 1.4$ 13. 20 15. -180

Page 216 Review Capsule for Section 8.4 1. $-3x^2 + 2x$ 2. $-3x^3 - 4x$ 3. $-6x^2 + 4x + 12$ 4. $-5x + 7$ 5. $x^2 + 3x - 5$

Page 217 P-1 a. $x^2 - x + 3$ b. $2x^2 + 4x - 5$ **P-2** a. $2x^2 + 4x + 3$ b. $x^3 - 2x^2 + 5x - 1$ **P-3** a. $4x^3$ b. $-2x^2$ c. $-x$ d. 8

Page 218 Classroom Exercises 1. $7x - 7$ 3. $-2x^2 - 5x - 1$ 5. $2x^2 + 4x + 1$ 7. $3x^2 + 3x - 3$ 9. $-x^2 - 4x$

Page 219 Written Exercises 1. $11x^2 - 2x - 8$ 3. $1.5x^2 + 2.9x + 0.8$ 5. $-2x^3 - 4x^2 - 3x + 9$ 7. $-2x^2 + 6x - 18$ 9. $3x^3 - 5x^2 - 3x + 2$ 11. $4x^2 - 7x + 15$ 13. $-3x^3 + 5x^2 - 3x + 7$ 15. $7x^3 - 7x^2 - 4x - 9$ 17. $2.1x^2 + 4.4x + 6.7$ 19. $12p + 3$ 21. $20y + 5$

Pages 220-221 P-1 a. $-2x^2$ b. $-15x$ c. 12 **P-2** a. 5 b. -12 c. 9 **P-3** a. $2x^2 + 9x + 4$ b. $6x^2 + 5x - 6$ c. $12x^2 - 23x + 5$ d. $2x^2 - 3x - 20$

Page 222 Classroom Exercises 1. $x^2 + 3x + 2$ 3. $x^2 + 2x - 3$ 5. $x^2 - 6x + 8$ 7. $2x^2 + 5x + 2$ 9. $4x^2 - 11x - 3$ 11. $2x^2 + 5x$ 13. $x^3 - 4x^2 + 7x - 6$

Pages 222-223 Written Exercises 1. $x^2 + 8x + 15$ 3. $x^2 - 2x - 8$ 5. $x^2 - 8x + 15$ 7. $x^2 + 3x - 40$ 9. $2x^2 + 13x + 15$ 11. $3x^2 + 11x - 4$ 13. $5x^2 - 17x + 6$ 15. $8x^2 - 2x - 15$ 17. $x^3 - 6x^2 + 11x - 12$ 19. $2x^3 + 13x^2 + 11x - 20$ 21. $x^3 + 6x^2 - 5x - 30$ 23. $6x^3 - 13x^2 + 27x - 14$ 25. $-20x^3 - 17x^2 - 6x - 20$ 27. $42x^4 + 57x^3 + 18x^2 + 30x + 15$ 29. $x^2 - 11x + 28$ 31. $3x^2 + 14x - 5$

Page 223 Review Capsule for Section 8.6 1. $-4x + 2$ 2. $-2x - 4$ 3. $8x - 3$ 4. $2x^2 + 8$

Pages 224-225 P-1 $6x^2$; $4x$; 2 **P-2** $2x^2$; $-x$

Pages 225-226 Classroom Exercises 1. $-5x^2$ 3. $3x^2 - 5$ 5. $x + 5$ 7. $3x + 2$

Page 226 Written Exercises 1. $2x + 3$ 3. $3 - 2xy$ 5. $3x + 4$ 7. $5x - 6$ 9. $3x - 4$ 11. $x^2 - 3x + 2$ 13. $2x^2 - 3x + 1$ 15. $x - 5$ 17. $x^2 + 3$ 19. $2x^2 - 5x + 10$ 21. $x^2 + 2x + 4$ 23. $x^3 + 3x^2 + 9x + 27$ 25. See diagram at right.

25.
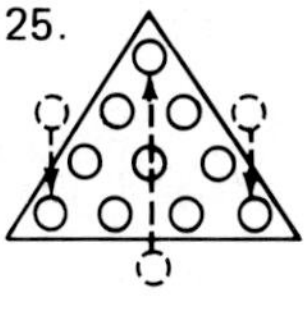

Pages 227-228 P-1 $3x$ **P-2** $-8x^3 + 10x^2 - 12x - 19$ **P-3** $2x^2$

Page 229 Classroom Exercises Note: In the answers for Ex. 1-5, Q stands for quotient and R stands for remainder. 1. Q: $x - 5$; R: 3 3. Q: $2x + 2$; R: -1 5. Q: $x + 1$; R: $-2x + 2$ 7. $10x^2 - 15x$; $6x - 9$ 9. $x^3 - 3x$; $2x^2 - 6$; $2x + 11$ 11. $x^3 - 3x^2 + 2x - 6$; $-3x^2 - 6$

Page 230 Written Exercises 1. $2x - 5 + \frac{3}{x + 4}$ 3. $2x + 7 - \frac{4}{2x - 3}$ 5. $3x + 4 - \frac{5}{5x - 6}$ 7. $5x - 6 - \frac{10}{2x + 5}$ 9. $6x - 5 + \frac{x}{2x^2 - 1}$ 11. $x^2 - 3x + 6 - \frac{8x - 8}{x^2 + x - 1}$ 13. $x - 5 - \frac{6}{x^2 + 8x + 2}$ 15. $x^2 + 2x - 3 + \frac{24}{2x^2 + 9x + 4}$ 17. $x^2 - y^2 - \frac{y^3}{x + y}$ 19. $x^2 - 2x + 4 - \frac{16}{x + 2}$ 21. $x^2 - 2xy + y^2 + \frac{y^3}{2x - 3y}$ 23. $4x^2 + 2x + 1$ 25. $a + 6$ 27. $x^2 - 5x + 29 - \frac{148}{x + 5}$ 29. $3a + 4$ 31. $4a^2 + 9a + 3$ 33. $x^2 - xy + y^2$ 35. $x - y$ 37. $x^3 + x^2 + x + 1$

Page 231 1. \$212.27 3. \$637.01 5. \$315.28

Pages 232-234 Chapter Review 1. Yes 3. No 5. -32 7. -12.8 9. $21t^6$ 11. $2a^4b^5$ 13. $-12t^2$ 15. $\frac{1}{6}x^9y^2z^2$ 17. $\frac{13}{4}x^2y^4$ 19. No 21. No 23. $7x^2 + 3x - 16$ 25. $-3x^3 - x^2 + 12x + 5$ 27. $4x^2 - 7x - 17$ 29. $5x^3 + 2x^2 - 11x - 5$ 31. $6x^2 - 7x - 5$ 33. $x^4 - 5x^3 - 2x^2 - 2x + 12$ 35. $6r + 4$ 37. $3n^2 + 8n + 8$ 39. $3x^2 + 4x - 32$ 41. $12x^2 - 64x + 45$ 43. $-16x^4 + 12x^3 - 20x^2 + 27x - 9$ 45. $6x^2 + 22x - 8$ 47. $3xy - 2$ 49. $5x + 8$ 51. $-x^2 + x - 5$ 53. $-3x + 4 - \frac{19}{6x - 14}$ 55. $x^2 + 4 - \frac{6}{2x^2 - 5}$

CHAPTER 9 FACTORING POLYNOMIALS

Pages 236-237 **P-1** $2x(2x - 3)$ **P-2** $4x^2$

Page 238 Classroom Exercises 1. Yes 3. No 5. No 7. Yes 9. $2(2x + 1)$ 11. $5(x + 3)$ 13. $2(x^2 + 5)$ 15. $-3(x - 2)$ 17. $5x(x^2 - 2)$ 19. $8(3 - x^2)$

Page 238 Written Exercises 1. $4(x - 1)$ 3. $5(x + 2)$ 5. $x(2x + 3)$ 7. $-3x(x - 3)$ 9. $4x^2(2x - 3)$ 11. $6x(2x^2 - 3)$ 13. $x(x^2 - 2x + 3)$ 15. $2x(x^3 - 2x - 5)$ 17. $-5x(x + 2)$ 19. $4x^2(2x^3 - 3x + 5)$ 21. $5(x^4 - 7x^2 - 2)$ 23. Prime polynomial

Page 238 Review Capsule for Section 9.2 1. $x^2 + 2x + 2x + 2 \cdot 2$ 2. $x^2 + 4x + 2x + 4 \cdot 2$ 3. $x^2 + 7x + 3x + 7 \cdot 3$ 4. $x^2 + 3x + 2x + 3 \cdot 2$ 5. $x^2 + 11x + 1x + 11 \cdot 1$ 6. $x^2 + 5x + 4x + 5 \cdot 4$

Pages 239-240 **P-1** a. Second b. First c. Zero d. Eighth **P-2** a. First b. Third c. Fourth **P-3** a. $(x + 7)(x + 3)$ b. $(x + 2)(x + 10)$ **P-4** 5; 2; 10 **P-5** 9 and 4

Page 240 Classroom Exercises 1. Second 3. First 5. Fifth 7. $(x + 4)(x + 3)$ 9. $(x + 3)(x + 5)$ 11. $(x + 5)(x + 4)$ 13. $(x + 4)(x + 8)$

Page 241 Written Exercises 1. $(x + 3)(x + 4)$ 3. $(x + 5)(x + 3)$ 5. $(x + 4)(x + 5)$ 7. $(x + 8)(x + 4)$ 9. $(x + 3)(x + 11)$ 11. $(x + 5)(x + 2)$ 13. $(x + 3)(x + 2)$ 15. $(x + 7)(x + 3)$ 17. $(x + 13)(x + 2)$ 19. $(x + 10)(x + 3)$ 21. $(x + 7)(x + 4)$ 23. $k = 12$ 25. $k = 18$

Page 241 Review Capsule for Section 9.3 1. $(x - 2)(x + 5)$ 2. $(x + 2)(x - 3)$ 3. $(x - 4)(x - 6)$ 4. $(x - 2)(x - 10)$ 5. $(x - 5)(x - 5)$ 6. $(x - 9)(x - 9)$

Pages 242-244 **P-1** 10; 16 **P-2** $r + s$ **P-3** rs **P-4** -7; 12 **P-5** -1; -20 **P-6** 5; 10

Page 244 **Classroom Exercises** 1. 3, 2 3. 6, 3 5. None 7. 4, −2 9. 2, 4 11. None 13. $x^2 + 4x + 3x + 12$ 15. $x^2 + 5x + 3x + 15$ 17. $x^2 - 3x - 2x + 6$ 19. $x^2 + 5x - x - 5$ 21. $x^2 + 6x - x - 6$ 23. $x^2 + 7x - x - 7$

Page 245 **Written Exercises** 1. 7; 12 3. 14; 33 5. 4; −32 7. −8; −48 9. −17; 72 11. −3; 2 13. $x^2 + 12x + x + 12$ 15. $x^2 + 5x + 6x + 30$ 17. $x^2 - 5x - 3x + 15$ 19. $x^2 - 8x - 3x + 24$ 21. $x^2 - 8x + 5x - 40$ 23. $x^2 + 7x - 3x - 21$ 25. $x^2 - 12x + 3x - 36$ 27. $x^2 - 6x - x + 6$ 29. $x^2 - 14x + 3x - 42$ 31. $(x + 1)(x + 1)$ 33. $(x + 5)(x + 7)$ 35. $(x - 4)(x - 3)$ 37. $(x - 6)(x - 5)$ 39. $(x - 4)(x + 3)$ 41. $(x - 6)(x + 5)$ 43. $(x + 6)(x - 5)$ 45. $(x - 7)(x - 2)$ 47. $(x + 7)(x - 2)$ 49. Prime polynomial 51. $(x - 5)(x + 3)$ 53. $(x + 17)(x + 2)$ 55. $(x - 24)(x + 2)$ 57. $(x + 2)(x + 1)$ 59. $(x - 8)(x + 7)$

Page 246 **P-1** a. 5 b. 1 c. −1

Page 248 **Classroom Exercises** 1. $r + s = 3$; $rs = 2$ 3. $r + s = 10$; $rs = 9$ 5. $r + s = 11$; $rs = 18$ 7. $r + s = 11$; $rs = 18$ 9. $r + s = 17$; $rs = 30$ 11. $r + s = -1$; $rs = -6$ 13. $r + s = -5$; $rs = -50$ 15. $r + s = -7$; $rs = -30$ 17. $r + s = -17$; $rs = 72$

Pages 248-249 **Written Exercises** 1. $2x^2 - 3x + 1$ 3. $3x^2 + x - 2$ 5. $(2x + 1)(x + 1)$ 7. $(3x + 1)(x + 3)$ 9. $(3x + 2)(x + 3)$ 11. $(3x + 1)(2x + 3)$ 13. $(3x + 1)(2x + 5)$ 15. $(3x + 2)(x - 1)$ 17. $(2x + 5)(x - 5)$ 19. $(5x + 1)(3x - 2)$ 21. $(2x - 3)(x - 2)$ 23. $(4x - 1)(2x - 3)$ 25. $(2x - 1)(x + 3)$ 27. $(5x - 2)(3x + 4)$ 29. $(9x + 7)(2x + 1)$ 31. $(12x + 3)(x - 3)$ or $3(4x + 1)(x - 3)$ 33. $(8x + 3)(x + 4)$ 35. 5994 37. 7 39. 3

Page 249 **Mid-Chapter Review** 1. $4x(3x - 1)$ 3. $3x(x^2 + 3x - 5)$ 5. $2xy(x + 3)$ 7. $5x^2y(x - 2 + 5y)$ 9. $(x + 7)(x + 5)$ 11. $(x + 10)(x + 5)$ 13. $(x + 8)(x - 5)$ 15. $(x - 11)(x + 8)$ 17. $(x + 12)(x - 6)$ 19. $(x - 11)(x - 9)$ 21. $(2x - 3)(x + 2)$ 23. $(5x - 2)(4x + 1)$

Page 249 **Review Capsule for Section 9.5** 1. $(x - 12)(x + 12)$ 2. $(x - 2y)(x + 2y)$ 3. $(2x - 3)(2x + 3)$

Page 251 1. 0 pure tall; 64 pure short; 64 hybrid tall 3. 64 pure tall; 0 pure short; 64 hybrid tall 5. 14 pure rough; 28 hybrid rough; 14 pure smooth 7. 28 pure rough; 28 hybrid rough; 0 pure smooth

Pages 252-253 **P-1** a and c **P-2** $(x + a)(x - a)$ **P-3** 11 **P-4** a. x^3 b. r^4 c. t^6

Page 253 **Classroom Exercises** 1. $(a + 1)(a - 1)$ 3. $(p + 3)(p - 3)$ 5. $(x + 5)(x - 5)$ 7. $(t + 7)(t - 7)$ 9. $(b + 9)(b - 9)$ 11. $(p + q)(p - q)$ 13. $(2x + y)(2x - y)$ 15. $(x^2 + 5)(x^2 - 5)$

Page 254 **Written Exercises** 1. $x^2 - 64$ 3. $x^2 - 121$ 5. $4x^2 - y^2$ 7. $x^2y^2 - 9$ 9. $25x^2 - 9$ 11. $x^4 - 4$ 13. $(a + 5)(a - 5)$ 15. $(2x + 1)(2x - 1)$ 17. $(3a + 4)(3a - 4)$ 19. $(x^2 + 2)(x^2 - 2)$ 21. $(2x^2 + 3)(2x^2 - 3)$ 23. $(2x + y)(2x - y)$ 25. $(ab + 1)(ab - 1)$ 27. $(4t + 9)(4t - 9)$ 29. $(xy + r)(xy - r)$ 31. $(4xy^2 + 3)(4xy^2 - 3)$ 33. $(x^3 + 2)(x^3 - 2)$ 35. $(x^5 + y^4)(x^5 - y^4)$ 37. 224 39. 165 41. 396 43. 884 45. 1591 47. 3599

Page 254 **Review Capsule for Section 9.6** 1. $x^2 + 12x + 36$ 2. $x^2 - 6x + 9$ 3. $4x^2 - 20x + 25$ 4. $25x^2 + 10x + 1$ 5. $4x^2 + 28x + 49$ 6. $x^2 - 2xy + y^2$ 7. $9x^2 + 12x + 4$ 8. $25x^2 - 10x + 1$

Pages 255-256 **P-1** 3; 9; 9 **P-2** The value of b is negative. **P-3** a. $4x^2 + 20x + 25$ b. $x^2 - 14x + 49$ c. $9x^2 - 6x + 1$ **P-4** a. $(x + 1)(x + 1)$ b. $(x - 6)(x - 6)$ c. $(3x - 4)(3x - 4)$

Page 256 **Classroom Exercises** 1. Yes 3. Yes 5. Yes 7. No 9. No 11. No 13. $(x + 2)(x + 2)$ 15. $(x - 3)(x - 3)$ 17. $(x + 8)(x + 8)$ 19. $(x - 12)(x - 12)$ 21. $(2x + 1)(2x + 1)$

Page 257 **Written Exercises** 1. No 3. Yes 5. No 7. No 9. No 11. Yes 13. $x^2 - 10x + 25$ 15. $x^2 + 14x + 49$ 17. $4x^2 + 12x + 9$ 19. $25x^2 - 10x + 1$ 21. $16x^2 + 40x + 25$ 23. $4x^2 - 4x + 1$

25. $(6x - 1)(6x - 1)$ 27. $(5x + 2)(5x + 2)$ 29. $(3x - 4)(3x - 4)$ 31. $(7x + 10)(7x + 10)$ 33. $(7x - 1)(7x - 1)$ 35. $(x + 2y)(x + 2y)$ 37. $(2x + 3y)(2x + 3y)$ 39. $(xy - 8)(xy - 8)$ 41. $(3xy + 7)(3xy + 7)$

Page 257 **Review Capsule for Section 9.7** 1. B 2. D 3. A 4. C

Page 258 **P-1** 2 **P-2** 3

Pages 259-260 **Classroom Exercises** 1. 4; $(x^2 - 3)$ 3. 4; $(x^2 - 5x + 7)$ 5. 5; $(3x^2 - x - 5)$ 7. y; $(x^2y - 3x + 5)$ 9. −4; $(x^2 + 5)$ 11. 3b; $(a^2b - 3a + 2)$ 13. 3y; $(x^2 - 10x + 25)$

Page 260 **Written Exercises** 1. $2(y + 3)(y - 3)$ 3. $2(x + 2)(x + 2)$ 5. $3(x + 7)(x + 3)$ 7. $4(x - 6)(x - 4)$ 9. $5(x - 5)(x + 4)$ 11. $(2y + 3)(2y - 3)$ 13. $a(x + 2)(x + 2)$ 15. $2(3x + 1)(x - 2)$ 17. $4(3x - 1)(x - 3)$ 19. $-3(x^2 + 4)$ 21. $a(2x + 3)(2x - 3)$ 23. $y(x - 5)(x - 5)$ 25. $3(x + 5)(x - 5)$ 27. $2(4 - d)(4 - d)$ 29. $3(x - 4)(x + 2)$ 31. $2(1 + 8x)(1 - 8x)$ 33. $a(b - 5)(b + 4)$ 35. $3(2a - 1)(4a - 3)$ 37. 6 39. 2

Pages 261-262 1. 16 3. −5 5. 7 7. 3 9. 3 11. 26 13. 8 15. −6 17. 30 19. 7.44 21. $\frac{1}{4}$ 23. 4 25. 2 27. 24 29. −17

Pages 263-264 **Chapter Review** 1. $2x(2x - 5)$ 3. $6ab(3a + 5ab - 7)$ 5. $abc(ac^2 - b + abc)$ 7. $(x + 7)(x + 3)$ 9. $(x + 5)(x + 10)$ 11. $(x + 7)(x + 3)$ 13. $(x + 6)(x - 3)$ 15. $(x - 6)(x - 3)$ 17. Prime 19. $(2x - 3)(x + 1)$ 21. $(4x - 1)(x + 2)$ 23. $(3x - 5)(x + 1)$ 25. $(3a + 4b)(3a - 4b)$ 27. $(rs + xy)(rs - xy)$ 29. $(x^2 + 13)(x^2 - 13)$ 31. Yes 33. No 35. No 37. $(x - 4)(x - 4)$ 39. $(2x - 3)(2x - 3)$ 41. $(x - 2y)(x - 2y)$ 43. $5(x + 2)(x - 2)$ 45. $3(x - 2)(x - 2)$ 47. $3(x^2 - 4x + 32)$

Pages 265-266 **Cumulative Review: Chapters 1-9** 1. −13 3. 28 5. $24x - 56$ 7. $-12n^2 - 6n$ 9. $x = 12$ 11. Let x = value of lot, then x + 4000 = value of house. x + 4000 = 20,000; x = $16,000 13. Let x = number of multi-family homes, then x + 80 = number of single-family homes. x + x + 80 = 188; x = 54, x + 80 = 134 15. $x = -4$ 17. On a number line, all points to the left of, and including, 2. 19. On a number line, all points to the right of, and not including, $-1\frac{1}{2}$. 21. {all real numbers greater than −2} 23. {all real numbers less than −2} 25. $\frac{1}{x^4y^4}$ 27. $\frac{25c^2d^4}{x^6}$ 29. $-\frac{3}{4}$ 31. 18 33. $2\sqrt[3]{3}$ 35. $2x^3y^4\sqrt{2xy}$ 37. $-\sqrt{3}$ 39. $2\sqrt{2p}$ 41. $\frac{a}{4}$ 43. $\frac{\sqrt{14mp}}{7p}$ 45. 12.247 47. 0.272 49. 16 inches

51. −2 53. $-7.2a^5$ 55. $-3c^5$ 57. $4a^2 - 4a - 12$ 59. $12w^2 + 17w - 5$ 61. $4x - 3$ 63. $5y^2(2y - 3)$ 65. $(x + 2)(x + 7)$ 67. $(g + 2)(g - 12)$ 69. $(3n - 1)(2n + 3)$ 71. $(2y + 5d)(2y - 5d)$ 73. $(3x + 6) \cdot (3x - 6)$ or $9(x - 2)(x + 2)$

CHAPTER 10 PRODUCTS AND QUOTIENTS OF RATIONAL EXPRESSIONS

Page 268 **P-1** $\frac{x^2 + x - 1}{1}$ **P-2** a. Yes b. Yes c. No d. Yes e. No f. Yes **P-3** −1

Page 269 **Classroom Exercises** 1. Yes 3. No 5. Yes 7. Yes 9. 0 11. 3 13. $\frac{1}{2}$ 15. 1, −1

Page 270 **Written Exercises** 1. Yes 3. Yes 5. Yes 7. No 9. No 11. Yes 13. $-\frac{5}{2}$ 15. $\frac{4}{3}$ 17. $\frac{4}{3}$ 19. 0 21. 0 23. 10 25. 0 27. 2, −2 29. $\frac{1}{3}$ 31. 1, −7

Page 270 **Review Capsule for Section 10.2** 1. $2 \cdot 3 \cdot 3 \cdot x \cdot x \cdot x$ 2. $2 \cdot 2 \cdot 3 \cdot a \cdot b \cdot b$ 3. $(x + 5)(x - 5)$ 4. $(c + 9)(c - 9)$ 5. $x(x + 7)$ 6. $(a + b)(a - b)$ 7. $(x - 3)(x - 3)$ 8. $3b(a + b)$ 9. $3a(a + 2b)$ 10. $a(5a - b)$ 11. $(x + 2)(x + 2)$ 12. $(x - y)(x + y)$

Page 272 **P-1** $x + 2, x - 2; x + 3, x - 2$

Page 273 **Classroom Exercises** 1. $\frac{1}{3}$ 3. $\frac{1}{x}$ 5. $\frac{3}{x}$ 7. $\frac{x+2}{x+3}$ 9. $\frac{1}{x}$

Page 273 **Written Exercises** 1. $\frac{x}{2}$ 3. $\frac{3}{7x}$ 5. $\frac{x}{5}$ 7. $\frac{2}{3}$ 9. $\frac{x+2}{x+5}$ 11. $\frac{4x-4}{3x+9}$ 13. $\frac{3}{5}$ 15. $\frac{x+3}{x+5}$ 17. $\frac{1}{x-3}$ 19. $\frac{x+4}{x}$ 21. $\frac{x}{x-2}$ 23. $\frac{x}{x-2}$ 25. $\frac{2x}{5y}$ 27. $\frac{a+b}{c}$ 29. $\frac{a-b}{3}$ 31. $\frac{5a-b}{3a+6b}$

Page 273 **Review Capsule for Section 10.3** 1. $2 \cdot 2 \cdot 3 \cdot x \cdot x \cdot y \cdot y \cdot y$ 2. $2 \cdot 2 \cdot 5 \cdot r \cdot r \cdot r \cdot s \cdot t \cdot t$ 3. $2 \cdot 3 \cdot 3 \cdot a \cdot a \cdot b \cdot b \cdot b \cdot b \cdot c \cdot c \cdot c$ 4. $2 \cdot 2 \cdot 3 \cdot 3 \cdot p \cdot p \cdot p \cdot p \cdot p \cdot q \cdot q \cdot r \cdot r \cdot r$

Pages 274-275 **P-1** a. 8 b. $\frac{4}{5}$ c. $1\frac{3}{4}$ d. $\frac{1}{20}$ **P-2** x^2 and 20 have no common factors.

Pages 275-276 **Classroom Exercises** 1. $\frac{2}{5}$ 3. 1 5. $\frac{2x}{3y}$ 7. $\frac{5}{6}$ 9. $\frac{3}{5a}$ 11. $\frac{2}{5}$ 13. $\frac{1}{15}$ 15. 6ab 17. $-\frac{x}{3y}$ 19. 1

Page 276 **Written Exercises** 1. $\frac{5}{24}$ 3. $\frac{5y}{3x}$ 5. $\frac{3}{5}$ 7. $\frac{14}{15}$ 9. 2x 11. $\frac{20y}{x}$ 13. $-\frac{21b^2}{5a}$ 15. $\frac{25xy}{4}$ 17. $6a^2bc$ 19. 1 21. $\frac{33a^2c}{50}$ 23. $-\frac{3ab}{2}$ 25. 10:53 A.M.

Pages 277-278 **P-1** $-a + b$ **P-2** a. $2 - x$ b. $t - 5$ c. $m - r$ **P-3** $2 - x$ **P-4** $x^2 + 3x + 2$; $3x - 9$

Page 279 **Classroom Exercises** 1. $\frac{2}{3}$ 3. $\frac{x^2+4x+3}{x^2+6x+8}$ 5. 1 7. $\frac{1}{x+2}$ 9. $-\frac{5}{6}$ 11. 1 13. $\frac{x+1}{x}$ 15. $-\frac{2}{3}$

Pages 279-280 **Written Exercises** 1. $\frac{11}{12}$ 3. $-\frac{14}{15}$ 5. $\frac{x-3}{x-1}$ 7. $\frac{x-3}{2y}$ 9. $x - 3$ 11. $\frac{2}{x+2}$ 13. $\frac{x^2-5x+6}{x^2+5x+6}$ 15. $-\frac{5}{x}$ 17. 1 19. $-\frac{1}{3}$ 21. $\frac{x}{x+3}$ 23. 0.6521739 25. 0.06034482

Page 280 **Mid-Chapter Review** 1. $-\frac{7}{5}$ 3. $\frac{9}{5}$ 5. $\frac{x}{2y}$ 7. $\frac{x+6}{2x}$ 9. $\frac{4}{15}$ 11. $-\frac{25rs}{6t^2}$ 13. $\frac{6}{5}$ 15. $2r - 3$

Page 280 **Review Capsule for Section 10.5** 1. $(x + 1)(x - 3)$ 2. $(x - 3)(x - 2)$ 3. $(2x - 1)(x + 3)$ 4. $(2x + 3)(2x - 1)$

Pages 281-282 **P-1** $x, x + 2; x + 2, x - 2$ **P-2** a. $2x - 1, x + 1$ b. $x + 6, x - 6$ c. $2x^2, 3x - 2$ **P-3** $x - 5, x + 2; x + 3, x - 3; x + 2, x + 2$

Page 283 **Classroom Exercises** 1. $\frac{x-2}{x+4}$ 3. $\frac{x+5}{x+2}$ 5. $\frac{x-1}{2}$ 7. $\frac{5x^2}{x^3-x^2-9x+9}$ 9. $\frac{x+1}{x}$

Page 283 **Written Exercises** 1. $\frac{x+2}{x}$ 3. $\frac{5x+10}{3}$ 5. $\frac{5x^2+10x}{x^2-25}$ 7. $\frac{x+1}{5}$ 9. $\frac{x+4}{4}$ 11. $\frac{-3x+6}{2}$ 13. $\frac{3}{2x^2+9x-5}$ 15. $\frac{2x^3-6x^2+12x-8}{3x-1}$ 17. $x = 2y$ 19. No; a and b can have any values such that the ratio $\frac{a}{b}$ is equivalent to $\frac{5}{3}$.

Pages 284-285 **P-1** a. $\frac{x-3}{x+2} \cdot \frac{x-2}{x}$ b. $\frac{3x}{5} \cdot \frac{9x}{4}$ c. $\frac{2x^2}{5} \cdot \frac{1}{x+1}$ **P-2** $\frac{x-2}{x} \cdot \frac{5}{x-2}$
P-3 $\frac{4(r-3)}{9pq}$

Pages 285-286 **Classroom Exercises** 1. $\frac{2}{3} \cdot \frac{6}{5}$ 3. $\frac{2x}{6} \cdot \frac{3}{x}$ 5. $\frac{2}{x-2} \cdot \frac{5}{x}$ 7. $\frac{x-7}{5x} \cdot \frac{4}{x+1}$
9. $\frac{1}{x+2} \cdot \frac{1}{x-3}$ 11. $\frac{x+1}{x} \cdot \frac{2x}{x-10}$ 13. $\frac{x+8}{x-5} \cdot \frac{x-10}{x+2}$

Page 286 **Written Exercises** 1. $\frac{2}{3}$ 3. 6 5. $\frac{3a}{5}$ 7. $\frac{1}{2}$ 9. $\frac{2xy}{3}$ 11. $\frac{5}{2xy}$ 13. $\frac{x-5}{5}$ 15. $\frac{2x^2+4}{x^2}$
17. $\frac{1}{2x+10}$ 19. $\frac{x^4+6x^2+9}{3}$ 21. $\frac{5x}{2}$ 23. $\frac{2x^2+4x+2}{3x^2-6x+3}$

Page 287 **P-1** $\frac{1}{x^2-25}$

Pages 288-289 **Classroom Exercises** 1. 2, $x-3$ 3. x, $3x-1$ 5. $x+2$, $x-2$ 7. 4, $x-4$
9. 3, $x-7$ 11. x, $x-7$ 13. $2x$, $x-5$ 15. $x+4$, $x-4$ 17. $3x$, $x-5$ 19. $5x$, $x+3$

Page 289 **Written Exercises** 1. $\frac{2}{5x^2+15x}$ 3. $\frac{1}{2}$ 5. $\frac{x-5}{2}$ 7. $\frac{x^2-5x+6}{x^2+5x+6}$
9. $\frac{x^2+10x+24}{3x^4-21x^3-18x^2+216x}$ 11. $\frac{xy-y^2}{2}$ 13. $\frac{2}{7}$ 15. $\frac{5a+18}{12}$ 17. $\frac{3a+6b}{10a+5b}$ 19. 1 21. $-\frac{5}{x+3}$

Page 290 1. 2300 volts 3. 1.625 amperes

Pages 291-292 **Chapter Review** 1. Yes 3. No 5. -3 7. $\frac{1}{2}$ 9. $\frac{5x}{6}$ 11. $\frac{1}{3}$ 13. $\frac{2}{x}$ 15. $\frac{2xy}{3ab^2}$
17. $\frac{1}{ax-a^2}$ 19. $\frac{x^2-2x}{7}$ 21. $-\frac{5}{6}$ 23. $\frac{x+1}{x+2}$ 25. 1 27. $-\frac{10x}{3a}$ 29. $\frac{2x-4}{x^2+2x}$ 31. $\frac{2x^2+8}{3x}$ 33. $\frac{13a}{25}$
35. $\frac{10a-12b}{5a+3b}$

CHAPTER 11 SUMS AND DIFFERENCES OF RATIONAL EXPRESSIONS

Page 294 **P-1** a. $\frac{5}{7}$ b. $-\frac{1}{2}$ **P-2** a. $\frac{9}{x}$ b. $\frac{7}{a}$ c. $\frac{x}{y}$ **P-3** 3

Pages 295-296 **Classroom Exercises** 1. $\frac{3}{5}$ 3. $\frac{1}{6}$ 5. $\frac{3}{x}$ 7. $\frac{7}{3x}$ 9. $\frac{11}{x+1}$ 11. $\frac{3x}{x+3}$ 13. $-\frac{4a}{a+1}$
15. $\frac{2x+3y}{x+y}$ 17. $\frac{3x}{11}$ 19. $\frac{2x+5}{y+1}$

Page 296 **Written Exercises** 1. $\frac{11}{13}$ 3. $-\frac{5}{17}$ 5. $\frac{16}{a}$ 7. $-\frac{5}{y}$ 9. $\frac{12}{x+3}$ 11. $-\frac{1}{a-3}$ 13. $\frac{7x}{x+2}$
15. $-\frac{4a}{a+1}$ 17. $\frac{a+b}{2a+b}$ 19. $\frac{3r-5s}{r-2s}$ 21. $-\frac{4y}{19}$ 23. $-\frac{2x}{3y}$ 25. $-\frac{4}{9t}$

Page 296 **Review Capsule for Section 11.2** 1. $x-3$ 2. $-t+6$ 3. $4m-16$ 4. r^2-4r-2
5. $3n^2+2n-12$ 6. $-2s^2+2s-3$

Page 297 **P-1** Their values cannot be the same. **P-2** The numerator and denominator have no common factors.

Page 298 **Classroom Exercises** 1. $\frac{x+5}{x+2}$ 3. $\frac{3y+3}{y+3}$ 5. $\frac{4b+1}{b+2}$ 7. 1 9. $\frac{2x+5}{x-3}$ 11. $\frac{2x+2}{x+y}$

Page 299 **Written Exercises** 1. $\frac{2x+5}{x+y}$ 3. $\frac{6r-1}{r+s}$ 5. $\frac{4x-7}{x+2}$ 7. $\frac{4a-5}{a+b}$ 9. 1 11. 2 13. x 15. 3a 17. $x-4$ 19. $y-4$ 21. 4 23. $-\frac{6}{x+5}$

Page 299 **Review Capsule for Section 11.3** 1. 2, 2, 2, 3, x, x, x, y, y 2. 2, 2, 2, 7, r, r, s, s, s, s, t, t, t 3. 2, 2, $(m-3)$, $(m+3)$ 4. $(x-5)$, $(x-1)$ 5. 3, 2, $(x+4)$, $(x-3)$ 6. 2, 2, $(a+2)$, $(a-2)$, $(a-2)$, $(a+1)$

Page 301 **P-1** $(x-y)(x+y)3xy$

Page 302 **Classroom Exercises** 1. $3 \cdot 7$ 3. $2 \cdot 3 \cdot x \cdot x \cdot y$ 5. $2 \cdot 3 \cdot 5 \cdot a \cdot b \cdot c$ 7. $(x-3)(x-5)$ 9. $(a+b)(a-b)$ 11. $3(a-b)$ 13. $(2x-y)(x+2y)$ 15. $3 \cdot 5 \cdot x \cdot y$ 17. $3 \cdot 5(a+b)(a-b)$

Page 302 **Written Exercises** 1. 30 3. $12x^2y$ 5. $60a^2b$ 7. $(x-3)(x+3)$ 9. $3(x-y)(x+y)$ 11. $6b(b+5)(b-6)$ 13. $x^2(x-5)(x+5)$ 15. $5(x+2)^2$ 17. $(x-4)^2(x+2)$ 19. $(x-2)(x-3)(x+3)$ 21. No. Other possible monomials are 6c, 12c, and 24c.

Pages 303-304 **P-1** Multiplying by these numbers gives the LCM, 6x, as the denominator for both fractions. **P-2** $40x^2y$ **P-3** There is no common factor. **P-4** $\frac{5r}{1}$

Pages 304-305 **Classroom Exercises** 1. $\frac{x}{x}$ 3. $\frac{2}{2}$ 5. $\frac{2x}{2x}$ 7. $\frac{3s}{3s}$ 9. $\frac{3ab}{3ab}$; 6ab 11. 3a; 10xy 13. $\frac{3x}{3x}$; $45x^2$ 15. $\frac{2y}{2y}$; $\frac{3x}{3x}$; 4y; $4y + 15x$

Pages 305-306 **Written Exercises** 1. $\frac{29}{35}$ 3. $\frac{5b+7a}{ab}$ 5. $-\frac{7}{2x}$ 7. $\frac{7}{10x}$ 9. $\frac{15b-5a}{12ab}$ 11. $\frac{2b+3a}{a^2b^2}$ 13. $\frac{15b^2-a^2}{6a^2b^2}$ 15. $\frac{2a^2+3b^2}{abc}$ 17. $\frac{2+15a^3b}{3a^2b}$ 19. $\frac{40c^2d-3d}{8c}$ 21. $c = 2d$ 23. $\frac{2x-2}{2x+3}$ 25. Answers will vary. One solution is shown.

1	3	5

27. 400 meters

Page 306 **Mid-Chapter Review** 1. $-\frac{4}{t}$ 3. $\frac{8m}{m+8}$ 5. $\frac{3a-3}{a+1}$ 7. $m-2$ 9. $6t(t+3)(t-3)$ 11. $\frac{2b-15a}{5ab}$ 13. $\frac{c^2+2d^2}{6cd}$

Page 306 **Review Capsule for Section 11.5** 1. $\frac{x-2}{x-1}$ 2. $\frac{5x}{x-3}$ 3. $\frac{2(x+1)}{x-4}$ 4. $\frac{2x}{3}$

Page 308 **P-1** $\frac{x+4}{x+4}$; $\frac{x-1}{x-1}$

Pages 308-309 **Classroom Exercises** 1. $x(x+1)$ 3. $(x+5)(x-3)$ 5. $r(r+4)$ 7. $(2x-1)(x+3)$ 9. $x(x-1)$ 11. $\frac{a+3}{a+3}$; $3a+9$ 13. $\frac{y-2}{y-2}$; $2y-4$ 15. $\frac{x-2}{x-2}$; x^2-3x+2 17. $\frac{x-2}{x-2}$; $x^2+3x-10$

Page 309 **Written Exercises** 1. $\frac{3x+2}{x(x+1)}$ 3. $\frac{2a+8}{a(a+2)}$ 5. $\frac{x^2+2x+25}{(x+5)(x-3)}$ 7. $\frac{a^2-2a+10}{a(a-5)}$ 9. $\frac{a^2+b^2}{(a+b)(a-b)}$ 11. $\frac{a^2+6a-7}{(a-3)(a+2)}$ 13. $\frac{-7x+14}{(2x-1)(x+3)}$ 15. $\frac{2(x^2+2x-5)}{(x-1)(x-2)}$ 17. $\frac{-4b+1}{(b-1)(b-2)}$ 19. $\frac{6x+1}{x(x-1)}$

Page 309 **Review Capsule for Section 11.6** 1. $6(2x-1)$ 2. $5(x+3)(x-3)$ 3. $x(x-2)(x+5)$

Pages 310-311 **P-1** 2, x + 2; 3, x + 2 **P-2** There are no common factors. **P-3** x, x + 1 **P-4** x + 1, x + 2 **P-5** x(x + 1)(x + 2)

Page 312 **Classroom Exercises** 1. x(x − 1)(x + 1) 3. $(x + 3)^2(x - 2)$ 5. (y + 2)(y − 3)(y + 1) 7. $2(x + 1)(x - 2)^2$ 9. 2, x − 3 11. 4, x + 3 13. x + 2, x − 1 15. a + 5, a − 2

Pages 312-313 **Written Exercises**

1. $\frac{3}{2x+2} + \frac{1}{x^2-1} = \frac{3}{2(x+1)} + \frac{1}{(x-1)(x+1)}$

$= \frac{3}{2(x+1)}\left(\frac{x-1}{x-1}\right) + \frac{1}{(x-1)(x+1)}\left(\frac{2}{2}\right)$

$= \frac{3(x-1)}{2(x+1)(x-1)} + \frac{1(2)}{2(x+1)(x-1)} = \frac{3x-1}{2(x+1)(x-1)}$

3. $\frac{1}{x^2+x-6} - \frac{2}{x^2-2x} = \frac{1}{(x-2)(x+3)} - \frac{2}{x(x-2)}$

$= \frac{1}{(x-2)(x+3)}\left(\frac{x}{x}\right) - \frac{2}{x(x-2)}\left(\frac{x+3}{x+3}\right)$

$= \frac{1(x)}{x(x+3)(x-2)} - \frac{2(x+3)}{x(x+3)(x-2)} = \frac{-x-6}{x(x+3)(x-2)}$

5. $\frac{2x-3}{3(x-3)(x-1)}$ 7. $\frac{x-7}{(x+2)(x-1)(x-3)}$ 9. $\frac{7x+5}{10(x-5)(x+5)}$ 11. $\frac{x^2+4x-2}{2(x-2)(x+2)}$

13. $\frac{-2a-49}{(a-3)(a+2)(a-7)}$ 15. $\frac{2(x^2+x-3)}{(x+1)(x-1)(x+3)}$ 17. $\frac{-a^2+5a+4}{(2a+1)(a-3)(a-1)}$

Page 313 **Review Capsule for Section 11.7** 1. 9r 2. t(t − 1) 3. 4(q + 2)

Pages 314-315 **P-1** 3 **P-2** 3x(x + 2)

Page 316 **Classroom Exercises** 1. 0, 3 3. 0, −1 5. $-\frac{1}{2}$ 7. 4(x − 3) 9. 2x(5x − 1) 11. 2x(x + 3)

Page 316 **Written Exercises** 1. x = 4 3. x = 0 5. x = 16 7. No solution 9. x = 15 11. x = −8 13. $x = -\frac{3}{4}$ 15. $x = \frac{2}{5}$ 17. 24

Pages 317-318 1. Tweeter Electronics 3. 600; 700; 900 5. No 7. No; the bars would not be as easily compared and the reader would be more likely to use the vertical scales. 9. The graph at the right 11. The graph at the right; this graph is a more impressive representation of the company's profits. 13. State: \$2000; Out-of-State: \$3200; Private: \$4800 15. No 17. No 19. No 21. No; since the scale begins with 0, the lengths of the bars can be compared directly.

Pages 319-320 **Chapter Review** 1. $\frac{8}{3x}$ 3. $\frac{a}{a+5}$ 5. 1 7. $\frac{2(a+2)}{5a-1}$ 9. 2 11. 6x(x − 2) 13. $y^2(y-2)(y-3)$ 15. 2(y + 2)(y − 2) 17. $\frac{5}{6a}$ 19. $\frac{9y-10x}{12xy}$ 21. $\frac{a^2+ab+a-b}{(a+b)(a-b)}$ 23. $\frac{4a^2-23a+3}{(a+7)(a-3)}$ 25. $\frac{2x^2-2x-7}{(x+1)(x+2)}$ 27. $\frac{3x-2}{3(x+2)}$ 29. $\frac{-9x+16}{(2x+3)(2x-3)}$ 31. $\frac{5x^2-4x+3}{x(x-1)^2(x+1)}$ 33. $x = \frac{11}{3}$ 35. $x = -\frac{15}{2}$ 37. No solution

CHAPTER 12 RELATIONS AND FUNCTIONS

Pages 322-323 **P-1** Right; Left **P-2** Up; Down **P-3** 3; −2 **P-4** They do not have the same location. **P-5** The first number is x. **P-6** The first coordinate, 5, is positive. **P-7** The second coordinate, 3, is positive. **P-8** Start at 0, and move left 2 units along the x axis; −2 **P-9** Down 4 units parallel to the y axis; −4 **P-10** (−2, −4)

Pages 323-324 Classroom Exercises 1. right 3, down 2 3. left 3, up 7 5. right 7, down 1 7. right 8, up 3 9. left 3, down 3 11. (2, 1) 13. (−3, −1) 15. (3, −2) 17. (2, −3) 19. (−4, −3)

Page 324 Written Exercises For Ex. 1-11, the moves described locate the given points. Always start at origin. 1. Move 1 unit right and 4 units up. 3. left 5, up 3 5. right 6, down 1 7. left 4, down 1 9. left 4, down 4 11. right 3, up 3 13. (−5, −1) 15. (−2, 3) 17. (−4, −2) 19. (4, 3) 21. (−1, 1) 23. (−2, −1)

Page 324 Review Capsule for Section 12.2 1. −4 and −3 2. 5 and 6 3. −1 and 0 4. −4 and −3 5. −4 and −3 6. −2 and −1

Pages 325-326 P-1 x negative, y negative; x positive, y negative **P-2** Start at 0 and move $2\frac{1}{2}$ units left. **P-3** Move $3\frac{1}{4}$ units up parallel to the y axis. **P-4** (0, 0); (−3, −4)

Page 327 Classroom Exercises 1. II 3. I 5. On the y axis, 5 units down from the origin 7. On the x axis, $5\frac{1}{2}$ units to the left of the origin

Page 327 Written Exercises For Ex. 1-19, each point may be located on a graph by the directions following each coordinate pair. Always start at (0, 0), the point of origin. Note in Ex. 5, 9, and 15 there is only one move to locate each point. 1. right $4\frac{1}{2}$ units, up 3 units 3. left 2, down $4\frac{1}{2}$ 5. up 4 7. left 3.5, up 2.2 9. left $3\frac{1}{2}$ 11. left 3, up 2.2 13. left 6, down 2 15. down 2 17. left 1, up 2 19. left 7, up 2 21. parallelogram

Page 327 Review Capsule for Section 12.3 1. $1\frac{1}{4}$ 2. $2\frac{3}{4}$ 3. −30 4. 2.1

Page 328 P-1 (2, 1), (0, 1), (−2, 1), (2, −1), (0, −1), (−2, −1)

Pages 329-330 Classroom Exercises For Ex. 1 and 3, sample answers are given. 1. (−3, 3), (0, 0), (3, 3) 3. (−2, 3), (0, 1), (3, −2)

5.

x	−2	−1	0	1	2
y	−2	−1	0	1	2

7.

x	−2	−1	0	1	2
y	4	1	0	1	4

9.

x	−2	−1	0	1	2
y	1	0	1	4	9

11. (1, 1), (−1, −1) 13. (1, 1), (−1, 1) 15. (1, 4), (−1, 0)

Pages 330-331 Written Exercises 1. {(1, 0), (2, 1), (3, 2), (4, 3)} 3. {(1, 1), (2, 4), (3, 9), (4, 16)} 5. {(1, 1), (2, 2), (3, 3), (4, 4)} 7. {(1, 1), (2, $\sqrt{2}$), (3, $\sqrt{3}$), (4, 2)} 9. {(1, 1), (2, 1), (3, 1), (4, 1)} 11. {(1, 1), (2, 3), (3, 7), (4, 13)} For Ex. 13-15, each consists of 4 points. To locate each point, start at origin. Move in the stated direction the stated number of units. 13. (1 right, 2 up), (1 right, 3 up), (3 left, 5 up), (3 left, $4\frac{1}{2}$ down) 15. (Note: $\pi \approx 3.1$); (origin), (up π), (down π), (right π) For Ex. 17-27, each graph consists of 4 points. Ordered pairs are given. 17. (−1, 0), (0, 1), (1, 2), (2, 3) 19. (−1, −6), (0, −7), (1, −6), (2, −3) 21. (−1, 1), (0, 0), (1, 1), (2, 2) 23. (−1, 1), (0, 1), (1, 1), (2, 1) 25. (−1, −1), (0, 0), (1, −1), (2, −2) 27. (−1, −1), (0, −1), (1, 1), (2, 5) 29. Loss; the 10% profit is $9 and the 10% loss is $11, resulting in a total loss of $2. 31. 888 + 88 + 8 + 8 + 8

Page 331 Mid-Chapter Review 1. (−4, −2) 3. (−2, 0) 5. (0, 3) 7. (−5, 3) For Ex. 9-13, each point may be located on a graph by the directions following each coordinate pair. Always start at (0, 0), the point of origin. 9. right 2 units, down 1 unit 11. left 2.5, down 3.5 13. right 4, down $3\frac{1}{4}$ 15. {(−2, 4), (−1, 3), (0, 2), (1, 3), (2, 4)}

Page 331 Review Capsule for Section 12.4 1. A(−2, 1), B(−2, −1), C(−4, 2), D(−4, −2) 2. A(−2, 3), B(2, 3), C(−1, 0), D(1, 0) 3. A(−3, −2), B(−1, −3), C(2, 1), D(2, −1)

Page 332 1. 8000 Btu's 3. 6000 Btu's 5. 3500 Btu's

Pages 333-334 P-1 −1, 0, 4, −3; 3, −2, 0 **P-2** Two x values are equal in A; no two x values are equal in B. **P-3** Show that no two x values of relation T are equal. **P-4** P and S **P-5** Yes; a vertical line will have only one point in common with the graph.

Page 335 Classroom Exercises 1. Domain: $\{3, -5, 4, 0\}$; Range: $\{7, 2, -1, 0\}$ 3. Domain: $\left\{3, \sqrt{2}, \pi, \frac{1}{3}\right\}$; Range: $\{8, -1, -2\}$ 5. Domain: $\{2, 3, 4, 5\}$; Range: $\{3, 4, 5, 6\}$ 7. Domain: $\left\{1, -1, 0, \frac{1}{2}, -\frac{1}{2}\right\}$; Range: $\left\{1, 0, \frac{1}{2}\right\}$ 9. Yes 11. Yes 13. Yes 15. Yes

Pages 335-336 Written Exercises 1. $D = \{5, -3, -1, 1\}$; $R = \{10, 2, 0, -3\}$ 3. $D = \left\{\frac{1}{2}, -1, -\frac{1}{2}, 2\right\}$ $R = \left\{2, \frac{1}{2}, -1\right\}$ 5. $D = \{-1, -2, \pi, 0\}$; $R = \{\sqrt{2}, \sqrt{3}, -1, -\pi\}$ 7. $D = \{0, 1, 1.5, -1, -\pi\}$; $R = \{0, 1, -1\}$ 9. $D = \left\{0, \frac{1}{2}, 1, 1\frac{1}{2}, -\frac{1}{2}, -1\right\}$; $R = \{0, 1, -1\}$ 11. Yes 13. Yes 15. No 17. No
For Ex. 19-25, apply the vertical line test to each graph. If the line has more than one point in common with a graph, the graph is not a function. 19. Yes 21. No 23. Yes 25. No

Page 336 Review Capsule for Section 12.5 1. (−2, 2), (1, 1), (7, −1) 2. (−2, −2), (1, 1), (0, 2) 3. (−3, 3), (−2, 2), (1, 1), (0, 0) 4. (−2, −2), (0, 0) 5. (−2, −2), (0, 0) 6. (0, 0)

Pages 337–338 P-1 −3 **P-2** {all real numbers greater than or equal to −1} **P-3** $\frac{5}{4}$

Pages 338-339 Classroom Exercises 1. 0, −2, 2, −4, 4 3. The vertical line does not cross the graph at more than one point. 5. {real numbers} 7. 0 9. {all real numbers greater than or equal to −1 and less than 3} 11. $1\frac{1}{2}$

Page 339 Written Exercises The graph is either a line, two lines, or a curve (parabola). Points on the graph are given. 1. Line; (0, 0), (2, −4) 3. Line; (0, 1), (2, −1) 5. Two lines (inverted V); (0, 0), (−2, −2), (2, −2) 7. Two lines (V-shaped); (−2, 1), (−1, 0), (0, 1), (2, 3) 9. Line parallel to the x axis; (0, −3), (2, −3) 11. Curve; (−2, −2), (−1, 1), (0, 2), (1, 1), (2, −2) 13. Curve; (−2, 4), $(-1, 1\frac{1}{2})$, (0, 0), $(1, -\frac{1}{2})$, (2, 0) 15. Curve; (−2, −6), (−1, −2), (0, 0), (1, 0), (2, −2)

Pages 341-342 Classroom Exercises 1. 20 meters 3. 75 meters 5. No 7. 18 kilometers 9. 30 kilometers

Page 342 Written Exercises 1. 7 meters 3. 10 meters 5. 38 meters 7. \$1.07 9. \$1.41 11. \$1.41

Page 344 1. \$240; Clue 2 only was used. 3. Can't tell. 5. 12 ounces; both Clue 1 and Clue 2 were used. 7. 5 miles per hour; both Clue 1 and Clue 2 were used. 9. 20; Clue 2 only was used.

Pages 345-346 Chapter Review Each point in Ex. 1-15 will be located by moving in the stated direction the stated number of units. 1. left 3, up 2 3. left 3, down 4 5. right 3, down 3 7. right 3, up 1 9. right 2, down $3\frac{1}{2}$ 11. right 2.25, up 1.75 13. up 3 15. left $2\frac{1}{2}$, down $3\frac{1}{2}$ For Ex. 17-21, each graph consists of four points, which are listed as a set of ordered pairs. 17. {(−1, −8), (0, −5),

(2, 1), (3, 4)} 19. {(−1, 0), (0, −1), (2, 3), (3, 8)} 21. {(−1, 2), (0, 1), (2, 1), (3, 2)} 23. D = $\left\{-\frac{1}{2}, 2, 4\right\}$; R = {0, 6, −3} 25. D = {−3, 0, −2}; R = $\left\{\frac{1}{2}, -5, 0, -4\right\}$ 27. Yes 29. No 31. The graph is a line through the points (0, −3), (1, −2), (2, −1), (3,0). 33. Two lines (V-shaped) through (−2,0), (0, −2), (2,0). 35. 30 feet 37. 60 feet 39. 83 feet

Pages 347-348 Cumulative Review for Chapters 1-12 1. Let x = cost of labor, then x − 500 = cost of parts. x + x − 500 = 2100; x = 1300, x − 500 = \$800 3. $y = \frac{21}{5}$ or $4\frac{1}{5}$ 5. $d < 16$ 7. $y > -\frac{23}{5}$ or $y > -4\frac{3}{5}$ 9. $-\frac{1}{125m^3}$ 11. $\frac{16x^{12}}{y^8}$ 13. 8 15. $2a^3n^4\sqrt{3n}$ 17. $2\sqrt{6}$ 19. $10\sqrt{2g}$ 21. −51 23. $-3c^3d^7e$ 25. 5x − 2 27. 3x(x − 4) 29. (g − 5)(g + 9) 31. (3x − 1)(2x − 5) 33. (2h − 5)(2h − 5) 35. 2(n − 3)(n + 6) 37. $-\frac{1}{2}$ 39. 2, −5 41. $\frac{3}{x-2}$ 43. $\frac{3}{ab}$ 45. $\frac{4x^2y}{3}$ 47. $-\frac{x-1}{x+9}$ 49. $\frac{x-1}{3}$ 51. $\frac{1}{x-2}$ 53. $\frac{17}{x+2}$ 55. $-\frac{1}{10y}$ 57. $\frac{7}{6(x-3)}$ 59. −2 61. 1 63. right 5, down 6 65. left $1\frac{1}{2}$, down 2 67. Yes 69. Two lines (V-shaped); {(−4, 2), (−2, 1), (2, 1), (4, 2)}

CHAPTER 13 LINEAR EQUATIONS

Pages 350-351 P-1 (3, 5), (−4, −9), etc. **P-2** Substitute 3 for x and 5 for y in y = 2x − 1. **P-3** m: −1, b: $\frac{1}{2}$; m: $\frac{2}{3}$, b: $-\frac{1}{4}$; m: −3, b: 10 **P-4** Functions a and b

Pages 351-352 Classroom Exercises 1. −6, −3, 0, 6 3. −49, −19, 1, 31 5. m: −3; b: 7 7. m: −1; b: −3 9. m: −7; b: 0 11. m: −0.5; b: 3.8

Page 352 Written Exercises For Ex. 1-11, four points are given. 1. (3, 0), (0, 3), (1, 2), (−2, 5) 3. (0, −1), (1, 2), (2, 5), (−1, −4) 5. (−1, −2), (2, 4), (0, 0), (3, 6) 7. (−3, −2), (1. −2), (0, −2), (2, −2) 9. (0, 3), (2, −5), (−2, 11), (3, −9) 11. (0, $\frac{1}{2}$), (2, $-\frac{1}{2}$), (−2, $\frac{3}{2}$), (3, −1) 13. Yes 15. Yes 17. Yes 19. Yes 21. No 23. Yes 25. 32 cm^2

Pages 353-354 P-1 In each the value of b is 2. **P-2** (0, 2) **P-3** x value: 0; y values: (A) 1, (B) −3, (C) 3, (D) −1 **P-4** a. $\frac{5}{2}$ or $2\frac{1}{2}$ b. 4

Pages 354–355 Classroom Exercises 1. 1 3. $\frac{2}{3}$ 5. $1\frac{1}{2}$ 7. $\frac{1}{2}$ 9. $-\frac{1}{3}$ 11. −1 13. 0 15. 6

Page 355 Written Exercises 1. 7 3. $\frac{7}{2}$ 5. $-\frac{5}{3}$ 7. 0.3 9. 6 11. 0 For Ex. 13-21, two points are given. 13. (0, 1), (−1, 0) 15. (0, −3), (−2, 1) 17. (0, 5), (4, 7) 19. (0, 0), (2, −6) 21. (0, $\frac{3}{2}$), (5, 4)

Page 355 Review Capsule for Section 13.3 In each exercies, two points are given. 1. (0,5), (2,6) 2. (0,3), (2,4) 3. (0,1), (2,2) 4. (0,0), (2,1) 5. (0, −3), (2, −2) 6. (0, −5), (2, −4)

Pages 356-357 P-1 2 **P-2** m **P-3** In all four the value of b is −1. **P-4** a. Slope: $-\frac{3}{4}$, y intercept: −2 b. Slope: 5, y intercept: $-\frac{1}{2}$ c. Slope: $-\frac{3}{2}$, y intercept: $\frac{5}{2}$ **P-5** I, II, and IV **P-6** I and III; II and IV

Pages 357-358 Classroom Exercises 1. Slope: 3; y intercept: 2 3. Slope: 1; y intercept: 5 5. Slope: $\frac{1}{3}$; y intercept: −6 7. y = 5x + 3 9. y = −1x + 0 or y = −x 11. y = 3x + 10

Page 358 **Written Exercises** 1. 1 3. $-\frac{1}{2}$ 5. 2 7. -3 9. $-\frac{1}{2}$ 11. -4 13. $y = -x + 5$ 15. $y = -\frac{1}{3}x - 3$ 17. $y = -x$ 19. I, III 21. II, IV 23. II, IV 25. I, III 27. II, IV 29. I, III

Page 358 **Review Capsule for Section 13.4** 1. $-\frac{3}{5}$ 2. $-\frac{3}{2}$ 3. 5 4. $\frac{5}{8}$ 5. $\frac{2}{3}$

Page 359 **P-1** $\frac{3}{2}$ **P-2** Vertical motion: 3 units; Horizontal motion: 2 units **P-3** 6 units; 4 units

Page 361 **Classroom Exercises** 1. $\frac{3}{5}$ 3. $\frac{3}{2}$ 5. $-\frac{1}{3}$ 7. -1 9. $\frac{1}{2}$ 11. -1 13. 2 15. $-\frac{1}{2}$ 17. 2

Pages 361-363 **Written Exercises** 1. 1 3. 0 5. $-\frac{3}{4}$ 7. $-\frac{7}{6}$ 9. -1 11. 0 13. $-\frac{3}{4}$ 15. $-\frac{3}{2}$ 17. $y = x + 5$ 19. $y = \frac{9}{4}x - 3$ 21. $y = \frac{8}{5}x$ 23. $y = -6$ 25. $y = -x + 5$ 27. $y = \frac{2}{3}x + 6$ In Ex. 29-33, a second point is given that can be found using the given slope. The line can be drawn using the two points. 29. (1, 3) 31. (6, −6) 33. (0, 3) 35. About 98 inches 37. About 27,000,000 miles

Page 363 **Mid-Chapter Review** For Exercises 1-5, 4 points are given. 1. (0, −3), (1, −2), (2, −1), (3, 0) 3. (−2, 5), (−1, 3), (0, 1), (1, −1) 5. (−2, −3), (0, −2), (2, −1), (4, 0) For Exercises 7 and 9, the y intercept and 2 ordered pairs are given. 7. y intercept: 3; (−1, 4), (1, 2) 9. y intercept: $\frac{5}{2}$; (1, 2), (3, 1) 11. $y = -\frac{1}{4}x - 1$ 13. -1 15. 0

Pages 364-365 **P-1** They appear to be parallel. **P-2** a and c **P-3** 0

Pages 365-366 **Classroom Exercises** 1. Yes 3. Yes 5. $y = 4x + 3$ 7. $y = \frac{3}{2}x + 10$ 9. $y = -5.3x + 0.6$

Page 366 **Written Exercises** In Ex. 1-5, each graph is a pair of parallel lines. Two points are given for each line. 1. (0, 3), (1, 1); (0, −1), (−1, 1) 3. (0, −2), (4, 4); (0, 1), (−2, −2) 5. (0, 3), (4, 8); (0, 0), (4, 5) In Ex. 7-9, the graph is a set of three parallel lines. Two points are given for each line. 7. (0, −3), (2, 1); (0, 3), (−2, −1); (0, 0), (2, 4) 9. (0, 3), (−2, 1); (0, −3), (2, −1); (0, 0), (3, 3) The graphs of Ex. 11 and 13 are horizontal lines. One point is given for each line. 11. (0, 5) 13. $(0, \frac{3}{2})$ 15. Yes; the two lines would have the same slope.

Page 366 **Review Capsule for Section 13.6** 1. $x = -2$ 2. $x = 2$ 3. $x = 12$ 4. $x = \frac{3}{4}$

Pages 367-368 **P-1** (−4, 0) **P-2** a. 5 b. $-\frac{1}{2}$ c. -4 **P-3** $\{-2\}$ **P-4** There is more than one y value for the x value, −2.

Page 369 **Classroom Exercises** 1. -3 3. 1 5. -2 7. $\frac{5}{2}$ 9. -5 11. 1 13. $4\frac{1}{2}$ 15. $x = -5$ 17. $x = -1$ 19. $x = 2\frac{1}{2}$

Page 369 **Written Exercises** 1. -6 3. 3 5. -2 7. 3 9. 4 11. -4 In Ex. 13-19, each graph is a straight line. Two ordered pairs are given for each line, one using the x intercept, the other using the y intercept. 13. (2, 0), (0, −4) 15. (6, 0), (0, −3) 17. (−3, 0), (0, 6) 19. $(\frac{5}{2}, 0)$, (0, −5) In Ex. 21 and 23, the graphs are vertical lines. One point is given for each line. 21. (5, 0) 23. $(2\frac{1}{2}, 0)$

Page 369 Review Capsule for Section 13.7 1. slope: 2; y intercept: 3 2. slope: 1.8; y intercept: 32 3. slope: π; y intercept: 0 4. slope: 1; y intercept: 273 5. slope: 2.5; y intercept: 3

Page 371 P-1 \$9

Page 371 Classroom Exercises 1. 38°C 3. −7°C 5. −29°C 7. 6 m 9. 7 m

Page 372 Written Exercises 1. See Ex. 2 on p. 357 ($y = 2x$). This graph is similar except that only the portion in the first quadrant is drawn. 3. The graph is a portion of a line and contains the points (−273, 0) and (0, 273). The graph is in quadrants I and II only. 5. 68°F 7. 104°F 9. −22°F 11. 18 km/hr 13. 54 km/hr 15. 0.6 cm 17. 1.3 cm

Page 373 1. 1102.5 meters 3. 314 meters

Pages 374-376 Chapter Review 1. No 3. Yes 5. The graph is a straight line. The coordinates of four points are given. (−1, 3), (0, 1), (1, −1), (2, −3) 7. −10 9. 3.8 11. $y = 4x - 1$ 13. $y = -2.8x + 2.9$ 15. II, IV 17. I, III 19. $-\frac{1}{2}$ 21. 0.8 23. −2 25. −3 27. $y = 2x - 2$ 29. $y = \frac{3}{4}x - \frac{5}{2}$ 31. The lines are parallel; two points are given. (0, −2), (4, 0); (0, 3), (4, 5); (0, 0), (4, 2) The graphs of Ex. 33 and 35 are horizontal lines. One point is given. 33. (0, −3) 35. (0, 3.5) 37. $-2\frac{1}{2}$ 39. 8 In Ex. 41 and 43, each graph is a straight line. Two ordered pairs are given for each, one using the x intercept, the other, the y intercept. 41. (0, 1), (−1, 0) 43. (0, −6), (2, 0) In Ex. 45 and 47, the graphs are vertical lines. One point is given for each line. 45. (−1, 0) 47. $(\frac{7}{2}, 0)$ 49. 19.6 m 51. 12.6 m

CHAPTER 14 SYSTEMS OF SENTENCES

Pages 378–379 P-1 The y intercept; The x intercept **P-2** $x > 2\frac{1}{2}$; $x \le 2\frac{1}{2}$

Page 380 Classroom Exercises

1.

x	0	2	4
y	2	0	−2

3.

x	−2	0	2
y	−1	0	1

5.

x	0	5	2
y	−5	0	−3

7.

x	0	5	−4
y	$-\frac{5}{3}$	0	−3

Page 380 Written Exercises In Ex. 1-7, each graph is a straight line. Three ordered pairs from the Classroom Exercises are shown for each exercise. 1. (0, 2), (2, 0), (4, −2) 3. (−2, −1), (0, 0), (2, 1) 5. (0, −5), (5, 0), (2, −3) 7. $(0, -\frac{5}{3})$, (5, 0), (−4, −3) In Ex. 9-17, each graph is a straight line. Two points are given for each line. 9. (1, −1), (3, 2) 11. (−2, −3), (2, −2) 13. (0, 0), (5, 4) 15. (−1, 0), (0, −1) 17. (0, −6), (2, 0)

Page 380 Review Capsule for Section 14.2 1. (−1, 2) 2. (−1, 1) 3. (−3, 14)

Page 381 P-1 (−3, −2)

Page 383 Classroom Exercises 1. No 3. Yes 5. No 7. Yes 9. Yes 11. {(−2, −1)} 13. $\{(-1\frac{1}{2}, 2)\}$

Page 384 Written Exercises In Ex. 1-17, each graph is a pair of straight lines. The solution set is given for each exercise. 1. {(1, 3)} 3. {(−1, 4)} 5. {(−3, 4)} 7. ϕ 9. {(5, −2)} 11. {(0, 0)} 13. {(0, 0)} 15. {(1, −3)} 17. {(1, 5)} 19. $k = 2$ 21. $k = -1$

Page 385 P-1 a. $2x = 9$ b. $3y = 12$ c. $5x = 10$ **P-2** a. $2y = 8$ b. $4x = 12$ c. $2x = -6$

Page 386 Classroom Exercises 1. $2x = 6$ 3. $5y = 15$ 5. $3x = 12$ 7. $x = 4$ 9. $x = 1$

Pages 386-387 **Written Exercises** 1. $\{(2, 4)\}$ 3. $\{(-4, 2)\}$ 5. $\{(-2, 1)\}$ 7. $\left\{(4, \frac{3}{2})\right\}$ 9. $\{(6, 0)\}$ 11. $\left\{(-1, \frac{3}{2})\right\}$ 13. $\{(-2, 1)\}$ 15. $\{(2, 7)\}$ 17. $\left\{(\frac{1}{8}, -\frac{5}{2})\right\}$

Page 387 **Mid-Chapter Review** For Ex. 1-5, each graph is a line. Two points on each line are given. 1. (0, 5), (5, 0) 3. (0, 1), (1, –3) 5. (0, 2), (4, 5) 7. $\{(-2, -3)\}$ 9. $\{(2, -2)\}$ 11. $\left\{(2, \frac{1}{3})\right\}$

Page 387 **Review Capsule for Section 14.4** 1. 3 2. $\frac{1}{2}$ 3. –4 4. 1 5. –1 6. $2x - 2y$ 7. $3x + 3y$ 8. $3x + 6y$ 9. $8x - 12y$ 10. $-9x + 15y$ 11. $-2x + 2y$ 12. $-4x + 2y + 6$ 13. $-15x + 3y + 15$

Pages 388-389 **P-1** Multiply each side of eq. 1 by: (a.) 3; (b.) –1; (c.) –7 **P-2** a. Multiply each side of equation 1 by 5 and each side of equation 2 by 8. b. Multiply each side of equation 1 by 4 and each side of equation 2 by 3. c. Multiply each side of equation 1 by 3 and each side of equation 2 by –2.

Pages 389-390 **Classroom Exercises** 1. Multiply equation 1 by 2. 3. Multiply equation 2 by –2. 5. Multiply equation 1 by 3 and equation 2 by 4.

Page 390 **Written Exercises** 1. $\{(4, 2)\}$ 3. $\{(0, 2)\}$ 5. $\left\{(\frac{1}{2}, 4)\right\}$ 7. $\{(2, -3)\}$ 9. $\left\{(\frac{9}{7}, -\frac{10}{7})\right\}$ 11. $\{(0, 0)\}$ 13. $\{(0, 0)\}$ 15. $\{(2, -2)\}$ 17. $(1, -\frac{1}{3})$ 19. No 21. Yes 23. No

Page 391 **Review Capsule for Section 14.5** 1. $x + y = -12$ 2. $y - x = 7$ 3. $x = y - 6$ 4. $y = x + 13$

Page 392 **Classroom Exercises** 1. First number: x; second number: y; $x + y = 25$ 3. Jerry's age: x; Mary's age: y; $x + y = 25$ 5. Width: x; length: y; $x + y = 51$ 7. First number: x; second number: y; $2x + 3y = 19$

Page 393 **Written Exercises** 1. Larger: x; smaller: y; $x + y = 28$; $x - y = 3$; Answer: $15\frac{1}{2}$, $12\frac{1}{2}$ 3. First: x; second: y; $x + y = -12$; $x - y = 42$; Answer: 15, –27 5. Smaller: x; greater: y; $2x + 2 = y$; $3x + 2y = 39$; Answer: 5, 12 7. Length: x; width: y; $x = 8 + y$; $2x + 2y = 34$; Answer: $4\frac{1}{2}$, $12\frac{1}{2}$ 9. Ann's age: x; John's age: y; $y + 7 = x$; $x + y = 45$; Answer: 26, 19 11. Sam's crate: x; Tina's crate: y; $x = y - 6$; $2y = 3x + 5$; Answer: 13, 7

Page 394 1. 29.6 grams 3. 21.9 grams

Pages 395-396 **Chapter Review** For Ex. 1-5, each graph is a straight line. Two points are given for each line. 1. (0, 3), (2, –1) 3. (–1, 0), (4, 1) 5. (0, –2), (3, 0) 7. $\{(-3, 2)\}$ 9. $\{(1, 2)\}$ 11. $\{(2, -1)\}$ 13. $\{(-4, 9)\}$ 15. $\{(9, -4)\}$ 17. ϕ 19. Greater: x; smaller: y; $x + y = 8$; $4x = 4y + 2$; Answer: $4\frac{1}{4}$, $3\frac{3}{4}$ 21. Width: x; length: y; $3x + 2y = 40$; $2x + 2y = 34$; Answer: 6, 11

CHAPTER 15 MORE ON SYSTEMS OF SENTENCES

Page 399 **P-1** $2x - 3$ **P-2** $\frac{1}{2}y + 2$

Page 400 **Classroom Exercises** In Ex. 1-9, other answers can also be given. 1. $x = 2y + 3$ 3. $x = 1 - 3y$ 5. $x = 3 - 3y$ 7. $x + (2x - 3) = 5$ 9. $-2(3y + 1) + 5y = 2$

Page 400 **Written Exercises** 1. $\{(6, 2)\}$ 3. $\{(14, 10)\}$ 5. $\{(10, 6)\}$ 7. ϕ 9. $\{(3, -4)\}$ 11. $\{(0, -2)\}$

Page 401 **P-1** Because xy represents the product of the digits, not the number.

Page 403 **Classroom Exercises** 1. $x + y = 12$ 3. $y = x - 5$ 5. $5x = 7y$ 7. $x + y = 2x - 1$ 9. $y - x = -8$

Page 404 **Written Exercises** 1. $y = x - 2$; $x + y = 12$; Ans: 75 3. $x = 2y$; $x = y + 4$; Ans: 84 5. $x + y = 15$; $2x = 3y - 5$; Ans: 87 7. $y = x + 1$; $10x + y = 8y + 2$; Ans: 34 9. $x - y = -4$; $10x + y = 2(x + y) + 3$; Ans: 15

Page 404 **Review Capsule for Section 15.3** 1. $x = 48$ 2. $x = 100$ 3. $x = 150$ 4. $x = 450$

Page 405 **P-1** 20 grams

Pages 406-407 **Classroom Exercises** 1. 8 3. 14 5. $0.06y$ 7. 120 9. 15; 485 11. 40; 30

Pages 407-408 **Written Exercises** 1. $0.08y$; $0.05(400)$; $x + y = 400$; $0.04x + 0.08y = 0.05(400)$; Ans: 4% solution: 300 g, 8% solution: 100 g 3. $0.12x$; $0.20(48)$; 15%; $0.15y$; $x + 48 = y$; $0.12x + 0.20(48) = 0.15y$; Ans: 12% solution: 80 g, 15% solution: 128 g 5. Let x = grams of 15% solution, then y = grams of 30% solution. $x + y = 150$; $0.15x + 0.30y = 0.18(150)$; Ans: 15% solution: 120 g, 30% solution: 30 g 7. Let x = grams of $2\frac{1}{2}$% solution, then y = grams of 4% solution. $x + 250 = y$; $0.025x + 0.10(250) = 0.04y$; Ans: $2\frac{1}{2}$% solution: 1000 g, 4% solution: 1250 g 9. 1806 11. Maria, Carol, Glenn, Luis

Page 409 **Mid-Chapter Review** 1. $\{(-1, -5)\}$ 3. $\{(3, 9)\}$ Let x = ten's digit and y = one's digit. $x = 2y + 2$; $10x + y = 9x + 11$; Ans: 83 7. Let x = grams of 2% solution, then y = grams of 3% solution. $x + y = 800$; $0.02x + 0.03y = 0.025(800)$; Ans: 400 g of 2% solution, 400 g of 3% solution

Page 411 1. 12 hours 3. 440 hours

Pages 412-413 **P-1** First row: π, -2; Second column: -2, $-\sqrt{3}$ **P-2** a. -15 b. 0 c. 2 **P-3** a. $2p - 3q$ b. $-5x$ c. $s^2 + t^2$

Page 414 **Classroom Exercises** 1. 2 3. 13 5. -3 7. 17 9. 7 11. -13 13. -48 15. 0 17. 6 19. -3

Page 414 **Written Exercises** 1. -28 3. 0 5. -2 7. 4 9. 0 11. -223 13. $\begin{vmatrix} 2 & -1 \\ 1 & -3 \end{vmatrix}$; -5 15. $\begin{vmatrix} -2 & 5 \\ 1 & -3 \end{vmatrix}$; 1 17. $\begin{vmatrix} 2 & -1 \\ 5 & -1 \end{vmatrix}$; 3 19. $\begin{vmatrix} 5 & 3 \\ -3 & 1 \end{vmatrix}$; 14 21. $\begin{vmatrix} 4 & 5 \\ 5 & 6 \end{vmatrix}$; -1

Page 415 **P-1** 13; -9; 7

Page 417 **Classroom Exercises** 1. $x = \dfrac{\begin{vmatrix} 2 & -1 \\ -1 & 2 \end{vmatrix}}{\begin{vmatrix} 3 & -1 \\ 1 & 2 \end{vmatrix}}$, $y = \dfrac{\begin{vmatrix} 3 & 2 \\ 1 & -1 \end{vmatrix}}{\begin{vmatrix} 3 & -1 \\ 1 & 2 \end{vmatrix}}$ 3. $x = \dfrac{\begin{vmatrix} 1.5 & -0.5 \\ -2.5 & 1.5 \end{vmatrix}}{\begin{vmatrix} 0.5 & -0.5 \\ -1.5 & 1.5 \end{vmatrix}}$, $y = \dfrac{\begin{vmatrix} 0.5 & 1.5 \\ -1.5 & -2.5 \end{vmatrix}}{\begin{vmatrix} 0.5 & -0.5 \\ -1.5 & 1.5 \end{vmatrix}}$

Page 417 **Written Exercises** 1. $\{(5, 1)\}$ 3. $\{(-2, 1)\}$ 5. $\left\{\left(-\frac{3}{23}, -\frac{42}{23}\right)\right\}$ 7. $\left\{\left(\frac{7}{4}, \frac{9}{4}\right)\right\}$ 9. $\left\{\left(0, \frac{1}{2}\right)\right\}$ 11. ϕ

Page 417 **Review Capsule for Section 15.6** 1. $5n$ 2. $100t$ 3. $5w$ 4. $12y$ 5. $850k$ 6. $75n$

Page 420 **Classroom Exercises** 1. $5x$; $10y$; 650; $x + y = 60$; $5x + 10y = 650$ 3. x; $325x$; y; $135y$; 55; $x + y = 55$; $325x + 135y = 11{,}415$ 5. x; $15x$; y; 25¢; $25y$; $x = y + 20$; $15x + 25y = 1500$

Page 421 **Written Exercises** 1. Records: x; tapes: y; $x + y = 9$; $6x + 7y = 58$; Answer: 5 records, 4 tapes 3. Tomatoes: x; beans: y; $x + y = 50$; $35x + 45y = 2030$; Answer: 22 cans of tomatoes, 28 cans of beans 5. Fives: x; tens: y; $x + y = 86$; $5x + 10y = 675$; Answer: 37 fives, 49 tens 7. Students: x; adults: y; $x + y = 82$; $35x + 50y = 3245$; Answer: 57 students, 25 adults

Pages 423-424 Classroom Exercises 1. $20x$ 3. $10(x + y)$ 5. $3x + 3y$

Pages 424-425 Written Exercises 1. 7; 7y; $x + y = 5$; $70x + 7y = 224$; Ans: 3 hours by bus, 2 hours hiking 3. 3x; 3; 3y; 3720; $x = 30y$; $3x + 3y = 3720$; Ans: speed of plane: 1200 km/hr, speed of carrier: 40 km/hr 5. 4500; 15; 4500; $18(x - y) = 4500$; $15(x + y) = 4500$; Ans: speed of runner: 275 m/min, speed of wind: 25 m/min 7. 6; 48; 4; 48; $6(x - y) = 48$; $4(x + y) = 48$; Ans: speed in still water: 10 km/hr, speed of current: 2 km/hr

Pages 426-427 1. B 3. A 5. B 7. B 9. C 11. A 13. B

Pages 428-430 Chapter Review 1. $\{(-3, 3)\}$ 3. $\{(2, -1)\}$ 5. Let x = ten's digit and y = one's digit; $x = 3y$; $x + y = 12$; Ans: 93 7. Let x = ten's digit and y = one's digit; $10x + y = 9y$; $x + y = 9$; Ans: 45 9. Let x = grams of 12% solution, then y = grams of 20% solution. $x + y = 800$; $0.12x + 0.20y = 0.17(800)$; Ans: 300 g of 12% solution, 500 g of 20% solution 11. 41 13. -1 15. $\{1, -1)\}$ 17. $\left\{\left(\frac{53}{11}, -\frac{19}{11}\right)\right\}$ 19. Adult: x; student: y; $x + y = 204$; $125x + 75y = 21500$; Ans: 124 adult tickets, 80 student tickets 21. 3; 3x; 2; 2y; 1000; $x = 6y$; $3x + 2y = 1000$; Ans: speed of plane: 300 km/hr, speed of jeep: 50 km/hr

Pages 431-432 Cumulative Review: Chapters 1-15 1. $x < -4$ 3. $x < -5$ 5. $\frac{1}{a^5b^5}$ 7. $-\frac{27x^3y^{12}}{125}$ 9. $\sqrt{15}$ 11. $2x^3y^4\sqrt{3y}$ 13. $8c^2d^6e^5$ 15. $(4n + 1)(3n - 5)$ 17. $3(x - 3)(x - 3)$ 19. $\frac{a^2}{a - 5}$ 21. 6 23. $\frac{x^2 - 6x + 13}{x^2 - 4}$ 25. 23 For Ex. 27 and 29, each graph is a line. Two points are given. 27. (0, 3), (1, 1) 29. (0, 0), (2, 8) 31. Slope: -1; y intercept: 7 33. $\frac{2}{5}$ 35. $-\frac{9}{4}$ 37. No 39. $\{(2, 0)\}$ 41. $\{(-2, -3)\}$ 43. $\{(-4, -3)\}$ 45. $\{(12, 4)\}$ 47. Width: x; length: y; $x = 2y - 7$; $2x + 2y = 19$; Ans: width: 4 meters, length: $5\frac{1}{2}$ meters 49. $\{(1, 2)\}$ 51. $\left\{\left(\frac{7}{5}, \frac{1}{5}\right)\right\}$ 53. Adult: x; student: y; $x + y = 520$; $3x + 2y = 1500$; Ans: 460 adult tickets, 60 student tickets

CHAPTER 16 SYSTEMS OF INEQUALITIES

Page 434 P-1 Many answers are possible; (0, -6) is one answer.

Pages 435-436 Classroom Exercises In Ex. 1-7, many other answers are possible. 1. $(-2, -3)$, (0, 0), (2, 2) 3. $(-2, 3)$, (0, 0), $(2, -2)$ 5. $(-2, 0)$, $(0, -1)$, $(2, -3)$ 7. $(-2, 4)$, (0, 4), (2, 6) 9. $y < 2x + 1$ 11. $y \le 3x + 2$ 13. $y < -x - 10$

Page 436 Written Exercises In Ex. 1-11, many other answers are possible. 1. $(0, -2)$, (1, 0), $(-1, -4)$, (2, 0), $(-2, -5)$ 3. (0, 1), (1, 0), (2, 0), $(-1, 4)$, $(-2, 9)$ 5. $(0, -4)$, $(2, -3)$, $(4, -2)$, $(-2, -7)$, $(-4, -6)$ 7. (0, 0), (1, 0), $(-1, -1)$, (2, 0), $(-2, 0)$ 9. (0, 2), (1, 15), $(-1, 0)$, $(-2, -18)$, (2, 25) 11. (0, 2), (1, 3), $(-1, 2)$, (2, 5), $(-2, 1)$ 13. $y > 3x + 5$ 15. $y \le -4x + 6$ 17. $y > x + 3$ 19. $y > 3x - 15$ 21. $y \ge -\frac{2}{3}x + 2$ 23. $y > 2x - \frac{5}{2}$ 25. $y \ge \frac{1}{2}x + \frac{3}{14}$ 27. $y < -\frac{5}{3}x + \frac{1}{3}$

Page 436 Review Capsule for Section 16.2 In Ex. 1-8, each graph is a straight line. Two points are given for each. 1. (0, 2); (2, 0) 2. $(0, -1)$; (1, 1) 3. $(0, -3)$; (2, 0) 4. (0, 2); (4, 7) 5. $(0, \frac{3}{2})$; (1, 2) 6. $(0, -1)$; $(1, -4)$ 7. $(0, -2)$; $(5, -2)$ 8. $(-3, 0)$; $(-3, -2)$

Page 437 **P-1** The region where $y > 2x - 1$. The line $y = 2x - 1$. The region where $y < 2x - 1$.

Page 439 **Classroom Exercises** 1. Yes 3. No 5. Yes 7. No 9. $y > -x$ 11. $y < -\frac{1}{3}x - 1$

Page 439 **Written Exercises** In Written Ex. 1-17, each graph is a region that does not include its linear boundary. The location of the region with respect to the boundary is indicated for each exercise. 1. right of the boundary 3. right 5. below 7. right 9. above 11. left 13. left 15. left 17. above

Page 439 **Review Capsule for Section 16.3** 1. $y \leq 2$ 2. $y \leq x - 5$ 3. $y \geq 2x$ 4. $y \geq \frac{1}{2} - 3x$ 5. $y \leq -x + 1$ 6. $y \geq 2 - x$

Page 440 **P-1** $-2.9, 0, 5, 17$ **P-2** -3 **P-3** $-3, -2.9, 5, 17$ **P-4** (2, 0), (7, –5), (0, 2), (1, 1); (–2, 2), (2, –1), (0, 1), (1, 0)

Page 442 **Classroom Exercises** 1. Yes 3. Yes 5. Yes 7. Yes 9. Yes 11. $y \geq -x$ 13. $y \leq \frac{1}{2}x + 1$

Page 443 **Written Exercises** In Ex. 1-17, each graph is a region that has a straight line as its boundary. The boundary is included in the graphs of Ex. 1-15 but not in the graph of Ex. 17. The location of the region with respect to its boundary is indicated for each exercise. 1. right of the boundary 3. right 5. right 7. below 9. right 11. below 13. above 15. left 17. right 19. $y \leq x + 1$ 21. $y \geq -2x - 2$ 23. $y \leq -\frac{1}{3}x + 1$ 25. Yes 27. Yes

Page 444 **Mid-Chapter Review** 1. $y < -5x + 2$ 3. $y < \frac{3}{2}x + \frac{5}{2}$ 5. $y < \frac{4}{3}x - \frac{1}{2}$ In Ex. 7-11, each graph is a region that does not include its linear boundary. The location of the region with respect to the boundary is indicated for each exercise. 7. left of the boundary 9. below 11. left In Ex. 13-17, each graph is a region that has a straight line as its only boundary. The boundary is included in each of the graphs. 13. left 15. left 17. right

Page 444 **Review Capsule for Section 16.4** 1. (–2, 3), (–4, 3) 2. (0, –1), (3, 5) 3. (2, –4) 4. (–4, 3)

Page 445 **P-1** 3.9, 1, –11 **P-2** –1.9, 0, 17 **P-3** –1.9, 0, 2, 3.8

Page 447 **Classroom Exercises** 1. IV 3. II

Page 447 **Written Exercises** In Ex. 1-11, each graph is a region with two straight lines as its boundaries. The boundaries are not included in the graphs. The quadrants that contain the region are indicated, as are the equations of the boundaries. 1. Quadrants: III, IV; Boundaries: Left: $y = x + 1$, Right: $y = -x - 1$ 3. Quadrants: I, II, and III; Boundaries: Bottom: $y = 2x + 1$, Top: $y = -x + 2$ 5. Quadrants: II, III; Boundaries: Top: $x + 2y = 4$, Bottom: $x - 2y = -4$ 7. Quadrants: III and IV; Boundaries: Left: $x - 3y = 3$, Right: $2x + y = -1$ 9. Quadrants: II and III; Boundaries: Right: $x = -1$, Top: $y = 2$ 11. Quadrants: I, III, and IV; Boundaries: Left: $x = -1$, Top: $y = 3x$

Page 450 **Classroom Exercises** 1. c 3. d

Page 450 **Written Exercises** In Ex. 1-15, each graph is a region with two straight lines as its boundaries. The quadrants that contain the region are indicated, as are the equations of the boundaries. 1. Quadrants: I and II; Boundaries: Left: $y = -x - 1$, Right: $y = x + 1$; Boundaries included 3. Quadrants: I, II, and III; Boundaries: Top: $y = -x + 1$, Bottom: $y = 2x$; Bottom boundary included 5. Quadrants: II and III; Boundaries: Top: $x + y = 4$, Bottom: $x - y = -4$; Boundaries included 7. Quadrants: III and IV; Boundaries: Left: $x - 3y = 2$, Right: $2x + y = -2$; Right boundary included 9. Quadrants: I and II;

Boundaries: Right: $y = x + 2$, Left: $y = -x - 2$; Right boundary included 11. Quadrant: III; Boundaries: Top: $y = x$; Right: $y = 3x$; Boundaries included 13. Quadrants: II and III; Boundaries: Top: $y = 3$, Right: $x = -2$; Right boundary included 15. Quadrants: I and IV; Boundaries: Top: $y = 3x$; Left: $x = 0$; Top boundary included 17. $y \geq x$; $y \geq -x$ 19. $y \leq x$; $y \geq -x$

Page 452 Table: Plant A: $\frac{x}{60}$, 1.30x; Plant B: $\frac{y}{40}$, 1.35y Plant A should produce 7200 calculators and Plant B should produce 3000 calculators. The cost would be $13,410.

Pages 453-454 **Chapter Review** In Ex. 1 and 3, many other answers are possible. 1. (0, −2), (1, 0), (2, −6), (−1, 0), (−2, 2) 3. (0, 1), (2, 5), (−2, −4), (4, 2), (−4, −10) 5. $y \geq 3x - 2$ 7. $y > \frac{1}{3}x + \frac{5}{3}$ In Ex. 9-13, each graph is a region that does not include its linear boundary. The location of the region with respect to the boundary is indicated for each exercise. 9. above the boundary 11. left 13. below In Ex. 15-19, each graph is a region that includes its linear boundary. The location of the region with respect to the boundary is indicated for each exercise. 15. above 17. above 19. above In Ex. 21 and 23, each graph is a region with two straight lines as its boundaries. The boundaries are not included in the graphs. The quadrants that contain the region are indicated, as are the equations of the boundaries. 21. Quadrants: II, III, and IV; Boundaries: Right: $y = -x - 1$, Top: $y = \frac{1}{2}x + 2$ 23. Quadrant: II; Boundaries: Top: $2x + y = -3$, Bottom: $x + y = 2$ In Ex. 25, the graph is a region with two straight lines as its boundaries. The quadrants that contain the region are indicated, as are the equations of the boundaries. 25. Quadrants: I and IV; Boundaries: Bottom: $x + 2y = 4$, Left: $-3x + y = -1$; Boundaries included

CHAPTER 17 QUADRATIC FUNCTIONS

Page 456 **P-1** a and c

Page 457 **Classroom Exercises** 1. Yes 3. No 5. No 7. No 9. No 11. Yes 13. Yes 15. No

Page 457 **Written Exercises** 1. No 3. Yes 5. No 7. Yes 9. No 11. Yes 13. (−3, −9), (−1, −1), (2, −4) 15. (−3, 4), (−1, 0), (2, 9) 17. (−3, 6), (−1, −2), (2, 1) 19. (−3, −9), (−1, 1), (2, −14)

Page 458 **P-1** a. $-1\frac{3}{4}$ b. $2\frac{1}{4}$ c. $-3\frac{3}{4}$ **P-2** {all real numbers greater than or equal to −4}

Pages 459-460 **Classroom Exercises** 1. y: 4, 1, 0, 1, 4 3. y: 6, 3, 2, 3, 6 5. y: 6, 1, 0, 3, 10 7. 2 9. $2\frac{1}{4}$ 11. 0, 2 13. $\frac{1}{2}$, $1\frac{1}{2}$

Page 460 **Written Exercises** In Ex. 1-5, each parabola turns at a high point (maximum), or low point (minimum). This point is given for each exercise. 1. Turns at (0, 1); low point 3. Turns at (1, 4); high point 5. Turns at (1, 2); low point In Ex. 7-11, the high or low point and three points on the parabola are given. 7. Turns at (0, −2); low point; (−1, −1); (0, −2); (1, −1) 9. Turns at $(-\frac{1}{2}, \frac{1}{4})$; high point; (−1, 0); $(-\frac{1}{2}, \frac{1}{4})$; (0, 0) 11. Turns at (3, 1); low point; (2, 2); (3, 1); (4, 2) 13. 7 + 1 = 8; 9 − 6 = 3; 4 × 5 = 20

Pages 461-462 **P-1** 1. $a = -\frac{5}{2}$; $b = 0$; $c = 0$ 2. $a = 3$; $b = -4$; $c = -1$ 3. $a = 1$; $b = 3$; $c = 0$ **P-2** $y = \frac{1}{2}x^2$; $y = x^2$; $y = 2x^2$ **P-3** $y = -\frac{1}{2}x^2$; $y = -x^2$; $y = -2x^2$

Pages 462-463 Classroom Exercises 1. The point with coordinates (0, 0) 3. Downward 5. {all real numbers greater than or equal to 0} 7. Downward 9. Downward 11. $y = 3x^2$ 13. $y = 0.09x^2$

Pages 463-464 Written Exercises In Ex. 1-9, the graphs are all parabolas opening upward or downward.

1.

x	−2	−1	0	1	2
y	4	1	0	1	4

, opens upward at (0, 0);

x	−4	−2	0	2	4
y	4	1	0	1	4

, opens upward at (0, 0)

3.

x	−3	−2	0	2	3
y	$-\frac{9}{2}$	−2	0	−2	$-\frac{9}{2}$

, opens downward at (0, 0);

x	−2	−1	0	1	2
y	−6	$-\frac{3}{2}$	0	$-\frac{3}{2}$	−6

, opens downward at (0, 0)

5.

x	−2	−1	0	1	2
y	−8	−2	0	−2	−8

, opens downward at (0, 0);

x	−2	−1	0	1	2
y	6	$\frac{3}{2}$	0	$\frac{3}{2}$	6

, opens upward at (0, 0)

7.

x	−2	−1	0	1	2
y	2	$\frac{1}{2}$	0	$\frac{1}{2}$	2

, opens upward at (0, 0);

x	−2	−1	0	1	2
y	0	$-\frac{3}{2}$	−2	$-\frac{3}{2}$	0

, opens upward at (0, −2)

9.

x	−4	−2	0	2	4
y	−12	−3	0	−3	−12

, opens downward at (0, 0);

x	−4	−2	0	2	4
y	−11	−2	0	−2	−11

, opens downward at (0, 1)

In Ex. 11 and 13, the parabolas open at the origin to the right or to the left of the y axis.

11.

x	1	1	0	4	4
y	1	−1	0	2	−2

, opens to the right

13.

x	2	2	0	8	8
y	1	−1	0	2	−2

, opens to the right

Page 464 Review Capsule for Section 17.4 1. $x^2 - 4x + 4$ 2. $x^2 + 6x + 9$ 3. $x^2 - 10x + 25$ 4. $x^2 + 12x + 36$ 5. $x^2 + 4.2x + 4.41$ 6. $x^2 - 5.4x + 7.29$

Page 465 P-1 $a = \frac{1}{2}$; $b = -2$; $c = 2$

Page 467 Classroom Exercises 1. 2 units right 3. 1 unit right 5. 2 units left 7. 10 units left 9. 7 units right 11. 6 units left

Pages 467-468 Written Exercises 1. 5 units right 3. 8 units left 5. $3\frac{1}{2}$ units left 7. 2.7 units right 9. Second parabola is 2 units to left of first. 11. Second parabola is 3 units to right of first. 13. Second parabola is $\frac{3}{2}$ units to left of first. 15. Second parabola is 2 units to right of first. 17. Second parabola is 1 unit to left of first. 19. 75 21. −149.36531

Page 468 Mid-Chapter Review 1. (−2, 4), $(-1, \frac{5}{2})$, (0, 0), $(1, -\frac{7}{2})$ 3. (−2, −1), (−1, −3), (0, −3), (1, −1) In Ex. 5, the high point and three points for the graph (parabola) are given. 5. High point: (0, 2); $(-1, 1\frac{1}{2})$, (0, 2), $(1, 1\frac{1}{2})$ In Ex. 7-11, three points are given for each curve (parabola). 7. $y = \frac{3}{4}x^2$: (0, 0), (2, 3), (−2, 3); $y = -\frac{3}{4}x^2$: (0, 0), (2, −3), (−2, −3) 9. $y = 2x^2$: (0, 0), (1, 2), (−1, 2); $y = \frac{1}{2}x^2$: (0, 0), (2, 2), (−2, 2) 11. $y = \frac{1}{2}x^2 + 1$: (0, 1), (2, 3), (−2, 3); $y = -2x^2 - 2$: (0, −2), (1, −4), (−1, −4) 13. 5 units right 15. 4 units left

Page 468 Review Capsule for Section 17.5 1. (−2, 9), (0, 3), (2, 1) 2. (−2, 7), (0, −3), (2, 3) 3. (−2, −5), (0, −5), (2, −21)

Pages 469-471 **P-1** Quadratic **P-2** 2 units up **P-3** Move 3 units right and 2 units up. **P-4** 3 units down **P-5** Move 4 units left and 3 units down. **P-6** If h is positive, it is to the right. If h is negative, it is to the left. **P-7** If k is positive, it is above. If k is negative, it is below.

Page 471 **Classroom Exercises** 1. Yes 3. No 5. Yes 7. a 9. k 11. 5 units right, 2 units down 13. 2 units left, 3 units down 15. 1 unit left, 1 unit down

Page 472 **Written Exercises** In Ex. 1-7, the ordered pairs for the first parabola and the motions to locate the second parabola are given. 1. $y = \frac{1}{2}x^2$: $(-2, 2), (-1, \frac{1}{2}), (0, 0), (1, \frac{1}{2}), (2, 2)$; $y = \frac{1}{2}(x - 2)^2 + 3$: move 3 units up, 2 units right 3. $y = -x^2$: $(-2, -4), (-1, -1), (0, 0), (1, -1), (2, -4)$; $y = -(x + 3)^2 + 2$: move 2 units up, 3 units left 5. $y = 2x^2$: $(-2, 8), (-1, 2), (0, 0), (1, 2), (2, 8)$; $y = 2(x + 1)^2 - 3$; move 3 units down, 1 unit left 7. $y = -\frac{3}{2}x^2$: $(-2, -6), (-1, -\frac{3}{2}), (0, 0), (1, -\frac{3}{2}), (2, -6)$; $y = -\frac{3}{2}(x - 3\frac{1}{2})^2 + 2\frac{1}{2}$: move $2\frac{1}{2}$ units up, $3\frac{1}{2}$ units right 9. $y = 2(x - 3)^2$ 11. $y = \frac{1}{2}(x + 2)^2 + 5$ 13. $y = \frac{1}{3}(x + 1)^2 - 5$ 15. $y = -(x - 3)^2 - 2$ 17. $y = 3x^2 - 5$

Page 472 **Review Capsule for Section 17.6** 1. 4; $(x - 2)(x - 2)$ 2. 9; $(x + 3)(x + 3)$ 3. $\frac{1}{4}$; $(x + \frac{1}{2})(x + \frac{1}{2})$ 4. $\frac{9}{4}$; $(x - \frac{3}{2})(x - \frac{3}{2})$ 5. 25; $(x + 5)(x + 5)$ 6. 81; $(x + 9)(x + 9)$ 7. $\frac{25}{4}$; $(x - \frac{5}{2})(x - \frac{5}{2})$ 8. $\frac{49}{4}$; $(x - \frac{7}{2})(x - \frac{7}{2})$

Page 473 **P-1** a. $a = -1, h = -3, k = -\frac{3}{4}$ b. $a = \frac{3}{2}, h = \frac{1}{4}, k = \frac{5}{4}$

Page 474 **Classroom Exercises** 1. $(x - (-1))^2$ 3. $(x - 3)^2$ 5. $(x - (-\frac{1}{2}))^2$ 7. $(x - \frac{5}{2})^2$

Page 474 **Written Exercises** 1. $x^2 - 6x + 11$ 3. $2x^2 - 4x - 5$ 5. $3x^2 + 12x + 15$ 7. $\frac{1}{2}x^2 - 2x + 8$ 9. $(x - (-1))^2 + 2$ 11. $(x - 2)^2 - 3$ 13. $(x - (-3))^2 - 14$ 15. $(x - 5)^2 - 26$ 17. $(x - \frac{1}{2})^2 + \frac{19}{4}$ 19. $(x - (-\frac{3}{2}))^2 - \frac{25}{4}$

Page 474 **Review Capsule for Section 17.7** 1. $a = 3$; $h = 2$; $k = 1$ 2. $a = 4$; $h = 5$; $k = -2$ 3. $a = \frac{1}{2}$; $h = -\frac{3}{2}$; $k = -5$

Pages 475-476 **P-1** Lowest point **P-2** $\frac{1}{2}$; 2; 3 **P-3** a. $(\frac{1}{2}, \frac{5}{2})$ b. $(-3, -7)$ **P-4** $x = 2$ **P-5** a. $h = 5$ b. $h = -3$ c. $h = \frac{1}{4}$ d. $h = -\frac{4}{3}$ **P-6** $(-3, -2)$

Pages 476-477 **Classroom Exercises** 1. (5, 2) 3. (−2, 6) 5. (0, 0) 7. $(\frac{1}{2}, \frac{3}{4})$ 9. (5, 0) 11. $x = -1$ 13. $x = \frac{1}{2}$ 15. $x = 0$ 17. $x = 0$

Page 477 **Written Exercises** 1. (1, 10) 3. (7, −2) 5. (−2, 3) 7. (−1, −8) 9. $(\frac{1}{2}, -\frac{2}{3})$ 11. $x = 5$ 13. $x = 3$ 15. $x = -1$ 17. $x = \frac{1}{3}$ 19. $x = -\frac{7}{2}$ 21. $x = -\frac{1}{10}$

Page 479 1. b. The median points are (20, 8) and (110, 58). d. About 36 hours 3. b. The median points are (64, 115) and (72, 180). d. About 148 pounds

Pages 480-482 Chapter Review 1. No 3. (−1, −5), (0, −1), (2, −11) In Ex. 5-9, the high or low point and three points on the parabola are given. 5. Turns at (0, −1); low point; (−1, 0); (0, −1); (1, 0) 7. Turns at $(1, \frac{1}{2})$; high point; (0, 0); $(1, \frac{1}{2})$; (2, 0) 9. Turns at (−1, −4); low point; (−2, −3); (−1, −4); (0, −3) In Ex. 11-17, three points are given for each parabola and the point at which it opens. 11. $(-1, -\frac{1}{2})$; (0, 0); $(1, -\frac{1}{2})$; downward at (0, 0). (−1, −2); (0, 0); (1, −2); downward at (0, 0). 13. $(-1, \frac{5}{2})$; (0, 0); $(1, \frac{5}{2})$; upward at (0, 0). $(-1, -\frac{5}{2})$; (0, 0); $(1, -\frac{5}{2})$; downward at (0, 0). 15. $(\frac{1}{2}, -1)$; (0, 0); $(\frac{1}{2}, 1)$; Turns right at (0, 0). 17. (−1, −1); (0, 0); (−1, 1); Turns left at (0, 0). 19. 3 units left 21. $\frac{5}{2}$ units right In Ex. 23 and 25, the ordered pairs for the first parabola and the motions to locate the second parabola are given. 23. $y = x^2$: (−2, 4), (−1, 1), (0, 0), (1, 1), (2, 4); $y = (x - 3)^2 - 2$; move 2 units down, 3 units right 25. $y = -\frac{1}{2}x^2$: (−2, −2), $(-1, -\frac{1}{2})$, (0, 0), $(1, -\frac{1}{2})$, (2, −2); $y = -\frac{1}{2}(x + 1)^2 + 3$: move 3 units up, 1 unit left 27. $y = -10(x + 5)^2$ 29. $y = \frac{5}{2}(x - 2)^2 - 3$ 31. $x^2 + 2x - 4$ 33. $-3x^2 + 12x - 11$ 35. $\frac{1}{2}x^2 + 4x$ 37. $(x - 3)^2 - 7$ 39. $(x - (-\frac{3}{2}))^2 - \frac{5}{2}$ 41. $(x - 1)^2 - \frac{1}{2}$ 43. (10, 5) 45. (6, 0) 47. x = 2 49. $x = -\frac{9}{2}$

CHAPTER 18 QUADRATIC EQUATIONS

Page 484 P-1 −1, 2

Pages 485-486 Classroom Exercises 1. $x^2 - 2x - 3 = 0$ 3. $3x^2 - 5x + 6 = 0$ 5. $x^2 + 2x + 4 = 0$ 7. 1, 3 9. 0, 4 11. 2

Page 486 Written Exercises 1. $x^2 - 5x + 3 = 0$ 3. $2x^2 - 5x + 10 = 0$ 5. $x^2 - 9 = 0$ 7. −3, 1 9. −3.4, 1.4 11. No real solutions 13. No real solutions 15. 2 17. 0.6, 3.4

Page 486 Review Capsule for Section 18.2 1. (x − 6)(x + 2) 2. (x − 5)(x − 2) 3. (x − 6)(x + 4) 4. (x + 8)(x − 7) 5. (2x − 1)(x + 4) 6. (3x − 2)(x − 1) 7. (7 + x)(1 − x) 8. (9x − 2)(x − 7)

Page 487 P-1 −5; 2

Page 489 Classroom Exercises 1. {3, −5} 3. {6, −6} 5. {−4, −7} 7. $\{\frac{1}{2}, -5\}$ 9. $\{-\frac{3}{2}, -1\}$

Page 489 Written Exercises 1. {0, −3} 3. {5, −5} 5. {0, 10} 7. {10, −10} 9. {−5, 1} 11. {−4, 1} 13. {−4} 15. {2} 17. {2, 6} 19. {3, 8} 21. {−9} 23. {−8, 9} 25. $\{-3, \frac{1}{3}\}$ 27. $\{-\frac{2}{3}, \frac{1}{2}\}$ 29. $\{-\frac{4}{3}, \frac{1}{4}\}$ 31. $\{-\frac{5}{4}, \frac{1}{2}\}$ 33. $\{-\frac{5}{2}, 9\}$ 35. $\{-1, \frac{3}{4}\}$

Pages 490-491 P-1 a. {4, −4} b. {7, −7} c. {10, −10} **P-2** −5 **P-3** 1; 16; 16; 16

Pages 491-492 Classroom Exercises 1. {1, −1} 3. {11, −11} 5. x − 6 = 6 or x − 6 = −6 7. $x - \frac{1}{2} = 4$ or $x - \frac{1}{2} = -4$ 9. 4; 2; No 11. 25; 25; Yes 13. $\frac{9}{4}$; $\frac{3}{4}$; No 15. x + 10, x + 10 17. $x - \frac{1}{4}, x - \frac{1}{4}$ 19. $x - \frac{3}{2}, x - \frac{3}{2}$

Page 492 Written Exercises 1. {3, −3} 3. {13, −13} 5. $\{\sqrt{7}, -\sqrt{7}\}$ 7. $\{5\sqrt{3}, -5\sqrt{3}\}$ 9. x − 5 = 6 or x − 5 = −6 11. x + 6 = 5 or x + 6 = −5 13. $x + \frac{1}{2} = \frac{1}{2}$ or $x + \frac{1}{2} = -\frac{1}{2}$ 15. {−12, 10}

17. $\{3+\sqrt{5}, 3-\sqrt{5}\}$ 19. $\{0, 1\}$ 21. $\left\{-\frac{2}{3}+\frac{\sqrt{5}}{3}, -\frac{2}{3}-\frac{\sqrt{5}}{3}\right\}$ 23. $\{-4\}$ 25. $\{5\}$ 27. $\left\{\frac{1}{4}\right\}$
29. $\left\{\frac{2}{5}\right\}$ 31. $\frac{9}{4}$ 33. 2.2967226 35. 4.53041 37. 1.4299625 39. 19.453883

Page 493 Mid-Chapter Review 1. $\{-4, 1\}$ 3. $\{1.3, -4.3\}$ 5. $\{0.6, -3.6\}$ 7. $\left\{\frac{1}{2}, -5\right\}$
9. $\left\{-\frac{1}{3}, 1\right\}$ 11. $\left\{-\frac{1}{3}, -\frac{1}{2}\right\}$ 13. $\{-9, 15\}$ 15. $\{-1, 2\}$ 17. $\{-2, -8\}$

Page 493 Review Capsule for Section 18.4 1. $x^2-6x=-3$ 2. $x^2+5x=4$ 3. $x^2-4x=-10$
4. $x^2-\frac{1}{2}x=-\frac{3}{2}$ 5. $2x^2-4x=-5$ 6. $2x^2+3x=10$

Page 496 Classroom Exercises 1. 4 3. 1 5. $\frac{1}{4}$ 7. $\frac{1}{36}$ 9. $\left\{\frac{\sqrt{3}}{2}, -\frac{\sqrt{3}}{2}\right\}$ 11. $\left\{\frac{\sqrt{5}}{2}, -\frac{\sqrt{5}}{2}\right\}$
13. $\left\{\frac{1}{2}+\frac{\sqrt{3}}{2}, \frac{1}{2}-\frac{\sqrt{3}}{2}\right\}$ 15. $x+9, x+9$ 17. $x+\frac{3}{2}, x+\frac{3}{2}$ 19. $x-\frac{1}{5}, x-\frac{1}{5}$

Page 496 Written Exercises 1. $\{-3, 5\}$ 3. $\{-5, 1\}$ 5. $\{-3, 2\}$ 7. $\{-1+\sqrt{6}, -1-\sqrt{6}\}$ 9. ϕ
11. $\{3+\sqrt{14}, 3-\sqrt{14}\}$ 13. $\{-4+\sqrt{3}, -4-\sqrt{3}\}$ 15. $\left\{\frac{3}{2}+\frac{\sqrt{13}}{2}, \frac{3}{2}-\frac{\sqrt{13}}{2}\right\}$ 17. $\left\{-\frac{3}{2}+\frac{\sqrt{29}}{2},\right.$
$\left.-\frac{3}{2}-\frac{\sqrt{29}}{2}\right\}$ 19. $\left\{-\frac{1}{5}+\frac{\sqrt{6}}{5}, -\frac{1}{5}-\frac{\sqrt{6}}{5}\right\}$ 21. $\left\{-\frac{3}{2}, 1\right\}$ 23. $\left\{-\frac{1}{3}, 2\right\}$ 25. $\left\{1+\frac{\sqrt{3}}{3}, 1-\frac{\sqrt{3}}{3}\right\}$
27. $\left\{\frac{1}{4}+\frac{\sqrt{41}}{4}, \frac{1}{4}-\frac{\sqrt{41}}{4}\right\}$

Page 496 Review Capsule for Section 18.5 1. $\frac{5}{2}$; -1 2. 2; -4 3. $\frac{2+\sqrt{6}}{4}$; $\frac{2-\sqrt{6}}{4}$

Page 499 Classroom Exercises 1. $a=1, b=3, c=-2$ 3. $a=1, b=5, c=0$ 5. $a=1, b=3, c=5$

Page 499 Written Exercises 1. $\{-6, 1\}$ 3. $\{-7, 1\}$ 5. $\{2, 5\}$ 7. $\{-2, 6\}$ 9. $\{3\}$ 11. $\left\{-\frac{2}{3}, \frac{1}{2}\right\}$
13. $\{\sqrt{3}, -\sqrt{3}\}$ 15. $\{0, 6\}$ 17. $\left\{\frac{3+\sqrt{29}}{10}, \frac{3-\sqrt{29}}{10}\right\}$ 19. $\{0.4, -2.4\}$ 21. $\{1.8, -0.3\}$
23. $\{2.4, 0.6\}$

Page 499 Review Capsule for Section 18.6 1. 33 2. −71 3. 44

Page 502 P-1 a. Two rational numbers b. Two irrational numbers c. No real numbers d. One rational number

Page 502 Classroom Exercises 1. Two rationals 3. One rational 5. Two irrationals 7. Two rationals 9. No reals 11. Two rationals 13. 9; Two rationals 15. 1; Two rationals 17. −64; No reals 19. 17; Two irrationals 21. −3; No reals

Page 502 Written Exercises 1. 25; Two rationals 3. 36; Two rationals 5. 17; Two irrationals 7. 0; One rational 9. 28; Two irrationals 11. −100; No reals 13. 49; Two rationals 15. 100; Two rationals 17 4; Two rationals 19. −24; No reals 21. 145; Two irrationals 23. 120; Two irrationals

Page 503 Review Capsule for Section 18.7 1. $x^2-4x-78=0$ 2. $2x^2+5x-102=0$ 3. $x^2-7x-1=0$ 4. $x^2+2x+4=0$ 5. $2x^2-x-128=0$ 6. $2x^2+x-10=0$

Page 505 Classroom Exercises 1. Consecutive integers: n, n + 1; Equation: $n(n+1)=210$
3. Integers: n, n + 3; Equation: $n(n+3)=108$ 5. Width: w, Length: w + 5; Equation: $w(w+5)=24$
7. Numbers: n, 3n; Equation: $n^2=(3n)^2-18$ or $n^2=9n^2-18$

Page 506 **Written Exercises** 1. Length: ℓ, Width: $\ell - 2$; Equation: $\ell(\ell - 2) = 80$; Ans: Length: 10, Width: 8 3. Consecutive integers: n, n + 1; Equation: $n(n + 1) = 30$; Ans: 5 and 6 or −6 and −5 5. Integers: n, n − 13; Equation: $n(n - 13) = -40$; Ans: −8 and 5 or −5 and 8 7. Equation: $(2x + 4)(3x) - (2x)(x) = 40$; Ans: Length: 8, Width: 6

Page 508 1. $\frac{3}{2}$; $\frac{5}{4}x$; $x^2 + \frac{5}{2}x$; $\frac{3}{2}$; $\frac{25}{16}$; $\frac{3}{2} + \frac{25}{16} = \frac{49}{16}$; $\frac{7}{4}$; $\frac{2}{4}$, or $\frac{1}{2}$ 3. $x = 1$ 5. $x = 2$ 7. $x = \frac{3}{2}$

Pages 509-510 **Chapter Review** 1. $x^2 + 5x - 2 = 0$ 3. $\{-1.7, 1.7\}$ 5. $\{-1, 10\}$ 7. $\{0, -5\}$ 9. $\{19, -19\}$ 11. $\{-2, 10\}$ 13. $\{-\frac{3}{4}, 1\}$ 15. $\{-\frac{1}{3}, \frac{1}{2}\}$ 17. 0; One rational 19. 76; Two irrationals 21. Width: w, Length: w + 4; Equation: $w(w + 4) = 60$; Ans: Width: 6 cm, Length: 10 cm

CHAPTER 19 PROBABILITY

Page 513 **Classroom Exercises** 1. 8 3. 24 5. 16

Page 514 **Written Exercises** 1. 18 3. 200 5. 676 7. 18

Page 515 **Review Capsule for Section 19.2** 1. $\frac{1}{4}$ 2. $\frac{2}{5}$ 3. $\frac{3}{7}$ 4. $\frac{5}{8}$ 5. $\frac{1}{4}$

Pages 517-518 **Classroom Exercises** 1. $\frac{1}{4}$ 3. $\frac{1}{4}$ 5. $\frac{1}{4}$ 7. $\frac{1}{2}$ 9. $\frac{3}{4}$ 11. 1 13. No; the two events do not include all possible outcomes.

Pages 518-519 **Written Exercises** 1. $\frac{1}{7}$ 3. $\frac{5}{14}$ 5. $\frac{1}{4}$ 7. $\frac{1}{2}$ 9. 1 11. No; the two events do not include all possible outcomes. 13. $\frac{1}{3}$ 15. $\frac{1}{4}$ 17. $\frac{1}{2}$ 19. 0 21. $\frac{3}{8}$ 23. $\frac{1}{2}$ 25. $\frac{1}{2}$ 27. Answers will vary.

Page 521 **P-1** a. $\frac{1}{18}$ b. $\frac{1}{4}$

Page 522 **Classroom Exercises** 1. $\frac{9}{16}$ 3. $\frac{1}{8}$ 5. $\frac{4}{7}$ 7. $\frac{3}{7}$

Pages 522-523 **Written Exercises** 1. $\frac{1}{36}$ 3. $\frac{5}{18}$ 5. $\frac{1}{6}$ 7. 0 9. $\frac{13}{20}$ 11. $\frac{9}{20}$ 13. $\frac{1}{2}$

Page 523 **Mid-Chapter Review** 1. 18 3. $\frac{1}{4}$ 5. 1 7. $\frac{1}{6}$ 9. $\frac{3}{5}$

Page 524 1. No; it is possible that the 8 plumbers surveyed are not representative of the larger population. 3. Yes; checking 10% of the pages will give a good indication of the book's quality. 5. No; those viewers of late night television who actually call in to vote are not a random sample of the community population.

Page 527 **Classroom Exercises** 1. $\frac{1}{9}$ 3. $\frac{1}{6}$ 5. $\frac{2}{9}$ 7. $\frac{1}{15}$

Pages 527-528 **Written Exercises** 1. $\frac{8}{45}$ 3. $\frac{1}{45}$ 5. $\frac{2}{15}$ 7. $\frac{7}{15}$ 9. $\frac{1}{45}$ 11. $\frac{7}{24}$ 13. $\frac{2}{45}$ 15. $\frac{1}{180}$ 17. They arrive at the same time. The total length of Latasha's paths to the right in the figure is the same as the length of Jerome's path to the right. Likewise, the total length of Latasha's paths down in the figure is the same as the length of Jerome's path down.

Page 528 **Review Capsule for Section 19.5** 1. 192 2. 82 3. 183 4. 195 5. 882 6. 798

Page 530 **Classroom Exercises** 1. $\frac{1}{50}$ 3. $\frac{33}{50}$ 5. 13 7. 429

Page 531 **Written Exercises** 1. $\frac{6}{25}$ 3. $\frac{2}{5}$ 5. $\frac{1}{25}$ 7. 276 9. 460 11. 46 13. 420

Pages 532-533 1. 30 3. 50 5. 21 7. 264 9. 12 students

Pages 534-535 **Chapter Review** 1. 45 3. 252 5. $\frac{3}{10}$ 7. $\frac{1}{5}$ 9. 0 11. $\frac{7}{10}$ 13. $\frac{3}{10}$ 15. $\frac{3}{25}$ 17. $\frac{1}{14}$ 19. $\frac{1}{5}$ 21. $\frac{9}{50}$ 23. 270

Pages 536-537 **Cumulative Review: Chapters 1-19** 1. $a = 2$ 3. $15e^9$ 5. $4x^2 - 49x + 55$ 7. $\frac{x+3}{2}$ 9. 2 11. $\frac{3x+19}{2(x-3)(x+3)}$ 13. $\{(2, -3)\}$ 15. $y > -3x + 1$ 17. $y \leq \frac{7}{2}x + 4$ In Ex. 19, the graph is a region with two straight lines as its boundaries. The quadrants that contain the region are indicated, as are the equations of the boundaries. 19. Quadrants: All; Boundaries: Right: $x = 3$, Bottom: $x - y = 3$; Boundaries included 21. $(-2, -4)$, $(0,0)$, $(3, -9)$ 23. $(-2,9)$, $(0,1)$, $(3,4)$ 25. Turns at $(\frac{1}{4}, -\frac{1}{8})$, low point In Ex. 27, three points are given for each curve (parabola). 27. $y = x^2$: $(2,4)$, $(0,0)$, $(-2, 4)$; $y = 2x^2$: $(1, 2)$, $(0, 0)$, $(-1, 2)$ 29. Move 5 units right. 31. $(x - (-2))^2 - 3$ 33. $(x - (-\frac{3}{2}))^2 - \frac{17}{4}$ 35. $x = 1$ 37. $\{2, 5\}$ 39. $\{-4 + \sqrt{17}, -4 - \sqrt{17}\}$ 41. $\{3, -\frac{1}{2}\}$ 43. Consecutive integers: n, $n + 1$; Equation: $n(n + 1) = 56$; Ans: 7 and 8 or -8 and -7 45. $\frac{1}{6}$ 47. $\frac{1}{36}$ 49. $\frac{6}{25}$

PICTURE CREDITS

Key: *(t)* top; *(b)* bottom; *(l)* left; *(r)* right; *(c)* center.

CHAPTER ONE: Page 1, *(l)*, IBM Corp., *(tr)*, IBM Corp., *(br)*, HBJ Photo/Jerry White; 24, HBJ Photo/Rodney Jones; 28, Frank Siteman/Taurus Photos; CHAPTER TWO: 29, *(all)*, E. R. Degginger; 52, Photri; CHAPTER THREE: 57, *(l)*, Jonathon Watts–Science Source Library/Photo Researchers, Inc., *(tr)*, J. Stewart/Bruce Coleman, Inc., *(br)*, John Shaw/Bruce Coleman, Inc.; 70, J. Gordon Miller/Shostal Associates; 79, © Richard Hutchings; 80, E. R. Degginger; 81, Jeff Apoian/Nawrocki Stock Photo; 85, Sullivan and Rogers/Bruce Coleman, Inc.; 86, *(r)*, Brian Brake–Rapho Division/Photo Researchers, Inc., *(bl)*, L. Druskis/Taurus Photos; CHAPTER FOUR: 89, *(l)*, E. R. Degginger, *(tr)*, Paul Silverman/Fundamental Photographs, *(br)*, E. R. Degginger; 92, Lester Sloan/Woodfin Camp & Associates; 93, James Drake for Sports Illustrated © Time, Inc.; 96, Warren Morgan/Focus on Sports; 98, Mark Antman/The Image Works; 99, Michal Heron/Woodfin Camp & Associates; 105, © Richard Hutchings; 107, Jon Feingersh/Tom Stack & Associates; 112, George Hunter/H. Armstrong Roberts; CHAPTER FIVE: 115, *(l)*, Jeff Hetler/Stock, Boston, *(r)*, Dave Forbert/Shostal Associates; 135, Craig Aurness/Woodfin Camp & Associates; CHAPTER SIX: 143, *(l)*, Dr. J. Lorre/Photo Researchers, Inc., *(tr)*, Leonard Lessin/Peter Arnold, Inc., *(br)*, M. I. Walker–Science Source Library/Photo Researchers, Inc.; 155, NASA; 170, John Eastcott–Yva Momatuck/The Image Works; CHAPTER SEVEN: 173, *(l)* and *(tr)*, Ronda Bishop–Contact Press Images/Woodfin Camp & Associates; 196, Mark Stevenson/Nawrocki Stock Photo; 203, Mark Segal/Click/Chicago; CHAPTER EIGHT: 207, *(all)*, NASA; 223, Rice Sumner Wagner; 231, HBJ Photo/Blaise Zito Associates; 234, B. Kulik/Photri; CHAPTER NINE: 235, *(all)*, HBJ Photos; 250, Lowell Georgia/Photo Researchers, Inc.; 251, Grant Heilman Photography; CHAPTER TEN: 267, *(l)*, Rafael Macia/Photo Researchers, Inc., *(tr)*, Steve Solum/Bruce Coleman, Inc., *(br)*, E. R. Degginger; 290, David York/The Stock Shop; CHAPTER ELEVEN: 293, *(l)*, Chuck O'Rear–West Light/Woodfin Camp & Associates, *(tr)*, Chuck O'Rear/Woodfin Camp & Associates, *(br)*, © Nelson Morris/Photo Researchers, Inc.; CHAPTER TWELVE: 321, *(l)*, NASA, *(tr)*, © The Science Museum, London, *(br)*, The Granger Collection; 332, Freda Leinwand/Monkmeyer; 340, Guy Sauvage–Agence Vandiptadt/Photo Researchers, Inc.; 341, The Granger Collection; 342, Elaine Wicks/Taurus Photos; 344, HBJ Photo/Blaise Zito Associates; CHAPTER THIRTEEN: 349, *(l)*, Donald Miller/Monkmeyer, *(tr)*, Hank Morgan/Photo Researchers, Inc.; *(br)*, Richard Megna/Fundamental Photographs; 373, Jerry Irwin/Shostal Associates; CHAPTER FOURTEEN: 377, *(l)*, Shostal Associates, *(tr)*, HBJ Photo/Blaise Zito Associates, *(br)*, © Mark Greenberg/Visions; 394, Martin Rother/Taurus Photos; CHAPTER FIFTEEN: 397, *(l)*, Jerry Ferrara/Photo Researchers, Inc., *(tr)*, Ellis Herwig/Taurus Photos, *(br)*, © Martin Rogers/Stock, Boston; 410, Mimi Forsyth/Monkmeyer; 411, R. Krubner/H. Armstrong Roberts; 419, Doug Lee/Tom Stack & Associates; 422, Donald Dietz/Stock, Boston; 422, Robert Chase; CHAPTER SIXTEEN: 433, *(l)*, E. R. Degginger, *(tr)*, Sinclair Stammers–Science Source Library/Photo Researchers, Inc., *br*, NASA; 451, Eric Carle/Shostal Associates; CHAPTER SEVENTEEN: 455, *(t)*, S. L. Craig/Bruce Coleman Inc.,*(c)*, Mickey Palmer/Focus on Sports, *(b)*, Farrell Grehan/FPG; 478, John Lei/Omni Photo Communications; CHAPTER EIGHTEEN: 483, *(l)*, Ben and Miriam Rose/The Image Bank, *(tr)*, Don Katchusky/Taurus Photos; CHAPTER NINETEEN: 511, *(l)*, IBM Corp.; 512, Elaine Wicks/Taurus Photos; 513, *(t)*, © Art Attack, *(b)*, Photo Researchers, Inc.; 514, © Richard Hutchings; 517, Fred Lyons/Photo Researchers, Inc.; 525, Sepp Seitz/Woodfin Camp & Associates; 527, Russ Kinne/Comstock; 531, Bohdan Hrynewych/Stock, Boston; 533, Chris Luneski/Photo Researchers, Inc.